THE MEAT
WE EAT

Fourteenth Edition

THE MEAT

John R. Romans

William J. Costello

WE EAT

C. Wendell Carlson

Marion L. Greaser

Kevin W. Jones

 INTERSTATE PUBLISHERS, INC. / Danville, Illinois

THE MEAT WE EAT

Fourteenth Edition

Prior Editions: 1943, 1948, 1949, 1952, 1953, 1954, 1958,
1962, 1966, 1974, 1977, 1985, 1994

Library of Congress Catalog Card No. 99-76342

ISBN 0–8134–3175–1

1 2 3 4 5 6 7 8 9 0   05 04 03 02 01 00

Order from

INTERSTATE PUBLISHERS, INC.

510 North Vermilion Street
P.O. Box 50
Danville, IL 61834–0050

Phone: (800) 843–4774
Fax: (217) 446–9706
Email: info-ipp@IPPINC.com

World Wide Web: http://www.IPPINC.com

COVER PHOTO: *Courtesy, Urschel Laboratories, Inc.*
Designers and Manufacturers of
Precision-Engineered Size Reduction Equipment

Preface

It was the best of times, it was the worst of times,
it was the age of wisdom, it was the age of foolishness,
it was the epoch of belief, it was the epoch of incredulity,
it was the season of light, it was the season of darkness,
it was the spring of hope, it was the winter of despair.

Thus begins Charles Dickens' great novel, *A Tale of Two Cities*. These same words aptly describe the current state of animal agriculture.[1]

Today, consumers receive nutritional advice and food safety warnings almost daily from the media. Governments and the worldwide meat industry are responding. As a result, all meat products are leaner, and efforts are being made to tell shoppers exactly what they are buying through complete nutritional labeling.

At the same time, more individuals are becoming concerned with the environment and animal welfare. The meat industry is working diligently to preserve our natural resources and to assure the humane harvest of animals. It is also making great strides in mechanizing and automating many of the procedures used to transform animals into healthful, safe food products.

The Meat We Eat, 14th Edition, reflects these concerns and the efforts of the meat industry to address them. The chapter on meat inspection and animal loss factors has been completely revised to address the Hazard Analysis Critical Control Point (HACCP) program. The text also reflects the movement away from the concept of slaughter to that of harvest when considering the conversion of live animals into carcasses and products.

The new edition also addresses the importance of websites. Several are listed, thereby reducing the number of long tables included in previous editions of the text. Examples are included on how the industry is using the Internet to market its products more effectively.

---

[1]Dr. Robert Easter, Head, Department of Animal Sciences, University of Illinois in his Postings (December 1999) letter.

Despite ravings in the popular press promoting vegetarianism, a recent report by the Council for Agricultural Science and Technology (http://www.cast-science.org) presents a very encouraging perspective on the growth in demand for animal products over the next 20 years.[2] Also, Dr. Allison A. Yates, Director, Food and Nutrition Board, Institute of Medicine, National Academy of Sciences, gave a balanced presentation entitled "Nutrition Today, and Implications for the Future."[3] Both of these works merit careful study by students of meat science.

By 1990, there were 23 billion cattle, sheep, goats, pigs, buffalo, chickens, ducks, and turkeys being raised for food, or about four for every man, woman, and child on the earth. This figure represents a tripling in the number of livestock in the world since the early 1960s, when there were only 7.5 billion. Each of these species is covered in this book.

Teachers and extension food specialists must relate current, sound dietary concepts so that students and consumers understand the principles and value of a balanced diet that includes muscle foods.

The Meat We Eat is the ideal text for anyone interested in meat science. As a college textbook, it presents the science involved in converting whole live animals into nutritious and appetizing food for human consumption. As a reference handbook, it provides a current source for a wealth of information affecting today's meat industry.

---

[2]Ibid.

[3]http://www.barc.usda.gov/bhnrc/foodsurvey/Ayates/html

About the Authors

John R. Romans worked in agricultural services and hog buying for John Morrell & Company before pursuing advanced degrees, which led to an academic position at the University of Illinois in meats extension, teaching, and research. There he received numerous teaching and extension awards, among them the Distinguished Extension-Industry Service Award from the American Meat Science Association.

In 1981, after 15 years at Illinois, he returned to South Dakota State University, where he had earlier received his M.S. and Ph.D. degrees, to assume the position as head of the Department of Animal and Range Sciences. After six years as an administrator, he returned to research and teaching in meat science. He was awarded the Distinguished and Dedicated Service Award by the South Dakota Pork Producers Council, the South Dakota State University Dean's Award for his contributions to the Meat Board's Lessons on Meat, and the Signal Service Award by the American Meat Science Association. In 1997, he retired from South Dakota State University and returned to the University of Illinois Animal Sciences Department as an adjunct professor.

Dr. Romans is a Charter Diplomate in the American College of Animal Food Science and an emeritus member of the American Meat Science Association. Research over a span of more than 30 years has focused on lipid composition, deposition mechanisms, location, and effects of sex condition in meat animals. He has judged state, regional, and national carcass and product shows and has served on official committees for Intercollegiate Meat Judging Contests. Dr. Romans has served on editorial boards of the *Journal of Animal Science* and *The Professional Animal Scientist*. He has held positions on the American Society of Animal Science board, was president of the Midwest Section, and was named a Fellow of ASAS. Dr. Romans earned his B.S. in Animal Husbandry and Agricultural Education from Iowa State University.

William J. Costello recently retired from the Department of Animal and Range Sciences at South Dakota State University. For more than 30 years, "Doc" taught meats courses to students of all ages. He shared coordinating the Animal and Range Sciences cooperative education / internship / field experience program and teaching animal science junior and senior seminar courses with other faculty.

Dr. Costello advised many undergraduates, served as thesis and major advisor for graduate students, and was the faculty advisor for a number of student organizations.

The SDSU Meat Laboratory, under Dr. Costello's supervision, supported teaching, research, and extension activities in the college and the department. His meat research utilized meat processing techniques and less-valued portions of carcasses in the production of structured and convenience meat foods. Dr. Costello provided meats expertise to youth activities, to the meat and food processing industry, to citizens, and to government agencies. He was identified by students as the College of Agriculture Teacher of the Year in 1974, 1986, and 1997 and received the Gamma Sigma Delta Teaching Award and the American Meat Science Association Distinguished Teaching Award. The student management of the 1999 SDSU Little I named "Doc" agriculturalist of the year. SDSU identified him as a Distinguished Professor in 1991 and more recently granted Dr. Costello the F. O. Butler Award for Service to Students at South Dakota State University.

Dr. Costello received his B.S. in Animal Husbandry from North Dakota State University and his M.S. and Ph.D. from Oklahoma State University. He was employed by John Morrell & Company in research and technical sales for three years before he joined the faculty of South Dakota State University in 1965.

C. Wendell Carlson served with the Food Inspection Service of the U.S. Army during World War II, after receiving his B.S. in Animal Husbandry from Colorado State University. He earned his M.S. and Ph.D. degrees from Cornell University and returned west to South Dakota State University in 1949. Except for a leave at Washington State University in 1961–1962 and a tour with the USDA in Washington, D.C., in 1975–1976, he made his career at South Dakota State University, retiring in 1985.

Dr. Carlson's research and teaching centered on animal and human nutrition, including food science. He was the leader of poultry research and extension and in charge of the Animal Science Graduate Program and the SDSU Feed Unit at the time of his retirement. He was awarded the National Turkey Federation Turkey Research Award by the Poultry Science Association in 1963 and was made a Fellow of the Poultry Science Association in 1973. He was president of that association in 1969–1970. He received the F. O. Butler Certificate of Excellence from South Dakota State University in 1984 and the Gamma Sigma Delta Research Award in 1988.

From 1989 through 1997, Dr. Carlson held a visiting professorship at the University of Minnesota, engaged in nutrition consulting, and continued to serve on the Nutrition Advisory Committee for the Minnesota Turkey Growers Association. He has held several leadership offices in the Midwest Poultry Federation. He served on five volunteer overseas assignments with the International Executive Service Corps and the Volunteers in Overseas Cooperative Assistance.

Marion L. Greaser is a widely recognized muscle biologist. His research in the 1960s involved studies on postmortem changes in the sarcoplasmic reticulum. He later purified and extensively characterized the subunits of troponin. His current research focuses on myofibril assembly mechanisms, the role of titin in muscle function, and postmortem changes in muscle proteins as they relate to meat tenderness.

Dr. Greaser received the Distinguished Research Award from the American Meat Science Association in 1981, the Outstanding Researcher Award from the American Heart Association–Wisconsin Chapter in 1985, and the Meat Research Award from the American Society of Animal Science in 2000. He obtained his B.S. in Animal Science from Iowa State University and his M.S. and Ph.D. degrees jointly in Biochemistry and Muscle Biology from the University of Wisconsin–Madison. Dr. Greaser was a Postdoctoral Fellow from 1968 to 1971 in the Department of Muscle Research, Boston Biomedical Research Institute, Boston, Massachusetts. He returned in 1971 to the University of Wisconsin–Madison and has served on the faculty in the Department of Meat and Animal Science since that time. He currently is director of the Muscle Biology Laboratory.

Kevin W. Jones is currently employed in private industry. He joined the faculty at South Dakota State University after completing his M.S. and Ph.D. degrees at the University of Nebraska. His research efforts at South Dakota State University focused on the nutritional quality of further processed meat products to meet the convenience demand of the modern consumer. He received several grants from the livestock industry to support this work.

While at South Dakota State University, he was honored by undergraduates for his teaching. Dr. Jones received a Meritorious Service Award from USDA Secretary Block in 1983 and the George Straethern Memorial Research Award from the California Beef Council in 1984. Upon leaving South Dakota State University, Dr. Jones served as the technical director of Johnsonville Foods from 1986 to 1990. He received his B.S. in Animal Science from Purdue University.

The current authors wish to recognize the extraordinary contributions of the original author of this text, the late **P. Thomas Ziegler.**

Classification of Meat Species[1]

Phylum — *Chordata* (internal skeleton and dorsal tubular nerve cord)

Subphylum — *Vertebrata* (segmented bony backbone or vertebral column)

Class — *Mammalia* (hair; mammary glands secrete milk)

Subclass — *Theria* (marsupials and placentals; nipples on mammary glands)

Infraclass — *Eutheria* (foetus develops entirely by means of placenta)

Order — *Artiodactyla* (even-toed ungulates)

Suborder — *Ruminantia* (stomach with three or four chambers; no upper incisors)

Infraorder — *Pecora* (stomach with four chambers; true ruminants)

Family — *Bovidae* (hollow-horned)

Genus — *Bos* (cattle)

Group — *Taurine* (of or like a bull)

Species — *B. taurus* (cattle)

B. indicus (humped cattle)

---

[1]Storer, T. R. 1951. General Zoology (2nd Ed.). McGraw-Hill Book Co., New York. With interpretations by South Dakota State University Professor Emeritus Ernest J. Hugghins.

Genus — *Bison*

Group — *Bison*

Genus — *Ovis* (sheep)

Group — *O. aries* (domestic)

O. canadensis (mountain, bighorn)

Genus — *Capra* (goats)

Group — *C. hirius* (domestic)

Genus — *Oreamnos montanus* (mountain goats)

Family — *Cervidae* (deer)

C. elaphus (European reindeer)

Genus — *Odocileus* (American deer)

Group — *O. virginianus* (white-tailed deer)

O. hemionus (black-tailed and mule deer)

Genus — *Alce*

Group — *A. americanus* (moose)

Genus — *Rangifer* (North American reindeer and caribou)

Family — *Antilocapridae* (pronghorn)

Genus — *Antilocapra*

Suborder — *Suiformes* (upper incisors, pointed molars; simple stomach)

Family — *Suidae* (true swine)

Genus — *Sus* — *Sus scrofa* (wild boars)

Group — *S. domesticus* (domesticated swine)

Order — *Perissodactyla* (odd-toed ungulates)

Family — *Equidae* (horse family)

Genus — *Equus* (horse family)

Group — *E. caballus* (horses)

E. asinus (asses, donkeys)

E. zebra (zebras)

Order — *Rodentia*

    Family — *Leporidae* (domesticated rabbits)

Class — *Aves* (feathered)

  Subclass — *Neornithes* (w/o teeth)

    Superorder — *Neognathae* (flying birds)

    Order — *Galliformes* (fowls)

      Family — *Phasianidae* (w/spurs)

        Genus — *Gallus* (combs)

          Species — *G. domesticus* (chickens)

        Genus — *Phasianus colchicus torquatus* (ring-necked pheasants)

        Genus — *Bonasa umbellus* (ruffed grouse)

      Family — *Numididae* (guineas)

      Family — *Meleagridae* (w/caruncles)

        Genus — *Meleagris* (dewbills)

          Species — *M. gallopavo* (turkeys)

    Order — *Anseriformes* (broadened bills)

      Family — *Anatidae* (web feet)

        Genus — *Anser*

          Species — *A. anser* (geese)

        Genus — *Anas*

          Species — *A. platyrhynchos* (ducks)

    Superorder — *Palaeognathae* (walking birds, usually flightless)

    Superorder — Ratitae (flat [no keel] breastbone) (ostrich, emu, moa, kiwi)

      Order — *Struthioniformes*

        Family — *Struthio camelus* (ostriches)

Superclass — *Pices* (paired fins, gills, and skin with scales)

  Class — *Osteichthyes* (bony fishes)

Contents

Preface v

About the Authors vii

Classification of Meat Species xi

1. Introduction 1

2. Meat Biotechnology and Microbiology 13

3. Meat Inspection and Animal Loss Factors 37

4. Preparations for Processing — Worker
and Equipment Safety 115

5. Hog Harvest 133

6. Cattle Harvest. 173

7. Sheep and Lamb Harvest 197

8. Veal and Calf Harvest 213

9. Poultry Processing 223

10. Game Processing 247

11. Packing House By-products 275

12. Federal Meat Grading and Its Interpretations 347

13. Meat Merchandising 455

14. Pork Identification and Fabrication 503

15. Beef Identification and Fabrication 541

16. Lamb Identification and Fabrication. 595

17. Veal and Calf Identification and Fabrication 629

18. Fresh Meat Processing 641

19. Preservation and Storage of Meat. 689

20. Meat Curing and Smoking 731

21. Sausages 779

22. Structure and Function of Muscle. 889

23. Meat as a Food 909

24. Preparing and Serving Meats. 929

25. Meat Judging and Evaluation 1003

Appendix 1039

Index 1085

1

Introduction

This chapter shows the relationships that exist between diet composition, meat consumption, population growth, the well-being of people, economic development, and the ecology and economy of the world.

Approximately two-thirds of the world's agricultural land is permanent pasture, range, or meadow; of this, at least 60 percent is unsuitable for producing crops that can be consumed directly by humans. In the United States, 44 percent of the total land area is composed of rangelands and forests that are used for grazing. This land produces cellulosic roughages in the form of grass and other vegetation that are digestible by grazing ruminant animals (cattle, sheep, goats, deer, and bison). Cellulose is the most abundant chemical constituent in the dry substance of plants, but it cannot be digested by humans. Ruminants harbor microorganisms in the rumen portion of their four-compartment stomachs. These microorganisms have the ability to utilize cellulose for energy and to synthesize essential nutrients, such as amino acids and certain vitamins. Furthermore, approximately 98 percent of the grain fed to animals in the United States consists of corn, sorghum, oats, and barley, only limited amounts of which are major sources of human food in this country.[1]

Ruminants utilize grasses and grains and convert them into a more suitable and concentrated food for humans as well as providing materials for clothing, pharmaceuticals, and many other valuable by-products. Figure 1–1 depicts the central role that ruminants play in human nutrition.

---

[1]Council for Agricultural Science and Technology (CAST). 1980. Foods from Animals: Quantity, Quality and Safety. CAST Report No. 82.

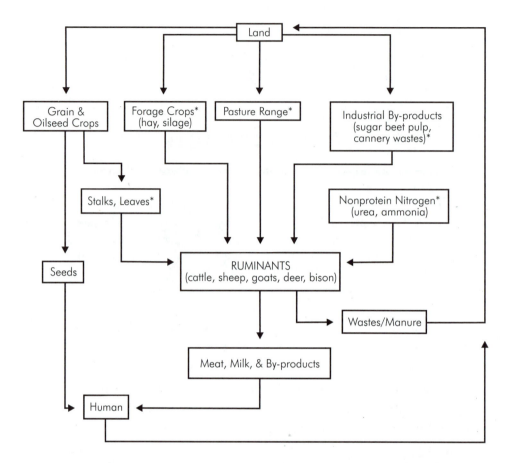

Fig. 1–1. The central role of ruminants in human nutrition. Items marked with an asterisk are converted by ruminants but are not eaten by humans. (From Coun. Agric. Sci. Tech. [CAST]. 1975. *Ruminants as Food Producers*, Spec. Publ. No. 4, p. 4.)

Because both swine and poultry possess a simple stomach (one pouch, like a human's), they cannot utilize roughage to the same degree as ruminants. However, they are very efficient in converting grain into meat. These conversion abilities have made herds and flocks of prime economic importance in the development of civilization.

The burgeoning world population has multiplied at an alarming rate (see Table 1–1). The rate of increase is beginning to slow, with the current 1.7 percent per year estimated to be 1 percent by 2025. Population experts predict that the world population may stabilize at somewhere between 10 and 14 billion people.

Low food consumption and undernutrition amidst global plenty will continue to persist in the near future. Nevertheless, for the first time in four

Table 1–1. World Population Trends[1]

Year	Approximate World Population	Intervals for 1 Billion Increase
1830	1 billion	
1930	2 billion	100 years
1960	3 billion	30 years
1978	4 billion	18 years
1988	5 billion	10 years
1998	Estimated 6 billion	10 years
2008	Estimated 7 billion	10 years
2019	Estimated 8 billion	11 years

[1]The United Nations Food and Agriculture Organization, New York, NY 10017.

decades, the United Nations Food and Agriculture Organization (UNFAO)[2] has reported a decline in the incidence of undernutrition in the developing world. As diets have become more diversified, including more animal products in developing economies, people tend to feel more positive about their future and are better able to plan for it. This relationship has been instrumental in slowing world population growth.

The importance of meat's contribution to a healthy, contemporary diet was recognized in a U.S. Senate resolution designating the first annual National Meat Week, January 22–28, 1984. The resolution stated that "meat is a wholesome and nutritious food, one of the most valuable sources of vitamins and minerals in the human diet, and a high quality source of protein" and that "the meat industry's annual sales of $70 billion make it the largest single component of U.S. agriculture." The size and impact of the meat industry has not changed appreciably since 1984, and since 1988, February has been designated National Meat Month. (The importance of including meat in a healthy diet is documented in Chapter 23.) Since 1964, February has also been designated by Congress as American Heart Month. The official designation of February as both Heart and Meat Month is appropriate, since today's lean meats fit naturally into the American Heart Association's "Prudent Diet."

Students interested in how we as humans developed in relation to our diet are encouraged to examine these relationships on the World Wide Web. Sug-

[2]New York, NY 10017.

gested Web sites include those of the Food and Nutrition Board; the Diet, Safety, and Health Economics Branch, Food and Rural Economics Division, Economic Research Service, USDA; and the USDA Food Surveys Research Group.

MEAT PRODUCTION, TRADE, AND CONSUMPTION

In any country, production, plus imports, minus exports, gives net meat disappearance or consumption. Table 1–2 provides an interesting, revealing history of muscle food consumption over the last 30 years. Pork consumption has maintained a relatively constant record, while poultry consumption has increased at an almost astronomical rate, and fish consumption has increased as well. Beef consumption has suffered the greatest decline, while veal and lamb consumption have also decreased. Current per capita meat and other food consumption data can be found by locating the United States Department of Agriculture (USDA), Economic Research Service (ERS), on the World Wide Web.

Table 1–2. Meat Consumption[1]

	Per Capita Consumption of Meat, Poultry, and Fish, Boneless Weight Equivalent								Per Capita Consumption of Meat, Poultry, and Fish, Retail Weight								
Year	Beef	Veal	Pork	Lamb	Chicken	Turkey	Fish	Total	Year	Beef	Veal	Pork	Lamb	Chicken	Turkey	Fish	Total
	- - - - - - - - - - - - - - - - - - (pounds) - - - - - - - - - - - - - - - - - -									- - - - - - - - - - - - - - - - - - (pounds) - - - - - - - - - - - - - - - - - -							
1970	79.6	2.0	48.6	2.1	25.2	6.4	11.8	175.7	1970	84.6	2.4	56.0	2.9	36.9	8.1	11.7	202.6
1971	79.0	1.9	53.0	2.1	25.1	6.6	11.5	179.2	1971	83.9	2.2	60.6	2.8	36.8	8.4	11.5	206.2
1972	80.7	1.6	48.1	2.2	26.2	7.1	12.5	178.4	1972	85.7	1.9	54.7	2.9	38.4	9.0	12.5	205.1
1973	75.9	1.2	43.4	1.7	25.3	6.7	12.8	167.0	1973	80.7	1.5	49.0	2.4	37.4	8.5	12.7	192.2
1974	80.6	1.6	47.1	1.5	25.2	6.9	12.1	175.0	1974	85.6	1.9	52.8	2.0	37.4	8.8	12.1	200.6
1975	83.0	2.8	38.5	1.3	24.9	6.5	12.2	169.2	1975	88.2	3.3	43.0	1.8	37.1	8.5	12.1	194.0
1976	88.9	2.7	41.0	1.2	27.2	7.0	12.9	180.9	1976	94.5	3.2	45.5	1.6	40.2	9.1	12.9	207.0
1977	86.2	2.6	42.6	1.1	27.6	7.2	12.7	180.0	1977	91.6	3.1	47.0	1.5	41.4	9.1	12.6	206.3
1978	82.3	2.0	42.8	1.0	29.2	6.9	13.4	177.6	1978	87.4	2.3	47.0	1.4	44.1	9.2	13.4	204.8
1979	73.5	1.4	49.1	1.0	31.6	7.3	13.0	176.9	1979	78.1	1.6	53.7	1.3	46.7	9.9	13.0	204.3
1980	72.1	1.3	52.6	1.0	31.5	8.1	12.8	179.4	1980	76.6	1.5	57.3	1.4	46.6	10.5	13.4	207.3
1981	72.7	1.3	50.4	1.0	32.2	8.4	12.9	178.9	1981	77.2	1.5	54.7	1.4	47.6	10.8	12.7	205.9
1982	72.4	1.4	45.3	1.1	32.4	8.4	12.3	173.3	1982	76.9	1.6	49.1	1.5	47.9	10.8	12.1	199.9
1983	73.8	1.4	47.7	1.1	32.7	8.7	13.1	178.5	1983	78.3	1.5	51.6	1.5	48.2	11.2	12.9	205.2
1984	73.6	1.5	47.5	1.1	34.0	8.7	13.7	180.1	1984	78.2	1.7	51.3	1.5	50.3	11.4	13.4	207.8

(Continued)

Table 1–2 (Continued)

	Per Capita Consumption of Meat, Poultry, and Fish, Boneless Weight Equivalent								Per Capita Consumption of Meat, Poultry, and Fish, Retail Weight								
Year	Beef	Veal	Pork	Lamb	Chicken	Turkey	Fish	Total	Year	Beef	Veal	Pork	Lamb	Chicken	Turkey	Fish	Total
						(pounds)								(pounds)			
1985	74.3	1.5	47.9	1.1	35.2	9.2	14.4	183.6	1985	78.9	1.8	51.7	1.4	52.1	12.1	14.3	212.3
1986	74.1	1.6	45.4	1.0	36.0	10.2	14.7	183.0	1986	78.5	1.8	48.8	1.4	53.2	14.4	14.5	212.6
1987	69.2	1.3	45.8	1.0	38.1	11.6	15.4	182.4	1987	73.5	1.5	49.0	1.3	56.3	15.2	15.5	212.3
1988	68.6	1.1	48.9	1.0	38.3	12.4	15.1	185.4	1988	72.3	1.4	52.2	1.4	56.4	15.9	15.0	214.7
1989	65.4	1.0	48.4	1.1	39.8	13.1	15.6	184.4	1989	69.4	1.2	51.7	1.5	58.7	17.1	15.6	215.2
1990	63.9	0.9	46.4	1.0	41.4	13.9	15.0	182.5	1990	67.4	1.1	49.8	1.5	60.9	18.4	15.0	214.1
1991	63.3	0.8	47.3	1.1	43.3	14.2	14.8	184.8	1991	66.8	1.0	50.4	1.5	65.6	18.0	18.8	217.1
1992	63.0	0.8	49.9	1.1	45.5	14.1	14.8	189.2	1992	66.5	1.0	53.1	1.4	68.2	18.0	14.8	223.0
1993	61.6	0.8	49.2	1.0	47.8	14.1	15.0	189.5	1993	65.1	0.9	52.3	1.3	70.5	17.8	15.0	222.9
1994	63.9	0.8	49.9	0.9	48.8	14.0	15.1	193.4	1994	67.0	0.9	53.1	1.2	71.4	17.8	15.1	226.5
1995	64.3	0.8	49.3	0.9	48.2	14.2	14.9	192.7	1995	67.4	1.2	49.1	1.2	71.9	18.0	15.1	223.9
1996	65.2	1.0	46.1	0.9	50.0	14.6	14.7	192.9	1996	68.2	1.2	49.1	1.1	71.7	18.5	15.1	224.9
1997	64.0	0.8	45.8	0.8	51.1	13.9	14.5	191.4	1997	66.9	1.0	48.7	1.1	71.9	17.6	15.0	223.4
1998	64.9	0.7	49.1	0.9	51.6	14.2	14.5	N/A[2]	1998	68.1	N/A	52.6	N/A	72.6	18.1	N/A	N/A
1999	62.3	0.6	49.1	0.8	45.0	14.0	N/A	N/A	1999	69.3	N/A	54.2	N/A	77.5	17.9	N/A	N/A
2000[3]	N/A	N/A	N/A	N/A	N/A	N/A	N/A	N/A	2000[3]	68.5	N/A	52.2	N/A	80.8	17.9	N/A	N/A

[1]USDA-ERS. (Ronald Gustafson.)

[2]N/A = not available.

[3]Preliminary.

Ready-to-cook (RTC) weight of poultry and fish is equal to the *carcass weight* of red meats for comparative purposes. *RTC weight* is a holdover term in the United States from the late 1970s and earlier when poultry and fish carcasses, because of their small size, were cooked whole or simply cut up into smaller pieces for cooking without significant weight loss. Therefore, all bones and fat in the carcasses were included in the RTC weight for poultry and fish. The amount of inedible material (bones, skin, fat) included in the carcass weight of red meats and the RTC weight of poultry and fish is relatively equitable among all the species. This is one basis for comparing disappearance (consumption) of the various species. Another is boneless weight eqquivalent (Table 1–2).

Obviously population has a great influence on per capita consumption, because total consumption is divided by population to determine per capita consumption. People of a country and an area of the world that lends itself to the economical production of a species will understandably consume more of the product produced by that species because of its availability and economy. The United States is clearly the leader in broilers, ranking first in production, exports, and total and per capita consumption. The United States leads in beef and veal production, imports, and total consumption, while it ranks third in exports and second (Argentina is the leader) in per capita consumption. China clearly leads the world in pork production and total consumption, but because of its large population, it ranks quite low in per capita pork consumption. Hungary and Poland lead in per capita pork consumption. Denmark, a member of the European Community, produces a large number of hogs and is very close to Hungary and Poland in per capita pork consumption. The European Community leads in lamb, mutton, and goat meat production, imports, and total consumption, while New Zealand and Australia clearly lead in per capita consumption. China, the former USSR, and Japan are the most successful at fishing, but they all rank below many smaller countries in per capita fish consumption.

The more highly developed countries of the world are those that are the most actively involved in meat production, trade, and consumption. In the United States, major sources of imported fresh beef are Australia and New Zealand; of processed beef, Argentina; of fresh pork, Canada and Denmark; of processed pork, Denmark, Canada, and Poland; of veal, New Zealand, Canada, and Australia; of mutton and lamb, Australia and New Zealand; of poultry, Canada, Denmark, Argentina, and Australia; and of edible fishery products, Canada, Thailand, and Ecuador. The major receivers of U.S. meat exports (percent of total) are Japan, 48; Mexico, 16; Canada, 11; the European Community, 9; South Korea, 4; and others, 12. U.S. poultry exports (percent of total) are more evenly divided among these: the former USSR and Eastern Europe, 22; Japan, 18; Hong Kong, 17; the Caribbean, 10; Mexico, 9; Canada, 9; the European Community, 5; and others, 10. Japan is clearly the largest recipient of U.S. fish exports, obtaining 56 percent of the total, while Canada (9 percent) and South Korea (9 percent) and many other countries (26 percent) get the remainder.

Within the United States, consumers can compare amounts of the various foodstuffs they purchase to eat by comparing the retail weight equivalents of the various foodstuffs, including meat. Retail weight equivalents are the only accurate way to compare the "consumption" of various foods that undergo different methods of preparation for serving. Because red meat carcasses are large, preparing them for retail sale and cooking involves removing considerable bone and some fat. When reporting the *retail weight* of red meats, the U.S. Department of Agriculture uses these factors to convert from *carcass weight:* beef × 0.705; veal × 0.83; lamb and mutton × 0.89; and pork ×

0.776. As noted previously, the RTC weight conversion for poultry (0.852) is comparable to the carcass weight of red meats.

In 1962, 87 percent of broilers were sold whole; however, recently less than 18 percent are sold whole. Processors now prepack broilers in packages containing individual cuts—for example, boneless, skinless breasts. The remaining bone and skin plus chicken backs and necks do not go into the human food chain at the former level; most are sold for pet food.

Purchases of all the muscle foods (red meats, poultry, and fish) should be expressed as "boneless, trimmed equivalent weight," which is less than carcass, ready-to-cook (RTC), and retail weights. When reporting the boneless, trimmed equivalent weight of red meats, the U.S. Department of Agriculture converts from carcass weight by using these factors: beef × .667; veal × .685; lamb and mutton × .658; and pork × .729. When converting to boneless, trimmed weight from RTC weight, the U.S. Department of Agriculture uses × .58 for chicken, × .785 for turkey, and × 1.00 for fish. The details of all these relationships can be found by contacting the USDA.

Consumption of red meat gradually declined from 130.2 pounds in 1970–1974 to 112.4 pounds in 1990, when the trend was reversed. Although beef consumption hit an all-time high of 82.8 pounds during the 1975–1979 period, since that time it has dropped to 66.1 pounds in 1999. Because beef is the largest component of the red meat group, it has the greatest effect on overall red meat consumption. Pork consumption has remained fairly constant. The consumption of lamb and mutton and of veal has decreased; however, the consumption of these red meats does not have a significant effect on the overall picture.

During this same period, the consumption of poultry meat has soared dramatically to 96.4 pounds. The consumption of chicken has gone up steadily from 27.4 pounds to 78.6 pounds. Turkey consumption has risen from 6.7 pounds to 17.8 pounds. More turkey and more chicken are being utilized in further processed meat items, as will be discussed in Chapters 9, 18, 19, and 20.

Over the last decade, average per capita fish and seafood consumption has remained relatively flat, at around 15 pounds, roughly 2 to 3 pounds less than turkey consumption. However, during this period, the source of seafood products has been shifting away from wild harvest and toward aquaculture. In 1997, U.S. production of processed catfish products was close to 1 pound per capita, imports of farm-raised shrimp were likely over 1 pound per capita, and the combination of farm-raised salmon, trout, tilapia, crawfish, and other aquaculture products probably added another pound. With about 20 percent of U.S. fish and seafood consumption now being farm-raised, aquaculture is becoming a recognized segment of the livestock complex, larger than veal, mutton, and lamb combined.[3]

In 1970–1974 total consumption of all muscle foods (red meats, poultry meats, and fish) was 176.4 pounds. This total has increased to 232 pounds in 1999.

The nutritional pyramid introduced by the USDA in 1992 (Figure 1–2) is beginning to be reflected in our food consumption trends. Note that the pyramid recommends 6 to 11 servings of bread, cereal, rice, and pasta per day. This food group forms the base of the pyramid. The next higher level of the pyramid is divided between vegetables (three to five servings per day) and fruits (two to four servings per day). One-half of the next higher, narrower level of the pyramid is occupied by milk, yogurt, and cheese, and the other

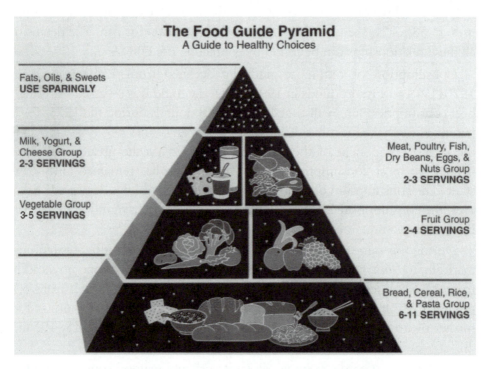

Fig. 1–2. The Food Guide Pyramid shows how to build a healthful diet by eating a variety of foods each day. (Courtesy, U.S. Department of Agriculture and U.S. Department of Health and Human Services)

---

[3]Aquaculture Outlook, March 1998, ERS-LDP-AQS-7. Aquaculture Outlook, a supplement to Livestock, Dairy, and Poultry, is published twice a year by the Economic Research Service, U.S. Department of Agriculture, Washington, DC 20036-5831.

half of this level is composed of *meat, poultry, fish,* dry beans, eggs, and nuts. Two to three servings per day from each group on this level are recommended. At the apex of the pyramid are fats, oils, and sweets, which should be used sparingly.

Food consumption data do not imply that each person actually eats the total quantities of foods they purchase. Spoilage, waste, and cooking loss must be taken into account. Dr. Burdette C. Breidenstein (now deceased), former Senior Vice President for Scientific Affairs, National Live Stock and Meat Board, Chicago, made computations showing that the retail weight of all red meat species averages only 82 percent of the carcass weight because of the bone and fat content and that actual consumption averages only 65 percent of the retail weight because of additional fat and bone trimming, cooking loss, and plate waste.[4]

Table 1–3 lists the conversion factors to change carcass weight to ingested weight. In addition to the consideration of bone and fat removal, cooking losses have also been estimated. The following cooked yields are used: fresh beef, 69.0 percent; ground beef, 70.25 percent; lamb and veal, 70.3 percent; and fresh pork, 71.6 percent. No consideration is given to cooking yields of processed meats since processed meats are precooked before sale and consumption and thus yield 100 percent. Seventy-five percent of the soft tissue of pork carcasses (muscle and fat) goes to processed meats, as does 15 percent of the lamb and mutton and 12 percent of the beef. Very little veal goes into processed meats. More and more poultry and increasing amounts of fish are being used in processed meats. *Surimi,* which is minced fish (usually Alaska pollack) that has been washed to remove fat, blood, pigments, and odorous substances, is becoming more popular and widely used. More than 50 percent of red meat ingestion involves processed meat (19.77 percent ground beef and 38.61 percent other), which is an indication that the U.S. "on-the-go" population values convenience highly.

Bergen and Merkel's estimate of per capita muscle food ingestion in the United States, which includes chicken and fish, is found in Table 1–4.[5] Note that these researchers have included variety meats but have not included processed meats. Be aware how food consumption data are presented and exactly what food products data are based on. Table 1–5 shows guidelines to convert live weight to carcass weight to retail weight to ingested weight.

---

[4]Food and Nutrition News. 1984. Vol. 56:3. May/June.

[5]Bergen, W. G., and R. A. Merkel. 1991. Body composition of animals treated with partitioning agents: implications for human health. FASEB J. 5:2951.

Table 1–3. U.S. Red Meat Ingestion per Day in 1999[1,2]

	Factor for Conversion from Carcass Weight	Oz.	Grams	Percent of Total Red Meat
Beef cuts	0.2778	0.97	27.6	36.0
Ground beef	0.1823	0.64	18.1	23.6
Total fresh beef		1.61	45.7	59.6
Fresh pork	0.1300	0.32	9.0	11.7
Lamb	0.3059	0.02	0.6	0.8
Veal	0.4524	0.02	0.6	0.8
Processed meats[3]	0.2017	0.73	20.7	27.0
Total red meat		2.70	76.6	99.9

[1]Consumption forecast by USDA. 1999. Economic Research Service.

[2]Conversions developed by Dr. Burdette C. Breidenstein, former Senior Vice President for Scientific Affairs, National Live Stock and Meat Board, Chicago.

[3]Composed of 79% pork and 21% beef; conversion factor used on total red meat carcass weight.

Table 1–4. Calculated per Capita Daily Ingestion of Meat, Poultry, and Fish Products[1,2]

Source	Ingested Portion[2]				Per Capita Daily Ingestion of		
	Capita/ Day	Fat[3]	Pro-tein[3]	Calories from Fat	Fat	Pro-tein	Choles-terol
	(g)	(%)	(%)	(%)	(g)	(g)	(mg)
Beef[4]	45.7	10.2	30.4	41.8	4.66	13.89	39.3
Pork[4]	25.4	9.7	29.5	43.2	2.46	7.49	21.8
Veal[4]	0.6	6.6	31.9	31.1	0.04	0.19	0.71
Lamb/mutton[4]	0.6	9.5	28.2	42.7	0.06	0.17	0.55
Variety meats[5]	9.1	11.1	18.6	49.3	1.01	1.67	3.6

(Continued)

Table 1–4   (Continued)

Source	Ingested Portion[2]				Per Capita Daily Ingestion of		
	Capita/ Day	Fat[3]	Pro-tein[3]	Calories from Fat	Fat	Pro-tein	Choles-terol
Poultry					0.29		
Broilers	51.1	7.4	28.9	36.0	3.78	14.77	48.0
Turkeys	12.6	7.4	28.9	36.0	0.93	3.64	9.6
Fish[6]	10.0	2.9	23.0	22.8		2.30	8.1
Total	155.1				13.23	44.12	128.4

[1]Bergen, W. G., and R. A. Merkel. 1991. Body composition of animals treated with partitioning agents: implications for human health. FASEB J. 5:2951. Adjusted to 1999 consumption forecast by USDA Economic Research Service for meat and poultry; USDA Food Surveys Research Group for Fish in 1994–1996.

[2]Ingested portion = raw product adjusted for raw edible portion minus cooking losses and plate wastes. Does not include eggs or milk products.

[3]Fat and protein content based on separable (cooked) lean because consumers will generally remove visible fat before ingestion.

[4]Retail composite.

[5]Variety meats; composite of one-third each of pork, beef, and poultry products.

[6]Composite of 50% whitefish, 25% tuna, 12.5% shrimp, and 12.5% lobsters/clams.

Table 1–5.   Conversion Guidelines

Live Weight (%)	Carcass Weight (% of Live)	Retail Weight (% of Carcass)	Ingested Weight	
			(% of Carcass)	(% of Retail)
Beef				
100 ≅	60 (page 177)[1]	64 (pages 587–589)	41	56
Pork				
100 ≅	70 (page 136)	83 (pages 533–540)	13	14 (fresh)
			53	64 (cured)
			66	78 (total pork)
Lamb				
100 ≅	50 (page 200)	76 (pages 625–626)	36	41

(Continued)

Table 1–5 (Continued)

Live Weight (%)	Carcass Weight (% of Live)	Retail Weight (% of Carcass)	Ingested Weight	
			(% of Carcass)	(% of Retail)
Veal				
100 ≅	60 (page 216)	82	47	57
Broilers				
100 ≅	74 (page 226)	65 (page 243)	46	70
Turkeys				
100 ≅	82 (page 226)	71 (page 243)	56	79
Fish				
100 ≅	74	66	85	100

[1]Reference page in *The Meat We Eat,* 14th ed.

We have discussed trends in U.S. food consumption. However, since these are national averages, they do not reflect changes in the composition of the population, such as the number of babies born and the number of people retired. Demography, the statistical study of the size, density, distribution, vital statistics (birth, death, marriage, divorce, health, and disease), age, and occupation of human populations, is important in interpreting food consumption data. For instance, the United States has a maturing population with changing life styles. Higher percentages of singles and dual-earner couples make up our population. This creates a stronger upscale market. More and stronger ethnic markets are developing as well.

A well-balanced agriculture provides its people with numerous types of food and consequently a better balanced diet. This type of agriculture also results in a more varied and stable source of income. By reviewing the data presented in this chapter, you can see how fortunate we are in the United States to have this vast array of wholesome food from which to build a balanced, healthful diet. U.S. consumers continue to spent only approximately 10.7 percent of their total disposable income on food (6.6 percent at home, 4.1 percent away from home). Of the food dollar, a little over $.51 is spent for food produced by animals, and of that, 69 percent, or $.35, is spent for meat. In no other country are people able to provide themselves with a balanced, nutritious diet this economically!

Meat is a staple that is part of a well-balanced diet for persons of all ages.

2

Meat Biotechnology and Microbiology

Microbiology and biotechnology are related to each other and to meat. Many important nutrients and growth factors make meat an ideal medium for microbiological growth. Knowing the characteristics of the specific organisms responsible for food poisoning and understanding how to control these organisms are of utmost importance in keeping meat wholesome. A meat plant must maintain the safety and wholesomeness of its meat products.

BIOTECHNOLOGY

A basic understanding of the principles of microbiology is necessary in order to conduct a meat processing operation that will assure meat safety.

Food microbiologists have taken the lead in introducing food biotechnology. Microorganisms are very useful as experimental systems in biotechnology because of their simplicity, which is desirable for genetic transformation, and their rapid growth rate. Under the proper conditions, some bacteria can double their numbers every 20 minutes. They generally grow well on a variety of substrates and produce a variety of products. The benefits that stem from controlling microorganisms and allowing them to contribute positively to meat and food product quality have long been known. Two of the most important are the formation of acid in summer sausage (fermentation), which gives summer sausage its characteristic tang, and the development of the "nutty" flavor of aged meat caused by other organisms. Also many dairy products, breads, alcohols, and pharmaceuticals are produced through the actions of friendly microorganisms. Beneficial bacteria are also found in the intestine, where they aid in digesting food.

Now through techniques used in food biotechnology, the genetic makeup of microbes may be altered so that they produce new products that improve meat aroma, flavor, and color. Enzymes for tenderizing meat that have unique properties, such as immobility and thermostability, are being developed. Through new concentrated starter cultures, fermentation can be intensified and more closely controlled. Newly developed microbes can detoxify food toxicants and reduce the use of stabilizers, such as antioxidants. These uses will enhance product quality and at the same time reduce the need for additives. Monoclonal antibodies[1] now being produced can detect other harmful and(or) beneficial organisms in meat.

In a broader sense, microorganisms can be used to improve the environment, which ultimately leads to safer meat products. Naturally occurring bacteria, known to have an appetite for hydrocarbons, were used to help clean up the 1989 Alaska oil spill.[2] Specific microorganisms that can detoxify potentially harmful industrial effluent, including packing house waste, are being grown. Biotechnology techniques can produce microbes for safe biodegradable pesticides and herbicides. Easily controlled antibiotics and economically produced vitamins may stem from growing genetically improved microorganisms for the livestock feeding industry. These are examples of how bacteria contribute to our general well-being. The application of these developments to meat processing will be covered in later chapters.

Students, consumers, and the population in general are exposed almost daily to the term *biotechnology*. In an article entitled "Biotechnology: A Boon to Food Safety," Frank E. Young, U.S. Commissioner of Food and Drugs,[3] defined *biotechnology* as "an umbrella term for the use of living organisms (or their by-products) to make or modify products, to improve plants or animals, or to develop microorganisms for specific uses." Young explains that "[b]iotechnology originated in practice over 8,000 years ago, when people learned to make beer with yeast. Farmers applied it through the centuries, as through selective breeding, they raised better strains of animals and plants. A basic breakthrough (though not a fundamental change in the essence of biotechnology) came in 1973 when scientists learned how to directly manipulate

---

[1]An antibody is a protein that reacts with a number of substances called antigens to cause a characteristic reaction that can be identified. A monoclonal antibody is one derived from a cell line.

[2]Crawford, M. 1989. Exxon bets on bugs in Alaska cleanup. Science. 247:704.

Crawford, M. 1990. Bacteria effective in Alaska cleanup. Science. 248:1537.

[3]An FDA Consumer Special Report. 1988. Safety first: protecting America's food supply. ISSN 00362–1332:50 (November).

the genetic material DNA.[4] Through cell fusion and genetic engineering, we can now induce improvements in food plants and animals rapidly with greater precision."

Meat scientists have been interested in the "cutting edge" of research for many years, and the discipline of meat science has grown out of both chemistry and nutrition. It is not unusual then that the leadership of The American Meat Science Association (see Appendix) invited presentations on biotechnology for several Reciprocal Meat Conferences during the 1980s.[5] These presentations provided the biotechnological groundwork for many meat scientists. Interest in biotechnology has been heightened by an active, gigantic federal project to map and sequence the human genome.[6] As a result, many scientists are writing about biotechnology and its application to agriculture and food.

Anyone studying the many aspects of animals and meat must keep as current as possible in related science and technology. Students can readily find the latest information on biotechnology — many conferences are being held on the subject, and many publications[7] and Web sites[8] are rapidly becoming available. The Biotechnology Information Center of the National Agriculture Library[9] is a good source for up-to-date information. Muscle foods are not being left out of the biotechnology thrust, but rather are at the center of it.

---

[4]DNA is deoxyribonucleic acid, a nucleic acid in which the sugar residues are deoxyribose. Macromolecules of DNA usually exist as double stranded polynucleotides held together by hydrogen bonds. DNA is the primary carrier of genetic information.

[5]Carlson, D. M. 1980. Recombinant DNA as a tool in animal research. Proc. Reciprocal Meat Conference. 33:107.

Wilson, G. A. 1982. Genetic engineering and its impact on the food industry. Proc. Reciprocal Meat Conference. 35:36.

Lewis, R. V. 1988. Biotechnology: what it can and cannot do. Proc. Reciprocal Meat Conference. 41:57.

Ivarie, R., J. Morris, and T. G. Clark. 1988. Gene regulation in muscle development and control of myogenic lineage by determination genes. Proc. Reciprocal Meat Conference. 41:61.

[6]McKusick, V. A. 1989. Mapping and sequencing the human genome. N. Engl. J. Med. 320:910.

[7]For example, A.S.M. 1998. Down on the animal pharm. Science. 282:2177.

Jaenisch, R. 1988. Transgenic animals. Science. 240:1468.

Knorr, D. 1987. Food Biotechnology. Marcell Dekker, Inc., New York.

Office of Public Liason. 1988. Agricultural biotechnology and the public, a report on four regional informational conferences. USDA.

Smith, L. M. 1989. DNA sequence analysis: past, present and future. Am. Biotech. Lab. Vol. 7 (No. 5):10.

Zimbelman, R. 1989. Biotechnology in the livestock animal: an overview of the advances in genetic engineering applied to food producing animals. Prof. An. Sci. 5:11.

[8]For example, the Illinois Biotech Resource Site (http://www.sdsnmr.com/ibrs/).

[9]Room 301, Beltsville, MD 20705, 301-344-3704 or 3218.

Support for this concept continues to build, as evidenced by this summary statement by Robert B. Shapiro, Chairman and CEO, The NutraSweet Company, "It's possible to envision a day when we can tailor molecules to give us the precise combinations of sensory, health and performance characteristics consumers want. We already have rudimentary capability to do this using molecular modeling and biotechnology. No one can predict how quickly we'll move into an age of designed foods, but the direction is clear; and I can tell you our company, together with others, large and small, [is] committed to using the most advanced scientific tools to create important new products."[10]

MICROBIOLOGY

Until a few years ago, muscle tissue in the living and growing animal was considered to be sterile, or nearly so, except for lymph nodes. However, researchers have found viable bacteria in the interior of muscle.[11] Even more so than in the past, maintaining a safe meat supply involves considering the general health of the live animal. In Chapter 3, animal loss factors, including parasites, will be discussed. When the process that converts living animal tissue into human food begins, chemical, physical, and biological degradation reactions also begin. If the knife used to sever the jugular vein and carotid arteries is not sterilized before use, bacteria may be introduced into all of the animal tissues, edible and non-edible, via the circulatory system. Sanitation is extremely important throughout the meat industry and will be covered thoroughly later in this chapter.

Meat microbiology is an integral part of the story of meat. Microbiology has an effect on all aspects of meat production, from raising the animal through its processing to providing food for the consumer. Only the most important points in meat microbiology are touched on here. You can consult excellent whole books and many Web sites on the subject for further study. A Report to the President on Food Safety,[12] an Institute of Food Technologists

---

[10]1989. Nutrition and health in America: the craze is not a phase. Food Marketing Institute Supermarket Industry Convention, Chicago (May 10).

[11]Johnson, J. L., M. P. Doyle, R. G. Cassens, and J. L. Schoeni. 1988. Fate of *Listeria monocytogenes* in tissues of experimentally infected cattle and in hard salami. Appl. Environ. Microbiol. 54:497.

Johnson, J. L., M. P. Doyle, and R. G. Cassens. 1990. Incidence of *Listeria ssp.* in retail meat roasts. J. Food Sci. 55:572.

[12]Food and Drug Administration, U.S. Department of Agriculture, U.S. Environmental Protection Agency, and Centers for Disease Control and Prevention. 1997. Food Safety from Farm to Table: A National Food Safety Initiative; A Report to the President.

Scientific Status Summary,[13] and a series of articles from *Science* magazine available on the World Wide Web[14] are all recommended for interested readers.

The prefix *micro* generally indicates that a microscope must be used to see the object, for it cannot be seen with the naked eye. Some people do not understand why they must wash their hands before eating or handling food because they cannot see microorganisms until a colony of them grows to a size that can be seen without a microscope. But long before this point, the microorganisms have done their damage. Consumed microorganisms (foodborne microorganisms) may precipitate health problems either by intoxication or by infection. Intoxication occurs when the microbe produces a toxin, which when ingested by the host (human), triggers sickness. There are several kinds of toxins produced, depending on the organism. An enterotoxin specifically affects the cells of the intestinal wall lining (mucosa), causing vomiting and diarrhea, the most common symptoms of food poisoning.

All toxins may be classified as either *exo*toxins or *endo*toxins (not to be confused with *entero*toxins). Exotoxins are found outside the bacterial cell, free in the medium. They are composed of proteins and can be destroyed by heat. Exotoxins are among the most poisonous substances known to humans. An example is the *Clostridium botulinum* toxin. Endotoxins are associated with the outer membranes of cells but are not released unless the cells are disrupted. Endotoxins are less specific, toxic, and potent. They are complex lipopolysaccharide (fat and carbohydrate) molecules that are not destroyed by heat. An example is the *Staphylococcus aureus* toxin.

Infection occurs when an organism is ingested by the host, then grows and by its presence disrupts the normal functioning of the system in which it is growing. Examples are *Salmonella* and *Listeria monocytogenes.*

There are several types of microorganisms that can grow on meat. Bacteria are clearly the most predominant and important. Bacteria are everywhere, on animals and people and in plants. A teaspoon of soil contains about 2 billion bacteria. The human body may carry up to 150 kinds of bacteria that total in number about 100,000 billion. Bacteria outnumber the body cells nearly 10 to 1. A laboratory test tube can hold 10 billion bacteria cells, twice the world population (see Table 1–1). (See Lederberg's essay on our relationship with

---

[13]Marth, E. H., and the Institute of Food Technologists Expert Panel on Food Safety and Nutrition. 1998. Extended shelf life refrigerated foods: microbiological quality and safety. Food Tech. 52:57.

[14]http://www.sciencemag.org

the microbes that share our world,[15] found in *Current Contents.*) Bacteria can be seen with the aid of a light microscope, and each averages approximately 1 micron (1 millionth of a meter) in length.

Molds and yeasts are fungi of some minor importance in meat. Molds are more apt to cause spoilage or to pose a health hazard in grains, cereals, flour, and nuts that have low water activity and in fruits that have a low pH. Yeasts may be involved in the spoilage of food products that contain significant amounts of sugar, such as fruits. Since meat has only approximately 1 percent sugar or carbohydrate, yeasts are generally not a problem. However, yeasts are very important in the baking industry and in the production of alcoholic beverages.

Viruses have the potential to be a leading cause of foodborne disease.[16] Other than raw and undercooked shellfish, meat is usually not contaminated by viruses. Problem foods that have been contaminated are ice, frostings, and salads. Viruses are inert and thus are unable to multiply outside a host cell. The particles of foodborne viruses are imperfectly spherical, with each particle ranging in diameter from 27 to 70 nanometers (nm = 1 billionth of a meter). These particles are 14 to 37 times smaller than bacteria and usually contain only ribonucleic acid (RNA) and protein. Because these particles cannot be seen with a light microscope, an electron microscope must be used (see Chapter 22).

A few parasites pose potential problems in meat. Parasite infection occurs in the live animal before it occurs in the human. (Parasites are discussed in Chapter 3.) *Trichinella spiralis,* a parasite long ago identified as capable of living in swine muscle, has nearly completely disappeared in the United States. *Toxoplasma gondii* is a small, intracellular protozoan that causes toxoplasmosis and occurs throughout the world. Infection has been observed in a wide range of birds and mammals. *Anisakis marina* is a roundworm parasite found only in fish.

Meat is an ideal culture medium for microbes; that is, they like it and thrive on it because meat is:

- High in moisture.

- Rich in nitrogenous foods of various degrees of complexity.

---

[15]Lederberg, J. 1989. Medical science, infectious disease, and the unity of humankind. Current Contents. 20 (No. 21):3.

[16]Cliver, D. O. 1988. Virus transmission via foods. Food Tech. Vol. 42 (No. 10):241.

- Plentifully supplied with minerals and accessory growth factors.

- In most instances, adequate in fermentable carbohydrates.

- At a fairly favorable pH (~5.6).

Factors Affecting Microorganism Growth on or in Meat

Temperature and Time

Some microbes grow well at temperatures of 32° to 68°F (0° to 20°C). They are known as *psychrophiles*. Examples are *Pseudomonas* bacteria and some yeasts and molds. These organisms are harmless and serve as a friendly warning that conditions are suitable for some microbial growth. The cliché that "Life begins at 40," meaning that meat is safe from harmful bacteria if it is kept under 40°F (4°C), was based on the observation that no harmful bacteria usually grew at temperatures under 40°F. Unfortunately, with more definitive techniques, scientists have recently found that the harmful bacteria *Listeria monocytogenes* and *Yersinia enterocolitica* can grow at temperatures as low as 34°F (2°C). Thus, the old familiar expression must be updated. In fact, the closer to 28°F (–2°C) meat is kept (its freezing point), during storage, transport, and display, the longer the shelf life, i.e., the maintenance of appearance, safety, and palatability.

Other microbes called *mesophiles* grow well at temperatures between 68° and 113°F (20° and 45°C). Most bacteria belong to this group. *Staphylococcus aureus* is an example.

A few microbes called *thermophiles* grow at higher temperatures between 113° and 150°F (45° and 65°C). An example is *Clostridium botulinum*.

The mismanagement of temperature is the most common reason for an outbreak of foodborne disease.[17] Closely related is time lapse (delay) at a critical temperature, which is the second most prevalent cause of an outbreak. Therefore, the combination temperature–time value is of utmost importance to assure meat safety. Cooling prevents most microorganisms from growing but does not kill them. Freezing for various lengths of time will kill some par-

---

[17]Bryan, F. L. 1988. Risks of practices, procedures and processes that lead to outbreaks of foodborne diseases. J. Food Prot. 51:663.

asites and will damage some bacteria; however, most merely become dormant and are revived when thawed.

A combined temperature–time value must be reached in order to kill microorganisms with heat. Attaining a given temperature is not as damaging to the microorganisms as holding a given temperature for some length of time. If you were to plot the number of surviving bacteria versus the time at a given temperature, you would observe a logarithmic (or exponential) rate of destruction. The time in minutes required at a given temperature to cause a decimal reduction (e.g., from 100 to 10 bacteria) is expressed by the letter "D." Thus, "D" represents a 90 percent reduction in the microbial population. In the food industry, heating time at a specific temperature is referred to as a 4D or 7D cook, etc., indicating the time required to kill 10,000 (4D) or 10,000,000 (7D) bacteria per gram. A subscript usually is attached to the "D," indicating the temperature of heating. For instance, the required temperature–time value for milk to be pasteurized is 161°F for 15 seconds (D_{161} = 0.25) or 143°F for 30 minutes (D_{143} = 30).

Meat can be kept safe when it is cold or hot, but not in between. Specifically, when meat is being cooked or cooled, it should pass through the temperature range from 40°F (4°C) to 140°F (60°C) or visa versa in four hours, or preferably less. Thus, temperature–time relationships must be monitored during each stage of the conversion process from live animals to ready-to-eat (RTE) meat products. Steam pasteurization (82.2°C for 6.5 seconds) effectively decreased the bacterial load on beef carcasses during harvest.[18]

At 140°F (60°C), most, but not all, microorganisms are killed. At this temperature, beef is rare (see the "Beef Steak Color Guide" in the color photo section). Rare beef has been served to millions of people for many years with no resultant or observed health problems. Consequently, it has been regarded as safe. As noted previously, the interior of muscle has long been considered to be sterile or nearly so. However, the outside of a piece of meat may have become contaminated through processing. When a piece of meat is cooked conventionally (not with a microwave oven), the outside gets hot first and normally reaches a higher final temperature than the interior. However, when the whole piece is thoroughly cooked through and through to minimize tem-

---

[18]Nutsch, A. L., R. K. Phebus, M. J. Reiman, J. S. Kotroloa, R. C. Wilson, J. E. Boyer, Jr., and T. L. Brown. 1998. Steam pasteurization of commercially slaughtered beef carcasses: evaluation of bacterial populations at five anatomical locations. J. Food Prot. 61:571.

perature gradations, steaks and chops are less palatable. Any organisms that might be present on the outside of a piece of meat are killed by the higher temperature of the outside of the cut, even though they might have survived 140°F (60°C). Nevertheless, because some school children became ill after eating incompletely cooked hamburgers, the U.S. Department of Agriculture[19] now recommends cooking meat to 160°F, which is medium (see the color photo section), to be completely safe. However, because of a more alkaline final pH, poultry should be cooked to 180°F, and if red meat is to be reheated, it should reach 165°F for maximum safety. Ground meats, as compared to intact meat cuts, are more apt to become contaminated because of increased surface area exposure and more processing steps by humans. Detailed discussions regarding temperature tolerances of individual microorganisms as well as temperature considerations when you are cooking and preserving meat will follow in this and later chapters.

Moisture

The water requirement for microbial growth is defined in terms of water activity (a_w) of the medium (solution that is meat or other food): a_w = vapor pressure of solution ÷ vapor pressure of pure solvent (H_2O).

Equilibrium relative humidity (ERH) is related to a_w as follows: ERH = $a_w \times 100$.

The procedures to calculate a_w are complex. In general, the dew point (which is equal to the ERH) above the sample in a closed container is determined.[20]

Most spoilage organisms require an a_w of 0.90 or greater for growth (see Table 19–3). Since fresh meat has an a_w of 0.99 or higher, it provides an ideal environment for growth. Some spoilage organisms can grow at a_w levels below 0.90. Those that may pose a problem in meat are noted in the discussions that follow and are listed in Table 19–3.

Moisture level can more easily be expressed as the percent of a substance that is water, and when expressed that way, most bacteria require at least 18 percent water in the medium to grow. Molds grow in a medium (tissue) that can be as low as 13 percent water. However, a_w and water content are not identical. Water activity describes the relative availability of water for many

---

[19]USDA, Food Safety and Inspection Service, Meat and Poultry Hot Line, 1-800-535-4555.

[20]Troller, J. 1983. Methods to measure water activity. J. Food Prot. 46:129.

actions and reactions in a food and may bear little or no relationship to the total amount of water present in the food (see footnote 20). Meat preservation by drying is covered in Chapter 19, and the effect of salt in lowering a_w is shown in Table 19–4.

Oxygen

Microbes that need oxygen present to grow are referred to as *aerobic*. These include yeasts, molds, and many bacteria. Those that cannot grow when oxygen is present are called *anaerobic*. Examples are *Clostridium* (which produces a deadly toxin) and the putrifiers (those that degrade proteins and form very strong-smelling gases). Anaerobic organisms are of particular concern to the producers of the recently developed precooked vacuum-packaged food products (see Chapter 19) and to the canning industry.

A third group of microorganisms is called *facultative,* indicating that although the organism may grow best under either aerobic or anaerobic conditions, growth can occur in both the presence and the absence of oxygen.

Oxidation–Reduction Potential

Oxidation and reduction processes are electron migrations between chemical compounds. Oxidation is the loss of electrons; reduction is the gain of electrons. When one substance gives up electrons, it is oxidized and the surrounding substance picks up the electrons and is thus reduced. Oxidation-reduction potential is referred to as *redox* and is indicated by the symbol E_h. In meat, which is nonhomogeneous, E_h is always changing. A high E_h occurs in aerobic conditions and may range up to $+300$ millivolts(mV), while low E_h occurs in anaerobic conditions and may range down to -420 mV. A high E_h will prevent anaerobes from growing, while a low E_h will prevent aerobes from growing. Thus, allowing or excluding oxygen through packaging will control E_h. The E_h of the interior of a piece of meat is different from the E_h of the exterior.

Degree of Acidity or Alkalinity (pH)

Most microorganisms thrive at a pH near neutrality (pH of 7.0). Most meat products fall in the pH range of 4.8 to 6.8. In general, microorganisms grow slower and less efficiently at a pH of 5.0 or lower. Because of this relationship, the acidity (low pH) that develops in summer sausage serves as a preservative as well as a flavor enhancer. Also, a dilute (1 to 2 percent) solution of an organic acid, such as lactic acid (a three-carbon acid produced naturally in muscle and by useful bacteria), may be sprayed onto the surface of carcasses soon after harvest to control microbial growth.[21]

Physical Properties of Meat

If the meat is in a form that provides a large surface area exposed to oxygen (air), the microbes have more room to grow. For example, a whole carcass has the minimum amount of surface area exposed. When it is cut into wholesale cuts, more area is exposed; when it is cut into retail cuts (steaks, etc.) still more area is exposed. When meat is ground, the maximum total surface area is exposed. Thus, the further meat is processed, the more vulnerable it is to microbial action.

Meat processors can control the factors that affect microbial growth and thus give meat a longer shelf life, meaning that the meat maintains wholesomeness and palatability longer. For example, a beef carcass could hang in a 33° to 35°F cooler for perhaps three weeks to age (not a standard practice), and depending on the relative humidity of the cooler, perhaps only a few molds and some harmless indicator bacteria, such as *Pseudomonas,* would grow on limited areas of the carcass. But if the carcass were to be cut into its cuts, or it were to be ground, or the temperature and(or) relative humidity of the cooler were to be raised, the above factors would be affected such that microbiological growth could flourish, and the product could become unwholesome.

The use of vacuum-packaging allows carcasses to be cut into primal, subprimal, or even retail cuts, after which each cut is placed in an airtight "envelope." In such a wrap, sanitation is maintained and contamination is prevented. Cuts may be held at the proper temperatures for extended time periods during marketing without the occurrence of damaging microbial growth.

---

[21]Prasai, R. K., G. R. Acuff, L. M. Lucia, D. S. Hale, J. W. Savell, and J. B. Morgan. 1991. Microbiological effects of acid decontamination of beef carcasses at various locations in processing. J. Food Prot. 54:868.

Competing Microorganisms

At any one time, all or any combination of the factors described previously in this chapter may exist. Organisms have different tolerances and optimal growing conditions. The presence and growth of helpful bacteria (lactic acid producers) retards the growth of some harmful organisms. All organisms that are present may not be able to grow at the same time. Those organisms that cannot grow under the conditions existing at any one time will not necessarily die, but most often they will remain dormant. If growing organisms are destroyed by changing any one or more of the conditions described previously [e.g., cooking to 140°F (60°C)], growth opportunities may be created for the dormant organisms. Such is the case when a cooked roast is left at room temperature for too long a time period (see the discussion on *Staphylococcus aureus*).

Organisms of Concern in Meat

The identification of causative microorganisms has been the basis for controlling foodborne disease for many years. However, recent studies have indicated that the genes carried by the microbe determine whether it will be harmless or potentially harmful.[22] Several organisms long thought to be harmless, now because of their genetic makeup, have been found to cause human disease. However, assigning the cause for foodborne illness to a particular organism should be done with caution. "In the vast majority of cases, **people** are responsible for foodborne disease outbreaks, either by abusing food or transmitting their disease to others via food. Basic rules of food preparation and hygienic practice are broken and disease results."[23] The Scientific Status Summary by the Institute of Food Technologists' Expert Panel on Food Safety and Nutrition provides a concise yet very complete summary of important bacteria associated with foodborne diseases.[24] An excellent and complete classification and summary of diseases transmitted by foods has also been compiled by Bryan.[25]

---

[22]Archer, D L. 1988. The true impact of foodborne infections. Food Tech. Vol. 42 (No. 7):53.

[23]Ibid.

[24]Institute of Food Technologists. 1988. Bacteria associated with foodborne diseases. Food Technol. Vol. 42 (No. 4):119.

[25]Bryan, F. L. 1982. Diseases transmitted by foods. HHS Publication No. (CDC) 81–8237.

Anyone who is interested in how widespread various diseases are, including those that are foodborne and described in this chapter, should consult the latest U.S. Department of Health and Human Services' Centers for Disease Control (CDC) Surveillance Summaries.[26] Those microorganisms that have the potential to cause disease and are known to grow on meat and meat products are discussed in detail in the following section.

Salmonella

Salmonella infection is caused by any one of a large number of species of that genus. Recall that a foodborne infection results from ingesting a high number of potentially infectious microorganisms or by allowing their growth and multiplication in the digestive tract. The bacteria must be living to cause infection. Thus, the onset of symptoms from a food infection will take longer to develop than symptoms from a food intoxication.

The salmonellae are gram-negative, nonspore-forming rods that ferment glucose, but not lactose or sucrose, usually with gas production. They grow best in nonacid foods at an optimum temperature of about 98.5°F but will grow well at room temperatures. Salmonellae are spread through the air, by direct contact with contaminated containers, and by handling during the manufacture and transportation of foods and food ingredients. It is considered nearly impossible to eliminate salmonellae altogether except in a sterile environment. Humans are contaminated by salmonellae through associations with animals and animal products. After ingestion, the incubation period will normally be 3 to 36 hours before digestive upsets occur. Symptoms may last from one to seven days.

Salmonellae are killed by pasteurization equivalent to that given milk (161°F for 15 seconds or 143°F for 30 minutes). The key control measures entail avoiding cross-contamination from raw meat to cooked food, or food which will be eaten raw, and maintaining high standards of food hygiene. Meat that is cooked to pasteurization temperatures is very palatable (see Chapter 24).

---

[26]U.S. Dept. of Health and Human Services, Public Health Service, Centers for Disease Control, Atlanta.

Escherichia coli

Virulent strains of *Escherichia coli* have been reported to be the causative agent in some food-poisoning outbreaks. *E. coli* is one of the most widespread types of microorganisms in nature; however, few virulent strains exist. One such strain is O157:H7, which is a very significant foodborne pathogen because of the severe consequences of infection that affects all age groups, its low infection dose, and its unusual acid tolerance[27]. This organism causes hemorrhagic colitis, a severe abdominal illness, with watery, bloody diarrhea and vomiting. The enterotoxin affects kidney function and the central nervous system. This can result in hemolytic uremic syndrome (HUS) and kidney failure, especially in young children. HUS can progress to thrombtic thrombocytopenia (TTP), which can result in blood clots in the brain, seizures, coma, and death.[28] Contamination of *E. coli* in meat usually occurs via improper harvest procedures and unsanitary handling of the cooked product or improper handling and cooking of the raw product.

The organism is facultative but is destroyed by cooking until the internal temperature of the meat product is at least 160°F.[29] Current safety recommendations include cooking hamburgers thoroughly and including a heating step that validates a 5D reduction of *E. coli* 0157:H7 in fermented sausage, such as Lebanon balogna and pepperoni.[30]

Campylobacter jejuni

C. jejuni is a gram-negative, motile, comma- to "S"-shaped organism that forms spirals. More foodborne infections are reported to be caused by this microorganism than any other. It has been estimated that because of its high body temperature and high pH, a large percentage of raw chicken is contaminated with *Campylobacter*.[31] *C. jejuni* will not grow at temperatures below

[27]Buchana, R. L., M. P. Doyle and the Institute of Food Technologists Expert Panel on Food Safety and Nutrition. 1997. Foodborne disease significance of *Escherichia coli* 0157:H7 and other enterohemorrhagic *E. coli*. J. Food Tech. 51:69.

[28]Brewer, M. S., and F. K. McKeith. 1999. HACCP Plan Development — Short Course Manual. The Board of Trustees, University of Illinois, Urbana, IL 61801.

[29]Kameswar, R. E., S. Doores, E. W. Mills, R. A. Wilson, R. C. Anantheswaran, and S. J. Knabel. 1998. Destruction of *Escherichia coli* 0157:H7 and *Salmonella typhimurium* in Lebanon bologna by interaction of fermentation pH, heating temperature and time. J. Food Prot. 61:162

[30]Ibid.

[31]Altekruse, S. F., N. J. Stern, P. I. Fields, and D. L. Swerdlow. 1999. *Campylobacter jejuni* — an emerging foodborne pathogen. Emerging Infectious Diseases. 5:28.

86°F, yet is destroyed by pasteurization. It grows slowly even under optimal conditions that include a low concentration of oxygen (5 to 10 percent). It does not compete well with other microflora and is sensitive to drying, storage at room temperature, acidic conditions, and disinfectants. If the organism nevertheless survives and is ingested by humans, the onset of symptoms usually occurs two to five days later. A very few microorganisms can elicit comparatively severe symptoms of profuse diarrhea (sometimes bloody), abdominal cramps, and nausea. Symptoms usually last only two to three days, but they can persist for weeks or months with complications. Control measures are the same as for *Salmonella,* namely good hygiene and avoidance of cross-contamination between raw or partially cooked food and ready-to-eat food.

Listeria monocytogenes

L. monocytogenes is a gram-positive, motile rod that causes the foodborne infection known as listeriosis. *Listeria* is everywhere — in soil, vegetation, and water. It is relatively resistant to drying, especially at lower temperatures. It survives and grows in sewage sludge because of its favorable E_h. Vegetable growers should not use manure or sewage sludge to fertilize vegetables that will be eaten raw because listeriosis has been traced to coleslaw. *Listeria* grows and multiplies at temperatures of 32°F (0°C) to 113°F (45°C) and survives freezing. It is salt-tolerant, surviving concentrations as high as 30.5 percent for 100 days at 39.2°F (4°C). Its survival in 30 percent salt can be shortened to five days when the temperature is raised to 98.6°F (37°C). It has a minimum water activity for growth of 0.93 to 0.92. It can grow in a pH range of 5.0 to 9.5. The presence of *Listeria* microorganisms in raw milk and in products made from raw milk is common, but these microorganisms are rarely found in pasteurized milk and milk products. *Listeria* has been found in raw poultry and red meat. It grows in damp environments found in meat plants.

Healthy people have very little risk from listeriosis. The carrier state, which many people might harbor, may be without symptoms, or so mild (diarrhea and mild fever) as to go unnoticed. If the disease proceeds, septicemia (blood poisoning) normally ensues. Most susceptible are very young infants, chronically ill people, elderly individuals, and pregnant women. The central nervous system is affected, with meningitis, encephalitis, or abscesses as the consequence. The disease also affects the heart and the eyes. The fetus of a pregnant woman is especially vulnerable. Listeriosis has been detected in newborns within 24 hours.

Control of *Listeria* in muscle foods is possible through good sanitation practices throughout processing. Strict elimination of cross-contamination between raw material, i.e., tissues that are subject to *Listeria* contamination (external and digestive), and internal tissues is essential. *Listeria* is about four times more heat-tolerant than *Salmonella*[32] and will survive milk pasteurization temperatures and time (see footnote 24). The D_{145} for *Salmonella* is 0.65 minutes and for *Listeria* 2.56 minutes (see footnote 32). Avoidance of post-cooking (heating) contamination is a very important control measure, because *Listeria* is a food safety hazard resonably likely to occur in ready-to-eat (RTE) products.

Staphylococcus aureus

S. aureus is gram-positive and occurs in masses like clusters of grapes or in pairs and short chains. It causes staphylococcal intoxication (ingestion of a food that contains a microbial toxin) by producing a very heat-stable toxin capable of precipitating severe gastrointestinal upsets. *S. aureus* is a facultative anaerobe, which grows best under aerobic conditions, but it can also grow in reduced concentrations of oxygen. It tolerates and grows in salt concentrations approaching saturation (10 to 20 percent), but it also tolerates nitrites fairly well. Thus, *S. aureus* can grow and multiply in cured meats, if other conditions are favorable. The cocci also can grow in lower pH solutions, below 5.0, more so under aerobic conditions than under anaerobic conditions. Growth may occur at a_w as low as 0.86 under aerobic conditions or a_w 0.90 under anaerobic conditions. The microorganism itself is somewhat heat-resistant, preferring temperatures around 95° to 98.6°F but with the ability to grow at temperatures as high as 118°F and as low as 44°F. The enterotoxin, however, is very heat-resistant.

More than half of all healthy adults in the United States will find *S. aureus* in their noses, throats, hair, and skin. Thus, any food that requires handling by humans is likely to become contaminated. The onset of illness resulting from this staphylococcal food poisoning occurs much more quickly than from any of the previously discussed infections because the active toxin is directly ingested. Symptoms of nausea, vomiting, retching, abdominal cramping,

---

[32]Wilson, G. D. 1988. *Listeria monocytogenes* — 1988. Proc. Reciprocal Meat Conference. 41:11.

sweating, chills, prostration, and subnormal body temperature may occur as early as 30 minutes or as late as 8 hours following the consumption of a product containing this organism. Most illnesses occur within 4 to 8 hours and last 24 to 48 hours.

This organism most often can cause a problem after a roast, either cured and smoked or fresh, properly cooked and served, remains at room temperature for a period of time before it is refrigerated. *S. aureus* if present, has a good opportunity to grow. Also, the previous cooking may have destroyed any competitive "friendly" microorganisms, thus giving *S. aureus,* because of its temperature tolerance, a chance to grow. Even if the leftover roast is thoroughly re-cooked before it is eaten, toxins formed during the period at room temperature will generally not be destroyed. Temperature control, i.e., adequate heat processing and cooking and proper cooling and refrigeration, will do much to alleviate the problem of staphylococcal food poisoning.

Clostridium perfringens

Bacteria from the family Bacillaceae, genus *Clostridium,* are spore formers. Spores are different from the vegetative cells in that they are more resistant to heat, radiation, chemicals, and other destructive action. Spores are the inactive or dormant state of the organisms. *C. perfringens* is a gram-positive, spore-forming, nonmotile, encapsulated short rod.

Food poisoning from *C. perfringens* is caused by the release of an enterotoxin during sporulation. Based upon their ability to produce various exotoxins, six types of *C. perfringens* are recognized: types A, B, C, D, E, and F. The food-poisoning strains belong mainly to the type A group.

C. perfringens is present in the soil and in the intestinal contents of almost all animals and humans. It is anaerobic and can withstand the adverse environmental conditions of drying and heating, as well as certain toxic compounds. This organism finds its way into meats either directly from the harvested animals or from subsequent contamination from containers, handling, or dust. The need to cook large amounts of food ahead of time, which necessitates heating, cooling, and reheating, makes food service establishments vulnerable sites for problems with this organism. The lowest reported temperature for growth is 68°F, while sporulation requires higher temperatures (98° to 99°F). Cooked meat should be kept above 140°F or below 40°F to inhibit growth of *C. perfringens.* Rapid, uniform cooling, especially of large pieces of cooked meat, to 50°F within two to three hours will prevent *C. perfringens* growth and sporulation. Cured meat products are rarely, if ever, responsible for outbreaks.

Sporulation (and subsequent toxin production) occurs in the gastrointestinal tract after the ingestion of large numbers (10^6 or more) of *C. perfringens* vegetative cells. Sporulation and toxin production may also occur in foods. Symptoms of acute abdominal pain and diarrhea develop 8 to 24 hours after ingestion and last 12 to 24 hours, but the food poisoning is rarely fatal. Knowledgeable regulation of temperatures is the most important factor in controlling food poisoning caused by *C. perfringens*.

Clostridium botulinum

Botulism, a food-borne intoxication is caused by various strains of *C. botulinum* that produce type A, B, E, and F toxins affecting the nervous system of humans. The organism is identified as a gram-positive, anaerobic, spore-forming rod, able to grow under relatively thermophilic conditions. Type A toxin is more deadly than the others and is destroyed by heating for 10 minutes at 176°F.

Clostridia are fairly widely distributed in soil, in marine and lake sediments, and in feces and carcasses of birds and animals. Conditions that favor growth and toxin production are relatively high moisture, low salt, low acid (pH greater than 4.6), no oxygen, and improper storage temperatures (temperatures above 38°F). Canned vegetables, meat, and fish are most often incriminated.

Sometime between 12 and 72 hours after the contaminated food has been consumed, the symptoms begin to be noticeable — e.g., nausea; vomiting; fatigue; dizziness; headache; dryness of skin, mouth, and throat; constipation; muscular weakness or paralysis; double vision; and difficulty in breathing. The toxin binds to the nerve endings, causing loss of those functions dependent on nerve action. The illness normally lasts 1 to 10 days, but approximately 10 percent of the cases are fatal. The nitrite component of a curing mixture has been shown to control the growth and toxin production of *C. botulinum* in cured meats (see Chapter 20).

Other Less Common Sources of Foodborne Infections and Food Poisoning

Numerous other microorganisms may be involved in food-poisoning outbreaks but are present in meat much less frequently than those bacteria discussed above in detail. *Yersinia enterocolitica* is an emerging food pathogen that grows at refrigeration temperatures, but it is not prevalent in meat.

Streptococcal food-poisoning outbreaks have been associated with cheese, turkey, turkey dressing, Vienna sausage, and barbecued beef. Specific organisms of concern are *Streptococcus faecalis, S. viridans,* and *S. pyogenes.*

Bacillus cereus causes a food-poisoning reaction similar to *C. perfringens;* however, the documented cases of food poisoning from *B. cereus* are quite low in the United States. Meat dishes that have been incriminated in outbreaks include meat loaf, minced meat, liver sausage, and soups.

Shigella bacteria are carried in the fecal matter of infected humans. No animal reservoir is known. The principal source of this pathogen is contamination of food by an infected food worker.

Vibrio parahaemolyticus food poisoning is contracted almost exclusively from seafoods. Symptoms consist of diarrhea, abdominal pain, nausea, vomiting, and chills. While its incidence is quite low in the United States, it is the leading cause of food poisoning in Japan.

Table 2–1 summarizes the characteristics of bacteria and the diseases they cause.

QUALITY CONTROL (QUALITY ASSURANCE)

Quality control (quality assurance) refers to a broad program that processors employ to ultimately assure consumers a safe food product. Quality control is emphasized throughout processing, but for it to work efficiently and to function for maximum benefit, Hazard Analysis Critical Control Points (HACCP) should be stressed. This means that certain points in the processing system are especially critical, and if these are not monitored closely, the whole process and all the products produced could be unacceptable. HACCP is the top priority of the USDA's Food Safety and Inspection Service (FSIS). The agency's goal is to make HACCP principles the foundation for meat and poultry inspection (see Chapter 3). Another term used in industry and government for this general procedure is Current Good Manufacturing Practices (CGMP). Quality control encompasses product composition, specifications, processing, packaging, storage, and distribution, as well as microbiological safety in relation to the plant's equipment, sanitation, and pest and rodent control. The use of computers and statistical process control is discussed in Chapter 18.

Table 2–1. Characteristics of Foodborne

Bacteria	Diarrhea	Vomit-ing	Nausea	Abdom-inal Cramps	Other Symptoms
Listeria monocytogenes	X				[1]
Clostridium perfringens	X			X (severe)	
Salmonella	X	X		X	
Escherichia coli serotype 0157:H7	X (bloody)		X	X (severe)	Urinary tract infection[2]
Campylobacter jejuni	X (profuse, sometimes containing blood)		X	X	Headache
Staphylococcus aureus		X	X	X	Sweating, chills, subnormal body temperature
Clostridium botulinum		X	X		Dizziness; headache; fatigue; dryness of skin, mouth, and throat; con-stipation; muscle weak-ness; paralysis; double-vision; breathing difficulty

[1]Normal, healthy people may unknowingly become carriers of *Listeria* with no further symptoms. Most vulnurable are elderly, weak, immunocompromised persons and unborn and newborn babies who may suffer central nervous system, heart, and eye damage because of advanced stages of listeriosis, resulting in meningitis, encephalitis, and other tissue-damaging diseases.

[2]Hemolytic uremic syndrome (HUS), a leading cause of acute kidney failure in children.

Diseases to Aid in Diagnosis

Time Until Symptoms Appear	Duration of Symptoms and Illness	How to Avoid
24 hours in newborn from infected mothers	Progresses suddenly in latter stages; 70% fatal	Sanitation; cook thoroughly.
8–24 hours	12–24 hours	Temperature control, adequate cooking, prompt and thorough cooling.
12–36 hours	1–7 days	Pasteurization (143°F for 30 minutes or 151°F for 15 minutes). Avoid cross-contamination between raw and ready-to-eat food.
3–4 days	2–9 days	Same as *Salmonella*
2–5 days	Usually 2–3 days, but can be weeks or months with complications	Same as *Salmonella*
As early as 30 minutes, normally 2–4 hours	24–48 hours	Same as *C. perfringens*
12–72 hours	1–10 days	Temperature control, use of nitrites. Be especially cautious with vacuum-packed foods.

Knowing that certain microorganisms thrive on the rich nutritional composition of meat should cause us to recognize that a quality control program with its associated rules and habits is very necessary and meaningful. Thus, a quality control program should be generated to improve the efficiency, service, and profitability of the meat operation, not merely to satisfy state or federal meat inspection regulations. Ultimately, of course, consumer satisfaction is the goal of all meat operations, and foremost among consumer concerns is the wholesomeness of the products that are eaten.

Management officials, supervisors, and all workers must be concerned about quality control, since it requires an all-out effort by all persons involved to produce wholesome, quality products.

Union Carbide Films Packaging Division has published a small brochure describing the need for quality control in plain, easy-to-understand terms: *keep it clean, cold, and covered to maintain product quality.*[33]

To keep meat *clean,* the producer must develop an effective sanitation program. The word *sanitation* comes from the Latin word *Sanitas,* meaning *health.* All equipment, buildings, people, wrapping materials, ingredients, and anything that comes in contact with the meat food must be kept clean. What looks clean may very well not be, since bacteria, yeasts, and molds (microorganisms) can be seen only through a microscope. It would take 25,000 bacteria laid end to end to make 1 inch.

A sanitation program must be well organized and very efficient. First, the contaminants must be identified in order to know what detergents will be needed to remove them, and what equipment should be used. The first step in equipment and floor cleaning is to physically remove the chunks and small particles of meat and offal with a dry squeegee, brush, or broom, and shovel. Using large amounts of water to move this material would be extremely wasteful and would eventually cause drains to clog and(or) septic tanks, cesspools, and various waste treatment facilities to become overloaded and fail prematurely. A basic (alkaline) cleanser should routinely be used to solubilize and release the remaining tiny meat residues. The water temperature should be 140°F or slightly higher to cut the fat residues. Periodically, once or twice a week, the basic detergent that is used on a daily basis should be followed immediately with an acidic cleanser to remove built-up mineral deposits.

---

[33]Union Carbide Films Packaging Division. 1974. The 3 C's of Plant Sanitation.

Repeated, constant use of a strong acid, however, will cause etching of metal equipment. After being cleaned, equipment and facilities should be sanitized with hot water (180°F) or an approved chlorine or iodine rinse. Finally, non-stainless steel metal equipment should be lightly sprayed with an edible mineral oil to prevent oxidation of the metal (rusting).

Workers must wash with approved germicidal soaps and warm water many times during the day, especially after breaks. At many plants, workers must change clothes completely when they go into the plants to work with meat. Workers in the ready-to-eat (RTE) and raw meat departments must not mingle or pass between departments without washing and changing clothes since cross-contamination from raw meat to RTE products is very possible. In large plants, workers wear color-coded clothing so that they can be readily spotted if they go into a department other that their own. Some RTE workers may wear masks like those worn by surgical doctors and nurses in hospital operating rooms.

Understandably, employees should not use tobacco, pick their noses, or scratch their heads while they are working with meat food. Furthermore, the water supply must be potable; i.e., it must be pure and non-contaminated. This basic requirement is discussed first in the FSIS regulations in Chapter 3.

Because proper sanitation is so important, just good common sense can go a long way toward establishing and maintaining satisfactory sanitation in any meat processing situation.

Cold was discussed adequately earlier in this chapter. It is one of the most important factors that affect microbial growth.

Excluding airborne microorganisms and vermin by using screens or other suitable barriers between processing areas is an important facet of keeping materials and equipment *covered*. Meats, carcasses, wholesale cuts, and primal cuts must be properly covered during transport. Retail cuts should be properly wrapped for display or for preservation by freezing. Obviously, workers should not cough, sneeze, or spit tobacco juice when they are handling meat food.

The factors that affect microbial growth, discussed earlier in this chapter, must be controlled in order for microbial growth to be restricted. Leistner's hurdle concept[34] consists of lining up the factors that control microbial

---

[34]Leistner, L. 1988. Shelf-stable edible meat. Proc. of Industry Day: part of the 34th International Congress of Meat Science and Technology, August 31, Livestock and Meat Authority of Queensland, Brisbane, QLD. 4000.

growth like a row of hurdles of various sizes, depending on the factor and its level. For microbes to grow, they must clear the whole line of hurdles.

The American Meat Institute has produced guidelines for quality control and is generally a good source for more detailed information in this area. Bacus discussed methods for microbial control in fresh and cured meats.[35] In addition to the factors emphasized in this chapter, Zaika described anti-microbial activity through the use of spices and herbs.[36] However, the addition of spices and herbs can also add microbial contamination.

Quality control, involving specifically the control of microbial growth for its positive benefit, is a most complex subject. This discussion gives the reader an exposure, as well as sources to seek for more detailed information.

---

[35]Bacus, J. 1988. Microbial control methods in fresh and processed meats. Proc. Reciprocal Meat Conference. 41:7.

[36]Zaika, L. A. 1988. Spices and herbs: their antimicrobial activity and its determination. J. Food Safety. 9:97.

3

Meat Inspection and Animal Loss Factors[1]

Government, educational institutions, and the livestock and meat industry are conscientiously working together to make the meat food items consumers purchase 100 percent wholesome and safe. It comes down to humans thinking and using common sense in order to maintain personal and facility sanitation practices. No meat purveyor wants to sell a product that is not completely safe to eat. Livestock producers must consider how their production system influences meat safety, meat processors must apply the HACCP system to monitor the many procedures involved in producing edible products, and consumers must take responsibility not to abuse meat products between the time they purchase them and the time they consume them.

The wholesomeness of U.S. meat and poultry is protected by the U.S. Department of Agriculture. Federal meat inspection dates back to June 30, 1906, with the passage of the Meat Inspection Act. Previous to that time, a limited form of federal inspection had been started in 1891, but this was only a voluntary inspection of cattle and hogs that were intended for export. Upton Sinclair's book *The Jungle*[2] has been credited with providing the impetus for the passage of the Meat Inspection Act of 1906, which has continually been evaluated and improved over the years. One major addition was the passage of the Wholesome Meat Act on December 15, 1967. The major thrust of this 1967 law, which still ties to the original 1906 act, was to make all of the

[1]The completeness and accuracy of this chapter would not have been possible without the generous assistance of Daniel L. Engeljohn, Ph.D., Director, Regulation Development and Analysis Division, Office of Policy, Program Development, and Evaluation, Food Safety and Inspection Service, U.S. Department of Agriculture. (202) 720–5627.

[2]Sinclair, U. 1906. The Jungle. Nal Penguin, Inc., New York.

various state inspection systems at least equal to the federal inspection system. Previous to this time, federal meat inspection applied only to meat and meat products in interstate or foreign commerce. Many states had their own inspection systems for meat and meat products moving within their own state borders, but others did not. Thus, consumers in the different states had varying levels of protection.

The 1967 law gave the states three years, until December 15, 1970, to inaugurate inspection systems equal to the federal system. In those states that could not or would not develop their own systems, the federal inspection system took over after December 1970. The federal government assists those states that continue to operate their own inspection systems (equal to federal) by assuming 50 percent of the cost of these state systems. One alternative, which is favored by some state legislatures and governors who are interested in finding new ways to balance their states' budgets, is to forego state inspection, turn it over to the federal system, and thus save all of the state cost. Talmadge-Aiken plants are federally inspected, but staffed by state employees.

On October 26, 1986, Congress passed Public Law 99–641, the Commodity Futures Act, which amended the Federal Meat Inspection Act to permit the U.S. Department of Agriculture to vary the intensity of inspection among plants. The passage of this act was based on the success of a Voluntary Total Quality Control Program that has been in effect since 1980, which gives processors incentives to develop and monitor their own quality control systems, provided they meet government inspection requirements. This program is called Discretionary Inspection (DI) and(or) Improved Processing Inspection (IPI) and is designed to put the greatest financial and human inspection input where the risk is greatest.

On July 25, 1996, the Final Rule entitled Pathogen Reduction; Hazard Analysis and Critical Control Point (HACCP) Systems, known throughout the meat industry as the "Mega Reg," was published in *The Federal Register* (Volume 61, Number 144). The purpose of this final regulation was to move meat inspection from command-and-control regulations to a science-based system. This development was suggested and supported by the National Academy of Sciences, the U.S. General Accounting Office, and inspection personnel.

ADMINISTRATION OF THE ACT

Meat inspection was first administered by the Bureau of Animal Industry, which became a part of the USDA's Agricultural Research Service. Later, meat inspection came under the USDA's Consumer and Marketing Service (C & MS). On April 2, 1972, the Secretary of Agriculture established the Animal and Plant Health Inspection Service (APHIS) and assigned to this new agency the meat and poultry inspection functions heretofore carried out by the C & MS.

On March 14, 1977, the Secretary of Agriculture established the Food Safety and Quality Service (FSQS) and assigned to it the responsibility for food inspection and grading. In addition to meat and poultry inspection, the new agency was responsible for the inspection and grading of dairy products, eggs, and fresh and processed fruits and vegetables. Another name change took place with a new administration in June 1981, when the meat grading and purchasing functions were transferred back to the Agricultural Marketing Service. The name *FSQS* was dropped, and meat, poultry, and egg inspection became the responsibility of the USDA's Food Safety and Inspection Service (FSIS).

Meat inspection is *not* synonymous with meat grading. Meat inspection guarantees meat wholesomeness, safety, and accurate labeling. All meat that is sold must by law be inspected. Every citizen, regardless of his or her meat-eating habits, pays for meat inspection through taxes. Conversely, meat grading designates expected eating quality (tenderness, flavor, and juiciness) and yield of carcasses and cuts. Meat grading is optional and is paid for by meat packers and processors and ultimately by the consumer, in the price of meat (see Chapter 12). The only way in which grading relates to the wholesomeness and safety of meat is in the regulation that meat must pass inspection before it can be graded. In some instances, the same properly trained and licensed USDA employee may perform both inspection and grading functions.

Quality is a factor not only in determining the palatability of meat as a food but also, through quality control and quality assurance, in determining the safety and composition of meat as a food. Caswell's bibliography is a valuable reference for anyone who wishes to investigate more thoroughly state and federal regulations governing food safety and quality.[3]

---

[3]Caswell, J. A. 1988 (September). Federal and State Regulations of Food Safety and Quality: A Selected, Partially Annotated Bibliography. NTIS stock number PB89-114342. Superintendent of Documents, U.S. Government Printing Office, Washington, DC 20402.

THE COST OF MEAT INSPECTION

Unlike any other food, meat is very much like the human body. Animals are subject to many of the same diseases as humans. For our protection, we must have rigid standards of inspection of the health of each animal and its products and the sanitary conditions under which it is processed. Who should pay for this inspection? If it is the packer, the added cost will be reflected in the price of the product, and consumers will pay by proxy. If it is the federal government's inspector who is in the plant to protect the public health, the public will pay through taxes.

Back in 1947, Congress included a provision in the USDA appropriation bill to pass along the cost of federal inspection to the packers and processors. This shift of the cost of meat inspection from the government to the packers was in effect until the passage of the Kem bill (S2256), when the cost reverted again to the U.S. Department of Agriculture, effective July 1, 1948. Since that time, the meat packing industry has been reimbursing the government only for overtime and special services. In addition to its mandatory inspection activities (paid for by taxes), FSIS provides voluntary, reimbursable inspection services in establishments. These may include plants that request exotic animal inspection, identification service, approved warehouse inspection, animal food inspection, and food inspection. Probably the main reason why the one-year trial in which the cost of all inspection was shifted to the packer did not succeed was that those who did not have federal inspection were able to undersell those who had to bear the cost of inspection.

The cost of meat inspection for each consumer remains amazingly low — approximately 21 cents per month. That's a tremendous bargain, considering that meat is a highly perishable food for which each consumer spends approximately 35 cents of his or her food dollar, which is about 11 percent of the average consumer's take-home pay.

Voluntary Inspection of Exotic Animals

Effective February 13, 1989, the FSIS adopted a regulation that provides voluntary antemortem (before harvest), postmortem (after harvest), and products inspection for American bison, elk, deer, antelope, reindeer, and water buffalo. This regulation allows for antemortem inspection at a producer's premises, on a transport vehicle, or at an official exotic animal meat plant. *It does not apply to exotic animals in the wild.* Every meat plant that is approved for federal inspection for any species receives a unique, individual identification number. If a plant is approved to harvest and process exotic animals, it has a specific identification number for exotic animals, which appears in the inspection stamp, so that the products can always be traced back to the plant of origin. The inspection stamp for exotic animals is triangular in shape (see Figure 3–5).

A packer or processor is permitted to harvest and process both exotic and common meat animals in the same plant, but it must be done at completely different times, the exotic animal meat must be kept completely separate and identified with a different identification number, and a complete, thorough cleanup and sanitization must be accomplished before the same facility can again be used for common species. Meat from exotic animals is permitted to be used in processed meat products as long as the final products contain at least 3 percent meat or 30 percent fat from inspected and processed cattle, sheep, swine, or goats. Other than these special provisions, all other inspection regulations remain the same as those described in the following material on meat animals.

This regulation is the result of requests from exotic animal producers; thus, it is classified as voluntary and is paid for by the producers and processors who receive the service.

Horse Meat

From prehistoric times until the fourth century, horse meat was eaten practically everywhere on Earth.[4] Then, beginning in the fourth century, its consumption by Christians was discouraged because the meat was used in pagan rituals. In 743, the pope decreed that Christians stop eating horse meat under penalty of death. The reluctance to consume horse meat continued until Napoleon's time, when one of his generals, seeing some of his troops starving, killed some of the horses and fed them to the troops. This changed the thinking about eating horse meat in Europe, but many people had already gone to the New World and held their dislike for horse meat.

In the United States, amid objections from horse lovers,[5] a revolutionary new product, canned horse meat, was introduced in pet shops in 1922 by

---

[4]Ron Corn, as reported by Patti Lewis in the Tri-State Livestock News, January 12, 1985.

[5]Dr. Jim Corbin, 1999, University of Illinois, personal communication.

P. M. Chappel of Rockford, Illinois. Chappel and his two brothers had already delivered 117,000 horses during World War I to various military governments and were well established in the horse trading business. When a French officer was inspecting some horses that were too small for war use, the officer commented, "What a pity these small animals cannot be used for human food in our country where our people need meat so badly." Mr. Chappel later implemented the use of horse meat as human food in Austria, Denmark, France, Belgium, Holland, Italy, Norway, Sweden, and several other countries. The Chappels were processing enough trimmings and small cuts into metal cans by the 1930s to feed "scientific, balanced" canned dog food to 500,000 dogs daily. It was estimated that one of every four families owned a dog. At that time the half-dozen major horse meat packing plants produced about 200 brands of canned horse meat.

Researchers at Clemson University[6] published results of a study in which they compared the palatability and proximate composition of patties using mixtures of ground beef, pork, and chevaline (horse meat). They found no difference in juiciness, flavor intensity, or flavor desirability, and the patties were similar in nutritive value. Readers will have no problem finding current horse and/or horse meat stories in the popular press.

The Horse Meat Act was approved July 24, 1919, making federal inspection necessary for horses harvested for interstate or foreign shipment, if the meat was to be sold for human consumption. The harvest establishment must be separate from those where other animals are harvested or where the meat products of other animals are handled. The inspection stamp on horse meat is applied with green ink rather than the purple found on other meat species. The act provides that such meat must be conspicuously labeled, marked, branded, or tagged "horse meat" or "horse meat product." Horse meat is also regulated by the federal Food, Drug, and Cosmetic Act, and the use of horse meat in every state is regulated by the U.S. Wholesome Meat Act. In recent years, there have been fewer than 20 USDA–inspected equine harvest plants in operation.

Accurate horse census figures are practically non-existent because the horse population is not large enough to support a major horse meat industry. Although human consumption of chevaline in the United States is very small,

---

[6]Eades, T. V., T. M. Moore, R. G. Godbee, E. Halpin, and G. C. Skelley. 1988. Comparison of patties using mixtures of ground meat from beef, pork, and chevaline. Meat Science. 24:127–132.

horse meat is considered a delicacy in the European and Japanese cultures. In recent years, U.S. horse meat has been one of the largest U.S. meat export items to the 12-nation European Community. As a result of the export demand, which exceeds the number of horses available for harvest, the price has risen dramatically. Previously, horse meat in the United States was used mainly in the manufacture of pet foods.

The meat is not the only source of revenue. Horse hides are used for leather; tallow for soap; hair, mane, and tail for hair goods; the glands for pharmaceuticals; and the offal as food for fish and carnivorous animals. Dead horses are converted into oils, glue, and fertilizer. By-products are discussed in Chapter 11.

FOOD SAFETY AND INSPECTION SERVICE ORGANIZATION AND RESPONSIBILITIES

The USDA Food Safety and Inspection Service (FSIS) verifies that meat and poultry products moving in interstate and foreign commerce for use as human food are safe, wholesome, and unadulterated and are accurately marked, labeled, and packaged. It also protects meat producers by insuring that no one will be able to gain an unfair economic advantage by putting unwholesome or misbranded products on the market.

The food safety strategy that FSIS outlined in the Pathogen Reduction/HACCP proposal[7] includes the following major elements:

- Provisions for systemic prevention of *biological, chemical,* and *physical* hazards through adoption by meat and poultry establishments of science-based process control systems.

- Targeted efforts to control and reduce harmful bacteria in raw meat and poultry products.

- Adoption of food safety performance standards that provide incentives for innovation to improve food safety and to provide a measure of accountability for achieving acceptable food safety results.

---

[7]Pathogen Reduction; Hazard Analysis and Critical Control Point (HACCP) Systems; Final Rule; Federal Register: July 25, 1996 (Vol. 61, No.144).

- Removal of unnecessary regulatory obstacles to innovation.

- Efforts to address hazards that arise throughout the food safety continuum from farm to table.

FSIS also stressed, as a central theme of its strategy, a need to clarify and strengthen the responsibilities of establishments *for maintaining effective sanitation.* Thus, establishing Sanitation Standard Operating Procedures (SSOPs) was the first requirement for processors, even before establishing a HACCP plan. There are many good references and sources covering SSOPs and HACCP plans that explain these topics in much more detail than we can cover in this textbook.[8, 9, 10, 11, 12]

The purpose of this book, and especially this chapter, is to relate government meat inspection regulations in a way that will promote understanding and also application, if appropriate. To quote a central Illinois bank's agricultural services newsletter[13] referring to government farm programs: "As in all areas where the government is involved, the rules, regulations and general hoops which must be jumped through can seem daunting and discourage participation." But be assured that unlike the farm program, if a meat processor decides to not participate in this HACCP program, that processor will no longer be in business. However, FSIS has moved away from command-and-control regulations where the exact same regulations are applied to each and every meat plant and poultry establishment. In this new HACCP program, establishments are to be afforded the flexibility to make establishment-specific decisions regarding food safety precautions. Another goal of this new program is to gain similarity and uniformity between FSIS-monitored livestock

---

[8]Brewer, M. S., and F. K. McKeith. 1999. HACCP Plan Development—Short Course Manual. The Board of Trustees, University of Illinois, Urbana, IL 61801.

[9]USDA-FSIS, Human Resources Development Staff. 1998. HACCP Regulatory Process for HACCP-Based Inspection Reference Guide. (This guide contains FSIS Directive 5000.1, which some industry leaders indicate is good information to have for you to get an idea of what your inspector is expecting.)

[10]International Meat and Poultry HACCP Alliance. Kerry Harris and Russell Cross, Texas A & M University, Center for Food Safety, College Station. 409-862-2036. <kharris@zeus.tamu.edu> and <russ-cross@idexx.com>.

[11]Silliker Laboratories. Pathogen Reduction and HACCP; Question and Answer Guide. Homewood, IL 60430. 708-957-8449. http://www.silliker.com

[12]National Meat Association, 1970 Broadway, Suite 825, Oakland, CA 94612. http://www.nmaonline.org/index/html

[13]FarmViews, 1999. Bank Illinois Agricultural Services, Vol. 7, No. 2 (June).

and poultry inspection procedures and between these and Food and
Drug Administration (FDA)–administered seafood inspection procedures.
James H. Hodges, President of the American Meat Institute Foundation, in
an update report to the Reciprocal Meat Conference,[14] stated: "HACCP is
working and will be successful if reasonable expectations prevail. . . . The
industry is taking more care than at any time in its history to ensure that only
clean, defect-free carcasses and birds emerge from slaughter, and processing is
conducted in a more hygienically controlled manner. . . . No meat and poultry processor wants a single illness or death from foodborne pathogens to
occur. . . . No other sector of the food industry operates with the degree of
regulatory oversight imposed on the meat and poultry industry. If other
industries were regulated in a similar manner, many more U.S. citizens would
be working as government inspectors."

Following the organization of most federal regulations, we will begin
with a listing of definitions.[15] That should help get this discussion going in
the right direction.

Adulterated. This term applies to any carcass, part thereof, meat or meat food
   product that bears or contains any such poisonous or deleterious substance
   which may render it injurious to health.

Anesthesia. Loss of sensation or feeling.

Animal food. Any article intended for use as food for dogs, cats, or other animals derived wholly, or in part, from the carcass or parts or products of the
   carcass of any livestock.

Artificial coloring. A coloring containing any dye or pigment, which dye or
   pigment was manufactured by a process of synthesis or other similar artifice, or a coloring which was manufactured by extracting a natural dye or
   natural pigment from a plant or other material in which such dye or pigment was naturally produced.

Artificial flavoring. A flavoring containing any sapid or aromatic constituent,
   which constituent was manufactured by a process of synthesis or other similar artifice.

Biological residue. Any substance, including metabolites, remaining in livestock at time of slaughter or in any of its tissues after slaughter as the result

---

[14]Hodges, J. H. 1999. Update on implementation of USDA's HACCP and pathogen reduction initia-
tives. Proc. 52nd Reciprocal Meat Conference of the American Meat Science Association. 52:97–98.

[15]Code of Federal Regulations, Title 9, CITE: 9CFR301.2. http://www.access.gpo.gov/nara

of treatment or exposure of the livestock to a pesticide, organic or inorganic compound, hormone, hormone-like substance, growth promoter, antibiotic, anthelmintic, tranquilizer, or other therapeutic or prophylactic agent.

Captive bolt. A stunning instrument which when activated drives a bolt out of a barrel for a limited distance.

Carbon dioxide. A gaseous form of the chemical formula CO_2.

Carbon dioxide concentration. Ratio of carbon dioxide gas and atmospheric air.

Carcass. All parts, including viscera, of any slaughtered livestock.

Chemical preservative. Any chemical that, when added to a meat or meat food product, tends to prevent or retard deterioration thereof but does not include common salt, sugars, vinegars, spices, or oils extracted from spices or substances added to meat and meat food products by exposure to wood smoke.

Consciousness. Responsiveness of the brain to the impressions made by the senses.

Cutting up. Any division of any carcass or part thereof, except that the trimming of carcasses or parts thereof to remove surface contaminants is not considered as cutting up.

Dead livestock. The body (cadaver) of livestock which has died otherwise than by slaughter.

Dying, diseased, or disabled livestock. Livestock which has or displays symptoms of having any of the following:

    (1)   Central nervous system disorder;

    (2)   Abnormal temperature (high or low);

    (3)   Difficult breathing;

    (4)   Abnormal swellings;

    (5)   Lack of muscular coordination;

    (6)   Inability to walk normally or stand;

    (7)   Any of the conditions for which livestock is required to be condemned on ante-mortem inspection in accordance with FSIS regulations.

Edible. Intended for use as human food.

Experimental animal. Any animal used in any research investigation involving the feeding or other administration of, or subjection to, an experimental biological product, drug, or chemical or any nonexperimental biological product, drug, or chemical used in a manner for which it was not intended.

Exposure time. The period of time an animal is exposed to an anesthesia-producing carbon dioxide concentration.

Federal Food, Drug, and Cosmetic Act. The Act so entitled, approved June 25, 1938 (52 Stat. 1040), and Acts amendatory thereof or supplementary thereto.

Food Safety and Inspection Service. The Food Safety and Inspection Service of the Department [of Agriculture].

Further processing. Smoking, cooking, canning, curing, refining, or rendering in an official establishment of product previously prepared in official establishments.

Inedible. Adulterated, uninspected, or not intended for use as human food.

Inhumane slaughter or handling in connection with slaughter. Slaughter or handling in connection with slaughter not in accordance with the Act of August 27, 1958 (72 Stat. 862; 7 U.S.C. 1901 through 1906, as amended by the Humane Methods of Slaughter Act of 1978, 92 Stat. 1069), and part 313 of . . . [Title 9] of the Code of Federal Regulations.

Inspected and Passed or U.S. Inspected and Passed or U.S. Inspected and Passed by Department [of Agriculture] **(or any authorized abbreviation thereof).** This term means that the product so identified has been inspected and passed under the regulations in . . . Title 9, and at the time it was inspected, passed, and identified, it was found to be not adulterated.

Label. A display of written, printed, or graphic matter upon the immediate container (not including package liners) of any article.

Labeling. All labels and other written, printed, or graphic matter:

(1)   Upon any article or any of its containers or wrappers, or

(2)   Accompanying such article.

Meat.   (1)   The part of the muscle of any cattle, sheep, swine or goats, which is skeletal or which is found in the tongue, in the diaphragm, in the heart, or in the esophagus, with or without the accompanying and overlying fat, and the portions of bone, skin, sinew nerve, and blood vessels which normally accompany the muscle tissue and which are not separated from it in the process of dressing. It does not include the muscle found in the lips, snout, or ears. This term, as applied to products of equines, shall have a meaning comparable to that provided in this paragraph with respect to cattle, sheep, swine, and goats.

(2) The product derived from the mechanical separation of the skeletal muscle tissue from the bones of livestock and poultry using the advances in mechanical meat/bone separation machinery and meat recovery systems that do not crush, grind, or pulverize bones, and from which the bones emerge comparable to those resulting from hand-deboning (i.e., essentially intact and in natural physical conformation such that they are recognizable bones, when they emerge from the machinery) which meets the criteria of no more than 0.15 percent or 150 mg/100 gm of product for calcium (as a measure of bone solids content) within a tolerance of 0.03 percent or 30 mg.

Meat broker. Any person engaged in the business of buying or selling carcasses, parts of carcasses, meat or meat food products of livestock on commission, or otherwise negotiating purchases or sales of such articles other than for his/her own account or as an employee of another person.

Meat by-product. Any part capable of use as human food, other than meat, which has been derived from one or more cattle, sheep, swine, or goats. This term, as applied to products of equines, shall have a meaning comparable to that provided in this paragraph with respect to cattle, sheep, swine, and goats.

Meat food product. Any article capable of use as human food which is made wholly or in part from any meat or other portion of the carcass of any cattle, sheep, swine, or goats. Exceptions are meat or other portions of such carcasses only in a relatively small proportion or historically not . . . been considered by consumers as products of the meat food industry as determined by specific labeling restrictions. This term, as applied to food products of equines, shall have a meaning comparable to that provided in this paragraph with respect to cattle, sheep, swine, and goats.

Misbranded. This term applies to any carcass, part thereof or meat food product if its labeling is false or misleading in any manner according to labeling restrictions. [See labeling discussion following.]

Nonfood compound. Any substance proposed for use in official establishments, the intended use of which will not result, directly or indirectly, in the substance becoming a component or otherwise affecting the characteristics of meat food and meat products, excluding labeling and packaging materials.

Official certificate. Any certificate prescribed by the regulations for issuance by an inspector or other person performing official functions under the Act.

Official establishment. Any slaughtering, cutting, boning, meat canning, curing, smoking, salting, packing, rendering, or similar establishment at which inspection is maintained under FSIS regulations.

Official inspection legend. Any symbol prescribed by the regulations in . . . [Title 9] showing that an article was inspected and passed in accordance with the Meat Inspection Act [Figure 3–5].

Official mark. The official inspection legend or any other symbol prescribed by FSIS regulations [Figure 3–5].

Packaging material. Any cloth, paper, plastic, metal, or other material used to form a container, wrapper, label, or cover for meat products.

Prepared. Slaughtered, canned, salted, rendered, boned, cut up, or otherwise manufactured or processed.

Renderer. Any person engaged in the business of rendering carcasses or parts or products of the carcasses of any livestock except rendering conducted under inspection. [See detailed discussion of rendering in Chapter 11.]

Shipping container. The outside container (box, bag, barrel, crate, or other receptacle or covering) containing or wholly or partly enclosing any product packed in one or more immediate containers.

Surgical anesthesia. A state of unconsciousness measured in conformity with accepted surgical practices.

U.S. Condemned. This term means that the livestock so identified has been inspected and found to be in a dying condition, or to be affected with any other condition or disease that would require condemnation of its carcass.

U.S. Inspected and Condemned (*or any authorized abbreviation thereof*). This term means that the carcass, viscera, other part of carcass, or other product so identified has been inspected, found to be adulterated, and condemned under FSIS regulations.

U.S. Passed for Cooking. This term means that the meat or meat by-product so identified has been inspected and passed on condition that it be cooked or rendered as prescribed by FSIS regulations. [See discussion in this chapter.]

U.S. Passed for Refrigeration. This term means that the meat or meat by-product so identified has been inspected and passed on condition that it be refrigerated or otherwise handled as prescribed by FSIS regulations.

U.S. Retained. This term means that the carcass, viscera, other part of carcass, or other product or article so identified is held for further examination by an inspector to determine its disposal.

U.S. Suspect. This term means that the livestock so identified is suspected of being affected with a disease or condition which may require its condemnation, in whole or in part, when slaughtered, and is subject to further examination by an inspector to determine its disposal.

Sanitation[16] and Sanitation SOPs[17]

Realizing that good sanitation is the critical foundation for HACCP, FSIS mandated that Sanitation Standard Operating Procedures (SSOPs) be in place before a HACCP plan is written and instituted. The term *SSOPs* refers strictly to procedures that are conducted daily, before and during operations, sufficient to prevent direct contamination or adulteration of product. Standard Operating Procedures (SOPs) and Good Manufacturing Practices (GMPs) are similar to each other and would generally include SSOPs. However, they would also include other steps in a process to assure consistent product content and quality.

The importance of sanitation was introduced in Chapter 2. This section is more specific in regard to FSIS sanitation requirements. Good sanitation means microbiological control. However, any amount of microbiological monitoring will not improve sanitation in a poor operation.

Operational sanitation must permit the production of wholesome products and product handling and processing without undue exposure to contaminants. Facilities and equipment must be properly cleaned at regular intervals. All employees must practice good personal hygiene, and management must provide the necessary equipment and materials to encourage such hygiene. Particular emphasis should be placed on products and product zones. The Environmental Sampling and Testing Recommendations (ESTRs) developed by the RTE group led by the National Meat Association (NMA) with the support of other organizations is a good place to start in designing an environmental testing program for pathogens in any system. The simple things, like personnel hygiene, managing the splash-back of high-pressure hoses before, during, and after operations, controlling water recycling systems, and ensuring proper sanitizing after thorough washing, can help tremendously. The harvest floor employee's rinsing off his or her apron

---

[16]Code of Federal Regulations, Title 9, Part 308—Sanitation. CITE 9CFR308. http://www.access.gpo.gov/nara

[17]Code of Federal Regulations, Title 9, Part 416—Sanitation. CITE 9CFR416. http://www.access.gpo.gov/nara

and letting the spray hit equipment is an easy problem to fix. The plant rolling stock that moves inter-departmentally can potentially transfer undesirable bacteria around the plant. Walk through the plant, and try to look at it as if you were a pathogen. Figure out how you would try to survive the process. The goal is to eliminate environments that are favorable to pathogens wherever possible and to manage their ability to spread from one area to another. Every plant is unique; no one solution works for all.

Microbiological monitoring involves the collection of samples by swabbing, the laboratory process of incubating the samples after they have been spread onto agar plates, and the interpretation of the plate counts. Clean equipment ready to use should have counts in the range of 0 to 30 colonies per square inch. Counts over 100 per square inch indicate a situation that is borderline and that could very likely cause problems in an operation.

The value and timing of microbiological testing are debatable among scientists in industry and government, so it is no wonder that consumers are confused about these issues. USDA-FSIS requires that harvest establishments regularly sample and test for generic *E. coli* to verify the adequacy of the organization's process control for the prevention and removal of fecal contamination and associated bacteria. In addition, *Salmonella* testing is required of harvesters and others providing raw ground products. Details of the sampling requirements and standards that must be met are available from USDA-FSIS (footnote 9).

Floors, Walls, Ceilings

Floors must be free of an accumulation of fats, blood, and other foreign material. Walls must be free of dirt, mold, blood, scaling paint, and other contaminants. Ceilings and overhead must be free of dust, scaling paint, scaling plaster, mold, rust, condensation, leaks, etc.

Equipment

All equipment must be in good condition and free from contaminants — rust, dust, dried blood, scrap meat, grease, etc. Each piece of equipment must be sanitized after use on each carcass.

- *Cattle* — Brisket saw, weasand rods, front shank tie-down chains, and dehorning equipment.

- *Calves* — Brisket saw or cleaver.

- *Swine* — Knife or other tools used to partly sever the head and a brisket-splitting device and a saw or cleaver, if carcass splitting occurs before viscera inspection is completed.

Personnel, Clothing, and Personal Equipment

- Establishment employees must wear clean and washable or disposable outer clothing. Street clothing should be properly covered. The wearing of sleeveless garments that would permit the underarm to be exposed should not be permitted. Suitable head coverings applicable to both sexes must be worn to cover the hair adequately. FSIS employees must appear neat and clean and demonstrate good working and sanitary practices that are expected of food inspection employees.

- All personnel must wash their hands as often as necessary to prevent product contamination, and always after they return from lavatory rooms. Special attention to fingertips, even with a fingernail brush, will cut down greatly on possible contamination. To insure complete washing and proper sanitation, some meat plants have installed automatic hand washing machines for their employees. The use of tobacco in any form — spitting or smoking — is not permitted in rooms where edible product is handled. Any practice that might be considered unsanitary should be prohibited.

- Personal equipment (knives, scabbards, steels, tool boxes, gloves, etc.) must be kept in a sanitary condition at all times.

Employee Welfare Facilities — Lunchroom, Locker, and Lavatory Facilities

Dressing rooms equipped with lockers or suitable alternate devices, lavatory rooms (showers in meat harvest plant), urinals, and wash basins (other than hand-operated ones) with soap and towels are required. Lavatory rooms and lavatory-room vestibules, which must be separated from adjoining dressing rooms, should have solid, self-closing doors. These areas must be free of odor, properly maintained, and kept clean at all times.

Coolers, Rails, Hooks, Drains, and Equipment

- Equipment in coolers must be clean and free of debris — corrosion, rust, dust, dried blood, scrap meat, and accumulation of fat. This equipment should also be kept in good operating condition. Overhead pipes, beams, and light fixtures, as well as ceilings and walls, must be free of contaminants, mold, and condensation.

- Rails must be clean and free of flaking paint, excessive oils and grease, rust, etc. Hooks must be clean and properly maintained.

- All drains and gutters should be properly installed with approved traps and vents.

Inedible and Condemned Rooms

The areas and equipment for inedible and condemned products handling should be adequate for the quantity of product. They must be separate from those for edible products and must be properly maintained. The areas must be properly constructed and vented, drained as required, and kept in good repair. An acceptable area for truck sanitizing should be available. Appropriate water connections for cleanup must be provided.

Offal Rooms and Coolers — Facilities and Equipment

Coolers must be free of condensation, and the floors, walls, and ceilings in the coolers must be free of the accumulation of dried blood, fat, scrap meat, mold, dirt and dust, and nuisances. Chutes, tables, pans, etc., must be constructed of rust-resistant materials, clean, and properly maintained.

Product Handling

- The product (carcass) must be handled in a clean and acceptable manner. Cooked, ready-to-eat, finished, perishable products must be kept completely separate from raw meat and raw meat emulsions. All products must be stored at a room temperature of 50°F or preferably lower (poultry, 40°F) and be accessible to inspection. Because of the metabolism of poultry, the ultimate (final) pH is higher (closer to neutral).

This is more favorable for microorganism growth. Thus, the lower holding temperature is necessary to compensate (see Chapter 2).

- Finished frozen products must be kept in a frozen state under non-fluctuating temperature, reasonably free of overhead frost, and accessible to inspection.

- Nonmeat materials (additives, etc.) must be approved by the federal Food and Drug Administration and properly identified. Product must be received only at a specified area. Nonmeat approval stickers must be applied.

- A suitable compartment or refrigerated area for holding return product pending disposition should be equipped for sealing in order to maintain security.

- Unpackaged custom products (not for sale, processed for the owner, inspection not required) must be held separately from inspected products (separate rail, racks, etc.).

- A thorough cleanup and sanitizing of equipment must be done after harvest and after the processing of the custom-exempt product before the handling of the inspected product can be resumed.

- Viscera separation and product handling must be conducted in a sanitary manner. Community baths are forbidden for all products. Paunches must be emptied without contaminating outer surfaces; however, in some instances, paunches can be sold full for immediate use as animal feed. The accumulation of full offal is not permitted. Pork hearts must be opened completely and all blood clots removed. Pork stomachs, chitterlings (intestines), and(or) ruffle fat (around intestines) must be clean and free of ingesta and(or) any other contaminants.

Carcass Cleanliness and Prevention of Contamination (Sanitary Dressing Procedures)

HEAD HANDLING

- *Cattle.* Head and corresponding carcass should be identified by duplicated numbered tags or by other acceptable means and removed in such a manner as to avoid soilage with rumen contents. Horns and all pieces of hide should be removed before the outer surfaces of the head are washed.

- *Sheep and goats.* Heads must be flushed and washed in a cabinet if they are being saved for edible purposes.

CARCASS PREPARATION

- *Swine.* After passing through the scalding and dehairing equipment, the carcass must be free of hair and scurf and the hind feet must be clean of hair and scurf before being gambrelled. If hogs are dipped in rosin, the nostrils and mouth must be closed with rubber bands or other acceptable means prior to being dipped. No shaving is permitted after the head is dropped.

- *Calves.* Before they are stunned, calves may be showered to aid in the washing of the hide; they must be washed clean before any incisions (except stick wounds) are made.

CARCASS SKINNING

- *Cattle.* The area of the skinning bed must be acceptably clean before the carcass is lowered. The head skin should be manipulated so that the neck is protected. The front and hind feet are removed before any other incision is made. The carcass must be removed from the skinning bed in such a way that will prevent contamination. Lactating udders must be removed in such a way that the udder contents do not soil the carcass. The supramammary lymph nodes must be left attached to the carcass until the inspection is completed. The dropping of bung should be made part of the rumping operations. The rectum must be tied, and the bladder must be tied or removed to prevent contamination.

  The FSIS permits the use of organic sprays on clean beef carcasses after they have been skinned, but prior to being eviscerated, to reduce bacteria on beef carcasses and to extend the shelf life of meat products. According to a 1989 AMI report,[18] microbial contaminants become firmly attached soon after contact; thus, prompt washing is important in removing them. The combination of 200 pounds per square inch

---

[18]American Meat Institute (AMI). 1989 (June). The Reduction of *Salmonella* on Beef and Pork Carcasses; A Research Report; A Pilot-Plant Scale Evaluation of Washing and Sanitizing Procedures.

(psi) with a flow rate of 3.36 gallons per minute (gpm) is most effective. The report also indicates that a mild salt solution (0.1 M KCl) will weaken the bacterial attachment. Organic acids (lactic more than acetic) do a good job of sanitizing and are more effective at 140°F than at ambient temperatures. Citric acid is also being tested. Washing with water alone may be counterproductive if it is followed with a sanitizer, because of sanitizer dilution. Dilute chlorine washes (50 parts per million chlorine), permitted by the FSIS, were somewhat less effective than the organic acids in this AMI study. For on-line sanitizing, the carcasses pass through cabinets in order to contain the mist and to evacuate the fumes and odors.

- *Calves.* The establishment has the responsibility for skinning and handling calf carcasses in a sanitary manner. When skinning operations start, the entire carcass should be skinned. **Note:** In cases where the establishment handles "hide-on" carcasses, the operation must be conducted in a sanitary manner. Hair-to-carcass contact is not permitted. Calf carcasses skinned after they have been chilled must be examined closely to detect injection lesions, foreign bodies, parasites, bruises, or other signs of pathology. All abnormal tissue must be removed.

- *Sheep, goats, and swine.* All operations in removing the pelt and skin must be done in such a manner that the carcass is not contaminated.

CARCASS EVISCERATION — Carcass evisceration must be done without contaminating carcasses or organs. The rectum must be tied to prevent soilage. Viscera should be presented in an orderly manner, with all the various parts and organs clearly visible, to facilitate inspection.

CARCASS WASHING — All carcasses must be thoroughly and properly washed. (See the preceding discussion on preevisceration washing and sanitation.) The final carcass wash with tepid water (90°F) may be followed with an organic acid rinse.

Harvest, Scalding, and Picking

POULTRY

- *Procedures.* A continuous intake of water must be sufficient to maintain acceptably clean scalding water and to provide a minimum overflow of 1 quart of water per bird per minute. Hair and feathers must be completely removed, and the bird must be given a final wash with potable water. "Singeing" to remove hairs is common.

- *Product washing.* The product must be effectively washed inside and out to remove excess blood, loose tissue particles, and any foreign material. Bacteria have been shown to adhere more tightly to poultry skin compared to other meat surfaces. Contamination of any tissue, other than the external skin surface, must be removed by trimming. All products must be clean before being chilled.

- Carcasses must be protected against possible contamination from floor cleanup and from any fixed objects.

Guidelines for Ready-to-Eat Products

National Meat Association (NMA) and microbiologists from Deibel Laboratories have provided these more specific guidelines for ready-to-eat (RTE) products, which will indicate the importance of detail in monitoring and maintaining a safe food environment: "Manufacturers of packaged fully cooked products should develop and maintain Clean Room Standard Operating Procedures (SOPs) to eliminate cross contamination. The Clean Room SOPs should include positive air movement in exposed clean rooms, restricted access to specific lines, sanitizing dips for hands and footwear, and the donning of sanitized clothing. In addition, equipment cleaning and sanitizing procedures should comprehensively address both product and nonproduct surfaces. A good preventive program incorporates microbial monitoring of the environment at an initial frequency of at least 20 samples per week. Microbial swabs should specifically target equipment contact surfaces, using a sponge technique and swabbing an area as large as possible for each sample. Consistently achieving negative test results may support reducing the sampling frequency to about 10 samples per week. Should environmental samples indicate a presumptive for *Listeria* species, begin a decontamination program that incorporates Ceiling-to-Floor strip cleaning and sanitizing immediately. Dismantle equipment to expose unsealed areas, and remove insulation from pipes and conduits, including plastic electrical ties and duct tape. Thoroughly wash and clean all equipment and environmental surfaces. Apply a peroxide solution to all surfaces. This application should then be followed by a fogging of the room with a 1000 PPM quaternary ammonia compound to insure that all crevices and micro cracks are coated with the fine droplets. After 30 minutes, rinse the equipment contact surfaces only, in preparation for use. It is important that the environment remain treated with the quaternary compound to prevent bacteria growth. Fogging should be repeated until environmental micro samples indicate that equipment surfaces are no longer contaminated with *Listeria*."

The HACCP Plan[19, 20, 21, 22, 23, 24]

Where Did HACCP Come From?

Rosemary Mucklow, National Meat Association (NMA) Executive Director, explains the origin of HACCP: "The pathway to the HACCP system started in 1959 when the Pillsbury Company was asked to produce a food that could be used under zero gravity conditions in space capsules. The challenge of such an endeavor is to create as close to 100% assurance as possible that no contamination is present, sickness in zero gravity situations being very dangerous. The concept and reduction to practice of the Hazard Analysis and Critical Control Point (HACCP) system was directly related to the Pillsbury Company's projects in food production and research for the space program. The basics were developed by the Pillsbury Company with the cooperation and participation of the National Aeronautics and Space Administration (NASA), the Natick Laboratories of the U.S. Army, and the U.S. Air Force Space Laboratory Project Group."

---

[19]Brewer, M. S., and F. K. McKeith. 1999. HACCP Plan Development—Short Course Manual. The Board of Trustees, University of Illinois, Urbana, IL 61801.

[20]USDA-FSIS, Human Resources Development Staff. 1998. HACCP Regulatory Process for HACCP-Based Inspection Reference Guide. (This guide contains FSIS Directive 5000.1, which some industry leaders indicate is good information to have for you to get an idea of what your inspector is expecting.)

[21]International Meat and Poultry HACCP Alliance. Kerry Harris and Russell Cross, Texas A & M University, Center for Food Safety, College Station. 409-862-2036. <kharris@zeus.tamu.edu> and <russ-cross@idexx.com>.

[22]Silliker Laboratories. Pathogen Reduction and HACCP; Question and Answer Guide. Homewood, IL 60430. 708-957-8449. http://www.silliker.com

[23]National Meat Association, 1970 Broadway, Suite 825, Oakland, CA 94612. http://www.nmaonline.org/index/html

[24]Code of Federal Regulations, Title 9, Part.417—Hazard Analysis and Critical Control Point (HACCP) Systems. CITE 9CFR417. http://www.access.gpo.gov/nara

[25]National Advisory Committee on Microbiological Criteria for Foods (NACMCF). 1992. Hazard Analysis and Critical Control Point System. Int. J. Food Micro. 16:1–23.

Developing a HACCP Plan

First, assemble all the pertinent information about your operation, such as a list of employees; the facility layout or blueprint; a list of equipment, including date purchased, manufacturer, etc.; a list of products produced; labels; a list of non-food items; water analysis; copies of any applicable government directives that might be used for references (cooking times, chill rates, etc.; see later section in this chapter); GMPs; SSOPs; and SOPs. *Second,* assemble a HACCP team. *Third,* describe your products or product groups. By grouping products that are produced by a similar process, you will cut down on the number of HACCP plans. You need a specific HACCP plan for each product or each group of products. *Fourth,* prepare a flow diagram.

Now you are ready to follow the HACCP principles. The seven HACCP principles are listed and discussed below.

- *Principle No. 1:* A *hazard analysis* of each process must be carried out. The purpose of the analysis is to identify and list the food safety hazards reasonably likely to occur in the production process for a particular product and the preventive measures necessary to control the hazards. A food safety hazard is any *biological, chemical,* or *physical* property that may cause a food to be adulterated or otherwise unsafe for human consumption. A *biological hazard,* if uncontrolled, will result in foodborne illness or even death. Bacteria are the most problematic biological hazard. A *chemical hazard,* such as excessive nitrite, may have an acute, toxic effect. A *physical hazard,* such as glass particles and metal shavings, can cause physical harm. A listed hazard must be of such a nature that its prevention, elimination, or reduction to acceptable levels is essential to the production of a safe food. Examples of questions to be considered in a hazard analysis are:

  1. What potential hazards may be present in the animals to be harvested or the raw materials to be processed?

  2. What are the avenues that might lead to contamination of finished product with pathogenic microorganisms, hazardous chemicals, or other potentially hazardous agents?

  3. What is the likelihood of such contamination, and what are the means of preventing it?

  4. Does the food contain any ingredient historically associated with a known microbiological hazard?

5. During processing, does the food permit survival or multiplication of pathogens or permit toxin formation?

6. Does the processing include a controllable step that destroys pathogens?

7. If it is likely that the food will contain pathogens, are they likely to increase during the times and conditions under which the food is normally stored before being consumed?

8. What product safety devices are used to enhance consumer safety (e.g., metal detectors, filters, thermocouples)?

9. Does the method of packaging affect the multiplication of pathogenic microorganisms and/or the formation of toxins?

10. Is the product epidemiologically linked to a foodborne disease?

- *Principle No. 2:* The *critical control points (CCPs)* of each process must be identified. A CCP is a point or step at which, or a procedure with which, control can be applied and a food safety hazard can be prevented, eliminated, or reduced to an acceptable level. All hazards identified during the hazard analysis must be addressed. The information developed during the hazard analysis should enable the establishment to identify which steps in its processes are CCPs. Identification of CCPs for controlling microbial hazards throughout the production process is particularly important because these hazards are the primary cause of foodborne illness. Readers may find the CCP decision tree in the manual by Brewer and McKeith (see footnote 19), originally published by the National Advisory Committee on Microbiological Criteria for Foods (NACMCF),[25] useful in the CCP identification process.

- *Principle No. 3:* The *critical limits* for preventive measures associated with each identified CCP must be established. A critical limit is the maximum or minimum value to which a process parameter must be controlled at a CCP to prevent, eliminate, or reduce to an acceptable level the identified physical, biological, or chemical food safety hazard. Critical limits are most often based on process parameters, such as temperature, time, physical dimensions, humidity, moisture level, water activity, pH, titratable acidity, salt concentration, available chlorine,

---

[25]National Advisory Committee on Microbiological Criteria for Foods (NACMCF). 1992. Hazard Analysis and Critical Control Point System. Int. J. Food Micro. 16:1–23.

viscosity, preservatives, or survival of target pathogens. Critical limits should be built upon applicable FSIS regulations or guidelines, FDA tolerances and action levels, scientific and technical literature, surveys, experimental studies, or the recommendations of recognized experts in the industry, in academia, or in trade associations. Establishments are encouraged to set critical limits more stringent than those now required by FSIS regulations or suggested by scientific data to ensure that regulatory requirements are routinely met, even when minor deviations occur.

- *Principle No. 4:* The **monitoring requirements** for CCPs must be determined. Monitoring is an integral part of HACCP and consists of observations or measurements to assess whether a CCP is within the established critical limit. Continuous monitoring is preferred, but when that is not feasible, monitoring frequencies must be sufficient to ensure that the CCP is under control. Assignment of the responsibility for monitoring each CCP is an important consideration. Personnel assigned the monitoring activities should be properly trained to accurately record all results, including any deviations, so that immediate corrective actions may be taken.

- *Principle No. 5:* The HACCP plan must include **corrective action** to be taken when monitoring indicates that there is a deviation from a critical limit at a critical control point. Although the process of developing a HACCP plan emphasizes organized and preventive thinking about what is occurring as the meat or poultry product is being manufactured, the existence of a HACCP plan does not guarantee that problems will not arise. For this reason, the identification of a planned set of activities to address deviations is an important part of a HACCP plan. When a deviation occurs, an action plan must be in place to determine the disposition of the potentially unsafe or noncompliant product and to identify and correct the cause of the deviation. The HACCP plan itself might require modification, perhaps in the form of a new critical limit or an additional CCP.

- *Principle No. 6: Effective record-keeping procedures* that document the entire HACCP system must be developed and maintained. A HACCP system will not work unless consistent, reliable records are generated during the operation of the plan and unless those records are maintained and available for review. One of the primary benefits of a

HACCP process control system to both industry and regulatory officials is the availability of objective, relevant data.

Here is what Teresa Frey, National Meat Association (NMA) Regulatory Aide, wrote about this principle: "HACCP without record-keeping (The Keystone HACCP Principle) is like a car without oil. A HACCP plan will not run properly and will eventually stall without proper records. Clear, concise recordkeeping can be an establishment's best asset. Confusing, sloppy and incomplete records may become a worst enemy of a plant. A plant may have the best sanitary practices, the most controlled CCP's and the most stringent monitoring procedures, but if it isn't recorded, it didn't happen."

Records should be made as "real-time" entries, and each entry should be dated and signed or initialed by the establishment employee making it. "Dry-lab" entries are not acceptable.

Prior to shipping product, a pre-shipment review of all records associated with the production of that product must be conducted to ensure that all critical limits were met and, if appropriate, corrective actions were taken. Where practical, this record check should be made by plant employees who did not make the original entries.

A plant may establish a record review system that best fits the needs of its facility. However, establishments must determine that all critical limits were met and that corrective actions were taken prior to product shipment.

Depending on the size of each plant, a lot identification system should be established and followed. In some operations, a lot may be the product manufactured from one cleanup until the next cleanup. Other operations may choose different time periods for their lot designation.

Records may be maintained on computers, provided appropriate controls are implemented to ensure the integrity of the electronic data and the signatures. A plant's written Hazard Analysis, decision-making documents supporting selection of CCPs and establishment of critical limits, and documents supporting both the monitoring and the verification procedures, as well as the frequency of those procedures, must be maintained onsite indefinitely. Records associated with monitoring, corrective actions, and verification procedures for harvest activities and refrigerated product must be maintained for at least one year. Records associated with monitoring, corrective actions, and verification procedures for frozen, preserved, or shelf-stable products must be maintained for at least two years. These must be kept on the premises of the establishment for a period of six months. After six months, the records may be relocated to offsite storage, provided they can be retrieved within 24 hours of an FSIS employee's request.

Periodic review of HACCP records can provide plant management with insight into processing trends. After all, FSIS will be periodically reviewing your records. Why shouldn't you? HACCP plans, procedures, and records required by the Final Rule should be available for inspection, review, and copying. It is important to note that although the inspector has the right of access and the right to copy records, the inspector does not have the right to take records. One form that a meat processor should avoid—and one that the processor will not be happy to see if it should appear—is the Non-Compliance Record (NR), on which the inspector documents a plant deficiency.

- *Principle No. 7:* HACCP systems must be *systematically verified.* After initial validation that the HACCP system can work correctly and effectively with respect to the hazards, the system must be verified periodically. Periodic verification involves the use of methods, procedures, or tests, in addition to those used for monitoring, to determine whether the HACCP system is in compliance with the HACCP plan and/or whether the HACCP plan needs modification and revalidation to achieve its food safety objective.

    Four processes are involved in the verification of the establishment's HACCP system. The establishment is responsible for the first three; FSIS is responsible for the fourth. The *first* is the scientific and technical process known as "validation" for determining that the CCPs and associated critical limits are adequate and sufficient to control likely hazards. The *second* is the process for ensuring, initially and on an ongoing basis, that the entire HACCP system functions properly. The *third* consists of documented, periodic reassessment of the HACCP plan. The *fourth* process defines FSIS's responsibility for certain actions (government verification) to ensure that the establishment's HACCP system is functioning adequately.

HACCP and the FSIS Food Safety Strategy

The food safety goal of FSIS's Pathogen Reduction/HACCP program is to reduce the risk of foodborne illness from meat and poultry products to the maximum extent possible by ensuring that appropriate and feasible preventive and corrective measures are taken at each stage of the food production process where food safety hazards occur. There is no single technological or regulatory solution to the problem of foodborne illness. Continuous efforts are required by industry and government to improve methods for identifying and preventing hazards and to minimize the risk of illness. FSIS's meat and poultry inspection program currently addresses and will continue to address many matters of importance to the safety and quality of the food supply, including supervision of industry compliance with sanitation standards, exclusion of diseased animals from the food supply, examination of carcasses for other visible defects that can affect safety and quality, and inspection for economic adulteration. These activities respond to some of the public's most basic expectations regarding the safety and quality of the food supply. Sanitation has been discussed, while the remaining topics will follow in this chapter.

Antemortem Inspection

Antemortem inspection procedures must be accomplished in such a manner that any animal that would produce unwholesome meat will be detected and removed. Animals that are down, disabled, diseased, or dead (4 *d*'s) are unquestionably unfit for human consumption. But other animals may be only "suspect," that is, suspected of being sick. The meat packer must provide a paved, drained, covered suspect pen for such animals plus some means of restraining these animals, e.g., a chute and nose tongs. The animal-holding area, where routine inspection is performed, must have a minimum light intensity of 10 foot-candles,[26] while the suspect area must have a minimum of 20 foot-candles. These light intensities are measured 3 feet above the floor (Table 3–1). The packer must also furnish a rectal thermometer, tags, and some means of securing the tags to the suspect animals, e.g., hog rings and a ringer.

The inspector observes the animals at rest in the pen and from each side, as the animals are slowly moved around. With poultry, the inspector also observes them on the day of harvest to the extent necessary to detect disease and(or) other harmful conditions.

Suspect animals are restrained, identified, temperatured, and held for a detailed examination. Normal rectal temperatures (°F) are 100°, horses; 101°,

---

[26]A foot-candle is a standard unit of illuminance on a surface that is everywhere 1 foot from a unit of luminous intensity equal to 1/60 of the luminous intensity of 1 square centimeter of a blackbody surface at the solidification temperature of platinum.

cattle; 102.3°, sheep; 102.5°, swine; 103.8°, goats; 104.9°, turkeys; and 107°, chickens. Animals to be condemned are either tagged "U.S. Condemned" and killed immediately by an establishment employee or released to approved authorities for treatment under official supervision.

Animals showing signs of disease must be separated from the other animals. A company employee must be available for moving, sorting, restraining, and identifying the animals. After an animal has had a detailed examination, it may be released without restriction or held after it has been harvested for further examination and disposition by the postmortem inspector.

Animal identity is established, enabling the postmortem inspector to assure that all animals being harvested have had antemortem inspection on the day of harvest. As of November 14, 1988, the U.S. Department of Agriculture has required that individual swine be identified in interstate commerce. However, in farrow-to-finish operations where the pigs are born and raised on one farm and marketed directly for harvest without being mixed with other hogs, and harvested one after the other as a group, the group can be identified at the plant. Efforts are being made to identify the origin of as many animals for harvest as possible in order to control illegal drug and chemical residues. Since the determination of residues must be accomplished after the animals have been harvested, live identification must be retained throughout the harvesting procedure.

Dead meat animals and dead poultry must be immediately tagged "U.S. Condemned" and tanked or effectively denatured under the supervision of an inspector. Denaturing involves completely marking by pouring a denaturant (dye), such as crude carbolic acid or finely powdered charcoal, over the animal or its parts.

Postmortem Inspection

Postmortem inspection procedures must be accomplished in such a manner that any unwholesome carcass, part, or organ will be detected and removed from human food channels. Lighting on the harvest floor is especially important so that inspectors can accurately appraise disease and contamination (Table 3–1). However, even under the best of conditions, inspectors cannot see microbial contamination. Thus, efforts are continuing to strengthen the scientific basis for postmortem inspection through the development of HACCP plans including rapid tests to check for contamination and adulteration. Image analysis techniques that use computer-assisted video, ultrasound, and reflectance photometry systems are being perfected.[27] These will aid in screening carcasses so that only the few carcasses with potential problems will be removed from the production lines for detailed examination and testing.

Head Inspection

Head inspection is very important for all species. All areas of head inspection and head washing for cattle, calves, hogs, and sheep must be illuminated to a minimum of 50 foot-candles (Table 3–1). In cattle, the head must be examined on all surfaces for pathological conditions and contamination; then, inspectors must properly incise and examine the mandibular, atlantal (if present), suprapharyngeal, and parotid lymph nodes (two each) for lesions of tuberculosis (page 104), actinomycosis (lumpy jaw), neoplasms (cancer), etc. The external and internal muscles of mastication must be incised in such a manner that will split the muscles in a plane parallel to the lower jawbone so that they can be inspected for tapeworm cysts (cysticercosis, pages 108–109). The tongue is observed and palpated along its entire length for bruises, hair sores, etc.

With calves, a visual inspection is made to determine if heads are free of hair, hide, horns, and contamination. Incisions for node inspection may be confined to the suprapharyngeal lymph nodes, unless there is reason to believe that other nodes should be examined.

In swine, both mandibular lymph nodes are sliced and examined for abnormalities, such as tuberculosis lesions (see Figure 5–31).

Viscera Inspection

CATTLE

- *Lungs.* Incision and inspection of the tissue of the right and left bronchial and anterior, middle, and posterior mediastinal lymph nodes; palpation of the parietal or curved surface; and observation of the ventral surface for tumor, abscess, or pneumonic conditions must be completed.

---

[27]Chen, Y. R. 1999. On-line Poultry Carcass Inspection System. Proc. 52nd Reciprocal Meat Conference of the American Meat Science Association. 52:49–50.

- *Heart.* The inner and outer surfaces of the heart must be examined, and the muscles of the left ventricle and interventricular septum must be incised and examined.

- *Liver.* Inspection includes surface observation and palpation for abscesses and other abnormalities, opening of the bile duct for liver flukes, and incision of the portal lymph nodes.

- *Spleen.* The spleen, along with the mesenteric lymph nodes and abdominal viscera, must be observed, and the ruminoreticular junction should be palpated. Mesenteric lymph nodes should be incised, if necessary.

CALVES — Lungs, heart, and liver must be observed and palpated, and the viscera, including paunch and intestines, must be carefully examined.

SWINE — Lung inspection includes observation of the parietal and ventral surfaces and palpation of the parietal surface, bronchial, and mediastinal lymph nodes. Both sides of the liver must be examined, and the parietal surfaces and portal lymph nodes should be palpated and incised, if necessary. The heart, spleen, and mesenteric lymph nodes should also be incised, if necessary.

SHEEP AND GOATS — The lungs and related lymph nodes, heart, spleen, and liver must be observed and palpated. The main bile duct must be opened by a meat plant worker and examined for parasites. The viscera must be carefully inspected.

Rail Inspection of Carcasses

CATTLE — All surfaces must be observed for pathology and cleanliness. The superficial inguinal (super-mammary), internal iliac, lumbar, and renal lymph nodes; exposed kidney and pillars (supporting tissues); and flat portion of the diaphragm must be examined and palpated. Incisions should be made, if necessary.

CALVES — There must be a visual inspection and observation of the carcass. The exposed kidneys and iliac nodes should be palpated, if necessary. The back of "hide-on" calves must be palpated to detect grubs and dirt.

SWINE — All parts of the carcass must be examined. Remnants of the liver and lungs, bruises, wounds, and other abnormalities must be removed by a meat plant worker before rail inspection can be completed. The kidneys must be palpated and observed for evidence of pathology, particularly kidney worms.

SHEEP AND GOATS — Both the internal and external surfaces of the entire carcass should be inspected. The prefemoral, superficial inguinal, popliteal, iliac lymph nodes and the diaphragm, kidneys, spleen (if present), and prescapular lymph nodes must be palpated.

POULTRY — The inspector must observe all external and internal surfaces of the cavity of each carcass, including the air sacs, kidneys, and sex organs, and must examine and palpate the legs, heart, liver, and spleen. The spleen in adult birds should be crushed. The inspector should signal the trimmer about removing defective parts and recording condemnations.

Disposition of Carcasses and Parts

Contaminated or diseased heads must be condemned or retained. When heads show pathological conditions, as previously discussed, the corresponding carcasses as well as the heads must be tagged for examination by a veterinarian. The veterinarian may permit the food inspector to dispose of viscera with localized lesions (abscessed livers, contaminated viscera, etc.). Carcasses that are contaminated or bruised or that have other abnormal conditions should be retained for trimming or for examination by a veterinarian before they are passed for final washing.

With poultry, questionable birds must be held for final postmortem inspection by a veterinarian.

Final Inspection Procedures and Dispositions

All organs, body cavities, and surfaces of the carcasses must be checked and examined. The lymph nodes must be exposed and incised, if necessary (TB, malignancies, etc.). With poultry, all carcasses and viscera retained for final inspection must be thoroughly examined by a veterinarian. For poultry, a minimum of 50 foot-candles at the lowest inspection point and the final postmortem areas is required (Table 3–1). All condemned as well as edible products must be carefully checked to assure that the postmortem line inspectors have made the proper dispositions. Disposition of carcasses and parts must be made in a professional manner on the basis of scientific training and reason and according to the meat inspection law and regulations. If necessary, samples should be sent to the laboratory for final pathological diagnosis to aid with dispositions.

Chilling and Moisture Control (Poultry)

Since poultry is often chilled with ice, certain restrictions apply. Immediately after harvest, evisceration, and washing, carcasses must be chilled by an approved method that will preclude adulteration. One such method is the use of potable ice (ice made from water suitable for drinking). An internal temperature of 40°F or less must be achieved. Moisture pickup must not exceed 8 percent. Efforts are underway to at least get the added moisture content listed on the label.

Control of Restricted Products, Animal Food Products, Condemned Carcasses, and Inedible Materials

- All *restricted products,* those meat products or ingredients which are undergoing tests for wholesomeness, must be under direct control or under lock or seal at all times until they are rendered acceptable for human consumption.

- *Animal food products,* those meat products suitable only for animals, must be under the inspector's control until they are packed and identified or denatured. Like products cannot be saved for human consumption and animal food simultaneously.

- *Condemned carcasses* or parts must be under the inspector's control until they are tanked (rendered safe through severe heat treatment) or properly denatured.

- *Inedible material* must be handled in a prompt, efficient manner by being placed in properly marked containers, under the inspector's supervision or control, until it is tanked or properly denatured, or until it is identified as food other than for humans.

Reinspection During Processing

Inspection and control of processed products (ham, bacon, sausages, etc.) must assure that only sound, wholesome products are distributed into human food channels. Even though animals and the meat they produce have already been inspected (both antemortem and postmortem), the meat, all added ingredients, and procedures are inspected again, thus, the term *reinspection*. Using only wholesome ingredients and approved quantities of acceptable chemicals, providing adequate protection during processing and storing, and controlling restricted products are procedures that are reinspected.

Facilities — Sufficient Lighting in All Areas

MEAT — All inspection areas (boning tables, grinders, bacon presses, slicers, choppers, etc.) must have 50 foot-candles of lighting. All other areas, except dry storage, where lighting sufficient for purpose is acceptable, must have at least 20 foot-candles. The lighting must be 10 foot-candles at the front shank level of the carcasses in the coolers. All lights must have protective coverings in processing rooms and areas where the product is exposed.

POULTRY — There must be 30 foot-candles of light in operating areas, 50 foot-candles at inspection stations, and 10 foot-candles in storage areas and coolers.

Table 3–1 summarizes inspection lighting requirements.

Table 3–1. Inspection Lighting Requirements

	Foot-Candles[1]
Antemortem	
Animal holding area (3 ft. above floor)	10
Suspect holding area (3 ft. above floor)	20
Postmortem[2]	
Heads (cattle, calves, hogs, sheep)	50
Poultry—lowest point and final	50

(Continued)

Table 3–1   (Continued)

	Foot-Candles[1]
Reinspection	
Boning tables, grinders, bacon presses, slicers, choppers, etc.	50
All other areas except dry storage	20
Front shank level for carcasses	10
Dry storage areas	"sufficient"
Poultry	
Operating areas.	30
Inspection stations	50
Storage areas & coolers	10

[1]A foot-candle is a standard unit of illuminance on a surface that is everywhere 1 foot from a unit of luminous intensity equal to $\frac{1}{60}$ of the luminous intensity of 1 square centimeter of a blackbody surface at the solidification temperature of platinum.

[2]Lights must have protective covering where product is exposed.

Management Controls — Products Received from Acceptable Sources

MEAT — Carcasses, cuts, and manufacturing meats must bear legible marks of inspection where they were harvested and(or) last processed.

POULTRY — Carcasses must be properly labeled with inspection legends.

NONMEAT — Supplies must be properly identified and labeled as required by the federal Food and Drug Administration.

Use and Handling — Storage of Raw Meat

MEAT — Raw meat, emulsions, and the finished perishable products must be stored at a room temperature of 50°F or lower, must be accessible to inspection, and must be handled in a manner that will avoid contamination.

POULTRY — Raw poultry meat, emulsions, and the finished perishable products must be held at an internal temperature of 40°F or less and handled in a manner to avoid contamination. The lower temperature requirement for poultry, compared to red meat, is based on a somewhat higher pH and more surface moisture, which makes poultry more susceptible to microbial growth.

Formulation Control and Identification

All ingredients, emulsions, mixtures, liquids, etc., must be identified through all phases of processing. All formulas and formulating procedures must be readily available for review by operating personnel and inspectors. The quantity of meat and nonmeat materials is controlled to produce a product in compliance with published standards and label declarations. Controls must be such that the inspector can evaluate the adequacy of the formulation.

Processing Controls for Curing, Pumping, and Smoking

CURING AND PUMPING — Restricted ingredients (nitrates, nitrites, phosphates, ascorbates, corn syrup) are to be used according to specific standards (see Chapters 20 and 21). All formulas for pickle and curing solutions and all curing and pumping procedures must be readily available for review by the plant manager and the inspector. The product must be uniformly cured and pumped.

SMOKING — Uniform procedures are used to shrink the product into compliance with applicable regulations (see Chapters 20 and 21). All pork products, except bacon, must be heated to a temperature not lower than 144°F for at least an instant to control trichinae. The method and control used must be known to insure such results. Bacon does not have to be heated to a regulated minimum temperature, since it is fried prior to being eaten. Cooked bacon, however, must be cooked sufficiently to control trichinae.

When cured and smoked poultry rolls are heated, they must reach an internal temperature of 155°F; all other poultry rolls must reach an internal temperature of at least 160°F.

Knowledge of Management Controls

The inspector must be knowledgeable of all procedures and controls used by management in the manufacture and formulation of all finished products.

Security of Brands, Certificates, and Seals

All brands and devices used for marking articles with the inspection legend, self-locking seals, official certificates, and other accountable items must be kept under adequate security, such as lock or seal, and an up-to-date inventory must be maintained of these security items (inventory does not include printed labels).

Potable Water

When water is used in areas where edible products are harvested, eviscerated, dressed, processed, handled, or stored, it must be potable.

To determine potability, the plant must have local authority certification on an analysis of samples taken from within the facility. The certification must be on an annual basis if the supply is from a municipal source, and if the supply is from a private well, cistern, spring, etc., the certification must be on a semiannual basis. Additional testing and certification are required when there is reason to believe that the water is being contaminated — cross-contamination of potable and nonpotable lines, back siphonage, surface drainage or ineffective drainage, floods, etc.

The usual test to determine potability is the total coliform count, a bacteriological test that, when positive, may indicate that the water has been contaminated with human or animal feces. However, coliforms occur naturally in soil, and most will grow in and on processing equipment in the presence of meat or related materials and moisture. The coliform group of bacteria is destroyed by normal heat treatment (pasteurization) and proper sanitation (see Chapter 2). Nitrate levels in the water are also checked. The meat industry is cooperating with the USDA–FSIS, the Food and Drug Administration (FDA), and the Environmental Protection Agency (EPA) to conserve water and to develop suitable methods to clean used water for further productive, safe use.

VACUUM BREAKERS — Where necessary, vacuum breakers of an acceptable type must be provided on waterlines connected to various pieces of equipment in order to prevent contamination of waterlines by back siphonage.

ICE — Ice must be made from potable water, certified by the appropriate local or state health agency, and handled and stored in a manner to avoid contamination. Block ice should be washed right before it is crushed.

NONPOTABLE WATER — The use of nonpotable water is permitted only in those instances when it cannot come in contact with edible products or potable water. Adequate identification of nonpotable lines is mandatory.

Sewage and Waste Disposal Control

Sewage and waste disposal systems must effectively remove sewage and waste materials — manure, paunch contents, trash, garbage, and paper. Such systems must also prevent undue accumulation or development of odors and must not harbor rodents and(or) insects. These systems must be approved by local or state health authorities for official plants. If there is no local or state agency with jurisdiction, or if the system is hooked directly into municipal lines, documentation must be provided.

ONSITE HANDLING — Conditions for onsite handling must be clean and sanitary. Sanitary problems created by the accumulation of objectionable materials and by the harboring of rodents or other nuisances are not acceptable.

Pest Control

The plant's pest control program must be capable of preventing or eliminating product contamination. Plant management must make reasonable efforts to prevent entry of rodents and other animals, as well as insects, into areas where products are handled, processed, or stored — including effective closures to outside openings (doors, screens, windows) — by using exterminating procedures, sprays, baits, etc. Only approved insecticides and rodenticides in the *List of Proprietary Substances and Non-food Compounds Authorized for Use Under USDA Inspection and Grading Programs,* Misc. Publ. 1419, may be used and must be applied in an approved manner.

Facilities and Equipment

The FSIS develops standards for facilities and equipment that will assure that the products produced in a plant will be sanitary and wholesome. Drawings and specifications of meat and poultry facilities and equipment must be approved by the FSIS prior to their use in federally inspected plants. Specific requirements may vary due to the size of the particular operation. For instance, the standard rail height specification to prevent carcasses from being contaminated from the floor is 11 feet, which requires very high ceilings for the support structures. A sizable operation needs the 11-foot-high rails throughout the plant for efficiency. The investment in the expensive construction is spread over a large volume of product, making it less costly to the operation. On the other hand, it would be practical for a small meat processor to build the high ceilings only for the harvest floor and chill cooler, where full carcasses and sides must hang intact, and build lower, less expensive ceilings and rails throughout the rest of the plant. In this type of situation, the operator would have to quarter (separate each side into forequarters and hindquarters) all beef carcasses to meet the floor clearance requirements in the rest of the plant, which might not be a problem for a small-volume operation. This way the facility sanitation requirements can be met with a lower investment.

Labels and Standards[28, 29]

As consumers have become more health-conscious, they have become increasingly aware of the value of the information on labels. The USDA–FSIS reviews and approves labels for all products that contain at least 2 percent poultry and 3 percent meat. Labels for other products are approved by the U.S. Food and Drug Administration. It is the task of label reviewers to make sure that each label is accurate and not misleading and that the product contains the appropriate ingredients. FSIS also develops formal product standards that specify the meat or poultry content and the ingredients of processed products.

When FSIS makes a major change in its regulations, this change affects a great number of processors and their approved labels. For instance, in 1990, two rule clarifications were made at the same time to minimize the cost of changing labels. One change reduced the number of substances that could be labeled simply as flavorings, and the second defined how much nonmeat protein could be used to calculate the permissible added water in sausage formulations. These two changes affected 2,500 plants and 45,000 labels.

---

[28]USDA–FSIS. 1987 (March). Meat and Poultry Labels Wrap It Up. Home and Garden Bulletin No. 238.

[29]Leddy, K. F. 1988. Labeling Policies for Meat and Poultry Products: Lean, Light, and Natural. Proc. Reciprocal Meat Conference. 41:21.

Components of a Label

A complete label must include these seven components:

1. Accurate product name

2. List of ingredients, in order from greatest to smallest amount

3. Name and place of business of packer, manufacturer, or person for whom product is prepared

4. Net weight

5. Official federal or state inspection stamp

6. Official plant number

7. Handling instructions, if the product is perishable (i.e., keep refrigerated, etc.)

Many manufacturers include more information on the label, such as nutritional details, which, although not yet required for all foods, is helpful to consumers who want to know how much of the Recommended Daily Allowance (RDA) they can expect from a serving. More and more consumer products are displaying this information on their labels. Although supplying dates is also optional, many manufacturers include dates on their products to guide consumers in safe storage and use. Cooking suggestions and menu tips are also helpful additions to a label. Figure 3–1 is an example of an approved label.

To be labeled with a particular name, a federally inspected meat or poultry product must be approved as meeting specific USDA product requirements so that consumers will get what they expect when they shop. A listing of standards for more than 200 popular meat and poultry products is available from the USDA–FSIS. The major products and their proper labels will be discussed in Chapters 18 through 21.

Special Label Claims

An increasing number of companies are interested in making nutrition-related claims on labels to capture the health-conscious market. The USDA–FSIS must approve and verify all such claims. A few examples are:

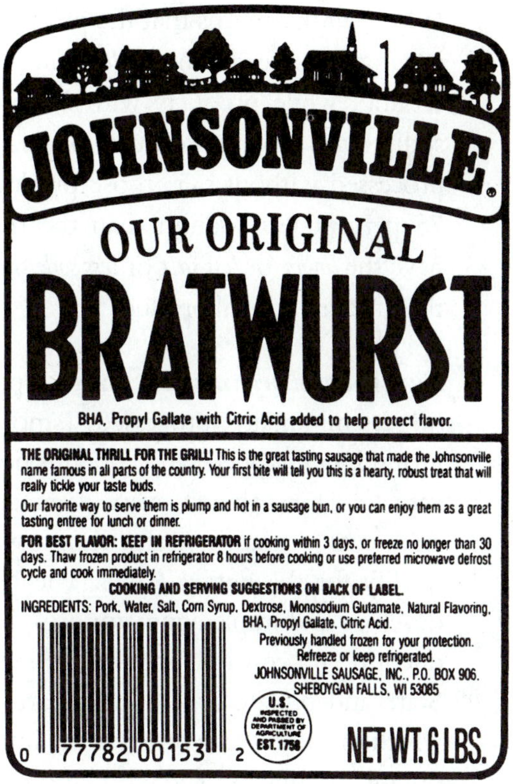

Fig. 3–1. Example of a label, Johnsonville Sausage Company, with the seven components.

- *Lean or low fat.* May be used on labels of meat or poultry products, *except ground beef and hamburger,* which possess no more then 10 percent fat by weight.

- *Extra lean.* As above, except no more than 5 percent fat by weight. *Ground beef and hamburger* may be labeled either *lean* or *extra lean,* provided the product contains no more than 22.5 percent fat. This figure represents a 25 percent reduction from the 30 percent fat previously permitted in these products (30 × 0.25 = 7.5; 30 − 7.5 = 22.5). In addition, the actual percentage of fat and lean must appear on the label.

- *Lite, light, lightly, etc.* All these terms refer to at least a 25 percent reduction in calories, fat, breading, or sodium from an approved source of comparison. They may also mean equal to or less than 40 cal/100 grams; 10 percent fat; 10 percent breading; 35 mg sodium/100 grams. Whenever *lite* is used on a label, it must be explained.

- *Sodium free* or *salt free; very low sodium; low sodium; unsalted* or *no salt added; reduced sodium;* and *lower* or *less salt* or *less sodium. Sodium free* or *salt free* means that the products contain 5 mg or less sodium per serving. Products with *very low sodium* have 35 mg or less of sodium per serving, while products with *low sodium* contain 140 mg or less of sodium per serving. Products labeled *unsalted* or *no salt added* have been processed without salt; those products labeled *reduced sodium* contain 75 percent less sodium than the traditional product; and those that have the *lower* or *less salt* or *less sodium* label have 25 percent less sodium than the traditional product.

- *Natural and organic.* The meat or poultry product that is labeled *natural* is minimally processed (smoked, cooked, frozen, fermented, ground, or chopped) and contains no artificial flavors, colors, or preservatives. Thus, almost every meat and poultry product, especially all fresh cuts — roasts, steaks, chops, whole poultry, and poultry parts — qualify for the *natural* label. The National Organic Program (NOP) develops and implements uniform federal standards related to the production, processing, and marketing of organically produced foods. State and private organizations are accredited through this program, which promotes and protects the interest of consumers, marketers, and organic producers. The NOP also coordinates the appointment of, and provides administrative support to, the National Organic Standards Board, an advisory board for the program, which was formed under provisions of the 1990 Farm Bill and is composed of scientists, handlers, farmers, environmentalists, consumer advocates, and a retailer. The board has considered the terms *natural* and *organic*.

  Private and state agencies certify products as *organic,* or the products are self-certified by producers. Each group uses a different set of standards, and there are no uniform national labeling regulations. Because of the current situation, the opportunity exists for mislabeling, fraudulent claims, and consumer confusion (http://www.ams.usda.gov/tmd/tmdnop.htm).

- *Extra* and *more than.* A particular component that is at least 10 percent more than the amount that is usually found in the regular product is labeled as *extra* or *more than.*

- *Exceeds USDA requirements for consumer protection.* When this appears on a label, it must be augmented with an additional statement that shows what requirement is exceeded, e.g., *protection from microbio-*

logical contamination. The company must prepare a control program that must be approved and monitored by the USDA–FSIS.

- *Imitation.* Products made to resemble or substitute for real products are labeled *imitation.* Such products may be lacking the specific ingredients, or they may contain some substituted ingredients that do not meet USDA–FSIS product standards.

- *Irradiation.* This is a preservation process (see Chapter 19) that must be indicated on the label by the word and the logo (see Figure 3–2).

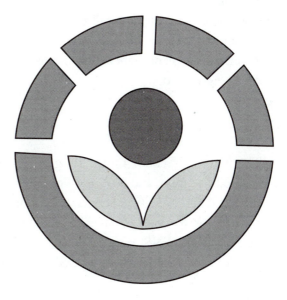

Fig. 3–2. Food irradiation symbol. *This symbol is required to be green and must be a part of the label on any food that has been irradiated.*

Nutritional Labeling Verification Procedures

A minimum of three laboratory analyses must be submitted with the label application. Once the label is approved, the product must be routinely verified as meeting the label statements. During each quarter of the first year after label approval, 12 random samples from 12 separate production shifts will be analyzed. If everything is satisfactory, during the second year and thereafter, samples will be taken and analyzed annually. This verification frequency will satisfy FSIS requirements because each company has continual, ongoing quality control programs that monitor the nutritional composition of its products.

Approved sources of comparison may be any of these three:

1. Regulatory standard (as with ground beef and hamburger)

2. Data from an appropriate reference source (e.g., USDA Agriculture Handbook No. 8[30])

3. Marketbasket survey

Verified Animal Production Control

Some livestock producers have received approved labels to sell meat from livestock that have been raised without the use of antibiotics or growth promotants. Producer testimonials and affidavits to substantiate the claims and protocol must list in detail the production practices on the farm, ranch, or feedlot and must be filed with the FSIS. Animals must be harvested in federally inspected plants where the carcasses can be positively identified and properly segregated.

Child Nutrition Program

The National School Lunch Program, School Breakfast Program, Child Care Food Program, and Summer Food Service Program are all administered by the Food and Nutrition Service (FNS) of the U.S. Department of Agriculture (http://www.usda.gov/). The FSIS maintains that establishments are complying with FNS requirements. Children must be served a specific amount of food from each of the USDA's 1992 five basic food groups:

1. Bread, cereal, rice, and pasta

2. Vegetables

3. Fruits

4. Milk, yogurt, and cheese

5. Meat, poultry, fish, dry beans, eggs, and nuts

The contribution of each individual food toward the total nutritional requirement is monitored through a special label (see Figure 3–3). In addition to the regular label requirements, the child nutrition label is printed with

---

[30]USDA, Human Nutrition Information Service. Agric. Handbook No. 8, Composition of Foods.

KEEP FROZEN

"Contains commodities donated by the United States Department of Agriculture."

SCHOOL LUNCH COMPONENT PACK FOR BEEF WITH TEXTURED VEGETABLE PROTEIN PRODUCT PIZZA

41.25 lbs. Component Pack for Cooked Beef and Textured Vegetable Protein Product Pizza: To each crust add 14.0 oz cooked beef topping, 10.00 oz. sauce and 16.00 oz. cheese/seasoning mixture and cut into 10 portions. Each portion provides 2½ oz. equivalent meat/meat alternate and 1½ servings of bread alternate and ⅛ cup vegetable for the Child Nutrition Meal Pattern Requirements. (Use of this logo and statement authorized by the Food and Nutrition Service, USDA 5-88.)

CONTAINS:

U.S. INSPECTED AND PASSED BY DEPARTMENT OF AGRICULTURE EST.

12 Crusts — 11.25 lbs.
Bag Cheese Topping —
11.85 lbs. cheese and
.15 lbs. of seasoning
Bag Beef and TYPP
Topping — 10.50 lbs.
Bag Sauce — 7.50 lbs.

CALCIUM PROPIONATE ADDED TO PRESERVE FRESHNESS OF CRUST INGREDIENTS: *TOPPING* (Chedder cheese, seasoning [onion, salt, spices, dehydrated romano cheese (made with cow's milk), garlic powder, parsley]). *CRUST* (Enriched flour, water, yeast, vegetable shortening, salt, dough conditioner [calcium sulfate, L-cysteine, hydrochloride, ascorbic acid], calcium propionate, sugar, wheat germ, dextrose, malted barley, calcium sulfate, ammonium chloride, potassium bromate). *BEEF AND TEXTURED VEGETABLE PROTEIN PRODUCT* (Ground beef [no more than 30% fat], water, textured vegetable protein product [soy flour, caramel color, zinc oxide, niacinamide, ferrous sulfate, copper gluconate, vitamin A palmitate, calcium pantothenate, thiamine mononitrate (B_1), pyridoxine hydrochloride (B_6), riboflavin (B_2), cyanocobalamin (B_{12})], salt, spices, dextrose, soy protein isolate, garlic powder, oleoresin paprika). *SAUCE* (Tomato purée, water, dextrose, modified food starch, salt, paprika, spices, citric acid, sugar, dehydrated romano cheese [made with cow's milk], garlic, onion, beef powder).

NET WT. 41.25 LBS.

Fig. 3–3. Label for products used in the Child Nutrition Program.

a distinct border, a six-digit identification number, and a statement of the product's contribution towards meal pattern requirements for the child nutrition programs.

Labels Track Time and Temperature

In an attempt to effect food safety, a new technology is being introduced into food labels. A special label can track a product for both time and temperature, helping prevent foodborne illness by warning when the product is at the danger point—that is, the point beyond which microorganisms may flourish. It is well known that after a product has been too warm for too long it may be contaminated, but there was previously no way to accurately track both times and temperatures, because each individual lot may experience a different amount of time inside and outside coolers. Readers may check the

NMA Web site (http://www.nmaonline.org) for information about these tracking labels and their sources.

Labeling Sausages Made with Natural Casings

Natural casings for meat and poultry sausages must be derived from the same species as indicated on the product label.

Substances Generally Recognized as Safe (GRAS) and Allergen Prevention

Meat processors should be aware of regulations governing food additives (Code of Federal Regulations 21CFR172) and those substances generally recognized as safe (GRAS) (Code of Federal Regulations 21CFR182). Processors should be sure to include on their product labels any additive that might cause allergic reactions in some consumers. Foods that commonly cause allergic reactions include peanuts, tree nuts (almonds, walnuts, hazelnuts, etc.), eggs, milk, soybeans, wheat, fish, and shellfish.

Consumer Participation in Setting Standards

New or revised standards and labeling rules are routinely being considered by the U.S. Department of Agriculture and the Food and Drug Administration. This information is readily available to news outlets. Consumers should express their views in writing.

Compliance

A group within FSIS monitors businesses engaged in interstate food marketing and distribution. This group investigates violations of the inspection laws; controls violative products through detentions, civil seizures, and voluntary recalls; and assures that appropriate criminal, administrative, and civil sanctions are carried out.

This compliance program operates between the processor and the consumer as a "second line of defense" against unfit products in marketing channels. Its aim is to prevent fraudulent or illegal practices once the product has left the processing plant. Compliance officers check for uninspected meat or poultry, counterfeit inspection stamps, inaccurate labels, and contamination or spoilage of products.

Enforcement actions are based on compliance reviews of meat and poultry handlers. Adjustable risk categories determine the frequency of scheduled reviews, and additional reviews are conducted randomly.

Pathology and Epidemiology

Pathology, epidemiology, and serology programs support meat and poultry inspection. Laboratory and investigative services study infectious agents associated with food and develop serological tests for infectious and toxic agents found in meat and poultry products.

The Meatborne Hazard Control Center investigates reports of potential health hazards. This group works with local, state, and federal public health agencies to control food-poisoning outbreaks by speeding up the identification of products responsible for human health hazards. The major cause for food poisoning has been found to be improper handling during the preparation of products at institutions, restaurants, and homes. Some samples of improper handling, as discussed in Chapter 2, are:

- Inadequate cooking.

- Storage at warm, "median" temperatures (between 40° and 140°F), which allows bacteria and other organisms to multiply rapidly.

- Failure to keep raw and cooked products separate during preparation.

- Contamination by human carriers of bacteria.

- Poor sanitation practices.

Residue Monitoring and Evaluation

Another important FSIS role is controlling unsafe drug and chemical residues that may occur in meat and poultry. There are two types of residues: (1) those present as naturally occurring components or contaminants in the environment and (2) those added by humans during the production, manufacture, or preparation of food products. Residue monitoring and surveillance is conducted in both the domestic and the import inspection programs. Animals, flocks, and herds suspected of illegal residues are sampled, and drug and pesticide enforcement agencies are alerted when violations are found. Residue Avoidance Programs (RAP), cooperative educational efforts involving the FSIS, producer organizations, and the USDA Extension Service are all instrumental in trying to prevent and control residues.

Among the FSIS testing capabilities is the enzyme-linked immunosorbent assay (ELISA), which can detect poultry, pork, beef, and horse antigens in meat products. Thus, this test is used to determine the species origin of meat products.

Federal-State Relations

Inspection is marked by continuing and expanding federal and state cooperation. The federal government offers technical, laboratory, and training aid. The federal-state relations staff provides this technical support and direction to state governments to assure that state inspection programs enforce requirements that are at least equal to those of the federal inspection laws. State-inspected plants may sell their products only within the state where they are located. However, as required by the 1996 Farm Bill, the Secretary of Agriculture had to submit recommendations to Congress for steps to achieve interstate shipment of state-inspected meat and poultry products. This is logical, if state and federal inspection are equal and if foreign meat can be sold anywhere in the United States (see next section). Perhaps it will become reality now, since all plants, both state and federal, are monitored by the HACCP program.

FSIS staff also give technical assistance to plants operating under the Talmadge-Aiken Act. State inspectors conduct federal inspections in these plants.

Foreign Programs

Imported meat and poultry products must meet the same standards as those produced in the United States. One measure of a country's inspection effectiveness is its body of laws, regulations, procedures, administration, and operations that must first meet U.S. standards. The FSIS evaluates importing country controls in these risk areas: disease, residues, contamination, processing, and economic fraud. It then approves the overall program. Individual plants within that country must then apply to their own government for certification to export to the United States. Each certified plant is subject to continuous inspection by inspectors of that country's government. This evaluation is supplemented by the work of FSIS veterinary medical officers with considerable experience in the domestic meat inspection system. They conduct periodic onsite reviews of certified foreign plants to assure the enforcement of the same standards of inspection as those in federally inspected U.S. plants.

The frequency of onsite review is determined by plant size, nature, and complexity of operations and by anticipated volume of exports to the United States. Plants that export large volumes, or those that are of special concern, are reviewed at least four times annually; other certified plants are reviewed at least once a year.

When imported meat and poultry products arrive in this country, they are reinspected by several federal agencies — U.S. Customs, USDA Animal and Plant Health Inspection Serivce (APHIS), and FSIS. They must bear prominent marking as to their country of origin. A meat-inspection certificate issued by the responsible official of the exporting country must accompany each shipment of meat offered for entry into the United States. The certificate identifies the product by origin, destination, shipping marks, and amounts. It certifies that the meat comes from veterinary antemortem and postmortem inspection; that it is wholesome, not adulterated or misbranded; and that it is otherwise in compliance with U.S. requirements.

To assure that the certifications made by foreign officials are valid, USDA inspectors, at the ports of entry and at destination points of inspection, inspect each lot of imported meat and poultry products.

A description of each lot arriving at U.S. ports is entered into the Automated Import Information System (AIIS) computer. This system centralizes inspection and shipping information from all ports and allows FSIS to set the inspection requirements based on the compliance history of each establishment. Information stored in the system includes:

- Amount and kinds of products offered from each country and establishment and the amount refused entry.

- Results of certification and labeling inspections.

- Results of organoleptic inspection for defects such as bone, hair, and cartilage.

- Results of laboratory samples tested for residues, proper cooking temperatures, economic and other adulterants. (An economic adulterant is a cheaper or less desirable ingredient that has been substituted for one that is required.)

To assure that representative samples are selected, statistical sampling plans are applied to each lot of product to be inspected. The sampling plans and criteria for acceptance or rejection of imports are the same as those used for U.S. federally inspected meat.

Although the sampling plans are generated by the AIIS to guide the inspection of imported lots, an inspector, whenever he or she thinks it is necessary, may hold a product and require additional samples or inspection procedures. All product not meeting USDA standards and found unfit for human consumption is directly controlled by the USDA import inspector from the time it is rejected until final disposition, a maximum of 45 days. Final disposition includes rendering, exporting, or processing as pet food. As a further check, imported meat that is subsequently used in domestic processed products receives additional examination in U.S. plants.

To assure that international regulations are standard, Codex Alimentarius, the international foods standards organization, was established in 1962. It is the major international organization responsible for protecting the health and economic interests of consumers, developing international food standards, and encouraging fair international trade in food. Codex is the joint food standards program of the Food and Agriculture Organization (FAO) and the World Health Organization (WHO) of the United Nations. The Commission is composed of members from 165-member nations, which represent 98 percent of the world's population.

Analytical and Scientific Support

The FSIS science program furnishes analytical support and scientific guidance for the meat and poultry inspection program. Laboratory analyses are designed to assure that meat and poultry products are safe from disease, microorganisms that cause food poisoning, harmful chemicals, and toxins.

The science group in the FSIS cooperates with other federal agencies, notably the Food and Drug Administration, the Environmental Protection Agency, and the Centers for Disease Control, and with state and local health authorities in carrying out its responsibilities. It develops and maintains close ties with national and international scientific communities in order to keep abreast of scientific and technological advances and to open new avenues for the exchange of scientific information.

FSIS chemists develop and improve practical, analytical procedures for detecting adulterants and chemical residues in meat and poultry products. They perform highly complex chemical analyses, coordinate an accredited laboratory program, and conduct onsite technical reviews of chemistry field-service laboratories to assure the quality and integrity of analytical results. In addition, they participate with the Food and Drug Administration in evaluating new animal drug applications.

Microbiologists provide analytical services to federal, state, and local agencies and advise others in the FSIS of the significance of laboratory results. They develop economical and efficient analytical screening methods for use in laboratories, in plants, and on the farm. They also carry out special investigations on the safety and quality of products and processes.

EXEMPTIONS FROM FEDERAL AND STATE MEAT INSPECTION

The requirements of the Wholesome Meat Act, the HACCP system, and the federal and state regulations for inspection do not apply to the following:

1. The harvesting by any individuals of livestock that they have raised, and the preparation by them, and the transportation of the carcasses or cuts, meat and meat food products of these livestock exclusively for use by them and the members of their households and their nonpaying guests and employees.

2. The harvesting and processing of an individual's livestock by another individual who has been paid for his or her services (custom harvesting and processing), in which case the same restrictions apply as in No. 1.

3. The harvesting by an individual of his or her own livestock and the delivery of the carcasses to a custom processor, in which case the same restrictions apply as in No. 1. Meat processors who only do custom harvesting and processing are known as *custom-exempt* processors and are subject to occasional inspections, perhaps biweekly or monthly, depending on the availability of inspectors. However, if custom operations are conducted in a plant that has inspection and sells meat (an official establishment), all of the provisions of the inspection act apply, including the following:

    a. Any products for sale are kept separate from the custom-prepared products at all times.

    b. The custom-prepared products are plainly marked "Not for Sale."

    c. If custom-exempt harvesting or other preparation of products is conducted in an official establishment, all facilities and equipment in the official establishment used for such custom operations must be thoroughly cleaned and sanitized before they can be used for preparing any products for sale.

    d. The custom-exempt, custom-prepared products must be prepared and handled in accordance with the provisions of the inspection act and cannot be adulterated.

4. The operations of types traditionally and usually conducted at retail stores and retail-type establishments for sale in normal retail quantities. However, many retail stores and restaurants are under city or county health department jurisdiction. The HACCP system should be used in all these outlets, but it remains practically voluntary in areas where inspection support is lacking. Large chain stores and restaurant chains are employing HACCP systems to assure food safety. It behooves consumers to seek these outlets for meat purchases. According to USDA–FSIS regulations, operations of types traditionally and usually conducted at retail stores and restaurants are the following:

    a. Cutting up, slicing, and trimming carcasses, halves, quarters, or wholesale cuts into retail cuts, such as steaks, chops, and roasts, and freezing such cuts.

    b. Grinding and freezing products made from meat.

    c. *Curing, cooking, smoking, or other types of preparation of products,*[31] except harvesting, rendering, or refining of livestock fat or the retort processing of canned products.

    d. Breaking bulk shipments of products.

    e. Wrapping or rewrapping products.

---

[31]A review of retail store inspection exemptions by the USDA is ongoing. The retail industry and the entire meat industry are continually changing. The italicized portion of item c is one critical area where uniform inspection regulations should be in force across the United States.

f. Consumer purchasing of any quantity of product from a particular retail supplier, which is deemed to be a normal retail quantity if it does not in the aggregate exceed one-half carcass. The following amounts of product will be accepted as representing one-half carcass of the species identified:

	One-Half Carcass (lb.)
Cattle	300.0
Calves	37.5
Sheep	27.5
Swine	100.0
Goats	25.0

g. Selling a product to consumers only. In terms of dollar value of total sales of a product, at least 75 percent represents sales to household consumers, and the dollar value limitations of total sales of product to consumers other than household consumers (i.e., hotels, restaurants, and institutions) are adjusted in accordance with Consumer Price Index changes when the amount of adjustment equals or exceeds $500.

VIOLATION OF WHOLESOME MEAT ACT

If any of the following conditions exist, the plant must be designated as endangering public health (EPH), and corrective action must be taken immediately.

- Use of nonpotable water in edible products departments.

- Improper sanitation that results in bacterial growth and development in or on the product, foreign matter entering the product, or the failure to control vermin and insects.

- Presence of carcasses or parts showing sufficient evidence to identify a systemic diseased condition or containing evidence of bearing a disease transmissible to humans.

- Use of unsound meat or poultry in processing meat or poultry food products.

- Presence of harmful chemicals and preservatives in excess of permitted tolerances.

- Failure to properly treat or destroy trichinae.

Inspectors should take immediate action to correct deficiencies in all phases of the operation within their purview. When a plant has been classified as endangering public health, it must be surveyed for corrective action after five working days. When a plant is deficient in one or more of the basic requirements but is not in an EPH category, it must be resurveyed no later than before the end of the succeeding quarter.

Each violation involving intent to defraud or any distribution or attempted distribution of an article that is adulterated shall make the person or persons representing a firm or corporation subject to imprisonment for not more than three years or to a fine of not more than $10,000. One of the largest fines levied against a company for violating the Federal Meat Inspection Act (FMIA) is $1,000,000 in a case in which company officials were found guilty of bribing federal meat inspectors and selling canned hams containing excessive amounts of water. This large fine resulted from the summation of many serious violations.

Before the FSIS turns over a suspected violation of the FMIA to the U.S. Department of Justice for prosecution, it notifies the suspected persons, firms, or corporations of its intent. The suspected violators have a limited time, depending on the seriousness of the violation, to respond. This advance notice will not be given if there is likelihood that the evidence might be destroyed or altered, persons might be injured or property damaged, the suspected violators might flee, undercover investigative techniques might be compromised, bribery or clandestine harvesting/processing operations might be involved, or if this violation is part of another criminal investigation.

OFFICE OF THE INSPECTOR GENERAL (OIG) AUDIT PLAN INCLUDES HACCP

The Office of the Inspector General work plan almost always includes audit investigations as follows:

- *Food Safety and Inspection Service HACCP Compliance:* OIG evaluates FSIS's and the industry's effectiveness in implementing, providing oversight of, and monitoring the HACCP Microbiological Testing of meat and poultry products.

- *OIG Assessment of FSIS's Controls over Meat and Poultry Production:* This assessment is to ensure:

    1. Adequate product sample selection and security.

    2. Reliable testing and reporting procedures.

    3. Sufficient enforcement actions, recall procedures, and reporting of product samples tested as positive for harmful pathogens.

- *OIG Review of FSIS's Food Safety Information Systems:* OIG determines whether FSIS has an effective automatic information system to schedule inspection tests, record inspection results, and track resolution of plant violations/deficiencies. OIG also determines whether the system is designed to develop plant profiles to identify problem plants, trends, or indications of widespread inspection system weakness/problems and, finally, to assess/monitor inspectors' performance.

THE MEAT INSPECTOR

FSIS personnel are divided into several classifications. Professional inspectors, administrators, laboratory scientists, etc., trained in veterinary science, chemistry, microbiology, meat science, muscle biology, and related disciplines, must pass required civil service examinations that qualify them for the GS 11–12 rank. Nonprofessional or lay inspectors must also pass a required civil service examination, designating them as GS 5–9.

In an effort to streamline the operation and costs of government regulatory agencies, the FSIS, the USDA division responsible for inspection, and the Agricultural Marketing Service (AMS), the USDA division responsible for grading, have signed an agreement to cross-utilize employees between the two agencies. Because persons will be properly trained and licensed in both inspecting and grading, double staffing can be eliminated in some plants.

THE MARKS OF FEDERAL AND STATE INSPECTION

Each establishment under federal or state inspection is granted an official number that appears on the inspection stamp and identifies the product wherever it is found (see Figures 3–4 and 3–5). For poultry plants, the inspection number is preceded with a "P."

Fig. 3–4. Examples of two state inspection stamps.

This mark is the federal stamp that is put on meat carcasses. It is only stamped on the major cuts of the carcass, so it may not appear on the roast or steak you buy.

This mark is put on every pre-packaged processed meat product — soups to spreads — that has been federally inspected.

This mark is stamped on federally inspected fresh or frozen poultry or processed poultry products.

This mark is stamped on federally inspected fresh or frozen exotic meat or processed exotic meat products.

Fig. 3–5. Examples of federal inspection stamps. Beside the triangle is printed "This mark is stamped on federally inspected fresh or frozen exotic meat or processed exotic meat products."

HUMANE HARVESTING

The federal Humane Slaughter Act went into effect on July 1, 1960, and was updated effective October 11, 1979. Under the act, all official establishments harvesting livestock under provision of the Federal Meat Inspection Act must employ humane methods of harvesting and handling livestock. Failure may result in temporary suspension of inspection, meaning operations must cease until the violation is corrected. Many states have enacted laws similar to the federal law. Ritual harvest methods, such as Kosher harvesting, are exempt from the act.

The federal act recognizes three methods of immobilization: mechanical, chemical, and electrical. Use of any one of the three methods, as described in Chapters 5 to 9, must produce complete unconsciousness with a minimum of excitement and discomfort.

THE KOSHER STAMP

The Humane Slaughter Act of 1960 does not apply to ritual harvest methods such as Kosher harvesting, a method in which the live animal is suspended (unstunned) and bled by an incision made by a specially trained rabbi or shohet across the throat (cut throat). The knife has a razor sharp, 14-inch blade. The throat of the animal must be washed free of any grit or foreign material so it will not nick the knife. The shohet makes no inspection of the carcass prior to placing the Kosher stamp (see Figure 3–6) in script or block letters on the carcass. The stamp carries no implication of the health or grade of the animal; only that it is proper, according to Hebrew law, and clean.

The *law* referred to is found in Leviticus 17:14, which states, "You shall not eat the blood of any creature, for the life of every creature is its blood; whoever eats it shall be cut off." The explanation given by those of the Jewish faith as to why they consume forequarter meat primarily is because it contains less blood and is more easily veined (blood vessels removed) than the hindquarter.

Fig. 3–6. Kosher stamp.

ANIMAL DISEASES AND
OTHER LOSS FACTORS

One of the major goals of the HACCP system is to have a completely safe source of food for the consuming public. Thus, this section on animal diseases and loss factors is doubly important to the overall food safety picture.

Anyone doing even a limited amount of harvesting, whether it be for home use or for subsequent sale, should be able to recognize unhealthy or unthrifty animals and should know something about the effect of the more common ailments on the wholesomeness, quality, and value of carcasses. Whenever there is any doubt concerning the health of an animal, a veterinarian should be consulted.

Workers at the University of Wisconsin–Madison have prepared a very complete series of films, videotapes, and fact sheets on this topic.[32] The diseases, parasites, and other loss factors covered in this series are those that directly affect the amount, quality, and wholesomeness of meat. Many other diseases and conditions influence animal health, growth, and thriftiness, but they are too numerous to be discussed here.

An additional, valuable reference for animal handling is a brochure prepared by the National Pork Producers Council.[33]

Pregnancy

Animals should not be harvested when they are in the advanced stages of pregnancy and obviously in the preparatory stage of parturition. This preparatory stage is characterized by dilation of the cervix and rhythmic contractions of the longitudinal and circular muscles of the uterus. During parturition, increased levels of the female hormones oxytocin and estrogen work to increase the irritability of the uterine musculature, allowing it to contract. The physiological condition of the female is thus disturbed, which may affect her whole musculature. Furthermore, her temperature most likely will have risen above normal. Following parturition and passage of the placenta, a female can be harvested, if desired, when her body temperature returns to normal (101°F).

---

[32]Departments of Agricultural Journalism and Meat and Animal Science, University of Wisconsin–Madison. 1988. Livestock and Carcass Abnormalities Series.

[33]Meisinger, D. 1999. A System of Assuring Pork Quality. National Pork Producers Council, in cooperation with the National Pork Board, P.O. Box 10383, Des Moines, IA 50306.

Accidental Death or Injury

A *healthy* animal that is killed or injured through accident, that suffocates from bloat, or that dies from a heart puncture caused by a nail or wire is fit for food, providing some knowledgeable person is there to cut the throat and bleed the animal. This should be done immediately in the case of death, and within minutes, or one or two hours, in the case of an injury, depending on the nature and severity of the injury. By no means is a dead or diseased animal suitable for human food. Once the formerly healthy animal is bled, the harvesting procedure should be completed and the carcass chilled as soon as possible, preferably within the hour.

Swine Porcine Stress Syndrome (PSS)

Pigs affected with PSS are unable to withstand the stress of management procedures that involve handling and crowding, transportation, or sudden environmental change. When these pigs are subjected to such stressful situations, they have a reaction that sometimes results in death.

The following sequence of events is typical, if a pig suffering from PSS is subjected to stress. The pig may show signs of trembling or muscle tremors and may be difficult to move. These initial symptoms are followed by irregular blotching of the skin of white pigs, labored breathing, and increased body temperature. The terminal stage of the syndrome is total collapse and a shock-like death of the animal. The condition is genetically controlled, so producers do well to select against the trait.

The PSS condition is usually associated with low-quality or pale, soft, exudative (PSE) pork. However, dark, firm, dry (DFD) pork may also result from the PSS condition. For further discussion, see Chapter 22.

Proper Livestock Shipping — Bruise and Death Loss Prevention

The growers of our meat supply face innumerable problems in raising animals for market. The fight against birth losses, disease, predatory animals, the elements — heat, cold, and drought — and nutrient and vitamin deficiencies is a continuous one. After producers have invested in feed, buildings, medications, equipment, and months of care, a shocking number of livestock never make it to market alive, or if they do, they are severely discounted for bruises and(or) broken limbs, etc. A bruise results from the hemorrhaging of a blood vessel under the hide. Bruised meat cannot be used for human food. The live-stock industry loses more than $50 million annually from bruises on cattle, hogs, and sheep.

The National Livestock Loss Prevention Board, organized in 1934, was reorganized in 1952 as the Livestock Conservation Institute[34] for the purpose of studying the causes of livestock losses on the farm and on the way to market. Two excellent videotapes available from this organization are "Cattle Handling and Transportation" and "Swine Handling and Transportation." The National Meat Association (NMA), in cooperation with Grandin Livestock Handling System, Inc.,[35] has developed films and videotapes for animal handling improvement and loss prevention. Most bruises occur during loading and unloading. In cattle, the major sites of bruising are in the valuable sirloin/hip area (31 percent of all cattle bruises), ribs (13 percent), and shoulders (36 percent). In swine, 66 percent of all bruises occur in the ham, followed by shoulders (10 percent) and loin (7 percent). In sheep, the major site is the legs (27 percent), followed by the loin (17 percent). Horns cause a large percentage of the bruises in cattle. Additional bruises result when animals are forced to turn short corners or forced through too narrow openings, such as when they are being loaded into the side of a semi-trailer. Ham damage in hogs results from human kicking and from slick floors that cause a hog to split or spread its legs violently out to the sides. A canvas slapper should be used for moving hogs. The most common cause of bruises in sheep is grabbing them by the wool or by the hindleg. When possible, a Judas goat (betrayer) or pet sheep should be used to lead them.

Loading facilities are very important. The recommended slope on loading chutes is 20°, although a 25° slope may be a reasonable compromise for maximum. Ramps should have solid sides to block out distractions from outside the chute that may spook the animals. A flat landing at the top of the chute is helpful for ease in getting animals to walk onto the truck.

Death loss in shipping is a special concern in hogs, although extra care in cattle shipments could result in reduced shrinkage. Shrink is the loss of animal weight (mostly stomach contents and water) between the farm or ranch and the market. It amounts to 5 and 8 percent in fed cattle hauled 50 and 750 miles respectively. Thus, most shrink comes in the early stages of a trip. All species shrink more in hot weather, and range livestock, not accustomed to

---

[34]Livestock Conservation Institute, 6414 Copps Ave., Suite 204, Madison, WI 53716. Phone 608-221-4848.

[35]Dr. Temple Grandin, Colorado State University, Boulder, CO. http://www.grandin.com

people, shrink more than other livestock. Hog transit deaths are much higher in the summer months, running from 48 to 60 per 100,000 hogs versus 33 to 38 per 100,000 during the other seasons. When livestock shipments are being planned, paying attention to the Livestock Weather Safety Index, which is widely disseminated over the media, can help alleviate death losses. Hot weather and high humidity are the cause of high mortality, especially in hogs. Sand should be used in the bottom of trucks and should be wetted down in hot weather and covered with straw in cold weather.

The Livestock Conservation Institute has provided space recommendations for shipping livestock. Hogs need more room in a truck during hot weather. Hogs weighing 200 pounds need a minimum of 3.5 square feet per animal. A 230-pound hog needs up to 4.4 square feet when the humidity is high and the temperature is over 75°F.

Table 3–2 provides a rule-of-thumb guide per running foot of truck floor (based on a 92-inch inside truck width) for varying weights of hogs when

Table 3–2. Truck Space Requirements for Hogs

Average Weight	Number of Hogs per Running Foot of Truck Floor (92-in. Truck Width)
(lbs.)	
100	3.3
150	2.6
200	2.2
250	1.8
300	1.6
350	1.4
400	1.2

Examples (250-lb. hogs):

44-ft trailer—44 × 1.8 = 79 hogs when temperature is mild. If the Livestock Weather Safety Index is in the Alert range, load 71. If the Weather Safety Index is in the Danger zone, ship only at night and load just 63 hogs.

44-ft double-deck trailer—88 × 1.8 = 158 during mild weather, 142 head if the Weather Safety Index is in the Alert range, and 126 head if it is in the Danger zone.

44-ft. possum-belly (two 10-ft front decks, three 25-ft. middle decks, and two 9-ft. rear decks, 113 ft. of floor space)—113 × 1.8 = 203 head in mild weather, 183 head when the Index is Alert, and 163 head if the Index says Danger.

temperatures are below 75°F. When the Livestock Weather Safety Index is in the Alert range, 10 percent fewer hogs should be loaded, and when it is in the Danger zone, 20 percent fewer hogs should be loaded.

Overloading of trucks is a major cause of bruises in loads of horned cattle or cattle which have had their horns tipped. Two or three too many horned 1,000-pound steers on a semitrailer can double the bruising. Tables 3–3 and 3–4 provide rule-of-thumb guides for the number of cattle and calves that can be loaded per running foot of truck floor in a truck with a standard 92-inch width. The recommendations apply to cows, range animals, and feedlot animals with horns or tipped horns. For feedlot steers or heifers without horns or with horn growths due to improper dehorning (scurs), the load can be increased by 5 percent.

Table 3–5 provides a rule-of-thumb guide for loading sheep.

Table 3–3. Truck Space Requirements for Cattle (Cows, range animals, or feedlot animals with horns or tipped horns; for feedlot steers and heifers without horns, increase by 5 percent)

Average Weight	Number of Cattle per Running Foot of Truck Floor (92-in. Truck Width)
(lbs.)	
600	0.9
800	0.7
1,000	0.6
1,200	0.5
1,400	0.4

Examples (1,000-lb cattle):

44-ft. single-deck trailer—44 × 0.6 = 26 head horned, 27 head polled.

44-ft. possum-belly (four compartments, 10-ft. front compartment; two middle double decks, 25 ft. each; 9-ft. rear compartment, total of 69 ft. of floor space)—69 × 0.6 = 41 head of horned cattle and 43 head of polled cattle.

Table 3–4. Truck Space Requirements for Calves (Applies to all animals in 200- to 450-pound weight range)

Average Weight	Number of Calves per Running Foot of Truck Floor (92-in. Truck Width)
(lbs.)	
200	2.2
250	1.8
300	1.6
350	1.4
400	1.2
450	1.1

Examples (450-lb. calves):

44-ft single-deck trailer—44 × 1.1 = 48 head.

44-ft. double-deck trailer—88 × 1.1 = 97 head.

Table 3–5. Truck Space Requirements for Sheep (Use for sheep being transported for harvest; load 5 percent fewer if sheep have heavy or wet fleeces)

Average Weight	Number of Sheep per Running Foot of Truck Floor (92-in. Truck Width)
(lbs.)	
60	3.6
80	3.0
100	2.7
120	2.4

Example (120-lb. sheep):

44-ft. triple-deck trailer—44 × 3 × 2.4 = 317 shorn sheep, 302 wooly sheep.

Swine Abscesses

In the early 1960s, the swine industry, including members of the Livestock Conservation Institute, became concerned about the incidence of losses of pork carcasses and carcass parts at packing plants due to swine abscesses. These losses began to increase in the late 1940s. A national survey conducted by the Livestock Conservation Institute indicated an annual product loss at the processing level of $12 million. These were largely jowl abscesses, although some packers found shoulder abscesses extending into the muscle as far as the needle was inserted when an earlier vaccination was given. After reaching a peak in the mid-1960s, the incidence of condemnations due to swine abscesses has declined steadily, largely because of the application of information obtained from research.

The Swine Abscess Committee of the Livestock Conservation Institute coordinated the findings of four state research groups and one USDA group and prepared and distributed a leaflet containing suggested control measures. The bacterial group *E. streptococcus* (GES) was found to be the principal cause of jowl abscesses. More recently, *Streptococcus equisimilis* (group C), one of the "streps" commonly associated with domestic animals, has been the major organism isolated from swine abscesses. These abscesses were 1 to 20 mm in diameter, mostly in intermuscular tissue, although a few were in the muscle itself, and had thick, fibrous walls, with yellowish, cheesy, odorless pus.

Abscesses induced in swine by GES occur primarily in the lymph nodes of the head and neck following infection via the oral or nasal routes. Involvement of lymph nodes in other areas of the body is unusual but may occur if GES enters the bloodstream. After ingestion, the organisms are quickly carried to the regional lymph nodes. Organisms have been found in mandibular and cervical lymph nodes as early as two hours after oral exposure. Microscopic abscesses begin forming in lymph nodes as early as 48 hours after exposure. These abscesses can be seen with the naked eye on cut surfaces of dissected lymph nodes as early as seven days after exposure and can usually be felt or seen externally within two weeks.

Susceptibility of swine to abscesses is influenced by age. The disease occurs primarily in post-weaning swine. Swine can become infected with GES by nasal contact with infected swine, via drinking water, and through feces. Abscesses that drain to the outside skin surface are also a source of infection. GES persists in pasture and swine pen soil for several weeks, especially at cold temperatures.

Swine can continue to transmit the disease as long as 2½ years after initial infection, but usually the carrier state lasts 10 months or less. Swine become immune to GES following recovery from the disease, and their serum will provide temporary passive immunity when it is injected into susceptible swine. Infected sows transmit passive immunity to their pigs via colostrum.

To cut down on costly abscesses resulting from various injections, the producer should make the injections just behind the pig's jawbone in the higher portion of the jowl (see Figure 3–7).

Injections at the site indicated seem to cause less tissue reactions, and if the needle is contaminated, any tissue reaction at this particular area causes less tissue trim than if injected higher or further back. Sanitary procedures must be observed while the animals are being vaccinated. Animals contaminated with fecal material should be cleaned before being injected. To prevent contamination of the bacterin in the vaccine bottle, producers should not insert the needle that was used for injecting a pig back in the bottle.

Fig. 3–7. Injections at the site indicated seem to cause less tissue reactions, and if the needle is contaminated, any tissue reaction at this particular area causes less tissue trim than if injected higher or further back. (Source: LCI83-8-14, *Vaccination Abscesses*)

CAUTION: Directions for withdrawal periods, as given on the label, should be carefully observed.

Beef Carcass Abscesses and Granulomas

The incidence of top sirloin abscesses and granulomas (inflamed tissue associated with an infective process) in beef cattle has lessened in recent years.

Some of the abscesses are so deep that they cannot be detected during harvest inspection. A deep abscess or granuloma, if present, does not become obvious until the top sirloin is cut into steaks. It and the surrounding tissue must be removed and discarded at that time. This is a serious problem, one that the industry is greatly concerned about.

The point in the production system when injections are made, the injection site on the animal, specific injected product composition, and injection hygiene are all important factors causing these abscesses.

To prevent this serious loss in one of the most valuable portions of the beef carcass, producers should strictly follow directions for any product they deem necessary to use, and if they give injections, they should make them in the less valuable neck area of the animal.

Cattle Liver Abscesses

Liver abscesses in feedlot cattle have been attributed to an abrupt increase in the intake of high-energy feeds.[36] A high-grain ration causes the rumen pH to become more acidic. Some cattle suffer from rumenitis (inflammation of the rumen wall) because of the acidic condition. The rumen wall becomes weaker and somewhat porous, allowing rumen bacteria (*Spherophorus necrophorus*) to pass out into the bloodstream.[37] These organisms eventually cause abscesses in the liver, which, in severe cases, can spread to other organs, sometimes resulting in adhesions. Economic losses can be as high as $60 per animal in organ and carcass damage, to say nothing of the cost in poor gains and efficiency.

Swine Arthritis[38]

Arthritis is commonly recognized as being a major factor in causing swine lameness. The disease is brought about by infection of the joints and the surrounding tissues by bacteria or *mycoplasmas* (nonmotile microorganisms without cell walls that are intermediate in some respects between viruses and bac-

---

[36]Brown, H., N. G. Elliston, J. W. McAskill, O. A. Muenster, and L. V. Tonkinson. 1973. Tylosin phosphate (TP) and tylosin urea adduct (TUA) for the prevention of liver abscesses, improved weight gains and feed efficiency in feedlot cattle. J. Anim. Sci. 37:1085.

[37]Smith, H. A. 1944. Ulcerative lesions of the bovine rumen and their possible relation to hepatic abscesses. Am. J. Vet. Res. 5:234.

[38]Pork Industry Handbook, PIH 36. Cooperative Extension Service.

teria). USDA meat inspection records indicate that trimming parts of pork carcasses and discarding whole carcasses because of arthritis are leading causes of loss at harvest. Greater loss probably occurs on the farm because of slower and less efficient gains and reduced performance by adult breeding stock and lactating sows. There are some instances of mortality, but death losses are not usually substantial.

One type of arthritis comes about as a result of erysipelas (diamond skin disease). This type can be prevented by vaccination. Streptococcal arthritis is another type that affects baby pigs. It can be treated effectively with penicillin if it is caught early enough. Two species of mycoplasmas cause arthritis in swine. Pigs under stressful conditions and practices and pigs weakened by disease or illness, such as pneumonia or diarrhea, are more susceptible to mycoplasma arthritis. Furthermore, breeding stock with good leg conformation and action should be selected from arthritis-free herds. General good management and sanitation are important in preventing arthritis.

In the acute stages of arthritis, joints show swelling, and pigs are lame and do poorly. The infection can spread throughout the body, not often resulting in death but often causing such carcasses to be condemned by FSIS personnel upon harvest. When the whole carcass is not infected, only the joint filled with excess fluid and the muscle surrounding that joint have to be removed and condemned. If the arthritic condition is localized and arthritis passes into the chronic stage before the animal is harvested, the affected joints become enlarged, stiffened, hard, and nonfunctional because connective tissue and bone have been deposited there. Such animals do not do well, but when harvested, they have less loss of product than those harvested in the acute stage.

Brucellosis

Brucellosis, caused by the bacteria *Brucella abortus,* inflicts a serious economic drain on the livestock industry. The cost to the producer resulting from cattle and swine abortions and delayed breeding is sizable, and many times it goes unrecognized. Packers are affected because brucellosis reduces the supply of animals for harvest and increases operating costs. Furthermore, because brucellosis is contagious, it can create a health hazard for livestock producers, veterinarians, and packing house workers. Since infected animals are normally culled from the herd and harvested, packing house workers are among those persons more apt to become exposed to the disease; however, the disease poses no threat to the wholesomeness of the meat processed from infected females.

Since the symptoms of human brucellosis — fatigue, fever, chills, weight loss, and body aches — are very similar to those of other common human infections, the disease is often overlooked or misdiagnosed. Chronic forms may cause more adverse effects, such as emotional disturbances, arthritis, and attacks on the central and optic nervous systems.

At the present time, there is no vaccine against brucellosis in humans, and relapses are common. Prompt diagnosis and treatment can shorten the course of the disease and prevent complications.

Livestock producers who assist in calving should wear protective gloves and scrub well afterwards as a precaution against contracting brucellosis. They should avoid touching their eyes, nose, and mouth after they have handled newborn calves or raw milk. They should burn or bury aborted fetuses and contaminated placental tissue and disinfect all contaminated areas, such as calving pens or areas where abortions have occurred. Finally, they should test all cattle suspected of brucellosis as well as any additions to their herds.

Tuberculosis

Various types of tubercle bacilli exist, and they most commonly infect a single species of animal. These bacilli are capable, however, of causing disease in a wide range of hosts — including humans and domestic animals. Three types — human, bovine, and avian — are of primary concern.

The human-type tubercle bacilli are usually associated with the familiar lung disease. They can also cause disease in other parts of the body and in other animals, including cattle, hogs, dogs, cats, canaries, and monkeys.

The bovine bacilli are usually associated with tuberculosis in cattle. Many other species, including humans, may also become afflicted — if they come in contact with infected cattle or drink unpasteurized milk from infected cattle.

The avian bacilli infect chickens primarily, but they may cause the disease in varying degrees in a wide variety of birds and mammals.

Hogs are susceptible to infection by all three types of tubercle bacilli. Odds are against hogs becoming infected with the human or bovine type in the United States, which is primarily due to an extensive campaign to eradicate bovine tuberculosis. Laws requiring the cooking of all garbage before it can be fed to hogs have played a large part in the eradication.

Avian tuberculosis, on the other hand, still causes condemnation of pork at harvest plants. In February 1972, the following additional meat inspection regulations pertaining to tuberculosis went into effect.

Carcasses Passed Without Restriction

- To be passed without restriction, a cattle carcass must be found to be free of tuberculosis lesions during postmortem inspection and must not be identified as a tuberculin test reactor.

- A swine carcass may be passed without restriction as long as the lesions are localized and limited to one primary site. Primary sites are cervical (neck area), mesenteric (abdominal area), and mediastinal (heart/lung area) lymph nodes.

Carcasses Passed for Cooking

In order for a carcass to pass for cooking, the packer must hold the product at 170°F for 30 minutes. The loss to the packer in lowered quality of this cooked meat is approximately two-thirds of its uncooked market value. The cooking requirement is more severe than that required for milk pasteurization (161°F for 15 seconds or 143°F for 30 minutes; see Table 2–1).

- A cattle carcass with a localized lesion of tuberculosis in any location must be passed for cooking.

- A swine carcass with a localized lesion of tuberculosis in any two primary sites must be passed for cooking. Therefore, a lesion in the head and one in the intestinal tract would result in that carcass being passed for cooking.

- Tuberculin test reactors must be passed for cooking, even if they are free of tuberculosis lesions or if they have lesions that are localized.

Condemnations

The pre–1972 regulations condemning carcasses are still in effect when the lesions of tuberculosis are generalized, i.e., distributed so widely that they could only happen by entry of the tubercle bacilli into the systemic circulation. Furthermore, lesions of tuberculosis in the hepatic lymph node, which drains the liver, indicate that the condition is generalized and that the carcass must be condemned.

Cattle Grubs

Cattle grubs are the immature forms of two species of warble flies, usually known as heel flies. The common cattle grub, *Hypoderma lineatum,* attacks cattle in all parts of the United States, but the northern cattle grub, *Hypoderma bovis,* is not usually found south of a line extending across northern New Mexico and northern Oklahoma. North of that line, untreated cattle are infested with both species, but the common cattle grub appears in the back of cattle two weeks to two months earlier than the northern cattle grub. Similarly, the adults of the common cattle grub are active a month or so earlier in any locality than those of the northern cattle grub.

Cattle growers are most apt to be aware of these parasites at two points in the life cycle of the grubs — first, when heel flies are chasing cattle in the spring and early summer and, then again, the following winter when grubs appear in the animals' backs after nine or more months inside their bodies.

The adult insects (heel flies) lay their eggs on the heels, legs, and other body parts of cattle. The eggs hatch into larvae (grubs) in three or four days. Soon after hatching, the young grubs burrow into the skin and slowly work their way through the animal's body. The common cattle grubs reach the gullet (throat or esophagus), while the northern cattle grubs reach the spinal canal, where they remain for several months before starting another migration, this time to the muscles in the animal's back.

When the grubs reach the animal's back, they settle just beneath the hide and cut breathing holes through it. At this time, the cattle grower may notice swellings, often called warbles or wolves, forming beneath the hide. The grubs remain in the animal's back for about six weeks. During this period, they gradually enlarge their breathing holes. When fully grown, the spiny grubs work their way out through the breathing holes and drop to the ground, where they change to pupae. Three to 10 weeks later, the time depending upon the temperature, the adult heel flies emerge from the pupal cases and are ready for mating and egg laying. The entire life cycle takes about a year, 8 to 11 months of which are spent as grubs in the bodies of cattle.

The losses begin when heel flies lay their eggs on the cattle. The heel flies cause no pain to cattle, but they frighten the animals and make them difficult to manage. When attacked, cattle run about wildly with their tails in the air and are often injured in this wild stampeding. Cattle find some relief from heel flies by standing for hours in deep shade or water, but while there, they are unable to eat.

The hide is damaged by the breathing holes cut by the grubs in the back, and these holes take two or more years to heal. Hides with five or more grub holes are downgraded, and hides with a large number of holes are virtually worthless (see Chapter 11).

The grubs produce abscesses on the inner surface of the skin, and these abscesses cause large, discolored swellings to form on the adjacent loin muscle and its covering. When the cattle are harvested, the yellowish-green, jelly-like substance must be trimmed from the carcass. The economic loss is substantial because the meat must be trimmed off and the carcass must be downgraded and sold at a lower price per pound. Total economic loss is difficult to determine because of the variation in losses from place to place and year to year.

The most commonly used and most practical control measure for cattle grubs is the application of a systemic insecticide (one that is distributed inside the body of the animal). The circulatory system carries the insecticide to the sites where the grubs are. Several pour-on chemicals have been successfully and extensively used for grub control. Cattle can also be sprayed or dipped with some of these chemicals. Also, there are commercial mineral mixes and feed additives that contain systemic insecticides, if the livestock owner prefers to use this method. Since the exact formulations often change, specific recommendations for the use of these systemic insecticides should be obtained each year from the local county agent, a veterinarian, a state university, or a product manufacturer.

Internal Parasites (Worms)

Many kinds of internal parasites (worms) cause damage and financial loss to the meat and livestock industry. The loss to the producer takes the form of retarded growth, failure to convert feed efficiently, and increased susceptibility to other diseases. Losses to the packer are a result of organ and carcass condemnations.

In swine, of the many worms that utilize the pig as a host, the two that cause the most organ and carcass damage are the large roundworm *(Ascaris suum)* and the kidney worm *(Stephanurus dentatus)*.[39] Large roundworms are

---

[39]Pork Industry Handbook, PIH 44. Cooperative Extension Service.

found in pigs of all ages — weanling pigs as well as mature sows and boars. The migrating larvae damage the liver and lungs and create conditions favorable for the development of bacterial and viral pneumonias, diarrhea, and occlusion of the intestine. Roundworms are the primary cause of liver condemnations in animals for harvest. Proper sanitary procedures aid in the control of all worms by removing the infective stage, but this lasts only a short time until the hogs are reinfected through manure, dirt clinging to sows, and even through the sow's first milk, colostrum. Employing a life-cycle deworming program that utilizes a broad spectrum dewormer, one that will eliminate several worm species in addition to the large roundworms, will control roundworms.

More of a problem in the southern United States, kidney worms generally appear in older pigs and breeding swine. These worms damage the kidneys, liver, tissues surrounding the kidneys, and ureters. Adhesions may also form between the liver and other organs. No treatment has been cleared for use against them. Thus, considerable damage can be done to the organs and the carcass.

In cattle, the internal parasite *Cysticercus bovis* (tapeworm) is a potential concern for humans.[40] Because of this, meat inspectors incise the external and internal masseter muscles of mastication (cheek meat), looking carefully for white tapeworm cysts that could be ingested by humans who eat contaminated, "rare" beef. The condition is known scientifically as cysticercosis, but the common designation is "beef measles," and the human form of the tapeworm is *Taenia saginata*. The cycle between humans and cattle is completed when birds, while drinking from open human sewage disposal plants, from effluent, or otherwise contaminated water, pick up tapeworm eggs and transmit them by defecating on feedstuffs that are later eaten by fattening cattle. The tapeworm eggs pass through the birds completely unaltered. After the human tapeworm egg is swallowed by a beef animal, it passes into the animal's intestine where a tiny larva hatches. The larva penetrates the intestinal wall and travels to all parts of the animal's body. Many larvae lodge in muscle tissue where they develop into ¼-inch white cysts, each of which contains the head of a new tapeworm. If these cysts are not discovered during post-harvest inspection, the life cycle of the tapeworm can be completed if infested beef is eaten "rare" by an unwitting human. "Well-done" cooking, as well as freezing the meat, destroys the cysts. In heavily infected carcasses, condemnation is the rule. (Also see the doneness chart in the color photo section.)

---

[40]Dunn, A. M. 1978. Veterinary Helminthology (2nd Ed.). William Heinemann Medical Books Ltd., London, pp. 276–279.

In infected humans, the tapeworm-containing cyst or cysticercus turns inside out and attaches itself to the wall of the small intestine. After the bladder or cyst is digested, the tapeworm begins producing segments that contain eggs. When egg production is completed within a segment, the segment may contain as many as 100,000 eggs. At intervals, the oldest segments detach themselves and pass out of an infested human through fecal material. An egg, about $\frac{1}{600}$ of an inch in size, can live for a year or longer under favorable environmental conditions. The human tapeworm has been known to reach a length of 70 feet, and, contrary to popular opinion, infected persons are usually obese because they are always hungry.

In the United States, the beef measles problem is mostly centered along the Mexican border — in California, Arizona, New Mexico, and Texas. Perhaps cattle have been infected prior to crossing the U.S.-Mexican border for finishing and ultimate harvest in this country. However, unsanitary toilet habits by feedlot personnel, campers, hikers, and picnickers all add to the tapeworm spread and threat. Humans are the sole host. More stringent efforts to eliminate or screen out birds around feedlots is recommended for tapeworm control.

Pork tapeworm (**Taenia solium**) *is being seen more frequently in the United States.* Although many cases may have been acquired in Mexico, where the incidence is higher, some people appear to have been infected by food handled by unhygienic food preparers who are carriers.

A large number of systemic control agents that can be administered by subcutaneous injection, orally, or by a slow-release implant are commercially available and are effective against a wide range of parasites. Check with your veterinarian, county agent, and(or) a commercial representative for selection and proper use of these products.

To view the "business end" of a tapeworm, check the Web site http://www.biosci.ohio-state.edu/~parasite/hymenolepis_nana.html and look for *Hymenolephis microstoma*.

Trichinosis[41, 42, 43]

The microscopic parasite *Trichinella spiralis,* or trichina, is not a widespread parasite in this country and can be controlled. The encapsulated larvae

---

[41]Livestock Conservation Institute, 6414 Copps Ave., Suite 204, Madison, WI 53716.

[42]USDA, Animal and Plant Health Inspection Service, Federal Building, Hyattsville, MD 20782.

[43]Gamble, R. H., and J. Keeton. 1998. Trichina. Facts. National Pork Producers Council and American Meat Science Association, P.O. Box 10383, Des Moines, IA 50306.

are found in the muscles of rats, dogs, cats, swine, horses, game (especially bear), and humans. Epidemics of human trichinosis have been traced back to raw horse meat in France and Italy. Humans acquire the parasite by eating the improperly prepared meat of an infected animal. Such animals become infected by eating infected rats or uncooked or partially cooked viscera or flesh of infected animals, generally through the medium of garbage.

The encysted larvae are liberated in the stomach of the host and pass on to the small intestine where they reach sexual maturity in a few days. After mating, the female penetrates the lining of the intestine and gives birth to young larvae that are carried by the bloodstream to the striated muscles where they attain maturity in several weeks. Encapsulation then takes place and, if the cycle is not repeated, the larvae in the cysts eventually die. About 250 of these minute parasites measure an inch.

In humans, the severity of the clinical disease is dependent upon the level of infection. A light infection of 1 to 10 larvae would probably not prompt a visit to the doctor. A moderate infection of 50 to 500 larvae would cause concern, while a severe infection of 1,000 or more larvae would be considered life-threatening. About 2 percent of humans who contract trichinosis die. The disease can mimic a wide variety of other illnesses. Most mild cases are misdiagnosed as influenza or other viral infections. The disease is characterized by fever, gastrointestinal symptoms, myositis (inflammation of the muscles), swollen eyelids, and an abnormal increase in the number of leukocytes in the blood (eosinophilia).

Because there has been widespread improvement of sanitary conditions in large hog-producing areas and because both state and federal legislation require all garbage to be cooked before it is fed to swine (Swine Health Protection Act, Public Law 96–468, 1980, amended 1984), the likelihood of any trichina infection is very small. The law prohibits the feeding of garbage to swine except when it is properly heat-treated at a licensed treatment facility. Garbage must be heated throughout by being boiled (212°F/100°C at sea level) for 30 minutes while being agitated. However, the law does not prevent the feeding of uncooked waste from ordinary household operations when it is fed directly to swine on the same premises where this household is located. Also, the federal law does not apply to garbage consisting of any of the following: rendered products, bakery waste, candy waste, eggs, domestic dairy products (including milk), fish from the Atlantic Ocean within 200 miles of the continental United States or Canada, or fish from the inland waters of the United States or Canada that do not flow into the Pacific Ocean.

The incidence of trichinosis in swine is greater in the United States than in many other countries (averaging 0.013 percent of all swine) and is a significant reason why pork does not have complete consumer acceptance in this country.

Infection can come from several sources, including wild animals, but in most of cases, the meat is known to have been eaten raw or inadequately cooked. Trichinae will not be present in beef muscle, since cattle are herbivores. However, if the processing of beef follows the processing of pork, and if the same equipment is used, and if strict cleanup and sanitation are not observed in the plant before the beef is processed, pork scraps may become mixed with the beef, thus creating a chance for cross-infection.

Methods of Detection

- A trichinoscope (microscope), which is able to detect one trichina per gram, the level that can induce trichinosis in humans, has been used in Germany since 1866 to determine the presence of trichinae in pork muscle.

- A digestion system, developed in the United States by Dr. William J. Zimmerman of Iowa State University, which can detect one trichina per 45 to 50 grams of diaphragm. With this system, the trichinae are freed from the surrounding muscle and connective tissue and thus are more easily seen and counted under a microscope.

Control

Pork products that are not customarily cooked in the home or elsewhere before they are consumed must be subjected to a USDA—FSIS approved treatment for the destruction of trichinae. The treatment consists of heating, controlled freezing, or curing, according to Section 318.10 of the FSIS regulations:

1. Heating to a temperature not lower than 144°F for at least an instant. (Note: Cooking recommendations are to heat fresh pork to at least 160°F as a safety margin.)

Temperature		Minimum Time
(°F)	(°C)	
120	49.0	21 hours
122	50.0	9.5 hours
124	51.1	4.5 hours
126	52.2	2 hours
128	53.4	1 hour
130	54.5	30 minutes
132	55.6	15 minutes
134	56.7	6 minutes
136	57.8	3 minutes
138	58.9	2 minutes
140	60.0	1 minute
142	61.1	1 minute
144	62.2	Instant

2. a. Freezing at temperature indicated. (Note: freezing will not destroy trichinae in bear meat.)

Temperature (°F)	6" Thick or Less (days)	More Than 6" but Less Than 27" Thick (days)
5	20	30
–10	10	20
–20	6	12

b. Freezing until the internal product temperature at the center of meat pieces is the following:

Temperature (°F)	Hours
0	106
– 5	82
–10	63
–15	48
–20	35
–25	22
–30	8
–35	½

3. Curing. Several methods for curing sausages, hams, and shoulders are spelled out in Meat and Poultry Regulations (USDA–FSIS). Generally, the combined effects of time, temperature, amount of salt, and diameter of product interact to destroy any trichinae that may be present. Small products and products containing salt that are heated to high temperatures take less processing time to become safe from trichinae. The substitution of potassium chloride for salt (sodium chloride) is permitted.

4. Irradiation. Low-level (1 KGy) irradiation treatment of pork carcasses to render the pork safe from trichinae is approved by the Food and Drug Administration. Spent fuel rods are used to provide gamma radiation from cesium 137 or cobalt 60. The USDA–FSIS has approved the procedure as an additive. An establishment with an approved Partial Quality Control (PQC) program can irradiate pork for the control of trichinae. (See Figure 3–2 for the radiation symbol used in labeling and Chapter 19 for a full discussion of irradiation.)

In addition to spreading trichinosis, garbage can serve as a means of transmission of numerous infectious or communicable foreign and domestic diseases of swine including, but not limited to, African swine fever (ASF), hog cholera, foot-and-mouth disease, and vesicular exanthema of swine. All of these diseases can be spread through infected meat scraps in improperly treated garbage that is fed to swine or through material that has been associated with such meat scraps. U.S. officials are conducting an intensified program to inspect meat and related products entering the United States, especially from countries in the Western Hemisphere with ASF. Complete surveillance is impossible considering the tremendous volume of international traffic, especially between the Caribbean Islands and the United States. A single contaminated meat product in garbage that reaches susceptible hogs could cause an outbreak. Thus, control of garbage is tremendously important to the U.S. swine industry.

Toxoplasmosis

Toxoplasma gondii is a small, intracellular protozoan parasite that reproduces only in a cat's intestine. Nevertheless, from 7 to 94 percent of certain human populations worldwide are infected. Toxoplasmosis is especially dangerous to pregnant females; it can cause miscarriage, still birth, or central nervous system problems, blindness, or mental retardation in babies who are born alive. In adults its symptoms are muscular weakness, weight loss, headaches, and diarrhea. *T. gondii* gets into the human system by contamination of food from cat feces. It survives and grows in raw and undercooked meat but is killed at 158°F. Thus, cleanliness and proper cooking temperatures are extremely important.

Anisakiasis

Eating raw fish (sushi) has become something of a delicacy in the United States. Besides the obvious microbiological problems persons could encounter by eating raw meat, by eating raw fish, persons expose themselves to the roundworm *Anisakis marina*. The *New England Journal of Medicine*[44] contains a graphic description (including a colored picture of the bright red, 1.75-inch-long nematode that was removed from a 24-year-old college student's abdomen) of a case mistaken for appendicitis. Heating fish at 140°F for 5 minutes or freezing fish at –4°F for 60 hours will kill *A. marina*.

HAPPY, HEALTHFUL EATING!

---

[44]Wittner M., J. W. Turner, G. Jacquette, L. R. Ash, M. P. Salgo, and H. B. Tanowitz. 1989. Eustrongylidiasis — a parasitic infection acquired by eating sushi. N. Eng. J. Med. 320:1124.

4

Preparations for Processing — Worker and Equipment Safety

Anthropologists have determined that our human ancestors were carving meat with sharp stones 2.5 million years ago.[1] However, nowadays, before we begin to process meat animals and meat products, worker safety must be thoroughly explained and understood. Having the right equipment available and in safe working order is of utmost importance. For example, the appropriate knife for the job must be sharp, and the necessary mechanical equipment must be correctly assembled and adjusted. Every employee must use proper safety equipment. The meat plant worker must be alert at all times; there is no room for carelessness.

One of the leading issues in the U.S. industrial complex during the 1980s was worker safety. Currently, almost every industrial publication and many newspapers routinely feature articles on worker safety. According to surveys conducted by the National Safety Council, meat packing has historically ranked high in the number of disabling injuries, but near the average of all industries in terms of severity of accidents. An injury is considered to be disabling if the injured person cannot return to the same work the next scheduled shift. Incidence of disabling injuries in the shipbuilding, anthracite mining, construction, and iron and steel industries normally ranks higher in frequency than in meat packing, while incidence in the communications, accounting, and automobile industries often ranks the lowest in frequency.

---

[1]White, T. 1999. Environment and behavior of 2.5-million-year-old Bouri hominids. *Science.* 284:625–629. (April 23.)

Gibbons, A. 2000. News this week: Chinese stone tools reveal high-tech *Homo erectus. Science.* 287:1556. (March 3.)

Unfortunately, the statistics for meat packing do not routinely differentiate between harvesting and processing or between the species harvested. Such a differentiation would be helpful in terms of applying appropriate regulatory activities, including worker's compensation rates, to the individual phases of the industry, since harvesting is more hazardous than wrapping meat, etc.

Ergonomics is a multi-disciplinary activity dealing with the interactions between people and their working environment, which includes atmosphere, heat, light, sound and all the tools and equipment of the workplace. One common affliction of assembly workers is *carpal tunnel syndrome* (CTS), which is caused by compression of the median nerve in the carpal tunnel. The medical name for the wrist is *carpus,* and the median nerve is the nerve closest to the thumb, i.e., on the inside of the wrist. The nerve runs down the arm to the hand through a tunnel in the wrist. When the wrist is repetitively turned unnaturally, and especially when under a work load, this nerve may be pinched, resulting in tingling, pain, or numbness in the thumb and first three fingers. A related ailment, called *tenosynovitis,* may develop. This happens when the sheath around the tendon that connects the muscles and bones in this area becomes inflamed. Both conditions reduce manipulative skills, particularly if the thumb is involved. Butchers and meat cutters are not the only workers potentially troubled with this condition, as it affects all who perform repetitive tasks, such as garment workers, grocery checkers, electronics assembly workers, typists, musicians, housekeepers/cooks, and carpenters.

The meat industry is investing large sums of money in safety programs to improve working conditions. "Good ergonomics is good economics" because a happy, healthy workforce is more efficient.

A publication entitled *Meat Industry Safety Guidelines,* produced by the National Safety Council,[2] and a unique, illustrated guide entitled *Management Strategies for Preventing Strains and Sprains: A Guide to Practical Ergonomics,* produced by the American Meat Institute[3] in cooperation with the National Safety Council, should be read by anyone who is responsible for meat plant safety. Each person working in a meat plant or with meat processing in any form or aspect should be responsible for safety, but of course, within any organization, large or small, certain key people, directed by top management, must assume major responsibility for implementing a strong, effective safety policy. It has been shown that positive reenforcement (the act of complimenting workers who continue to work safely) is much more effec-

---

[2]444 North Michigan Avenue, Chicago, IL 60611.

[3]P.O. Box 3556, Washington, DC 20007.

tive than negative criticism. In one study, unsafe acts and conditions were reduced 40 percent, while productivity, quality, and yield of product increased when this positive approach was implemented.

Other appropriate references that provide additional details about safety programs are: *Health and Safety Guide for Meat Packing, Poultry Dressing and Sausage Manufacturing Plants,*[4] *OSHA Handbook for Small Businesses,*[5] and *What to Do About OSHA.*[6]

Since many accidents and injuries are the result of a combination of environmental and personal causes, controlling the working conditions and controlling the actions of people — their behavior — will help prevent many of these needless, careless accidents and injuries.

KNIFE SAFETY

Because knife accidents are the most prominent cause of disabling injuries in the meat industry, knowing how to select, sharpen, and use knives is fundamental.

Knife Selection

Proper knife selection is basic to knife safety. Many different types of knives are available, due to the wide variation of functions performed with knives. Therefore, careful selection of knives is important. Selecting the proper knife for a given task involves knowing the parts of a knife. The two main parts are the handle and the blade. The handle may be made of rosewood, beechwood, hard rubber, plastic, or any of various other composition materials. Important considerations in selecting a knife handle are size, safety grip, balance, ease of sanitation, and resistance to strong detergents and the action of dishwashers. Beechwood is soft and not as satisfactory as rosewood, since the latter is more attractive and will not readily absorb grease. Hard rubber and plastic handles will not shrink, warp, splinter, or crack, but they may lose some luster after repeated washings, as will most wooden handles. The

---

[4]U.S. Dept. of Health and Human Services, Public Health Service, National Institute for Occupational Safety and Health (NIOSH), NTIS, Springfield, VA 22161. 1979. Publ. PB 274–194.

[5]U.S. Dept. of Labor, Occupational Safety and Health Administration (OSHA), Room N–3641, 3rd and Constitution Avenue NW, Washington, DC 20210. 1979. OSHA 2209.

[6]Chamber of Commerce of the United States, 1615 H Street NW, Washington, DC 20062. 1978. ISBN No. 0–9834–002–0.

shape and size of the handle is a personal matter; for example, persons with large hands favor a thick handle.

In all knives, the blade is fastened to the handle by the portion of the blade called the "tang." The tang may extend to the butt end of the handle or only part way. The handle is fastened to the tang, usually with three rivets or steel pins. Wooden handles are attached this way. The pins must fit tightly and remain so in order to prevent meat particles and moisture from accumulating, which could cause microbial growth and the swelling and cracking of wooden handles. Plastic handles are usually molded directly around the tang to prevent this. In attempts to eliminate carpal tunnel syndrome (mentioned previously), some knife manufacturers are attaching the handle to the blade at an angle which allows for a more natural positioning of the wrist when the handle is grasped, compared to the straight attachment on the knives shown in Figure 4–1. In fact, some knives are now manufactured with the handle at a right (90°) angle to the blade, which puts the wrist at its most natural position for work. The "old-time" meat cutters may at first have difficulty using these knives.

Knife blades vary in length, width, thickness, stiffness, curve, and design (see Figure 4–1). The blade of a knife should be a good bit longer than the meat portion to be cut. Thus, boning knives and skinning knives are generally made in 5- and 6-inch blade lengths, while steak knives, breaking knives, slicing knives, and butcher knives come in 8-, 10-, and 12-inch blade lengths. Personal preferences exist among butchers and meat cutters. Some prefer straight boners, others prefer curved; some prefer flexible, others prefer stiff; some prefer narrow, others prefer wide. A wide, thick blade will wear longer, but it will drag against the meat. A thin, flexible, narrow blade is usually preferred by boners because it turns easily in tight places, can be sharpened more easily than a thick blade, and moves through meat with less resistance.

The steel in almost all knife blades is approximately 80 percent iron and 1 percent carbon. The remainder is composed of 14 to 17 percent chromium in stainless steel blades, with other metals, such as manganese, silicon, molybdenum, nickel, tungsten, and vanadium, entering into the various alloys that compose particular knife blades. The more carbon in the steel, the tougher the steel, which makes it more difficult to work into a blade. The measure of hardness of steel is known as the "Rockwell number." An American metallurgist named Rockwell developed a machine with a diamond-pointed cone which is pressed to a standard depth into the metal, registering the force required on a dial. Rockwell 57 is somewhat standard, while Rockwell 67–68 definitely indicates a metal too tough for sharpening.

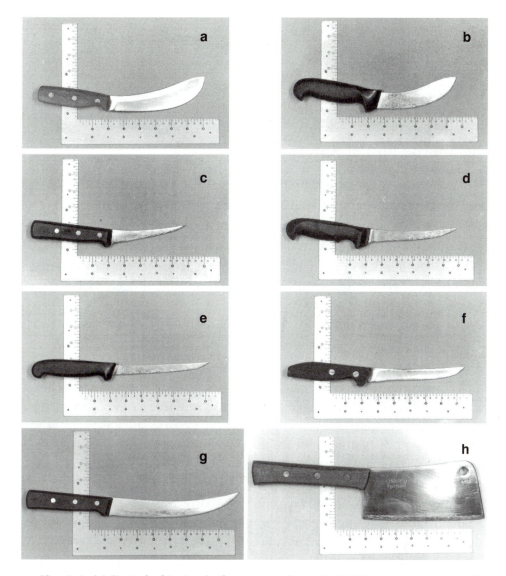

Fig. 4–1. (a) Six-inch skinning knife; rosewood handle. (b) Five-inch skinning knife; molded plastic handle. (c) Five-inch curved boner; medium flex, rosewood riveted handle. (d) Six-inch curved boner; medium flex, molded plastic handle. (e) Six-inch straight boner; flexible, narrow, molded plastic handle. (f) Six-inch straight boner; stiff, wide, hard rubber riveted handle. (g) Eight-inch carving knife and breaker; stiff, wide, rosewood riveted handle. (h) Eight-inch stainless steel cleaver; rosewood riveted handle. This tool is used less frequently in modern meat processing.

One interested person once visited the research lab of one of our largest steel producers and asked three highly skilled research metallurgists to recommend a steel for knife blades. He received three different answers. Even the experts cannot agree on the best alloy for knife blades. Theoretically, the perfect blade would never rust, never break, and never have to be resharpened. There really is not such a blade. However, modern knife manufacturers have formulated special steels that come close to the perfect blade described. There are many reputable knife manufacturers from which dealers for the butchers and meat cutters may purchase knives.

The Sharpening of Knives

A dull knife is inefficient and ineffective except for cutting the person using it. Those of us who marvel at the speed and dexterity of the workers in the harvest and cutting rooms of large and small packing houses must remember that they work with sharp tools and must be experts to hold their jobs.

The conventional steps of grinding, honing, and steeling are described here, followed by a description of a diamond-imbedded hand tool that can replace all three of these steps.

Grinding

Manufacturers of knives may or may not market them sharpened for immediate use. Grinding is done to get extra thinness to the cutting edge. Some knives may need no further grinding on a coarse stone because of the type of grind (V, hollow, concave, or diamond-sharpened) already on the blades (see Figure 4–2). A sand stone, an emery stone, or a diamond sharpener may be used for grinding. The stones should be water- or oil-cooled to avoid heating the knife and to float the metal particles away.

The blade usually need not be ground back more than ¼ inch from the edge, forming what is known as a bevel. This bevel should be the same on

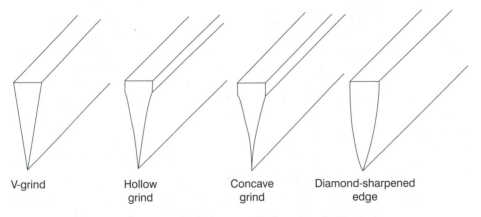

| V-grind | Hollow grind | Concave grind | Diamond-sharpened edge |

Fig. 4–2. Types of grinds normally found on knife blades.

both sides of a skinning knife so that the operator may use it with either hand in siding (removing the hide from the side of a carcass). If the bevel is only on one side of the knife, the bevel side must be next to the hide when it is being skinned.

It is advisable to grind the knife by running the stone with the edge of the knife (see Figure 4–3), rather than at a right angle against the edge. This method is safer, and the user is less likely to scar the blade any further back than the actual bevel itself. Since the smooth finish of a knife is less likely to rust than the ground surface, this technique is rather important in maintaining the neat appearance of the blade. A sand stone of medium fine grade is preferable to an emery stone for grinding knives. The use of an electric emery belt grinder can be disastrous to the inexperienced meat cutter, who may inadvertently grind a good portion of the blade away before realizing what is happening.

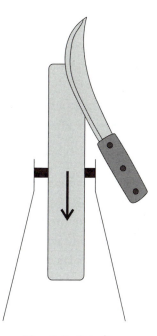

Fig. 4–3. Grinding.

Honing

Honing puts a fine, smooth edge on a blade. All knives must be honed initially and periodically thereafter as needed. Honing may be done on a flat, fine carborundum stone on which water, liquid soap, or oil is used to maintain a clean, abrasive surface (see Figure 4–4).

The flat stone should be set in a block of wood or placed on a damp cloth to keep it from sliding (see Figure 4–4). The handle of the knife is grasped with one hand, and the heel of the knife blade is placed on the end of the stone nearest to the operator. The blade of the knife should be tilted up enough to make the bevel lie flat with the stone. The angle of tilt can vary from almost flat to no more than 20°, exactly the same as the bevel on the knife. The operator should place the finger tips of his or her other hand on the flat side of the blade near the back edge to exert pressure on the blade.

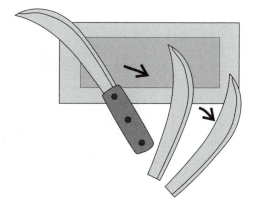

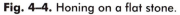

Fig. 4–4. Honing on a flat stone.

With a sweeping motion toward the end of the stone, the operator draws the knife across the stone and at the same time hones the full length of the edge. Each pass across the stone should traverse the length of the stone and the length of the edge.

The operator then turns the knife over in the palm of his or her hand by twisting the thumb and index finger and draws the knife across the stone toward himself or herself. Honing against the edge of the knife avoids the formation of a wire edge. To finish the sharpening process, the operator should run the edge of the blade lightly over the flat of his or her thumbnail to test the knife for sharpness and smoothness of edge. If the knife slides easily, it lacks the proper sharpness. A sharp edge will pull on the nail, and a rough or wire edge will rasp the nail. Another method is to move the ball of the thumb lightly over the edge, but this is not recommended for inexperienced users.

The technique for using the round, hand-held hone, which resembles a steel, except that it is a carborundum or ceramic stone (see Figure 4–5), is

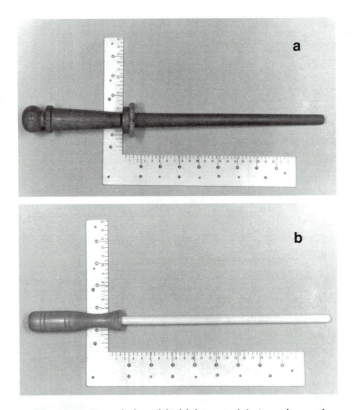

Fig. 4–5. Round, hand-held hones. (a) A carborundum hone that will break if dropped. (b) A ceramic hone known as a "fiddlestick" that is somewhat more resistant to damage if it is dropped.

similar to that for steeling a knife blade, a description of which follows. A word of caution: "When using the hand-held carborundum hone, ***don't drop it;*** it will easily break into several pieces, rendering it useless." The ceramic stone is somewhat tougher and more resistant to breakage.

Steeling

Steeling lines up the tiny microscopic teeth on the blade and straightens a wire edge that may have developed during honing.

There are steels of various types on the market, which are adapted to certain uses. The carborundum and ribbed steels are primarily for kitchen use, where knives need not be razor sharp. The mirror smooth steel for a razor sharp edge is the one best suited for harvest plant, processing, and retail meat dealer needs. Steels range in length from 8 to 14 inches, the 10- and 12-inch lengths being the most popular (see Figure 4–6). With the proper wrist and

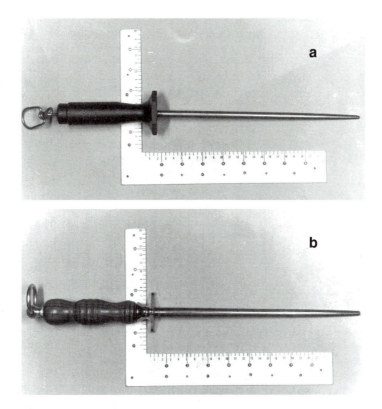

Fig. 4–6. (a) Ten-inch steel; ⅜-inch diameter, smooth polish, molded handle with guard. Light weight and easily carried on waist chain opposite scabbard. (b) Same style and size steel as (a) except that it has a stylish rosewood handle and a larger diameter steel tapering to ⅜ inch; thus, it is heavier and more suitable for home and demonstration carving and cutting.

elbow action, about 7 or 8 inches of the steel is all that is used, except when a steak knife is being steeled.

Rather than using a "surrounding grip" (see Figure 4–7), the operator should hold the steel firmly in the palm of his or her hand, with the thumb in line with the fingers in a position almost diagonal to the body but with a slight upward tilt. This is important as it permits the free movement of the knife across the steel without drawing it too close to the hand holding the steel. The heel of the blade is placed against either the near side or the far side of the tip of the steel at an angle ranging from almost flat to no more than 20°, exactly the same as the bevel, and the blade is brought down across the steel toward the other hand with a quick swinging motion of the wrist and forearm. The entire blade should pass lightly over the steel. Then the knife is brought into position on the opposite side of the steel, and the same motion is repeated. Once the single stroke, just explained, is mastered, the double stroke will come easily. This consists of placing the knife on the steel for the backward stroke in addition to the downward stroke. Less than a dozen strokes of the knife should be sufficient to realign the edge on a knife that is not very dull.

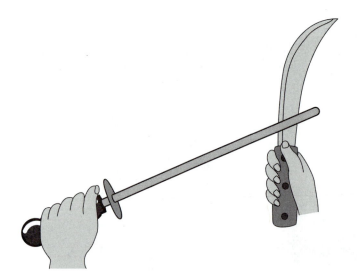

Fig. 4–7. Steeling a knife. Proper way to grip steel. Note how thumb is in line with fingers and thus protected.

Sharpening any tool requires the removal of metal. Diamond, the hardest known substance, can work against even the hardest metals with little effect on its own structure. A patented process by which diamond particles are permanently attached to a strong metal-backing surface to keep the diamonds in precise alignment for sharpening has produced a handy tool that can replace grinder, hone, and steel. The diamond sharpener (see Figure 4–8) is strictly a

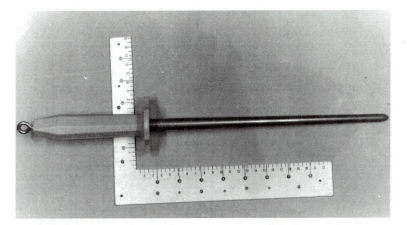

Fig. 4–8. Diamond® sharpener.

hand tool, but with it a person can sharpen (cut metal) as well as or better than with a conventional hand or mechanical knife sharpener.

To sharpen with the Diamond® sharpener,[7] follow these steps:

- Lay the knife nearly flat on the sharpener. The preferred angle is approximately 7°, although 85 percent of all new knives have angles varying between 10° and 40°.

- Alternating between the sides, rub in circles with moderate to heavy pressure to establish the edge (see Figure 4–9). Figure 4–10 gives

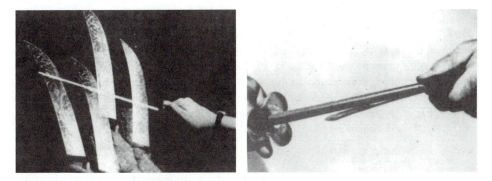

Fig. 4–9. *(Left)* The rotary motion and angularity the operator should use while sharpening with the Diamond® sharpener. The angle should be no steeper than 5° to 7° between the sharpener and blade. *(Right)* Alternately, the sharpener may be used as a file if the knife is held still and the sharpener is rubbed on the knife. (Courtesy, EZE–LAP Diamond Products, Westminster, California)

---

[7]EZE–LAP Diamond Products, Box 2229, Westminster, CA 92683.

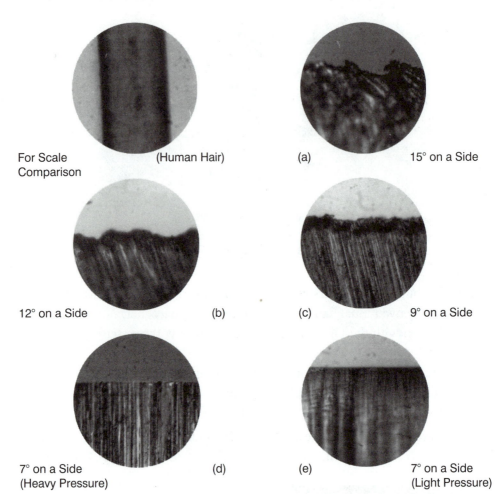

For Scale Comparison   (Human Hair)   (a)   15° on a Side

12° on a Side   (b)   (c)   9° on a Side

7° on a Side (Heavy Pressure)   (d)   (e)   7° on a Side (Light Pressure)

Fig. 4–10. The results of sharpening at several angles. A human hair is included for size-scale comparison. (a) Metal particle pullout, which will occur with any abrasive during sharpening a large angle. In (b), (c), and (d), the edge improves as the sharpening angle is progressively reduced down to approximately 7°. The edges shown in (d) and (e) were made with a single EZE–LAP® sharpener, with a lightening of pressure to achieve the razor edge. (Courtesy, EZE–LAP Diamond Products, Westminster, California)

micro-photo comparisons of knife edges sharpened by this method at different angles.

- To finish, use the same procedure, but with light to very light pressure. Taking a few final light strokes into the cutting edge at the same flat angle creates micro-sawteeth aligned in one direction. These micro-sawteeth hook into the material being cut, just as a hand saw's teeth hook into wood.

The natural wobble in hand and arm motion will give the perfect angle. The operator should mark each side of the cutting edge with a black, non-permanent felt-tip marker. After the knife has been sharpened, as instructed above, the user will notice that the ink has been ground away with the metal all along the area behind and on the cutting edge. The technique really works.

This circular sharpening technique eliminates one of the biggest problems in edge maintenance, the thickness developed immediately behind the cutting edge, normally called "the shoulder" of the knife, created by using the age-old technique of steeling.

It is generally known that standard steels do not remove metal, but only realign the cutting edge. The steel actually may dull because after each stroke the fine microscopic cutting edge is bent from side to side. The cutting edge breaks off just like a piece of fine wire bent back and forth between the fingers. The result is an edge that must be reshaped by grinding. Laying the sharpened blade perfectly flat against another blade and then rubbing the blades together in small circles may eliminate the fine sawtooth or wire edge. There should be absolutely no angle between the knife blades. The wobble from the circular motion creates a perfect angle to eliminate any wire edge. A steel is never really required.

Utilization of Protective Equipment

Various types of protective equipment should be used, depending on the circumstances and speed of work required. Plated or stainless steel mesh gloves for the thumb and two fingers or for the whole hand are available. Fibrous gloves with stainless wire woven in are lighter and cheaper than the metal mesh gloves. Other items of protection are properly engineered arm guards made of metal, plastic, or neoprene and abdominal protectors made of stainless steel, aluminum mesh, neoprene, or suede leather with steel studs. Knife guards, built into the handle or blade, prevent the user's hand from sliding over the handle onto the blade.

Safe Handling of Knives
When Not in Use

Instead of placing the knife when it is not in use on the table or in table slots, which have proven to be unsafe, the worker should carry the knife in a scabbard worn around his or her waist. The scabbard must be composed of a material that can be routinely cleaned and sterilized (see Figure 4–11).

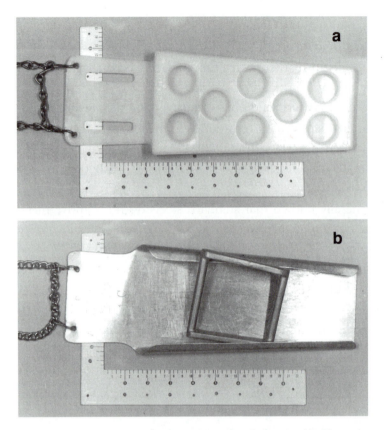

Fig. 4–11. (a) Plastic knife scabbard and (b) metal knife scabbard. Note that both scabbards are held to the worker's body by metal chains for ease of cleaning and sanitizing each entire unit.

Training in Using a Knife Safely

It is important to take time at the onset to train a neophyte knife handler. All of the elements mentioned above are important to knife safety, but workers need time to practice and guidance in the proper grip and motion to be used in the various knife functions.

One of the nation's largest hog harvesters, which in two shifts every 24 hours harvests some 15,000 hogs, has a special knife-sharpening crew to sharpen knives. A cart of sharp knives is taken to the hog harvest for dispersion among the meat cutters. A good, experienced meat cutter working hard and fast for eight hours will use one knife per day. A beginner, in the same situation, will dull five or six knives per day. Thus, the industry places proper training in knife use as a high priority.

POWER TOOLS AND EQUIPMENT

Many power tools and much power equipment are used in the meat industry. Each tool and machine must be properly guarded, and every operator must be properly trained in its use. OSHA's "Lock Out/Tag Out" rule specifies that if a machine is shut down for cleaning or repair, the power to the machine must be *locked* off, not just turned off, and tagged. A significant number of industrial injuries and deaths have been caused by machinery that was inadvertently turned on while service or other personnel occupied dangerous positions in or under the equipment.

Even though mechanization has been slow to come to the meat packing industry because of the wide variety of shapes and sizes of animals and animal products and because mechanization is expensive, as time goes on, more and more operations, especially in large plants, are becoming mechanized. For example, water knives for cutting meat are being developed. The water comes in pulsating, high-pressure jet streams that cut easily. This process has been used in other industries. An AMI report[8] evaluates the status of mechanical vision, which allows machines to see, and mechanical assists, which help workers do physical jobs.

Materials Handling

Lifting is a major activity in the meat industry. There is a proper way for workers to lift without injuring themselves. The National Institute for Occupational Safety and Health (NIOSH) has developed a lifting guide that shows how an ideal weight limit based on the number of lifts required per minute and the length of time involved in each lift can be calculated. Using information from this report, the meat industry has redesigned work stations to bring loads closer to the body, thereby reducing strain. At the same time, more operations are being mechanized to eliminate human lifting. One example is a mechanical leaf fat puller. (See Figure 5–50 for an example of the hand pulling methods.) More hand lifting is being replaced by mechanical lifting devices, power trucks, conveyors, and elevators. Workers must know exactly how to operate and control each new machine that comes on line. Thus, worker training in safe machine use is essential.

---

[8]American Meat Institute. 1989 (January). Applications for New Technology in Reducing Repetitive Trauma Injuries in the Meat Packing Industry.

Robotics in the Meat Industry

Meat processors are turning to robotics for their flexibility, ruggedness, and repeatability. Robots can be programmed to do multiple tasks. They can withstand sanitation washdown procedures and are unaffected by differing or constant ambient temperature. Pallatizing and lifting in general are two of the main uses of robots. Robot use cuts down considerably on worker injuries, resulting in improved employee mental attitude and physical health.

OTHER ASPECTS OF SAFETY

Buildings and equipment must be engineered so that electrical hazards are avoided, and railings, stairways, floor surfaces, and aisles are safe. Power transmission equipment — lifts and hoists, chains and cables — must have safety guards in place. Many other items of equipment — water hoses, steak-tenderizing machines, frozen meat slicers, meat grinders, stuffers, patty-forming machines, etc. — must be properly installed and used safely.

OSHA requires that every company inform its employees of all possible hazards involved in the workplace. NIOSH regularly publishes recommendations for limits of exposure to potential hazardous substances or conditions in the workplace.[9] One area of possible concern in the meat industry is fumes from the solvent dimethylformamide (DMF) used in the leather tanning process. Fumes and gases must be closely monitored. Workers must be advised to avoid closed spaces that could be harmful to them. Workers must know the location of respiratory equipment and how to use it properly. They must also know emergency and first-aid and medical procedures. All workers must have their hearing and vision checked annually. Every company must strictly adhere to the noise standards established by OSHA.

An orderly arrangement of operations, tools, equipment, storage facilities, and supplies allows for proper cleaning which is required.

Only those chemicals authorized by the USDA–FSIS can be used as cleaning agents. They must be used properly and in the correct amounts so that alkali, acid, and germicide hazards can be avoided.

---

[9]NIOSH Recommendations for Occupational Safety and Health Standards 1988. 4676 Columbia Parkway, Cincinnati, OH 45226.

Proper maintenance involves not only keeping equipment and facilities in functional condition but also properly adjusting personal safety devices. Maintenance workers must practice using ladders and portable tools correctly, and when working with refrigeration, other electrical, and welding equipment, they must exercise caution.

The factors that cause fires should be eliminated or at least minimized. All personnel must be aware of fire protection rules at all times. All workers should be instructed in the proper use of fire prevention and extinguishing equipment and in personal protection techniques in the event of fire.

Dealing with and Reporting Injuries

From the National Meat Association Risk Management page comes this warning: Most accidents can be prevented, but sooner or later someone in your organization will have an accident. Knowing exactly what to do immediately will help prevent serious repercussions and additional physical trauma for the unfortunate employee who experienced the accident. The main objective is to be sure this person gets the best medical attention as soon as possible. Be sure that all employees understand what should be done and the proper sequence for activities. Do not forget that the accident must be reported to company and government officials.

5

Hog Harvest

In 1866 the American Society for the Prevention of Cruelty to Animals was incorporated. We now refer to the conversion of live animals into carcasses, meat cuts, and meat products as *harvesting.* The meat industry is concerned that animals, which provide us with the most nutritious, well-balanced food available, be treated kindly at all times and not suffer pain.

This chapter deals with hog harvest. Zooarchaeologists now estimate that the pig was perhaps the first farm animal domesticated—before sheep or cattle—about 11,500 years ago.[1] This new estimation is earlier than previous estimations of domestication of meat animal species.

Pigs are more like humans than they are like cattle or sheep. Readers who have worked with pigs know that each pig has its own "mind" and resists coercion from any direction to change it. Pigs are monogastrics and are beneficial in human research. In Iowa, where pigs outnumber people nearly 5 to 1, residents especially value the pig and feature it in displays in several metropolitan settings. In the summer of 2000, decorated fiberglass pigs "hogged the streets" of Cincinatti and of cities in northern Kentucky.

It is of utmost importance to meat quality and animal welfare that pigs be treated properly before, during, and immediately after harvest. Although many readers will not be involved in the actual harvest of pigs, all persons should know the procedures involved in converting a live, contented pig into delicious pork chops and ham.

In 1998 there were 101,029,300 hogs harvested commercially in the United States, excluding those relatively few harvested on the farm.[2] Table 5–1 lists the states in the order of commercial pork harvest volume in 1998.

---

[1]Archaeology Special Section. 1998. Reading the signs of ancient animal domestication. 1998. Science. 282:1448. http://www.sciencemag.org

[2]USDA, National Agricultural Statistics Service, Agricultural Statistics Board. March 1999. Livestock Slaughter—1998 Summary. (Mike Easdale.)

Table 5–1. Commercial Hog Harvest Ranked by States, 1998[1,2]

Rank	State or Region	Thousand Head
1	Iowa	28,905.6
2	North Carolina	9,786.4
3	Illinois	9,197.1
4	Minnesota	7,940.0
5	Indiana	6,326.8
6	Nebraska	6,285.8
7	South Dakota	4,452.3
8	Virginia	4,172.7
9	Oklahoma	4,119.2
10	Missouri	3,926.5
11	Kentucky	2,530.4
12	Pennsylvania	2,427.9
13	California	2,212.4
14	Ohio	1,261.4
15	Wisconsin	375.1
16	Texas	324.2
17	Georgia	296.5
18	Delaware–Maryland	278.5
19	Oregon	188.5
20	Florida	141.2
21	New Jersey	96.3
22	Alabama	90.7
23	North Dakota	72.4
24	New York	52.7
25	Colorado	41.2
26	Louisiana	37.7
27	Hawaii	36.5
28	Montana	28.1
29	Utah	27.3
30	New England[3]	26.4
31	West Virginia	11.0
32	Arizona	9.2
33	Wyoming	5.9
34	New Mexico	2.7
	Total United States[4]	101,029.3

[1]USDA, National Agricultural Statistics Service, Agricultural Statistics Board. March 1999. Livestock Slaughter—1998 Summary. (Mike Easdale.)

[2]Includes animals harvested in federally inspected and other plants, but excludes those harvested on the farm.

[3]New England includes Connecticut, Maine, Massachusetts, New Hampshire, Rhode Island, and Vermont.

[4]States with no data are still included in the U.S. total; individual data are not printed to avoid disclosing individual operations.

For current information on what method of marketing (direct, terminal, sale barn, grade and yield) was used for these hogs, you may wish to contact the USDA Agricultural Marketing Service (http://www.ams.usda.gov/).

Before proceeding to harvest a hog, be sure that you have proper and safe equipment; that your equipment is ready to use (sharp and clean); and that the animal you are about to harvest is healthy. In other words, you should study and master the first four chapters before applying the material you will learn in this chapter. A review fact sheet covers principles to assure pork quality before, during, and following harvest.[3]

FAST (DRIFT) PRIOR TO HARVEST

Hogs should be held off feed for 16 to 24 hours prior to harvest to assure completeness of bleeding and to ease evisceration. Furthermore, a time of fasting will reduce the carbohydrate available for postmortem conversion to lactic acid, resulting in higher ultimate muscle pH, a more desirable muscle color and increased water-holding capacity; thus improved pork quality.

In regard to food safety, the intestinal mucosa (mucous lining of the intestine) normally prevents bacteria of the gut from penetrating the bloodstream and thus infecting the meat. However, if the intestines are distended with feed when the pig encounters the stress and excitement of harvest, the mucosa is put under exceptional pressure and may "give way" to the organisms. Also, when the intestinal tract contains a heavy load of ingesta, it is simply harder to handle during evisceration and is more likely to burst or be accidentally cut, thus contaminating the carcass.

WEIGHING AND RECORDING[4]

The live hog or a group (draft) of hogs should be weighed on a balanced scale prior to harvest in order that yield (dressing percent) can be accurately accounted for. The sex, breed, and ID number, along with the live harvest weight, should be recorded. Dressing percent is calculated as follows: Carcass

---

[3]National Pork Producers Council. 1998. Critical Points Affecting Fresh Pork Quality Within the Packing Plant. Fact Sheet 04328 — 9/98.

[4]Engeljohn, D., and J. R. Romans. 1979. Pork Slaughter in Small Plants. Dehairing and Skinning Methods. Slide set E 70. College of Agriculture Instructional Resources, University of Illinois at Urbana–Champaign.

weight ÷ live weight × 100 = dressing percent (yield). Routinely, carcass weight is "hot," or "warm," meaning it was taken prior to carcass chilling. Chilling causes carcasses to lose approximately 2 percent because of the moisture evaporation; thus, the dressing percent would be lower if chilled carcass weight were used. Furthermore, animals weigh less immediately prior to harvest than at purchase, since animals "shrink" in transit. Animals standing in yards awaiting harvest rarely gain back to their purchase weight, even though they have feed and water before them. Thus, a dressing percentage calculated on harvest weight would be higher than if it were calculated on purchase weight. Realistically, purchase weight should be used in order to compute financial returns, but harvest weight should be recorded as well to be able to determine what shrink the animals have undergone between purchase and harvest. Hogs that are scalded and have their hair removed, in general, dress approximately 70 percent. Heavier hogs will dress higher in the 70s. Excessively filled, light weight hogs will dress lower, in the high 60s. Most packers avoid purchasing hogs weighing less than 210 pounds because the yield of carcass and salable cuts is less and because the cost is the same to harvest a light weight hog as it is to harvest a heavier hog. Therefore, most packers will expect a dressing percent in the 70s. Muscle is more dense than fat, so if hogs that are equal in weight, fatness, and fill are compared, the heavier muscled hogs will dress higher. Nevertheless, across the industry and throughout the many droves composed of individual hogs with different amounts of muscling, fatness, and fill, the fatter droves usually average out with a higher dressing percent because the fatter hogs are usually heavier. Thus, fat, fill, muscling, and weight are four important factors that affect dressing percent. Hogs that are skinned and have their feet removed will dress approximately 8–10 percent less than scalded hogs with feet on. The skin accounts for 6 percent and the feet 2–3 percent.

IMMOBILIZATION

Hogs must first be restrained. A small confinement pen with a squeeze gate is sufficient in a small operation. However, in large commercial operations, a series of chutes and restrainer conveyers move the hogs into position for stunning. The V restrainer/conveyer shown in Figure 5–1, or a similar system, is used in most large hog harvesting operations. The facility and equipment should be designed to allow as quiet and gentle handling as is possible so that the hogs will produce meat with acceptable quality. A prototype double rail restraining system to be used with an automatic electric stunner for hogs is under development by Grandin Livestock Handling System, Inc.[5] With this system, hogs ride on the double rail to the immobilization station. Double rail systems are being used in cattle (see Figure 6–27), sheep (see Figure 7–19) and calf (see Figure 8–11) harvesting operations. The main advantage of the double rail system is that animals ride to the immobilization point in a more comfortable, natural position, rather than with their legs and bodies squeezed in the V conveyer. The incidence of minute hemorrhages at the skin surface is reduced in pigs restrained through this system. Furthermore, the person responsible for immobilization is able to safely stand closer to the animal and thus accomplish the task more quickly and accurately with less effort and less trauma to the animal.

Hogs must be stunned with a federally acceptable device (electrical, mechanical, or chemical) so that they are rendered unconscious prior to being "shackled, hoisted, thrown, cast, or cut." The use of a commercial electric stunner is shown in Figure 5–1. The normal settings for electrical stunning are 70 to 90 volts and 0.3 amps for 2 to 10 seconds (minimum 1.25 amps for 2 to 3 seconds). (See the Appendix section on electricity, which explains the various electrical terms and relationships.) Because most stunners have variable settings, the directions on electric stunners should be checked before they are used. Cardiac arrest stunning will render an animal insensible to pain and sensation prior to hoisting and bleeding more effectively than conventional electric

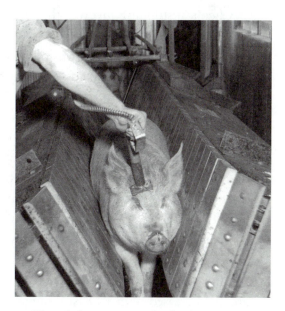

Fig. 5–1. A commercial electric stunner. Hogs are driven into a squeeze chute that has moving sides to propel the hogs to the stunner. (Courtesy, Cincinnati Butchers' Supply Co.)

---

[5]2918 Silver Plume Drive, No. C-3, Ft. Collins, CO 80526. http://www.grandin.com

A complete discussion of this system, including pictures and diagrams, is available from Dr. Temple Grandin.

Grandin, T. 1988. Double rail restrainer conveyer for livestock handling. J. Agric. Eng. Res. 41: 327–338.

stunning. In cardiac arrest stunning, an electric current is passed through both the brain and the heart to produce permanent insensibility. This allows more time, if needed, between stunning and bleeding (sticking) with no possibility of the animal becoming conscious again. Cardiac arrest stunning is recommended for sheep, calves, and poultry, as well as for pigs.

Mechanical stunning involves the use of a compression bolt with either a mushroom head or a penetrating head. The force may be provided with compressed air (see Figure 7–1) or a cartridge (see Figure 8–1). Placement of the bolt on the forehead is most effective, although some prefer placement behind the poll. Mechanical stunning of hogs is largely confined to smaller operations. The use of a .22 caliber rifle is approved in some locals, but extreme care must be exercised for human safety. Furthermore, the brain from a hog that has been shot with a rifle is not edible because of possible lead or steel contamination by the rifle shell particles.

Chemical stunning involves the use of CO_2, which makes the blood oxygen levels of the animals drop, causing them to become anesthetized. Workers at the Danish Meat Institute and the Swedish University of Agricultural Sciences have completed some interesting studies on CO_2 stunning.[6] In Denmark, 90 percent of the 15 million hogs harvested annually are CO_2 stunned. The European researchers have shown that when pigs are introduced into an atmosphere of 30 to 40 percent CO_2 before being subjected to 70 to 80 percent CO_2 for final anesthetization, the extent of blood splashing, shoulder and backbone fractures, and pale, soft, exudative (PSE) pork (see Chapter 22) is dramatically reduced compared to electrical stunning. Hogs go to "sleep" relatively peacefully because they lose sensitivity before they are completely unconscious. Although they go through a brief period of excitation and uncontrolled muscular activity, they do not feel any pain. Because of less noise and a calmer atmosphere, human worker safety and comfort are greatly increased with CO_2 stunning. In the Danish system, a hog enters a cage, which is lowered in stages into the two levels of CO_2. The Danish system can handle a maximum of 600 hogs per hour.

In the United States, CO_2 stunning was popular around 1950 but in the ensuing years more packers have used mechanical and electrical stunning

---

[6]Rust, R., and G. McCain. 1991. CO_2 Stunning Revisited. Meat and Poultry, 37:100 (No. 9, September); also Patricia Barton Gade, Danish Meat Research Institute, Roskilde, Denmark; and butina ApS, Holbaek, Denmark <info@butina.dk>.

methods. In the older U.S. systems, the CO_2 was confined to a chamber lower than the surrounding surface because it is heavier than air. The hogs were driven and carried by a conveyer into the lower chamber. In order for this new system to be used, U.S. engineers would have to design one that could handle the 1,000 to 1,100 hogs per hour that are harvested by U.S. packers.

STICKING (EXSANGUINATION)

Whether hogs are stunned by the use of one of the three methods previously discussed, or whether they are harvested by cardiac arrest, the time interval between stunning and sticking should be minimized to prevent blood splashes in the muscle. Blood splashes are caused by the rupture of capillaries as a result of increased blood pressure in the constricted capillaries while the heart continues to pump. After stunning, the quicker the pressure is released through sticking, the less chance there is of capillary rupture. Sticking the stunned hog while it is in a prone position, rather than in a suspended position, cuts down this time interval and thus often prevents blood splashes. When employing cardiac arrest, blood splashes are fewer.

A 6-inch sticking knife, sharpened on both sides of the tip, is large enough for the ordinary hog. A straight, rigid, 6-inch boning knife will work as a sticking knife. For large hogs (400 to 600 pounds), the 7-inch blade is preferable. If the hog is suspended, the butcher should steady it by placing the flat of his or her hand on the hog's shoulder (never by grasping a leg). After the cranial (front) tip of the sternum bone has been located, the butcher should insert the knife straight into the soft tissue in front of the sternum initially, then with the point of the knife between the sternum and the backbone and directed toward the tail (this is very important), give an upward thrust toward the anus, dipping the point until it strikes the backbone, thereby severing the carotid arteries and jugular veins. Then, the knife should be withdrawn. The stick hole should be no wider than the knife blade to avoid contamination from the scalding tank water. The knife should be kept midway between the shoulders to avoid cutting into shoulder muscles, which would expose them to blood flow from the jugular veins and carotid arteries. This is known as a shoulder stick. No twisting or cross cutting of the knife is necessary. If the knife is inserted too far cranially to (in front of) the sternum bone, the animal will not bleed freely and will die slowly.

When the hog is stuck properly (see Figure 5–2), blood will gush out of the opening in a steady stream (see Figure 5–3) because the carotid arteries (bright red blood) and the jugular veins (dark blood) have been severed. A proper stick means a fast bleed.

Blood should not be allowed to run into the regular sewer, since it creates a high biological oxygen debt (BOD), meaning it would retard complete sewage degradation by microbes in a sewage system. Rather, blood should be saved through a special drain or removed from the animals into vats or barrels (see Figure 5–3), to be processed and used for animal feed or fertilizer. Only if it is kept completely sterile by removal from the animals through tubes or syringes can it be

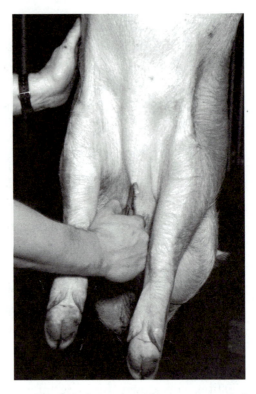

Fig. 5–2. Properly sticking a suspended hog.

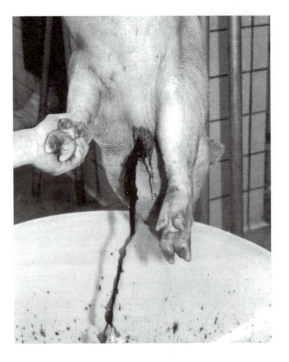

Fig. 5–3. The blood gushing out of the small opening is caught in a large, funneled vat.

used for human food (see Chapter 11). Blood volume in animals may approximate 6 to 8 percent of live weight, with considerable individual variation. A good stick results in removal of approximately 50 percent of the total blood in an animal.

SCALDING

Sufficient water of the right temperature (143±2°F; 61±1°C) and removal of the hog from the water when the hair slips easily will assure a good, quick job. A safe scalding temperature is 135° to 160°F, the lower temperature requiring more time. Water up to 180°F can be used but is not recommended since the hog must be withdrawn as soon as the hair slips easily. Initial heating causes the protein in the hair follicles to denature, thus loosening the hair. Overscalding is referred to as "setting the hair" because it causes the skin to contract around the base of the bristles, holding them tight. Even further overscalding begins to cook the skin, and it may deteriorate, allowing the contaminated scalding water to enter the meat. Severely "cooked" hogs must be condemned. A rule of thumb (or finger) for determining the proper scalding temperature, if a thermometer is not readily available, is to carefully place your finger in the water and begin slowly counting — "one-thousand-one, one-thousand-two, one-thousand-three" — and then if you must remove your finger at about "one-thousand-four" or "one-thousand-five," the temperature is approximately 143°F. Although this four- to five-second rule works for most people, it is not completely foolproof because some people are more or less heat tolerant; however, it does give a good safety range to avoid overscalding a hog.

Once the hog is in the water, it should be kept under water and continually moved and turned to get uniform scalding. In large plants, hogs enter the scalding tub and are moved through the tub by a conveyer set at the proper speed to allow the proper scalding time. Thus, if the water temperature is right, everything is "automatic" (see Figure 5–4). These large plants may run the temperature at 139°F, which would allow 4½ minutes before overscalding, during the normal season. During the hard-hair season (September–November), the water temperature should be 139° to 140°F and the immersion period 4 to 4½ minutes, while in the easy-hair season (February and March), a temperature of 136°F for 4 minutes is preferable. However, in small plants without the automation, hair condition must be checked periodi-

Fig. 5–4. Hogs sliding into a scalding tub. From there they are elevated into the dehairing machine. (Courtesy, G. A. Hormel & Co.)

cally during the scalding period. The ham, flank, belly, and head regions should be checked. Once the worker determines that the hair is loose, the hog should immediately be taken out of the water. At 145°F, scalding time may be two to three minutes, but red and black hogs may take longer.

The procedure for scalding hogs in barrels on farms is to insert a hook in the side of the mouth (for light hogs) or between the lower jaw bones and then scald the rear half of the hog. After this half has been scraped and the hind feet shaved clean, the tendons should be opened and a gambrel should be inserted to manipulate the hog. (Figures 5–20 and 5–22 show a gambrel in use.) The front half should be immersed, scalded, and scraped, then the carcass should be hoisted, the remainder of the hair shaved from the carcass, and the carcass rinsed.

HAIR AND SCURF REMOVAL

Various dehairing machines, sometimes called "polishers," are manufactured to remove hair from scalded hogs. The large daily harvest of hogs by packers has made it necessary to devise mechanical equipment to handle large numbers of hogs per hour. Today, a single dehairing machine will handle from 150 to 500 or more hogs per hour, depending upon its length. Large plants are equipped with twin machines that will handle up to 1,000 hogs per hour. These machines are constructed of heavy V-shaped bars, a heavy steel frame, and two shafts to which belt scrapers with metal tips are attached. The lower shaft runs from 55 to 60 revolutions per minute (rpm), and the upper shaft runs around 100 rpm. Both shafts run in the same direction. Hot water, 140°F, is sprayed on the hogs as they pass through the dehairer toward the discharge end.

Medium and small plants may have smaller dehairing machines that handle one or two hogs at a time. These machines have belt scrapers attached to rotating bars which cause the hogs to rotate while the scrapers knock and pull the hair out (see Figure 5–5). The action of the flippers and the turning of the hogs causes the hair to be removed, provided the hogs were properly scalded. After about 15 to 30 seconds, most of the hair should be removed. A machine that combines the scalding and dehairing (polishing) operations is now available for small plants (see Figure 5–6).

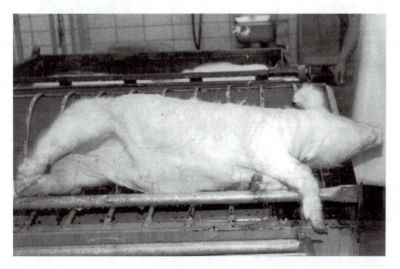

Fig. 5–5. In this small dehairing machine, the action of the flippers and the turning of the hog cause the hair to be removed.

After dehairing, and before the hog cools down, the toenails should be removed with a toenail puller. If possible, the butcher should engage both hooks of the puller at the top of the nail (see Figure 5–7) and then pull the nail. The butcher should ***not*** hold the leg with one hand while he or she pulls with the other — if the puller slips off the toenail, it could become buried in the butcher's forearm. Once the animal cools off, the nails are harder to pull. In the case of tough, hard-to-remove toenails, a pruning shear may be used to clip off the toes.

Fig. 5–6. A combination scalding and dehairing machine. (Courtesy, E-Zuber Engineering & Sales, Inc., Minneapolis)

Fig. 5–7. Toenails are removed with a toenail puller. ***Do not hold the leg with your nonpulling hand*** — if the puller were to slip off the toenail, it could become buried in your forearm.

Once the nails have been removed from the front and hind feet, the skin and hair are removed from between the toes on all four feet. The two tendons in the hind foot must be exposed so that the trolley hook or gambrel can be inserted to hang the hog on the rail. Two cuts should be made as deep as possible on the back side of the hind foot. One cut should be off to the left next to the dewclaw, and the other cut should be off to the right of the center of the foot next to the other dewclaw (see Figure 5–8). Cutting to the left and to the right of the center will avoid cutting the tendons. If the cut is properly executed, the butcher will be able to place a finger between the bone and the two tendons.

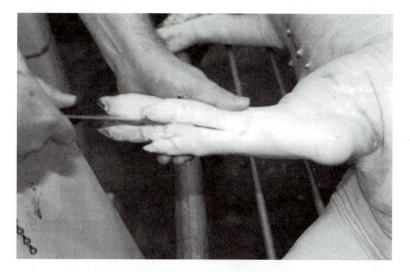

Fig. 5–8. Loosening the tendons. The small and large tendons on the back of the hocks are used for supporting the carcass on the gambrel.

A bell-type scraper that is fairly sharp is a useful tool in the effective removal of scurf. The round working surface permits rapid dehairing, and plenty of pressure can be applied to remove the bristles and dirt (see Figure 5–9). All evidence of hair, including eyebrows and hair on the lips and inner ear, must be removed before head removal and evisceration can take place. In some cases it may be more expedient to trim away tissues, rather than to clean them.

If the head is free of hair and debris and if the animal has been stunned in an approved manner so that no part of the head is contaminated (a bullet or penetrating captive bolt would contaminate the head), the head can be utilized for human consumption. A hog's inner ear contains wax and dirt, which would contaminate any meat product made from the hog's head, so it must be removed. To remove the inner ear, the butcher holds the ear with one hand and inserts the knife with the other hand, cutting in a circle around the

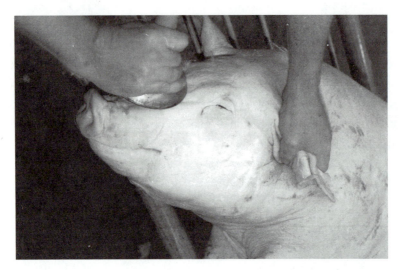

Fig. 5–9. A bell scraper with its sharp, rounded working edge removes bristles and dirt.

inside of the ear, as though coring an apple (see Figure 5–10). Once this circle has been cut around the inner ear, the butcher pulls this section with one hand and cuts off the inner section at the base in order to clean the ear of contaminates.

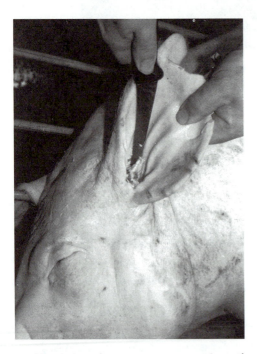

Fig. 5–10. The inner ear, contaminated with wax and dirt, must be removed if the head is to be saved for human consumption.

Next, the hair and skin surrounding the eye, including the eyelid, are removed. At a point directly above the eye, make a cut as deep as you can, until you hit the skull. Then, pull the skin surrounding the cut above the eye, and continue cutting towards the eye (see Figure 5–11). You may cut directly over the eye, or you may cut out the eye. Continue cutting with one hand and pulling with the other until the skin surrounding the eye is removed. Try to remove only the skin, not the lean tissue.

To remove the lips, first place a knife directly above the corner of the mouth, and then make a cut as deep as possible until you hit the jawbone. By placing your finger in the cut you just

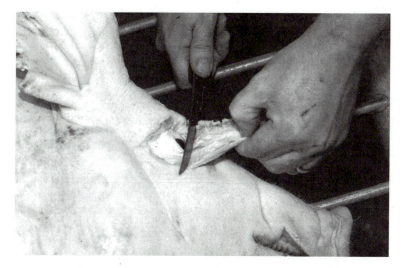

Fig. 5–11. The hair and skin surrounding the eye, including the eyelid, must be removed.

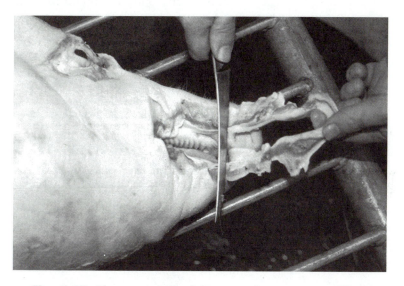

Fig. 5–12. Lips are removed because they are very difficult to clean completely.

made, pull the lips towards you and continue cutting them (see Figure 5–12). You should be able to remove both the upper and the lower lips at the same time. Again, as with the eyebrows, try to remove only the skin, and leave the lean tissue intact.

The rest of the carcass must be scraped in order to remove the remaining hair. Use your skinning knife as if it were a razor blade. Rub your hand back and forth on the skin surface to detect areas you may have missed. Be sure to scrape under the legs. If the hair is hard to remove or if a bruise or callus is evident, remove that area by cutting it out with your knife.

When the hog is completely cleaned, it is ready to be hoisted up and landed onto the rail. Place a gambrel in the tendons that were previously opened. The hog should now be scraped once more, starting at the ham end and working down, to remove clinging water and loose hair. Small, fine, and tough hairs that remain can be removed by singeing them with a torch fueled by propane or another suitable fuel. The burning action of the torch causes the hair that was missed to char and become evident. Be sure to singe the hind feet area well, as the hair is hard to see there. The head region is also difficult to clean; therefore, while singeing, be sure to lift the front legs and ears. However, holding the flame in one area too long will burn the skin.

Once you have singed the remaining hair on the entire carcass, wash and scrape off the charred hair. Scrape as if you were shaving the animal. You can feel the brittle ends of the hair if you miss a spot. Before finishing the scraping process, be sure to lift up the legs and clean around them and the head. Once the hog is free of singed hair and brittle ends of hair, wash the hog thoroughly from the hind feet to the head.

DEPILATING HOG CARCASSES

At one time the practice of dipping each carcass after it was dehaired into a hot solution (250° to 300°F) of rosin and cottonseed oil for a period of six to eight seconds to remove hair, stubble, and roots was quite popular. When the rosin coating plasticized after cooling, it was stripped by pull-rolling it down the carcass, taking with it the remaining hair, stubble, and roots. However, in recent years, many packers have discontinued its use because of the shortages of trees, wood, and rosin and have turned instead to mechanical brushes and torches to completely clean dehaired pork carcasses.

PULLING HOG SKINS IN A SMALL PLANT[7]

The following method of skinning a hog can be used when there is a scarcity of labor or equipment. It is a one-person job and requires neither hot

---

[7]Method perfected by Bill Condradt, Plains Processing, Shiocton, WI, and Wieland Kayser, Kayser's Butchering — Lena Maid Meats, Lena, IL. Slide set E 71, College of Agriculture Instructional Resources, University of Illinois at Urbana–Champaign.

water nor scalding equipment. However, conventional hand skinning involves considerable knife skill. Because the fat is soft and easily cut, the knife must be kept tight against the skin at all times. Deep gashes made by uncontrolled strokes of the knife do considerable damage to hams and bacons that are to be cured. Furthermore, the soft, thin skin is easily cut and, if cut, is worthless as garment material and must be sold at low prices for manufacture into animal protein feeds.

An easier method of removing the pigskin from the carcass by pulling results in a salable pigskin and, at the same time, produces a nice, smooth carcass free of gouges. The light area shown in Figure 5–13 is the area which is to be hand skinned. Figure 5–17 shows a hog "pre-skinned" ready for pulling. Novices may be inclined to skin too far back into each side with a knife, as can be seen by comparing Figures 5–13 and 5–17. Too much hand skinning may result in a damaged hide and a scored carcass.

The hog is humanely stunned and bled in the conventional manner. After bleeding is completed, the hog is placed on the skinning rack or cradle. Depending upon what use is to be made of the pig's feet, they may be skinned out (see Figure 5–14), with only the hooves and dewclaws discarded and the feet

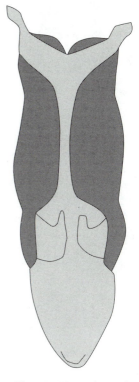

Fig. 5–13. When the skin is pulled, the lighter area is the only area that should be hand skinned. Compare to Fig. 5–17.

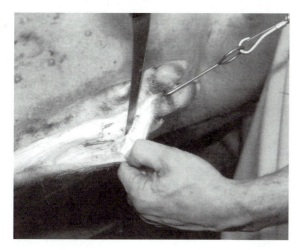

Fig. 5–14. Feet may be skinned out more easily by holding them taut with a bent shroud pin attached to a scabbard chain.

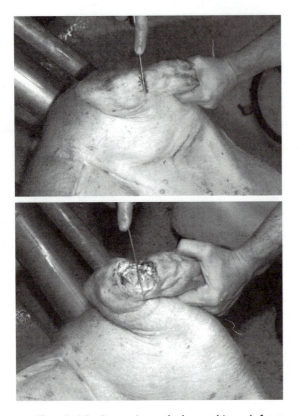

Fig. 5–15. Removing whole unskinned foot. *(Top)* Applying down pressure to locate the flat joint. *(Bottom)* Retaining tendon anchor for carcass hanging.

saved for human food, or they may be completely cut off without skinning (see Figure 5–15) and discarded into the inedible barrel to be processed into animal feed. When removing the hind foot, be sure to cut below the hock joint itself at the flat joint to avoid destroying the tendon anchor used to hang the carcass. The value of pigs' feet is not standard in all areas, so it may not pay to spend the time and effort to clean them properly for human food. Feet with the skin on will average approximately 1 pound each. At any rate, the feet must be removed or skinned before the hide can be removed in order to prevent contamination of the skinned carcass with particles of dirt and(or) manure dropped from the feet.

Once the feet have been skinned or removed, make an opening in the hide down the belly from the tip of the jaw to the tail. Also make an opening down the rear of each hindleg, meeting in the center several inches ventral to the anus. Start skinning the inside of the ham. Follow up forward just past the teat line toward the shoulder, skinning part of the hollow dimple behind the front leg. Begin skinning under the jowl, where the original stick opening was made. Skin down on the jowl and head as far as possible before turning the hog to complete head skinning in the cradle later. Skin down the shoulder to the point of the shoulder. Then start on the other side. Skin out as much of the head as possible. Skin down on the outside of the shoulder to the point and past the dimple behind the front leg. Be careful here *not* to cut through the skin. Skin toward the other ham just to clear the teat line and then skin the inside of the ham. An electric or air-powered skinning knife is useful;

however, avoid doing too much skinning — just clear the teat line and leave the skin on the rear of the hams.

To fully prepare the carcass for skin pulling, you must completely skin out the head. This is a critical area, and care should be taken **not** to cut through the skin. Turn the pig in the cradle to gain access to the top of the head and neck area. Skin past the eyes and ears. Around the ears the skin is very thin and close to the flesh. Skin past the hump of fat on the crown of the skull and around the ear area just far enough to get past the lean connecting tissues. Skin only until you get through the lean and you are back into the fat again. Figure 5–16 shows the head properly skinned. As an alternate method for skinning the head, insert a stainless steel meat hook into the lower jawbone, raising the head part way by hoist in order to gain easier access to skin out the back of the skull.

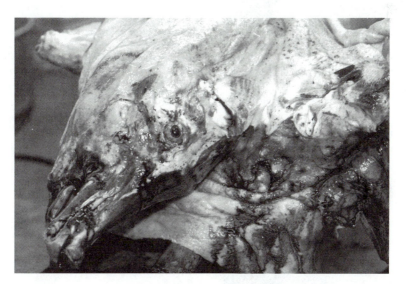

Fig. 5–16. The head has been properly skinned prior to pulling.

Figure 5–17 shows the rear view of a hog properly prepared for hide pulling.

To position the hog for pulling its skin, insert a gambrel in the rear legs, and then raise the hog. If two hoists are available, one may be used to keep tension on the carcass while the second pulls the skin off. A ring should be securely fastened in the floor directly under the point of the jaw. Insert a strong meat hook into the ring in the floor and into the lower jawbone of the

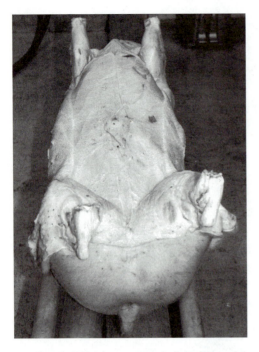

Fig. 5–17. A hog properly "preskinned," ready for hide pulling. Compare to Fig. 5–13.

Fig. 5–18. A trucker's load tightener can be used to maintain tension on the carcass as the skin is being pulled.

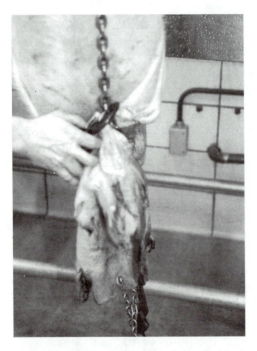

Fig. 5–19. Loose head and shank skin is gathered within the hoist chain for pulling.

hog. Raise the first hoist to take up the slack, but do not exert too much tension as you might pull the head right off. An alternate method is one in which the carcass is landed on a rail, freeing that hoist for pulling the skin and thus using only one hoist per hog. A chain is cinched from the S-hook to the floor ring and tightened by hand, or with a cinching device, such as a trucker's load tightener (see Figure 5–18).

Gather and roll together as much of the loose head and shank skin as possible, and place a short piece of chain around it, or use the hoist chain itself (see Figure 5–19).

Raise the hoist slowly at first to put pressure onto the pigskin, and check that the skin is starting to pull off properly (see Figure 5–20). If it is pulling off into the fat, stop and carefully use your skinning knife to release the skin. Proceed to pull the rest of the skin up and off the carcass (see Figure 5–21).

Figure 5–22 shows the smooth carcass free of gouges that remains after this method of pulling the skin. In Figure 5–23 a carcass that has had the skin pulled this way is compared with one that has been conventionally hand skinned by an experienced butcher.

All hogs do not pull the same, just as they do not scald the same. Even with patience and the desire to change over to a new system after doing it for years another way, a person may be frustrated at first.

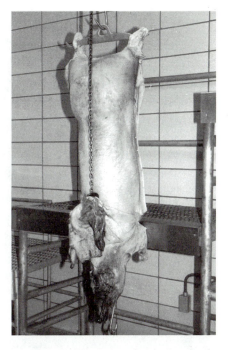

Fig. 5–20. Raise the hoist slowly and somewhat tentatively at first to check the tension and to make sure the skin is starting to pull off properly.

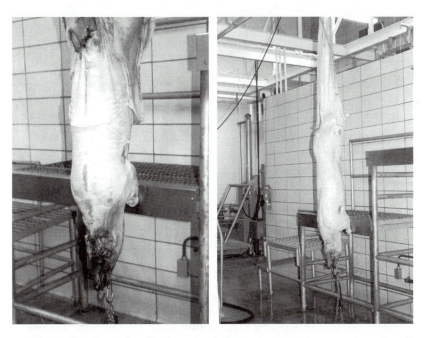

Fig. 5–21. Once the skin is properly started, maintain a steady pull until the skin is completely pulled from the carcass.

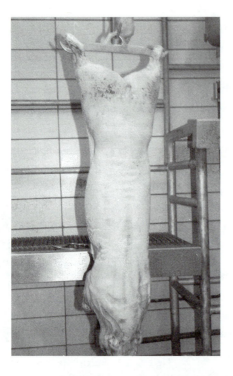

Fig. 5–22. A smooth, gouge-free carcass following skin pulling. The tiny dark spots on the ham are blood from fat surface capillaries that have ruptured when the skin pulled off. These spots have no effect on the meat quality since they are only on the exterior fat surface which is always trimmed off. Even so, the fat is completely edible with the tiny blood spots present, because the blood and fat are wholesome.

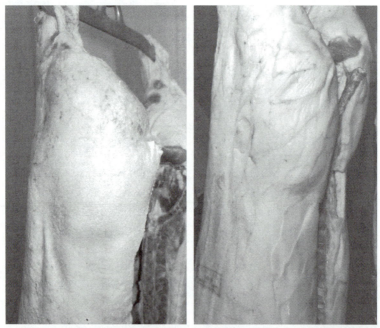

Fig. 5–23. (Left) Closeup of a carcass from which the skin was pulled by the method described under "Pulling Hog Skins in a Small Plant." (Right) Closeup of a carcass from which the skin was removed by an experienced meat cutter. The advantage in carcass smoothness is clearly evident.

The first couple of days are hard on the nervous system. One fault most butchers may have is that of overskinning by hand, especially through the belly region out past the teat line.

The skin represents 6 percent of the weight of the carcass. For other than home consumption, skinning has not previously been economical because skin removal means weight loss, because the market for skins has been poor, and also because the process is slower than dehairing in large commercial machines. Pork skin or rind is not necessary either in the curing or in the subsequent keeping of the cured meat. Curing tests conducted at The Pennsylvania State University have shown that skinned pork cuts take the cure faster and keep as well as unskinned cuts. Other advantages are that the task can be done by one person, no hot water is required, and no rinding (removal of skin) of pork fat for rendering is necessary.

Pigskins must be prepared for storage until they are picked up by the purchaser. Immediately after the skin is pulled from the carcass, trim off the head skin right behind the ears and trim out the tail and anus at the other end.

After placing the skin, with the flesh side up, on a piece of plywood, salt it with a liberal application of ice cream salt or coarse sack or stock salt. Salt and fold so as to hold the salt on the hide, and then place the hide in a 50-gallon drum. When the drum is full, sprinkle a little salt on top. Salted pigskins do not smell or cause a fly problem. Pigskins should be salted as soon as they are available. They should never be held over unsalted.

In recent years, as new equipment has been developed, a number of commercial hog skinning operations have begun in the United States. Among them is the Jimmy Dean Meat Co., which skins hogs for the production of its whole-hog pork sausage (see Figures 5–24 to 5–27).

The USDA–FSIS now permits the injection of compressed air under the skin in the abdominal area to facilitate skin-

Fig. 5–24. Power clippers are used to remove hind feet. The beginning of the skinning operation is the folding over of skin on the ham region. (Courtesy, *The National Provisioner* and the Jimmy Dean Meat Co.)

Fig. 5–25. Using a power knife, continue folding skin down to the chest area. Skinning continues down across the chest. The USDA–FSIS now permits the injection of compressed air into the area between the skin and the body wall to aid in skin removal. The front feet are removed, and the head and shoulders are completely skinned out. (Courtesy, *The National Provisioner* and the *Jimmy Dean Meat Co.*)

Fig. 5–26. Carcass at right shows the loose skin around the head, which is gathered up in a chain. Note the carcass anchors between the jawbone and the platform. The next carcass is in the process of being skinned, and the process has been completed on the remaining carcasses. (Courtesy, *The National Provisioner* and the Jimmy Dean Meat Co.)

Fig. 5–27. Skin take-off, at this instant, is at the point of bung area. Note the large machines used for this commercial operation. (Courtesy, *The National Provisioner* and the Jimmy Dean Meat Co.)

ning. The practice of skinning hogs has increased, since a majority of hams and bellies are now merchandised skinless. Furthermore, the hide is more valuable when removed intact before it is cured, as portions are used for human burn treatments and clothing (see Chapter 11).

HEAD REMOVAL

With practice, removing the head is quite simple; however, for the beginner, it tends to be frustrating and complicated. A sharp knife is especially important for the process. Enter just behind the ears at the natural seam, and follow the seam down both sides of the face, leaving the jowls completely on the carcass. Make a cut directly above the ears (see Figure 5–28). Grasp one ear with one hand and push the head down. This helps separate the opening. Next, position your knife straight up and down, and cut into the seam, outlining the jowl (see Figure 5–29). Cut as close to the jawbone as possible, leaving as much of the jowl on the carcass as you can. Do this to both sides of the head.

Now that you have freed the head of the jowl and tissues in front of the backbone, you are ready to actually cut the joint that holds the head. This is the atlas joint (the first cervical vertebra-skull connection), which is shaped somewhat like a "U." Your knife must follow this path ⁓⁓ to sever the

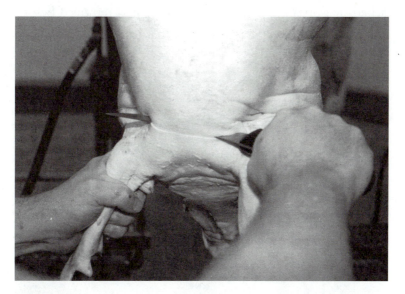

Fig. 5–28. When beginning to remove the head, grasp one ear and pull it down while making an inward cut directly above the ears to the backbone.

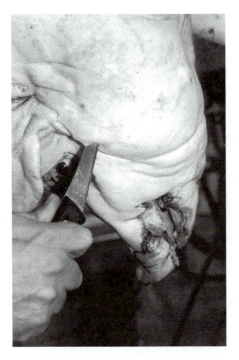

Fig. 5–29. Cutting vertically, follow the natural seam down both sides of the face, separating the head from the jowl and carcass.

joint. To be sure that you are located near this joint, stick your knife directly into the center of the joint. If the blade goes in as far as the handle, you are in the right place. If it does not, you are probably one joint too high. You can sever the joint either by inserting your knife inside the center of the joint and cutting both directions from the inside out or by locating the opening and cutting upward and then downward from the outside in (see Figure 5–30). One side of the atlas joint is shown in Figure 5–31.

Once you are through the atlas joint, cut through the esophagus and trachea, allowing the head to drop further. To avoid cutting through the *mandibular* lymph nodes needed for inspection, sever the windpipe and esophagus directly above *(caudal)* the epiglottis (Adam's Apple), leaving it with the head. This facilitates evisceration later. Leave the head attached by skin and remaining muscle. If it is removed before inspection is completed, keep proper identification. The salivary glands are rather

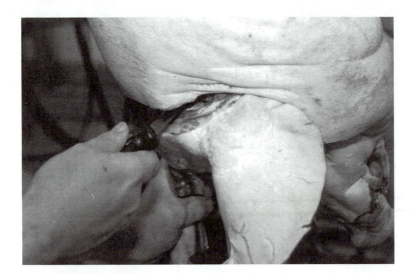

Fig. 5–30. Severing the atlas joint.

large and are important to identify because they are located directly above the mandibular lymph nodes, which the inspector must examine (see Figure 5–31) to check for the grainy tuberculosis (TB) condition. Clean, clear glands indicate a healthy pig.

After the head has been inspected and passed by the inspector, the tongue must be "dropped." With the hog head resting on a rack, place the fingers of one hand into the windpipe and pull toward you. With your knife in your other hand, begin cutting around the mouth cavity. Be sure to cut between the salivary gland and the jawbone so that the salivary gland can be removed. Once all the connective tissue and muscle have been freed, the tongue will drop, or you can pull it partly out. Since the salivary glands and epiglottis are present on the end of the tongue, they must be removed. Cut directly across the end of the tongue. After the head is cleaned

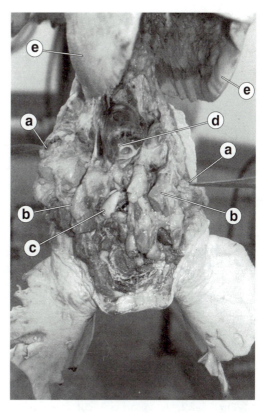

Fig. 5–31. Pig head hanging by skin attachments at the jowl. (a) *Mandibular* lymph nodes (sliced). (b) Salivary glands (right gland sliced approximately in half). (c) One side of atlas joint. (d) Epiglottis (Adam's apple). (e) Jowls.

and washed, the inspection stamp must be applied to the forehead. Many edible products, including brains and head meat for sausage, come from the head (see Chapter 11). Head weights from forty-two 218-pound University of Illinois hogs used for teaching averaged almost 11 pounds each, or 5 percent of the live weight.

EVISCERATION

The first step in the evisceration of a barrow is to remove the pizzle, or penis. Make a superficial cut between the hams on the underline (see Figure 5–32). Continue cutting down towards the pizzle, cutting a little deeper.

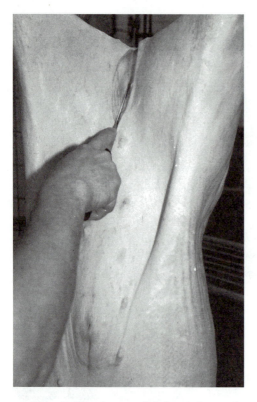

Fig. 5–32. A superficial cut is made between the hams.

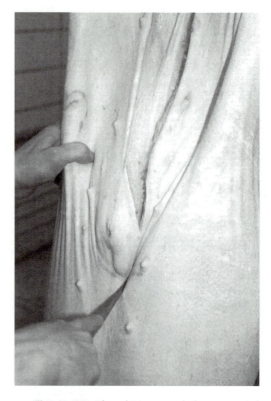

Fig. 5–33. The skin around the preputial pouch is cut.

As you cut, you should notice a white cord. This cord is part of the pizzle. As you approach the preputial pouch, angle off the midline to cut around and under it (see Figure 5–33).

Now cut a little deeper in the opening you made so that the pizzle cord will be exposed. Remain outside the abdominal wall but beneath the pizzle. You should cut deeply enough so that you can completely free the pizzle (see Figure 5–34). When loosening the pizzle, do not cut too deeply because the intestines are located directly dorsal to the pizzle.

When the whole pizzle is free, except at the point of origin which is next to the aitch bone, pull the pizzle back towards the anus and cut away the lean and connective tissue that adheres to the penis. Do not cut off the pizzle. Just loosen the connections around it until you can tell when you are at its point of attachment. If you cut off the pizzle, the bladder could possibly leak urine and contaminate the carcass.

A gilt, of course, has no pizzle to remove; thus, these steps are omitted if you are working on a gilt.

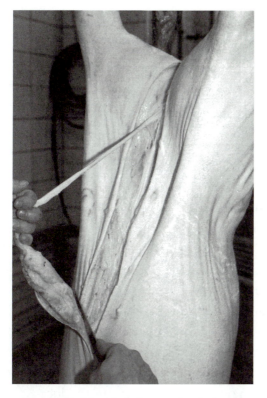

Fig. 5–34. The pizzle is completely freed.

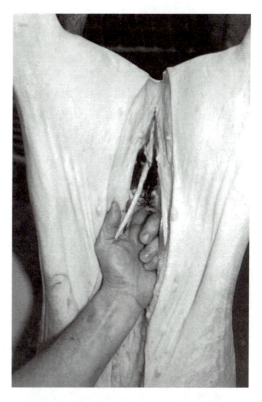

Fig. 5–35. After opening the abdominal wall, you can feel the ridge where the aitch (pelvic) bone is joined in cartilage.

Make a small opening through the abdominal wall in the flank area. Be very careful not to puncture any part of the intestine. The aitch bone is located directly between the two hams. Separate the hams at the midline, indicated by a silvery membrane. In males, this is exactly at the center of the pizzle root. You can feel the exact center of the aitch bone by placing your finger on it (see Figure 5–35).

The aitch bone can be split at its midpoint where it is joined by cartilage. In Figure 5–36, the knife is positioned in the center of the aitch bone. Press carefully and firmly with a sharp knife. The aitch bone will split easily in a young pig if you are exactly on the center cartilaginous seam. If you are unable to split the aitch bone, check your location. Do not be overzealous in applying force on your knife because it may slip and sever the large intestine, which lies immediately behind the aitch bone. The aitch bone is just a few inches long.

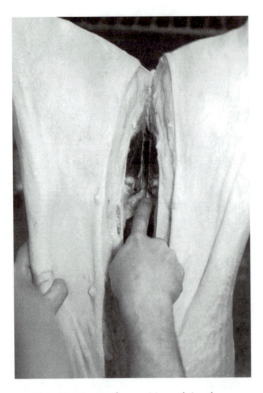

Fig. 5–36. Knife positioned in the very center of the aitch bone where the bone splits easily through the cartilaginous joint.

Bunging, or loosening the anus and the reproductive tract, is the next step. It may be easier for a beginner to bung from the backside of the animal where the detail of the anus can be easily seen; however, experienced butchers bung from the abdominal side of the carcass. They have learned to feel the critical portions of the anus and to cut around them without seeing exactly where the knife is.

Holding your knife vertically, cut to the backbone on both sides of the anus (see Figure 5–37). The anus is attached to the backbone; therefore, it is necessary to cut behind the bung in order to avoid cutting the intestinal wall (see Figure 5–38).

You should be aware of the location of the knife tip so that you do not puncture the bung (large intestine). When all the connective tissues of the bung are severed, you should be able to pull the bung up and out to the ventral side, where you can examine carefully for any punctures which should *not* be present. Although the *sphincter* muscle prevents fecal matter from

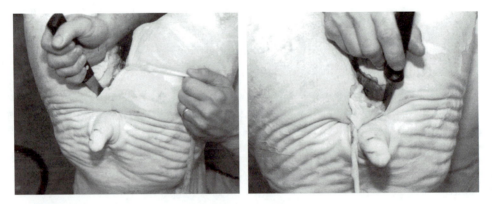

Fig. 5–37. In bunging, a cut is made on first one side and then on the other side of the anus.

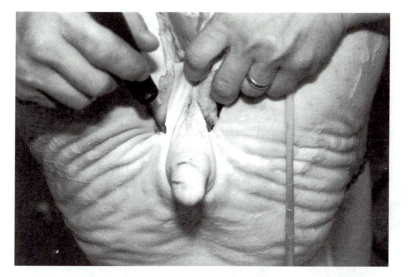

Fig. 5–38. Separate the anus from the backbone by cutting, *being careful not to puncture the intestinal wall.*

escaping, you may tie off the bung for extra safety. (String should always be available on the harvest floor.) Figure 5–39 illustrates the properly loosened bung, including penis (pizzle) and preputial pouch.

Splitting the sternum is the next step. First, open the skin from the stick hole to the bottom of the jowl (see Figure 5–40). Be sure to make only one opening. Then, extend the stick hole (the original opening). The sternum bone extends from the stick hole to approximately the first teat. Cut through the skin and fat to the bone (see Figure 5–41).

You may split the sternum by using a hand saw held with the blade facing up (see Figure 5–42). You must saw on the center of the sternum bone, being careful not to puncture the intestines with the saw

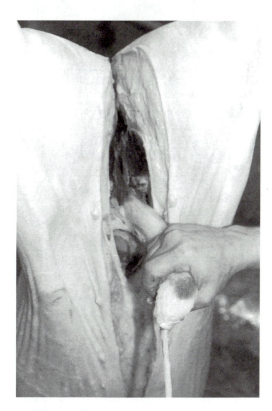

Fig. 5–39. A properly loosened bung, including penis and preputial pouch.

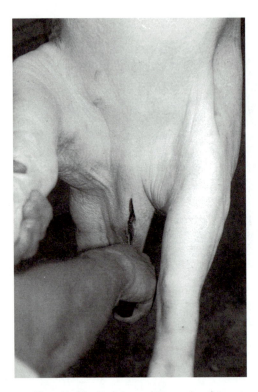

Fig. 5–40. Before splitting the breast-bone, make an opening by extending the stick opening. *Never make a second opening,* as it will expose the meat unnecessarily.

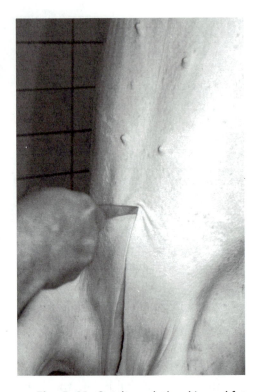

Fig. 5–41. Cut through the skin and fat to the bone, and continue to a point near the first teat.

at the top end of the sternum. Alternately, you may split the sternum with a knife; however, you must cut off-center to get through the cartilaginous connections of the ribs, thus making unequal spareribs on the two sides, with one side containing all the sternum bone.

Now you are ready to open the ventral side. It is very important to protect the intestines from the edge of the knife. Place your thumb and index finger at the heel of the knife blade. When cutting, separate the knife edge from the intestines with your fingers. Insert your knife, handle first, into the upper opening where you had split the hams, and with the blade protruding outward, cut through the abdominal wall for the entire ventral length of the carcass. Keep your fingers next to the wall of the abdomen and in front of the intestines. Put pressure on your knife, and cut in *one smooth, continuous motion* down through the opened sternum (see Figure 5–43). ***Once you start, do not stop!*** If you stop, the intestines will begin to fall out of the cavity, and you will

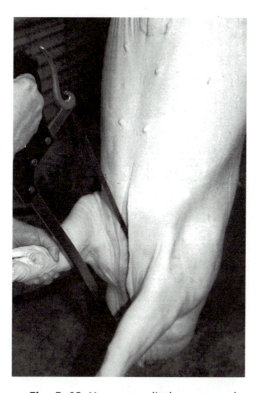

Fig. 5–42. You may split the sternum by using a hand saw held with the blade facing up.

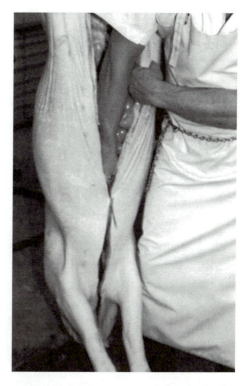

Fig. 5–43. In opening the ventral side, keep your fingers next to the abdominal wall and in front of the intestines. Cut in one smooth, continuous motion.

have difficulty finishing opening the ventral side without puncturing the intestines.

Once you have completed the ventral side opening, the intestines will fall out only partially. Grasp the bung and pull firmly. You will be able to pull the bung a little; however, you will notice two white cords that are connected to the kidneys. These cords are the *ureters,* which are connections between the kidneys and the bladder. Cut the ureters, allowing the bung to be easily pulled (see Figure 5–44).

A silvery membrane which is part of the diaphragm separating the abdominal cavity from the thoracic cavity must be cut from the top to the bottom. The membrane goes with the intestines, and the diaphragm muscle should be left in the carcass to be saved for human food (see Figure 5–45).

Once both sides of the silvery membrane have been cut, you must sever the membranes and connective tissues that are connected to the backbone. Make a cut just above the liver all the way to the backbone. The kidneys and

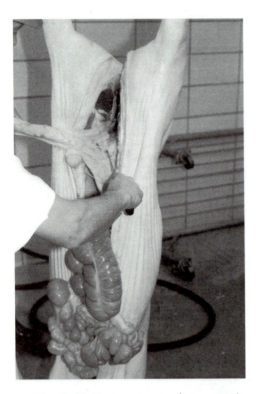

Fig. 5–44. You must sever the ureters in order to successfully pull the bung out of the body cavity.

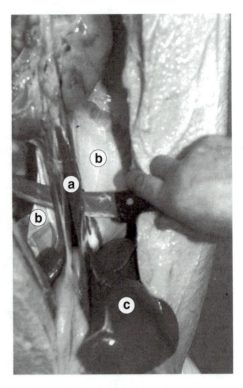

Fig. 5–45. Cutting the diaphragm to release the viscera. (a) Central diaphragm muscle. (b) Silvery diaphragm membrane (the portion to the left has been cut). (c) Liver. Care must be taken to avoid cutting the esophagus, which extends through the diaphragm. Here the knife is posed several inches above the esophagus.

kidney fat are still present and remain in the carcass at this point. As soon as you make this cut, all the internal organs, except the kidneys, will fall out. Therefore, you must be prepared to catch and hold all the intestines and organs. A firm grasp of the stomach and part of the intestines will assure you of not dropping the remaining viscera (see Figure 5–46).

Generally, the pluck, or contents of the thoracic cavity, will not come out easily. Therefore, you must cut to your and the pig's left, inside the thoracic cavity, cutting the *pericardium* (heart sack). Then push down. By firmly pushing down on the windpipe, you will free the remaining organs. Place the entire tract on a table or a tray (*not on the floor*) for inspection. The liver must be removed from the rest of the viscera after it passes inspection. The pale

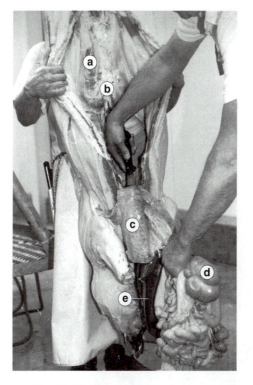

Fig. 5–46. Completing evisceration. A firm grasp on the stomach and intestines is necessary to prevent the viscera from falling to the floor while the remaining support tissues are cut. (a) Tenderloin (psoas) showing through under the leaf fat. (b) Kidney. (c) Lungs. (d) Blind gut (caecum). (e) Liver.

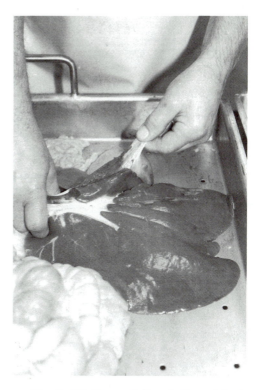

Fig. 5–47. The pork liver is three-lobed. Removing the gall bladder involves loosening the duct away from the bladder and then pulling steadily on it.

green bag is the gall bladder, which contains bile salts in solution. To remove the gall bladder, first sever the white bile duct as far away from the gall bladder as possible by making only a superficial cut across the bile duct. Next, pull the bile duct and the gall bladder off the liver. The duct and gall bladder pull off quite easily, if you free the duct and pull firmly and steadily on it (see Figure 5–47). *Do not puncture or spill the gall bladder contents.*

Once the gall bladder is free, either empty the gall to be saved (see Chapter 11) or discard it in the inedible barrel if no market exists, and then wash the liver. If the esophagus (weasand) is saved for human consumption, the thyroid gland, if present, must be removed because it is not safe for human consumption (see Chapter 11).

SPLITTING, INSPECTION, WEIGHING, AND REFRIGERATION

The carcass must be split in half in order for it to adequately chill and for ease of cutting and fabricating cuts. A power circular saw may be used by experienced, qualified workers, as long as it is equipped with an automatic brake in the blade. A safer, small, portable Wellsaw may also be used (see Figure 5–48).

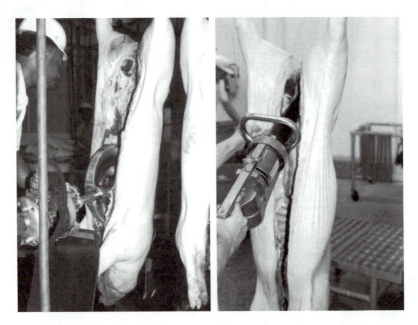

Fig. 5–48. *(Left)* Halving the carcass with an electric disc saw, such as packers use. *(Right)* A small Wellsaw is often used in small plants to split hog carcasses. (Left, University of Illinois photo)

The safest procedure for students is to use the hand saw (see Figure 5–49). Begin at the tail bone and work down. Lay the saw on the sacrum so that it will be at a 45° angle. Keep this angle as you saw through the backbone. The carcass should be split precisely in half. Watch for the spinal cord as you are sawing, since this will show that you are in the center as you split each vertebra. When completing the splitting of the shoulder, be especially careful. Leaving the skin and fat slightly attached between the sides will prevent the two separated sides from slipping off the gambrel ends.

After the carcass has been split, remove the glands and blood clots that hang in the neck and jowl regions.

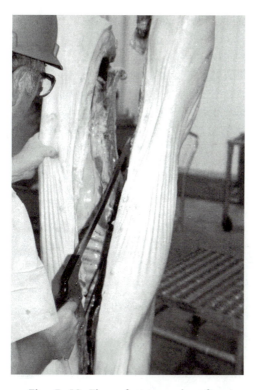

Fig. 5–49. The safest procedure for students is to use the hand saw for splitting.

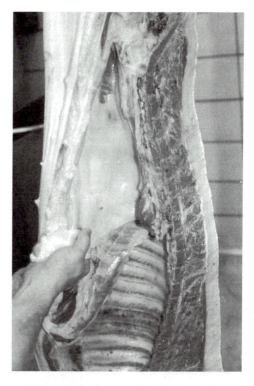

Fig. 5–50. Remove leaf (kidney) fat by starting where it joins the diaphragm muscle and then pulling up and away.

Next, remove the leaf fat and kidneys. The leaf fat is a high-quality fat lining the abdominal wall, commonly called kidney fat in beef and sheep. A strong person can easily remove the leaf fat. First, pull the end of the fat from the belly at a point just above the diaphragm muscle (see Figure 5–50). Continue pulling up. You can also pull the kidneys and kidney fat out with the leaf fat.

At the top near the ham, be careful not to damage the *femoral* artery, which may be used later to cure the hams (see Figure 5–51).

Facing the ham will be the next step. This process decreases the carcass weight and thus dressing percentage; however, it increases the attractiveness and lowers the fat trim on the carcass. Begin cutting a strip of skin and fat directly above the last teat. Continue pulling with one hand and cutting with your knife (see Figure 5–52). The skin and fat may be removed back to the tail bone.

Finally, wash the carcass, starting at the top and spraying downward towards the jowl. The purpose of washing is, of course, to remove any bacterial contamination that might have become associated with the carcass during

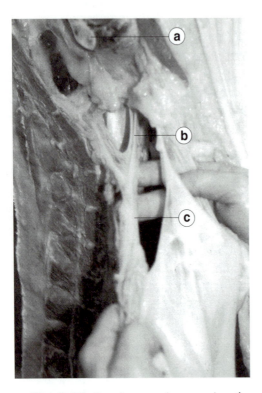

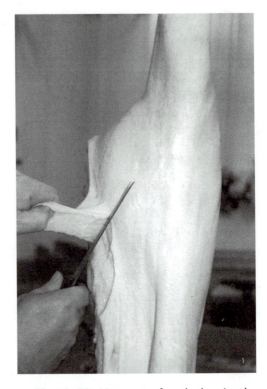

Fig. 5–51. Exercise care in removing the leaf fat so that the femoral artery (b) is not damaged. (a) Aitch bone. (c) *Dorsal aorta*.

Fig. 5–52. Hams are faced; that is, the fat collar is removed to improve appearance.

harvest. A mild salt solution (0.1 M KCl) weakens the bacterial attachment to carcasses and makes the bacteria more accessible to sanitization procedure, especially if the solution is applied promptly.[8] Pressure of 200 pounds per square inch (psi) and a flow rate of 3.36 gallons per minute (gpm) are most effective. Dilute organic acids (2 percent lactic acid and 3 percent acetic acid) are good sanitizers. All washes will destroy bacteria at ambient temperatures. In industry, carcass washing is mechanized. As carcasses pass through cabinets on the harvest line, the proper solutions are applied at the most effective pressure.

Once the carcass has been washed (and sanitized), the inspector will examine it one final time. The inspector will be looking for hair and other contaminants, as well as abscesses and other abnormal growths. After the inspector has examined the carcass, weigh it, subtracting the weight of the trolley

[8]American Meat Institute. 1989. A Research Report: A Pilot-Plant Scale Evaluation of Washing and Sanitizing Procedures. P.O. Box 3556, Washington, DC 20007.

(taring). At the direction of the inspector, apply the inspection stamp to each wholesale cut. Blot the surface to be stamped with a paper towel to prevent the ink from running. The stamp is the buyer's insurance that the carcass has been inspected for wholesomeness. Notice the shape and the information contained on the stamp, including the establishment number unique to the plant where the pig was harvested (see Figure 5–53). All the wholesale cuts, except the Boston butt, are stamped. The Boston butt cannot be stamped because the outer surface for stamping is under the clear plate. The loin and spareribs should be stamped on the inside with a smaller-sized stamp, where the ink will not be trimmed off when the carcass is cut. The carcass should be marked with an ID number.

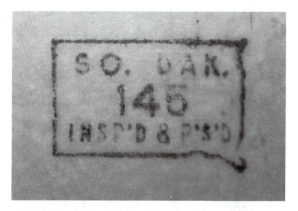

Fig. 5–53. The inspection stamp (in this case, the state stamp) is applied to each wholesale cut.

Place the carcass in a 32° to 34°F (0° to 1°C) cooler with air velocity typically 5 to 15 miles-per-hour (mph) equating to 23°F wind chill, for a 24-hour chill period. However, with accelerated (hot) processing, the carcass may be tempered (held at an intermediate temperature [60°F]) for several hours or boned immediately (see Chapter 22). When large numbers of warm carcasses are handled, the chill room is normally precooled to a temperature several degrees below freezing (27°F), bringing the wind chill to 16°F to compensate for the heat from the carcasses, which raises the cooler temperature considerably. (See the Appendix for the formula to calculate wind chills.)

The cooler shrink on a 24-hour chill will average between 2 and 3 percent, depending upon the humidity of the cooler. However, spray chilling is permitted by the USDA–FSIS, and the water spray solution may contain up to 5 parts per million (ppm) available chlorine which acts as a sanitizer.[9] Canadian workers demonstrated that spray chilling pork carcasses during the first 24

---

[9]Association of Official Analytical Chemists (AOAC). 1984. Official Methods of Analysis.

hours postmortem for 1 minute every 15 minutes for 10 hours significantly reduced carcass shrinkage from 2.6 percent for conventional chilling to 0.5 percent.[10] The spray-chilled carcasses were slightly tougher. For thorough chilling, the inside temperature of the ham should reach at least 37°F.

At the John Morrell and Co. plant in Sioux Falls, South Dakota, hot pork carcasses directly off the harvest floor pass on line into and through a freezer. Company officials report customers are very happy with the bright colored fresh pork that is firm and lacking excessive surface moisture, which sometimes detracts from the appearance and quality of fresh pork. Furthermore, there have been no reports of any toughening, and this process, when compared to conventional chilling, reduces carcass shrink.

WILTSHIRE SIDE

A Wiltshire is an English style side of pork that must conform to rather rigid specifications in order to satisfy the English market. It consists of the entire side, or half, of a hog carcass, minus the head, feet, aitch bone, backbone, tenderloin, and skirt (diaphragm). The spareribs and neck rib (bones) are left in the carcass.

The ideal side weighs approximately 60 pounds, a minimum of 40 pounds and a maximum of 80 pounds. The ideal length from the fore part of the first rib to the fore end of the aitch bone is 29 inches, with a minimum of 26 inches and a maximum of 32 inches. Wiltshires are cured and then packed, either in bales or in boxes, for export. They are smoked after they reach their destinations.

COMPUTERIZED HARVEST AND ROBOTICS

It is extremely difficult to automate harvest procedures because of the biological variability in animals, their carcasses, and cuts. Nevertheless, in some parts of the world much progress is being made in computerizing harvest procedures and installing robots to accomplish the most repetitive tasks. These innovations have usually been introduced and used in low-volume plants.

---

[10]Jones, S.D.M., A. C. Murray, and W. M. Robertson. 1988. The effects of spray chilling pork carcasses on the shrinkage and quality of pork. Can. Inst. Food Sci. Technol. J. 21:102.

6

Cattle Harvest

In 1866 the American Society for the Prevention of Cruelty to Animals was incorporated. We now refer to the conversion of live animals into carcasses, meat cuts, and meat products as **harvesting.** The meat industry is concerned that animals, which provide us with the most nutritious, well-balanced food available, be treated kindly at all times and not suffer pain.

This chapter deals with cattle harvest. Cattle were probably tamed about 9,000 years ago. They were used for draft and meat but were not raised for milk until 3,000 years later.[1] Cattle ownership was an indication of wealth. The word *cattle* is related to the words *chattel* and *capital,* which relate to goods or property.

Cattle are not native to the United States. Columbus brought over cattle from Europe, and later Portuguese traders brought over more. The English brought over many cattle in the early days of our history; these cattle are still known today as English breeds. In the front of this textbook, note the classification system showing two species of cattle — *Bos taurus* and *Bos indicus.* The latter are humped cattle.

There are now approximately 1.1 billion cattle in the world and 100 million cattle in the United States. The mystique of the cow has pervaded our major cities. Both Chicago, in the summer of 1999, and New York, in the summer of 2000, featured fiberglass cow statues in public places and buildings. Such cow art displays are planned for other major cities of the world.

Although many readers will not be involved in the actual harvest of cattle, all persons should know the procedures involved in converting a live, contented bovine into delicious steaks, roasts, and ground beef.

---

[1]Archaeology Special Section. 1998. Reading the signs of ancient animal domestication. 1998. Science. 282:1448. http://www.sciencemag.org

In 1998 there were 35,465,000 cattle harvested commercially in the United States, excluding those relatively few harvested on the farm.[2] Table 6–1 lists the states in the order of commercial cattle harvest volume in 1998. For current information on what method of marketing (direct, terminal, sale barn, grade and yield) was used for these cattle, you may wish to contact the USDA Agricultural Marketing Service (http://www.ams.usda.gov/).

----

[2]USDA, National Agricultural Statistics Service, Agricultural Statistics Board. March 1999. Livestock Slaughter — 1998 Summary. (Mike Easdale.)

Table 6–1. Commercial Cattle Harvest Ranked by States, 1998[1,2]

Rank	State or Region	Thousand Head
1	Kansas	7,540.8
2	Nebraska	7,300.4
3	Texas	6,767.0
4	Colorado	2,417.2
5	Wisconsin	1,649.7
6	California	1,019.1
7	Illinois	1,019.0
8	Pennsylvania	958.9
9	Iowa	932.2
10	Washington	924.9
11	Idaho	743.2
12	Minnesota	685.3
13	Arizona	463.0
14	Michigan	394.7
15	Georgia	359.1
16	South Dakota	249.3
17	Alabama	228.6
18	North Carolina	162.0
19	Ohio	150.3
20	New York	86.6
21	Missouri	75.9
22	Florida	50.7
23	Kentucky	49.8

(Continued)

Table 6–1   (Continued)

Rank	State or Region	Thousand Head
24	Indiana.	43.3
25	Delaware–Maryland	36.3
26	Oklahoma	34.7
27	New England[3].	27.5
28	Louisiana	25.5
29	New Jersey	24.4
30	Montana.	21.9
31	Arkansas.	20.7
32	New Mexico	20.6
33	Virginia.	19.3
34	Hawaii	18.2
35	Oregon	15.0
36	West Virginia.	14.3
37	Wyoming	6.4
38	Nevada	1.0
	Total United States[4]	35,465.0

[1]USDA, National Agricultural Statistics Service, Agricultural Statistics Board. March 1999. Livestock Slaughter—1998 Summary. (Mike Easdale.)

[2]Includes animals harvested in federally inspected and other plants, but excludes those harvested on the farm.

[3]New England includes Connecticut, Maine, Massachusetts, New Hampshire, Rhode Island, and Vermont.

[4]States with no data are still included in the U.S. total; individual data are not printed to avoid disclosing individual operations.

In 1962, direct purchases of cattle by packers accounted for only 43 percent of the total purchases, while more than half were bought in terminal markets. Since that time, the number of cattle bought in terminal markets has declined, while the number of direct purchases has risen. The U.S. Department of Agriculture no longer reports terminal and auction purchases separately. Carcass grade and weight purchases are a form of direct purchase based on actual carcass grade and weight. Direct sources for steers and heifers are higher (over 92 percent) than the total average, while public market sources are more important (nearly 60 percent) for cows and bulls.

Before proceeding to harvest a beef animal, be sure that you have proper and safe equipment; that your equipment is ready to use (sharp and clean);

and that the animal you are about to harvest is healthy. In other words, you should study and master the first four chapters before applying the material you will learn in this chapter.

FAST (DRIFT) PRIOR TO HARVEST

Cattle will lose from 3 to 4 percent of their weight if they are kept off feed for 24 hours. This is referred to as "shrink," or "drift." It has become a general practice for buyers to demand this shrink, either as a mathematical deduction from full-fed weight or as an actual off-feed practice. If reputation means anything at all to feeders, they will never salt or fill their cattle before sale time. A cattle buyer is seldom fooled and then only once by the same person. The carcass yield will betray the perpetrator.

Thus, cattle should be kept off feed at least 24 hours previous to harvest. Results of a test conducted at the Pennsylvania Agricultural Experiment Station to determine the effects of fasting beef animals for periods ranging from 24 hours to 48 hours on the yield and appearance of the carcasses were definitely in favor of the longer fast period. The fasted animals bled out more thoroughly and were easier to dress, and the carcasses were brighter in appearance than those from cattle allowed feed up to the time of harvest. (Note previous discussion on hogs in Chapter 5.) Undue rough handling or excitement causes the blood to be forced to the outermost capillaries, from which it will not drain as thoroughly as it would under normal heart action. This results in a fiery carcass (pink tinge to the fat), which has lower keeping qualities because of the retained blood.

WEIGHING AND RECORDING

The live beef animal should be weighed on a balanced scale prior to harvest so that yield (dressing percent) can be accurately determined. Carcass yield is often confused with USDA Yield Grade, but the two terms signify distinctly different aspects of animal value. *USDA Yield Grade* refers to the percent of the carcass that is available for sale as retail product, while *carcass yield* refers only to the weight of the entire carcass (muscle, fat, and bone), expressed as a percent of the animal's live weight. *Carcass yield* is more clearly designated as dressing percent. Hot dressing percent (hot carcass weight ÷ live weight × 100) is a function largely of fill and fat, but weight and muscling are important too. Fatter, heavier cattle will usually dress higher. However, muscle is denser than fat, so a lean, heavily muscled animal may dress higher than a fat, less-muscled animal. In any case, the animal with more fill will dress lower. All cattle will average around 60 percent. Heifers run slightly lower in dressed yield, mainly because they have a greater amount of internal fat surrounding the intestines, which does not stay with the carcass. Buyers discount prices offered for open heifers as insurance against pregnancy. Pregnancy lowers dressing percent considerably, in some cases as much as 10 percent.

STUNNING

Various types of compression guns can be used on properly restrained animals. They are made with either long or short handles and are of the penetrating or the nonpenetrating type. The nonpenetrating type have mushroom heads. These guns operate on the forehead or behind the poll. Several different types of stunners and the locations of stunning are shown in Figures 6–1, 6–2, 6–3, and 6–27.

Electrical harvest and stunning are permitted and used in some parts of the world. (Note previous discussion about stunning hogs in Chapter 5.)

Fig. 6–1. A power-actuated stunner. It uses .22 caliber rim fire power loads in five graded strengths for effective stunning of all weights of cattle, calves, hogs, and sheep. This packing house tool meets all legislative and humanitarian requirements and delivers effective stunning blows to pates (crowns) or backs of heads. Automatic penetrator rod retraction, lever-type trigger, sleeve-style bolt, and light weight and compact design are some of the features of the stunner, made for easy portability and comfortable, one-hand operation. (Courtesy, Remington Arms Co., Bridgeport, Connecticut)

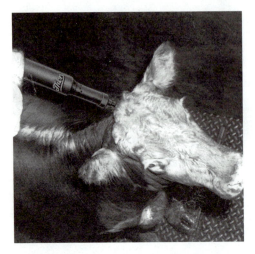

Fig. 6–2. A compression stunner for cattle. In this case, the stunner (penetrating type) is aimed at the *medulla oblongata*. (Courtesy, Thor Power Tool Co., Aurora, Illinois)

Fig. 6–3. Stun the forehead at the spot marked with the *x*, which is the point at which imaginary lines from each eye to the opposite horn root, or poll, cross.

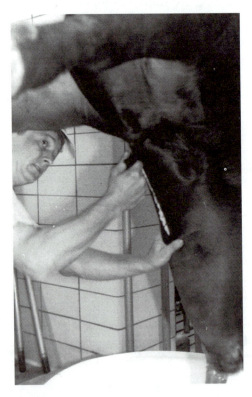

Fig. 6–4. Make an incision through the hide only between the jaw and the point of the brisket.

STICKING (EXSANGUINATION)

Standard Method

After the animal has been hoisted, make an incision through the hide only, between the point of the brisket and the jaw (see Figure 6–4). Insert the knife in front of the brisket at a 45° angle (see Figure 6–5), and sever the carotid arteries and jugular vein (see Figure 6–6).

Kosher Method

The rabbi, or sholet, cuts across the throat in one continuous stroke with a 16-inch knife (calif) without stunning the animal. If it is properly done, the animal loses consciousness within three seconds.

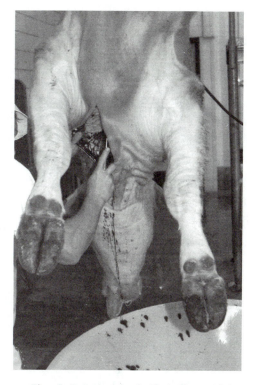

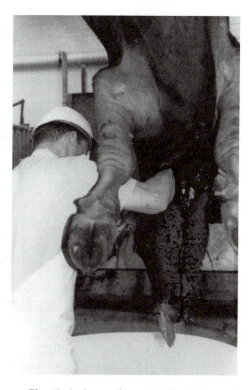

Fig. 6–5. Insert the knife in front of the brisket at a 45° angle.

Fig. 6–6. Sever the *carotid* arteries and *jugular* vein. Blood is caught in a large, funneled vat.

ELECTRICAL STIMULATION

Electrical stimulation refers to the passage of an electric current through a carcass at some point during the harvesting process. The electrical stimulation of pre-rigor muscle causes the carcass to undergo a rapid series of muscle contractions and relaxations. This rapid series of contractions and relaxations accelerates the rigor process (stiffening of the carcass) by speeding up normal biochemical and physical events that occur in muscle after the animal's death, e.g., depletion of energy stores and decline in muscle pH. (See Chapter 22 for a more detailed discussion of the effects of electrical stimulation on postmortem muscle changes and meat quality.)

In 1951, electrical stimulation was first recognized for its tenderizing effect on meat (Harsham and Deatherage).[3] However, there was limited com-

---

[3]Harsham, A., and F. E. Deatherage. 1951. Tenderization of Meat. U.S. Patent 2,544,581.

mercial interest in the process (in the United States) until scientists from the Texas Agricultural Experiment Station determined that the use of stimulation had a beneficial effect on the *quality grade* designation for beef carcasses (see Chapters 12 and 22). Since the U.S. marketing system for beef carcasses is highly dependent on the quality grade a carcass receives, this finding provided the economic incentive for the production of commercial electrical stimulation equipment. Thus, early in 1978, the LeFiell Company developed the first commercial stimulator. Thereafter, this technology was rapidly adopted within the beef packing industry and is now widely practiced.

The benefit of improved tenderness is no longer recognized as the primary benefit of electrical stimulation. In fact, improvements in tenderness are generally so little that they are often not detected by the human palate. There are, however, a number of other advantages for electrically stimulating beef carcasses. These advantages include better dissipation of heat ring; improved lean color, firmness, texture, and marbling score (see Chapters 12 and 22); improved bleeding of the carcass; and greater ease of hide pulling.

The two recognized methods of electrical stimulation are high-voltage stimulation (greater than 500 volts) and low-voltage stimulation (less than 75 volts). The high-voltage stimulation equipment is designed for larger packers with continuous chain operations that can handle up to 350 head per hour. These systems are considerably more expensive than low-voltage systems and have complex safety mechanisms built in to prevent accidental electrocution of workers. Because complex safety equipment is not needed in low-voltage electrical stimulation systems, these systems are much cheaper. However, low-voltage systems are considerably slower and have not been well adapted to high-speed, continuous chain operations. Low-voltage stimulation also operates on a principle that is slightly different from that of high-voltage stimulation. Low-voltage stimulation must be completed no later than 15 minutes after stunning because this method relies on the still functioning nervous system, which triggers muscle contractions, whereas high voltage directly stimulates the muscle itself.

Figure 6–7 illustrates a steer being stimulated with a low-voltage stimulator. Note that stimulation is done immediately after exsanguination in this case. Violent muscle contractions cause the forelegs to point upward, the tail and ears to become erect, and the hindlegs to spread apart. Excess blood is "pumped" because of the rigorous muscle contractions, resulting in a more complete bleed.

Fig. 6–7. Low-voltage electrical stimulation of beef. *(Left)* Steer in relaxed state with no electric current applied. *(Right)* Steer in state of contraction with 75 volts (1.5 amps) of electricity applied. Electricity is generally "pulsed" through carcasses at 0.5- to 1.5-second intervals. The electricity travels from the two probes (one inserted in each hock) to the nose clamp (ground). Note upward contracted state of forelegs, spreading of hindlegs, erect tail and ears in the steer in the right photo. The steer's forelegs were secured with a rope to prevent them from violently swinging about.

RODDING THE WEASAND

The esophagus must be separated from the trachea and other connective tissue attachments in the thorax to facilitate evisceration. The separation, referred to as "rodding the weasand," is done at some point following bleeding. A metal rod, with a handle on one end and formed into several spiral loops on the other end, is threaded onto the esophagus just behind the Adam's apple and forced toward the rumen, thus separating the esophagus from the trachea (see Figure 6–8).

As an alternative, this step can take place on the skinning cradle after the brisket has been split (see Figure 6–19). A knot is made in the esophagus, or a string or rubber band is put around it to prevent spillage of rumen contents.

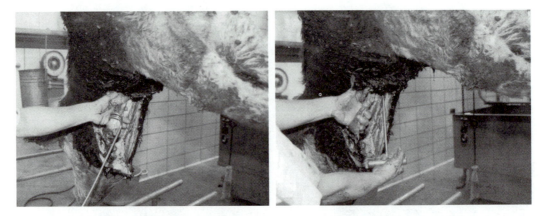

Fig. 6–8. Rodding the weasand. *(Left)* Metal rod with spiral loops is threaded onto the esophagus and *(right)* forced toward the rumen, thus separating the esophagus from the trachea.

HEADING

Open the hide from the poll down to one nostril. Continue the opening from the stick opening down through the center of the jaw and skin out one side of the face (see Figure 6–9). Skin out the front of the face (see Figure 6–10). Turn the head and skin to the opposite side (see Figure 6–11). Grasp the jaw in one hand, bend the head back on its poll, and remove the head by cutting through the Adam's apple and the atlas joint (see Figure 6–12).

The USDA–FSIS permits injection of properly filtered compressed air under the hide in the region of the head to aid in hide removal. Head weights on forty-three 1,037-pound University of Illinois steers used for teaching averaged 29 pounds and ranged from 1.5 percent to 4.5 percent of live weight.

Beef heads furnish several edible products — cheek meat (*masseter* muscles), the brain, and the tongue (see Chapters 11 and 18).

SHANKING OR LEGGING

Place the skinning rack under the withers and then lower the animal onto the rack. Prior to opening up the hide, remove the feet to prevent contamination of the carcass with manure and dirt dropped from the hooves. Open the

Fig. 6-9. Heading. Skinning out one side of the face.

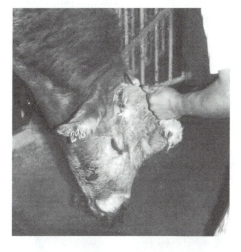

Fig. 6-10. Skinning out the front of the face.

Fig. 6-11. Skinning out the other side of the face.

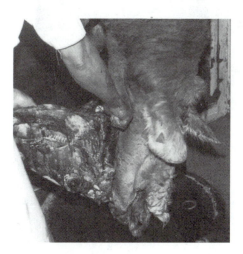

Fig. 6-12. Removing the head after severing the atlas joint.

hide on the rear of the foreshank and the rear of the hindshank by placing the tip of your knife under the skin and then cutting outward. The cut should begin at the dewclaws on the hindshank and continue down the leg, over the hock joint to end at the midline, slightly below the anus. On the front shank sever the tendon to release tension by cutting across the shank. Then, use the same procedure, opening from the dewclaws, or from the point of removal of the shank if it has already been removed, to the elbow, and then across to the midline on the brisket.

Remove the shanks at the articulating joint between the carpal bones (foreshank) and the tarsal bones (hindshank) located below the knee and hock, respectively. Skinning out the shanks will help you to see the joints more clearly. On the front legs, this joint is at the enlargement, about 1 inch below the knee joint, just where it tapers down to the cannon bone (see Figure 6–13). You have found the correct joint if you expose two flat bones in each joint face. If you expose three rounded bones in each joint, you are too high and into the knee joint itself. Your knife should slip through this joint, unless the animal is matured, in which case the joint may be calcified, and you may have to use a saw.

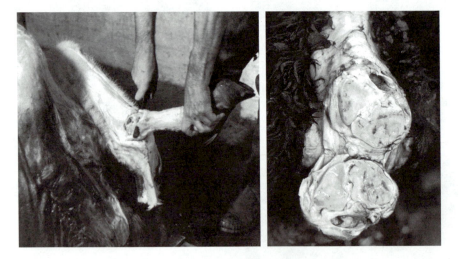

Fig. 6–13. *(Left)* Removing the foreshank. *(Right)* Closeup of flat joint, proper location of foreshank removal. Note the two flat bones in each joint face.

On the rear hocks, the proper spot is about 1 inch toward the hooves from where the taper begins. A decided groove is evident when the knife rests at the proper spot. It is especially important to remove the hindshank at the proper location below the hock joint itself in order to preserve the tendon anchors for carcass hanging. Cut around to either side, and then grasp the shank near the foot and give a sharp thrust downward and outward from the stifle joint (see Figure 6–14).

The USDA–FSIS permits the injection of properly filtered air under the skin of the shanks to aid in sanitary removal for human consumption.

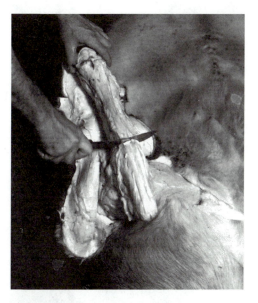

Fig. 6–14. Removing the hindshank.

SIDING

Open the hide down the middle from the stick opening to the bung. Siding can be started at the fore end or rear end (see Figure 6–15). To side, grasp

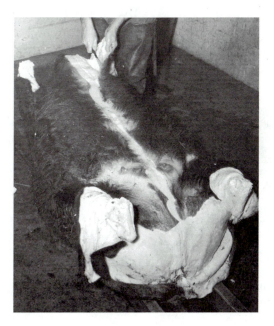

Fig. 6–15. Siding can be started at the fore end or rear end. This beef is held on its back by means of a skinning rack.

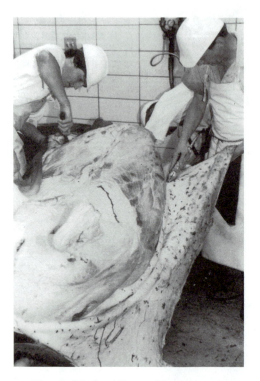

Fig. 6–16. In siding, with either a conventional hand knife or a power rotary skinner, hold the hide tightly, and keep the bevel of your knife flat to the hide in order to avoid making cuts or scores in the hide.

the hide firmly with an upward pull, and with long, smooth strokes of the skinning knife, remove the hide down over the sides. Air or electrically powered rotary skinning knives are popular in many plants because they make the job of skinning much easier. The bevel of either type must be flat to the hide to avoid making cuts or scores (see Figure 6–16). This is one of the most difficult tasks in the skinning operation and requires considerable experience before satisfactory progress can be made.

Note the proper hide pattern in Figure 6–17. Attempt to avoid scoring or cutting the hide, as this decreases its value for leather (see Chapter 11).

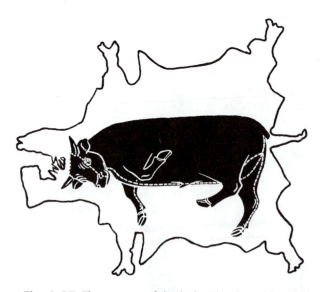

Fig. 6–17. The pattern of the hide. The dotted line indicates where the opening is made in the hide (see also Fig. 11–3). (Courtesy, USDA)

OPENING

After the hide has been sided down to the rail on the skinning rack on each side, open the brisket. Cut through the fat and muscle of the brisket at its center with a knife, and then saw through the sternum (see Figure 6–18).

Once the chest cavity is opened, rod the weasand; i.e., separate the esophagus from the trachea (see Figure 6–19), unless you did this earlier when the animal was suspended (see Figure 6–8).

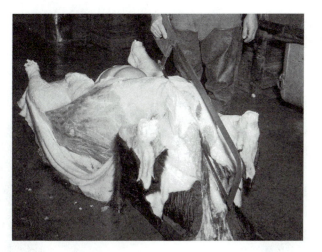

Fig. 6–18. Splitting the brisket.

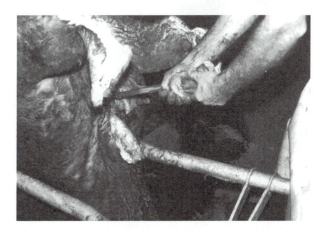

Fig. 6–19. Rodding the weasand. The esophagus (weasand) has been separated from the trachea and is tied in a knot to prevent rumen spillage on the carcass. This operation may have been done earlier (see Fig. 6–8).

A beef tree (a large, spreading gambrel) or beef trolley, designed to operate in conjunction with a power lift and spreader, may be inserted in the hocks so that you can raise the carcass to a height that is convenient for splitting the aitch bone, for bunging and tailing, and for removing the hide from the round and rump. In a male, the pizzle must be removed first. Cut it loose from the belly wall and back to its origin at the pelvic junction. Do not cut it off, but move the loose end back out of the way. With the pizzle out of the way, proceed to separate the rounds with your knife. Each top (inside) round muscle is covered with a tough membrane, and where the two join over the high point of the pelvis (aitch bone), they form a decidedly heavy, white-appearing membrane. Follow the membrane with your knife, and avoid cutting into the muscle until you reach the pelvis. It is possible to split the aitch bone on very young cattle with your knife, but to do it successfully, you must be exactly on the cartilaginous center. Also, you run the risk of breaking your knife blade and(or) puncturing an intestine. Thus, using a saw is quicker, safer, and more accurate, especially for beginners (see Figure 6–20).

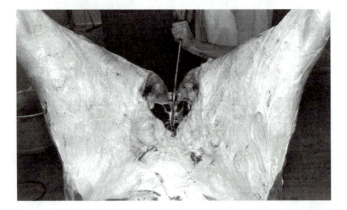

Fig. 6–20. Splitting the aitch bone with a saw, while the carcass is raised to a convenient height.

In bunging a male, pull the pizzle taut posteriorly, and loosen the anus by cutting completely around it (see Figure 6–21). While severing all supporting connections, be extremely careful that you do not puncture the intestine itself. When bunging a female, you must grasp the anus itself, in lieu of the penis, to hold it steady for bunging. Once loosened, the bung is tied with a cord and pushed through into the abdominal cavity, where it can be reached from the belly side.

To remove the tail, skin around its base and split the hide on the ventral midline for the entire length of the tail. Sever the tail two joints from the body, and skin entirely around its base. By placing a paper towel over the skinned base stub, you can easily pull out the tail the remainder of the way (see Figure 6–22).

Some butchers prefer to start splitting the carcass at this half-hanging stage. Because the carcass is more stable in this position, splitting the sacral

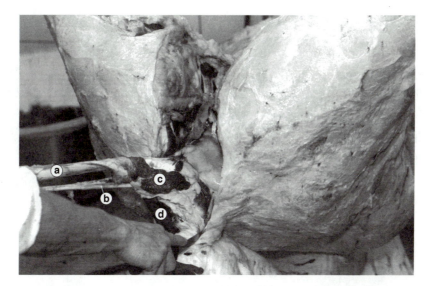

Fig. 6–21. In bunging a male, while holding tension on the pizzle, cut completely around the anus (bung gut). While severing all supporting connections, *be very careful not to puncture the intestine.* (a) Penis. (b) Retractor penis muscle. (c) Thin layer of *gracilis* muscle from inside the round, adhering to the bung. (d) Anus.

Fig. 6–22. Removing the tail.

vertebrae section of the backbone can be done more accurately, if the cutter is splitting with a hand saw.

After freeing the bung and removing the tail, raise the carcass to a hanging position to complete hide removal and evisceration. Removal of the hide from the back is called backing. This operation consists of running the knife around the back between the hide and the carcass and then letting the hide drop of its own weight (see Figure 6–23). When completing the siding operation, you may also remove the hide by cutting in from either side.

Fig. 6–23. Backing, i.e., letting the hide drop of its own weight after the connective tissue attachment between the hide and the carcass has been lightly cut.

Fig. 6–24. Opening the belly cavity with one continuous motion. Insert the knife handle into the opening at the cod/udder area, pointing the blade upward and outward, while at the same time using your fist and the knife handle to protect the intestines and the rumen.

EVISCERATING

Insert the handle of the knife in the abdominal cavity at the cod/udder area, with the blade leaning upward and outward and with your fist protecting the intestines and rumen. Open the belly cavity, using one continuous motion (see Figure 6–24).

Loosen the fat and membranes that hold the bung gut and bladder to the backbone, and then cut the white ureters that connect the kidneys to the bladder (see Figure 6–25). The kidneys and kidney fat remain in a beef carcass, largely because of custom; however, the kidney fat does protect the valuable tenderloin muscle from drying out and darkening during aging.

Loosen the liver with your hands, and then sever it from the backbone with a knife. Place the liver on an inspection tray for observation by the inspector. Following its inspection, remove the gall bladder by cutting across the top of the bile duct at the center of the liver and peeling it, rather than cutting it out (see Figure 5–47).

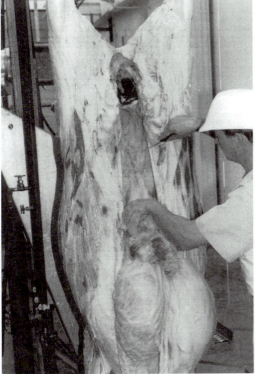

Fig. 6–25. A few well-placed cuts of the supporting membranes and ureters connecting the bladder and the kidneys will allow the paunch and the intestines to drop into the gut cart.

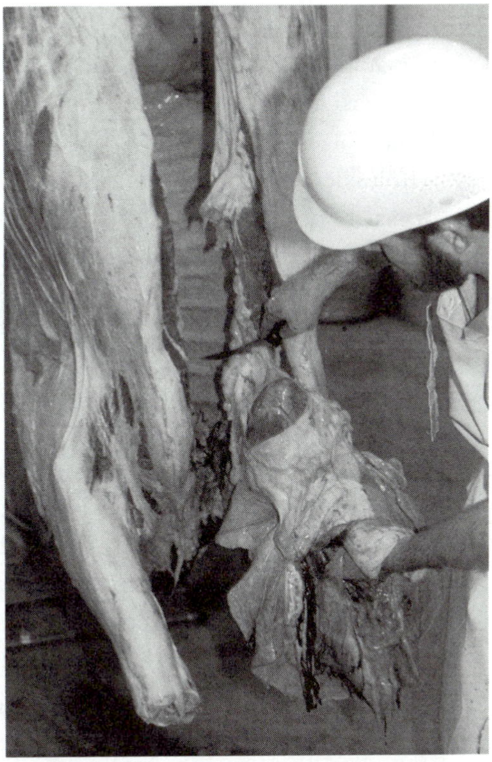

Fig. 6–26. Removing the pluck, consisting of the heart, lungs, and windpipe. The diaphragm muscle remains in the carcass.

Pull the previously loosened esophagus up through the diaphragm to allow the abdominal viscera to fall freely into the inspection cart.

The diaphragm separates the abdominal and thoracic cavities. It consists of the diaphragm muscle and the membrane joining the muscle. Cut out only the membrane, as the diaphragm muscle is good, edible meat and is commonly known as the skirt or hanging tenderloin. The organs that lie in the thoracic cavity are called the pluck and consist of the heart, lungs, and windpipe (see Figure 6–26).

HALVING (SPLITTING)

Using a hand or an electric beef-splitting saw, saw through the exact center of the backbone to split the beef into sides. Figure 6–33 shows an electric saw in use in a commercial plant.

Standing on the belly side of the carcass, use a hand saw to saw through the caudal (tail) vertebrae to the sacrum and through the sacrum and the lumbar vertebrae. Make sure you split each superior spinous process (feather bone) of each vertebra because splitting each of these affects the weight of wholesale and retail loin, rib, and chuck cuts. Since the feather bones in the dorsal region (rib area) of the backbone are quite long and narrow, stay on center by pointing the saw toward the tail and then sawing at the same approximate 45° angle that you started when sawing the sacrum.

WASHING

All blood should be washed off both the inside and the outside of the carcass. Pump the shoulders by working the shanks up and down to get thorough blood removal. Then, wash the carcass with cold or lukewarm water. (Review the discussion on carcass washing in Chapter 5.) After the final rinse, the inspector must make a final examination.

Shrouds made from unbleached duck cloth, immersed in warm water, stretched tightly, and pinned over the outside of the sides of warm beef before they were moved into the cooler were once widely used in the industry. The shrouds improved the appearance of the carcasses for the potential buyers by absorbing the blood, smoothing the external fat covering, and making the fat to appear white and dense. The shrouds, or cloths, were removed after the carcasses had chilled. With the advent of increased fabrication of beef carcasses into wholesale and retail cuts in packing plants, most packers have discontinued the use of shrouds in an effort to save unnecessary costs.

CARCASS CHILLING

A thorough chilling during the first 24 hours is essential, otherwise the carcasses may sour. This occurs first at the hip joint — a deep-seated joint from which heat is slow to escape. A desirable temperature for chilling warm beef carcasses is 32°F. Since a group of warm carcasses will raise the temperature of a chill room considerably, it is good practice to lower the temperature of the room to 5° below freezing (27°F) before the carcasses are moved in. Temperatures more severe than this can cause cold shortening, an intense shortening of muscle fibers, which brings about toughening (see Chapter 22).

Cooler shrinkage with conventional chilling averages from 1 to 2 percent in the first 48 hours and from 3 to 6 percent if the carcasses are aged for 2 weeks. The amount of cooler shrinkage is dependent upon cooler humidity and air velocity (typically 5 to 15 miles per hour equating to a 12°F wind chill effect). Fatter carcasses will shrink less and be less likely to have cold shortening because of the external insulation. Kansas State University workers measured the effects of spray chilling beef carcasses with 37°F water in a 32° cooler for 90 seconds every 15 minutes during the first 8 hours of chilling and found that spray-chilled carcasses shrunk only 0.3 percent compared to 1.1 percent for conventionally chilled carcasses.[4] (Review spray chilling of hog carcasses in Chapter 5.)

COMMERCIAL BEEF HARVESTING OPERATIONS

It is not unusual to find operations in the United States harvesting nearly 3,200 cattle in an eight-hour day, or 400 per hour. To accomplish such an endeavor, these operations use equipment that is very mechanized, as can be noted in Figures 6-27 to 6-33.

Canada Packers, Ltd., designed a 14-station system of on-the-rail dressing of beef that has eliminated the necessity of operators having to stoop while working. Many large packers use a system similar to this. In this system, driving into the knocking pen and knocking the beef are performed by one person at station 1. Then, two people shackle the animal, also at station 1. Sticking and scalping (heading) are performed

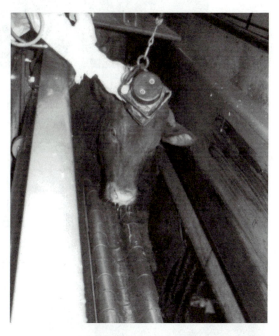

Fig. 6–27. Easily and quickly, this worker places the air-powered mechanical stunner on the head of a restrained animal in a double rail conveyer. (Courtesy, Grandin Livestock Handling Systems, Inc., and *The National Provisioner*)

---

[4]Allen, D. M., M. C. Hunt, A. Luchiari Filho, R. J. Danler, and S. J. Goll. 1987. Effects of spray chilling and carcass spacing on beef carcass cooler shrink and grade factors. J. Anim. Sci. 64:165.

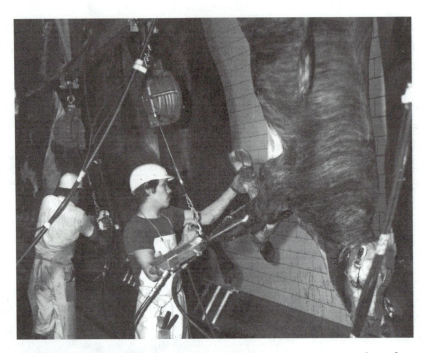

Fig. 6–28. Making full use of power tools, employees sever front feet and horns, if present, from bled carcasses. (Courtesy, *The National Provisioner*)

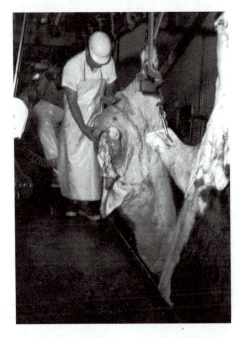

Fig. 6–29. At the second transfer station, carcasses are shown being taken up as the employee performs butting operation on the next carcass. (Courtesy, *The National Provisioner*)

at station 2. Another worker legs, butts, inserts trolley, and removes the shackle at station 3, while a fourth worker legs, butts, and inserts trolley in the second leg at station 4. At station 5, one worker removes the front foot, performs work on the brisket, dehorns, and removes the head. One worker rims (opens the hide over the brisket and plate), clears the shanks, and works on the chuck and neck at station 6. At station 7, work is performed by a single worker on the rump; the bung is dropped, and the tail is pulled. The rosette (cutaneous omobrachialis muscle on the surface of the outside chuck) is cleared with an air-operated skinning tool, flanking is performed, and the hide is pulled by the machine operated by one worker at station 8. The brisket is sawed, and the carcass is eviscerated at station 9, and the hide pull is completed at station 10. In the pull off, the hide drops directly into a chute. The splitter and scriber at station 11 operates on an elevating bench, using a foot switch. Trimming is done at station 12, weighing at station 13, and shrouding at station 14.

Several newer hide-pulling machines, some of which pull down the hide, are on the market. One advantage of a down pull is that hide-borne contaminants, such as loose hair, dirt, and manure, either fall to the floor or drop progressively away from the carcass to the lower portions of the hide.

Steam pasteurization is USDA-FSIS approved as an anti-microbial treatment for beef carcasses. Systems in place utilize a three-part pasteurization chamber on line in beef harvest plants. Surface water is removed from a carcass in the first compartment, the carcass surface is steamed in the second compartment, and chilled water is applied in the third compartment to reduce carcass surface temperature to 36°F. Research is being conducted on the feasibility of

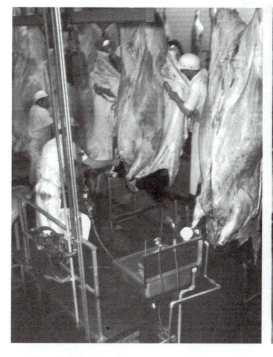

Fig. 6–30. The final step in hide preparation for strip-off includes freeing the small of the back (loin) as shown. In the foreground the worker is removing the hide from the brisket. (Courtesy, *The National Provisioner*)

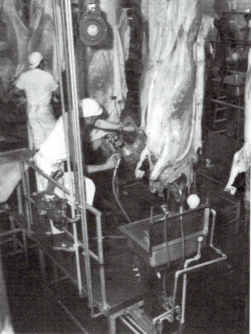

Fig. 6–31. With a power tool, the employee at the hide removal station opens the brisket. (Courtesy, *The National Provisioner*)

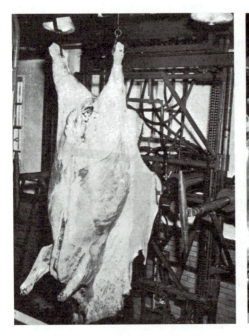

Fig. 6–32. Can-Pak® hide puller.

Fig. 6–33. At this station, a cantilever platform moves the meat cutter downward and forward as he splits the carcass. Note this commercial operator is splitting from the back side because this method fits into the company's completely mechanized operation. (Courtesy, ConAgra, Inc. [Monfort], Grand Island, Nebraska, and *The National Provisioner*)

chemically removing the hair and attached material from beef animals shortly after sticking. This is another effort to improve a Critical Control Point in the harvest of beef animals. For an update on this procedure, contact Dr. Gary Smith at Colorado State University <gsmith@ceres.agsci.colostate.edu>.

7

Sheep and
Lamb Harvest

In 1866 the American Society for the Prevention of Cruelty to Animals was incorporated. We now refer to the conversion of live animals into carcasses, meat cuts, and meat products as **harvesting.** The meat industry is concerned that animals, which provide us with the most nutritious, well-balanced food available, be treated kindly at all times and not suffer pain.

This chapter deals with sheep and lamb harvest. Sheep had long been thought to be the first farm animal to be domesticated. However, zooarchaeologists now estimate that the pig was perhaps the first to be domesticated — before sheep or cattle — about 11,500 years ago.[1] But it was perhaps less than 1,000 years later that sheep and goats were domesticated.

Sheep are gregarious, meaning that they are "flockbound," even to excess. The fact that sheep will follow is well known in the meat industry, because almost every meat packer has a goat to lead the lambs to harvest. The goat is usually most reliable, returning time after time to lead the lambs to their demise. The goat that fills this role is known as the Judas goat for its deceiving ways.

It is of utmost importance to meat quality and animal welfare that sheep and lambs be treated properly before, during, and immediately after harvest. Although many readers will not be involved in the actual harvest of sheep, all persons should know the procedures involved in converting a live, contented lamb into delicious chops and roasts.

---

[1]Archaeology Special Section. 1998. Reading the signs of ancient animal domestication. 1998. Science. 282:1448. http://www.sciencemag.org

In 1998 there were 3,803,700 sheep and lambs harvested commercially in the United States, excluding those relatively few harvested on the farm.[2] Table 7–1 lists the states in the order of commercial sheep and lamb harvest volume in 1998. Many areas of the country are no longer served by either adequate live markets or harvest facilities for sheep and lambs. For current information on what method of marketing (direct, terminal, sale barn, grade and yield) was used for these sheep and lambs, you may wish to contact the USDA Agricultural Marketing Service (http://www.ams.usda.gov/).

Before proceeding to harvest a sheep or lamb, be sure that you have proper and safe equipment; that your equipment is ready to use (sharp and clean); and that the animal you are about to harvest is healthy. That is, you should study and master the first four chapters before applying the material you will learn in this chapter.

FAST (DRIFT) PRIOR TO HARVEST

A 24-hour fast previous to harvest is probably of greater importance with sheep than with other livestock because the digestive tract composes a higher percentage of the live weight than is the case with cattle and hogs. A fast not only facilitates the eviscerating process significantly but also adds to the bright appearance of the carcass. (See the previous discussion on hogs in Chapter 5 and cattle in Chapter 6.) The removal of the pelt is made somewhat easier by a limited fasting period; however, water must be present during the fasting to avoid dehydration, which results in tissue shrink and difficulty in pelt removal.

HANDLING

Lifting a sheep by grasping the fleece will cause surface bruises on the carcass. Instead, grasp it by the leg. Place one hand under the jaw and the other at the dock and lead the lamb. (See Figure 7–19 for a commercial sheep handling system.)

---

[2]USDA, National Agricultural Statistics Service, Agricultural Statistics Board. March 1999. Livestock Slaughter — 1998 Summary. (Mike Easdale.)

Table 7–1.   Commercial Sheep and Lamb Harvest Ranked by States, 1998[1,2]

Rank	State or Region	Thousand Head
1	Colorado	1,288.9
2	Iowa	540.2
3	Michigan	210.1
4	South Dakota	95.2
5	New Jersey	93.8
6	Pennsylvania	70.4
7	New Mexico	36.8
8	Indiana	35.8
9	New York	30.0
10	New England[3]	22.5
11	Virginia	17.0
12	Ohio	10.1
13	Wisconsin	9.2
14	Kentucky	7.8
15	Oregon	5.9
16	Idaho	4.0
17	Georgia	3.4
18	Montana	3.2
19	Tennessee	3.1
20	Louisiana	2.9
21T	Kansas	2.1
21T	North Carolina	2.1
23	Nebraska	1.6
24	Wyoming	1.4
25	Arizona	1.0
	Total United States[4]	3,803.7

[1]USDA, National Agricultural Statistics Service, Agricultural Statistics Board. March 1999. Livestock Slaughter—1998 Summary. (Mike Easdale.)

[2]Includes animals harvested in federally inspected and other plants, but excludes those harvested on the farm.

[3]New England includes Connecticut, Maine, Massachusetts, New Hampshire, Rhode Island, and Vermont.

[4]States with no data are still included in the U.S. total; individual data are not printed to avoid disclosing individual operations.

YIELD (DRESSING PERCENT)

Normally, dressing percent is based on the hot carcass weight rather than the chilled carcass weight and is compared to the purchase weight of the sheep. Dressing percent is calculated as follows: Carcass weight ÷ live weight × 100 = dressing percent. Most sheep and lambs will average in the 50 percent range. Heavy-muscled and fatter lambs can go up to 54 to 55 percent, depending on weighing conditions. Lambs that are full will dress less, as will thin ewes, light weight, unfinished lambs, and sheep that are carrying fleece. For instance, an average fleece weight of 6 pounds would decrease the dressing percent approximately 3 percent (see Chapter 11).

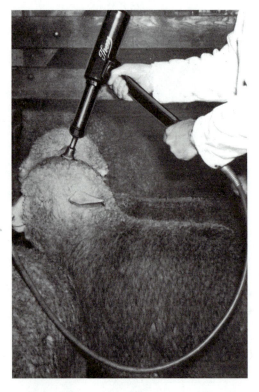

Fig. 7–1. A compression stunner. This particular model uses compressed air — no cartridges. Its high velocity blow is effective in painlessly stunning cattle, hogs, and sheep. It is furnished with either a penetrating or a nonpenetrating head. This stunner illustrates the mushroom, or nonpenetrating, type head used for veal and lamb. (Courtesy, Thor Power Tool Co., Aurora, Illinois)

TOOLS AND EQUIPMENT

A 5-inch curved boning knife or a thin well-ground skinning knife works well for pelting a sheep.

For holding a sheep on its back, a table or platform may be used. However, a trough-like skinning rack on legs about 18 inches high is very handy. The trough is 6 inches wide at the bottom with sloping sides 6 inches high. This type of rack is called a lamb or veal cradle.

STUNNING AND STICKING (EXSANGUINATION)

A sharp blow on top of the poll will stun sheep that do not have horns. A captive bolt powered by cartridges or air with the mushroom head (see Figure 7–1) or an electric stunner (see Figure 7–2; also see Figure 5–1) is recommended for stunning.

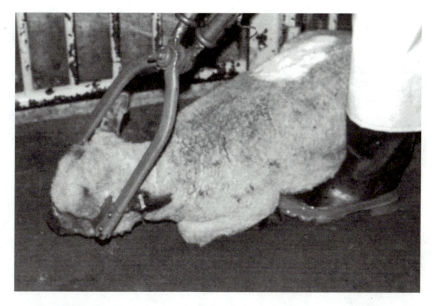

Fig. 7–2. An inexpensive, handy electric stunner suitable for use in smaller operations.

Cardiac arrest was discussed in Chapter 5. This method is very acceptable for sheep and lambs. New Zealand workers used 400 volts and 0.8 amperes for three seconds applied to three points (two on the head and one on a foreleg) to cause cardiac dysfunction (complete unconsciousness).[3] In a study conducted in Germany, cardiac arrest that brought about heart fibrillation, electrical stunning of lower voltage that did not cause the heart to stop, and captive bolt stunning were compared.[4] Less blood was removed from lambs in the cardiac arrest group than from those that were stunned with low voltage and captive bolt (1.7, 3.6, and 3.5 percent of live weight, respectively). However, the additional blood left in the lambs from the cardiac arrest group was not evident and would be beneficial in terms of yield and nutritive value.

Hoist or place the stunned sheep on a table or on a sheep and veal rack. Grasp the jaw or ear with one hand, insert the knife behind the jaw, blade edge outward (see Figure 7–3), and draw the knife out through the pelt (see Figure 7–4), severing the jugular veins and carotid arteries. Blood yield for lambs will average approximately 3 percent of live weight.

---

[3]Blackmore, D. K., and C. V. Petersen. 1981. Stunning and slaughter of sheep and calves in New Zealand. N.Z. Vet. J. 29:99.

[4]Paulick, C., F. A. Stolle, and G. von Mickwitz. 1989. The influence of different stunning methods on the quality of sheep meat. Fleischwirtsch. 69:227.

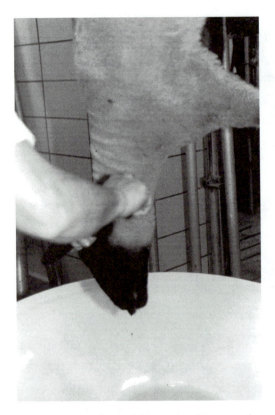

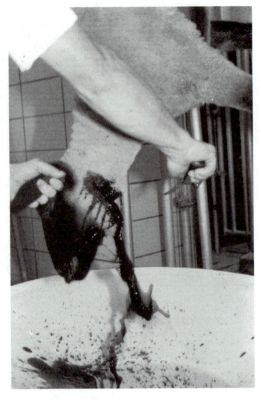

Fig. 7–3. When sticking a lamb, grasp the jaw or ear with one hand, and with the other hand, insert the knife behind the jaw, blade edge outward.

Fig. 7–4. Draw out the knife through the pelt, severing the jugular veins and carotid arteries. Note that blood is being caught, rather than being washed down the drain.

PELTING

With the sheep lying on its back on the rack, grasp a foreleg, or secure the foreleg to a scabbard chain with a bent shroud pin, and then open the pelt down the front of the leg from the break joint to the breast (see Figure 7–5). Do the same on the other foreleg, having the two cuts meet in a point in front of the breast.

Skin out the forelegs at this time. The front foot is removed at the break joint in lambs (see Figure 7–6). The break joint is recognized by a swelling in the cannon bone at its lower extremity just above the hoof (wrist or spool joint). In yearling mutton, and mutton older than 15 months, the break joint is ossified, and the front foot must be removed at the spool joint. Figure 7–7 shows a spool joint and a break joint.

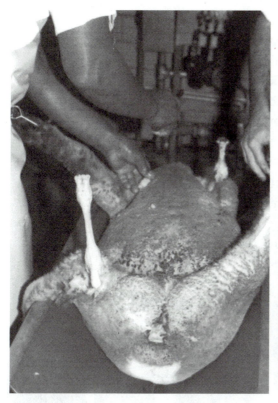
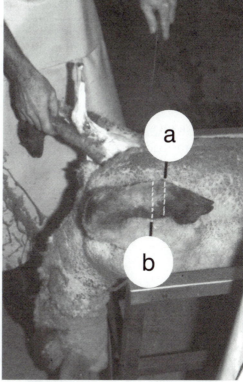

Fig. 7–5. Technique used to skin out forelegs and hindshanks. While skinning, meat cutters hold legs taut with shroud pin hook. Note break joint on foreleg, spool joint on hindleg. (Courtesy, University of Illinois)

Fig. 7–6. Skinning out a foreleg. (a) Location of spool or mutton joint. (b) Location of lamb or break joint. (Courtesy, University of Illinois)

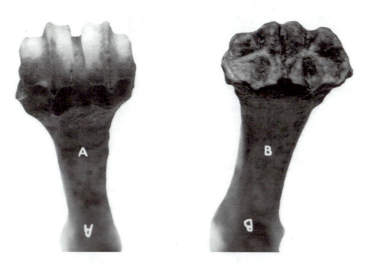

Fig. 7–7. (A) A spool or mutton joint. (B) A lamb or break joint.

Grasp a hindleg and then open it down the back from the hoof to the bung. In order to avoid cutting the tendon and the fell (the thin, colorless, connective tissue membrane just under the pelt that separates the pelt from the carcass and that should always be left intact on the carcass) or exposing the muscle, hold the knife fairly flat to the carcass and leg when you make the opening. Remove the foot at the lowest possible joint (see Figure 7–8) so that an intact tendon anchor is left for hanging the carcass. Loosen the tendon over the back of the hock, and then proceed to skin out the opposite hindleg.

Grasp the cut edge of the pelt at the flank (cod or udder area). Pull and at the same time use the fist of your other hand to loosen the pelt (see Figure 7–9). Repeat the process coming in over the breast. Then, fist the pelt over the belly by turning and pushing your fist against the pelt, not against the carcass.

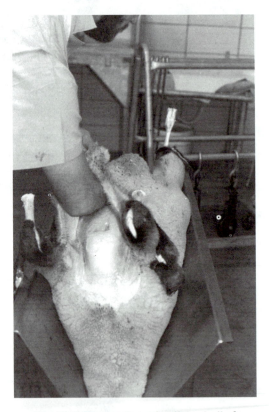

Fig. 7–8. Disjointing the foot on the hindleg. The dotted line across marks the break joint. If the foot were removed at this joint, the anchorage of the tendons (A) would be weakened, and the legs would have to be tied together at the hocks instead of the tendons.

Fig. 7–9. Fisting the pelt off the belly from the breast end while the sheep remains on the cradle.

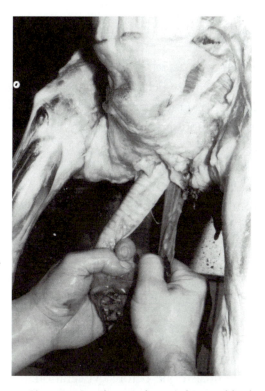

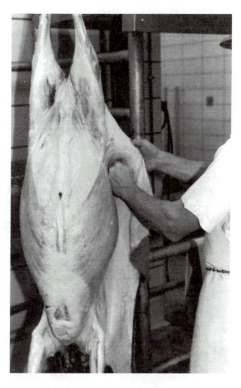

Fig. 7–10. The trachea (white, ribbed "tube" in butcher's left hand) and the esophagus (dark, smooth "tube" in his right hand) are separated for ease of evisceration.

Fig. 7–11. Fisting a lamb on the rail, being careful not to break the fell.

The *trachea* (windpipe) and esophagus must be separated (see Figure 7–10) to allow the abdominal cavity organs and contents to be removed separately from the thoracic cavity organs during evisceration. The loosened esophagus (properly tied off or knotted) slips easily out of the thoracic cavity while still securely fastened to the rumen. If this prior separation has not been completed, you must split the breast prior to evisceration to remove both the abdominal and the thoracic cavity offal at the same time.

You may use a strong cord to tie the tendons of both hindlegs together, or you may insert an S-hook through the tendons of both legs before you hang a lamb on an overhead trolley or rack. In order to prevent back and groin injuries to yourself, use a mechanical hoist to lift the lamb to the overhead trolley or rack. If a mechanical hoist is not available, get help. The job should be done by two persons.

Open the pelt down the center of the belly, and fist it loose around the side and up the leg, being careful to avoid breaking the fell (see Figure 7–11). It is safer to fist up the leg than to pull the pelt down the leg. Unless the skin is started exactly right, pulling the pelt down the leg may tear the protective fell and expose the muscle.

Sever the bung by cutting across it where it is attached to the pelt, and then pull and fist the pelt from the tail. Next, fist the pelt over the shoulder and pull it off the back and neck. When fisting, you must have clean hands to avoid contaminating the lamb carcass with wool and dirt from the pelt.

Sever the head at the atlas joint. In many areas there is no market for edible sheep heads and tongues, so the heads are simply removed and made into animal feed and fertilizer. If there is such a market, then skin out the head and remove the tongue. Head weights from forty-two 118-pound University of Illinois lambs used for teaching averaged almost 6 pounds and ranged from 4.5 to 5.5 percent of live weight.

EVISCERATING

Just as with hogs and beef cattle, splitting the breast is best done with a saw. If the esophagus has been previously separated from its thoracic cavity attachments, it is not necessary to split the breastbone before evisceration. You must avoid breaking the digestive tract and contaminating the carcass during evisceration.

Since the lamb carcass is not split during harvest, the pelvic bone is not split, because the small size allows complete chilling of the intact carcasses.

Taking a position to the rear of the carcass, cut around the bung, and then loosen it (see Figure 7–12).

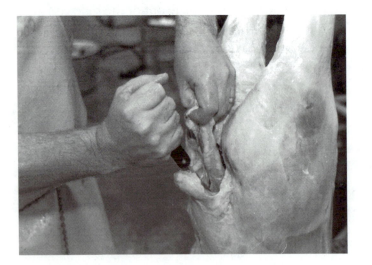

Fig. 7–12. Bung the sheep by carefully cutting around the anus.

On a male carcass, loosen the pizzle and remove it from the surface of the abdomen back to its anchor at the pelvic junction (aitch bone). Holding it taut to keep it out of the way, make an opening at the cod (see Figure 7–13). Cut off the pizzle as near as possible to its root deep in the crotch.

Insert the first and second fingers to guard the point of the knife, or insert the handle into the abdominal cavity the same way as in pork and beef, and continue the opening to the breast (see Figure 7–14).

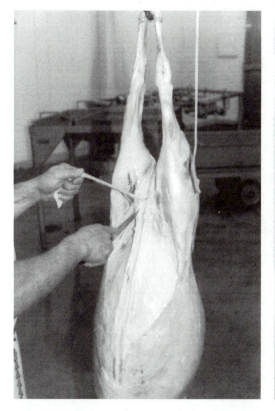

Fig. 7–13. Hold the loosened pizzle tight to keep it out of the way while making a small opening at the cod (udder). Then cut off the pizzle as close as possible to its root, deep in the crotch.

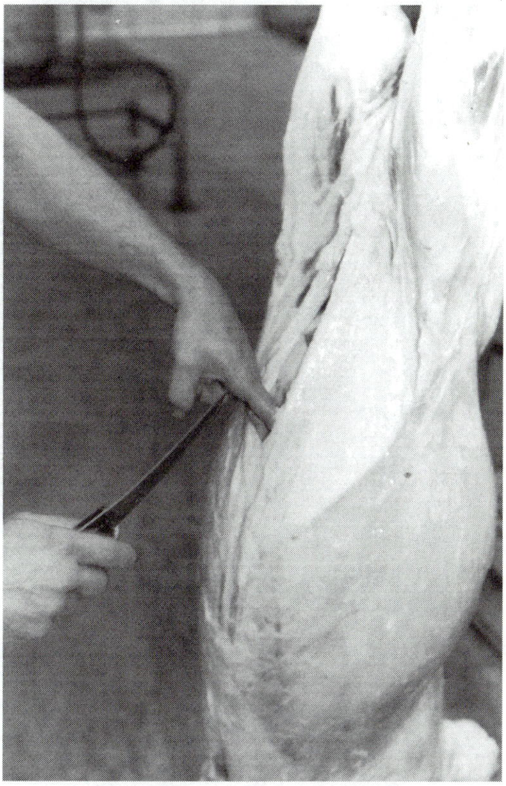

Fig. 7–14. As you open the belly, insert your first and second fingers to guard the point of your knife.

Grasp the loose bung. Use the knife to sever the ureters that lead to the kidneys. These are strong and will tear out the kidneys and the kidney fat if they are not cut (see Figure 7–15). The large four-compartment stomach and intestines can now be easily pulled out.

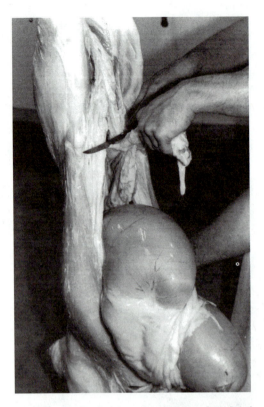

Fig. 7–15. Cutting the ureters that lead from the kidneys to the bladder, and other supporting tissues, in order to release the intestinal tract.

The liver is removed from the abdominal tract, and the bile duct is removed from the liver. Split the breast with a saw or a knife, if this has not been done previously, and remove the pluck. Packers do not remove the spleen (melts).

Wash both the inside and the outside of the carcass, and especially wash the neck and chest cavity to remove any traces of blood (see Figure 7–16).

Trim all scraggly ends off the neck, and double the foreshank up against the arm, using a tendon from the foreshank (see Figure 7–17) to hold it. This operation plumps the shoulder, keeps the foreshank out of the way in crowded lamb coolers, and makes the break joint easily visible to the grader.

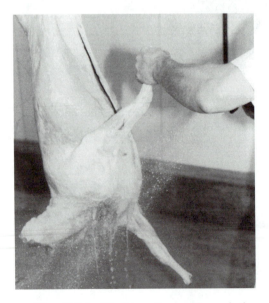

Fig. 7–16. Washing the neck and the breast.

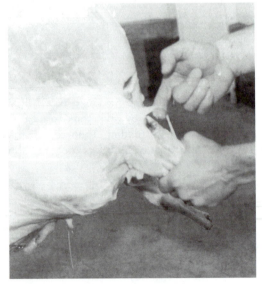

Fig. 7–17. Using the loosened tendon to hold the doubled-up foreshank.

Because they are small, lamb carcasses chill out rapidly and are usually merchandised early the next morning after harvest.

HOTHOUSE LAMBS

Hothouse lambs are rated by epicureans as being the most delectable of the lamb age groups. They are dropped during the months of October, November, December, and January and are marketed between the ages of 6 and 10 weeks. Although the name *hothouse* is rather ambiguous, it indicates that these lambs have been housed in barns or sheds where they are protected from the cold weather.

Hothouse lambs may be defined as "lambs that are dropped out of the regular lambing season and marketed at live weights ranging from 25 to 60 pounds." This makes the hog-dressed weight, which is about 70 percent of the live weight, range between 18 and 42 pounds. *Hog dressed* means dressed with head and pelt on but with feet and viscera removed (see Figure 7–18). The object of this method of dressing is to hold down shrinkage and thus aid

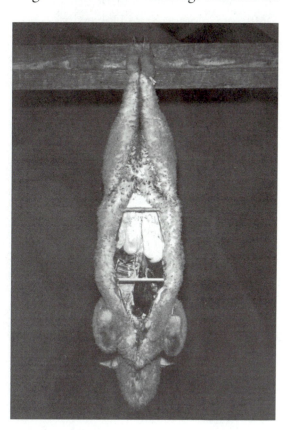

Fig. 7–18. Hothouse lamb (hog dressed). This lamb graded "extra fancy" on the New York City market. It was sired by a purebred Southdown ram and was out of a Hampshire–Dorset–Merino cross-bred ewe. It had an abundance of kidney fat, which is one of the factors in judging the finish and determining the grade. Hothouse lambs are dressed "pluck in." The pluck, in this case, consists of the liver, heart, lungs, gullet, and windpipe. This lamb weighed 39 pounds alive and 28 pounds, hog dressed. Hothouse lambs are still popular in New York City.

in maintaining the pink color of baby lambs. The full-dressed weight is 48 to 55 percent of the live weight. New York City prefers lambs weighing 30 to 40 pounds hog dressed.

COMMERCIAL LAMB HARVEST

Most sheep and lamb harvest takes place in large, highly mechanized harvest plants, each capable of handling 200,000 to 700,000 head per year. The rate of harvest generally ranges from 250 to 350 animals per hour in a plant with automated equipment, which includes a continuously or intermittently moving rail. A glimpse of such a mechanized operation is provided in Figures 7–19 through 7–23.

Monfort Lamb, of Greeley, Colorado, has installed a New Zealand inverted chain system for lamb harvest. Rather than lambs being suspended by their hind legs, with this inverted system the lambs are suspended by their front legs—thus, the term *inverted*. This system appears to be more efficient and allows for closer control of carcass cleanliness, according to Elizabeth Duffy, of Colorado State University.

Fig. 7–19. This double rail restraining system permits sheep to ride on the double rail to the immobilization station. The main advantage of this double rail system is that animals ride to the immobilization point in a comfortable, natural position. The person responsible for immobilization is able to safely stand closer to the animal, and thus accomplish the task more quickly and accurately with less effort and trauma to the animal. (Courtesy, Grandin Livestock Handling Systems, Inc.)

Fig. 7–20. Lambs are suspended by all four limbs to position for initial legging. (Courtesy, John Morrell & Co.)

Fig. 7–21. Pelts are pulled by hand as lambs move at approximately 200 per hour. (Courtesy, John Morrell & Co.)

Fig. 7–22. Federal inspector (in white) examines each carcass following evisceration. (Courtesy, John Morrell & Co.)

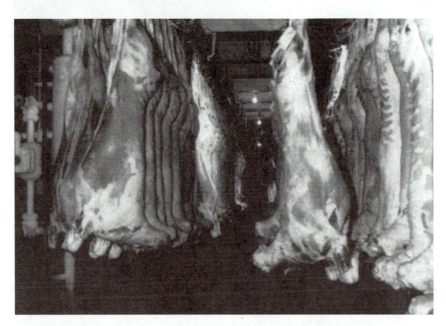

Fig. 7–23. Partial view of some 2,000 to 3,000 lamb carcasses in the cooler containing the day's harvest. They are crowded but adequately spaced for chilling. (Courtesy, John Morrell & Co.)

8

Veal and Calf Harvest

In 1866 the American Society for the Prevention of Cruelty to Animals was incorporated. We now refer to the conversion of live animals into carcasses, meat cuts, and meat products as **harvest-ing.** The meat industry is concerned that animals, which provide us with the most nutritious, well-balanced food available, be treated kindly at all times and not suffer pain.

This chapter deals with the harvest of vealers and calves. It is most important to make sure that these young animals are fed and managed properly as they approach the age for harvest and that they are handled with extreme care just prior to and during harvest. The largest percentage of vealers and calves are young males resulting from the dairy industry. That is why veal and calf harvest is centered in the dairy states (see Table 8–1). Also, U.S. consumers do not eat much veal (see Chapter 1).

It is of utmost importance to meat quality and animal welfare that vealers and calves be treated properly before, during, and immediately after harvest. Although many readers will not be involved in the actual harvest of vealers and calves, all persons should know the procedures involved in converting a live, contented young bovine into delicious chops and roasts.

In 1998 there were 1,457,800 vealers and calves harvested commercially in the United States, excluding those relatively few harvested on the farm.[1] Table 8–1 lists the states in the order of commercial veal and calf harvest volume in 1998. For current information on what method of marketing (direct, terminal, sale barn, grade and yield) was used for these vealers and calves, you may wish to contact the USDA Agricultural Marketing Service (http://www.ams.usda.gov/).

---

[1]USDA, National Agricultural Statistics Service, Agricultural Statistics Board. March 1999. Livestock Slaughter — 1998 Summary. (Mike Easdale.)

Table 8–1. Commercial Veal and Calf Harvest Ranked by States, 1998[1, 2]

Rank	State or Region	Thousand Head
1	Wisconsin	210.7
2	New York	200.5
3	Pennsylvania	193.4
4	California	190.7
5	New Jersey	129.7
6	Illinois	123.8
7	Minnesota.	56.3
8	New England[3]	52.4
9	Ohio	51.5
10	Texas	43.0
11	Louisiana	20.5
12	Delaware–Maryland	3.8
13T	Idaho	2.1
13T	Missouri	2.1
15	Mississippi	1.6
16	Oklahoma	1.1
17T	Arkansas.	1.0
17T	Virginia.	1.0
19	Kentucky.	0.9
	Total United States[4]	1,457.8

[1]USDA, National Agricultural Statistics Service, Agricultural Statistics Board. March 1999. Livestock Slaughter—1998 Summary. (Mike Easdale.)

[2]Includes animals harvested in federally inspected and other plants, but excludes those harvested on the farm.

[3]New England includes Connecticut, Maine, Massachusetts, New Hampshire, Rhode Island, and Vermont.

[4]States with no data are still included in the U.S. total; individual data are not printed to avoid disclosing individual operations.

Considerable confusion is evident in circles outside the livestock industry as to what constitutes the distinction between a vealer and a calf. The U.S. Department of Agriculture defines a *vealer* as an immature bovine animal, usually not over three months of age, that has subsisted largely on milk or milk replacers. The color of its carcass lean is light grayish pink. Such veal has the characteristic trimness of middle associated with limited paunch development. A *calf* is defined as an immature bovine animal between three and eight months of age, which, for a considerable period of time, has subsisted in part or entirely on feeds other than milk and has thus developed a heavier middle. Grayish red is the typical color of calf carcass lean.

Since it is rather difficult to determine the age of a vealer or calf, no set age can be given as a definite dividing line. Weight and conformation are used more as a basis for determining their classification, with weight being the determining price factor among vealers and calves of equal conformation, finish, and quality.

VEALERS

There is no sex classification made for vealers, since they are not old enough for sex conditions to have had much influence on their physical characteristics. The greatest supply of vealers comes from dairy farms during the spring and fall months. The large market centers for veal are New York City, Buffalo, Chicago, Detroit, Milwaukee, and South St. Paul.

Immature Veal

The practice on many dairy farms that do not have purebred stock is to allow the calf to suckle the dam for several days to remove the colostrum milk. The calf is then sold to a dealer for harvest, or it may be sold to a farmer or dealer who keeps some cows for the purpose of vealing calves. The carcasses of these immature vealers are usually designated as *bob veal*. To discourage the sale of immature veal, most states have legislation regulating the legal age at which veal can be harvested. Bob or immature veal, although not unwholesome, is not an economical buy because of (1) the high moisture content, (2) the large proportion of bone to lean, and (3) the low quality.

The skins from bob or immature veal are called "deacon skins" and generally weigh under 9 pounds.

The skins from stillborn calves are called "slunk skins" and have short, fine hair. Slunk skins are tanned with hair on and are used for apparel items such as jackets and vests and for linings for other leather articles (see Chapter 11).

The term *vealing calves* refers to the feeding of young calves, either by hand or by letting them suckle nurse cows, and supplementing the milk ration with a grain gruel. The most recent calf feeding practice is to special feed the calves with milk replacers. When such feeding has produced calves of the desired weight, the calves are sent for harvest.

CALVES FOR HARVEST

Sex conditions cause some changes in the physical characteristics of calves over 3 months of age that are not evident in vealers; hence, the market classifies calves as to sex. Size and weight are important in the selection of calves, either for harvest or for further feeding.

Before proceeding to harvest a vealer or a calf, be sure you have proper and safe equipment, that your equipment is ready to use (sharp and clean), and that the animal you are about to harvest is healthy. That is, you should study and master the first four chapters before applying the material you will learn in this chapter.

METHODS OF DRESSING
VEAL AND CALVES

Veal is dressed with either "skin off" or "skin on" (hog dressed). In the past, the "hog style" carcass was popular because it prevented the outer surface of the carcass from becoming dark and dry, and it is still being used in some parts of the United States. However, a problem arises in the cooler because these carcasses lose hair, which can become affixed to other meat and thus contaminate it. So more and more plants are removing the skins on the harvest floor and wrapping the carcasses in pliofilm wrap or bags to prevent dehydration and darkening and to preserve the fresh appearance and bloom.

Special-fed veal, weighing 360 to 370 pounds live, will dress about 70 percent with the hide on. The hide will affect the dressing percent about 10 percent, i.e., these calves will dress 60 percent hide off. Southern calves, those just off grass, and bob veal will dress about 10 percent lower than special fed veal. Calf carcasses are dressed generally in the same manner as beef, with skin off and split into sides.

HANDLING, STUNNING, AND STICKING

Vealers and calves should be kept off feed for 18 hours before they are harvested. They should be handled with care to avoid bruises and undue excitement.

Any of the mechanical stunners (see Figures 8–1 and 8–11) can be used.

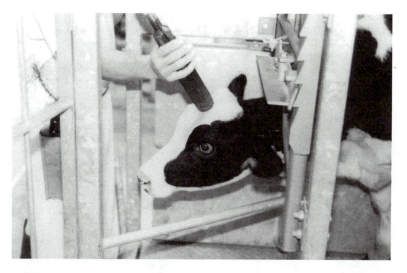

Fig. 8–1. Stunning a calf with a light load in a captive bolt stunner.

Two methods of sticking are common. One is to "Kosher stick," or cut the throat just back of the jaw. The other is to stick in front of the brisket, as is done in beef (see Figure 8–2).

RODDING THE WEASAND

The esophagus (weasand) must be separated from the trachea in order for the abdominal cavity organs and contents to be removed during evisceration. This is called "rodding," because a rod is used (see Figure 8–3). Once the rod has encircled the weasand, it is pushed until the end reaches the rumen (see Figure 8–4) so that the esophagus is separated from the trachea (windpipe) and other attachments along its entire length. The weasand is then tied off to prevent spillage of rumen contents.

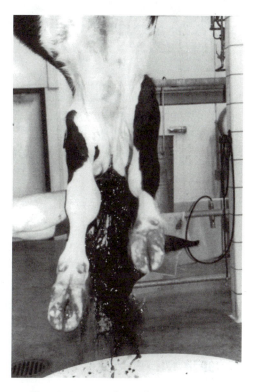

Fig. 8–2. In sticking, first open the hide from the jaw to the brisket, as in beef. Enter the opening in front of the brisket, and sever the jugular vein and carotid arteries. Note that blood is being caught in a funneled vat rather than being allowed to run down the floor drain.

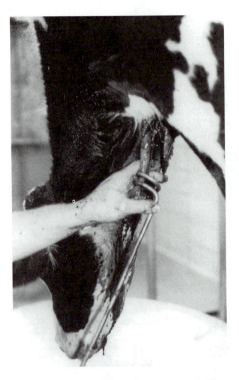

Fig. 8–3. Rodding the weasand. The loops of the rod have been threaded onto the esophagus (weasand).

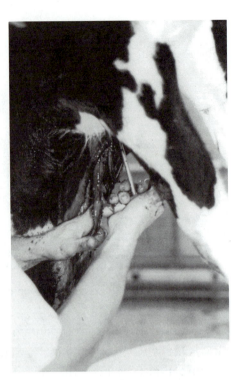

Fig. 8–4. As the rod is pushed upward toward the stomach (rumen), it separates the esophagus and the trachea.

DRESSING

Skin Off

After the head has been removed, the calf or veal is placed in a cradle, and the method of removing the feet and opening the skin is the same as in beef (see Figure 8–5).

Since a calf skin is thinner and softer and more readily scored or cut with a knife than is a beef hide, fist (see Figure 8–6) or pull (see Figure 8–7) the skin off the sides and back of the carcass.

Skin On (Hog Dressed)

Open the skin from the hoof to the knee on the foreshank and to the hock on the hindshank. Skin out the foreshank and hindshank, and then remove them at the articulating joint as in beef. Skin out the head, and then remove it at the atlas joint. Split the skin and carcass over the median line of the belly from the back end of the brisket to the cod or udder.

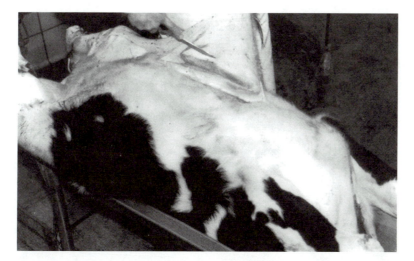

Fig. 8–5. The calf has been placed in a cradle, the feet removed, hide opened down the centerline, and siding begun.

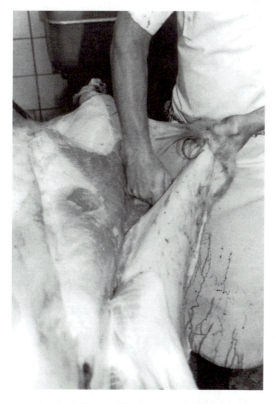

Fig. 8–6. The calf hide may be fisted free of the carcass, since it is thin and soft. This method may not be possible on older calves, but when used, it prevents nicks and gouges in the hide and the carcass.

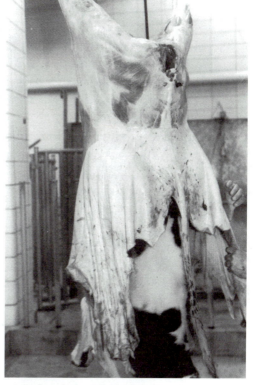

Fig. 8–7. After landing on the rail, the hide may be pulled off the back.

EVISCERATION

Cut around the bung, and let it drop into the abdominal cavity (see Figure 8–8). Then remove the entrails from the abdominal cavity (see Figure 8–9), but leave the liver in the carcass. However, remove the gall bladder from the liver.

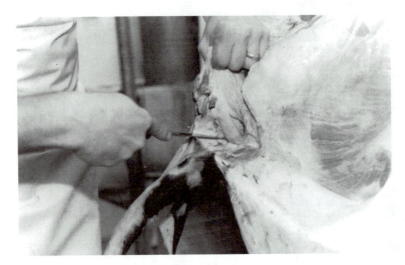

Fig. 8–8. Loosening the bung involves carefully cutting around both sides and finally cutting the attachment to the backbone.

Cut the diaphragm, and remove the pluck (see Figure 8–10). In this operation be extremely careful not to mutilate the *thymus* gland (sweetbread). The sweetbread and liver are considered part of a veal carcass and are weighed with the carcass. (They are removed from calf carcasses.) The *thyroid* gland must be excluded from any trim used from the throat area, since the gland can be harmful to some humans.

HALVING

Because of their small size, veal are not split. Calves are split for easier handling and thorough cooling.

COMMERCIAL VEAL AND CALF HARVEST

A modern veal and calf processing plant, utilizing a continuous, mechanically powered rail system, is depicted in Figures 8–11, 8–12, and 8–13.

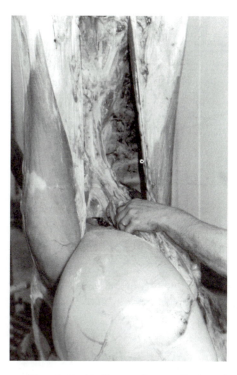

Fig. 8–9. Making a few well-placed cuts through the supporting tissues will remove the abdominal cavity contents, including the rumen in the foreground.

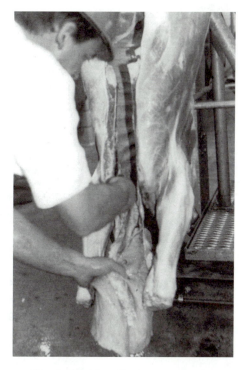

Fig. 8–10. The diaphragm membrane is cut, leaving the muscle intact. This allows the pluck, composed of the heart, sweetbreads (thymus gland), trachea, and lungs to be removed.

Fig. 8–11. This double rail restraining system permits vealers and calves to ride on the double rail to the immobilization station. The main advantage of this double rail system is that animals ride to the immobilization point in a comfortable, natural position. The person responsible for applying the air-powered mechanical stunner is able to safely stand closer to the animal, and thus accomplish the task more quickly and accurately with less effort and trauma. (Courtesy, Grandin Livestock Handling Systems, Inc.)

Fig. 8–12. A general view of a well-illuminated, compact veal and calf dressing department. Note hearts and livers on rack in right foreground. (Courtesy, *Meat Processing*)

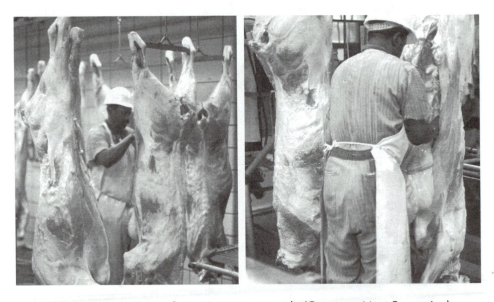

Fig. 8–13. Two views of an eviscerator at work. (Courtesy, *Meat Processing*)

9

Poultry Processing

Review the dramatic increase in poultry meat consumption in recent years noted in Chapter 1. Also, compare the concentration of harvesting plants between the various species of meat animals and poultry by referring to Tables 5–1, 6–1, 7–1, 8–1, 9–1, and 9–2. Poultry harvesting is concentrated in fewer plants across fewer states than is the case with the other species. Therefore, probably very few readers will have an opportunity to actually harvest a chicken or a turkey. More readers may have opportunities to process birds into retail products. The whole picture of harvesting and processing is presented in this chapter.

Also, some consumers may have interest in ostriches and emus. These flightless birds are members of the family Ratitae (see the classification system in the front of this textbook). A brief discussion of their production is included.

Since 1925 there has been a spectacular rise in the production of commercial broilers in the United States. Broiler production started to expand in the Delmarva section of Delaware, Maryland, and Virginia and increased from 34 million (4 percent of the total chicken meat supply) in 1934 to approximately 5.8 billion in 1990. Broiler production in the United States has expanded south and, to a lesser extent, west from the original Delmarva area. Table 9–1 lists the numbers of commercial, federally inspected young chickens harvested in the United States in 1998. Table 9–2 presents comparable information for young turkeys.

Broiler meat consumption accounts for over 96 percent of all chicken meat consumption. (See Chapter 1 for a full discussion on food production and consumption.)

Highly commercialized operations for harvesting, dressing, eviscerating, cutting up, packaging, and transporting chickens provide a constant supply of ready-to-cook poultry to supermarkets across the United States. So rapid and

Table 9–1. Commercial Young Chicken Harvest Ranked by States, 1998[1, 2]

Rank	State or Region	Thousand Head
1	Arkansas.	1,151,310
2	Georgia	1,129,588
3	Alabama	854,456
4	Mississippi	681,989
5	North Carolina	624,596
6	Texas	506,515
7	Virginia.	356,710
8	Missouri	355,823
9	Delaware	272,285
10	California.	246,375
11	South Carolina	213,358
12	Maryland	209,706
13	Tennessee.	180,795
14	Louisiana	155,169
15	Florida	131,374
16	Pennsylvania.	109,003
Total United States[3]		7,822,899

[1]USDA, National Agricultural Statistics Service, Agricultural Statistics Board. March 1999. Livestock Slaughter—1998 Summary. (John Lange.)

[2]Includes chickens harvested in federally inspected and other plants, but excludes those harvested on the farm.

[3]States with no data printed are still included in the U.S. total; individual data are not printed to avoid disclosing individual operations.

Table 9–2. Commercial Young Turkey Harvest Ranked by States, 1998[1, 2]

Rank	State or Region	Thousand Head
1	Minnesota.	43,060
2	North Carolina	38,111
3	Virginia.	29,154
4	Arkansas.	25,933
5	Missouri	23,460
6	California.	18,881
7	Iowa.	14,472
8	Pennsylvania.	1,188
Total United States[3]		271,469

[1]USDA, National Agricultural Statistics Service, Agricultural Statistics Board. March 1999. Livestock Slaughter—1998 Summary. (John Lange.)

[2]Includes turkeys harvested in federally inspected and other plants, but excludes those harvested on the farm.

[3]States with no data printed are still included in the U.S. total; individual data are not printed to avoid disclosing individual operations.

constant are the present supply channels that less than 10 percent of this product is sold frozen, the major portion being sold fresh chilled. Turkeys are somewhat seasonal, with 60 percent being produced during the last half of the year and 80 percent being frozen for future consumption. The corresponding figures for ducks are 70 and 40 percent respectively. Almost 100 percent of the geese are processed from October to January, and most of them are marketed frozen.

DRESSING POULTRY

Selection of Poultry for Dressing

The value of poultry for meat varies considerably with the strain and the breed. Some are thin-meated, some are deep-sided and rangy, and others are thick-meated and particularly full-breasted. The thick-meated, full-breasted, well-finished birds will be top grade on any market, provided they have been properly dressed. To avoid having too many birds of the lower grades, growers should not market thin chickens but retain them on a finishing ration if they are of the quality that can be finished. Turkeys should be fed until enough fat has been deposited under the skin over the breast and legs so that the skin no longer appears blue. "Produce what the market demands, not what is convenient at the time" is good advice to follow.

HANDLING PREVIOUS TO DRESSING

Poultry, like animals, should be handled carefully to avoid bruises, abrasions, and broken limbs. Most bruises occur within 24 hours of harvest. The birds should not be subjected to temperature extremes in the crates or holding pens, which should be well ventilated. The birds should have free access to water. Water is a heat regulator and helps the birds to eliminate waste products. If birds are denied water for too long a time, they will lose weight and dress poorly. Feed should be withheld for 8 to 10 hours before the chickens are harvested, because those full of feed will not bleed as well and will be harder to eviscerate. However, withholding feed for more than 12 hours will cause a marked loss in dressing yield.

A bird should be caught by the leg below the thigh in such a manner that it does not strike its breast on a hard surface. Holding one wing while you pick up the bird by the shank will prevent it from struggling. Over-heated or over-excited birds will bleed poorly, producing carcasses of higher blood content and lower keeping quality. Such stock often die during transport. Hauling live birds over 50 miles is conducive to excessive losses in numbers and quality and thus should be avoided if at all possible.

Dressing Percent

The overnight fasting shrink has been shown to vary from as little as 2 percent in chickens to as much as 7 percent in heavy turkeys. The dressing loss, whether blood-and-feather dressed or full-dressed, i.e., bleeding followed by feather removal or the further removal of head, feet, and viscera, depends upon the weight and condition of the birds. Chickens under 5 pounds will lose an average of 11 percent blood-and-feather dressed and 27 percent full-dressed. Chickens over 5 pounds will average 9 and 25 percent loss, respectively. Male turkeys weighing between 13 and 17 pounds will lose an average of 10 percent blood-and-feather dressed and those over 20 pounds will lose about 8 percent. The same birds full-dressed will average a loss of 20 and 18 percent, respectively. Blood-and-feather–dressed female turkeys weighing under 10 pounds average 10 percent loss, those weighing between 10 and 12 pounds average 9 percent, and those weighing between 12 and 15 pounds average 7.5 percent loss. These same birds will show a full-dressed loss of approximately 19 to 21 percent, or slightly more than the toms.[1]

TOOLS AND EQUIPMENT

Commercial dressing practices of both animals and poultry have been streamlined and mechanized. In fact, the mechanical processing of poultry is very similar to the methods employed by the meat packing industry in the harvest of hogs. The commercial poultry processing plant has overhead tracks (much lighter in structure and lower than those used for larger animals), which take the form of belt chains to which oval link chains are suspended to hold the shackles for suspending the birds. These belt chains move at a controlled speed, dependent on the ability of the personnel and equipment to handle the product.

---

[1]NCM–46 Regional Technical Committee. 1975. Factors Affecting Poultry Meat Yields. Bull. 630. S.D.S.U. Agric. Exp. Sta.

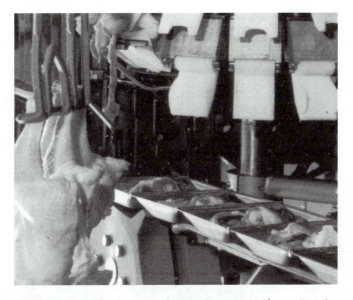

Fig. 9–1. The Maestro System® removes the entire visceral package in the eviscerator and deposits it in a pan. The pan is then presented at the inspection station, along with the bird from which the package was removed. The bird is automatically tilted for ease of inspection, with is visceral package just underneath. This system reduces contamination and thus the number of people required for reprocessing. It conforms to the HACCP principle of removing the visceral package at the earliest possible point in the harvesting process. (Courtesy, Meyn Poultry Processing, Oostzaan-Holland, and Cantrell Machine Co., Inc., Gainesville, Georgia)

Up to 6,000 birds per hour can be processed with a single system of machines. Stunners, harvesters, scalders, pickers, hock cutters, eviseration machines (see Figure 9–1), gizzard processing systems, lung removers, neck breakers, head cutters and neck skin cutters, continuous chill systems, giblet wrappers, sizing systems, and cut-up machines (see Figure 9–2) are all automated. However, for the most part, hock pickers, outside bird washers, eviscerating troughs, oil sac cutters, opening cut machines, combination washers, continuous giblet chillers, and other forms of equipment are not automated. The most modern plants will usually also have one or more automatic deboning machines. There are at least two types: one which macerates the total carcass, separating the bone from muscle by sieve and centrifugal action, and the other which simulates hand action and retains some of the muscles intact. The yields are usually greater than with hand deboning (see Chapter 18).

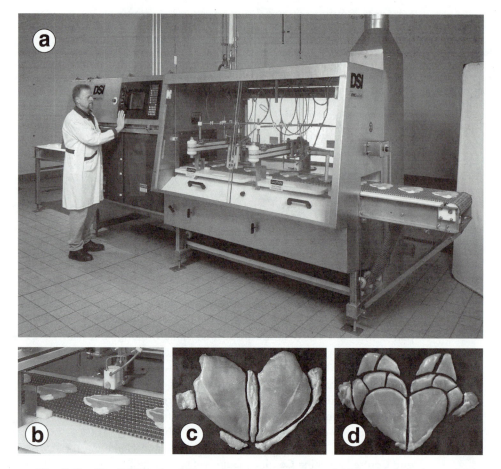

Fig. 9–2. The DSI Portioner® poultry cutting machine. (a) The single lane model — compact, high value. (b) Closeup of the machine, showing its ability to cut on the X and Y axes. (c) Chicken breast separated from breast bone and fat. (d) Chicken breast cut into nuggets. (Courtesy, Stein, Inc., Sandusky, Ohio; Richard A. Kish, Marketing Manager)

Extensive modification of procedures is now underway as the result of HACCP (see Chapter 3). The main reason for this is to improve the quality of the product, i.e., reduce bacterial contamination. Because of the characteristics of HACCP, each processor may have its own set of procedural guidelines — the result of the federal inspector's concurrence with guidelines proposed by the processor to minimize contamination. Since HACCP's inception in January 1998, when the larger processors were required to participate, contamination has been drastically reduced — up to 80 percent in some cases. The goal of HACCP is to reduce contamination to zero. Some have said that HACCP is the most important thing to have ever happened to the meat processing industry.

When only a few birds are dressed, the equipment may consist of a scalding tub, a shackle for holding the bird, and probably a bleeding cup. The knives necessary in each case are a sticking knife (a 3-inch blade for chickens and a 4-inch blade for turkeys), a pinning knife (a paring knife will do), several sizes of boning knives (3½- to 6-inch blades) for eviscerating, and a linoleum knife for splitting the back (or a power meat saw, if available). A thermometer for testing water temperature and bone shears for severing the head, neck, and shanks are desirable. Although the blades of the knives used for sticking and braining are narrow (¼ inch wide), the handles should be of standard grip.

BLEEDING PRACTICES

Immobilization

Birds are suspended by the legs with shackles made of heavy wire, which not only hold the feet in the V-shaped vise but also spread the legs. Some form of stunning, which is required by law in the United States, should immediately precede cutting the throat. For turkeys and some larger chickens, the use of an electric stunning knife or automated stunning device is recommended for aiding in feather removal and improved bleeding. However, this does not accomplish electrical stimulation, which is described later. Immobilization with carbon dioxide gas has been tested and would be acceptable, and in fact would allow for better bleeding, but the difficulty in providing gas chambers has precluded its use.

Cutting the Veins

Cutting the veins inside the mouth is the most widely practiced method for bleeding (see Figure 9–3 A2). Grasp the head, and hold it firmly in the left hand (if you are right-handed), pressing the thumb and forefinger on both sides of the junction of the upper and lower beaks. This forces the mouth open so that the point of the sticking knife can be inserted, sharp edge downward, to the base of the skull. Press the point of the knife into the flesh, lift the handle upward, and cut downward and to the right, severing the jugular veins. If a good bleed does not result, try again until there is free bleeding. The feet of the bird should be level with the eyes of the worker for convenience of operation.

DRY-PICKING AND DEBRAINING

If the birds are to be dry-picked, i.e., the feathers are removed without water immersion, debraining is essential, and of some advantage even with scalded fowl. Hold the head of the bird in the same manner and position as for bleeding. Insert the knife blade (blade edge up) into the cleft in the roof of the mouth and force it through to the rear lobe of the brain *(medulla oblongata)* (see Figure 9–3 B1). The chicken gives a peculiar squawk if properly debrained; whereas the turkey relaxes its wings and spreads out its main tail feathers in the shape of a fan. The puncturing of the brain relaxes the feather muscles, causing the feathers to become loose and more easily plucked. This condition lasts for two to three minutes before the muscles begin to tighten up because of rigor mortis, thus necessitating rapid and orderly plucking.

In dry-picking, "roughing" the bird before the muscles reset is the process of removing the major part of the plumage in the order in which the parts of the bird bleed out. It consists of twisting out the tail and main wing feathers and then plucking the breast, neck, back, thighs, and legs. This is followed by the more tedious task of pinning, i.e., removing tiny, immature feathers called pinfeathers. The object is to pluck and handle the bird so that the outside layer of skin is free from tears, abrasions, or bruise spots and maintains its nat-

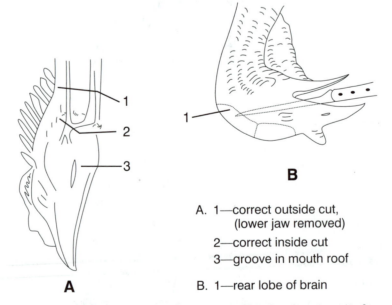

A. 1—correct outside cut,
   (lower jaw removed)
   2—correct inside cut
   3—groove in mouth roof

B. 1—rear lobe of brain

Fig. 9–3. Bleeding and debraining a turkey. (A) Bleeding locations (1–from outside, 2–from inside). (B) Debraining. The turkey is properly suspended by its legs, and the knife is penetrating the rear lobe *(medulla oblongata)* of the brain.

ural bloom. Dry-picking has largely been replaced by semi-scalding, except for some Kosher-type processing.

SCALDING PRACTICES FOR CHICKENS AND TURKEYS

Hard, or Hot, Scalding

Hard, or hot, scalding (water temperature in excess of 155°F) was one of the earliest methods used for the quick removal of the feathers. It is still a common practice employed in home dressing birds of all kinds — it is not practiced on birds destined for the commercial market. The speed at which feathers are loosened is dependent upon the temperature of the water and the period of immersion. The warmer the water, the quicker the feathers are loosened. But the temperature of the water to use also is dependent upon the age and nature of the bird to be scalded. Young birds with tender skins should not be scalded in water over 150°F, whereas mature birds scald well at 155° to 160°F. Mature birds can be scalded in water around 185°F, but the immersion period must be short to avoid cooking the skin. When high scalding temperatures are used, it is well to immerse the scalded birds in cold water as soon as the feathers are loosened. This stops further scalding action.

The hot scald ($\geq$ 160°F) works well on birds having a large number of pinfeathers. The operator must be careful not to overscald. Overscalding causes the skin to tear and discolor and gives the bird a cooked appearance, producing a carcass that lacks bloom and turns brown rapidly (or bright red upon freezing).

Fat birds will hold their natural color longer because the melted fat forms a film over the skin, excludes the air, and retards desiccation. When a deep yellow color is desired on fat birds of the yellow-skinned variety, one practice is to dip the dressed birds into boiling water and then douse them immediately into cold water. The hot water melts the fat and draws it, along with the yellow pigment, to the surface of the skin. The cold water causes the fat to harden and the color to set in the fat.

Semi-scalding (Slack Scalding)

The semi-scalding method was developed in the late 1920s and is *now universally used* in all the large poultry packing plants. It lends itself well to mechanization and has the advantage of lower labor costs. It improves the appearance of the birds because they do not turn red or brown but keep a natural bloom. This, incidentally, allows more birds to be retained in the higher grades, resulting in greater financial return.

The temperature of the water should be 125° to 126°F for young, tender-skinned birds, 127° to 128°F for roasters and young turkeys, and 130° to 132°F for aged birds. These temperatures do not loosen the feathers as much or as rapidly as the hot scald temperatures, nor do they cook or cause the outer skin to peel. The period of immersion varies from 20 to 30 seconds for broilers and up to 60 seconds for older and larger birds.

The scalding vats should·be equipped with thermostatically controlled steam jets in order to keep the desired water temperature. A detergent added to the water may shorten the scalding time.

Because the skin is not weakened in strength, mechanical pickers can be used with this type of scalding. An automated picking machine consists of a drum upon whose circumference are mounted innumerable fingers made of rubber. As the drum revolves, the semi-scalded birds are held or suspended against its outer surface, then when a shackled bird passes between two drums, the feathers are rubbed off.

If plucked by hand, the birds also should be suspended from shackles. The plucking cannot be done by rubbing as can be done with the hot-scald method. The feathers must be pulled out as with dry-picking.

WAX PICKING

Wax picking works very well and is ordinarily used in combination with the semi-scalding method for ducks and geese. After the birds are roughed on the picking machine, they pass through a drying machine because the wax will stick to the dry feathers and stubs, which are very short broken feathers, more tightly than it will to wet feathers.

The dried birds are then dipped, by hand or automatically, into a special preparation of melted wax (patented) with a melting point of around 120°F. Suspended by the head and feet, the birds move along a processing rail. The bucket containing the heated wax moves up and envelops each bird at a temperature of 125° to 130°F for a period of 30 to 60 seconds. The birds then pass under a cold water spray or through a cool air blast to harden the wax. Frequently the birds are wax-dipped a second time. The hardened wax is then pulled from the birds with the feathers, pinfeathers, hair, and scale encased in it, thus producing an attractively dressed product. In most waterfowl dressing operations, the water is boiled out (if water spray was used for cooling) and strained off the feathers to renovate this wax for reuse.

CHILLING

Subjecting the birds to a temperature of 32° to 36°F is highly essential for the immediate removal of body heat. Birds should not be frozen before the animal heat has been removed because "cold shortening" (see Chapter 22) may occur, thus making the meat less tender as the actin and myosin filaments of the muscle fibers slide and lock together. The most common procedure is to place the birds in tanks of ice slush containing phosphates. Phosphates aid in plumping the birds, chilling them, absorbing water, and reducing cooking losses. The maximum water uptake by this means is limited by practice to 5 percent, even though the USDA–FSIS permits up to 8 percent, for when the flesh is cooked, it tends to become mushy if an excessive amount (over 5 percent) has been used (see Chapter 3). In-line chillers with prior electrical stimulation greatly reduce the chilling time.

In former years, when birds were thoroughly chilled, their heads were wrapped with paper, their wings were folded against their bodies, and they were packed in paper-lined boxes or barrels for shipment. Birds processed to this point were designated as blood-and-feather dressed, or New York–dressed. This technique is no longer practiced. Off-flavors were a common defect in birds handled in this manner.

EVISCERATING

The processing procedure is largely completed when the birds are eviscerated. Mechanization of eviscerating equipment continues to occur, making it possible, for the most part, to eliminate hand work, except for hanging the carcass (see Figure 9–1). However, only hand techniques will be discussed here.

The different operations in their proper order are as follows:

1. Remove the head. Using the bone shears, cut through the back of the skin on the neck, peel down the skin, and sever the neck close to the shoulders and from the head at the other end. The neck is sold with the carcass.

2. Remove the crop and trachea. Hook the short gullet (between crop and gullet) with your index finger and peel the crop loose from the skin by working it forward, cutting at the lower end of the gullet. (see Figure 9–4).

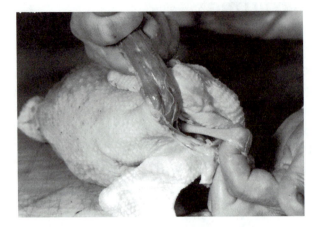

Fig. 9–4. Remove the crop and the trachea.

3. Remove the lungs. Place your index finger between the ribs and the lungs to loosen the lungs from the chest wall.

4. Remove the feet at the hock joint.

5. Make an incision from the rear end of the keel bone to the rectum, and cut around the rectum (see Figure 9–5).

Fig. 9–5. Cut around the vent with the opening made to the rear of the keel bone.

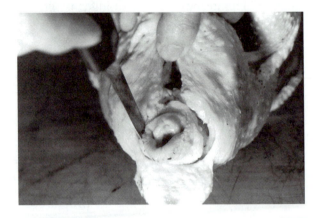

6. Draw the intestinal tract, the heart, the lungs, and the liver through this opening (see Figure 9–6). Chilled or partially chilled birds are easier to draw than warm ones.

7. Remove the bile sac from the liver, separating it from the intestines.

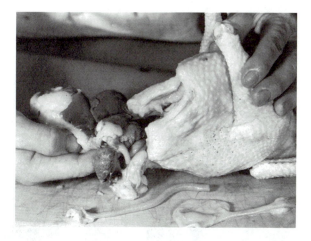

Fig. 9–6. Remove the intestinal tract, the heart, the lungs, and the liver.

8. Cut away the gizzard from the intestines, and split along the edge of the fleshy part of the gizzard sufficiently deep to cut the muscle but not the inner lining (see Figure 9–7). Proper pressure of both thumbs pulling the halves apart should permit peeling without breaking the lining and spilling the contents of the gizzard (see Figure 9–8). Gizzards are easier to peel if they have been partially chilled in ice water.

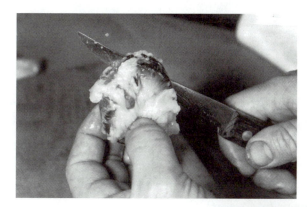

Fig. 9–7. Make the opening cut on the fleshy side of the gizzard.

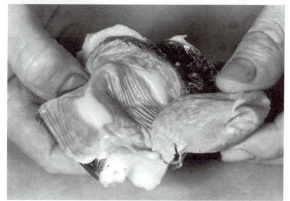

Fig. 9–8. Peel the gizzard.

Fryers or broilers may be eviscerated in a different manner. Use a linoleum trimming knife to cut along either side of the backbone, beginning at the rear and cutting forward. This leaves the backbone and neck in one piece. Remove the neck with bone shears. The two halves are laid open sufficiently that the entrails can be removed (see Figure 9–9). Split through the breast with a cleaver (in industry a power saw or cleaver/cutter is used) to halve the bird. The halves can then be quartered, if desired.

Fig. 9–9. Fryers or broilers can be eviscerated in a different manner. Use a linoleum trimming knife to open the back, splitting along either side of the backbone. The side can be quartered, if desired.

Electrical stimulation is frequently used after evisceration to more rapidly induce rigor and tenderness, thus greatly reducing chilling time. With electrical stimulation, poultry carcasses can be further processed the same day instead of having to be chilled overnight. Exact techniques are patented, but the principles are similar to those for beef (see Chapters 6 and 22). In one study, when electrical stimulation was accompanied with brine chilling[2] or high temperature conditioning,[3] tender broiler meat portions were produced. The 20 minutes of brine chilling at 34°F (1°C) was preceded with 10 minutes of brine emersion at 60°F (20°C). The high temperature conditioning involved electrical stimulation prior to a 103°F (39°C) water bath for one hour postmortem, which was followed by ice chilling.

For maximum attractiveness, the birds should be shaped to give them a plump, compact appearance. The wings are compressed against the sides of the carcass, and the legs are brought together at the vent. Birds for roasting

---

[2]Dawson, P. L., D. M. Janky, M. G. Dukes, L. D. Thompson, and S. A. Woodward. 1988. Tenderness of hot-boned broiler breast fillets chilled in either water or brine using two different chilling regimens. Poult. Sci. 67:1545.

[3]Sams, A. R. 1989. Electrical stimulation in high temperature conditioning of broiler carcasses. Poult. Sci. 68 (Suppl. 1):129 (Abstr.).

are trussed. A length of cord or wire is drawn over the fore part of the breast and over the wings and crossed over the back (see Figure 9–10), brought over the ends of the drumsticks, and tied tightly at the back of the rump (see Figure 9–11).

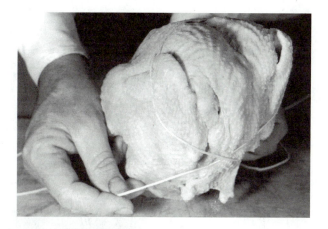

Fig. 9–10. The first step in trussing. A length of cord is drawn over the forepart of the breast and over the wings and crossed over the back.

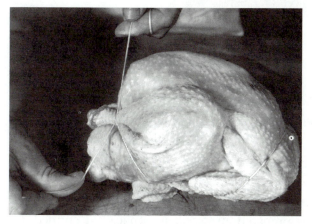

Fig. 9–11. Completing the trussing. The cord is brought over the ends of the drumsticks and tied tightly at the back of the rump.

CHICKEN PARTS

Broilers are being sold as parts in ever-increasing proportions. Some consumers prefer white meat, so an all-breast pack best meets their need. Others prefer dark meat, so thighs, drumsticks, or both leg parts are most suitable for them. Parts are very adaptable for specialty dishes and frequently for barbecues. The demand for white meat exceeds that for dark, so processors are often faced with the dilemma of cut-rate pricing for dark meat. Because of this price difference, most of the dark meat goes into processed meat, sausages, bologna, frankfurters, etc. (see Chapter 21).

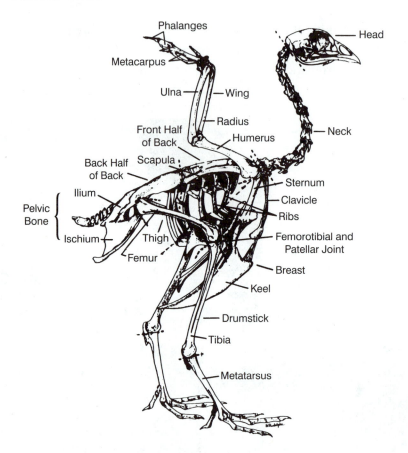

Fig. 9–12. Avian anatomy.

Most of the parts are made by cuts at certain joints, as can be seen by examining the skeleton (see Figure 9–12) and relating to the parts shown. The North American Meat Processors Association (NAMP) (http://www.namp.com) has prepared an excellent publication entitled *The Poultry Buyers Guide* that contains very complete information (including colored pictures) on food-service cuts for poultry.

Although grade standards are dependent on the carcass grade given before the bird is cut into parts, the parts also must conform to the following specifications:

Wings

Wings include the entire wings (*humerus, radius, ulna,* and *metacarpus*) with all muscle and skin tissue intact. The wing tips (*phalanges*) may be removed (see Figures 9–12 and 9–13). The *radius,* the *ulna,* and the *metacarpus/phalanges* are often sold at retail bars or restaurants as buffalo wings (barbeques).

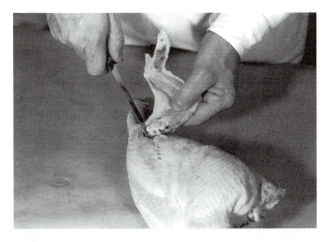

Fig. 9–13. Cut up chicken. First, remove the wing.

Legs, or Thighs and Drumsticks

A leg includes the thigh *(femur)* and drumstick *(tibia)* as removed from the hip joint with the *metatarsus* removed at the hock joint. Back skin or pelvic bones are not included, but the pelvic meat may be kept. A drumstick may be separated from the thigh by a cut made through the knee joint *(femorotibial* and *patellar* joint) (see Figures 9–12 and 9–14).

Greater demand for breast meat parts often causes leg parts to be in excess, and so additional processed items are often produced from deboned thigh and drumstick flesh. An automatic deboning machine may be used, or the muscle may be removed by hand, especially when muscle particles are desired for products such as turkey hams (see Chapter 20).

Fresh frozen deboned thighs are a popular item as well as intact frozen thighs.

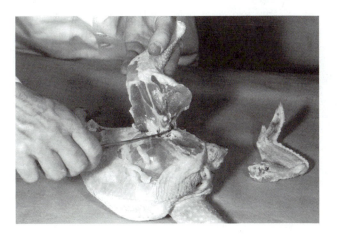

Fig. 9–14. Second, remove the leg.

Breasts

A breast is separated from the back at the shoulders by a cut from that point along the junction of the vertebral and sternal ribs (see Figures 9–12 and 9–15). Ribs may or may not be removed, and the breast may be cut

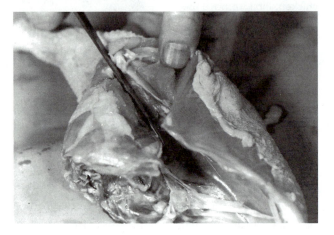

Fig. 9–15. Third, remove the tail piece (not shown). Fourth, separate the rib and neck piece from the breast.

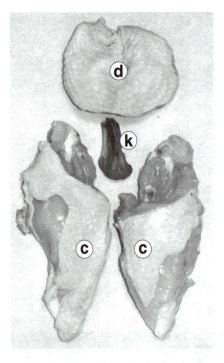

Fig. 9–16. Fifth, divide the breast into three sections: (c) half breast, (d) wishbone breast, and (k) sternum (discarded). (See also Fig. 9–2.)

along the breastbone into halves after the sternum or calcified portion has been snapped out and removed. Breasts may be labeled "with ribs" if the ribs were not removed.

Producing a third part, that containing the wishbone or *clavicle,* entails severing it from the breast halfway between the front point of the breastbone and the end of the clavicle to a point where the clavicle joins the shoulder. Neck skin is not included (see Figure 9–16).

Complete deboning provides a boneless breast or a product to be used in further processed items. Frozen boneless breasts, either whole or in the form of "chicken tenders" from the underlying minor muscle, are most popular among consumers.

Meat and Poultry Magazine, in June 1999, reported these forecast prices for the last quarter of 1999:

> Beef ribeye (IMPS 112A, 11 lbs. and up) . . . . $455.87/cwt.
>
> Chicken breats (bonelss, skinless) . . . . $163.67/cwt.

This comparison of boneless muscle value from these two species makes obvious one reason for the popularity of chicken breast muscle.

Backs and Necks

Most backs and necks are mechanically deboned (separated), and the resulting muscle is used in processed meat products (see Chapters 18 and 21). Intact backs include the pelvic bones and all the vertebrae posterior to the shoulder joint. They may be cut into two parts as shown in Figure 9–17.

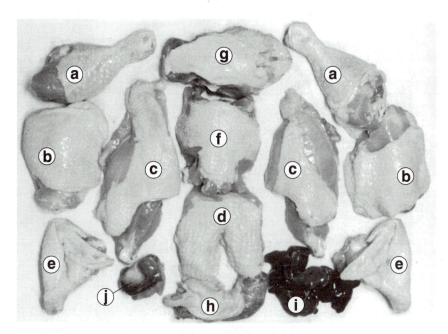

Fig. 9–17. The cut-up chicken. (a) Drumstick, (b) thigh, (c) half breast, (d) wishbone breast, (e) wing, (f) rib back, (g) sacral back, (h) neck, (i) liver, and (j) gizzard. (See also Fig. 9–2.)

Intact backs that meet all the applicable quality standards must include the meat on the *ilium* (oyster muscle), pelvic meat and skin, and vertebral ribs and *scapula,* with meat and skin, except for any one of these meat portions, which may be eliminated. The neck should not have the skin removed.

The cuts in order of their monetary value are breasts, legs (or thighs and drumsticks), wings, backs, necks, and gizzards (see Figure 9–17). The livers and hearts may bring a good price on some markets.

In the December 23, 1998, newsletter of the National Chicken Council, formerly known as the National Broiler Council, the organization released the results of its marketing survey. The compiled data were provided by 30 broiler processors and 22 distributors and present a good industry overview of 1997. The National Meat Association (NMA) published this chicken marketing survey in its January 11, 1999, *Lean Trimmings* and *Herd on the Hill.*

Share of Total Marketings (%)	Market Channel	Quantity (billions of lbs.)
29.1	Retail grocery	7.86
17.6	Export	4.75
15.5	Fast food	4.19
11.0	Pet food / renderers	2.97
9.2	Other food service	2.48
7.8	Further processor	2.11
4.9	Brokers/traders	1.33
2.7	Club store	0.73
1.0	Unclassified	0.27
0.6	Government	0.16
0.6	Institutions	0.16

FABRICATED OR FURTHER-PROCESSED TURKEY

The utilization of large tom turkeys for roasting is not practical in many homes in the United States. The chief outlet is to the hotel and restaurant trade. However, by devising several variations of turkey processing, which are presently in use in the industry, investigators have widened the market for large toms, thus keeping their price more in line with hen turkeys, which are more desirable for the average family.

The binding of pieces or chunks of meat together to form rolls or loaves has received considerable attention by the poultry industry, especially in the utilization of turkey meat. Over 60 percent of all turkey meat produced is used in the production of convenience items.

Breast meat may be made into turkey rolls or steaks, and thigh meat may be formed, cured, and smoked to simulate ham. Drumsticks, wings, and necks may be mechanically deboned, and the resultant product used for turkey burgers, turkey patties, turkey frankfurters, or turkey sausages. Cornell University workers summarized information on roll manufacture and indicated that the addition of dried egg albumen, wheat gluten (vital), or gliadin, all protein materials, aided significantly toward binding meat chunks together into a simulated roast to be easily sliced and served to the consumer.[4] This technique has been modified with similar or other binders being used.

Half and Quarter Turkey

Halving a bird is done by sawing through the back and breast. Quartering a large bird is accomplished by making the division midway between the wing and leg.

The sale of cut-up turkey has been somewhat less in recent years because some consumers have the impression that turkey must be roasted whole. However, the roasting of half turkey with the dressing underneath has been well received, and the roasting of quarter turkey is convenient for small families.

Edible Yield

Edible yield from the full-dressed carcass in terms of muscle and skin will average from 65 percent for fryers or broilers to 71 percent for larger turkeys, with bone losses of 35 percent and 29 percent, respectively.[5]

DUCKS AND GEESE

Waterfowl are very tight-feathered, which makes them difficult to scald; consequently, much higher temperature conditions must be used. One good

---

[4]Vadehra, D. V., and R. C. Baker. 1970. The mechanism of heat initiated binding of poultry meat. Food Technol. 24:766.

[5]Matthews, R. H., and Y. G. Garrison. 1975. Food yields summarized by different stages of preparation. Agric. Handbook No. 102. USDA, Washington, DC.

method is to steam them. This is accomplished by churning them in water that is near the boiling point or by wrapping burlap sacks around them and then immersing them in hot water. Water temperatures as low as 160°F can be used, but the scalding time is considerably longer (up to two to three minutes) than with higher temperatures.

Properly bled and debrained waterfowl may be readily dry-picked, but the fine down feathers must be removed by scalding or waxing. Because waxing limits the value of the down feathers, the birds are often not waxed until after most of the down has been otherwise removed. Waxing usually involves a double-dip procedure, i.e., building up a sufficient coat of wax so that when the birds have been chilled, the wax can easily be pulled off with the pinfeathers. (See earlier discussion on waxing.)

How to produce geese with minimal amounts of pinfeathers is a major unsolved problem. Geese could be grown to market weight in less than 10 weeks, but to produce them with minimal pinfeathers still requires over 20 weeks. Temperature changes during growth also seem to be a factor in that a frost tends to minimize the pinfeathers.

Long Island for many years was a large production center for so-called "green" ducks (8 to 12 weeks of age). This production has now largely moved to the Midwest, especially Michigan, Indiana, and Wisconsin, but the term "Long Island Style Ducklings" remains a legend among connoisseurs of fine food, and so green ducklings have retained that name.

Although geese are not listed in Table 9–1 or 9–2, South Dakota, the longtime home of the chapter's author, has held the No. 1 position for many years, harvesting approximately 200,000 geese annually.

Except for limited processing and sale as fresh items around holidays, most waterfowl are frozen and held as such until sold to consumers.

GUINEA FOWL AND PHEASANTS

Most guineas are dressed in the same manner as chickens — dry-picked or semi-scalded. The young birds are marketed at dressed weights of 2 pounds or under.

Pheasants produced commercially are probably one of the meatiest birds for their size of any bird processed for human consumption. They are generally semi-scalded before they are picked, eviscerated, frozen, and marketed as full-dressed birds. They weigh from 1 to 2 pounds ready-to-cook, and the major portion of the meat is on the breast. The bones are quite small in proportion to the amount of edible meat present. Game birds, including pheasants, may be skinned, as described in Chapter 10.

OSTRICHES AND EMUS

Both ostriches and emus are flightless birds that belong to the family Ratitae. These birds have the same genetic base as chickens and turkeys but when mature will weigh about 300 pounds. They were traditionally farmed in Africa and Australia for the production of feathers, high-quality leather, and fat. The feathers are used in making hats, boas (long, fluffy scarfs), and other fine clothing and accessories. The hide is processed into expensive leathers for boots, shoes, purses, and the like. The $3^1/_2$- to 4-pound eggs that do not hatch may be worth around $40 each to painters and collectors.

Production units have been established in many other parts of the world. The first ostriches reached the United States in 1989. Most flocks are in the Southwest, where the climate is favorable.

There are now several ostrich and emu processing plants in the United States producing steaks, ground meat, patties, sausages, and jerky. Meat is no longer a by-product, but still only a marginal fraction of the meat is used for human consumption. It is often served in metropolitan restaurants. The intense red color of the meat, due to a high pigment content, is very sensitive to oxidation, thus limiting the storage of fresh meat under aerobic conditions to short periods.

An ostrich's eye is bigger than its brain. The bird is often seen with its head in the sand. For further information, simply enter the words *ostrich, emu,* or *ratite* in your Internet search engine. Be sure to use only information from refereed journals.

STORAGE

Almost all broiler or fryer-type chicken is sold fresh, primarily to avoid the bone darkening that occurs after frozen birds are thawed. When the birds are older, as with roasters and turkeys, there is limited tendency for this to occur. The younger chickens are usually packed in ice for shipment and short-time holding. After being vacuum packed in flexible film, most other poultry are first exposed to a hard chill (–20°F) with air blast or brine and then frozen completely at 0°F. Afterwards, the carcasses may be boxed individually or in groups of four, six, or more. They are then held at a constant temperature ($0°$ ± 5°F) for up to six months, after which the quality diminishes significantly (see Chapter 19).

It should be noted that, according to USDA-FSIS regulations, the term *fresh* cannot appear on the label of raw poultry whose internal temperature has ever been below 26°F.

10

Game Processing

Game meat has sometimes been touted to consumers as being more healthful than meat from our tame meat species. Perhaps, due to diet, some game is leaner than beef, pork, or lamb, and this may be an advantage. However, all game must be handled very carefully immediately after harvest. Review Chapter 2, "Meat Biotechnology and Microbiology," if you have any doubts about the effects of temperature and time relationships on microbiological growth. Sanitation is a key factor as well. Some wild game species — for example, buffalo, rabbits, and channel catfish — have been "tamed" and are being raised in production systems. Harvesting these confined animals is more conducive to maintaining food safety.

DEER AND OTHER BIG GAME

The conservation of game by legislation, which restricts the period when such game may be legally shot, coupled with greater widespread awareness and concern for the welfare of wild animals, is making venison and other game animals more abundant. In some families, game is an attractive addition to the family larder. The commercial production of captive deer and elk is a growing industry. State and provincial regulations regarding the hunting of deer and other big game animals vary considerably throughout the United States and Canada. It is the hunter's obligation to know and abide by these regulations. Every effort should be made to make a quick and efficient harvest for optimal meat quality and to minimize the suffering of the animal, and every effort should be made to prevent the flagrant waste of the carcass that has sometimes been evident in its preservation and utilization. Although the following discussion pertains to deer, the same principles apply to other big game, such as antelope, elk, moose, and bighorn sheep. However, one must realize that it is far easier to harvest, clean, and cool a 150-pound white-tail deer than a 600-pound elk, or an even larger moose.

Precautions at Time of Harvest[1]

Make sure your deer is dead! Many hunters have leaned their guns against a tree and prepared to field dress their deer only to see it spring up and bound off. A good rule of thumb is to wait 15 minutes before approaching a downed deer and 30 minutes before trailing a wounded deer. A wounded deer will generally travel less than 1 mile before lying down if it feels that it is not being followed. Given ample time (the 30-minute wait), the deer will often die where it has bedded down, or at least it can be approached more easily for a second shot as it becomes less wary. However, when a wounded deer is trailed immediately, it will generally keep running (sometimes 5 miles or more), which significantly reduces the hunter's chance of finding it. The good hunter will know his or her capabilities in placing a shot for a quick, clean harvest and will avoid wounding game animals.

The deer should be approached with caution from its tail end. Male deer (bucks) sometimes turn very aggressive in the mating season and have been known to gore and trample humans to death. With your gun ready for another shot, nudge the deer with your foot. It may have been wounded and simply dropped from exhaustion. If it is still alive, you will know it. Keep out of reach of its legs until you are certain it is dead. Adhere to the state laws regarding tagging and reporting the harvest.

Bleeding

It is not necessary to bleed your deer unless it has been shot in the head, and often not even then. A shotgun slug in the neck or body cavity does a good job of bleeding it for you. If you want to bleed your deer, and if it happens to be a corn-fed trophy buck, do not ruin the head by slashing the neck. Instead, make a small cut at the base of the neck. Insert the blade of the hunting knife (4½- to 5½-inch blade) several inches in front of the point of the breast, with the point of the blade aimed toward the tail. Plunge it up to the hilt, press the blade downward to the backbone, and withdraw it with a slicing motion. Elevate the rear portion of the deer to permit the blood to drain by gravity.

---

[1]Excerpts from Field Dressing Your Deer. Illinois Dept. of Conservation, Springfield, IL 62706.

Field Dressing and Handling

Many hunters think that they should cut the glands from a deer's legs. This is not really necessary. Deer have glands not only on their legs but also between their toes and at the corners of their eyes. All these glands are inactive after the animal's death. Carry a small sharp knife for field dressing. Unless you plan on bagging a deer in hand-to-hoof combat, a big knife is a handicap. Those Bowies with the 10- and 12-inch blades may look good swinging from your belt, and they may be just the tools for Green Berets or commandoes, but when it comes to field dressing a deer, they are extremely awkward. A clasp knife or a sheath knife with a sharp 4- or 6-inch blade is all you need.

A 6- to 8-foot section of good clothesline is handy to have with you. With it you can tie the deer's leg to a tree to give yourself working space. If you do not have a rope, you can spread the deer's hindlegs by inserting a small branch about 3 feet long between them. With the animal on its back, block it on either side with logs or rocks to keep it in place.

If you have shot a buck, cut the genitals from the body. Carefully cut the hide in the abdominal area. Minimize the opening, for when dragging or carrying the deer, you will expose it to weeds, soil, and insects, which contaminate it. Open the body cavity slightly. Tilt the animal sideways and let the blood drain from the body cavity. Do not split the aitch bone now, but cut the bung loose as you would with a lamb (see Chapter 7), by cutting around the bung and pulling it back through the abdominal opening. Split the aitch (pelvis) bone when you get the deer home where you can work under more sanitary conditions. Having loosened the rectum, remove all of the intestinal tract, being extremely careful to minimize internal fecal contamination of the carcass. Remove the pluck (heart, lungs, gullet, and windpipe) and liver by cutting the gullet (esophagus) and windpipe as far forward in the chest cavity as possible. Carry a plastic bag with you to place the liver and heart in, as they will make a very nutritious and palatable camp meal.

In parts of the country, warm temperatures and insects can be a problem. Flies often become a problem in southern deer ranges and western antelope ranges due to the earlier seasons. Wrapping the carcass in cheese cloth to prevent the entry of flies into the opened body cavity will remedy fly and other insect problems. Some hunters will rub ground black pepper on the inside of the carcass to further distract flies from laying eggs. In warm weather, meat spoilage can ruin an otherwise successful hunt. Getting the carcass to a locker operator with a refrigerated cooler within a day of the harvest will prevent meat spoilage. If this is not possible, and borderline spoilage temperatures exist (above 40°F), hang the carcass in a well-shaded spot, and prop the body cavity open with a clean stick to facilitate chilling.

Tagging and Transporting

Many hunters prefer to tag their white-tailed deer as soon as they find them. Others prefer to tag them after they are dressed. Just be sure that you tag your deer before you load it into your car. There is no easy way to get the deer out of the timber into your car, but one method seems to require the least effort. Tie the front legs of your deer to its head and drag it. Larger elk and antelope may need to be split or even quartered to be moved. Remember, normally, you must check in your deer at the check station specified for your county on the same day it was harvested. Investigate your local laws.

Corn-fed white-tailed deer are fine eating, but their carcasses must be kept as cool as possible. Some hunters transport field-dressed deer attached to the outside of their vehicles (they must not rest over the hoods where they will absorb engine heat) to give mute evidence of the hunters' good fortune and to serve as refrigeration in transit.

Disposition of the Carcass

At the journey's end, after being checked in, the deer should be skinned, unless it is to be hung and aged in the cold for a week. Leaving the skin on the carcass during aging holds down shrinkage and avoids discoloration. Proper aging should improve tenderness and overall palatability. The aging temperature should be 32° to 38°F. If you intend to process the carcass yourself, you must be especially aware that the temperature of the carcass should not exceed 40°F while it is aging.

If you intend to deposit the carcass with a locker operator/meat processor for processing, be sure to make previous arrangements. Strict federal and state regulations govern the processing of game in meat processing establishments. These regulations are:

- Wild game carcasses must be dressed prior to entering the processing or refrigerated areas of the licensed establishment.

- Wild game carcasses stored in the refrigerated areas of the licensed establishment must be contained and handled in a manner that will assure complete separation of wild game from domestic meat and meat products. This may be accomplished by, but not limited to, the following: (1) the use of separate coolers, (2) the enclosing of game in metal cages, and (3) the complete enclosing of game carcasses with plastic or shrouds.

• A written request should be made by the establishment to the responsible government agency for a listing of the days and time of day wild game carcasses may be processed.

• All equipment that comes in contact with wild game must be thoroughly cleaned and sanitized before it can again be used on domestic animal or poultry carcasses.

Skinning

Open the skin over the rear of the hock and down the back of the leg to the rectum. Skin around the hock, and remove the leg at the break joint on the lower part of the hock. Make an opening between the tendon and the hock, and insert a hog gambrel or a heavy stick that will hold the hocks apart. Raise the carcass until the haunches are at shoulder height. Remove the forelegs at the smooth joint (just below the knee joint). Very little knife work is necessary, since the pelt can be pulled and fisted from the carcass. If the head is to be mounted, the skin on the neck (cape) should be opened on the topside of the neck and behind the shoulder (see Figure 10–1). A properly caped deer (see Figure 10–2) should have sufficient shoulder skin for the taxidermist to work with. It is far better to leave too much skin with the cape than too little.

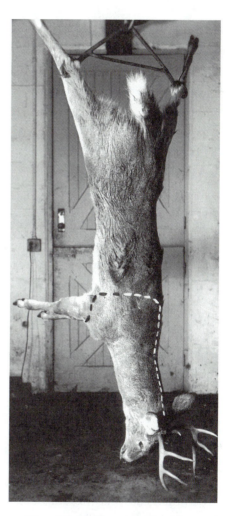

Fig. 10–1. The dotted white and black line indicates where you should make the opening when you are skinning out the cape for a head that is to be mounted. (Courtesy, The Pennsylvania State University)

Fig. 10–2. *Do not ruin your trophy deer by cutting the cape too short.* Note how the prominence of the shoulder adds to the beauty and balance of the finished mount. (Photo by Kevin W. Jones)

Proceed to pull the pelt from the rounds (see Figure 10–3). Use your fist to remove the pelt from the sides, and continue to pull it down the back.

The head is removed at the atlas joint after the neck has been caped out such that the cape and head can be removed in one piece. After the pelt and head have been removed, split the underside of the neck, and remove the gullet (esophagus) and windpipe and the remainder of the pluck, if this has not been done previously. Use a stiff brush and plenty of clean water to wash the hair and soil from inside the carcass. Refrigerate the carcass. Figure 10–4 shows a full-dressed carcass.

Pelts

Care of Hide and Head

Rub the skin side with a liberal amount of fine salt, and apply plenty of salt to the head. Let the salt be absorbed for 24 to 48 hours, then fold the pelt, hair side out, and tie securely with strong cord. Tag it according to law, and ship it to a taxidermist for mounting and tanning, unless you wish to attempt to tan it yourself (see Chapter 11). Use clean table salt to avoid mineral stains, particularly if the pelt is to be made into buckskin. Save time and money by discarding badly torn or scored pelts.

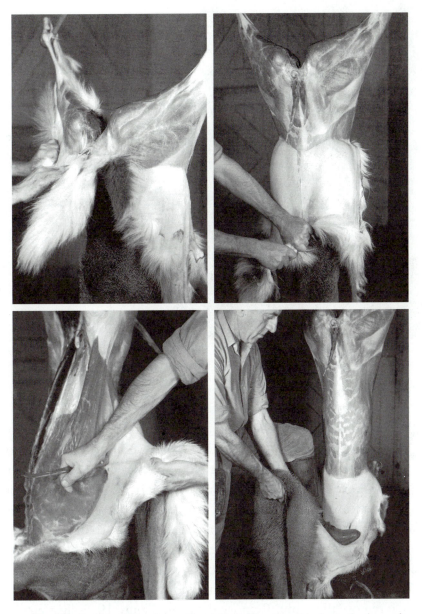

Fig. 10–3. Skinning a deer. *(Top left)* Pulling the skin from the haunch. *(Top right)* Pulling the pelt from the loin. *(Bottom left)* Fisting over the side. *(Bottom right)* Using body weight to pull the pelt over the back and shoulder. (Courtesy, The Pennsylvania State University)

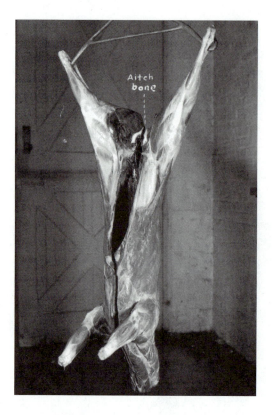

Fig. 10–4. The full-dressed carcass showing the aitch bone. (Courtesy, The Pennsylvania State University)

Cutting the Carcass

Split the carcass through the center of the backbone, dividing it into two sides; however, if the neck is to be used for pot roast or neck slices, remove it before you split the carcass. Place the side of venison on the table, inside down, and remove the hindleg by cutting in front of and close to the hipbone *(ilium)*. Move forward to the shoulder, and remove it by cutting between the fourth and fifth ribs. The breast that is attached to the back must be removed. Cut across the ribs about 3 inches from the backbone on the blade end to the loin end. Separate the ribs from the loin by cutting directly behind the last rib. The leg is placed on the table, aitch bone on top. Cut parallel to the aitch bone, and remove the rump. Remove the flank. Figure 10–5 illustrates where these cuts should be made.

Venison rib chops, round steaks, and rolled shoulder roasts are shown in Figure 10–6. The shanks, breast, and flank are usually boned and ground into deer burgers or incorporated with pork for sausage. The heel and neck portions (see Figure 10–5, items 2 and 8, respectively) of the carcass are also often used for sausage or ground venison.

Fig. 10–5. One method of cutting a venison carcass: (1) rear shank; (2) heel; (3) round steak; (4) rump and sirloin; (5) loin chop or roast; (6) rib chop or roast; (7) shoulder blade; (8) neck pot roast; (8A) neck slices; (9) arm roast; (10) foreshank; (11) breast; (12) flank. The letters in white indicate the line where wholesale cuts are made. This carcass weighed 92 pounds and was in excellent finish. (Courtesy, The Pennsylvania State University)

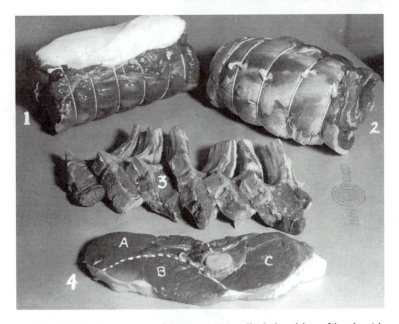

Fig. 10–6. Some cuts of venison: (1) rolled shoulder of buck with slice of fresh pork fat for self-basting; (2) rolled shoulder of fat doe (it has sufficient fat); (3) rib chop; (4) venison round steak indicating (A) inside round, (B) outside round, and (C) round tip. (Courtesy, The Pennsylvania State University)

Boneless Cuts

An alternate method of cutting is to make all cuts boneless, including the rib and loin chops. Butterfly chops from these tender cuts are highly desirable. Furthermore, by making boneless cuts, you eliminate the need for a saw, which is a nuisance in home cutting. Many of the principles in Chapter 16 on lamb cutting can be applied to cutting a venison carcass. Figures 10–7 to 10–12 depict a venison carcass and the boneless sub-primal cuts obtained from it.

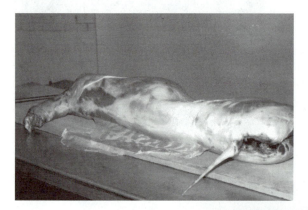

Fig. 10–7. A 96-pound field-dressed doe yielded this 73.5-pound carcass, with a loin eye of 4.35 square inches. The thin exterior muscle (cutaneus trunci) was removed prior to the removal of the longissimus muscle.

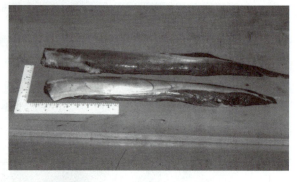

Fig. 10–8. The doe's boneless loin–rib muscles weighed 5.1 pounds. The thin connective tissue cover *(facia)* must be removed for palatable chops.

Fig. 10–9. The tenderloins weighed 0.6 pound.

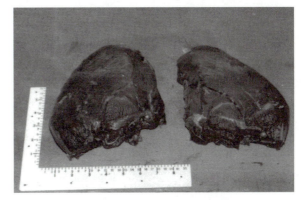

Fig. 10–10. The top round roasts totaled 4.3 pounds.

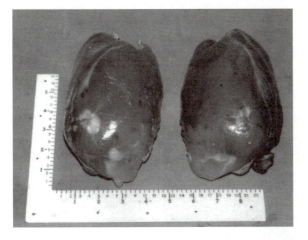

Fig. 10–11. The round (sirloin) tip roasts totaled 3.8 pounds. These are ideal for roasts and quite satisfactory for steaks.

Fig. 10–12. The bottom round roasts weighed 4.1 pounds. These roasts can be ground with the shoulder muscles and the remaining usable trim from the less desirable thin rough cuts (flank, plate, shank) to be mixed with pork trim for fresh sausage or processed into summer sausage. The bottom also makes a delicious pot roast.

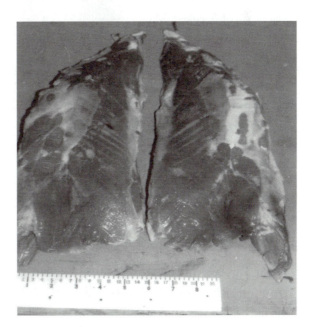

A University of Illinois study that involved seven deer (three does, four bucks) had the following results:

Average for Seven Deer

	Pounds	
Field-dressed weight	112	
Estimated age (years)	1.6	
Carcass weight to cut (Figure 10–7)	87	
Dressing percent	78	
Loin-eye area at 12th rib (sq. in.)	4.5	

		% of Carcass
Lean trim (excluding mutilated)	27.4	31.5
Boneless loin and rib (Figure 10–8)	7.0	8.0
Tenderloin (Figure 10–9)	1.0	1.1
Boneless top round (Figure 10–10)	5.3	6.1
Boneless round tip (Figure 10–11)	4.6	5.3
Boneless bottom round (Figure 10–12)	4.4	5.2
Edible meat	49.7	57.2
Mutilated (not usable)	9.8	11.2
Bone	22.7	26.1
Fat	3.6	4.1
Nonedible	36.1	41.4
Total (edible and nonedible)	85.8	98.6

The loin-rib muscle (*longissimus*) and tenderloin were made into 1¼-inch chops and the top round into 1¼-inch steaks. Panel members tasted the chops and steaks with and without the thinly adhering fat cover. No preference was shown in regard to fat cover or animal age, but doe meat was slightly preferred over buck meat.

Mutilated or Bloodshot Areas

If a large area of the carcass is affected by the shot, portions of it may be salvaged. First, wash it free from hair. Then, make a weak brine by dissolving ½ pound of salt in 1 gallon of water. Soak the carcass in the brine solution overnight. The salt will draw out most of the blood, and the best meat will be suitable for grinding or stewing. Badly mutilated meat can be used for dog food, if all bone chips and slugs or shot are removed.

Sausages

A variety of tasty sausages and processed products can dramatically improve the palatability of the trimmings that would otherwise be used for ground venison. In cases where older bucks are taken, particularly those which have not had access to grain diets, the entire carcass can be better utilized in sausage. This will help reduce the extreme "gaminess" of the meat that is often present in such carcasses. One taste of sausage made from the less-tender cuts of venison ground with fat pork trimmings will assure anyone that no part of the venison carcass should be wasted. Some creative ideas on how to utilize this type of venison are given in the following recipes. If you are intent on home processing your deer and making sausage, you will need some equipment and supplies. Your local locker operator/meat processor might provide some assistance in locating a supplier.

Fresh Venison Sausage

10 lb. lean venison	175 g salt (9 tbsp.)
10 lb. 50 percent fat pork trim	10 g coriander (2 tsp.)
12 g ground sage (1 tsp.)	20 g ground black pepper (4$^{1}/_{8}$ tsp.)
10 g mace (2 tsp.)	10 g hot mustard (2 tsp.)

NOTE: For conversions: 1.0 pound = 454 grams

1.0 ounce = 28.4 grams

Grind lean venison and fat pork through a $^{3}/_{8}$-inch coarse grinder plate. Mix uniformly with salt and spices. Regrind sausage through a $^{1}/_{8}$-inch fine grinder plate, and stuff into sheep or hog casings. This sausage can also be wrapped in freezer paper in bulk form and subsequently made into patties before it is cooked, if desired.

Unlike the other sausages described here, fresh venison sausage is not cured or smoked and is thus the easiest to make. This sausage will keep for only three to five days (in a refrigerator); thus, it should be sharp frozen until ready for consumption.

Polish Venison Sausage

10 lb. lean venison

10 lb. 50 percent fat pork trim

17 g ground coriander (3½ tsp.)

10 g garlic powder (2 tsp.)

1 lb. cold water (1 pt.)

15 g paprika (3 tsp.)

200 g salt (¾ cup)

1 g ground nutmeg (2¼ tsp.)

40 g sugar (8¼ tsp.)

11 g ground ginger (2¼ tsp.)

22 g certified cure (4½ tsp.)[2]

6 g ground caraway (1¼ tsp.)

30 g ground black pepper (6¼ tsp.)

Grind venison through a ³/₈-inch coarse grinder plate, and then regrind through a ¼-inch fine plate. Blend venison with salt, water, and cure in a mixer for two to three minutes. Next, grind pork trim through the coarse and the fine grinder plates, add to mixer, and mix an additional four minutes while adding the remaining ingredients. Stuff into sheep or hog casings, and link casings by twisting in opposite directions every 6 inches (see Figure 21–5). Hang the linked polish sausage on smokehouse sticks and cook at 130°F for one hour. After this drying period, smoke the product for two to three hours, while gradually increasing the temperature to 155°F. Increase the temperature of the smokehouse to 165°F, and cook the sausages until the internal temperature reaches 152°F.

Venison Summer Sausage

10 lb. lean venison

10 lb. 50 percent fat pork trim

30 g ground black pepper (6¼ tsp.)

16 g ground coriander (3¼ tsp.)

2 lb. cold water (1 qt.)

4 g ground nutmeg (³/₈ tsp.)

225 g salt (¾ cup)

4 g ground red pepper (¾ tsp.)

100 g dextrose (5 tbsp.) or 4 g whole mustard seed (¾ tsp.)

200 g corn syrup (1 cup)

0.5 g ground ginger (¹/₈ tsp.)

22 g certified cure (4½ tsp.)[3]

Starter culture (if available)

---

[2]Certified cure is a nitrite cure that when used according to directions, usually 1 ounce per 25 pounds of meat, will give the correct nitrite level of 150 ppm (parts per million). *It is very important that you weigh out this ingredient accurately and that you do not use more than the amount called for.* Certified cure cannot be purchased through a grocery store or a drug store but rather through a local locker operator, who deals with a spice company.

[3]Ibid.

Grind venison through a $^3/_8$-inch coarse grinder plate, then regrind through a $^1/_8$-inch fine plate. Mix ground venison with salt, water, and cure for three minutes. Add pork trim that has been ground through a $^1/_8$-inch fine grinder plate, and add the remaining nonmeat ingredients while mixing. The summer sausage should be mixed an additional four minutes after all the ingredients have been added. Stuff the product into 3-inch fibrous casings, then clip or tie the ends of the casings. Care should be taken to avoid air pockets within the stuffed casings. If a commercial starter culture has been used, the product can go straight into the smokehouse and be fermented for 12 to 18 hours at 80° to 100°F. If no starter culture has been used, hold the product in a 38°F cooler for three days before following the same fermentation schedule. (For more information on fermentation processes and the principles behind fermentation, see the section in Chapter 21 on dry and semi-dry sausages.) The product should be smoked during the fermentation process. After the fermentation process, gradually increase the smokehouse temperature to 165°F, and cook the product to an internal temperature of 152°F. Immediately after cooking, rinse with cold water to remove the greasy film on the product surface and to shrink the casing to prevent the formation of wrinkles, and then place the product in a 34° to 38°F cooler. If a "tangier" summer sausage is desired, increase the level of dextrose in the formulation, and ferment for a longer period of time.

Venison Jerky

3 lb. lean venison round muscle
10 g Accént (2 tsp.) or monosodium
   glutamate (MSG)
250 g water (1 cup)
100 g soy sauce (¾ cup)
6 g garlic powder (1¼ tsp.)

20 g seasoned salt or barbecue salt
   (4$^1/_8$ tsp.)
6 g black pepper (1¼ tsp.)
6 g Worcestershire sauce (1¼ tsp.)
2 g liquid smoke (optional) (½ tsp.)

The round muscle should be in one piece and free of any fat. Semi-freeze the meat, and slice it with the grain into $^1/_8$-inch slices, and cut into strips of the desired size. Mix the remaining ingredients together to form the marinade. Pour the sauce over the meat. Marinate for 24 hours, stirring occasionally. Lay the venison strips on oven racks in the smokehouse, and then smoke for six to eight hours at 110° to 120°F, or until the desired degree of chewiness is achieved. The jerky can also be dried in an oven at the lowest setting, with the door ajar.

Venison Smoky Sausage Sticks

20 lb. lean venison

10 g ground red pepper (2 tsp.)

250 g salt (¾ cup + 1 tbsp.)

4 g ground coriander (¾ tsp.)

22 g certified cure (4½ tsp.)[4]

4 g garlic powder (¾ tsp.)

35 g ground black pepper (7¼ tsp.)

Grind venison through a ³/₈-inch coarse grinder plate, and then regrind through a ¹/₈-inch fine plate. Mix the ground venison with salt and then cure until the meat is very tacky (three to five minutes). Add the remaining spices, and then mix for an additional four to five minutes. Stuff the sausage into the smallest diameter *edible* casings available (see Chapter 21). (Sheep casings or collagen casings that are ½-inch diameter will work.) There is no need to link this sausage, since it will be cut into 6-inch "sticks" when it is finished. Place the stuffed sausage in a cooler at 38°F for three days. After the three days, put the sausages on smokehouse sticks in a smokehouse that has wet or damp sawdust. Smoke them for six hours at 80° to 100°F. After that, raise the smokehouse temperature to 120°F for an additional two hours, and then finish the product at 150°F for two more hours. Cut into 6-inch sticks when the sausages have cooled.

For more information on home sausage production and the principles behind the various steps outlined here, see Chapter 21.

Antelope (Pronghorn), Elk, and Moose

Antelope (pronghorn) are generally smaller than deer and thus are easier to handle. Elk and moose are also members of the deer family. Elk are larger than deer, and moose are larger than elk. Procedures in the field vary depending on the size of the harvest. Remember, it is essential to cool out the carcass as completely as possible, as soon as possible after the harvest. Larger carcasses (elk and moose) may need to be halved and(or) quartered to get them out of the field and to hasten cooling. Principles covered in the preceding chapters on harvesting and in the later chapters on cutting domestic species apply to the handling and processing of game animals.

---

[4]Ibid.

Bighorn Sheep

The hunting of bighorn sheep is permitted only during a restricted season, which is limited to a few states. South Dakota has such a season. Only a very few licenses are permitted, so authorities may hold either a lottery or an auction to determine the license holders. The price of the permits can escalate into the tens of thousands of dollars in the United States. International mountain sheep permits have been known to cost $25,000.[5] Check with the local authorities in your area. If you succeed in securing a license and bagging a bighorn sheep, handle it in much the same way as you would a deer, using the techniques described in this chapter and in Chapters 7 and 16 — with one exception. Be extremely careful in removing the head and pelt because these are your "trophies."

Buffalo (Bison)

The hunting of buffalo in the United States is controlled. There are several herds in state and national parks, as well as many private herds. More private herd owners are encouraging hunting, with the cost running from $700 up to $3,000 or more per hunt, depending on the location. For this price the hunter will probably be guaranteed a buffalo and will sometimes be given the hide and the head. However, the hunter often must pay for the meat at a price several times higher than the price of equivalent cut(s) of beef. Buffalo meat, particularly buffalo burgers, is becoming more popular because of its perceived health benefits. (See Chapter 23 for a comparison of the nutritional value of all muscle foods.) Buffalo are very difficult to handle and move. Seven-foot heavy fences are required simply to hold them for growing, since they are capable of jumping 6 feet high or more. For moving them in confined areas, such as in corrals and chutes, even more sturdy closed, curved fencing is needed (see Chapter 3).

There are approximately one-quarter of a million buffalo in the United States, and the number is growing. It has been reported[6] that buffalo meat is the best seller of game meat even though it is two or three times more expensive than beef. The high price arises from the difficulty in raising the animal, as well as its relative scarcity (there are nearly seven times as many cattle har-

---

[5]Marshall, E. 1990. Mountain sheep experts draw hunter's fire. Science 248 (No. 4954, April 27):437.

[6]Deli Business, May 1999, as reported by the National Meat Association Lean Trimmings, June 28, 1999.

vested in a day as there are buffalo harvested in a year). Buffalo is merchandised in the same retail cuts as beef (see Chapter 15).

If you contemplate going into the buffalo business, you should consult with an experienced producer in your area before beginning. Not too many meat processors are eager to "harvest" buffalo because they are so hard to handle and will break a lot of equipment. Once buffalo are harvested, disassembly proceeds similar to beef, except extreme care must be taken with the head and hide removal, because the head and the hide are two of the most valuable "trophy" by-products.

GAME BIRDS

Pheasant, quail, grouse, partridge, and wild turkeys account for the majority of game birds taken by hunters. Turtle doves, although very small, are also popular among hunters in many regions of the country. With the exception of the wild turkey, game birds are generally cleaned by removing the skin with the feathers intact, since this is much easier and faster than plucking.

Pheasants

Pheasants are probably the most popular of the upland game birds and, for their size, are one of the meatiest birds used for human consumption. These birds commonly will have a 1- to 2-pound carcass, with the majority of the meat on the breast. State hunting regulations regarding season and bag and possession limits vary considerably. *It is illegal to shoot a hen pheasant in most states.* For this reason, many states require the head or feet to be left on the carcass when it is being transported. It is the obligation of the hunter to know and abide by the regulations of the state in which he or she is hunting.

There are two popular methods of cleaning pheasants. After the bird has been skinned, follow the same procedures outlined for dressing poultry (see Chapter 9). An alternative method of processing pheasants is faster and neater and preferred by many hunters because they do not have to handle the entrails. However, this procedure may not be legal if the heads are required to be left on the carcasses that are transported across state lines. This alternative dressing procedure is as follows:

- Skin the bird by making a small lateral incision on the underside of the breast and then pulling the skin and feathers off the carcass.

- From the top (dorsal) side, cut down both sides of the back, starting at the cranial end and cutting through to the last rib.

- Separate the carcass by pulling the breast apart from the neck, back, and legs. The intestinal tract, heart, lungs, and liver will remain attached to the back portion.

- Remove the feet and lower legs at the joint below the drumsticks.

- Cut off each leg from the back portion by cutting immediately adjacent to the back and through the ball and socket joint.

Fig. 10–13. The three-piece dressing method for pheasants yields one breast-wing portion and two leg (thigh) pieces. This method is preferred by many experienced pheasant hunters because it is faster and cleaner than the conventional method.

Dressing pheasants in this manner will produce three pieces (two thigh and drumstick pieces and one breast) from each bird (see Figure 10–13).

The back and neck are discarded, since they contain very little meat (see Figure 10–14).

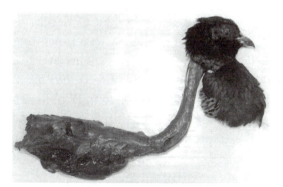

Fig. 10–14. The pheasant's neck, back, and viscera remain in one piece and are discarded. There is very little salvageable meat on a pheasant's neck and back.

Quail, Grouse, Partridge, Doves

These smaller game birds, like pheasants, have the majority of their carcass weight in the breast. Most hunters prefer removing the skin with the feathers rather than performing the more tedious task of dry-picking. These birds are generally dressed by removing the entrails from the whole carcass. In the case of doves (and sometimes quail) only the breast portion is kept for human consumption because of the small size of these birds and the limited value of the remaining portions.

Waterfowl

Ducks and geese are among the most difficult of birds to clean. If the skin is removed, the meat will dry out during roasting. For this reason, ducks and geese are usually scalded or dry-picked and dipped in wax to remove the feathers (see Chapter 9).

RABBITS

Wild rabbits, taken as part of the hunter's bag, can be dressed in the same manner as domestic rabbits. Wild rabbits will exhibit a darker meat that is often tougher than the meat of domesticated rabbits.

A communicable disease known as *tularemia* is prevalent in wild rabbits. While dressing rabbits that have the disease, humans can become infected through abrasions in their skin. Tularemia is transmitted from one rabbit to another by the rabbit louse or tick. The disease has not been observed in domestic rabbits, and the disease organism is destroyed when the rabbit is cooked.

Domestic rabbits are one of the most efficient meat producers of all animals. They have very high reproductive rates, averaging 48 to 64 offspring per year. Young rabbits can reach 4- to 5-pound market weights at eight weeks of age. Rabbits are also highly efficient converters of feed into meat, requiring approximately 2½ pounds to produce 1 pound of gain. Rabbits can be fed high-forage, low-grain diets, which are not competitive with human food needs, and still maintain high levels of production efficiency.

A symposium entitled "A New Look at Rabbits in the World" was held at the 91st Annual Meeting of the American Society of Animal Science in Indianapolis on July 22, 1999. At this symposium, one of the speakers, Pat Lamar, of the American Rabbit Breeders Association (http://www.arba.net/), provided this list of Web sites for those interested in raising rabbits for meat:

Rabbit Ranching	http://www.ardeng.net/rabbitranch/
Rabbit Farming	http://www.rabbitfarming.com/

Commercial Rabbit Industries   http://www.3-cities.com/~fuzyfarm

The Rabbit Web   http://www.rabbitweb.net

Another speaker at this symposium, Dr. Steven D. Lukefahr, of Texas A & M University–Kingsville <steven.lukefahr@tamuk>, gave a very informative presentation entitled "Unique Opportunities for Meat Rabbit Projects in Developing Countries." Dr. Lukefahr is coauthor of the textbook *Rabbit Production*.[7]

Consumption of rabbit meat in the United States remains very low, probably because many people perceive the rabbit to be a pet animal rather than a meat animal.

Inspection

Rabbit inspection is identical in scope and completeness to poultry inspection, which is discussed in detail in Chapter 3. The mark of inspection is identical for both poultry and rabbits and is illustrated in Chapter 3.

Handling

Never lift a rabbit by the ears or the legs. Grasp a fold of skin over the rabbit's shoulders, support the rump with your free hand, and hold the back of the rabbit against your body.

Dressing Rabbits

The method of harvesting rabbits consists of the following:

- Give a sharp blow to the top of the rabbit's head to stun it.
- Make an incision at the rear of the hock between the bone and the tendon.
- Suspend the carcass by hanging it on a hook through the hock.
- Sever the head at the atlas joint.
- Remove the free rear leg at the hock joint.

---

[7]J. I. McNitt, N. M. Patton, S. D. Lukefahr, and P. R. Cheeke. 2000. Rabbit Production, 8th ed. Interstate Publishers, Inc., Danville, Illinois.

- Remove the tail and forelegs (knee joint).
- Cut the skin on the rear of the loose leg to the base of the tail and up the rear of the suspended leg (see Figure 10–15).

Fig. 10–15. The procedure followed in skinning a rabbit. (Courtesy, U.S. Fish and Wildlife Service)

- Pull the edges of the cut skin away from the flesh and down over the carcass. Make no other cuts in the skin.
- Eviscerate by opening the median line of the belly, leaving the heart, liver, and kidneys in the carcass.
- Remove the suspended rear leg, and then rinse the carcass in cold water to remove any hair or blood.
- Joint the carcass by removing the forelegs and hindlegs, cutting the loin in one piece, and separating the shoulders.

Pelts

Rabbit pelts have a fur skin value. A small pelt should be stretched on a thin board or wire stretcher 24 inches long and 4 inches wide at the narrow end and 7 inches wide at the base. The skins of 10- and 12-pound rabbits need a board 30 inches long, 4 inches wide at the narrow end, and 9 inches wide at the base. Stretch the warm skin on the board with the fore part over the narrow end, smoothing out the wrinkles. Place both front legs on one side of the shaping board. Remove any surplus fat, and make sure that the skin dries flat. Do not dry the skin in the sun or in artificial heat, and do not use salt. When dry, the skin can be stored in a tight box, but each layer should be sprinkled with naphtha soap flakes or moth balls to ward off moths.

FISH

Young people especially enjoy fishing, and once "hooked" on fishing, an individual may have the urge to go fishing the rest of his or her life. Fish constitute the largest and most diverse class of vertebrates. Figure 10–16 shows the nomenclature of the main fish parts. The over 20,000 species create an enormous spectra of ideal living conditions, life style, size, and longevity. Most people who fish crave a limited number of these many species.

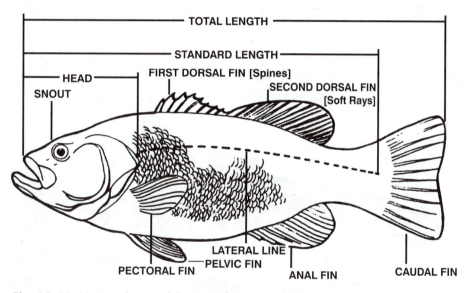

Fig. 10–16. Nomenclature of the main fish parts. (Courtesy, Reynolds and Tainter)

Figure 10–17 shows the identifying features of a channel catfish, a species that has become very popular because of its adaptability to farm production (aquaculture). Once confined to the rural south, this firm-fleshed fish can taste muddy unless farmed. Almost all catfish consumed in the United States is farm raised and is served in posh restaurants. According to the Catfish Institute, catfish is now the fifth most popular fish in the United States, behind tuna, pollack (used to make imitation lobster and crab meat), salmon, and cod.

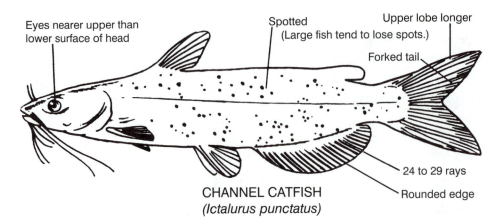

CHANNEL CATFISH
(Ictalurus punctatus)

Fig. 10–17. Identifying features of a channel catfish. Reproductive length, 12 to 15 inches; number of eggs per female, 2,500 to 70,000; number of days to hatch, 6 to 10; time of spawning, spring; type of reproduction, nest; average life span, eight years; average size, 1 to 2 pounds; maximum size, 50 pounds; principal food eaten, insects; most suitable habitat, rivers. (From *What Fish Is This?* Illinois Department of Natural Resources, Division of Fisheries, 3078118-50M-5/99, Springfield, IL 62701-1787)

Figure 10–18 shows the identifying features of the Atlantic salmon, also known as landlocked salmon. Pounds of U.S. farm-raised salmon (a carnivore) is much less than pounds of U.S. farm-raised catfish.

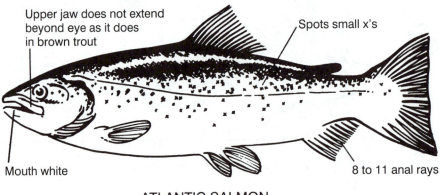

ATLANTIC SALMON
(Salmo salar)

Fig. 10–18. Identifying features of the Atlantic salmon, also known as landlocked salmon. Reproductive age, three years; number of eggs per female, 2,000; number of days to hatch, 90 to 120; time of spawning, fall; type of reproduction, eggs buried in gravel; average life span, seven years; average size, 4 pounds; maximum size, 100 pounds; principal food eaten, fish; most suitable habitat, Lake Michigan. (From *What Fish Is This?* Illinois Department of Natural Resources, Division of Fisheries, 3078118-50M-5/99, Springfield, IL 62701-1787)

Fish farms are found in all 50 states. Nevertheless, probably more than half of the fish on the U.S. menu is imported. The Bering Sea, which separates Russia and Alaska, is the most productive source of pollack, and many countries covet this source. The Economic Research Service (ERS) of the USDA (http://usda.mannlib.cornell.edu/reports/livestock/ldp-aqs/) publishes *Aquaculture Outlook* twice a year.

Fishing seasons are more plentiful and timely than hunting seasons, so your favorite fish are probably available several times during the year.

Fish consumption has increased in recent years (see Chapter 1) primarily because "fish farming" has increased, making fish more available, and because eating fish is perceived by more people as being good for their health.[8] If fish are properly handled throughout the entire time period between being caught and eaten by humans, they will indeed be a healthful food product.

The Food and Drug Administration (FDA), not the USDA-FSIS, protects the microbiological safety of seafood. This is confusing to consumers and lends credence to the idea of having one organization responsible for all food safety.

In an attempt to make government food inspection consistent, FDA regulations do require a HACCP plan for seafood processors. That is why some chefs and consumers prefer a "fresh" farm-raised fish rather than a wild, imported fish that has been sitting around an airport or seaport for an unknown period.

Handling and Cleaning[9]

As soon as fish are caught, their quality begins to decline. They should be handled carefully because they bruise easily. They should be kept out of the sun to retard deterioration. They should be cleaned, dressed, and washed; packed in ice; and then refrigerated.

The following are the basic steps for cleaning fish.

1. Wash the fish in clean, drinking-quality water. Figure 10-19a shows a round (whole) fish.

---

[8]S. Specht-Overholt, J. R. Romans, M. J. Marchello, R. S. Izard, M. G. Crews, D. M. Simon, W. J. Costello, and P. D. Evenson. 1997. Fatty acid composition of commercially manufactured omega-3 enriched pork products, haddock, and mackerel. J. Anim. Sci. 75:2335-2343.

[9]Reynolds, A. E., and S. Tainter. 1978. Freshwater Fish Preservation. Coop. Ext. Serv. Publ. No. E–1180. Michigan State Univ., East Lansing.

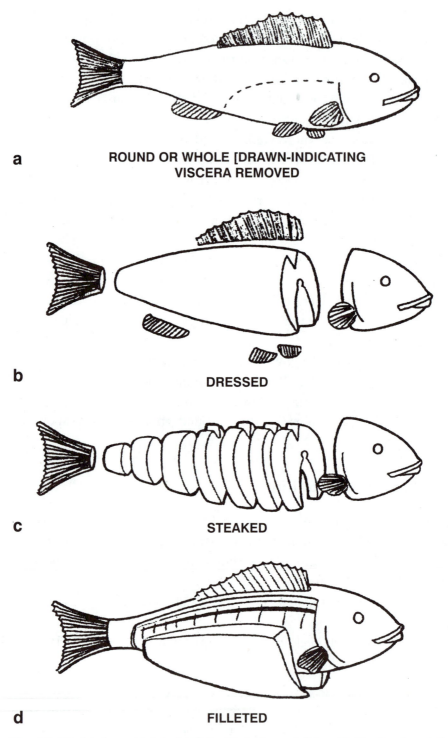

a   **ROUND OR WHOLE [DRAWN-INDICATING VISCERA REMOVED**

b   **DRESSED**

c   **STEAKED**

d   **FILLETED**

Fig. 10–19. (a) A round (whole) fish. (b) A dressed fish with the head and fins removed. (c) Large fish may be steaked. Cut fish crosswise into portions about 1 inch thick. (d) Fish may be filleted. (Courtesy, Reynolds and Tainter)

2. Scale or skin the fish. To remove the scales, scrape with the dull edge of a knife from the tail to the head. If you skin the fish, scaling is not necessary. To skin, split down the back, and loosen around the fins. Then remove the skin with pliers, pulling from the head to the tail.

3. Cut the entire length of the belly from vent to head. Cut out the viscera and gills if the head is not removed.

4. To remove the head, cut to the backbone above the collarbone. Break the backbone over the edge of a table or a cutting board, then cut any tissue holding the head to the body.

5. Cut along each side of the dorsal (large back) fin and then pull out the fin. The fins should never be trimmed off because the bones at the base of the fins will be left in the fish. Figure 10-19b shows a dressed fish with the head and fins removed.

6. Wash the fish thoroughly in cold running water.

7. To steak large fish, cut the fish crosswise into portions about 1 inch thick (see Figure 10–19c).

8. To fillet fish, cut down the back of the fish from the tail to the head with a sharp knife (see Figure 10–19d). You do not have to eviscerate the fish before you fillet it. Then, cut down the backbone, angling the knife to cut away the flesh from the backbone and running the knife over the rib bones. Lift off the side piece, which will free the fillet at the tail. After turning over the fish, cut the fillet from the other side. If you wish to skin the fillet, lay it skin-side down on a cutting board, and holding the tail end with your fingers, cut through the flesh to the skin. While holding the free end of the skin firmly between your fingers, flatten the knife on the skin, and run the knife forward to cut the flesh away from the skin.

9. Refrigerate the fish until it is cooked or preserved. The techniques for preserving and cooking fish are described later in this text (see Chapters 19 and 24, respectively).

A small, soft-covered book entitled *Going Wild*[10] makes an entertaining and useful addition to the library of avid hunters and fishers.

---

[10]Urban, G., and M. Marchello. 1987. Going Wild. Mateb Marketing, Inc., Sartell, MN 56377.

11

Packing House By-products

In many cases, the use and development of by-products by the meat industry means the difference between profit and loss for meat processors. Also, without meat by-products, consumers would lack a vast array of usable products that make their lives more comfortable.

Throughout history, animals have been utilized as a food source. Early peoples also made use of many animal products such as skins for clothing and shelter; bones and horns for tools; tendons and intestines for weapons, tools, and bindings; teeth, claws, feathers, and hair for ornaments; stomachs, bladders, and skins for containers. Nothing was wasted following a successful hunt for meat. Modern society learned well from those ancestors, for today's meat industry still utilizes the nonmuscle portions of livestock. Mr. Tony Javurek, who guided two tours daily at John Morrell's Sioux Falls packing plant for more than 20 years, expressed the meat industry philosophy about by-products during each tour. Tony always said, "We use all parts of the pig except the squeal and the curl in its tail."

Meat harvest by-products (*offal*) consist of all parts of the animal that are not included in the carcass. Cutting and processing of the carcass results in additional nonmuscle by-products, namely, fat, bone, and other connective tissues.

By-products play a dual role as both assets and liabilities in the meat industry. Modern living is enhanced by many nonmeat products, which originate from the meat industry. However, environmental concerns, energy costs, and increased volume have combined to make by-product handling a major economic and management problem in the meat industry. A hog-harvest operation with a 1,000-head-per-hour capacity must be able to process approximately 72,000 pounds of by-product material per hour. The by-product varies from edible muscle and organ meats to hair and manure, therefore requiring a myriad of physical movement, sanitation, and processing technologies.

Meat industry by-product handling practices have changed dramatically since 1896, when *The National Provisioner* reported that a man in South Omaha made $10,000 yearly by skimming grease from the sewers that carried refuse from the big packing houses into the Missouri River. That same year, the Butchers Board of Trade of San Francisco was considering a sanitary improvement in the form of a sewer that would empty far out into the bay. Offal from the harvest plants was being dropped into the bay near shore and was drifting to and fro with the tide, causing unpleasant odors.[1]

Today, economics, modern technology, and industrial concern for the environment result in maximum salvage and utilization of all by-product materials. In many communities, the air and water effluent (flow out) from meat operations must be as clean as or cleaner than the water and air entering the plant.

Processed by-products have been a significant source of income to the meat processing industry. Edible offal contributes 4.9 percent of the total value of beef to the packer, inedible offal returns 8.44 percent (over 50 percent of this is from the sale of the hide; see Table 11–14), and the rendered products provide 1.73 percent (half is dried blood value), for a total return from by-products of 15.07 percent of the total sale value of the products of a harvested beef animal.[2] Beef harvest operations cover the harvest costs, profits, and part of the purchase cost of the livestock with sales of by-product material (see Chapter 13). The meat industry continues to seek new uses and markets for the many nonmeat items that result from the disassembly lines.

Rendered products are used in four major sectors of today's economy[3]:

1. The human food industry (edible tallow, gelatin, defatted meat tissue, etc.).

---

[1]Meat for the multitudes II. July 4, 1981. The Natl. Prov. 185 (1):60.

[2]Pietraszek, G. April 22, 1989. Review of edible meat by-products. The Natl. Prov. Pearson, A. M., and T. R. Dutson (Ed.). 200 (16):17.

[3]Halpin, K. M., J. E. Sullivan, R. Bradfield, and Q. Liu. 1997. Expanding the meat–by-product market with pet foods. In: Proceedings of the 1997 Meat Industry Research Conference, Chicago, p. 51.

2. The livestock, poultry, and pet food industries.

3. Soaps and personal-care products.

4. Industrial production of metals, rubber, pesticides, and fertilizers.

It is difficult to cut production and distribution expenses for the majority of commodities; hence, one of the most important opportunities for gaining a competitive advantage, or even for enabling an industry or individual business to maintain its position in this new competition, is to reduce its manufacturing expense by creating new credits for products previously unmarketable. For some individual businesses, this manufacture of by-products has turned waste into such a source of revenue that the by-products have become more profitable per pound than the main product.

Some of the uses of various meat by-products and the techniques by which these by-products are transformed to serve these uses are discussed in the remainder of this chapter.

EDIBLE BY-PRODUCTS/VARIETY MEATS

Edible by-products, sometimes referred to as "variety meats," include livers, hearts, tongues, brains, sweetbreads (thymus and/or pancreas), lamb and calf fries (testicles, often called mountain oysters), kidneys, oxtails, tripe (stomach), chitterlings and natural casings (intestines), and, in some societies, blood. Meat trimmings derived from heads are used in processed meats and are also characterized as "edible offal" or "edible by-product items." Edible fats removed during harvest include pork leaf fat (kidney fat) and the caul fat shrouding the rumen and stomach. The beef grading regulations (see Chapter 12) may permit significant amounts of surface (subcutaneous) fat and internal fats to be removed during the harvest process. This procedure, called "hot fat trimming," may facilitate rapid chilling, seam fat removal, and peripheral muscle removal, as well as achieve labor efficiencies.

All edible by-products must be wholesome and must be segregated, chilled, and processed as specified by the FSIS (see Chapter 3). Several edible offal items are illustrated in Figure 11–1. The yields of some edible offal items for beef and pork are listed in Table 11–1.

Variety meats are economical sources of valuable nutrients (see Tables 11–2, 23–4, and 23–7). Liver is the most nutritious of all meat items. The *nutrient density* of liver exceeds that of muscle meats, which is high. One

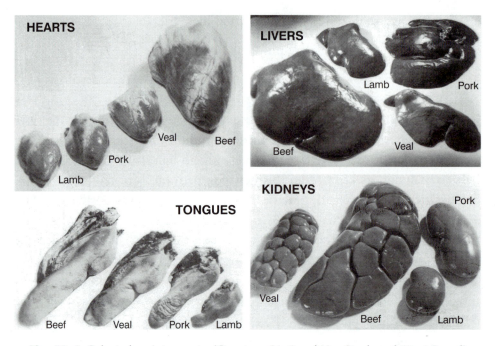

Fig. 11–1. Selected variety meats. (Courtesy, National Live Stock and Meat Board)

Table 11–1.   By-product Yields

By-products	1,150-Pound Beef				230-Pound Hog	
	(lbs.)	*(%)*[1]	*(lbs.)*	*(%)*[2]	*(lbs.)*	*(%)*[1]
Blood	46.0	4.0	28.75	2.5	7.0	3.04
Brain	1.3	0.11		NA	0.25	0.11
Edible harvest fat	40.3	3.5		NA	8.0	3.5
Hooves/feet		NA	19.5	1.7	5.0	2.17
Head and cheek meat	5.2	0.45	6.9	0.6	1.38	0.6
Heart	5.8	0.50	5.8	0.5	0.6	0.25
Hide or hair	80.5	7.0	86.2	7.5[3]	2.0	0.87
Intestine		NA	74.8	6.5	4.2	1.83
Kidney	1.4	0.12	2.3	0.2	0.5	0.22
Liver	12.6	1.10	16.1	1.4	3.25	1.4
Lungs	6.9	0.60	10.4	0.9	1.0	0.43
Spleen	1.9	0.17	2.2	0.19		NA
Sweetbread (thymus)	0.3	0.03	0.8	0.07		NA

(Continued)

Table 11–1   (Continued)

By-products	1,150-Pound Beef				230-Pound Hog	
	(lbs.)	(%)[1]	(lbs.)	(%)[2]	(lbs.)	(%)[1]
Tail	1.7	0.15	3.0	0.26	0.25	0.1
Tongue	4.0	0.35	4.1	0.36[3]	0.75	0.3
Tripe (stomach)	23.0	2.0	23.0	2.0	1.5	0.65
Weasand (esophagus)	0.5	0.04	0.3	0.03	0.13	0.05
Manure	51.8	4.5		NA	4.0	1.74
Inedible raw material	149.5	13.0		NA	23.0	10.0
Rendered fat, edible	28.8	2.5		NA	28.0	12.17
Rendered fat, inedible	46.0	4.0		NA	6.0	2.61
Cracklings	34.5	3.0		NA	5.0	2.17

[1]Packers Engineering and Equipment Co., Inc., catalog, and Sisson and Grossman. 1953. Anatomy of Domestic Animals. W. B. Saunders Co.

[2]Terry, et. al. 1990. Yields of by-products from different cattle types. J. Anim. Sci. 68:4200.

[3]Trimmed.

3½-ounce (100-gram) serving of beef liver provides 0.85 to 0.9 ounce (24 to 25 grams) of high-quality protein, while contributing only approximately 160 calories to the diet. An excellent source of readily digested heme iron, liver also provides B vitamins, particularly B_{12}, as well as vitamin A to consumers who enjoy its unique flavor. (Heme is the O_2 carrying component of hemoglobin and myoglobin and is a source of iron more readily absorbed by the human digestive system than most dietary iron forms.) Variety meats are also relatively high in protein, with the exception of brain and tripe. More extensive use of meat animal by-products for human food has been proposed as one means of solving world nutrition problems. In addition to enhancing human nutrition, new developments in meat by-product utilization would increase the overall efficiency of livestock production.

Although many consumers in the United States purchase and use variety meats in their meal planning, larger amounts of some by-products are utilized in processed meats, principally sausage (see Chapter 21). FSIS regulations require that the label on all processed meats containing variety meats indicate in the ingredient list each specific variety meat that is included in the formulation.

People in other societies of the world, particularly those in European countries and in China and Japan, consume a much larger amount of variety

Table 11-2. Proximate Protein, Fat, and Calorie Content of 100 Grams of Cooked Variety Meats

Cooked Variety Meats	Protein	Fat	Calories
	----------- (grams) -----------		
Veal[1]			
Brain...............	10.5	7.4	111
Heart	26.3	4.5	153
Kidney.............	26.3	5.9	165
Liver	21.5	7.6	160
Lung	18.8	2.6	104
Pancreas	29.1	14.6	256
Spleen.............	23.9	2.6	125
Sweetbread (thymus)	18.4	2.9	105
Tongue	26.2	8.3	187
Lamb[1]			
Brain...............	12.7	9.2	137
Heart	21.7	5.2	140
Kidney.............	23.1	3.4	129
Liver	23.7	10.9	199
Lung	20.9	3.0	116
Pancreas	23.3	9.6	186
Spleen.............	27.3	3.8	150
Tongue	21.5	20.5	277
Turkey[2]			
Gizzard	29.4	3.9	163
Heart	26.8	6.1	177
Liver	24.0	5.9	169
Neck (meat only)	26.8	7.3	180

[1]B. Bloch. The Meat Board Meat Book. 1977. National Live Stock and Meat Board.
[2]USDA. 1979. Composition of Foods, Poultry. Agric. Hdbk. No. 8–5.

meats per capita than do people in the United States. (For U.S. consumption, see Table 1–4.) In fact, whole pig heads are popular in China. Therefore, the U.S. meat industry has developed overseas markets for those nutritious by-products of animal agriculture. Table 11–3 indicates that edible offal items (variety meats) produced a positive international trade balance of $400 to $600 million for the United States in recent years.

Table 11–3. Value of U.S. Imports and Exports of Meat By-products, 1990–1998[1]

By-products	Imports				Exports			
	1990	1996	1997	1998	1990	1996	1997	1998
	- ($1,000,000) -							
Variety meats	75.7	108.9	141.7	124.8	361.0	714.2	595.2	595.6
Tallow, grease, and lard	72.3	15.2	21.9	15.7	401.3	488.0	437.9	545.5
Casings	53.3	65.2	72.5	76.7	32.3	50.2	59.2	61.7
Hides, skins, and leather	86.8	139.3	133.4	104.9	2,237.4	1,308.8	1,326.9	1,018.8
Wool, mohair, and other hair	182.1	152.0	154.2	126.8	106.3	79.1	61.4	46.7
Total	470.2	480.6	523.7	448.9	3,138.3	2,640.3	2,480.6	2,268.3

[1]**Livestock, meat, and related products — share of total export value, 1998:** red meats, 39%; variety meats, 6%; live animals, 6%; poultry and poultry meats, 24%; animal fats, 6%; hides and skins, 11%; other, 8%.
Source: Meat and Poultry Facts. 1991, 1999. American Meat Institute.

Variety meats are generally more perishable than other meats and should be cooked and served soon after purchase. Precooking fresh variety meats after purchase will extend storage times. Variety meats are processed under the same rigorous meat inspection criteria as other meat items (see Chapter 3). A very brief description of some preparation techniques for variety meats is included here.

Liver

Liver is most often sliced, relatively thin, after the outer connective tissue membrane has been removed. The slices may be cooked by a variety of methods, such as frying, broiling, sauteing, and braising. Liver may be ground or chopped and added as an ingredient to loaves, sausages, spreads, or other dishes.

Heart

Heart is generally less tender than liver and requires moist heat and long-term cookery. Heart cavities that have been slashed open for inspection may be filled with a dressing and stitched shut and then roasted (in the presence of moisture) like a turkey. The meat has an excellent flavor, and the dressing takes on the "hearty" flavor of the meat.

Tongue

After braising, both heart and tongue (see Figure 11–2), thinly sliced, are excellent cold sandwich meats. Garnishes, mustard, horseradish, or other dressings make those cold sandwiches truly unique snacks or lunch fare. Removal of the tough outer membrane of the tongue is facilitated by blanching (short exposure to boiling water) prior to long-term, moist-heat cookery. Tongue may be available cured or pickled in some market areas.

Fig. 11–2. Beef tongues and brains are individually bagged, placed in shipping boxes, and frozen after a postmortem chill period. (Courtesy, IBP Corp., Dakota City, Nebraska)

Brains, Sweetbreads, and Fries

Brains, sweetbreads (beef and veal thymus and pork pancreas glands), and fries (testicles, sometimes known as mountain oysters, or Rocky Mountain oysters) are often sliced thinly and dipped in batter or flour (breaded) before being deep fried. Precooking aids in removing the outer membranes of sweetbreads and brains (see Figure 11–2) as well as "setting" the soft, delicate products for easier slicing. These variety meats may be broken up into small pieces and mixed with scrambled eggs.

Kidneys

Kidneys are sometimes left in lamb and veal loins to become part of the loin or "kidney chops." Kidneys may be included as one of the ingredients in meat casseroles, stews, and pies. Lamb and veal kidneys may be broiled or skewered because they are more tender than beef kidneys.

Oxtail

Oxtail is an unsurpassed source of rich, meaty flavor and texture for soups and stews. The tail sections should be browned (not essential), then allowed to simmer until the meat is tender and separated from the bone. The bone portions can be removed prior to serving. The stock (with or without the meat) may be added to other soup ingredients. Bones located throughout the carcass are also suitable, better if they have some adhering meat, for the preparation of soup stock.

Tripe

Tripe is the name applied to edible stomach tissue. Tripe is derived from the first (rumen) and second (reticulum) compartments of the four-part stomach of cattle and the stomach of hogs (maws). Following evisceration, the organs are emptied, thoroughly washed, and rapidly chilled. The inner surface layer (membrane) is also removed in the processing of stomachs for tripe. Tripe is sold by packers fresh, partially cooked, or fully cooked. It is less tender and has a firmer texture than other variety meats. Best cooked with moist heat, it may be served along with various sauces and dressings; it may be the main item in various meat dishes, as indicated for kidneys; and it may be added to spicy soups.

Chitterlings (Intestines)

One source identifies chitterlings as beef or pork entrails used as food.[4] Pork intestines are thoroughly cleaned and cooked with sauces to be served as chitterlings. After being precooked in water with seasonings, the intestines may be cut into small pieces, breaded with raw egg and crumbs, and then deep fat fried.

Sausage Materials

Intestines, as well as other portions of the digestive tract, are used as casings for sausage products (see Chapter 21). Many specialty sausages are derived largely from edible offal items. Jellied products, such as headcheese, souse, and scrapple, may rely on meat sources with high collagen content, such as pork skin trimmings and head trims, for unique identifying characteristics. *Haggis* is a Scottish food made from the hearts, lungs, and livers of sheep or calves, which is highly seasoned, mixed with oatmeal, and then boiled in a sheep stomach.

Blood

According to the USDA–FSIS, no blood that comes in contact with the surface of the body of an animal or is otherwise contaminated can be used for food purposes. Blood contains around 17 percent protein. Only blood from inspected animals may be used for meat food products. The defibrination of blood intended for food purposes cannot be performed with the hands. In some European countries and in the former USSR, blood proteins are utilized as food ingredients to a greater degree than they are in countries in the Western Hemisphere. Those countries have developed collection systems for edible blood, which in many cases comply with U.S. meat inspection requirements. Some of these systems utilize cannula-like funnel devices for blood removal, and some draw the blood directly from the animals' vascular system into sterilized vacuumized containers.

Blood proteins can be utilized as emulsifiers in human foods, but the procedures involved in doing this and the products that result are not widely uti-

---

[4]Shannon, E. 1962. Dictionary of Culinary Terms. A. S. Barnes, San Diego, p. 46.

lized in U.S. industry. In areas of the former USSR, 40 percent of all harvest blood is recovered for human food. The use of blood protein as an ingredient in foods is under investigation in the United States. As evidence of this, a cake recipe in which ingredients derived from blood are used received nationwide coverage.

ANIMAL FATS

Animal fats have long served as a food energy source; however, our ancestors found many other uses for them. Melted fat, or grease, rubbed into the hides made the tepee (*tipi* in Sioux) leather supple and waterproof. Bear grease, or other grease, provided a sheen and a unique odor to the hair of our ancestors for many generations. Today, the lanolin in skin preparations and cosmetics of all kinds is a component of the lipid extracted from wool. The rendering (extraction) process originated with the discovery of soap making techniques, which involved boiling fat and ashes together. Animal and fish oils provided the fuel for the oil lamps that preceded the kerosene, gasoline, propane, and electric lamps of modern years. At one time during the twentieth century, the value of the fat derived from hog harvest and cutting was equal to the meat value. Of course, hogs were fatter and meat was cheaper in those days than they are now. In California's early history, cattle were harvested for the hide and tallow, and the meat was left to rot.

Fats derived from animal sources and plant sources differ, mainly, in melting point and degree of saturation. Fats and oils are combinations of saturated and unsaturated, long-chain fatty acids that are chemically bonded to the three-carbon alcohol, glycerol, to form triglycerides (three fatty acids bound to one glycerol). No double bonds occur between the carbon atoms (from 4 to 22+) that form saturated fatty acid molecules. Contrastingly, double bonds are found between some of the carbons of unsaturated fatty acid molecules. The double bonds are less stable than single bonds and are responsible for reducing the melting point of soft fats (containing unsaturated fatty acids) compared to firmer fats (made, principally, of saturated fatty acids). Because many of the commercially available lipids derived from plants and fish have a greater degree of unsaturation, they are liquid at room temperatures and are referred to as oils. Animal-derived lipids are usually solid at room temperature because of the greater degree of saturation and are called fats. (See also Chapter 23.)

Rendering

In the meat industry, the fat is extracted from almost all by-products that have no other use or market potential. The extraction technique used is called *rendering*. Rendering usually involves the heating or cooking of the raw material to liquefy the fat and to break down membranes or other structures that may hold the fat. The melted fat can then be drawn off, leaving the solid or semi-solid nonfat materials. In animal tissue, the nonfat material is mainly protein and mineral. Rendering serves two purposes: it separates fat and protein, and it cooks the animal tissue. The separated fat and protein have greater value than the fresh raw material. Cooking significantly increases the storage stability of the fat and protein products by killing the microorganisms that were present in the raw material and by removing most of the moisture.

The rendering process has evolved along with all other aspects of the meat industry. Early rendering was accomplished by direct application of heat to containers holding the animal tissue. The next step in the evolution of the rendering process was the injection of live (pressurized) steam into the product. The steam increased the rate of temperature rise and speeded up the process, but it also added moisture and resulted in overheating some portions of the product. The tank water that accumulated in wet rendering was evaporated and reduced to the consistency of molasses. That fluid, referred to as "stick" or "stick liquor," was mixed with the tankage (precipitated solids) and dried.

Confinement of the steam in a jacket around the vessel, which contained the constantly stirred product, was a further improvement. Closing the vessel and drawing a vacuum on the inside of the steam-jacketed chamber reduced the boiling point of the moisture, and the moisture vapor was evacuated. The rendering process was therefore further shortened, and the time and(or) temperatures required were reduced. Dry rendering was used extensively in the manufacture of lard and in the extraction of edible (for humans) and inedible fats and oils. Additional advantages over wet rendering (live–steam injection into the product) are that it eliminates the expense and problems of handling tank water and it reduces odor problems.

Many rendering operations utilize large, steam-jacketed tanks, which can be charged with thousands of pounds of fresh raw materials, closed, and vacuumized. The product is constantly agitated during the cooking process. At the end of the cycle, the liquid fat is removed, and the solid protein material is mechanically pressed to squeeze out as much fat as possible. The pressed solid is ground and sold as high-protein meal. The areas of the meat plant where the rendering takes place are traditionally referred to as "tank houses" or "tank rooms." The product destined to be rendered is to be "tanked."

Although rendering has historically been a batch process, continuous rendering systems are now in use. The particle size of the raw material is reduced, and the product is heated rapidly in a variety of cookers, depending upon the system. The product moves through the cooker as heat is transferred to it from steam pipes and(or) steam-heated agitators inside the cooking chamber. The cooked mass flows or is pumped through "drainers" or "percolators" where the fluid fat runs off. The solids travel on to presses that remove the remaining fat, then through driers and grinders to be sold as meal. Some systems preheat the raw product to permit it to be pumped through a heat exchanger, which increases the product temperature in a short time interval. The liquefied material is separated into solids, water, and fat by centrifugation. The fat temperature may be reduced through another heat exchanger, minimizing the time–temperature impact on fat quality. Not only is the fat produced from this system of exceptional quality, but also the minimal heat denaturation of the protein fraction results in a material that may be utilized in some edible meat products, if edible raw fats were processed. The edible protein material is called "partially defatted beef fatty tissue (PDBFT)" or, in the case of pork, "PDPFT." The remaining fat in that material may also be mechanically extracted (pressed) from the protein material for the production of meat meal. Other "low-temperature" continuous systems press the raw product first, removing about 45 percent of the water, and subsequently heat the dried raw material by using lower temperatures in long cookers and adding hot fat or water to the cold mass. The melted mix is then separated through percolators and(or) centrifuges. The protein fraction is pressed and dried. Continuous systems are used to process both edible and inedible raw materials. A large meat processor has introduced a low-fat beef product that has been separated from high fat content trimmings. It can be used as an ingredient in either fresh or processed meat products. Solvent extraction systems for inedible fatty materials have also been used successfully in the meat and rendering industries.

Meat animal harvest and processing operations usually have two completely separate rendering operations within their facilities, one for edible (for humans) raw materials and one for inedible raw materials. Edible raw materials must be segregated, chilled, and handled as designated by FSIS during the harvest and cutting procedures that generate them. The edible rendering facilities must be physically separate from any areas of the plant where inedible raw materials and products are processed or stored. However, some meat plants may only process edible materials onsite and will send the inedible by-products to another renderer.

A major segment of the animal products processing industry is made up of rendering operations that are not operated by members of the meat industry and that render a wide variety of material, including dead animals, meat retail trimmings and bone, as well as restaurant and garbage greases. Renderers recycle highly perishable waste material, much of it originating with the livestock and meat industry, into inedible, rendered lipid and protein products. For further information about the rendering industry, check these Web page addresses provided by *Render Magazine:*

American Feed Industry Association	http://www.afia.org
American Meat Institute	http://www.meatami.org
Animal and Plant Health Inspection Services	http://www.aphis.usda.gov
Animal Protein Producers Industry	http://www.animalprotein.org
Center for Veterinary Medicine	http://www.fda.gov/cvm
Fats and Proteins Research Foundation	http://www.fprf.org
Food and Drug Administration	http://www.fda.gov
Food Safety and Inspection Service	http://www.fsis.usda.gov
National Cattlemen's Beef Association	http://www.beef.org
National Renderers Association	http://www.renderers.org
Occupational Safety and Health Administration	http://www.osha.gov
Render Magazine	http://www.rendermagazine.com
U.S. Department of Agriculture	http://www.usda.gov

LARD

Trends in Lard Production and Consumption

Lard production and consumption steadily declined (see Table 11–4) until 1985, when production stabilized at 10 to 11 pounds per hog and per capita consumption stabilized at 1 to 2 pounds per year. A market-weight hog grading U.S. No. 1 yields about 7 pounds of lard per 100 pounds live weight, whereas the same weight hog grading U.S. No. 3 yields about 14 pounds (Chapters 12 and 14).

Table 11–4. Trend in U.S. Lard Production and Consumption

Year	Total Hog Harvest	Total Lard Production	Lard Production per Hog[1]	Consumption[2] Total	Consumption[2] per Person	Civilian Population July 1
	(1,000 hd.)	(mil. lbs.)	(lbs.)	(mil. lbs.)	(lbs.)	(mil.)
1950	79,263	2,631	33.2	1,891	12.6	150.2
1955	81,051	2,660	32.6	1,639	10.1	162.3
1960	84,150	2,562	30.4	1,358	7.6	178.1
1965	76,462	2,045	26.7	1,225	6.4	191.5
1970	87,012	1,913	22.0	939	4.7	201.9
1975	69,823	1,012	14.5	633	3.0	213.8
1980	97,174	1,217	12.5	594	2.6	225.7
1985	84,938	929	10.9	422	1.8	236.2
1986	79,956	874	10.9	417	1.7	238.4
1987	81,422	862	10.6	441	1.8	240.6
1988	88,136	938	10.6	433	1.8	242.9
1989	89,007	935	9.9	442	1.8	245.1
1990	85,481	919	10.1	468	1.2	249.4
1991	88,445	952	10.8	429	1.7	252.1
1992	95,157	1,025	10.8	426	1.7	255.1
1993	93,296	1,005	10.8	449	1.7	257.7
1994	95,905	1,034	10.8	598	2.3	260.3
1995	96,535	1,040	10.8	585	2.2	262.8
1996	92,569	998	10.8	606	2.3	265.2
1997	92,125	993	10.8	615	2.3	267.7

[1]Computed by dividing total lard production by total hog harvest.

[2]As lard direct only and not including lard used in manufactured foodstuff.

Source: Agricultural Statistics. 1999. USDA. (Mike Easdale.)

The decrease in production, coupled with increased competition from fats of vegetable origin and consumer health/fat concerns, has brought about a decrease in consumption since 1950. However, lard does provide a source of energy and an essential fatty acid (linoleic) for the human consumer. Lard is an easily digested fat because its melting point is near body temperature. Because the fatty acids in lard vary in hardness, the cook or baker is able to take advantage of its wider "plastic" range at lower temperatures than some other fats provide. The value of lard as a shortening is indicated by the light, flaky crusts that result from its use. The shortening value of a pastry product ingredient is measured by its ability to reduce the force required to break a standard pie crust or cracker. Edible lard is, therefore, useful as a cooking fat, as a shortening, as a flavor ingredient in many foods, and as a nutritional source.

Official Definitions[5]

Lard, Leaf Lard

(a) Lard is the fat rendered from clean and sound edible tissues from swine. The tissues may be fresh, frozen, cooked, or prepared by other USDA–approved processes that will not result in the adulteration or misbranding of the lard. These tissues shall be reasonably free from blood and shall not include stomachs, livers, spleens, kidneys, and brains or settlings and skimmings. "Leaf lard" is lard prepared from fresh leaf (abdominal) fat. (*Leaf fat* is the meat industry terminology for pork kidney fat that is removed from the carcass during harvest.)

(b) Lard (when properly labeled) may be hardened by the use of lard stearin or hydrogenated lard or both and may contain refined lard and deodorized lard, but the labels should state such facts as applicable.

(c) Products labeled "lard" or "leaf lard" must have the following identity and quality characteristics to insure good color, odor, and taste of the finished product:

(1) *Color*: White when solid. Maximum 3 red units in a 5¼-inch cell on the Lovibond scale.

---

[5]Meat and Poultry Inspection Regulations. USDA.

(2) *Odor and taste:* Characteristic and free from foreign odors and flavors.

(3) *Free fatty acid:* Maximum 0.5 percent (as oleic) or 1.0 acid value, as milligrams KOH per gram of sample.

(4) *Peroxide value:* Maximum 5 (as milliequivalent of peroxide per kilogram fat).

(5) *Moisture and volatile matter:* Maximum 0.2 percent.

(6) *Insoluble impurities:* By appearance of liquid, fat, or maximum 0.05 percent.

(d) Products found upon inspection not to have the characteristics specified in paragraph (c) . . . but found to be otherwise sound and in compliance with paragraph (a) . . . may be further processed for the purpose of achieving such characteristics.

Rendered Animal Fat or Mixture Thereof

"Rendered animal fat," or any mixture of fats containing edible rendered animal fat, shall contain no added water, except that "puff pastry shortening," may contain not more than 10 percent water.

Stearin and Lard Oil

If rendered pork fats are held at a temperature of 90° to 100°F, the product will separate into liquid and solid material. Mechanical separation, by pressing, of the material will produce two commercially important products of animal fats. The solid material is called stearin, and the liquid is identified as lard oil (olein). Stearin is composed of glycerol esters of higher melting point, saturated fatty acids, principally stearic acid, whereas lard oil is made up of the glycerol esters of lower melting point, less saturated fatty acids, mainly oleic acid. Lard oil is used in the manufacture of margarine, as a burning oil, and as an industrial lubricant.

Antioxidants for Animal Fats

Animal fats, particularly those fats and oils that include unsaturated fatty acids, react chemically with oxygen. The reaction, oxidation, results in flavor and odor changes that are commonly referred to as rancidity. Exposure to oxygen, heat, light, salt, iron or other elements stimulates oxidative changes in fats.

Antioxidants are additives that reduce the rate of oxidative changes in fats or products containing fats without causing undesirable changes in the fat or fat-containing product. All antioxidants used in edible fats must be approved for use by the USDA–FSIS (see Chapter 3). Approved antioxidants must be tasteless, odorless, and non-toxic to humans, and they must stabilize the fat when present at a low concentration. Approved antioxidants include BHA (butylated hydroxyanisole), BHT (butylated hydroxytoluene), glycine, propyl gallate, resin guaiac, TBHQ (tertiary butylhydroquinone), and tocopherols. Commercial combinations of approved antioxidants, such as Griffith Laboratories G–4™ and Eastman Chemical Products, Inc.'s Tennox™ products, are effective not only because they stabilize stored fats but also because they "carry through" to protect products made from those fats.

Rendering Pork Fat in the Home

Fresh pork fat resulting from home harvest (see Chapter 5) or available from local meat processing operations may be rendered in the kitchen. Most supermarket meat departments are unable to supply the consumer with raw pork fat because they receive fresh pork as closely trimmed cuts. Leaf fat (kidney fat removed during harvest); subcutaneous fat, trimmed from the surface of pork wholesale and retail cuts; and intermuscular fat, found between muscle systems of the carcass, may be rendered.

Lard quality will be improved if the skin is removed from the subcutaneous fat (called *rinding*) before rendering. The fat should be cut or ground into uniform pieces no more than 1 inch square. Smaller pieces speed the rendering and reduce the heat damage to the products.

The fat should be placed in a cooking vessel either on the stove or in the oven. The fat should be stirred as it warms to avoid scorching. Oven rendering has some advantages, especially if the temperature controls on the oven are accurate. The temperature of the lard should not exceed 240°F. Boiling will occur after the fat has liquefied. Water is being evaporated, and the fat will not exceed the boiling temperature until most of the water has been removed. After the boiling stops, the temperature should be monitored closely. The fat should be allowed to cook until the solid material, called *cracklings,* attains a golden or amber color. Further cooking may result in overcooking and will probably not increase the yield of lard.

Whenever lard or grease is heated, during rendering or cooking, be extremely careful to prevent spilling, especially when removing heavy, lard-filled vessels from the oven, because hot grease clings to exposed skin surfaces and clothing, resulting in severe, deep burns.

Chilling and Storing Rendered Lard

The hot lard should be transferred by filtering through several thickness of cheese cloth, or the equivalent, to clean, dry, non-metallic, heat stable storage containers. A lard press may be used to squeeze the remaining lard from the hot solids, or cracklings. The press should be clean and uncontaminated by oxidized meat or fat. The cracklings may be wrapped in the cheese cloth filter and placed in the chamber of the press, which has an outlet for the liquid and a lever or gear and ratchet mechanism to exert pressure on the mass. Small hand-operated sausage stuffers may serve double duty as lard presses. Hand pressing the cracklings should result in a lard yield of 80 to 85 percent of the weight of the skinless raw fat.

Rapid cooling of the lard produces a firm, smooth-textured lard and reduces free fatty acid formation. To further enhance texture, stir the product well as it reaches the creamy stage in cooling. If the lard is unstirred or chilled slowly, the lard oil may separate or a grainy texture may develop. Lard should be stored at 40°F or less. Lard may be frozen, but it should be packaged in airtight containers and used within six months or less to minimize oxidative changes in flavor and aroma.

The stability of home-rendered fats may be extended by adding approved antioxidants or by mixing 2 to 3 pounds of commercially prepared hydrogenated vegetable shortening with 50 pounds of the hot fat before packaging. The antioxidant in the added product will provide "carry-through" protection to the lard.

Overcooking lard, during rendering, results in more pronounced flavors and odors, darkens the color, reduces the keeping quality, and increases the free fatty acid content, which in turn lowers the temperature at which the lard will give off smoke during cooking (smoke point, 250° to 425°F, depending upon free fatty acid content). Skin or meat particles remaining on the raw fat will also cause discoloration of the finished lard. Metallic ions, especially copper, stimulate oxidation of fatty acids; therefore, copper or brass containers or utensils should not be used in processing fat, particularly when it is hot. Stainless steel rendering vessels are preferable, but aluminum and rust-free iron kettles are also acceptable.

Cracklings

Cracklings are the protein-rich solids that remain after the lard has been pressed from rendered pork fat. The edible hot cracklings should also be cooled rapidly. Some people add salt to cracklings and enjoy them as a snack food. Cracklings may also be used as an ingredient in products such as "crackling bread."[6] Pork skin pieces that were left on the fat are often included in the cracklings and may also be used as snack items. Fresh pork skins that are fresh, clean, and free of harmful microorganisms may be cooked to produce a crisp product much like crackling skins. This can be done in a microwave oven by heating the skin portions until they puff. However, the rapid cooking process may not be sufficient to eliminate microorganisms on the skin surface. The skin pieces may also be cooked in deep fat.

TALLOWS AND GREASES

Classification and Grading

Table 11–5 lists the standards for tallows and greases, as recognized by the industry and marketers of those products. The criteria for the classes and grades of animal fat products are based upon certain chemical and physical characteristics. The source of the raw material and the processing procedure may influence these characteristics.

The major classifications, based on human consumption, are *edible* and *inedible*. Edible fats may be produced only from edible carcass parts maintained under conditions approved by USDA–FSIS meat inspection regulations. The beef and pork fats utilized in the manufacture of margarine are examples of edible fats, as is lard. Nearly all other fats are inedible.

- *Titer:* Minimum titer refers to the temperature developed as the liquid fat cools (congeals) to a solid. A higher titer is more desirable. Fats with titers equal to or over 104°F (40°C) are *tallows*, and those with titers under 104°F (40°C) are *greases*.

---

[6]Lustig, L. S., S. C. Sondheim, and S. Rensel. 1939. Southern Cookbook. Culinary Arts Press, Reading, Pennsylvania, p. 31.

- *FFA (free fatty acid) content:* A lower percentage of free fatty acid is preferable.

- *MIU (moisture, impurities, and unsaponifiables):* A lower percentage of MIU, indicating less "contamination" with useless material, is preferred.

Table 11–5. Trading Standards[1] for Tallows and Greases

Tallows and Greases	Minimum Titer[2]	Maximum FFA[3]	MIU[4]	FAC Color[5]
	(°C)	(%)	(%)	(score)
Tallows				
Edible	41.0	0.75	[6]	3
Top white	41.0	2.0	1.0	5
All beef packer.	42.0	2.0	1.0	none
Extra fancy.	41.0	3.0	1.0	5
Fancy	40.5	4.0	1.0	7
Bleachable fancy	40.5	4.0	1.0	none
Prime	40.5	6.0	1.0	13–11B
Special	40.0	10.0	1.0	21
No. 2	40.0	35.0	2.0	none
"A"	39.0	15.0	2.0	39
Greases				
Lard (edible).	38.0	0.5	[6]	[7]
Choice white	36.0	4.0	1.0	13–11B
Yellow.	[8]	15.0	2.0	39

[1]Approved by American Fats and Oils Association, effective January, 1992.
[2]Crystalization temperature.
[3]Free fatty acid percentage.
[4]Moisture, impurities, and unsaponifiables.
[5]Fat Analysis Committee — standards are matched (white).
[6]Moisture maximum 0.20 percent. Insoluble impurities maximum 0.05 percent.
[7]Lovibond color maximum 1.5 red. Lard peroxide value 4.0 ME/K max.
[8]Titer minimum, when required, to be negotiated contract-by-contract basis.
Source: Render, The National Magazine of Rendering. April, 1992. 21:42.

- *FAC (Fat Analysis Committee) color:* Color designation varies in descending desirability from the lightest and whitest to the darkest as follows: Extra Fancy, Fancy, Bleachable Fancy, Prime and Choice White, Choice White A, Choice White B, Yellow, House (intermediate between Yellow and Brown), and Brown.

- *Lovibond color:* The Lovibond color system uses a set of glass standards to match the color of the product. The fats are graded lower as color intensity increases.

Edible Fats

Edible tallows and lard are used in oleomargarine (margarine), shortenings, and cooking fats. The original name *oleomargarine* is derived from *oleo,* which designates beef fat in the industry terminology. Edible lamb tallow may be utilized in the manufacture of oleomargarine and other shortenings, while inedible lamb fats may be utilized in the manufacture of feeds, soaps, and lubricants. Although early margarines were made largely with animal fat, most margarines on the market today are blends of vegetable oil, water, milk, salt, preservatives, and sometimes diacetyl, to import a buttery flavor. Still, in recent years, approximately 30 million pounds of edible animal fats were merchandised and consumed in the form of margarine (see Table 11–6).

The meat inspection regulations concerning margarine production are rather extensive, due to the wide variety of ingredients and the extensive blending and processing required.

The shortening products, used in pastries and many other foods and as a cooking medium for frying, represent a larger market for edible fats (see Table 11–7). Whereas only 1.5 to 2 percent of the margarine produced during a recent period was animal fat, the shortenings produced in the same period were 4.5 to 5.5 percent lard and 5 to 6 percent beef fat.

Table 11–4 demonstrates how the reduction in the amount of lard produced per hog has brought about a decline in the total lard production in the industry. Conversely, edible tallow production (see Table 11–8) has increased significantly (from approximately 800 million pounds in 1977 to an average 1,500 million pounds recently), while the number of cattle harvested declined from 42 million head in 1977 to an average of 35 million head in 1998. Since fat content (per carcass) has been decreasing, the explanation for the increase in tallow production seems to be that the transition of the industry to the cen-

Table 11–6. Margarine: Fats and Oils Used in U.S. Manufacture, 1975–1997

| Year | Selected Reported Fats and Oils Consumed in Margarine Manufacture | | | |
	Animal Fats[1]	Soybean Oil	Corn Oil	Total[2]
	---------------------- *(million lbs.)* ----------------------			
1975	52	1,568	188	1,917
1980	104	1,653	223	2,016
1985	65	1,628	210	1,946
1986	48	1,741	190	2,041
1987	22	1,615	248	1,931
1988	35	1,619	210	1,895
1989	32	1,573	214	1,875
1990	35	1,749	208	2,102
1991	43	1,853	196	2,160
1992	37	1,926	176	2,174
1993	31	2,013	161	2,239
1994	42	1,793	NA[3]	2,003
1995	41	1,684	NA	1,847
1996	28	1,694	77	1,816
1997	14	1,650	61	1,733

[1]Lard and beef fats.

[2]Includes small quantities of cottonseed, peanut, coconut, palm, and sunflower oils.

[3]NA = not available.

Source: Agricultural Statistics. 1999. USDA.

Table 11–7. Shortening: Fats and Oils Used in U.S. Manufacture, 1975–1997

Year	Selected Reported Fats and Oils Consumed in Shortening Manufacture				
	Lard	Edible Tallow	Soybean Oil	Cottonseed Oil	Total[1]
	- (million lbs.) - - - - - - - - - - - - - - - - - - -				
1975	166	602	2,025	154	3,728
1980	378	673	2,660	189	4,200
1985	289	1,015	3,625	173	5,564
1986	274	973	3,375	182	5,453
1987	224	890	3,434	136	5,303
1988	265	840	3,562	170	5,377
1989	295	752	3,509	192	5,338
1990	264	637	4,004	252	5,684
1991	274	462	4,152	260	5,767
1992	310	427	4,150	241	5,761
1993	296	404	4,951	266	6,544
1994	287	405	4,929	216	6,365
1995	325	374	4,673	212	6,031
1996	284	320	4,690	237	5,935
1997	272	312	4,517	257	5,679

[1]Includes small quantities of coconut, palm, corn, peanut, safflower, and sunflower oils.
Source: Agricultural Statistics. 1999. USDA. (Mike Easdale.)

tralized breakdown of beef into primals and sub-primals (boxed beef) has resulted in increased salvage and edible processing of beef-cutting fats. These beef-cutting fats were formerly discarded totally or converted to inedible material by local retail cutting operations and renderers.

Export marketing and prices, both world and domestic, are influenced to a greater extent by fluctuations in seed oil supply than in animal fat supply. That fact helps explain the failure of export sales tonnages to reflect the production levels of animal fats. Generally, the world has an oversupply of fats and oils, which keeps all prices low.

**Table 11–8. Production, Export,[1] and Utilization
of Edible Lard and Tallows**

	Lard			Tallows		
Year	Production	Export	Domestic Use	Production	Export	Domestic Use
	------------------------------- (million lbs.) ------------------------------					
1977	966	132	861	795	18	768
1980	1,159	144	1,023	1,122	137	995
1985	899	108	810	1,625	44	1,595
1986	853	103	745	1,278	59	1,244
1987	916	120	787	1,300	112	1,184
1988	940	127	811	1,201	246	981
1989	909	85	845	1,193	324	816
1990	915	110	807	1,202	262	955
1991	987	131	856	1,515	346	1,197
1992	1,011	129	886	1,414	295	1,103
1993	1,015	118	890	1,499	301	1,204
1994	1,063	140	924	1,542	259	1,268
1995	1,013	94	922	1,505	220	1,303
1996	970	103	871	1,390	176	1,200
1997	1,065	125	917	1,500	236	1,275
1998	1,100	110	990	1,415	210	1,202

[1]Exports are exports net of imports.

Source: Oil Crops Outlook and Situation, October, 1998. Economic Research Service, USDA. (Mike Easdale.)

Inedible Fats

Meat harvest and processing plants that have rendering facilities must have two separate rendering units. The two units must be separated physically to prevent any intertransfer of raw material, product, or contamination from the inedible-rendering area to the edible-rendering area. Equipment used to handle, contain, or process inedibles must be clearly marked "Inedible," and any edible product that comes in contact with any so marked pans, trucks, or conveyors immediately is designated as being inedible. In meat and poultry operations, all soft tissue, some bones, sweepings, scrapings, and skimmings that are not classified as edible or do not have other uses, are cooked and processed into inedible fat and meal by rendering. In most cases, with the exception of the beef hide, dead or condemned animal products are processed by inedible rendering. Table 11–9 lists the production statistics and gives some information on the usage of inedible tallows and greases in the United States.

Table 11–9. Inedible Tallows and Grease: U.S. Supply and Disposition

	Supply				Disposition		
					Factory Consumption[1]		
Year	Stocks Jan. 1	Production	Total	Exports	Total	Use in Soap	Use in Feed
				(million lbs.)			
1977	355	6,106	6,465	2,885	3,180	737	1,330
1982	451	6,026	6,477	3,035	2,898	545	1,418
1987	316	5,602	5,918	2,491	3,082	571	1,727
1988	407	6,158	6,565	2,807	3,147	461	1,864
1989	399	5,848	6,247	2,679	3,194	368	1,919
1990	374	5,217	6,097	2,267	3,061	402	2,000
1991	357	5,759	6,116	1,936	2,949	392	1,748
1992	349	5,768	6,117	2,279	3,050	334	1,954
1993	309	6,621	6,930	2,117	3,018	300	1,995
1994	320	6,712	7,032	2,167	3,190	301	2,102
1995	348	6,745	7,093	2,663	3,223	264	2,166
1996	373	6,376	6,749	1,993	3,289	245	2,253
1997	262	6,249	6,511	1,685	3,399	245	2,401

[1]Also for lubricants, fatty acids, etc.

Source: Agricultural Statistics. 1999. USDA. (Mike Easdale.)

Use in Animal Feeds

A major domestic use of tallows and greases is in animal feeds (see Table 11–9). These fats are usually stabilized with an approved antioxidant to prevent rancidity, thereby maintaining feed palatability. Fats are the richest food nutrient in terms of energy, and as such have been used successfully in cattle, poultry, swine, and pet feeds since the 1940s.

In addition to their energy value, fats reduce the dust, improve the color and texture, enhance the palatability, increase pelleting efficiency, and reduce machinery wear in the production of animal feeds.

Pet Foods

"The pet food industry represents a significant outlet for meat by-products, including rendered fats and protein meals. Roughly 25 percent of all meat by-products produced in the U.S. are utilized in pet foods. The inclusion rate of meat by-products in specific cat and dog foods may be as high as 50 and 40 percent respectively. The 67.9 million cats in the United States slightly exceed the 55.8 million dogs in number. However, total cat and dog food sales in 1996 were estimated at 3.95 and 5.41 billion dollars respectively"[7] and in 1998 totaled 11 billion dollars.[8]

In the early 1990s, the retail value of products produced by the U.S. pet food industry totaled approximately $7 billion per year. This compares with $5.7 billion (9 billion pounds of dog and cat food) during 1987, $1.6 billion in the mid–1970s, and $350 million in 1958.[9] An informative history of dog food development is presented by Corbin (see footnote 8).

International sales are a fast-growing component of the U.S. pet food industry. Although most pet foods are cereal-based, animal fats are used as energy sources in pet foods. Meat meal is also used extensively in some pet foods for protein and mineral sources as well as for palatability enhancement. Fresh by-products (uncooked) may be ingredients of canned and fresh frozen pet and special animal foods (zoo foods, mink food, racing and guard dog food, and fox food).

Fatty Acids

Fatty acids are obtained from animal fats through a process referred to as splitting and are being used in ever-increasing quantities in the manufacture of scores of products. Splitting involves the separation of the fatty acids from

---

[7]Halpin, K. M., J. E. Sullivan, R. Bradfield, and Q. Liu. 1997. Expanding the meat–by-product market with pet foods. In: Proceedings of the 1997 Meat Industry Research Conference, Chicago, p. 51.

[8]Corbin, J. 1998. Dog foods development — how they came to be; at Penn State University Students at Dads Products, Inc., Meadville, Pennsylvania, November 23–24.

[9]Britzman, D. February, 1988. Gone to the dogs. Harv. States J. 8 (2):5; and personal communication, 1992.

glycerine, the market term for glycerol. This hydrolysis of fats is accomplished with the use of steam, high temperature (up to 480°F/250°C), pressure, and, sometimes, catalysts. Lipase enzyme systems that make the fat separations with less energy input and produce improved products have recently been proposed. Both processes occur in continuous systems. The list of uses for fatty acids and other derivatives of natural fats, animal, plant, and marine (fish), by the chemical industry is impressive.[10]

- *Uses of fatty acids and derivatives:* Plastics; metal soaps; washing and cleaning agents; soaps; cosmetics; alkyd resins; dyestuffs; textile, leather, and paper industry products; rubbers; lubricants.

- *Uses of methyl esters of fatty acids:* Cosmetics; washing and cleaning agents.

- *Uses of glycerol and derivatives:* Cosmetics; toothpastes; pharmaceuticals; foodstuffs; lacquers; plastics; synthetic resins; tobacco; explosives; cellulose processing.

- *Uses of fatty alcohols and derivatives:* Washing and cleaning agents; cosmetics; textile, leather, and paper industry products; mineral oil additives.

- *Uses of fatty amines and derivatives:* Fabric conditioners; mining materials; road-making materials; biocides; textile and fiber industry products; mineral oil additives.

- *Uses of drying oils, neutral oil, and derivatives:* Lacquers; dyestuffs; varnishes; linoleum; soaps.

Oils

Lard oil, made from white A grease, is used in the manufacturing of a high-grade lubricant for delicate running machine parts. The oil from white B grease is sometimes called "extra neat's-foot oil" and is used in giving viscosity to mineral oils. The oils made from brown grease are used in compounding

---

[10]Baumann, H., M. Buhler, H. Fochem, F. Hirsinger, H. Zoebelein, and J. Falbe. 1988. Natural fats and oils — renewable raw materials for the chemical industry. Angew. Chem. Int. Ed. Engl. 27:41.

cutting oils, heavy lubricating oils, special leather oils, and illuminating oils. They are also combined with paraffin in making candles.

Pure neat's-foot stock is made from the shin bones and feet of cattle. This stock is grained and pressed to secure the pure neat's-foot oil, which is used in the leather and textile industries.

Soap

The fatty raw materials for the manufacture of soap include animal fats (lard oil and tallows) and vegetable oils (coconut, olive, corn, palm, palm kernel, cottonseed, soybean, peanut, linseed, rapeseed, and sesame). Other greases used for making soap for special purposes are extracted from garbage, wool scourings, bone boiling, glue making, and hide trimmings. The garbage grease is changed into fatty acids and then distilled, resulting in a very white fat of good odor that is used by practically all large soap manufacturers.

Prior to the 1960s, the greatest utilization of tallows and greases was in soap making. However, with the increased use of phosphate-based detergent powders and liquids, soap production declined significantly. Furthermore, the domestic use of animal fats in soap has been significantly replaced by the use of plant lipid sources. The development of detergents and other cleaning agents and the replacement by vegetable fats posed serious threats to the market for the major products of the rendering industry. The loss of a portion of the soap market stimulated the renderers to expand markets and to diversify the products of the industry. New markets were researched and developed in the feed industry, in the export market, and in the application of fatty acids to industrial uses.

When fats and alkali (sodium or potassium hydroxide) are cooked together, soap is the result. The chemical term for the reaction involved is *saponification*. The glycerine is separated from the fatty acids in the triglyceride structure, and the fatty acids combine with the sodium or potassium to form the soap. Water and glycerine are the other products of the reaction. The water is removed by evaporation, and the syrupy red glycerine settles out of the system and is drawn off. The demand for glycerine as a raw material for the production of explosives, including nitroglycerine, varies with the level of peace in the world. The glycerine extracted from 4 pounds of animal fat will produce about 1 pound of nitroglycerine. Contrastingly, glycerine is also used as a means of applying external medications.

Modern soap making occurs in continuous processing systems that utilize fatty acids "split" or stripped from raw fats (as described earlier). Soap making originated and remained a batch process involving more time and less technology until the middle of the twentieth century. The writings of a curious Roman, Pliny the Elder, who died in A.D. 79, contain early references to soap prepared from goat tallow and wood ashes.[11]

The many different types of soaps available today result from variations in the raw materials used, the processing procedures, and a wide range of additives to enhance the odor, texture, water softening ability, and pH. Soft soaps retain some of the glycerine and water, and thus remain fluid. Floating soaps include minute air spaces and contain more water that non-floating milled soaps. Cleansers may contain abrasives and alkaline salts to enhance their effectiveness.

Because soap is biodegradable, it has an advantage over the phosphate-based detergents that replaced it in many areas of the household and market place some years ago. Phosphates tend to accumulate in the water supply and are responsible, in part, for the stimulation of algae growth and oxygen depletion in lakes and streams. Tallow-based cleaning products with detergent-like traits, effective in hard and cold water, have been developed and are in use in parts of the world. Environmental concerns and fat utilization research may help fat-based cleansing materials gain a larger share of future cleanser markets.

MEAT MEALS

The dry, defatted, high-protein material that results from rendering may vary, depending on the raw materials used and on the processing technique employed. The protein products of rendering may be utilized in a number of ways but are marketed most extensively as animal feeds (see Table 11–10). Because animal nutritionists have detailed knowledge of specific nutrient requirements and use computers to balance rations for specific amino acids and micronutrients, the nutrient content and the nutrient availability of feed ingredients should be standardized. The animal protein sources incorporated into livestock rations have not been well standardized, but the U.S. rendering industry is installing rendering systems that result in less heat damage to nutrients and is improving the quality control and handling of raw materials

---

[11]Burnham, F. 1978. Rendering, The Invisible Industry. Aero Publishers, Inc., Fall River, Massachusetts, p. 9.

**Table 11–10. Use of High-Protein Animal By-products
in Commercial Feeds**

Year Beginning October	Tankage and Meat Meal	Fish Meal	Dried Milk[1]	Total
	- - - - - - - - - - - - - - - - - - - (1,000 tons) -			
1980	2,458	379	305	3,142
1985	2,800	511	412	3,724
1988	2,567	292	446	3,305
1989	2,557	357	461	3,375
1990	2,527	275	458	3,260
1991	2,541	257	470	3,268
1992	2,372	528	464	3,364
1993	2,446	719	470	3,636
1994	2,551	334	463	3,349
1995	2,536	290	420	3,246
1996	2,783	308	428	3,519
1997	2,501	488	413	3,402

[1]Includes dried skim milk and whey for feed, but not milk fed on farms.
Source: Agricultural Statistics. 1999. USDA. (Mike Easdale.)

in an effort to standardize rendered protein products. In addition to reducing the variation in nutrient content and availability, the improved methods will lessen the potential microbial contamination of valuable feed ingredients.

Many non-fat products of rendering may be utilized as organic fertilizers. Animal proteins, especially blood meal and dried blood, are the base materials of some adhesives. Bone meal is marketed to the producers of china, instrument keys, steel alloys, glass, water-filtering agents, and enamels.

The major non-fat products of rendering are described as feedstuffs, since that is the principal use of the materials listed. The International Feed Number (IFN), which is an identification system for feed ingredients,[12] is indicated for each product.

---

[12]Association of American Feed Control Officials. 1999. Official Publication, pp. 189–195.

IFN 5-00-385   Animal meat meal, rendered

*Meat meal** is the rendered product from mammal tissues, exclusive of any added blood, hair, hoof, horn, hide trimmings, manure, or stomach and rumen contents except in such amounts as may occur unavoidably in good processing practices. It shall not contain added extraneous materials not provided for by this definition. The calcium (Ca) level shall not exceed the actual level of phosphorus (P) by more than 2.2 times. Meat meal shall not contain more than 12 percent pepsin indigestible residue, and not more than 9 percent of the crude protein in the product shall be pepsin indigestible. The label shall include guarantees for minimum crude protein, minimum crude fat, maximum crude fiber, minimum phosphorus (P), and minimum and maximum calcium (Ca). If the product bears a name descriptive of its kind, composition, or origin, it must correspond to that name.

IFN 5-00-386   Animal tankage meal, rendered

*Meat meal tankage** is similar to meat meal (IFN 5-00-385) except that it may contain added blood or blood meal. However, it shall not contain any other added extraneous materials not provided for by this definition.

IFN 5-00-387   Animal tankage with bone, rendered

*Meat and bone meal tankage** is similar to meat meal tankage (IFN 5-00-386) except that it includes bone and shall contain a minimum of 4 percent phosphorus (P).

IFN 5-00-388   Animal meat with bone, rendered

*Meat and bone meal** is similar to meat meal (IFN 5-00-385) except that it includes bone and shall contain a minimum of 4 percent phosphorus (P). Don A. Franco, D.V.M., M.P.H., Director, Scientific Services, National Renderers Association,[13] explained the nutritional value of meat and bone

---

*Use of this ingredient, from mammalian origins, is restricted to nonruminant feeds unless specifically exempted by 21 Code of Federal Regulations (CFR) 589.2000. Because of the possibility of the transmission of the bovine spongiform encephalopathy (BSE) pathological prion protein through the feeding to other ruminants, feeds containing prohibited material must bear the following label statement: *"Do not feed to cattle or other ruminants."*

[13]Franco, D. A. 1999. Marketing meat and bone meal for its mineral values: A perspective. Render. 28:16–21.

meal. The main difference between by-product feeds IFN 5-00-385, -386, -387, and -388 is the inclusion or exclusion of bone and blood.

IFN 5-01-977   Fish meal, mechanically extracted

Fish meal is the clean, dried, ground tissue of undecomposed whole fish or fish cuttings, either or both, with or without the extraction of part of the oil. It must contain not more than 10 percent moisture. If it contains more than 3 percent salt (NaCl), the amount of salt must constitute a part of the brand name, provided that in no case must the salt content of this product exceed 7 percent.

IFN 5-03-795   Poultry feathers meal, hydrolyzed

Hydrolyzed poultry feathers meal is the product resulting from the treatment, under pressure, of clean, undecomposed feathers from harvested poultry, free of additives and/or accelerators. Not less than 75 percent of this product's crude protein content must be digestible by the pepsin digestibility method.

IFN 5-03-798   Poultry by-product meal, rendered

Poultry by-product meal consists of the ground, rendered, clean parts of the carcasses of harvested poultry, such as necks, feet, undeveloped eggs, and intestines, exclusive of feathers, except in such amounts as might occur unavoidably in good processing practices. The label shall include guarantees for minimum crude protein, minimum crude fat, maximum crude fiber, minimum phosphorus (P), and minimum and maximum calcium (Ca). The calcium (Ca) level shall not exceed the actual level of phosphorus (P) by more than 2.2 times.

IFN 5-26-005   Animal blood meal, conventional cooker dehydrated
IFN _____   Animal blood meal, steamed, dehydrated
IFN _____   Animal blood meal, hydrolyzed, dehydrated

Blood meal _____ is produced from clean, fresh animal blood, exclusive of all extraneous materials, such as hair, stomach belchings, and urine, except in such amounts as might occur unavoidably in good processing practices. The process used must be cited as a part of the product name, such as conventional cooker dried, steamed, or hydrolyzed. The product usually has a black-like color and is rather insoluble in water.

IFN 6-00-400   Animal bone meal, steamed

Steamed bone meal is the dried and ground product sterilized by cooking undecomposed bones with steam under pressure. Grease, gelatin, and meat fiber may or may not be removed. The product must be labeled with guarantees for phosphorus (P) and calcium (Ca). The term "steamed bone meal" must be used in all labeling.

HIDES AND PELTS

Classification of Beef Hides

Hides, pelts, and skins are classified according to (1) the species and class from which they come, (2) their weight, (3) the nature and extent of branding, and (4) the type of packer producing them.

The general term *hide* refers to a conventional beef "skin" weighing more than 36 pounds (29 pounds if trimmed and fleshed). Hides are designated as steer, cow and heifer, or bull and may be classified by weight into extra-light (often called ex-light), 36½ to 57 (29½ to 48) pounds; light, 57 to 69 (48 to 58) pounds; and heavy, 70 (59) pounds and up. Lighter weight units are classified as skins and are subdivided into calf skins (the lightest) and kipskins (intermediate between calf skins and hides).

Brands and their locations affect the value of hides, since those branded on the side and butt (rear quarter) cause damaging scar tissue in this most valuable area when they are processed into leather. The term *Colorado* or *Texas* is applied to those hides branded on the side, with the Texas hides generally being plump and close-grained. Butt-branded hides are often indicated as such in the market. Unbranded hides are known as "natives" (see Table 11–11).

Hides are further classified into major packer, small packer, and country hides. The principal differences are in the manner of take off and subsequent handling. Major packer hides are uniform in pattern, have a minimum of cuts and scores, are free from dung, and have been cured, sorted, and stored under standard conditions. Small packer and country hides generally are not uniform in take off (pattern), contain more cuts and scores, and because of the various systems of handling by different individuals, they are sometimes undersalted, showing hair slips and maggot infestation. Hides are graded 1 (no deep scores or cuts), 2 (up to four cuts), or 3 (five or more cuts with some loss of hair), with Grade 1 being the highest quality hide. Hides that have been removed from dead animals (those from rendering plants, those that have been range salvaged, etc.) are called "renderer" or "murrain" hides.

Table 11–11. Beef Hide Classifications, Based on Sex, Weight, and Brand Location

Selection	Cured Weight	Green Weight
	------------- (lbs.) -------------	
Cows		
Light native cows	30¾–53	36½–63
Heavy native cows	54 and up	64 and up
Branded cows	30¾ and up	36½ and up
Steers		
Extra-light native steers	30¾–48	36½–57
Light native steers	49–58	57–69
Heavy native steers	59 and up	70 and up
Light and extra-light butts	30¾–58	36½–69
Heavy butts.	59 and up	70 and up
Light and extra-light Colorados	30¾–58	36½–69
Heavy Colorados	59 and up	70 and up
Light and extra-light Texas	30¾–58	36½–69
Heavy Texas	59 and up	70 and up
Bulls		
Native and branded bulls	59 and up	70 and up
Calves		
Light calves.	9½ and down	11 and down
Heavy calves.	9½–15	11–18
Kips	15–25	18–28
Overweight.	25–30	28–36

Note: All green weights above are net (tared for mud and manure), as are the cured weights (tared for mud, manure, and salt).

Source: Minnock, J., and S. R. Minnock (Ed.). 1979. Hides and Skins, National Hide Assn.

Handling Hides and Pelts

Proper handling of the most valuable by-product of the bovine species, the hide, begins when the calf is born. Good livestock management involves proper housing, effective pest and parasite control, and good nutrition, among other factors. Injuries that tear the skin, caused by projecting nails, wires, or other fencing materials, horns, and poorly maintained chutes and vehicles, result in scars or irregularities in finished leather produced from hides and pelts. Branding reduces the value of the hide, but the degree of value impact may be influenced by brand location and branding technique. Pests and disease, such as ticks, lice, grubs, mange, scabies, ringworm, pox, and warts, produce skin lesions that carry over to the leather. Proper animal health management and medication can prevent the causative infections and infestations.

Attention to hide removal pattern (see Figure 11–3), and prevention of cuts or scores in the hide during the skinning process will maintain the value of a healthy hide. Trimming thick edges and flesh, washing away manure and dirt, and preserving hides immediately after skinning are essential to prevent microbial deterioration of skin and leather quality. Packer hides receive premium prices over country hides because of the greater emphasis on skinning skills and hide management practices as well as the larger volume offered by the meat packer.

Skinning and hide, skin, and pelt handling procedures described in Chapters 6, 7, and 10 will result in realization of the greatest possible value from small lots or individual hides from home-harvested or salvaged casualty animals. Folded, cured hides, skins, and pelts should never be tied with wire, as the rusted metal will produce a stain or mar on the leather. Polyester (plastic) twines should not be used on woolen pelts because poly fragments cannot be separated from wool and will cause fabric problems. Cool, dry storage and prompt marketing of salted hides, skins, and pelts will also enhance their value.

A Voluntary Hide Trim Standard

A beef hide trim that is designed to encourage economy in shipping hides and to improve the quality of leather products was developed at the University of Cincinnati and became effective April 4, 1965 (see Figure 11–3). Its use has been widely accepted, and it is currently a voluntary standard used within the industry. The trim is as follows:

- *Heads:* Ears, ear butts, snouts and lips, fat, and muscle tissue should be removed. The pate should be removed from the pate (crown) side of the head by cutting through the eye hole. The narrow side of the head should be trimmed through the eye in a similar manner. All ragged edges should be removed.

- *Kosher heads:* These should be removed by cutting across at the top of the Kosher cut. Headless Kosher hides are to be put into their respective weight classifications by lowering the testing weights 3 pounds for cow hides and ex-light steer hides and 5 pounds for light and heavy steer hides.

- *Shanks:* Foreshanks should be trimmed straight across through the knee at a point that will eliminate the cup. Hindshanks should be trimmed one-third the distance from dewclaw holes to the wide flare of the shank.

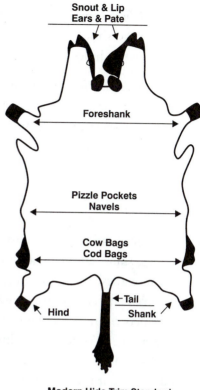

Modern Hide Trim Standard

Fig. 11–3. The dark portions must be trimmed off in order to conform to the trim standard (see also Fig. 6–17). (Developed at the University of Cincinnati).

- *Cow bags, teats, cow navels, and cod bags:* Cow bags, teats, cow navels, and cod bags should be removed straight with the belly line. Cow hides that have the bags trimmed out because of government regulations are not considered as being off-pattern if they are properly shaped in other respects and are deliverable as No. 1 hides if they conform to other trade standards of that designation.

- *Pizzle pockets:* Pizzle pockets should be split through the center but left on the hides for identification.

- *Tails:* Maximum tail length should be no more than 4 inches cured, measured from the root. (Removal of the switch is necessary because it is difficult to cure. The presence of a poorly cured appendage will cause a hide proper to spoil.)

- *Cheek brands:* When a hide has only a cheek brand, that portion of the head should be trimmed off, and the hide should be placed in with the native hides.

- *Wire in hides:* Green hides should be examined carefully, and before they are salted, all wire hog-rings should be removed because their presence can be damaging to hide processing and tanning machinery and to leather.

HIDE PROCESSING

Deterioration of all parts of the animal begins at the time of harvest. Hides, skins, and pelts are more susceptible to deterioration than the carcass because they are highly contaminated with bacteria, mud, blood, and fecal material and are often wet when removed from the carcass. As with all other products of meat animal harvest, controlling the rate of hide deterioration is imperative in order to maximize the value of the final product. Hide processing, therefore, serves two purposes. It prevents or reduces the rate of deterioration, and it transforms the raw hide, skin, or pelt into a usable product. The steps in hide processing are (1) preserving the trimmed, washed hide temporarily (usually by curing); (2) removing the wool or the hair (dehairing), fat, and other tissue (fleshing); (3) preparing the fleshed, dehaired hide, chemically, for the tanning process; and (4) tanning the hide into leather.

Hide Preservation

Hide preservation has traditionally been accomplished by curing, but other techniques may be equally effective while avoiding some of the problems encountered with salt curing. Research has been conducted by at least two groups in Canada on systems and facilities to preserve fresh hides using radiation.[14] Immediately after flaying (skinning), the hides are trimmed according to the standards described earlier, washed, demanured, and fleshed. After being dipped in a patented chemical precursor, the hides, in one continuous system, are squeezed, bagged individually, laid out flat on a conveyor,

---

[14]Pietraszek, G. May 23, 1987. Areas of concern faced by domestic tanning industries. The Natl. Prov. 196 (19):16.

and passed under an electron beam. The treatment sterilizes the hides and does not reduce the strength of the leather. No nuclear material is required since the ionizing irradiation is generated from electricity.

Curing hides, skins, and pelts is accomplished in much the same manner as curing the meat. Salt is used to extract some of the moisture from the hide (reduce water activity, see Chapter 2) and is absorbed by the hide to achieve salt levels toxic to many bacteria (see Chapter 2). A green beef hide will be almost two-thirds water (60 to 65 percent); whereas after the hide has been cured, the water content should be no more than 45 to 48 percent, and the salt level should be at least 14.5 percent. The hide should be approximately 85 percent salt-saturated to prevent spoilage. Hide curing salt mixes and brines often include chlorinated phenols to prevent damage to the product by salt-tolerant organisms. Other bactericides[15] that are less difficult to handle in the effluent waters at the curing plants and the receiving tanneries are being sought. Curing is more effective if the hides have been washed to cool them and to remove dirt and manure before curing is initiated. Some packers trim, flesh, and demanure the hides prior to curing, which results in greater curing uniformity and efficiency. A properly cured hide can be stored and shipped without deterioration. Curing is a preservative process only and is not a necessary step in leather tanning.

Salt Pack Curing

Largely a process of the past, salt packs are still used to produce cured hides in this country and around the world. In salt pack curing, hides are stacked hair side down in the hide cellar or hide room with salt uniformly spread over the flesh side of each hide. If the hides are arranged properly, with edges folded, the stack (several feet high) will be able to hold the brine, which is the solution produced as the salt extracts moisture from the hide within the stack. The salt solution penetrates the hides and cures them. One reason why salt pack curing has been replaced by other processes is the 30-day curing time requirement. The cost of maintaining the inventory, in addition to the labor required and materials handling problems, has contributed to the replacement of salt packs in modern high-volume harvest operations.

---

[15]Bactericides extend hide life. January 9, 1988. The Natl. Prov. 200 (2):8.

Brining Hides

The more commonly used and rapid methods of curing hides have resulted from dissolving the salt to produce a saturated solution (brine) into which the hides are introduced. The pack-curing procedure required that the salt first extract water from the hide. Agitating the hides immersed in a saturated curing solution further speeds up salt penetration. Brine curing can be done in large vats, raceways (large oval vats with a center island, much like a race track), revolving drums, or hide processors. Depending upon conditions, hides may be satisfactorily brine cured overnight in agitated vats and raceways and, routinely, within 24 hours. Many of the salt-cured beef hides produced in the United States are raceway (see Figure 11–4) or hide processor cured.

Fig. 11–4. A raceway hide-curing system. Hides are being unloaded from the raceway and hooked to the conveyor to be transferred to the grading area. (Courtesy, IBP Corp., Dakota City, Nebraska)

Hide Processor Curing

Hide processors are large stainless steel, inclined axis, stationary, concrete mixer–shaped units or wooden drum units. The capacities range from 250 to 400 beef hides or more. Hides direct from the hide pullers are dumped into

Fig. 11–5. Hide grading and palletizing area. Note hides in transit to processing area on overhead conveyors. Fleshing machine operators are on the balcony. Hides are spread on the tables in the background for trimming, grading, and folding. Grades are sorted on different pallets. (Courtesy, IBP Corp., Dakota City, Nebraska)

the mixers for an initial chilling/washing cycle. Some systems water-cool the hides to set up the soft tissues, run the hides through the fleshing and demanureing machine, and then transfer them to the processor. After the wash water has been removed from the processor drum, the hides are dry cured by fresh salt, which is added at the rate of 20 to 24 percent of the total hide weight. The processing schedules vary among different plants, but with the hide processor technology, curing may be completed in six to seven hours. The use of brine instead of dry salt improves the appearance of the cured hides but does not influence the final leather quality. Brine curing, when compared to dry salt curing in the hide processor, increases the time required to completely cure the hides. Cured hides are tied and palletized (see Figure 11–5).

Further Preparation of the Hides

Leather has many uses, even in today's "synthetic world." One reason why leather is so versatile is because it has a variety of physical and chemical characteristics. Stated another way, leather is many different products. The selection and preparation of hides, skins, and pelts for tanning and the tanning procedure itself involve numerous combinations of physical and chemical processes, determined, in part, by the purpose of the leather to be produced.

The preserved hide (cured, in most cases) must be hydrated, cleaned, dehaired, and prepared chemically to receive the tanning agents before the actual tanning is initiated. Much of the preparatory treatment may be accomplished before the hide leaves the harvest plant, or it may all be done by the tanning company.

If not trimmed previously, the hide is *trimmed* according to the guidelines listed earlier. Preserved hides and skins, which have been dehydrated by salting or by drying, must be rehydrated by water *soaking* to reinstate flexibility and to allow penetration of the various chemicals used in subsequent treatments. Soaking is not required if the hides were preserved by irradiation or arrived fresh. The hides must be *washed* to remove manure, blood, loose tissue, soil, and remaining salt and *fleshed* (scraped) to remove the remaining fat, connective tissue, and muscle from the underside of the hides if they were received unwashed and unfleshed. Machines which *demanure* and flesh in one operation are used. Nearly all cattle hides and some skins are *dehaired*. Similar procedures are used to pull wool from most sheep pelts and skins. The outer skin layer, or epidermis, and hair or wool, which is mainly keratin, a protein which is degraded by strong bases (alkalies), are removed. The hides, skins, or pelts may be immersed in vats or revolving drums, or painted with a solution or paste of hydrated lime (calcium hydroxide) to remove the hair or wool (liming). If the hair or wool is saved, the hides, skins, or pelts are removed from the lime solution and passed through dehairing machines or wool pullers when the hairs become loosened in the follicles. In some operations, the hair may be digested (burned) by the base. Sulfides are often added to enhance the effectiveness of the dehairing operation. The major protein in the hide tissue, collagen, is not sensitive to bases.

Dehaired hides, even after being washed to remove the hair, are high pH (basic) systems. Acid baths are used to neutralize (delime) the basic conditions and to produce conditions in the hide that will enhance the action of bating enzymes. The liming treatment also opens up the collagen structure, which exposes the subsurface material to the action of the pancreatic enzymes that are used to "bate" the hide. *Bating* is a step in the transition from hide to leather during which protein degradatory enzymes derived from pancreatic tissue and(or) microbial sources are introduced. The enzymes digest the non-collagenous components of the hide, freeing it from the collagen protein. The loosened non-collagenous material, along with epidermal and follicular remnants of the liming process, are termed scud. Therefore, the mechanical scraping action that forces those materials out of the hide structure and off the surface is called *scudding*. The liming treatment, followed by acid neutralization and enzymatic digestion (bating), leave the collagen loosely organized and swollen. The structural changes result in a thickening of the hide referred to as *plumping*.

The plump hide is able to retain significant amounts of moisture from the washings that follow bating and scudding. The plump structure is weaker, may be less likely to diffuse tanning agents through the collagen matrix, has a basic pH, and is subject to deterioration during the storage period that may occur between hide preparation and initiation of the tanning process. *Salt pickling*, immersion or drumming in a salt/acid solution, shrinks the collagen, reduces the pH, dries the hide, and stabilizes it for storage. Pickling with salt is an essential preparatory action for some tanning processes. The pickled hide is run through rollers to wring the water and excess pickle from it. The pickled hide may be stored, shipped, or tanned.

At some point during the preparation of the hide, it may be sided or cut longitudinally down the midline into two sides. This is often called *splitting* the hide. Hides may be processed as intact hides or as sides, depending on the preference of the organization. Sides are easier to handle in the hide processors and drums and permit larger batches to be processed per machine.

The preceding steps were seldom accomplished by a meat packer years ago. In those days, the hide-processing steps beyond salt curing were accomplished by the tanner or by an intermediary operation called a "hide processor." Although packers today do ship cured hides, some are taking hides through the pickling procedure and some even through the first chrome-tanning stage to produce what the industry refers to as "wet blue" hides.[16] Hides that are immediately processed through the pickled or wet blue stages (see Figure 11–6) by the packer are not cured, since there is no need to preserve fresh hides that will be shipped as wet blues within the week after they are *flayed* (removed from the carcasses).

---

[16]Minnoch, J., and S. R. Minnoch (Ed.). 1979. Hides and Skins. Distributed by National Hide Assn., printed by Eakin Press, P.O. Box 178, Burnet, TX 78611.

Fig. 11–6. Trimmed, fleshed hides, delivered directly from the harvest area of one U.S. packer are placed into very large wooden drums where all of the processing procedures to produce chrome-tanned "wet blue" hides are accomplished. (Courtesy, IBP Corp., Dakota City, Nebraska)

The advantages to the industry of further processing include not only reduced shipping weights and costs due to trimming and moisture loss at the site of production (see Figure 11–7) but also the salvage of trimmings and fleshings through the packing plant rendering operation. Pollutants resulting

Fig. 11–7. The chrome-tanned "wet blue" hides marketed by packers and hide processors are more stable and contain less moisture and contaminants than salt-cured hides. The hides are spread for evaluation and folding after passing through the rollers behind the worker. (Courtesy, IBP Corp., Dakota City, Nebraska)

from dehairing, bating, and, particularly, the chrome process are produced in a more remote area and within an operation well versed in dealing with pollution. With more processing being done at the packer level, the added value at a high-volume position in the process should have a positive economic impact on the whole industry.

Commercial Tanning

Tanners produce finished leather, whether they start with fresh, cured, pickled, or wet blue hides. Some tanners may also fabricate consumer products from the leather. Several steps are necessary to produce finished leather, and the procedures and conditions vary widely, depending on the traits required in the finished product. Multiple volume encyclopedias on the tanning process have been written.[17] The coverage here will be very brief.

Tanning

Tanning is the incorporation of organic or inorganic compounds into the properly prepared hide or skin collagenous structure to produce a stable, flexible, sheet material called leather. The common tanning processes may be grouped into vegetable, chrome, and miscellaneous on the basis of the source of the incorporated compounds.

VEGETABLE TANNING — *Tannins,* the term applied to a group of compounds that have been known to transform hide into leather for at least 5000 years,[18] are extracted from various portions of many different plants. The wood and(or) bark of woody plants (trees) provide the source of some of the tannin material used commercially; however, other plant parts may also be sources of tannins. The wood of a South American tree, the quebracho, provides raw material for much of the commercially used tannin. Among the many other plant sources are mimosa or black wattle tree, chestnut, valonia, mangrove, oak, hemlock, and spruce. The composition of the tannin varies, depending upon the source; therefore, the tanner selects the tannin source based upon the leather to be produced and the operating conditions.

---

[17] A Modern Course in Leather Technology, a multi-volume series. 1966. The Leather Science Division of the Commonwealth Library. Pergamon Press, New York.

[18] Reed, R. (Ed.). 1966. Science for Students of Leather Technology. Pergamon Press, New York, p. 218.

Although one of the oldest known materials processing industries, tanning remains a combination of art, skill, chemistry, and, in some cases, luck.

Traditionally, the prepared hide or skin material is advanced through a series of increasingly concentrated solutions of the tannin and other agents in large vats or pits. Tanning may be accomplished in rotating, closed containers (drums) with the various solutions pumped in and out. The temperature, pH, and solution strength all have an impact upon the rate of diffusion of the tanning agents through the hide collagen fibers and upon the chemical bonding of fibers and tannin. If too much tannin bonds to the surface collagen fibers early, diffusion of tannin deeper into the hide structure is prevented. This sealing of the structure, called *case hardening,* may be stimulated by placing hides directly into relatively concentrated tannin solutions. The kind of acid used to achieve the desired pH may have a significant effect on the rate of tanning at different stages. Vegetable tanning is a long-term process, requiring varying numbers of days, depending upon the system and the product.

CHROME TANNING — Chromium salts are also effective in cross linking the collagen fibers in animal skins to produce leather. Chrome-tanned leather is stronger and tougher than vegetable-tanned leather. However, because the chrome tanning process blocks the ability of the tanned collagen to swell and close its structure when wetted, chrome-tanned leather does not turn water, but lets it pass through more readily. Whereas tannin-tanned leather is natural or light tan in color, chrome-tanned leathers are green to blue.

Chrome tanning is faster than vegetable tanning and is more likely to be accomplished in a drum. Packers who sell chrome-tanned blue hides place as many as 400 fresh, cooled, trimmed, fleshed, and washed hides in one drum (see Figure 11–6). The hides are dehaired (limed), washed, delimed, bated, washed again, salt pickled, and chrome tanned before they are removed two or three days later. The chemistry and control of the tanning environment is as important to success with chrome tanning as it is with vegetable tanning. Blue stock, which is properly prepared through chrome tanning, can produce almost any type of leather.

Aluminum salts can also be used in processing hides, but the collagen cross link that forms is less stable to moisture and the "tan" can be literally washed out of aluminum-tanned leather. Chrome and aluminum tanning are sometimes used together to produce the white leathers that result from aluminum tanning. Zirconium is a mineral tanning agent that results in light leather but is very expensive to use for that purpose. The zirconium tan is more resistant to water than the aluminum tan.

Both hide and leather processing result in effluent materials that may impact the environment. Protection of the environment is an important facet of chrome-tanning plant design, in particular. The industry must invest heavily in reclaiming systems and in facilities to clean and monitor effluent to assure a safe environment.

MISCELLANEOUS TANNAGES — Other products that are sometimes referred to as *syntans* (synthetic tannins) can be used to produce leather from hides and skins. Formaldehyde, which is a tanning agent too, is involved in the production of some syntans. Phenols in conjunction with formaldehyde and sulphones can be formulated to produce tanning abilities. The results are not equal to those derived from vegetable tannins but are adequate for many leather uses. Syntans are often used with vegetable tans to speed the tanning or to produce color or other physical alterations in the leather.

Chrome-tanned leather may be retanned with vegetable tannin and vice versa to result in leather that combines the qualities of both processes.

Post-tan Processing

Whatever the tanning process, subsequent treatments are required to finish the leather for almost any purpose. The final leather thickness, pliability, water resistance, and surface properties must be established after the tanning is completed. Since the tanning is done in solutions, the tanned hide will contain water that must be removed. One process is termed *setting out*. The leather is mechanically pressed or passed through tight rollers that squeeze the moisture and remaining tanning residue from the tanned collagen (see Figure 11–7). Uniform drying of heavy leather requires experience and skill. Oil may be used to control the movement of moisture from the deeper layers and the rate of surface drying.

The thickness of leather is altered by *splitting* the sheets into two or three layers on special machines. (Do not confuse this process with separating a hide into two sides, which is also called splitting.) The upper (hair) surface is called the *grain* layer and the lower surface, the *flesh* layer. Splitting may also occur before the tanning process to permit the layers to be used for purposes that require different tanning procedures. Flesh layers of sheep skins are used for chamois; thus, they may be tanned and further processed to maximize flexibility. Uniformly thick leather of a specific depth is the result of *shaving*. The shaving machine trims away more of the thicker portions of the skin and also produces a more finished surface. The shavings may be pressed and bound into sheets to be used as an economical leather product, sometimes called *shoddy leather*.

Tanners use special formulations and combinations of oil, soap, or fat to *fatliquor* leather. The flexibility of the finished leather, the resistance of the leather to the effects of moisture, and the appearance of the leather are all improved by this incorporation of oil or fat between and on the surface of the fibers. In some cases leather may be *stuffed* with oils and(or) other materials such as fine clay or sugars to provide special characteristics.

In the early days of leather processing in the home, tanners softened skins by making a bend across the length of the leather and by biting along the length of the fold. Another fold was made adjacent to the first, and the process was repeated until the entire skin was softened. One baby bunting skin could take hours to soften. In the tannery, machines flex and work the leather to assure softness and flexibility. This *staking* operation is done as the leather is drying and in conjunction with fatliquoring. It is probably called staking because hides were pulled and flexed over stakes or posts before the days of mechanical staking.

Finished leather has surface finishes and textures that vary from glistening patent to coarse suedes. The patent leather has been *finished* with varnish to produce the unique shine. The suede has probably been *buffed* or sandpapered to produce the rough surface. By *plating,* special textures or designs may be embossed on the grain surface. Leather is bonded or pressed tightly against the surface of an etched plate while it is drying, much as a seal is applied to paper or ink is applied to newsprint. *Dyeing*, special finishing, and *drying* techniques are all used to enhance the product's utility, appearance, and longevity in these final phases of the transition from animal skin to quality leather.

CALF SKINS

Skins are the pelts of small animals, wild or domestic, such as calves, sheep, goats, muskrats, foxes, and minks.

Calf skins and kipskins produce leather that some experts refer to as the "elite of the industry."[19] These fragile hides must be processed carefully to insure the attainment of leather quality. Curing must be accomplished quickly

---

[19]Minnoch, J., and S. R. Minnoch (Ed.). Hides and Skins.

and adequately, but excessive salting should be avoided. Trimming a calf hide is very important, due to its small size and the importance of a proper pattern (see Figure 11–3). Storage intervals of calf skins and kipskins should be minimized to reduce the probability of deterioration. The skins of near-term unborn calves are termed slunk skins and command high prices because of their scarcity and high leather quality.

PIGSKINS

As indicated in Chapter 5, most U.S. pork processors have used in the immediate past and are presently using a dehairing technique that leaves the skin on pork carcasses until they are processed into wholesale or retail cuts. With such a system, the only fresh skin available, of any consequence, is that resulting from the fatback and the hams, and this skin is largely used in gelatin production. Most whole pigskins for tanning are imported. Pigskin is used as leather for gloves, wallets, handbags, brief cases, toiletry cases, tobacco pouches, bookbindings, and leggings.

Wolverine World Wide Incorporated has developed a machine to remove scalded skin from the belly and backfat portions of the carcass.[20] Single-side units are capable of removing a 3- to 3½-square-foot piece of skin extending from the ham cut-off line to the shoulder cut-off line, while double-side units can remove the skin from the middle section of the whole carcass in one piece. The pieces are processed and tanned, and much of the leather is merchandised in the form of sueded footwear.

The pigskin leather is tough and produces scuff-resistant footwear. The hog bristle (hair) is unique in that it grows through the skin from the follicle in the subcutaneous fat layer. The holes, or pores, through which the hairs pass, result in a naturally "air-conditioned" type of leather.[21] Wolverine uses significant amounts of pigskins, particularly for sueded leathers.

Scalded skins can be pulled with skinning equipment, but the leather is 10 percent thinner and has less tensile strength than leather from unscalded skins. The heat damage incurred during scalding makes many skins unsuitable for use as leather; however, with the introduction of the mechanical pig skinner (see Chapter 5), several processors are gearing up for hog-skinning operations. Although scalding is more labor-efficient during the harvest phase,

---

[20]Ibid.

[21]Ibid.

labor is required to remove the skins at later stages of processing. Curing procedures for pork skins utilize the hide-processor technique in some plants. In some cases, skins may be transferred fresh to commercial processors or directly to a tanning company, such as Wolverine. Defleshers (fleshing machines) smaller than the ones used for beef hides are available. Many are constructed of stainless steel because the pork fleshings are edible and can be converted to edible lard, if processed through FSIS–approved equipment and procedures.

Specially selected and treated hog skins, because of their similarity to human skin, are used in the treatment of humans suffering from massive burns. Other injuries that have removed large areas of skin and persistent skin ulcers may respond to hog skin treatment as well. Hog skins are cut into strips or patches, shaved to remove the hair, split to 0.008 to 0.020 inch in thickness, and then cleansed, sanitized, and packaged. The skins are then applied directly to the injured areas to decrease pain, inhibit infection, and prevent loss of body fluids. Closely adhering "porcine" dressings help prepare the patient for permanent skin grafting by promoting the development of granulation tissue, so essential before skin grafting can begin. They also ease the flexing of joints and the stretching of scar tissue early in the treatment procedure, a vital factor in the patient's return to full physical capability.[22]

Skins for porcine dressings are sent from source plants to Genetic Laboratories, Inc., of Minneapolis. Upon arrival there, the skins are shaved, split, and packaged in a variety of forms in surgically clean conditions and chill-stored or frozen to await shipment to hospitals and burn-treatment centers.

In explaining the rapidity of this process, the director of the Skin Bank at Genetic Laboratories said that within 24 hours from the time a hide is removed from a hog, the porcine skin made from the hide can be used as a dressing on a burn victim. It has been estimated that these porcine skin dressings save at least one life per day in this country, and they alleviate much pain and shorten the length of hospitalization for other severe burn victims.

HAIR

A substitute for camel's-hair brushes is made from the delicate hairs on the inside of the ears of cattle. Hog bristles for making brushes were formerly

---

[22]Hog Is Man's Best Friend. 1978. National Live Stock and Meat Board, Chicago.

imported from China but are now being produced in the United States in increasing amounts. It requires considerable hand labor to collect the proper length hair that is found over the shoulder and back of the hog. The fine hair of the bulk of our domestic hogs is not suitable for brush making and is processed along with other inedibles through the tank house (rendering department) and sold as meal.

SHEEP PELTS

Sheep pelts are classified by the length of wool they carry as follows: Full wool, 1½ inches or more; Fall shorn, 1 to 1½ inches; Shearling No. 1, ½ to 1 inch; Shearling No. 2, ¼ to ½ inch; Shearling No. 3, ⅛ to ¼ inch; and Shearling No. 4, ⅛ inch.

Sheep pelts are further classified as to their origin as follows: River — Missouri River area; Southwest — Texas, Oklahoma, Arizona area; Northern — Minnesota and the Dakotas area.

Sheep pelts are normally cured with the wool left on and are then sent to a "pulling" operation that removes the wool. The freshly removed pelts are rubbed with fine salt to preserve them until the time they are sold. In packing houses, a depilatory paste made of lime and sodium sulfide is spread on the skin side of pelts, and the next day the wool can be pulled. Depending on the wool length, the wool is used in a wide variety of ways. The resulting skin is processed into leather, with techniques similar to those described for beef hides. The leather produced from sheep skins is used for suedes, shoe uppers, linings, garments, and accessories. Some sheep skins are processed and tanned with the wool left on to provide coat and boot linings, as well as rugs, decorative pelts, and seat covers. A great deal of sheep skin is used for bookbinding, hat sweat bands, shoe linings, gloves, and chamois skins. Because they are larger and wear better, goat skins are more valuable than sheep skins.

Sheep skins of long-wooled breeds from which the wool has been pulled supply the best wearing leather. Of the long-wooled breeds, skins of the Merinos supply the poorest leather.

Bed pads made of special glutaraldehyde-tanned shearling (sheep skins with the wool evenly clipped) are effective nursing aids for preventing and healing bedsores. The painful sores develop in patients who cannot move and must lie for long hours in one position. The wool is placed in direct contact with the patient's skin. This allows for free circulation of air and absorption of perspiration. It also minimizes skin abrasion, thereby aiding in preventing and in healing the bedsores. The pads are resilient and distribute the weight of the patients evenly. The wool is highly flame-retardant. Conventionally tanned shearlings are hardened and shrunken by a few launderings, whereas the glutaraldehyde-tanned pelts retain usefulness after 50 or more launderings.[23]

WOOL

Pulled wools constitute a significant portion of all the wool produced in the United States (2 to 8 pounds of pulled wool per head). A typical 8- to 10-pound fleece (approximately 4.5 pounds of scoured wool) will make about 3 yards of worsted fabric, which is sufficient for the production of one suit. The degreasing of wool removes an oil that may be 15 percent of the weight of the wool. This oil, when treated, produces some valuable products. They include lanolin, which is used as a base for ointments and cosmetics, leather dressings, and fiber lubricants. The potassium carbonate removed in the wash water in wool cleaning represents a significant part of the by-product value of degreasing.

HOME TANNING

Although the available equipment and the number and kind of hides, skins, or pelts to be processed will determine the technique to some degree, a few suggestions for the home tanner who may be dealing with anything from a buffalo bull hide to a bunny pelt are offered. One procedure for the preparation of fresh sheep skins for tanning would be to start by soaking the pelts for 10 to 15 minutes in a tub while working out the blood and manure. It may take more than one cycle to complete the task. The pelts should be trimmed of residual fat and muscle tissue and ragged shank skin or cut areas. The flesh surface should be scraped (fleshed) of all remaining non-hide tissue. The pelts could then be placed in a washing machine, if one is available, and run through the complete wash cycle with a mild dish soap. When the pelts are removed from the washer, the small particles of dirt, hair, and loose tissue can be wiped, carded, or combed off the skin surface. Thorough hand scrubbing,

---

[23]USDA. December, 1965. Agric. Res., p. 8.

with a brush, soap, and warm water, is necessary if a washing machine is not available. Careful and complete rinsing will minimize the residual soil and soap. Hand wringing or a washing machine wringer will remove the water. The skins are then salted and cured up to 30 days. The procedure just described could also be used for salted pelts after they have been softened by a water soak.

Salt Alum[24] Tanning

Soak the salted sheep skins in water until they are soft, and then place them on tables or beams where they can be trimmed and fleshed. Wash the pelts, using the procedures indicated in the preceding section. Place the pelts on tables, fleece side down, and rub them thoroughly with a mixture of one part of powdered alum to two parts of common salt (4 ounces of alum and 8 ounces of salt will tan the average-sized sheep pelt). Leave the pelts in this position overnight; the next morning, hang them over a rail, skin side up. Allow them to hang in this position for 24 hours, then sponge the skins to remove the unabsorbed salt, and rub 1 to 2 ounces of neat's-foot oil or glycerine into the damp, soft skins. As the skins dry, during the next two days, stake (page 322) them several times to keep them from becoming hard. When they are dry, buff them, using a coarse-grade sandpaper fastened over a block of wood. Card the wool, using an ordinary wool card. The pelts are ready for use.

Alum Tanning

This method is suitable for fur pelts and skins, since the aluminum sulfate does not color the fur or the skin. Dissolve 1 pound of aluminum sulfate in 1 gallon of water, and dissolve 4 ounces of crystallized sodium carbonate (soda ash) and ½ pound of salt in ½ gallon of water and then pour this slowly into the alum solution, stirring it vigorously. Place the prepared skins (soaked, fleshed, and washed) in the solution for two to four days, depending upon their thickness, and then rinse them and put them through a wringer. Rub the damp skins with oil or glycerine, and as they dry, stake and then buff them.

---

[24]Alum is (a) potassium aluminum sulfate — $KAl(SO_4)_2 \cdot 12\ H_2O$ or (b) ammonium aluminum sulfate — $NH_4\ Al(SO_4)_2 \cdot 12\ H_2O$ or (c) aluminum sulfate — $Al_2(SO_4)_3$.

Before the skins are rubbed with oil or glycerine, retanning them with 1 pound of Leukanol (a syntan made by Rohm and Haas Company), dissolved in 1 gallon of water, will make them tougher and softer without discoloration.

Vegetable Alum Tanning

Dissolve ½ pound of aluminum sulfate and ½ pound of salt in a small quantity of water. Dissolve 2 ounces of gambier or Terra Japonica[25] in a little boiling water. Mix the two solutions, and add sufficient water to make 1 gallon. Add sufficient flour with the 1 gallon of tanning liquor to make a moderately thin paste. Take a properly prepared pelt (soaked, trimmed, fleshed, and washed) and apply three coatings, about ¼ inch thick, at two-day intervals, to the skin side of the pelt, removing each previous coating before applying the next. When the pelt is practically dry, rinse it in warm water containing some borax, and then rinse in fresh water. Squeeze out the water and then slick the skin with a dull knife. After applying a coating of glycerine, hang the pelt up to dry. Stake the pelt several times while it is drying and then buff with coarse sandpaper.

This method produces a yellow skin of good tensile strength.

Salt Acid Tanning

Make up a solution of 3 ounces (fluid) commercial sulfuric acid and 2 pounds of common salt per gallon of soft water in a wooden or other non-metallic container and in sufficient volume to immerse the prepared hides (soaked, fleshed, and washed). Small, thin (calf) skins should remain in the tanning solution 12 to 24 hours. Cow or steer hides may require two to seven days, depending upon the size and thickness of the hides. The hides should be stirred or moved in the solution every few hours to assure even tanning. Wash the hides in cold water to remove excess acid and salt, and then stretch the hides to drain. Sheep pelts should be spread wool down, and the tanning solution daubed on with a rag or a handful of wool. The application should be repeated several times to assure that the hides are thoroughly treated with the solution. After 12 hours, wipe off any excess tanning solution, and then

---

[25]A yellowish, dry, resinous, astringent substance obtained from a Malayan woody vine.

wash the surface with a damp cloth. When the hides are nearly dry, work them (staking), and apply a coat of neat's-foot oil. Allow them to dry, then moisten thoroughly and work them while they are drying. This technique is said to produce a strong white leather.

Hides can be tanned with the hair left on in most of the procedures described, or they can be dehaired prior to tanning by being immersed in a solution of lime water (2 pounds of lime dissolved in 5 gallons of water). The hides should remain in the solution until the hair slips, which should occur in three to five days. After that, the hair should be scraped off with a fleshing knife, and all the lime water should be washed from the hides. The dehaired hides are then ready for tanning.

LEATHER, PAST AND FUTURE

At one time, the greatest use for leather in the United States was to manufacture harnesses for the many horses that provided power for work and transportation. With the advent of mechanized power and transportation, the need for harness leather decreased rapidly. Leather was then used extensively to transfer the horsepower from engine pulleys to machine pulleys by leather belts. But again, technology found new and better methods. There remained, however, a brisk demand for leather in shoe manufacturing, until synthetic composition soles gradually began to take over — by 1960, only 35 percent of the shoes manufactured in the United States had leather soles. The composition soles could be produced cheaper. However, the resilient, pliable wearing qualities of leather are still utilized extensively in the shoe uppers, if not in the soles. The hides from many cattle are required every year for leather accessories in the sports field. Footballs, commonly referred to as pigskins, are covered with cowhide today.

U.S. hide producers are exploring new frontiers in domestic hide consumption and at the same time eyeing the growing demand for hides to export. To do this, they are making greater use of research and technology. The United States is a prime source of raw materials for countries that manufacture leather (Italy, Japan, South Korea, for example). Table 11–12 shows that approximately equal numbers of U.S. hides are being exported as are being domestically processed each year.

The use of leather in the United States includes 15 to 20 percent for the manufacture of garments and gloves and 5 to 10 percent for miscellaneous products, in addition to the 75 to 80 percent for footwear. Leather garments

Table 11–12.   U.S. Production of Beef Hides, Skins, and Leather and Exports of Hides

Year	Hide and Skin Production		Leather Production[1]	Export Cattle Hides	
	Cattle	Calves	Number	Number	Value
	- - - - - - - - - - - - - - - (1,000 pieces) - - - - - - - - - - - - - - -				($1,000,000)
1983	36,649	3,077	18,470	21,281	728.4
1984	37,582	3,297	17,000	25,029	1,066.6
1985	36,293	3,385	15,280	24,956	991.9
1986	37,288	3,408	14,510	26,336	1,195.5
1987	35,647	2,815	14,750	23,953	1,306.2
1988	35,079	2,504	13,300	24,349	1,465.1
1989	33,917	2,172	12,932	23,234	1,296.9
1990	33,242	1,805	14,820	21,342	1,286.6
1991	32,690	1,436	14,800	20,017	1,073.6
1992	33,069	1,420	15,900	19,098	1,038.9
1993	33,504	1,242	18,057	18,226	926.6
1994	34,376	1,315	18,842	17,911	1,059.2
1995	35,817	1,477	18,092	20,044	1,224.6
1996	36,760	1,815	18,769	20,289	1,125.1
1997	36,492	1,619	19,592	18,991	1,134.3
1998	35,637	1,501	20,297	17,805	871.1

[1]"Cattle Hide Equivalents" accounting for cattle, calf, kip, goat, sheep, lamb, and horse leathers produced.

Source: Compiled from official USDA and Department of Commerce statistics by U.S. Hide, Skin, and Leather Assn. (Jerry Breiter.)

are popular and more practical than in the past, as a result of improvements in the resistance of leather to the effects of drycleaning solvents and the production of light weight, supple cowhide garment leathers. International markets for U.S. leather have remained strong or have improved (doubled in the European Community between 1985 and 1986), partly due to the activities of The Leather Industries of America, a trade organization of U.S. tanners, that has been promoting "American Lifestyle Leather" at trade shows worldwide.

Hides and skins exported to other countries (see Table 11–12) return to the United States as leather, or as consumer products made of leather (see Table 11–13). Tanning and leather fabrication are labor-intensive activities. Countries with large, low-cost labor forces can utilize raw materials obtained from the United States to profitably produce products sold in the United States at prices lower than U.S. manufacturers can produce the products. The imports of leather into the United States usually approximately equal the exports of leather (Table 11–13).

Table 11–13.   Leather: U.S. Foreign Trade, 1974–97

Year	Imports	Exports
	- - - - - - - - - - - - - - - ($1,000) - - - - - - - - - - - - - - -	
1974	124,513	102,116
1975	87,953	141,715
1980	217,306	271,944
1985	394,124	283,704
1986	406,485	313,408
1987	562,744	395,483
1988	748,363	506,483
1989	743,615	624,925
1990	682,987	750,836
1991	570,845	680,348
1992	631,360	705,038
1993	735,793	764,120
1994	959,090	811,951
1995	1,088,959	870,247
1996	1,139,426	950,510
1997	1,375,730	1,145,664
1998	1,571,196	1,289,547

Source: U.S. Hide, Skin, and Leather Assn.

Although technological advances continue to produce materials that replace leather in current uses, new uses, as well as the improvement of traditional leather products, such as the garment cowhide, also result from technological advances. New domestic uses for hides are being developed, as evidenced by the previous discussion concerning the medical uses of pigskins and lamb shearlings, or expanded, as indicated by the increased use of leather-covered upholstery in homes and autos. Is there a leather future? Indications are that the single most valuable by-product of beef harvest (see Table 11–14) will remain so.

Table 11–14. Percent of USDA's Estimated Hide and Offal Value for Harvested Beef Accounted for by Hide Value, 1982–1998

Year	Est. Hide and Offal Value[1]	From Hide
		(%)
1982	$5.58	44.6
1983	5.91	49.3
1984	6.68	53.4
1985	5.84	54.1
1986	6.28	60.9
1987	7.79	62.0
1988	8.56	59.6
1989	8.10	65.2
1990	8.41	66.5
1991	7.41	58.7
1992	7.56	60.2
1993	7.75	58.8
1994	8.36	59.2
1995	8.89	59.3
1996	8.87	57.5
1997	8.83	57.5
1998	7.03	56.0

[1]Value estimated weekly by USDA on basis of 100 lbs. live weight of steers for harvest.

Source: Compiled from USDA statistics by U.S. Hide, Skin, and Leather Assn.

PHARMACEUTICAL BY-PRODUCTS AND XENOTRANSPLANTATION

Animal products have been used in the healing process for centuries. In fact, some animal products have held "magical" healing powers for certain societies throughout history. Similarly, the minute portions of certain animal extractives that many humans use each day can literally be the difference between life and death for these people.

Scattered through various parts of the animal body are a number of internally secreting, ductless endocrine glands. The substance secreted by each exercises some specific control over the conduct, character, and development of the body. The functions of these glands are so interrelated that undersecretion or oversecretion of any one of several of the glands will cause abnormalities. Many of the magical products derived from animal tissues saved by the meat industry and extracted, purified, and prepared by the pharmaceutical industry are hormones. Enzymes and other types of chemicals are also derived from animal harvest by-products.

Much of the following information about the pharmaceutical by-products has been extracted from two National Live Stock and Meat Board publications: *Hog Is Man's Best Friend* and *The Good Things We Get from Cattle Besides Beef.* Even though many of these pharmaceutical by-products are now produced synthetically through biotechnological procedures (see Chapter 2), some people in the United States and around the word still demand animal-derived pharmaceuticals.

Xenotransplantation is the transplantation of cells, tissues, and organs from one species into another.[26] There is a tremendous backlog of humans waiting for organ transplants. Officials from the Food and Drug Administration (FDA), the Centers for Disease Control and Prevention (CDC), and the National Institutes of Health (NIH) agreed to allow research to continue under stringent safeguards. With pig-to-human transplantation, for example, the ongoing concern is that even if the human immune system does not reject pig organs, the organs might cause humans to become infected with pig diseases. Transgenic pigs have been produced by injecting human genetic material into pig embryos. The human immune system will

---

[26]Vogel, G. 1998. No moratorium on clinical trials. Science. 279:648.

not reject organs from these pigs. Disease transmission from the pig tissue to humans may not be as serious as originally proposed.[27, 28]

Adrenals

The adrenals, also called the suprarenal glands, are two in number. They are bean-shaped in the sheep, measure approximately 1 inch by ½ inch, and may be found some 2 inches from the kidneys. In the ox, they are located near the center of the animal (medial), anterior (towards the front) of the kidneys, and are triangular or heart-shaped. In the pig, they are long and narrow and lie on the medial border of the kidney. In humans, they rest astride the kidneys and are larger than most endocrine glands. They are reddish-brown in color and are somewhat bean-shaped. The cortex (outer portion) produces steroid secretions essential to life maintenance. The medulla (inner portion) of the glands produces epinephrine. It requires the adrenals of 13,000 head of cattle to produce 1 pound of epinephrine. Each adrenal gland weighs approximately ½ ounce (14 grams).

Until recently, hog adrenal glands were an important source of many different hormones that physicians used to treat illnesses or chemical imbalances in the human body. Now many adrenal compounds are being made synthetically.

Corticosteroids, from the cortex or outer shell of the hog's adrenal glands, influence and regulate the human body's utilization of minerals and nutrients. They regulate water, nitrogen, potassium, and sodium (salt) balance in the body. They are used in treating shock, deficiencies of the adrenal glands, and Addison's disease — a debilitating ailment brought on by malfunction of the human adrenal cortex, which causes a person to lose weight and become thin, to develop dark-skin pigmentation, to lose strength, and to lose sodium in the body.

Cortisone, one of the corticosteroids, influences fat, sugar, carbohydrate, and water metabolism (utilization of food and other materials within the body), improves muscle tone, and reduces pain caused by calcium deposits in humans. It has many therapeutic uses, such as treatment for shock, arthritis, and asthma. Cortisone for pharmaceutical uses is now produced synthetically.

The hormones epinephrine and norepinephrine, produced by the adrenal medulla, are also manufactured synthetically. Epinephrine, also called adrena-

[27]Weiss, R. A. 1999. Xenografts and retroviruses. Science. 285:1221–1222.

[28]Paradis, K., G. Langford, Z. Long, W. Heneine, P. Sandstrom, W. M. Switzer, L. E. Chapman, C. Lockey, D. Onions, The XEN 111 Study Group, and E. Otto. 1999. Search for cross-species transmission of porcine retrovirus in patients treated with living pig tissue. Science. 285:1236–1241.

lin, stimulates body processes in the utilization of food. This drug is used to relieve some of the symptoms of hay fever, asthma, and some forms of allergies affecting the mucous membrane of the nasal passages. It may be used to shrink blood vessels during certain types of surgery, control bronchial asthma spasms, and reduce inner-eye pressure during glaucoma treatment. It is also used to restore heartbeats in cases of cardiac arrest. Dentists use it to prolong the effects of local anesthetics.

Norepinephrine helps shrink blood vessels in humans, reducing the flow of blood through the body and slowing the rate of rapid heartbeats.

Blood

Blood albumin from meat animals is used in human blood Rh factor typing. Blood fibrin extract from hog blood is used to make amino acids that are part of parenteral (infused as intravenous) solutions for nourishing certain types of surgical patients.

Fetal pig plasma is important in the manufacture of vaccines and in tissue culture media. Fetal pig blood contains no antibodies and therefore is unlikely to stimulate immune reactions.

Plasmin, a hog blood enzyme that has the unique ability to digest fibrin in blood clots, is used to treat patients who have suffered heart attacks.

Thrombin helps create significant blood coagulation. It aids in the treatment of wounds, particularly in cases in which the injury is in an inaccessible part of the body, such as the brain, bones, or gastrointestinal tract (as in the case of peptic ulcers). In skin grafting, thrombin helps to keep the graft in place and to "cement" gaps where tissues have been surgically removed.

Fibrinolysin combined with deoxyribonuclease from the pancreas helps remove dead tissue that results from certain vaginal infections. It is a valuable cleansing agent for infected wounds or clotted blood and can speed up the healing of skin damaged by ulcers or burns.

Blood is also used in cancer research, microbiological media, cell cultures, protein hair conditioners, fertilizers (see page 346), animal feed (see pages 307–308), and human food (see pages 284–285).

Brain

Hog brains are a potential source of cholesterol, the raw material from which vitamin D_3 is made. Vitamin D_3 is necessary in building bones and teeth and is used to treat rickets in children and premature infants.

The hypothalamus produces hormone-releasing hormones, relatively small molecules that cause the release of various hormones from the pituitary gland.

Thromboplastin, made from the brains of cattle, is used as a blood coagulant in surgery.

Gall Bladder

Chenodeoxycholic acid, once derived from hog bile acid, to be given to humans to dissolve gallstones, is now produced chemically at half the cost of extracting it from animal tissue.

Ox bile extract from liver bile or a component (dehydrocholic acid) is used in the treatment of indigestion, constipation, and bile tract disorders that stem from disease or surgery.

A unique market has developed in China for gallstones, where they are thought to have mystical values. The price for gallstones in the United States may be as high as $600 per ounce. One large beef packer, harvesting more than 100 cattle per hour, accumulated less than 1 pound of gallstones in two years. However, a trade company in Hong Kong indicates that even 30 kg (67 pounds) per year from any country is worthwhile.

Heart

Hog heart valves, from young pigs to full-sized market hogs, are specially preserved and treated and surgically implanted in humans to replace heart valves that have been weakened or injured by rheumatic fever or by birth defects.

Hog valves are superior to mechanical valves in several ways. For one, the vast majority of patients with mechanical valves usually require constant infusion of anticoagulant drugs to prevent the valves from sticking, and in many patients, the use of anticoagulants over long periods of time may cause many types of undesirable side effects. In contrast, the hog valve is a naturally formed and functioning organ from a living animal that continues to function normally in the human body as an inanimate object, generally with fewer anticoagulants required. If a problem develops with a hog valve, early warning symptoms alert patient and physician in time for surgery. Also, anticoagulants are generally not prescribed for children because they tend to retard physical growth. Hog valves implanted in children with a minimum of anticoagulants usually correct congenital heart defects without disturbing physical development.

Since the first operation in 1971, many thousands of heart valves have been implanted in adults and children ranging in ages from less than 1 year to more than 70 years. Since hog valves do not grow while implanted in the human heart, youngsters, as they grow and develop, may have to undergo three surgeries, with progressively larger valve replacements each time.

Bovine pericardial tissue (the membrane enclosing and attaching the heart within the chest cavity) is processed and utilized to "patch" the pericardial tissue of patients after bypass or other heart surgery. Use of the treated bovine tissue prevents adhesions of the pericardium to the sternum, which often follow cardiac surgery.

Intestines

Heparin, classed as one of the "essential" pharmaceuticals and obtained almost exclusively from the inner (mucosa) lining of the hog's small intestine and from lungs, is a natural anticoagulant that thins the blood and dissolves, prevents, or retards clotting during surgery, especially surgeries involving organ transplants. Heparin is also used as a gangrene preventative in cases of frostbite and as a burn treatment. A smaller form of heparin, used in Europe, has been depolymerized enzymatically or chemically and is more effective and specific in its action.

Enterogastrone, a hormone taken from the hog duodenum (beginning of the small intestine), regulates gastric secretions in the stomach. It is also being used experimentally to speed the emptying time of the stomach.

Secretin hormone, also from the duodenum, stimulates pancreas glands to produce pancreatic juices. It is injected in humans to test for disease of the pancreas.

Much of the "cat gut" used for surgical suture is derived from sheep and other meat animal intestines.

Liver

In 1926, researchers reported that patients with pernicious anemia showed marked improvement when lightly cooked animal liver — a source of vitamin B_{12} — was included in their diets. Funded by a grant from the National Live Stock and Meat Board in 1924, Dr. G. H. Wipple of the University of Rochester researched the importance of liver as a treatment for pernicious anemia. In 1926, Dr. George Minot of Harvard prescribed the practice of eating raw liver. Today, patients may be given vitamin B_{12} by injection. Desiccated liver, containing added nutrients, is often used as a nutritional supplement. Liver extract was sometimes combined with folic acid and injected into the blood-stream to treat various types of anemia, including pernicious anemia. Liver injections were also used to treat sprue (a long-term condition associated with diarrhea), weakness, emaciation, and anemia.

Cholic acid has been purified from bile. It is an intermediate in the formation of chenodeoxycholic acid and ursodeoxycholic acid, which are also derived from bile to be used in the treatment or prevention of gallstones.

Catalase is an enzyme that is used in dairy processing, mainly cheese making. Hydrogen peroxide is added to milk to sterilize the milk to eliminate harmful bacteria. Those microorganisms that are beneficial to the cheese-making process are not sensitive to hydrogen peroxide. After the treatment, the excess hydrogen peroxide is decomposed by the addition of catalase. Then chymosin, which has been produced from extracted calf stomach tissue, is added to curdle the milk.

Lung

Lungs may be used as a source of heparin, but intestinal mucosal heparin extractions are more easily purified. Lung tissue is a source of a pancreatitis treatment product called aprotinin.

Ovaries

Hog ovaries are a source of progesterone and estrogen used to treat various reproduction problems in humans. Sow ovaries are the major source of relaxin, a hormone often used during childbirth. The harvest of 145 female hogs will produce 1 pound of fresh ovaries from which corpus luteum and ovarian extracts are prepared.

Pancreas

The pancreas is more commonly known as the pork sweetbread but should not be confused with the commercial veal sweetbread (thymus gland). The pancreas has both internal and external secretions, the latter passing into the small intestines to aid in the digestion of starch, protein, and fat. The internal secretion (insulin) regulates sugar metabolism. Failure of the pancreas to supply the insulin to regulate sugar metabolism results in the disease known as diabetes mellitus. Diabetes in humans was a killer disease before it was discovered that animal insulin could be used in humans.

Insulin, first isolated by Drs. Banting and Best, is secured from specialized groups of cells in the pancreas known as the Islets of Langerhans. Dr. Sanger and associates were able to establish the primary structure of insulin, that is, they determined the number and sequence of amino acids that make up the insulin protein molecule. Thus, it was theoretically possible to synthesize insulin in a "test-tube," and such a task has been accomplished. With the use of biotechnology (see Chapter 2), a product referred to as *humulin* is being produced and is replacing animal sources of insulin. Two types of bacteria have been engineered to produce each of the two chains in insulin's structure. The chains are then combined to form the final product. Although the human form may be higher priced, more diabetics are using humulin all the time. All newly diagnosed patients are put on the new product, and many former animal insulin users are being converted.

Hog pancreas glands are still a source of insulin hormone. Hog insulin is especially important because its chemical structure most nearly resembles that of humans, which is significant because approximately 5 percent of all diabetics are allergic to insulin from other animals and can tolerate only insulin from hogs.

Glucagon is a hormone given to raise the blood sugar level of alcoholics who have suffered a low blood sugar episode and to treat insulin overdoses in diabetics. It has a specialized use in the treatment of some mental disorders.

Kallikrein, also called kininogenase, is a proteolytic enzyme. It catalyzes a hydrolysis that forms kallidin. Kallidin and bradykinin, which is formed by the action of trypsin, are kinins. Kinins dilate vascular smooth muscle tissue and thus have the ability to reduce blood pressure. Kallikrein is sold as a hypertensive agent in the Japanese market.

Chymotrypsin is an enzyme used to cleanse wounds and to remove dead tissue where ulcers and infections occur. It can be used in treating serious injury or following surgery when localized inflammation and swelling result due to excess fluids. Chymotrypsin is generally used for the removal of devitalized tissue in eye surgery.

LPH (lipotropic hormone) is used as a digestive aid and is important in the digestion and absorption of fats and oils.

Pancreatin is a mixture of pancreatic enzymes used to treat faulty digestion in humans. Because of its high-fat digestive capability, pancreatin is also used in the treatment of cystic fibrosis, a disease afflicting approximately 4 million people in this country.

Trypsin helps break down food by aiding in the hydrolysis of protein in the upper part of the small intestine. Both trypsin and chymotrypsin are prescribed to remove dead and diseased tissue from wounds and to speed healing after surgery or injury.

Young animals have higher levels of insulin than older animals, but the opposite is true of the enzyme levels, which are higher in more mature animals. If the pancreas is collected to extract insulin, young animals are the preferred source, but if the products of interest are enzymatic, packers that are harvesting mature animals are more likely to be sought as sources.

Because the pancreas glands have been a source of insulin and other hormones and enzymes, many meat harvest operations have been saving them. Tissues saved for pharmaceutical use must be properly handled to maintain levels of the extractives and to prevent contamination with organisms that could degrade the tissues and their products. For that reason, the pharmaceutical companies that buy the glandular by-products have guidelines and specifications for the collection and storage of the raw materials. They also have employees who travel to plants to train persons to collect and handle the hormone- and enzyme-rich materials.

Parathyroids

Parathyroids consist of four small glands the size of a grain of wheat, which are located close to the thyroid gland. Their secretions regulate the calcium content of the bloodstream and maintain the tone of the nervous system. The removal of all the parathyroids will cause death within a few weeks. To secure 1 pound of parathyroid extract would require the harvest of approximately 3,600 animals.

The parathyroid hormone is used to compensate for the human body's inability to naturally produce this hormone. Parathyroid deficiency can result in convulsions, painful muscular spasms, loss of calcium from the bones, abnormal tooth development, and cataracts.

Pineal

The pineal gland is located in a brain cavity behind and just above the pituitary. It is about one-third the size of the pituitary and is reddish in color. Its secretion regulates child growth — hastening or retarding puberty and maturity.

The hog's pineal gland secretes the hormone melatonin, which is used in the treatment of personality and mental disorders. It also affects the color of the skin and the formation of freckles.

Pituitary

Located at the base of the brain and well-protected in a separate bone cavity, the pituitary gland is about the size of a pea and is grayish yellow in color. It is made up of an anterior and a posterior lobe that have distinct functions. The anterior lobe is known to produce (1) the growth-promoting hormone (GH), (2) the thyroid-stimulating hormone (TSH), (3) the mammary-stimulating hormone or prolactin, (4) the gonad-stimulating hormone, and (5) the adrenal-cortex–stimulating hormone (ACTH). The posterior lobe excretes principles that (1) control blood pressure and pulse rate (ADH or vasopressin), (2) regulate the contractile organs of the body, and (3) govern energy metabolism.

The growth-promoting hormone (GH) has shown great potential for increasing animal production; however, most of it is now being produced by genetic engineering techniques.

TSH (thyroid-stimulating hormone, or thyrotropic hormone) is used in conjunction with isotopes, etc., to locate small particles of thyroid cancer that may have spread to other parts of the body. Thyrotropin (TSH) is a hormone that stimulates the thyroid gland. It is used as a diagnostic tool to determine the cause of hypothyroidism in a patient, whether it stems from anterior pituitary failure or from complete failure of the thyroid gland. (In the event of anterior pituitary failure, the drug will stimulate proper functioning of the gland.)

Prolactin stimulates milk secretion in the mammary glands, and it may play a role in the future treatment of breast cancer.

Oxytocin hormone is used to treat obstetrical complications, induce labor, increase uterine muscle contractions, cause milk release by mammary glands, and lower the body's blood pressure to control uterine bleeding at childbirth. Oxytocin has also been used as a wound-closer by physicians attending professional boxers.

Corticotropin (ACTH) is a valuable diagnostic tool. Its most important medical use is to assess the operation of the adrenal glands. It can also be used in the treatment of psoriasis, the control of severe allergic reactions (rhinitis and bronchial asthma), eye inflammation due to allergies, certain respiratory diseases, anemia, arthritis, rheumatism, infectious mononucleosis, and multiple myeloma, a terminal form of leukemia. It would take the pituitary glands from 10,000 cattle to produce 1 pound of ACTH, and 1,800 hogs are needed to produce 1 pound of pituitary glands, which contain only ¾ gram (about $1/40$ ounce) of ACTH.

ADH (antidiuretic hormone, or vasopressin) helps regulate the body's water losses by the kidneys through urine production. Failure of the body to produce ADH results in excessive water loss in urine — a disease called diabetes insipidus. ADH also influences blood pressure by contracting the smooth muscle that reduces the diameter of arteries.

ADH is used in testing for renal functions. It is also employed to stimulate proper movement of material through the intestinal tract following operations and to dispel "gas shadows" that occur when abdominal X–rays are being made.

Thus, pituitary glands, primarily from hogs, produce a great number of hormones used to control human growth and metabolism problems and to regulate the activity of the body's other endocrine glands.

Skin

Gelatin from hog skin collagen is used for coating pills and making capsules. Gelatin is taken orally, theoretically to improve fingernail strength. (See the discussion of pork skins used as burn bandages earlier in this chapter.) A porcine collagenous product has been developed to stimulate clotting during surgery. The product is applied directly to the surface of the bleeding tissue.

Spleen

Splenic fluid affects capillary permeability and blood clotting time and speeds up recovery from inflammatory conditions (redness and swelling).

Stomach

Linings of hog stomachs contain proteins and enzymes used in many commercially produced digestive aids and antacids.

The pyloric lining of the hog stomach is rich in "intrinsic factors," or unidentified factors that must be present before the human body can utilize vitamin B_{12} from food or vitamin preparations to relieve or prevent pernicious anemia.

The pink mucous lining of the hog stomach is the richest natural source of pepsin, which is used in the treatment of achylia gastrica, a condition in which the stomach fails to produce gastric juices (acid and pepsin). Achylia gastrica is often present in cases of pernicious anemia and stomach cancer accompanied by achlorhydria, or lack of hydrochloric acid.

Mucin is used to treat peptic and duodenal ulcers. It also lubricates food movement through the digestive tract and is considered a valuable adjunct to many digestive products.

Rennet (rennin) or chymosin, a mild enzyme, is used to help infants digest milk and is used in cheese making. Chymosin is an extract of the fourth portion of the milk-fed calf stomach (abomasum). Chymosin is a more selective enzyme than pepsin or other enzymes that are used to curdle milk for cheese. Pepsin continues to hydrolyze other bonds that may soften the curd or produce bitter flavors. Thus, chymosin is used when high-quality products are desired.

Testes

Hyaluronidase, an enzyme that attacks the complex glycoprotein, hyaluronic acid, found in joints and other connective tissues, is derived from testes. Hyaluronidase is used to aid drug dispersion in connective tissues and other tissues.

Thymus

In veal, the thymus gland has a commercial food value. It is cream-colored, is located in the neck near the chest cavity, and has two lobes, the second lying within the chest cavity. The principal function of the thymus gland may be related to the inhibition of the activity of the gonads (glands related to sexual development and function) in young animals. Atrophy of the thymus gland following puberty is thought to be evidence of that function. The thymus is also considered to be a source of factors affecting the ability of the body to resist infections and to react to the presence of foreign protein and other introduced elements.

Thyroid

In the sheep, the thyroid gland is dark with a long, ellipsoidal outline measuring about 2 inches by ½ inch. It is located on the first five or six rings of the trachea. In the ox, the gland has two lateral lobes connected by an isthmus and is located just below the larynx (expanded portion of the respiratory tract containing the vocal cords). In a calf, the thyroid is about 3 inches long and dark in color. In swine, it is triangular-shaped, is about 2 inches across, may be located some distance from the larynx, has no isthmus, and somewhat adjoins the esophagus. It is smaller in cattle than in humans, and its secretion is an iodine-containing compound termed thyroxin. In the young, a deficiency of thyroid tissue causes a condition known as *cretinism,* which results in physical deformity and defective mentality, or idiocy; in the adult, it causes *myxedema,* defined as "severe thyroid deficiency" (hypothyroidism), characterized by dry skin and hair and loss of physical and mental vigor.

A deficiency of iodine in the diet or the water supply may cause a simple enlargement of the thyroid, termed a goiter. Goiter in humans and animals can be treated by supplying the necessary iodine. Oversecretion of the thyroid increases basal metabolism, causing the afflicted humans and animals to become nervous and thin. The action of thyroid secretions is interrelated with other glands.

Desiccated thyroid is used extensively in keeping hypothyroid patients from the slow-moving, slow-talking, inactive existence they would otherwise lead. It is one of the few glandular substances that is effective when taken orally. Because of the effect when orally administered, thyroids are ***not allowed to enter human food products*** (they are declared inedible at inspected harvest points). Their use in animal food without processing may result in physiological effects also. It requires 40 beef thyroids to make a pound (14 to 21 grams per gland). Hog thyroids are equally valuable.

Thyroglobulin, obtained exclusively from hogs, is given as a supplement to persons with underactive thyroid.

Calcitonin is given to lower calcium and phosphate levels in the blood and to regulate the heartbeat. It is also used in the treatment of Paget's disease, a painful malady of the bone.

Nervous System

Cholesterol comes from the spinal cord. It is essential in the synthesis of sex hormones that may be administered to humans when natural development of sex characteristics does not occur. These hormones are also used to treat menopausal syndromes and to prevent swelling of breasts and milk production when a mother does not nurse a new-born baby.

Bone Cartilage

Plastic surgeons may use the cartilage from the breastbone of young cattle to replace flat bones, such as facial bones. The specially processed xiphoid or xiphisternal cartilage (see Figure 15–1) permits bone damage repair.

OTHER BY-PRODUCTS

Glue and Gelatin

The raw materials used to produce glue and gelatin are high in collagen. They are connective tissue; skin or hide trimmings; sinews, horn piths, lips, ear tubes, pizzles, and cartilage; beef and calf bones; mammary glands; heads of cattle, calves, and sheep; and knuckles and feet. Pigskins are a good source of gelatin. Both glue and gelatin are colloidal proteins. They are chemically and physically similar and differ mainly in that gelatin is made from clean, edible materials prepared under sanitary conditions to make it edible.

Glue Stocks

The three main types of glue are hide glue, bone glue, and blood albumin glue. The latter is water-resistant and is used widely in the manufacture of plywood.

The oldest and widest use for glue is in the furniture and veneer industry. Glue has so many varied uses that it has been said that glue holds the world together. It is used in sizing paper; in the manufacture of wool, silk, and other fabrics; in sizing straw hats; in sizing walls that are to be painted; in sizing barrels or casks that are to contain liquids; on the heads of matches to make airtight caps over the phosphorus; in the manufacture of sand and emery paper to hold the abrasive on the paper; in the manufacture of dolls, toys, and ornaments; in the making of picture frames, mirror frames, rosettes, billiard balls, composition cork, imitation hard rubber, printing rolls, mother-of-pearl, gummed tape, paper boxes, calcimine, automobile bodies, caskets, leather goods, and bookbinding; and many other products.

Gelatin Stocks

The two types of gelatin according to their source are hide gelatin and bone gelatin.

Gelatin finds wide use in the manufacture of ice cream; in the making of certain pharmaceutical preparations and capsules for medicine; in the coating of pills; in the making of mayonnaise dressings and emulsion flavors; in the clarifying of wine, beer, and vinegar; in the making of court plaster; in photography; in electroplating; as a bacteria culture medium; and for various other uses.

A large percentage of gelatin comes from the bones of veal. The heads of veal calves find favor with some people who cook them and use the head meat and broth with noodles or as gelled meat.

Blood

As stated earlier, no blood which comes in contact with the surface of the body of an animal or is otherwise contaminated can be used for food purposes. Blood contains around 17 percent protein. If the blood is allowed to coagulate, the gelled portion contains fibrin and cellular proteins, whereas the blood albumin remains in the fluid serum. The fibrin portion may be sold as dried blood in tankage or fertilizer. The serum is clarified and dried and sold as blood albumin. Blood albumin is used in certain malt extracts and in fixing pigment colors in cloth, in finishing leather, in clarifying liquors, and in manufacturing glue, as noted earlier.

Pharmaceutical and medical products derived from blood were discussed earlier.

One hundred pounds of beef blood treated with an anticoagulant and centrifuged will yield about 40 pounds of solid (cells) material and 60 pounds of plasma, or 16 pounds of dried solids, 3.4 pounds of dried serum, and 3.5 pounds of wet fibrin.

Blood is also used in the manufacture of buttons and imitation tortoise shell articles.

Bones, Hooves, and Horns

The shin bones of cattle, with the knuckles removed, are cooked to remove the meat and neat's-foot oil and then washed and air-dried. They are then sawed into flat slabs from which crochet needles, bone teething rings, pipestems, dice, chess pieces, electrical bushings, washers, collar buttons, flat buttons, knife handles, and many other articles are made. Some other uses for bone are in the case hardening of steel; in the manufacture of bone black used as a bleach for oils, fats, waxes, sugar, or pharmaceutical preparations; as a feed ingredient (ground bone meal, steamed bone meal); and as a fertilizer.

White hooves are used for making imitation ivory products. Black hooves find use in the manufacture of potassium cyanide for extracting gold.

Cattle horns can be split into thin strips, pressed in heated molds of various patterns, and colored to make imitation tortoise shell. Horns are used for making napkin rings, goblets, tobacco boxes, knife and umbrella handles, and many other articles.

12

Federal Meat Grading and Its Interpretations[1]

Meat grading is completely optional and completely different from meat inspection, which is required by law. Many consumers confuse these two functions of our federal government. Chapter 3 covered meat inspection.

Meat grading is very detailed, with particular requirements for each species. This chapter lists the main factors and discusses them in some detail. However, the reader who is serious about a future as a meat grader should use the many references listed in this chapter to get more detailed information.

As we move into the new millennium, technology is constantly helping make our meat grading more objective and more accurate. Nevertheless, we are trying to categorize a population of biological organisms that, by their very nature, are diverse. As biotechnical gene splicing continues to develop (refer to Chapter 2), our meat-producing species will become more uniform. However, there will always be need for subjective grading by talented human beings.

FEDERAL MEAT GRADING

The Federal Meat Grading Service was established by the Sixty-eighth Congress of the United States on February 10, 1925; however, tentative standards were formulated for grades of dressed beef in 1916. They provided the

[1]The completeness and accuracy of this chapter would not have been possible without the generous assistance of Dr. Jimmy Wise, Mr. Dean Jensen, and Mr. Marty O'Connor, Meat Marketing Specialists, USDA–AMS, Livestock and Seed Division, Livestock and Meat Standardization Branch, 14th and Independence Avenue SW, Washington, DC 20250, and Dr. Tom Carr, University of Illinois.

basis for uniformly reporting the dressed beef markets according to grades, which became a voluntary grading service early in 1917. The grade specifications were improved from time to time, as experience gained through their use indicated that changes were necessary. They were published first in mimeographed form in June 1923. After slight changes, they were included in Department Bulletin No. 1246, *Market Classes and Grades of Dressed Beef,* which was published in August 1924, but did not become effective until July 1, 1926. The meat grading service was officially started with beef on May 2, 1927. Grade standards for veal and calf carcasses and for vealers and calves for harvest were introduced in 1928. The official standards for grades of lamb and mutton carcasses became effective on February 16, 1931. Tentative standards for grades of pork carcasses and fresh pork cuts were issued by the U.S. Department of Agriculture in 1931. These tentative standards were slightly revised in 1933. Presently, beef grading accounts for over 95 percent of the approximately 13 billion pounds of meat graded annually.

Federal meat grading was administered by the Livestock Division of the Agricultural Marketing Service (AMS) of the U.S. Department of Agriculture until 1977, when it was placed under the USDA's new Food Safety and Quality Service (FSQS). In 1981, the FSQS was abolished, and the grading service was returned to the AMS to be administered by the new Livestock and Seed Division.

A comparison of the Canadian and U.S. beef grading systems can be found in *The National Provisioner,* Vol. 201, No. 19, November 6, 1989.

The Purpose of the Act

The purpose of meat grading is to segregate the large population of carcasses from meat-producing animals into classes and grades with similar meat characteristics. An unbiased employee of the federal government performs this task before the carcasses are broken into wholesale and retail cuts, when all the parts of the carcass necessary to predict the different meat characteristics are still intact.

The federal grade stamp on meat provides consumers with a reliable guide to eating quality and quantity. Each quality grade name is associated with a specific combination of carcass traits that indicates to producers, processors, retailers, and consumers the use for which the meat product is best suited. Federally graded meat is widely found in retail stores. Most retailers sell only the grade or grades that their customers request or that they deem meets the needs of their customers. The quality grade name appearing on the grade stamp can be used to serve the consumer in two important ways: as a guide to eating quality and as a guide to preparation. Meat of each grade will provide a satisfactory dish if it is appropriately cooked.

Consumers can learn, either by study or from experience, what government grade names mean and represent. Eating quality (tenderness, juiciness, and flavor) in meat is quite variable and difficult for the average person to recognize in the retail cut. Thus, the consumer has come to depend on items bearing brand names. If a particular brand meets a consumer's approval, a new customer has been added — until some different graded product is tried and accepted. The manufacturer of a food product may be unknown to the public, but the brand name of the product, if good, is on every tongue. The terminology used by the U.S. grading service should help the consumer who may be perplexed by the vast number of brand names.

Yield grades, which indicate the quantity of edible meat in a carcass, are more meaningful to the retailer than to the ordinary consumer. The retailer uses yield grades to indicate the expected yield of salable meat from a carcass and its various wholesale cuts. However, when a consumer buys a side or a quarter of beef, the yield grade is extremely important, since it will indicate an expected yield of edible meat from the hanging carcass. The yield grade does not provide any guidelines for eating quality.

Inspection Requirements

Products, to be eligible for grading service, must be prepared under federal inspection or any inspection system acceptable to the U.S. Department of Agriculture. Recall that inspection guarantees *wholesomeness* (see Chapter 3) and is paid for with tax dollars. Not all persons in each of the many facets of the meat industry (producers, processors, retailers, and consumers) agree with every aspect of the U.S. meat grading system. Thus, meat grading is voluntary, and those who value its contributions use it and pay for it.

Types of Service

Federal meat grading is conducted by two divisions of the AMS: the Livestock and Seed Division and the Poultry Division. The Meat Grading and Certification Branch of the Livestock and Seed Division grades beef, veal, pork, lamb, and mutton for sale through regular commercial channels. The

Poultry Division of the AMS grades all poultry and rabbits for the same purpose. Both divisions have additional responsibilities that involve the examination and certification, for conformance with specifications for grade and other factors, of meats offered for delivery to federal, state, county, and municipal institutions that purchase meat on the basis of contract awards. This latter service covers all kinds of meats, meat products, and by-products.

BEEF CARCASS GRADING

Development of the Standards[2]

The official standards were first formulated in 1916, first published in 1923, and slightly changed in 1924. In May 1927, the voluntary beef grading and stamping service began. In July, 1939, the official standards were amended to provide a single standard for the grading and labeling of steer, heifer, and cow beef according to similar inherent quality characteristics. The amendment also changed certain grade terms for steer, heifer, and cow beef from *Medium, Common,* and *Low Cutter* to *Commercial, Utility,* and *Canner,* respectively.

Similar changes in the grade terms for bull and steer beef were made in an amendment in November 1941, which established the following grade terminology for all beef: *Prime, Choice, Good, Commercial, Utility, Cutter,* and *Canner.*

Compulsory grading by the Office of Price Administration (OPA), necessitated by World War II, was in effect from 1941 to 1946.

An amendment in October 1949 eliminated all references to the color of fat.

In December 1950, the official standards for grades of steer, heifer, and cow beef were amended. The Prime and Choice grades were combined and designated as *Prime,* the Good grade was renamed *Choice,* and the Commercial grade was divided into two grades: *Good,* which designated the beef produced from young animals included in the top half of the grade, and *Commercial,* which designated the remainder of the beef in that grade.

---

[2]USDA–AMS. 1997. United States Standards for Grades of Carcass Beef. Title 7, Chp. I, Pt. 54, Sec. 54.102–54.107, Code of Federal Regulations. (January 31)

In June 1956, the official standards for grades of steer, heifer, and cow beef were again amended. The Commercial grade was divided into two grades strictly on the basis of maturity, with beef produced from young animals being designated as *Standard,* while *Commercial* was retained as the grade name for beef produced from mature animals.

In June 1965, these changes were instituted:

- The official standards for grades of steer, heifer, and cow beef were revised to place less emphasis on changes in maturity in the Prime, Choice, Good, and Standard grades.

- The rate of increase in required marbling was changed to offset increasing maturity, and the minimum marbling permitted was reduced for more mature carcasses by as much as 1½ degrees in Prime, 1 degree in Choice, and ¾ degree in Good and Standard. Consideration of the two degrees of marbling in excess of that described as abundant was eliminated.

- The manner of evaluating conformation was also clarified — carcasses could meet the conformation requirements for a grade either through a specified development of muscling or through a specified development of muscling and fat combined.

- The requirement was established that all carcasses must be ribbed prior to being graded.

- Standards for yield grades of carcasses and certain wholesale cuts of all classes of beef were also established.

In July 1973, the official standards were again revised to provide separate quality grades for beef from young (A maturity) bulls in a class designated as *Bullock.* The quality grade standards for bullock were the same as those for steer, heifer, and young cow beef. *Bull* was retained as the class designation for beef from more mature bulls, but the quality grades for bull and stag beef were eliminated, leaving only the yield grade standards to apply. *Stag* beef was redesignated as *Bullock* or *Bull,* depending on its evidence of maturity.

In February 1976, compensation for increasing maturity within A maturity by an accompanying increasing marbling requirement was eliminated. As a result, the minimum marbling requirement was increased one-half degree for the youngest Good but reduced one-half degree at the A/B maturity line for Good and reduced one degree at the A/B line for Prime, Choice, and

Standard (see Figure 12–6). The maximum maturity permitted in Good and Standard was reduced to the same as that permitted in Prime and Choice. Conformation was eliminated as a quality grade factor. All carcasses graded were required to be identified for both quality and yield grades.

In October 1980, these changes were made:

- Removal of the yield grade stamp from an officially graded beef carcass would be allowed, provided the fat thickness did not exceed ¾ inch.

- Carcasses with the characteristics of the rib eye or thickness of fat over the rib eye altered would not be eligible for grading.

- Grading would only be done on beef that was in carcass form, in the plant where the animal was harvested, and not before at least a 10-minute period has elapsed following ribbing.

On November 23, 1987, the name *Good* was changed to *Select*. The standards were not changed.

On April 9, 1989, quality grading and yield grading were uncoupled. This rescinded the rule requiring carcasses to carry both a quality and a yield grade.

On January 31, 1997, the official standards were revised to restrict the Select grade to A maturity only and to raise the marbling degree required for Choice to minimum modest throughout B maturity.

Classes of Beef Carcasses

The first step in beef carcass grading should be to determine the class.

Class determination of beef carcasses is based on evidences of maturity and apparent sex condition at the time of harvest. The classes of beef carcasses are steers, bullocks, bulls, heifers, and cows. Carcasses from males — steers, bullocks, and bulls — are distinguished from carcasses from females — heifers and cows — as follows:

Steer, bullock, and bull carcasses have a "pizzle muscle" (attachment of the penis) and related "pizzle eye" adjacent to the posterior end of the aitch bone (see Figures 12–1 and 12–3).

Steer, bullock, and bull carcasses have, if present, rather rough, irregular fat in the region of the cod. In heifer and cow carcasses, the fat in this region, if present, is much smoother (see Figures 12–1 and 12–2).

In steer, bullock, and bull carcasses, the area of lean exposed immediately ventral to the aitch bone is much smaller than in heifer and cow carcasses (see Figures 12–1, 12–2, and 12–3).

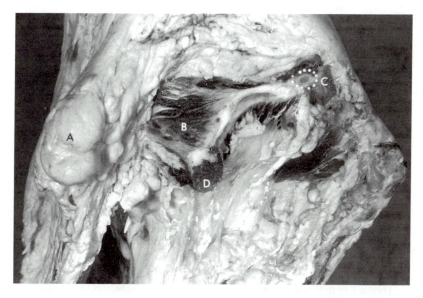

Fig. 12–1. Steer carcass. The cod fat (A) to the left, the half-closed face of the *gracilis* muscle (B), and the pizzle eye (C) at the right of the aitch (pelvic) bone (D) identify this as a male.

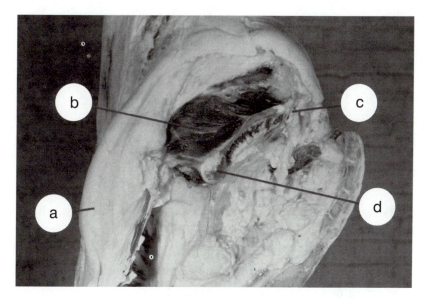

Fig. 12–2. Heifer carcass. Note the presence of the udder (a), the exposed face of the *gracilis* muscle (b), and the lack of a pizzle eye (c) at the right of the aitch bone (d).

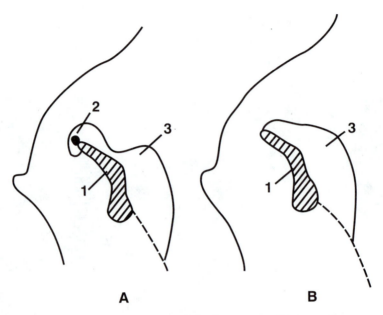

Fig. 12–3. (A) The contour of the lean area of the *gracilis* muscle on a steer carcass showing (1) aitch bone, (2) pizzle eye, and (3) lean area of *gracilis*. (B) The same area on a heifer carcass. This shows the method of identifying a steer round from a heifer round when the rump and the shank have been removed.

Steer, bullock, and bull carcasses are distinguished by the following:

- In steer carcasses, the pizzle muscle is relatively small, light red in color, and fine in texture, and the related pizzle eye is relatively small.

- In bullock and bull carcasses, the pizzle muscle is relatively large, dark red in color, and coarse in texture, and the related pizzle eye is relatively large.

- Bullock and bull carcasses usually have a noticeable crest, caused by a shortening and bunching of the neck muscles. Male hormones cause this secondary sex characteristic to become evident as bulls mature.

- Bullock and bull carcasses, in contrast to steer carcasses, also usually have a noticeably highly developed round muscle (*gluteus medius*) adjacent to the hipbone, commonly referred to as the "jump muscle." However, in carcasses with a considerable amount of external fat, the development of this muscle may be obscured.

- Although the development of secondary sex characteristics is given primary consideration in distinguishing steer carcasses from bullock or bull carcasses, this differentiation is also facilitated by consideration of the color and texture of the lean. In bullock and bull carcasses, the lean is frequently at least dark red in color with a dull, "muddy" appearance, and in some cases it may have an iridescent sheen. Also, it is frequently coarse and has an "open" texture.

- The distinction between bullock and bull carcasses is based solely on the evidence of their skeletal maturity. Carcasses with the maximum maturity permitted in the Bullock class must still qualify for A maturity and have slightly red and slightly soft chine bones; the cartilages on the ends of the *thoracic* vertebrae should have some evidence of ossification; the *sacral* vertebrae should be completely fused; the cartilages on the ends of the *lumbar* vertebrae should be nearly completely ossified; and the rib bones should be slightly wide and slightly flat. Bull carcasses have evidence of more advanced maturity.

Heifer and cow carcasses are distinguished by the following:

- Heifer carcasses have a relatively small pelvic cavity and a slightly curved aitch bone. In cow carcasses, the pelvic cavity is relatively large, and the aitch bone is nearly straight.

- In heifer carcasses, the udder will usually be present. In cow carcasses, the udder will usually have been removed. However, neither of these is a requirement.

Two Considerations — Quality and Yield

The grade of a steer, bullock, heifer, or cow carcass is based on separate evaluations of two general considerations: (1) The palatability-indicating characteristics of the lean, referred to as the *quality* grade, and (2) the indicated percent of trimmed, boneless major retail cuts to be derived from the carcass, referred to as the *yield* grade. When graded by a federal meat grader, the grade of a steer, bullock, heifer, or cow carcass may consist of both the quality grade and the yield grade, or either grade only. The grade of a bull carcass consists of the yield grade only.

The carcass beef grade standards are written so that the quality and yield grade standards are contained in separate sections. Eight quality grade designations — Prime, Choice, Select, Standard, Commercial, Utility, Cutter, and Canner — are applicable to steer and heifer carcasses. Except for Prime, the same designations apply to cow carcasses. The quality grade designations for bullock carcasses are Prime, Choice, Select, Standard, and Utility. There are five yield grades applicable to all classes of beef, denoted by numbers 1 through 5, with Yield Grade 1 representing the highest degree of major retail cut yield.

When officially graded, bullock and bull beef will be further identified for their sex condition; steer, heifer, and cow beef will not be so identified. The designated grades of bullock beef are not necessarily comparable in quality or yield with similarly designated grades of beef from steers, heifers, or cows. Neither is the yield of a designated yield grade of bull beef necessarily comparable to a similarly designated yield grade of steer, heifer, cow, or bullock beef.

Ribbing the Carcass

In order for the quality grade or yield grade of a carcass to be determined, it must be split equally down the back into two sides, and one side must be partially separated into a hindquarter and a forequarter. Insofar as practicable, it should be cut with a saw and a knife as follows: A saw cut should be made perpendicular to both the long axis and the split surface of the vertebral column across the twelfth thoracic vertebra at a point which leaves not more than one-half of this vertebra on the hindquarter. The knife cut across the rib eye muscle starts — or terminates — opposite the above-described saw cut. From that point it extends across the rib eye muscle perpendicular to the outside skin surface of the carcass at an angle toward the hindquarter, which is slightly greater (more nearly horizontal) than the angle made by the thirteenth rib with the vertebral column of the hindquarter posterior to that point. As a result of this cut, the outer end of the cut surface of the rib eye muscle is closer to the twelfth rib than is the end next to the chine bone. Thus, the cut surface of the rib eye muscle is perpendicular to its long axis. Beyond the rib eye, the knife cut continues between the twelfth and thirteenth ribs to a point that adequately exposes the distribution of fat and lean in this area. The knife cut may be made prior to or following the saw cut, but it must be smooth and even, such as would result from a single stroke of a very sharp knife.

Other methods of ribbing may prevent an accurate evaluation of quality grade and yield grade determining characteristics. Therefore, carcasses ribbed by other methods are only eligible for grading under the standards if an accurate grade determination can be made by the official grader.

Beveling of the fat over the rib eye, application of pressure, or any other influences that would alter the area of the rib eye or the thickness of fat over the rib eye may prevent an accurate yield grade determination. Therefore, carcasses subjected to such influences may not be eligible for a yield grade determination. Also carcasses with more than minor amounts of lean removed from the major sections of the round, loin, rib, or chuck are not eligible for a yield grade determination.

See Figure 15–3 for illustration and further discussion.

Conformation

Conformation, the manner of formation of the carcass, refers to the thickness of muscling and to the overall degree of thickness and fullness of the carcass. Conformation has been eliminated as a beef quality grade factor, but it is still a major factor in comparative carcass judging and evaluation. Some segments of the modern meat industry still value conformation quite highly, and some consumers prefer shapely, symmetrical top loin or rib steaks rather than those that are long and narrow.

Conformation is evaluated by averaging the conformation of the various parts of the carcass, considering not only the proportion that each part is of the carcass weight but also the general value of each part as compared to the other parts. Thus, although the chuck and round are nearly the same percentage of carcass weight, the round is considered the more valuable cut. Therefore, development of the round is given more consideration than development of the chuck in the evaluation of overall conformation of a carcass. Similarly, since the loin is a greater percentage of carcass weight and also generally a more valuable cut than the rib, its conformation receives more consideration than the conformation of the rib. Superior conformation indicates a high proportion of meat to bone and a high proportion of the weight of the carcass in the more valuable parts. It is reflected in carcasses that are very thickly muscled; that are very full and thick in relation to their length; and that have a very plump, full, and well-rounded appearance. Inferior conformation indicates a low proportion of meat to bone and a low proportion of the weight of the carcass in the more valuable parts. It is reflected in carcasses that are very thinly muscled; that are very narrow and thin in relation to their length; and that have a very angular, thin, and sunken appearance.

Quality — Maturity

After class determination, the next step in quality grading should be the determination of *maturity.* The maturity of the carcass is determined by evaluating the size, shape, and ossification of the bones and cartilages — especially the split chine bones — and the color and the texture of the lean flesh. In the split chine bones, ossification changes occur at an earlier stage of maturity in the posterior portion of the vertebral column (*sacral* vertebrae) and at progressively later stages of maturity in the *lumbar* and *thoracic* vertebrae. The ossification changes that occur in the cartilages on the ends of the split thoracic vertebrae are especially useful in evaluating maturity, and these vertebrae are referred to frequently in the carcass beef standards. Unless otherwise specified in the standards, whenever the ossification of cartilages on the thoracic vertebrae is referred to, this is construed to mean the cartilages attached to the thoracic vertebrae at the posterior end of the forequarter (see Table 12–1).

The size and the shape of the rib bones are also important considerations in evaluating differences in maturity. In the very youngest carcasses considered as beef (A– maturity), the cartilages on the ends of the dorsal processes of the chine bones (feather bones) show no ossification, cartilage is evident on all of the vertebrae of the spinal column, and the sacral vertebrae show distinct separation. In addition, the split vertebrae usually are soft and porous and very red in color. In such carcasses, the rib bones have only a slight tendency toward flatness. In progressively more mature carcasses, ossification changes become evident first in the bones and cartilages of the *sacral* vertebrae, then in the *lumbar* vertebrae, and still later in the *thoracic* vertebrae. In beef that is very advanced in maturity (E+ maturity), all the split vertebrae will be devoid of red color and very hard and flinty, and the cartilages on the ends of all the vertebrae will be entirely ossified. Likewise, with advancing maturity, the rib bones will become progressively wider and flatter; thus, in beef from very mature animals, the ribs will be very wide and flat.

In steer, heifer, and cow carcasses, the range of maturity permitted within each of the grades varies considerably. The Prime, Choice, Select, and Standard grades are restricted to beef from young cattle; the Commercial grade pertains to beef from cattle too mature for Select or Choice; and the Utility, Cutter, and Canner grades include beef from animals of all ages. By definition, bullock carcasses are restricted to those in which evidence of maturity does not exceed A maturity.

Table 12–1 summarizes the information that is necessary for determining maturity groupings. In addition to the rib eye muscle, the entire visible skeleton, including the rib cage, should be observed and studied before a composite final evaluation of maturity is made. The extremes at both ends of the maturity span are A minus (A–) at the youngest end (referred to as A^0 when

Table 12–1. Maturity Descriptions

Approx. Chron. Age	Maturity Group	Vertebral Ossification			Ribs	Rib Eye Muscle
		Sacral	Lumbar	Thoracic		
9–10 months	A – (A⁰)	Distinct separation	Cartilage evident on all vertebrae	Cartilage evident on all vertebrae; soft, porous, and very red chine bones	Slight tendency toward flatness	Light grayish in color; very fine in texture
30 months	A + /B – (A¹⁰⁰/B⁰)	Completely fused	Nearly completely ossified	Cartilage has some evidence of ossification; slightly red and slightly soft chine bones	Slightly wide; slightly flat	Light red in color; fine in texture
42 months	B + (B¹⁰⁰)	Completely fused	Completely ossified	Cartilage partially ossified; chine bones tinged with red	Slightly wide; slightly flat	Tends to be fine in texture
43 months	C – (C⁰)	Completely fused	Completely ossified	Cartilage moderately (20–30%) ossified; chine bones tinged with red	Slightly wide; slightly flat	Tends to be fine in texture

(Continued)

Table 12–1 (Continued)

| Approx. Chron. Age | Maturity Group | Vertebral Ossification | | | Ribs | Rib Eye Muscle |
		Sacral	Lumbar	Thoracic		
60 months	$C + /D - (C^{100}/D^0)$	Completely fused	Completely ossified	Cartilage shows consider-able ossi-fication, but out-lines are plainly vis-ible; mod-erately hard, rather white chine bones	Moderately wide and flat	Moderately dark in color; slightly coarse in texture
90 months	$D + /E - (D^{100}/E^0)$	Completely fused	Completely fused	Cartilage barely vis-ible; hard, white chine bones	Wide and flat	Dark red and coarse in texture
Oldest	$E + (E^{100})$	Completely fused	Completely fused	Completely ossified	Very wide and flat	Very dark red and coarse in texture

the width of the A maturity area is considered in the grading chart, Figure 12–6, in terms of percentages) and E plus (E+) or E^{100} at the extreme oldest end of the maturity span. Both ends of the span are relatively easy to identify, but to determine the exact location within the span is more difficult. Two critical points exist in the span: (1) *the A^{100}/B^{0} junction,* because the Select grade is restricted to A maturity only and because a modest degree of marbling is required to balance B maturity in the Choice grade, and (2) *the B^{100}/C^{0} junction,* after which cattle are no longer eligible for the Prime, Choice, Select, or Standard grade.

If any one skeletal area is more important in determining maturity grouping, it would be the *thoracic* area; thus, diagrams of that area are included in Table 12–1. Nevertheless, a composite evaluation must be made, including an appraisal of conformation, since angularity implies maturity. In Table 12–1, descriptions are listed for points of merger between the maturity groupings, so by comparing such points, you can locate a carcass within the maturity span and estimate the percentage of distance between the points of merger. Also included in Table 12–1 is an approximation of chronological age of cattle at which carcasses of each given maturity will be produced.

Determining a subjective trait such as carcass maturity takes a good deal of judgment and at least some experience. By using a percentage system, you can evaluate the separate factors as listed in Table 12–1 and then put them together for a composite evaluation of maturity. When you are undecided, skeletal development takes precedence over muscle firmness, texture, and color.

Quality — Color and Texture

In steer, heifer, and cow beef, the color and the texture of the lean flesh also undergo progressive changes with advancing maturity. In the very youngest carcasses considered as beef, the lean flesh will be very fine in texture and light grayish red in color. As carcass maturity increases, the texture of the lean will become progressively coarser and the color of the lean will become progressively darker red. In very mature beef, the lean flesh will be very coarse in texture and very dark red in color. Slightly more emphasis is placed on the characteristics of the bones and cartilages than on the characteristics of the lean in determining the maturity of a carcass in which the skeletal evidences of maturity are different from those indicated by the color and texture of the lean. In no case can the overall maturity of the carcass be considered more than one full maturity group different from that indicated by its bones and cartilages.

References to the color of lean in the standards for steer, heifer, and cow beef pertain to colors associated only with changes in maturity. They are not intended to apply to colors of lean associated with so-called *dark-cutting beef*. Dark-cutting beef results from a reduced glycogen (starch) content of the lean at the time of harvest. As a result, this condition does not have the same significance in grading as do the darker shades of red associated with advancing maturity. The dark color of the lean associated with dark-cutting beef is present in varying degrees — from being barely evident to actually nearly black and usually with a gummy texture *(black cutters)*. The dark-cutting condition does affect the palatability of beef by causing a decrease in tenderness and flavor desirability.[3]

The condition favors microbial growth due to the elevated pH. It is considered in grading because of its effect on retail acceptability and value. (A more complete discussion of this phenomenon is contained in Chapter 22.)

Depending on the degree to which this characteristic is developed, the final grade of carcasses that would otherwise qualify for the Prime, Choice, or Select grade may be reduced as much as *one full grade*. In beef otherwise eligible for the Standard or Commercial grade, the final grade may be reduced as much as *one-half a grade*. In the Utility, Cutter, and Canner grades, this condition is not considered.

Bullock carcasses that have darker muscle color than expected for the grade under consideration will first be graded according to skeletal characteristics only. The final grade is then determined in accordance with the procedures specified in the standards for grading dark-cutting beef.

Quality — Marbling (Intramuscular Fat)

For steer, heifer, and cow beef, the *marbling* and *firmness* that are observed in a cut surface, in relation to the apparent maturity of the animal from which the carcass was produced, are the factors to be considered in the evaluation of *quality* of lean. A certain level of marbling is necessary to assure optimum pal-

---

[3]Wulf, D. M., J. K. Page, T. R. Schwotzer, and G. R. Dunlap. 1998. Using measurements of muscle color/pH/waterholding capacity to augment the current USDA beef carcass quality grading standards and improve the accuracy and precision of sorting beef carcasses into palatability groups. Final Report to the National Cattlemen's Beef Association, Denver.

atability, especially in terms of juiciness and flavor. The degrees of marbling, in order of descending quantity, are abundant (Ab), moderately abundant (MA), slightly abundant (SA), moderate (Md), modest (Mt), small (Sm), slight (Sl), traces (Tr), and practically devoid (PD). Illustrations of the lower limits of eight of the nine degrees of marbling considered in grading beef appear in Figure 12–4. Reproductions of rib eyes containing SA, Md, Mt, Sm, Sl, and Tr amounts of marbling are shown in the color photo section.

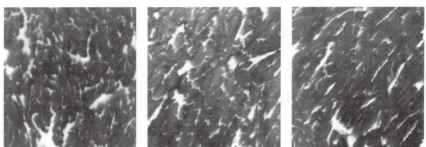

Abundant (Ab)   Moderately Abundant (MA)   Slightly Abundant (SA)

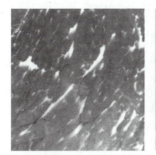

Moderate (Md)   Modest (Mt)

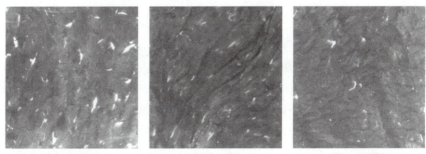

Small (Sm)   Slight (Sl)   Traces (Tr)

Fig. 12–4. Degrees of marbling. This series of black and white photos was used as the USDA standard until 1981, when the National Live Stock and Meat Board published a set of colored pictures that now serve as the USDA standards. See color section for reproductions of these standards. (Courtesy, USDA)

Figure 12–5 shows the positive relationship between marbling level and the percent fat extracted from the fresh *longissimus* muscle with diethyl ether, a common chemical fat solvent (see Chapter 18).

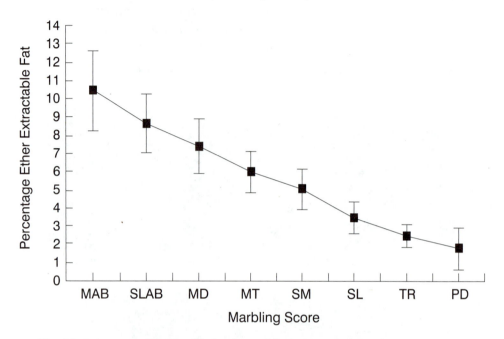

Fig. 12–5. Intramuscular fat (as determined by extraction with ether) plotted against marbling level. The brackets around each marbling point indicate the variation in fat content of top loin steaks within each marbling score among the 518 steer and heifer beef carcasses used in this study. Two-thirds of the fat determinations lie within the brackets, which indicate one standard deviation. Note that the higher marbling levels have the largest variation, except for practically devoid. (From *Designing Foods, Animal Options in the Marketplace.* 1988. National Research Council, National Academy Press, Washington, D.C., p. 352, based on research by Savell, Cross, and Smith. 1986. *J. Food Sci.* 51:838.)

Except for the youngest maturity group (A), within any specified grade, the requirements for marbling and firmness increase progressively with evidence of advancing maturity. However, firmness is seldom a limiting grade factor. To make it easier to equalize advancing maturity with higher levels of marbling, the standards recognize five different maturity groups (see previous section) and nine degrees of marbling. Marbling can be scored in percentages equally as convenient as maturity. Note the depth of the marbling segments on the beef-grading chart (see Figure 12–6), and compare the marbling photos with the rib eye in question. For instance, if the marbling level is higher than the minimum small pictured, but less than the minimum modest, the

RELATIONSHIP BETWEEN MARBLING, MATURITY, AND CARCASS QUALITY GRADE*

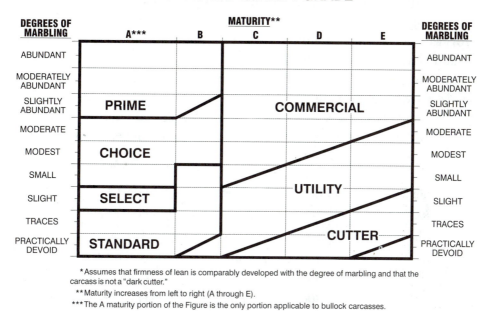

* Assumes that firmness of lean is comparably developed with the degree of marbling and that the carcass is not a "dark cutter."

** Maturity increases from left to right (A through E).

*** The A maturity portion of the Figure is the only portion applicable to bullock carcasses.

Fig. 12–6. Relationship between marbling, maturity, and quality.

marbling level is somewhere between Sm^0 and Sm^{100}, perhaps 50 percent of the way to modest; thus, the marbling level is Sm^{50} or typical small. Percentages should be designated in no smaller units than 10, and as such will make combination with maturity for final grade very straightforward.

Relationship Between Marbling, Maturity, and Quality

The relationship between marbling, maturity, and quality is shown in Figure 12–6. For instance, the minimum marbling requirement for Choice does not increase from a minimum small amount for the A maturity carcasses, but it does increase to a minimum modest amount for carcasses having the maximum maturity permitted in Choice. Likewise, in the Commercial grade, the minimum marbling requirement varies from a minimum small amount in beef from animals with the minimum maturity permitted to a maximum moderate amount in beef from very mature animals.

The marbling and other lean flesh characteristics specified for the various grades are based on their appearance in the rib eye muscles of properly chilled carcasses that are ribbed between the twelfth and thirteenth ribs.

In certain instances (e.g., meat-grading contests and carcass-evaluation demonstrations), it is necessary to determine the grade to the nearest third. Combining maturity and marbling by using percentages is straightforward, provided a few rules are followed, which are determined by which grades and maturity levels are under consideration.

In A maturity (and for low Choice in B maturity), the same marbling degree is required throughout the maturity group to qualify for a given grade. The *Prime* and *Choice* grades are each *three* marbling levels deep in A maturity (see Figure 12–6) or 300 percent for the full grade in terms of marbling. Thus,

Sm^{90}	=	90/300	=	$Choice^{30}$	=	Low Choice
SA^{90}	=	90/300	=	$Prime^{30}$	=	Low Prime
Mt^{80}	=	180/300	=	$Choice^{60}$	=	Av Choice
MA^{80}	=	180/300	=	$Prime^{60}$	=	Av Prime
Md^{70}	=	270/300	=	$Choice^{90}$	=	High Choice
Ab^{70}	=	270/300	=	$Prime^{90}$	=	High Prime

The *Select* grade is *one* marbling level deep; thus, percentage slight marbling equals percentage grade; i.e., Sl^{50} = $Select^{50}$ = Av Select.

The *Standard* grade is *two* or *three* marbling levels deep or 200 to 300 percent for the full grade in terms of marbling. Thus in A maturity:

PD^{60}	=	60/200	=	$Standard^{30}$	=	Low Standard
Tr^{50}	=	150/200	=	$Standard^{75}$	=	High Standard

In B (except in the Select grade), C, D, and E maturity, as maturity increases, an equal increase in marbling is required in order to maintain a given grade (see Figure 12–6). The logical steps are:

1. Set up a minimum marbling requirement for the grade called for by the maturity and marbling combination in question.

2. Subtract the minimum marbling percent from the actual marbling percent, giving the excess marbling above the minimum requirement.

3. Adjust the excess marbling percent by dividing by the number of marbling levels associated with the grade in question. In the Select grade (one marbling level deep), no adjustment is needed as this equals position in grade, but in the Standard grade (two marbling levels

deep) and all other grades (three marbling levels deep), the excess percent must be adjusted by dividing as was done for A maturity carcasses. *Caution:* Use only 10 percent increments.

Examples:

Maturity	Marbling	Grade	Minimum Marbling for Grade	Marbling Excess	Adjustment for Marbling Depth	Quality Grade
B^{10}	Mt^{10}	Ch	Mt^{10}	0	$0/300 = Ch^0 =$ Low Choice	
B^{10}	Sm^0	St	PD^{10}	300	$300/300 = St^{100} =$ High Standard	
B^{50}	Mt^{70}	Ch	Mt^0	70	$70/300 = Ch^{23} =$ Low Choice	
D^{20}	Tr^{20}	Ut	Tr^{20}	0	$0/300 = Ut^0 =$ Low Utility	
E^{50}	Md^{50}	Comm	Md^{50}	0	$0/300 = Comm^0 =$ Low Commercial	
E^{40}	SA^{60}	Comm	Md^{40}	120	$120/300 = Comm^{40} =$ Av Commercial	
B^{20}	PD^{60}	St	PD^{20}	40	$40/200 = St^{20} =$ Low Standard	
B^{30}	Sl^{50}	St	PD^{30}	120	$120/200 = St^{60} =$ High Standard	
C^{50}	Sl^{60}	Ut	PD^{50}	210	$210/300 = Ut^{70} =$ High Utility	
D^{30}	Md^{20}	Comm	Mt^{30}	90	$90/300 = Comm^{30} =$ Low Commercial	

Specifications for Bullocks, Bulls, and Stags

Bullocks, by definition, include only carcasses within the A maturity range. Thus, the specifications for bullock beef are identical to those for steer and heifer beef in the Prime, Choice, Select, Standard, and Utility grades when spelled out for A maturity.

Bulls and stags (all exceeding A maturity) *are not quality graded,* but they may be yield graded.

Specifications for Official U.S. Standards for Quality Grades of Carcass Beef (Steer, Heifer, Cow)

Prime

As the name implies, beef of this grade is highly acceptable and palatable. Prime grade beef is produced from young and well-fed steers and heifers — cow beef is *not* eligible for the Prime grade. The youth of the cattle and careful intensive feeding combine to produce very high-quality cuts of beef. Such cuts have liberal quantities of fat interspersed within the lean (marbling). These characteristics contribute to the juiciness and the flavor of the meat. Rib roasts and loin steaks of this grade are consistently tender, flavorful, and juicy, and cuts from the round and chuck should also be highly satisfactory.

Maturity groups: A and B

Marbling minimum requirements:
A maturity minimum marbling is constant from A^0 to A^{100}
B maturity minimum marbling increases equally as B^0 increases to B^{100}

Grade	Minimum Marbling	FOR	Maturity A^{0-100}	B^{0-100}
Low Prime (P^{0-33})	Slightly abundant		SA^0	SA^{0-100}
Av Prime (P^{34-66})	Moderately abundant		MA^0	MA^{0-100}
High Prime (P^{67-100})	Abundant		Ab^0	Ab^{0-100}

Examples for quality grade computations were given previously.

Choice

This grade is preferred by most consumers because it is of high quality but usually has less fat than beef of the Prime grade. More of this grade of beef than of any other grade is stamped. Roasts and steaks from the loin and rib are tender and juicy, and other cuts, such as those from the round or chuck, which are more suitable for braising and pot roasting, should be tender with a well-developed flavor.

Maturity groups: A and B

Marbling minimum requirements:

A maturity minimum marbling is constant Small from A^0 to A^{100} and constant Modest in B maturity as B^0 increases to B^{100}

Grade	Minimum Marbling	FOR	Maturity A^{0-100}	B^{0-100}
Low Choice (Ch^{0-33})	Small		Sm^0	Not eligible
Av Choice (Ch^{34-66})	Modest		Mt^0	Mt^0
High Choice (Ch^{67-100})	Moderate		Md^0	Md^{0-100}

Select

This grade pleases economy-conscious consumers who seek beef with little fat but with an acceptable degree of quality. Although cuts of this grade may lack the juiciness and flavor associated with a higher degree of fatness, their relative tenderness and high proportion of lean to fat make them the preference of some people.

Maturity groups: A and B

Marbling minimum requirements:

A maturity minimum marbling is constant from A^0 to A^{100}
B maturity not eligible

Grade	Minimum Marbling	FOR	Maturity A^{0-100}	B^{0-100}
Low Select (Se^{0-33})	Slight		Sl^0	Not eligible
Av Select (Se^{34-66})	Slight, small		Sl^{34}	Not eligible
High Select (Se^{67-100})	Slight, small		Sl^{67}	Not eligible

Because of the narrowness of this grade, it is often divided in halves rather than thirds for grading contests; the grade is divided at Sl^{50} in A^{0-100} maturity.

Standard

Standard grade beef has a very thin covering of fat and appeals to consumers who are primarily concerned with having a high proportion of lean. When properly prepared, such beef is usually relatively tender because it is youthful. It is mild in flavor and lacks the juiciness usually found in beef with more marbling. However, less than one-tenth of 1 percent of the beef graded is Standard. Consumers rarely see this grade at retail markets, for most of it is included with the some 20 percent ungraded beef that is further processed for use in the hotel and restaurant trade (see Chapters 13 and 18).

Maturity groups: A and B

Marbling minimum requirements:
   A maturity minimum marbling is constant from A^0 to A^{100}
   B maturity minimum marbling increases equally as B^0 increases to B^{100}

Grade	Minimum Marbling	FOR	Maturity A^{0-100}	B^{0-100}
Low Standard (St^{0-33})	Practically devoid		PD^0	PD^{0-100}
Av Standard (St^{34-66})	Practically devoid, traces		PD^{66}	$PD^{66-Tr66}$
High Standard (St^{67-100})	Traces, slight		Tr^{33}	$Tr^{33-Sl33}$

Commercial

Beef that is graded Commercial is produced from older cattle and usually lacks the tenderness of the higher grades. Cuts from this grade, if carefully prepared, can be made into satisfactory and economical meat dishes. Most of these cuts require long, slow cooking with moist heat to make them tender and to develop the rich, full beef flavor characteristic of mature beef. This grade is also rarely seen at retail markets because the ribs and loins are tenderized for hotel and restaurant use as less expensive steaks, and the chucks, rounds, and rough cuts are processed into sausages, ground beef, and related products (see Chapters 18, 21, and 22).

Maturity groups: C, D, and E

Marbling minimum requirements: increase equally as maturity increases in all three maturity groups.

Grade	Minimum Marbling	FOR	Maturity		
		C^{0-100}	D^{0-100}	E^{0-100}	
Low Commercial (Co$^{0-33}$)	Small, modest, moderate	Sm^{0-100}	Mt^{0-100}	Md^{0-100}	
Av Commercial (Co$^{34-36}$)	Modest, moderate, slightly abundant	Mt^{0-100}	Md^{0-100}	SA^{0-100}	
High Commercial (Co$^{67-100}$)	Moderate, slightly abundant, moderately abundant	Md^{0-100}	SA^{0-100}	MA^{0-100}	

Utility

Beef of this grade is produced mostly from cattle somewhat advanced in age and is usually lacking in natural tenderness and juiciness. Cuts from this grade carry very little fat but provide a palatable, economical source of lean meat for pot roasting, stewing, and boiling or for ground-meat dishes. For satisfactory results, long, slow cooking by moist heat is essential. The cuts of this grade rarely appear in the retail markets because the ribs and loins are tenderized for hotel and fast-food restaurant use as less expensive steaks, and the chucks, rounds, and rough cuts are processed into sausages, ground beef, and related products (see Chapters 18, 21, and 22).

Maturity groups: A, B, C, D, and E

Marbling minimum requirements: increase equally as maturity increases in all five maturity groups.

Grade	Minimum Marbling	FOR	Maturity			
		A^{0-100}	B^{0-100}	C^{0-100}	D^{0-100}	E^{0-100}
Low Utility (Ut$^{0-33}$)	Devoid, practically devoid, traces, slight	Dev^0	Dev^{0-100}	PD^{0-100}	Tr^{0-100}	Sl^{0-100}
Av Utility (Ut$^{34-66}$)	Traces, slight, small	Std not Ut		Tr^{0-100}	Sl^{0-100}	Sm^{0-100}
High Utility (Ut$^{67-100}$)	Slight, small, modest	Std not Ut		Sl^{0-100}	Sm^{0-100}	Mt^{0-100}

Cutter

In all groups, the rib eye muscle is soft and watery and dark red in color. The marbling requirement increases from practically devoid at the B/C junction to traces at E^{100}. This grade of beef is not presented to the buying public in retail markets; instead, it is processed into frankfurters, bologna, and hamburger.

Canner

This grade includes only those carcasses that are inferior to the minimum requirements specified for the Cutter grade and are utilized in the same manner.

Live Animal and Carcass Illustrations of Beef Quality Grades

Beef carcass quality grades are based on marbling and maturity (see previous discussion). (Figures 12–7 to 12–18, provided by the University of Illinois, South Dakota State University, and the National Live Stock and Meat Board, depict live animal and carcass beef quality grades.[4]) In live cattle, the trait of maturity (age) is somewhat easily recognized by a person who has had some training and experience. Marbling, however, is very difficult to estimate on a live animal without the use of supplemental information, such as the length of time on feed, composition of ration, and genetic background. Thus, many cattle are bought on "reputation," that is, buyers learn the way various feeders feed a certain kind of cattle in order to produce carcasses of a given grade.

Some characteristics relating to overall fatness and therefore indirectly to marbling can be observed on live cattle. However, some fat cattle do not marble, and some lean cattle do marble. Therefore, the pictures depicting especially the live grades of Prime, Choice, and Select must be recognized as *typical* examples only. Certainly breeds and types other than those shown for each grade qualify daily in the United States for the various grades. Furthermore, the final carcass grade is not known until the carcass is ribbed and the marbling evaluated.

---

[4]Carcass pictures showing beef quality grades are shown in color in the Meat Evaluation Handbook, published by the National Livestock and Meat Board in 1988. This handbook is highly recommended for the serious student of meat judging and grading.

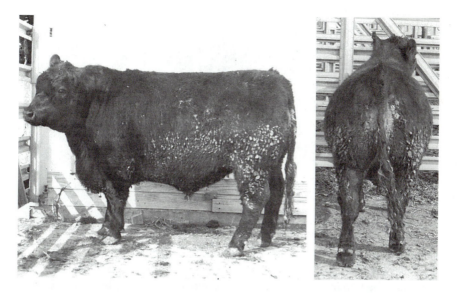

Fig. 12–7. A 1,090-pound Prime steer. From the side, the fullness of his brisket and foreflanks and rear flanks is evidence of his having been fed. Fullness here represents fat. From the rear, his width through his middle and fullness of twist again indicate feed, fat, and the potential to marble.

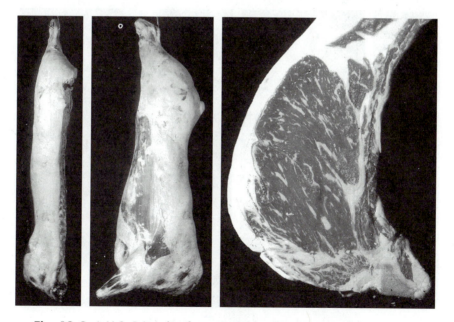

Fig. 12–8. A U.S. Prime beef carcass. This typical A maturity carcass has a light red color of lean; red, porous chine bones; large cartilaginous "buttons" on the *thoracic* vertebrae; and nearly completely fused *sacral* vertebrae, all which cannot be seen in these photos. The moderately abundant marbling and firm rib eye muscle combine with A maturity to indicate Prime quality, an Average Prime carcass. The carcass is wide and thick in relation to its length and is thickly muscled throughout — note the plumpness in the round and the thickness in the loin, rib, and chuck.

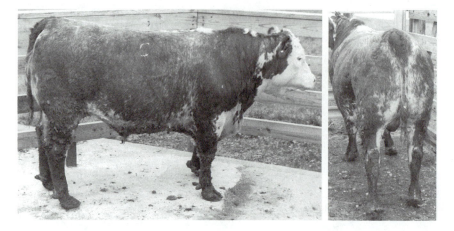

Fig. 12-9. A 1,150-pound Choice steer. From the side, his brisket and foreflank are moderately full, indicating some fatness. From the rear, he appears to be trim in the twist and lower round, but he shows evidence of some fatness over the edge of his loin.

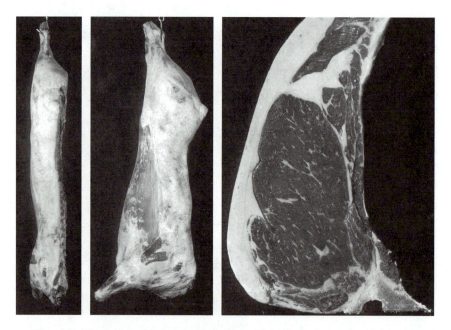

Fig. 12-10. A U.S. Choice beef carcass. The rib eye muscle of this typical A maturity carcass displays Average Choice quality. It has a weak moderate (moderate[30]) amount of marbling and slightly firm lean. This quality, irrespective of the average muscling evidenced by a moderately plump round and a moderately thick loin, rib, and chuck, results in an Average Choice grade carcass.

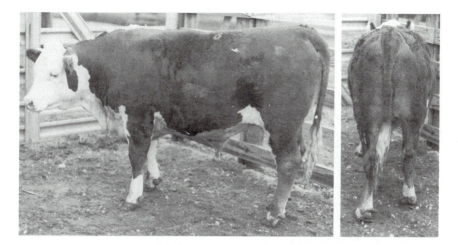

Fig. 12–11. A 980-pound Select steer. The steer's light weight and his trimness about the brisket and flanks as well as behind the shoulders indicate a short time on feed. This steer is narrow behind, lacking muscling and bulge to the round.

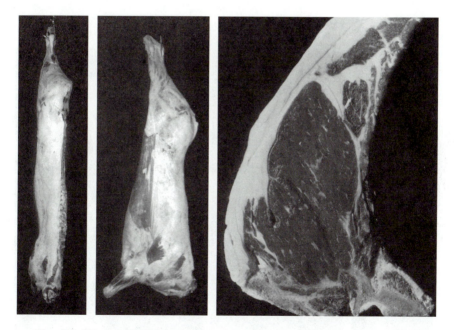

Fig. 12–12. A U.S. Select beef carcass. This very young (A– maturity) carcass has distinct separation of the sacral vertebrae; red, porous chine bones; and a very light red color of lean, none of which can be seen in these illustrations. The slight amount of marbling in the moderately soft rib eye muscle qualified it for Average Select. This carcass has slightly plump rounds and slightly thick and full loins, ribs, and chucks.

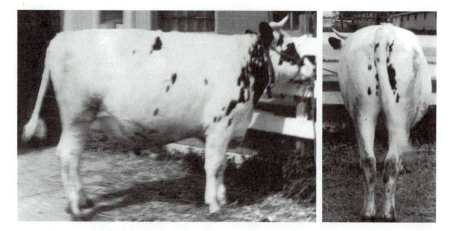

Fig. 12–13. A 1,125-pound Standard steer. This steer is narrow behind and shows little evidence of external finish. Young steers of dairy breeding predominate here, although young cattle of any breed that are somewhat underfinished or that do not marble readily because of genetics qualify for the Standard grade. On the other hand, many dairy steers move up to the Good and Choice grades.

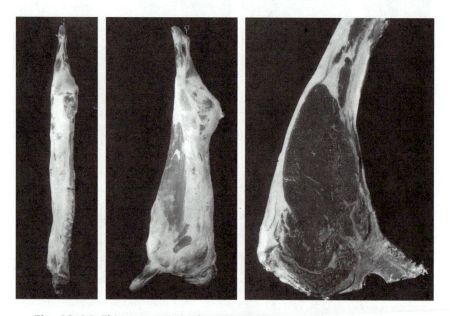

Fig. 12–14. This carcass is in the youngest maturity group (A). The rib eye muscle, which is moderately soft and has traces of marbling, is typical of Average Standard grade. This carcass is slightly thinly fleshed throughout.

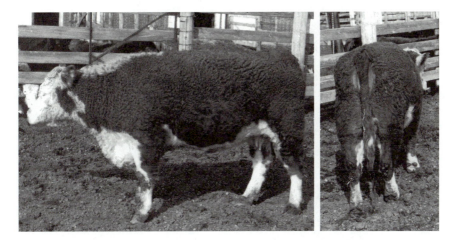

Fig. 12–15. A 1,120-pound Commercial cow. The advanced age of this cow is apparent from her angular conformation. The fullness in her brisket and behind her shoulders indicates a well-finished cow.

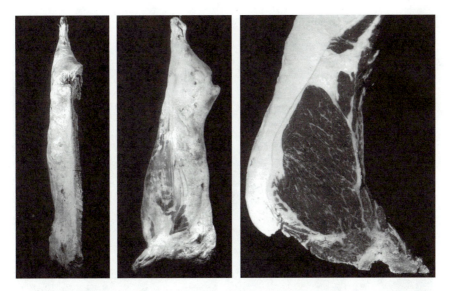

Fig. 12–16. A U.S. Commercial beef carcass. This very hard-boned carcass with no cartilages on the ends of the chine bones because they are completely ossified and flinty is E+ maturity. This is also indicated by the dark red, coarse-textured rib eye muscle. The moderately abundant amount of marbling indicates Average Commercial grade quality.

Fig. 12–17. An 840-pound Utility cow. From the side, her extreme angularity from front to rear is obvious. Note her protruding ribs, indicating practically a complete void of cover. From the rear, concave rounds and sunken sirloin indicate a lack of muscling. Normally, such cows are dry before going to market. This cow was harvested to obtain an active udder as part of a research project, thus her obviously preponderant udder.

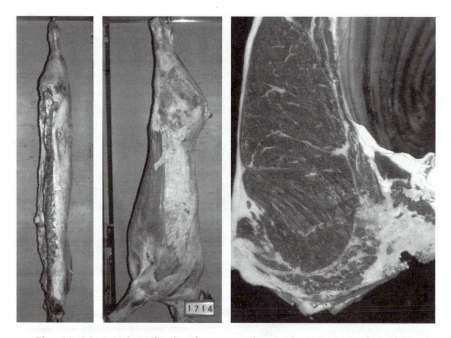

Fig. 12–18. A U.S. Utility beef carcass. The modest amount of marbling in the dark red, coarse-textured rib eye muscle of this E+ maturity carcass qualified it for Average Utility. This carcass has slightly concave rounds, slightly thin sunken loins, and flat, thinly fleshed ribs.

Yield Grades

Yield grades indicate the quantity of meat, that is, the amount of retail, consumer-ready, ready-to-cook, or edible meat, that a carcass contains. The net amount available for eating is extremely important economically. Yield grade should not be confused with dressing percent, which is often called *yield*. Yield grade refers to the amount of *edible product from the carcass*. Dressing percent (yield) refers to the amount of *carcass from a live animal*.

The lower the number of the yield grade, the higher the percent of the carcass in total retail product, including all roasts, steaks, and lean trim (ground beef). However, yield grades were originally determined by measuring only the boneless, closely trimmed roasts and steaks from the round, loin, rib, and chuck (see Table 12–4).

Determination of a Yield Grade

Beef yield grades are determined by a regression equation that is based on four carcass traits: (1) the amount of external fat; (2) the amount of kidney, pelvic, and heart (KPH) fat; (3) the area of the rib eye muscle (REA); and (4) the hot carcass weight (HCW).

The amount of *external fat* on a carcass is measured perpendicular to the outside fat surface at a point three-fourths of the length of the rib eye from its chine bone end (see Figure 12–19). This measurement may be adjusted, as necessary, to reflect unusual amounts of fat on other parts of the carcass, or if it is evident that fat has been removed from the measurement area during the harvest/hide removal procedure. Determining the amount of this adjustment requires that particular attention be given to the amount of fat in the brisket, plate, flank, cod or udder, inside round, rump, and hips in relation to the actual thickness of fat over the rib eye. Thus, in a carcass that is fatter over other areas than is indicated by the fat measurement over the rib eye, the measurement is adjusted upward. Conversely, in a carcass having less fat over the other areas than is indicated by the fat measurement over the rib eye, the measurement is adjusted downward. In many carcasses, no such adjustment is necessary; however, an adjustment in the thickness of fat measurement of 0.1 or 0.2 inch is not uncommon. In some carcasses, a greater adjustment may be necessary. As the amount of external fat increases, the percent of retail cuts decreases. Each 0.1-inch change in adjusted fat thickness over the rib eye changes the yield grade by 25 percent of a yield grade.

The amount of *kidney, pelvic, and heart fat* considered in determining the yield grade includes the kidney knob (kidney and surrounding fat), the lumbar and pelvic fat in the loin and round, and the heart fat in the chuck and brisket area, which are removed in closely trimmed retail cuts (see Chapter 15). The amount of these fats is evaluated subjectively and expressed as a percent of the carcass weight. As the amount of kidney, pelvic, and heart fat increases, the percent of retail cuts decreases. A change of 1 percent of the carcass weight in these fats changes the yield grade by 20 percent of a yield grade.

The *area of the rib eye* is the area considered where this muscle is exposed by ribbing between the twelfth and thirteenth ribs (see Figure 12–19). This area is usually estimated subjectively; however, it may be measured. Area of rib eye measurements may be made by means of a grid calibrated in tenths of a square inch, by a compensating planimeter measurement of an acetate tracing or by a computerized digital tracing. An increase in the area of rib eye increases the percent of retail cuts. A change of 1 square inch in the area of rib eye changes the yield grade by 32 percent of a yield grade.

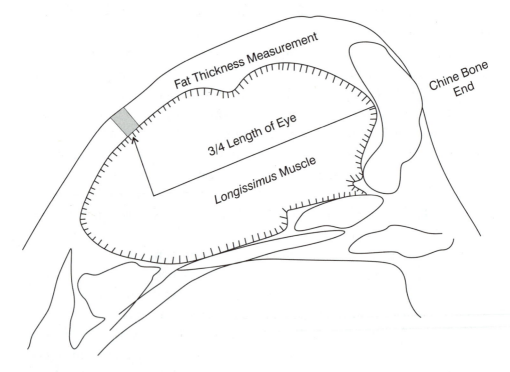

Fig. 12–19. Area where fat thickness and rib eye are measured. See page 356 for details on ribbing the carcass. (Courtesy, American Meat Science Association)

Hot carcass weight (or chilled carcass weight × 102 percent) is used in determining the yield grade. As the carcass weight increases, the percent of retail cuts in general decreases, because a heavier carcass is usually fatter. A change of 100 pounds in hot carcass weight changes the yield grade by approximately 38 percent of a yield grade.

The standards include a mathematical equation for determining yield grade group. This group is expressed as a whole number; any fractional part of a designation is always dropped. For example, if the computation results in a designation of 3.9, the final yield grade is 3 — it is not rounded to 4. *Equation*: 2.50 + (2.50 × adjusted fat thickness, inches) + (0.20 × percent kidney, pelvic, and heart fat) + (0.0038 × hot carcass weight, pounds) – (0.32 × rib eye area, square inches).

The following are examples of calculations of yield grades.

- A 610-pound (hot weight) Prime grade carcass has 0.7 inch of fat thickness over the rib eye (other external fat covering normal); 4.5 percent of its carcass weight in kidney, pelvic, and heart fat; and 10.4 square inches of rib eye. This is an old-fashioned steer but one still seen too many times. Thus, the equation 2.50 + (2.50 × 0.7) + (0.20 × 4.5) + (0.0038 × 610) – (0.32 × 10.4) = 2.50 + 1.75 + 0.9 + 2.32 – 3.33 = 4.14, resulting in Yield Grade 4.

- A modern steer having a 700-pound Low-Choice carcass with 0.2-inch fat thickness; 2 percent kidney, pelvic, and heart fat; and 14 square inches in rib eye area calculates: 2.50 + (2.50 × 0.2) + (0.2 × 2.0) + (0.0038 × 700) – (0.32 × 14.0) = 2.50 + 0.5 + 0.4 + 2.66 – 4.48 = 1.58, resulting in Yield Grade 1.

In actual practice, USDA and company graders, other industry personnel, and teachers and students preparing for intercollegiate grading competition use a simplified or shortcut system such as described here.

- From the fat thickness measurement, determine a *preliminary yield grade* (PYG) by including any adjustments for unusual fat deposition as described earlier. (See Table 12–2 for the relationship between fat thickness and preliminary yield grade.)

Table 12–2. Preliminary Beef Yield Grade by Fat Thickness

Thickness of Fat over Rib Eye	Preliminary Yield Grade
(in.)	
0.0	2.0
0.2	2.5
0.4	3.0
0.6	3.5
0.8	4.0
1.0	4.5
1.2	5.0
1.4	5.5
1.6	6.0

- Because the rib eye area is largely dependent on the hot carcass weight, every carcass has an expected rib eye area depending on its weight. The relationship between hot carcass weight and expected rib eye area is shown in Table 12–3. Note that the rib eye area increases and decreases at the rate of 0.3 square inch for each 25 pounds of hot carcass weight. The hot carcass weight is determined as the carcass leaves the harvest floor and is recorded on a carcass tag that is visible to the grader. The preliminary yield grade should be changed according to how much over or under the estimated or measured rib eye area is from the expected rib eye area. Each 1 inch over or under the expected rib eye area causes an adjustment of 0.3 of a yield grade as discussed earlier. All adjustments are rounded to the nearest tenth of a grade.

Table 12–3 gives an expected rib eye area according to carcass weight. Heavily muscled carcasses may have rib eye areas equal to or exceeding 2 square inches per 100 pounds of carcass weight.

- Determine the *final yield grade* (1 to 5) by further modifying the preliminary yield grade, as necessary, for variations in the kidney, pelvic,

Table 12–3. Beef Carcass Weight — Area of Rib Eye Schedule

Hot Carcass Weight	Area of Rib Eye	Hot Carcass Weight	Area of Rib Eye
(lb.)	(sq. in.)	(lb.)	(sq. in.)
		675	11.9
350	8.0	700	12.2
375	8.3	725	12.5
400	8.6	750	12.8
425	8.9	775	13.1
450	9.2	800	13.4
475	9.5	825	13.7
500	9.8	850	14.0
525	10.1	875	14.3
550	10.4	900	14.6
575	10.7	925	14.9
600	11.0	950	15.2
625	11.3	975	15.5
650	11.6	1,000	15.8

Guide to rib eye area for other weights of carcasses between the 25-pound graduations:

		600–608 lb. 11.0 sq. in.
600	11.0	609–616 lb. 11.1 sq. in.
625	11.3	617–624 lb. 11.2 sq. in.
		625–633 lb. 11.3 sq. in.

and heart fat from 3.5 percent. For each percent of kidney, pelvic, and heart fat more than 3.5 percent, add 0.2 of a grade to the preliminary yield grade. For each percent of kidney, pelvic, and heart fat less than 3.5 percent, subtract 0.2 of a grade from the Preliminary Yield Grade.

For example, use this simplified shortcut system to yield grade the two carcasses yield graded earlier.

	Carcass A	*Carcass B*
Fat thickness, in.	0.7	0.2
PYG (Table 12–2)	3.75 = 3.8	2.5
HCW, lb.	610	700
Expected REA, sq. in.	11.1	12.2
REA, sq. in.	10.4	14.0
REA adjustment	+0.21 = 0.2 (0.3 × 0.7)	−0.54 = 0.5 (0.3 × 1.8)
KPH fat, %	4.5	2.0
KPH adjustment	+0.2 (0.2 × 1)	−0.3 (0.2 × 1.5)
YG	4.2	1.7

For live animal evaluation, a similar simplified shortcut system is sometimes used.

- Start with a preliminary yield grade based on your estimate of the animal's carcass fat thickness and overall fatness (see Figures 12–23 to 12–32).

- Adjust the preliminary yield grade for hot carcass weight, rib eye area, and kidney, pelvic, and heart fat a constant 0.1 unit of variation from a base as follows:

	Base	*Amount to Equal 0.1 YG*
HCW	600 lb.	25 lb.
REA	11.00 sq. in.	0.33 sq. in.
KPH fat	3.5%	0.5%

Round off all estimates and calculations to 0.1. Example (using carcasses graded by equation on previous pages):

	Carcass A	*Carcass B*
Fat thickness, in.	0.7	0.2
HCW, lb.	610	700
REA, sq. in.	10.4	14.0
KPH fat, %	4.5	2.0
PYG	3.8	2.5

Adjusted for

HCW   +0.04 = 0   +0.4

$(610 - 600 = 10/25 = 0.4 \times 0.1 = 0.04)(700 - 600 = 100/25 = 4 \times 0.1)$

REA   +0.2   −0.9

$(11 - 10.4 = 0.6/0.33 = 1.8 \times 0.1)(14 - 11 = 3/0.33 = 9.1 \times 0.1)$

KPH   +0.2   −0.3

$(4.5 - 3.5 = 1/0.5 = 2 \times 0.1)(3.5 - 2 = 1.5/0.5 = 3 \times 0.1)$

Yield grade   4.2   1.7

The equation is the most accurate, while the simplified shortcut method is off a few hundredths due to rounding. The method used by graders works most efficiently and quickly because graders know the requirements verbatim and sometimes must apply them and a quality grade to beef carcasses passing on a moving chain that may run at a speed in excess of 350 carcasses per hour.

Determination of Percent of Boneless Closely Trimmed Retail Cuts from the Round, Loin, Rib, and Chuck

The same four carcass traits that are used to determine yield grades are used to determine the percentage yield of a beef carcass in boneless closely trimmed retail cuts from the round, loin, rib, and chuck. This percentage is low because it does not include all the lean trim and additional cuts that contribute to the edible portion of a beef carcass (see Chapter 15). The equation for the percent of boneless closely trimmed roasts and steaks from the round, loin, rib, and chuck = 51.34 − (5.784 × adjusted fat thickness) − (0.0093 × hot carcass weight) − (0.462 × percent kidney, pelvic, and heart fat) + (0.740 × rib eye area). The relationship between these percentages and yield grades is presented in Table 12–4.

Table 12–4. Beef Yield Grade Boneless Primal Roast and Steak Yield Equivalents

Yield Grade	Percent of Carcass in Closely Trimmed Roasts and Steaks from Rounds, Loins, Ribs, and Chucks
1	52.6–54.6
2	50.3–52.3
3	48.0–50.0
4	45.7–47.7
5	43.5–45.4

Specifications for Official U.S. Standards for Yield Grades of Carcass Beef

The following descriptions of carcass characteristics in each yield grade provide a guide that can help in the subjective determination of yield grades. The yield grade for most beef carcasses can be determined accurately on the basis of a visual appraisal by a trained and experienced grader.

Yield Grade 1

Carcasses have a thin layer of external fat over the ribs, loins, rumps, and clods and slight deposits of fat in the flanks and cods or udders; they have a very thin layer of fat over the outside of the rounds and over the tops of the shoulders and necks.

Muscles are usually visible through the fat in many areas of the carcass.

The following descriptions are of two different weight carcasses near the borderline of Yield Grades 1 and 2.

- A 500-pound carcass might have 0.3 inch of fat over the rib eye, 11.5 square inches of rib eye, and 2.5 percent of its weight in kidney, pelvic, and heart fat.

- An 800-pound carcass might have 0.4 inch of fat over the rib eye, 16 square inches of rib eye, and 2.5 percent of its weight in kidney, pelvic, and heart fat.

Yield Grade 2

Carcasses are completely covered with fat, but the lean is plainly visible through the fat over the outside of the rounds, the tops of the shoulders, and the necks.

There is a slightly thin layer of fat over the loins, ribs, and inside rounds, with a slightly thick layer of fat over the rumps, hips, and clods.

Small deposits of fat occur in the flanks and cods or udders.

- A 500-pound carcass (near the borderline of Yield Grades 2 and 3) might have 0.5 inch of fat over the rib eye, 10.5 square inches of rib eye, and 3.5 percent of its weight in kidney, pelvic, and heart fat.

- An 800-pound carcass might have 0.6 inch of fat over the rib eye, 15 square inches of rib eye, and 3.5 percent of its weight in kidney, pelvic, and heart fat.

Yield Grade 3

Carcasses are usually completely covered with fat, and the lean is usually visible through the fat only on the necks and lower part of the outside of the rounds.

There is a slightly thick layer of fat over the loins, ribs, and inside rounds, with a moderately thick layer over the rumps, hips, and clods.

Slightly large deposits of fat occur in the flanks and cods or udders.

- A 500-pound carcass (near the borderline of Yield Grades 3 and 4) might have 0.7 inch of fat over the rib eye, 9.5 square inches of rib eye, and 4 percent of its weight in kidney, pelvic, and heart fat.

- An 800-pound carcass might have 0.8 inch of fat over the rib eye, 14 square inches of rib eye, and 4.5 percent of its weight in kidney, pelvic, and heart fat.

Yield Grade 4

Carcasses in this group are usually completely covered with fat. The only muscles visible are those on the shanks and over the outside of the plates and flanks.

There is a moderately thick layer of fat over the loins, ribs, and inside rounds and a thick covering of fat over the rumps, hips, and clods.

Large deposits of fat occur in the flanks and cods or udders.

- A 500-pound carcass (near the borderline of Yield Grades 4 and 5) might have 1 inch of fat over the rib eye, 9 square inches of rib eye, and 4.5 percent of its carcass weight in kidney, pelvic, and heart fat.

- An 800-pound carcass might have 1.1 inches of fat over the rib eye, 13.5 square inches of rib eye, and 5 percent of its weight in kidney, pelvic, and heart fat.

Yield Grade 5

A carcass in this group has more fat on all the various parts, a smaller area of rib eye, and more kidney, pelvic, and heart fat than a carcass in Yield Grade 4.

Yields and Value Differences Between Yield Grades

Value differences between yield grades naturally fluctuate from week to week and month to month because of overall meat price fluctuations and variations in cutting and trimming methods. Further discussion on yield of cuts and value differences between carcasses is found in Chapters 13 and 15.

Applying the Grade Stamp

The federal meat grader weighs all the factors that have been discussed, decides upon the grade, and makes an identifying mark with a stamp designating the grade: "xxx" for Prime, "xx" for Choice, etc. After the grading task is concluded, the grader or an assistant applies the ribbon grade stamp over the entire length of the carcass with a roller that includes the grader's initials (see Figures 12–20 and 12–21) and at such other points that all principal

Fig. 12–20. Applying the grade stamp. Note the round inspection stamp adjacent to the rolled grade stamp. (See also Figs. 12–21 and 12–22.) (Courtesy, Livestock and Seed Division, Agricultural Marketing Service, USDA)

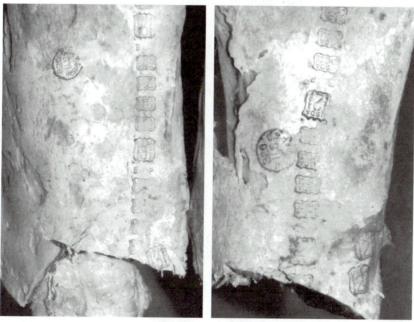

Fig. 12–21. *(Left)* Closeup of a hindquarter showing the combination yield, quality roll, and inspection stamp. "E P" are the initials of the grader who graded the carcass. Note the normal presence of kidney knob. *(Right)* Closeup of a hindquarter from which the kidney knob has been removed, making this carcass a Yield Grade 2. To indicate that kidney, pelvic, and heart fat has been removed, the grade is rolled on upside down. (Verified, 1999, by Martin ("Marty") E. O'Connor, Agricultural Marketing Specialist, USDA)

retail cuts will bear the grade label. That is why federally graded beef and lamb are often referred to as "rolled." The grade designations are in full grades. The Cutter and Canner beef grades are *not* rolled. If the buyer wishes to know whether the rolled beef of the grade being bought is in the top, middle, or low bracket of the grade, he or she can ask for a grading certificate that will so indicate.

The grade stamp has been rolled on beef and lamb carcasses for many years. However, with the added emphasis on lean meat, subcutaneous fat is usually trimmed away before meat is seen by consumers. In this instance, the grade roll comes off with the fat unless special precautions are taken to leave the strip of fat displaying the grade roll. Also, because most packers have discontinued using shrouds (see Chapter 6), the fat surface on carcasses is not smooth, and the grade roll is often difficult to read. Alternate methods, other than roller branding, to identify the USDA grade on carcasses and cuts are being explored. One prototype system involves using grade-labeled plastic bags to maintain the grade identity of each individual cut. The bags are sealed in a way that retains the integrity of the product. Also the grade can be stamped on each box or package of product.

Do not confuse grading with inspection or the grading stamp with the inspection stamp. By law, ***every animal that is harvested for meat sale must be inspected.*** The federal inspection stamp is round and appears on each wholesale cut (see Figures 12–21 and 12–22). Grading is ***not required*** by law.

Fig. 12–22. Two round inspection stamps and the grade stamp showing the combined quality and yield roll. Two inspection stamps indicate that the animal was harvested in one plant, cut into a wholesale rib in a second plant, and shipped to a third plant or store.

Live Animal and Carcass Illustrations of Beef Yield Grades

Beef carcass yield grades are based on four factors (see previous discussion): (1) carcass weight; (2) rib eye area; (3) external fat thickness; and (4) percent kidney, pelvic, and heart fat. Since these final yield grade factors are more objective (that is, may be actually measured) than are those for the quality grades, live appraisal for yield grade is somewhat more straightforward than live appraisal for quality grades.

For instance, certain key areas in the live animal can be evaluated for indications of trimness and muscling. Those animals that are deep and full in the twist, flanks, and brisket and those that are wide and flat over the top of the back will be fat, because those places fill up with fat as animals get overfat. Muscling indicators in the live animal are width through the lower round in the area of the stifle joint, width through the chest, leg placement well spaced on all four corners, evidence of forearm muscling, and a rounded, muscular appearance over the loin and back. Muscles are rounded, not square. Squareness over the back indicates fatness, not muscling.

Fatness and muscling are the two important yield grade factors, and they can be reasonably appraised. Carcass weight is, of course, a function of live weight and can easily be considered in the evaluation of the live animal. Kidney, pelvic, and heart fat is somewhat difficult to estimate. Cattle with wasty middles may have excessive internal fats, but this is easily confused with fill (feed in the digestive tract, primarily in the rumen). Thus, an estimate of kidney, pelvic, and heart fat based on overall fatness is probably the most useful method. Dairy animals are known for carrying heavier than normal amounts of kidney fat.

In actual evaluations, movement of the animal is extremely helpful in ascertaining muscling and trimness. Obviously, these pictures (see Figures 12–23 to 12–32), provided by the University of Illinois, South Dakota State University, and the National Live Stock and Meat Board, do not provide movement, and thus serve only as a general guide to evaluation. When evaluating, the student should always observe animals in motion. Oftentimes, judges like to feel an animal to determine the fatness, but such an opportunity is seldom available during the cattle marketing sequence. Thus, top cattle buyers perfect the "eye ball" method of evaluation.

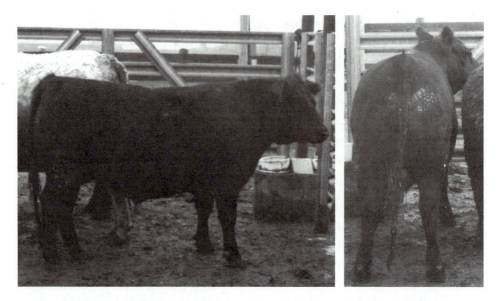

Fig. 12–23. A 1,020-pound Yield Grade 1 steer. From the side, you can see the trimness in his flank and brisket. From this view, you can also observe muscling in the shoulder and bulge to the lower round. From the rear, note the leg placement squarely on the corners. He lacks depth of twist, indicating trimness, but he is wide and thick through the lower round, indicating muscling.

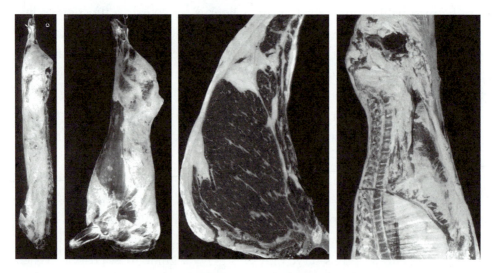

Fig. 12–24. Yield Grade 1 beef carcass.

Carcass weight . . . . 645 pounds
External fat thickness. . . . . 0.2 inch
Rib eye area . . . . 13.9 square inches
Kidney, pelvic, and heart fat 2.5%
Yield grade. 1.5
Quality grade . . . . High Choice

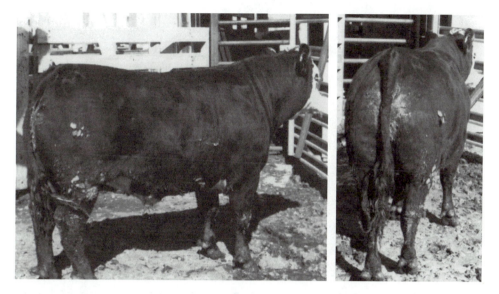

Fig. 12–25. A 975-pound Yield Grade 2 steer. From the side, this steer shows considerable forearm muscling and is moderately trim in the rear flank. His stance from behind is somewhat closed, although he does show width through the center of his rounds and a muscular turn to his loin.

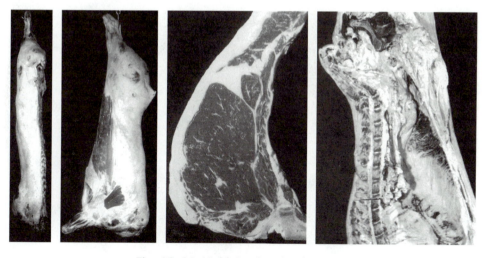

Fig. 12–26. Yield Grade 2 beef carcass.

Carcass weight . . . . 605 pounds
External fat thickness. . . . . 0.4 inch
Rib eye area . . . . 12.3 square inches
Kidney, pelvic, and heart fat 3.0%
Yield grade. 2.5
Quality grade . . . . Average Choice

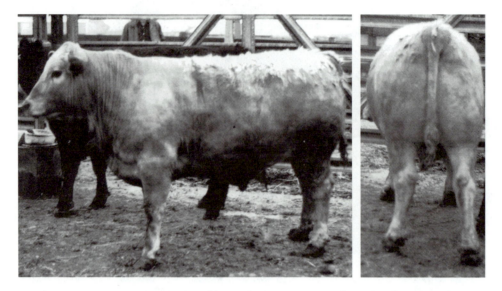

Fig. 12–27. A 1,120-pound Yield Grade 3 steer. From the side, this steer appears upstanding but does show some flesh in his brisket. Forearm and rear quarter muscling are evident. From the rear, excess fat is evident around the tail head, and his turn of loin indicates possible excess finish. Muscling from this view appears adequate but not exceptional.

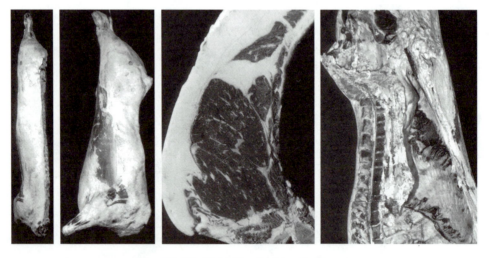

Fig. 12–28. Yield Grade 3 beef carcass.

Carcass weight . . . . 700 pounds
External fat thickness. . . . . 0.6 inch
Rib eye area . . . . 11.8 square inches
Kidney, pelvic, and heart fat 3.5%
Yield grade. 3.6
Quality grade . . . . Average Choice

Fig. 12–29. A 1,060-pound Yield Grade 4 steer. From the side, excessive development of the brisket is apparent. The steer appears very heavy in the front quarter and lacks muscle in the valuable hindquarter. From the rear, his broad, flat-topped back indicates excessive finish.

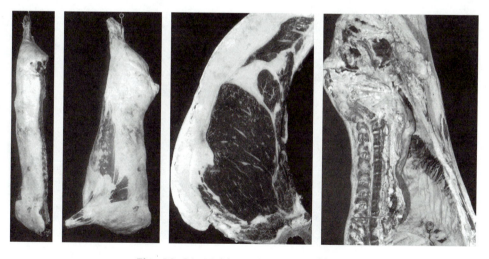

Fig. 12–30. Yield Grade 4 beef carcass.

Carcass weight . . . . 665 pounds
External fat thickness. . . . . 0.9 inch
Rib eye area . . . . 10.5 square inches
Kidney, pelvic, and heart fat 3.5%
Yield grade. 4.6
Quality grade . . . . Average Choice

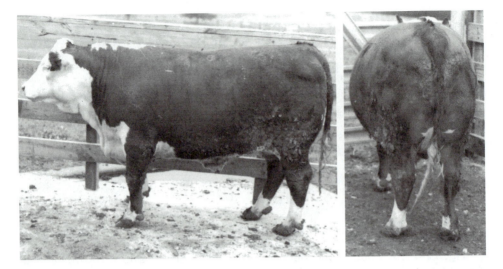

Fig. 12–31. A 1,200-pound Yield Grade 5 steer. This big steer from the side appears to have adequate muscling, but excessive fat is especially evident over the tail head and sirloin. From the rear, he is much wider in his middle and up over his back than he is through his rounds, indicating excessive fat. His rear stance is closed, giving evidence of poor muscling.

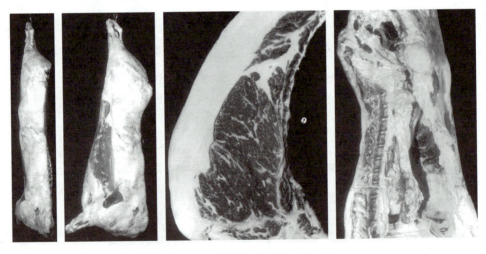

Fig. 12–32. Yield Grade 5 beef carcass.

Carcass weight . . . . 750 pounds
External fat thickness. . . . . 1.1 inches
Rib eye area . . . . 10.9 square inches
Kidney, pelvic, and heart fat 5.0%
Yield grade. 5.6
Quality grade . . . . High Choice

National Summary of Beef Graded

See Table 12–5 for a summary of the total number of beef carcasses graded, the numbers in each grade for quality and yield for 1998, and the Web site for updating this information.

Table 12–5. National Summary of Beef Graded, 1998 (Number of Head)[1]

Quality Grade	Yield Grade					Quality Only	Total Quality Graded This Pd.	% of All Quality Graded	% of Total Steer and Heifer Harvest
	#1	#2	#3	#4	#5				
Prime	11,604	188,456	385,472	29,228	2,116	249,955	866,831	3.3	3.0
Choice	892,131	6,246,505	6,438,225	285,475	23,269	1,709,274	15,594,879	59.7	54.4
Select	1,585,714	4,724,085	2,259,542	45,514	3,120	1,036,354	9,654,229	36.9	33.7
Standard	185	294	77	0	0	948	1,504	0.0	0.0
Commercial	6	474	522	1	0	53	1,056	0.0	
Utility	143	5,926	3,887	0	0	69	10,025	0.0	
Cutter	0	0	0	0	0	0	0	0.0	
Canner	0	0	0	0	0	0	0	0.0	
Yield only	429,478	521,914	326,137	19,737	1,460				
Total yield graded this period	2,919,261	11,687,654	9,413,862	379,955	29,965		Total bull/bullock 14		
% of all yield graded	11.9	47.8	38.5	1.5	0.1				

95.7% of federally inspected steer and heifer harvest represents total steers and heifers graded.

0.2% of federally inspected cow harvest represents total cows graded.

83.2% of federally inspected beef harvest represents total beef (steers, heifers, and cows) graded.

[1]Information provided by U.S. Department of Agriculture, Agricultural Marketing Service, Livestock and Seed Division, Meat Grading and Certification Branch. http://www.ams.usda.gov/lsg/mgc/Reports/mn-cy98.pdf

VEAL AND CALF CARCASS GRADING

The high demand for beef has caused the onset of the practice of castrating and feeding male dairy calves for subsequent harvest as beef steers. Most of the more thickly muscled calves are selected for further feeding, leaving calves with less well-developed conformation for the veal and calf market.

Grading standards have been continually updated to reflect industry changes. The most current change allows for grading only in carcass form after the hide has been removed and only in the establishment where the hide removal occurs.[5] In 1971, changes were made that (1) increased the emphasis placed on the color of lean in classifying veal; (2) reduced the conformation requirements one full grade in both veal and calf; (3) reduced the quality requirements by varying degrees for both veal and calves; and (4) eliminated the Cull grade for both veal and calves, leaving Prime, Choice, Good, Standard, and Utility as the carcass grades for both classes.

Changes in the standards for grades of veal and calf carcasses previously made effective March 10, 1951, coincided with revisions made in the live grades and (1) combined the former Choice and Prime grades under the name *Prime;* (2) renamed the Good grade *Choice;* (3) established a new grade called *Good,* which included meat from the top half of the former Commercial grade; (4) continued the remainder of the Commercial grade as *Commercial;* and (5) left the Utility and Cull grades unchanged.

A further revision was made in October 1956, changing the Commercial grade to *Standard* and making certain changes in the phrasing of the standards designed to facilitate their interpretations.

Differentiation Between Veal, Calf, and Beef Carcasses

Differentiation between veal, calf, and beef carcasses is made on the basis of the color and texture of the lean; character of the fat; color, shape, size, and ossification of the bones and cartilages; and general contour of the carcass. Typical veal carcasses have a *grayish pink* color of lean that is very smooth and velvety in texture. They also have a slightly soft, pliable character of fat and narrow, very red rib bones. By contrast, typical calf carcasses have a *grayish red* color of lean, a flakier type of fat, and somewhat wider rib bones with less pronounced evidence of red color. Carcasses that show *any* indication of ossification in the *sacral* vertebrae are classified as beef.

Classes of Veal and Calf Carcasses

Class determination is based on the apparent sex condition of the animal at the time of harvest. Hence, there are three classes of veal and calf carcasses —

---

[5]USDA–AMS. 1980. Official United States Standards for Grades of Veal and Calf Carcasses. Title 7, Chp. I, Pt. 53, Sec. 53.107–53.111, Code of Federal Regulations. (October 6).

steers, heifers, and bulls. While recognition may sometimes be given to these different classes on the market, especially calf carcasses from bulls that are approaching beef in maturity, the characteristics of such carcasses are not sufficiently different from those of steers and heifers to warrant the development of separate standards for them. Therefore, the grade standards that follow are equally applicable to all classes of veal and calf carcasses.

Application of Standards

Veal and calf carcasses are graded on a composite evaluation of *conformation* and *quality. Conformation,* or the manner of formation of the carcass, refers to its thickness or fullness.

Quality of lean — in all veal carcasses, all unribbed calf carcasses, and ribbed calf carcasses in which their degree of marbling is not a consideration — usually can be evaluated with a high degree of accuracy if equal consideration is given to the following factors, as available: The amount of feathering (fat intermingled within the lean between the ribs) and the quantity of fat streakings within and upon the inside flank muscles. (In making these evaluations, the grader considers the amounts of feathering and flank fat streakings in relation to color [veal] and maturity [calf].) In addition, however, consideration also may be given to other factors if, in the opinion of the grader, a more accurate quality assessment will result. Examples of such other factors include firmness of the lean, distribution of feathering, amount of fat covering over the diaphragm or skirt, and amount and character of the external and kidney and pelvic fat.

When ribbed calf carcasses are graded, the quality evaluation of the lean is based entirely on characteristics that are exposed in the cut surface. Figure 12–33 illustrates how the factors are combined to arrive at the quality grades shown in Figure 12–34.

The *final grade* of a carcass is based on a composite evaluation of its conformation and quality. Conformation and quality often are not developed to the same degree in a carcass, and it is obvious that each grade will include various combinations of development of these two characteristics. The principles governing the compensations of variations in the development of quality and conformation are as follows: In each of the grades, a superior development of quality is permitted to compensate, without limit, for a deficient development of conformation. In this instance, the rate of compensation in all grades is on an equal basis — a given degree of superior quality compensates for the same degree of deficient conformation. The reverse type of compensation — a

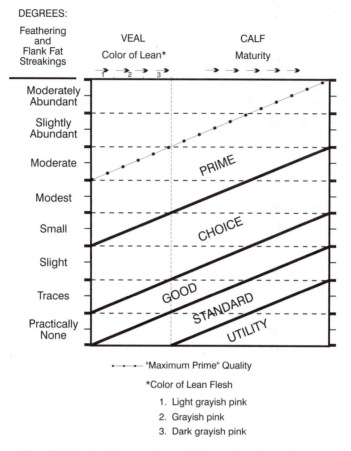

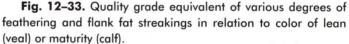

Fig. 12–33. Quality grade equivalent of various degrees of feathering and flank fat streakings in relation to color of lean (veal) or maturity (calf).

superior development of conformation for an inferior development of quality — is not permitted in the Prime and Choice grades. In all other grades, this type of compensation is allowed, but only to the extent of one-third of a grade of deficient quality. The rate of this type of compensation is also on an equal basis — a given degree of superior conformation compensates for the same degree of deficient quality.

National Summary of Veal and Calf Graded

See Table 12–6 for a summary of the total number of veal and calf carcasses quality graded, the numbers in each quality grade for 1998, and the Web site for updating this information.

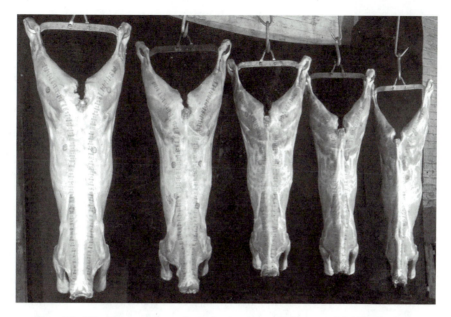

Fig. 12–34. Five grades of veal carcasses *(left to right)*: Prime, Choice, Good, Standard, and Utility. *(Note:* the grade name *Good* is retained in veal and calf grading.)

Table 12–6. National Summary of Veal and Calf Graded, 1998 (Number of Head)[1]

Quality Grade	Total This Pd.	%
Prime	38,855	11.8
Choice	285,875	87.2
Good	3,023	0.9
Standard	0	0.0
Utility	0	0.0
Total	327,753	

22.8% of federally inspected veal and calf harvest represents total veal and calf graded.

[1]Information provided by U.S. Department of Agriculture. Agricultural Marketing Service, Livestock and Seed Division, Meat Grading and Certification Branch. http://www.ams.usda.gov/lsg/mgc/Reports/mn-cy98.pdf

LAMB, YEARLING MUTTON, AND MUTTON CARCASS GRADING

Development of the Standards[6]

- In February 1931, grades for lamb and mutton carcasses became effective.

- In October 1940, the standards were amended so as to change the grade designations Medium and Common to *Commercial* and *Utility*, respectively.

- In April 1951, the official standards were again amended when Prime and Choice grades were combined and designated as *Prime*. The Good grade was renamed *Choice*, which also became the highest grade for carcasses of mutton older than yearlings.

  The top two-thirds of the Commercial grade was designated as *Good*. The lower one-third of the Commercial grade was combined with the top two-thirds of the Utility grade and designated as *Utility*, thereby eliminating the Commercial grade name. The lower one-third of the Utility grade was combined with the Cull grade and designated as *Cull*.

- In February 1957, the standards for grades of lamb carcasses were amended. The quality requirements for Prime and Choice grade carcasses were reduced from those for more mature lambs. The quality requirements for the Good grade were increased slightly, particularly for carcasses from very young lambs.

  Carcasses with quality indications equivalent to the lower limit of the upper third of the Good grade were permitted to be graded *Choice*, provided they had a development of conformation equivalent to the mid-point of the Choice grade or better.

  Practically all references to quantity of external and kidney and pelvic fat were eliminated.

---

[6]USDA–AMS. 1992. Standards for Grades of Lamb, Yearling Mutton and Mutton Carcasses and Standards for Grades of Slaughter Lambs, Yearlings, and Sheep; Rule 7, Code of Federal Regulations, Pts. 53 and 54. Federal Register 57, No. 97:21338–21345, May 9.

- In March 1960, both the conformation and the quality requirements for the Prime and Choice grades were reduced about one-half grade. In addition, a minimum degree of external fat covering was prescribed for the Prime and Choice grades.

    The emphasis placed on internal factors considered in the evaluation of quality was decreased — the emphasis on feathering between the ribs was reduced, the consideration of overflow fat was eliminated, and the emphasis on firmness of fat and lean was increased.

- On March 1, 1969, yield grades were adopted for lamb, yearling mutton, and mutton carcasses.

- In October 1980, the standards were changed to require grading only in carcass form and only in the establishment where the animals were harvested or initially chilled.

- In October 1982, the standards were revised to allow a carcass with one or two break joints to be classed as lamb. Feathering was dropped as a quality grade criterion.

- On July 6, 1992, the standards were changed to require the application of the yield grade for lamb, yearling mutton, and mutton when it was quality graded and the removal of kidney/pelvic fat >1 percent of carcass weight prior to grading. Leg conformation was dropped as a yield grade criterion.

Differentiation Between Lamb, Yearling Mutton, and Mutton Carcasses

Differentiation between lamb, yearling mutton, and mutton carcasses is made on the basis of differences that occur in the development of their muscular and skeletal systems (see Table 12–7). Typical lamb carcasses must have at least one front shank break joint. Yearling mutton carcasses may have either break joints or spool joints on their front shanks (see Figures 7–6 and 7–7). Typical mutton carcasses always have spool joints on their front shanks.

**Table 12–7. Lamb, Yearling Mutton, and
Mutton Maturity Indicators**

	Young Lambs	Mature Lambs	Yearling Mutton	Mutton
	(A maturity)	(B maturity)		
Rib bones	Moderately narrow, slightly flat	Slightly wide, moderately flat	Moderately wide, tend to be flat	Wide, flat
Break joints	Moderately red, moist, and porous	Slightly red, dry, and hard	May or may not be present	Spool joints
Color of inside flank muscles	Slightly dark pink	Light red	Slightly dark red	Dark red
Texture of lean	Fine	Moderately fine	Coarse	Coarse

Two Considerations — Quality and Yield

The grade of a lamb, yearling mutton, or mutton carcass is based on separate evaluations of two general considerations: (1) the palatability-indicating characteristics of the lean and conformation, referred to as the *quality grade* and (2) the indicated percent of trimmed, boneless major retail cuts to be derived from the carcass, referred to as the *yield grade*. When graded by a federal meat grader, the grade of a lamb, yearling mutton, or mutton carcass may consist of the quality grade, the yield grade, or a combination of both the quality grade and the yield grade.

The terms *quality grade* and *quality* are used throughout the standards. The term *quality* is used to refer to only the palatability-indicating characteristics of the lean. As such, it is the major consideration in determining the quality grade. Although the term *quality grade* is used to refer to an overall evaluation of an *ovine* carcass based on both its quality and its conformation, this usage is not intended to imply that variations in conformation are either directly or indirectly related to differences in palatability, since research has shown no such relationship.

Conformation is the manner of formation of the carcass with particular reference to the relative development of the muscular and skeletal systems, although it is also influenced, to some extent, by the quantity and distribution of external finish.

Quality of the lean flesh is best evaluated from the observation of a cut surface to consider the texture, firmness, and marbling of the lean in relation to the apparent maturity of the animal from which the carcass was produced. However, in grading carcasses, direct observation of these characteristics is not always possible. Therefore, the quality of the lean is evaluated indirectly. Equal consideration is given to both the streaking of fat within and upon the inside flank muscles and the firmness of the fat and lean in relation to the apparent evidence of maturity.

The apparent *sex condition* of the animal at the time of harvest is not normally considered in ovine carcass grading. However, carcasses with thick, heavy necks and shoulders, typical of uncastrated males, are discounted in grade in accord with the extent to which these characteristics are developed. Such discounts may vary from less than one-half grade in carcasses from young lambs, in which such characteristics are barely noticeable, to as much as two full grades in carcasses from mature rams, in which such characteristics are very pronounced.

Lamb Carcass Quality Grades

Prime

Carcasses possessing the minimum qualifications for the Prime grade are moderately wide and thick in relation to their length and have moderately wide and thick backs, moderately plump and full legs, and moderately thick and full shoulders.

Requirements for firmness of lean and fat vary with changes in maturity.

- Young lambs — small quantity of fat streaking on inside flank muscle.
- Mature lambs — modest amount of fat streaking on inside flank muscle.

The lean flesh and the exterior finish must not be less than tending to be moderately firm. The minimum external fat requirements for Prime are at least a very thin covering of external fat over the top of the shoulders and the outside of the center parts of the legs, and at least a thin covering of fat over the back; that is, the muscles of the back must be plainly visible through the fat.

The development of quality that is superior to that specified as minimum for Prime may compensate, on an equal basis, for the development of conformation that is inferior to that specified as minimum for Prime. For example, a carcass that has evidence of quality equivalent to the midpoint of the Prime grade may have conformation equivalent to the midpoint of the Choice grade and remain eligible for Prime. However, in no instance may a carcass that has a conformation inferior to that specified as minimum for the Choice grade be graded Prime.

Choice

In the description of the Choice grade, the word *slightly* may be substituted for the word *moderately* that is used in the description of the Prime grade. A carcass that has conformation equivalent to at least the midpoint of the Choice grade may have quality equivalent to the minimum for the upper third of the Good grade and remain eligible for Choice. Superior quality may compensate on an equal basis for a conformation that is inferior to that specified as minimum for Choice. For example, top Choice quality in a carcass with top Good conformation makes the carcass eligible for the Choice grade. However, in no instance may a carcass that has a conformation inferior to that specified for the Good grade be graded Choice.

Good

Lamb carcasses possessing minimum conformation qualifications for the Good grade are slightly thin-muscled throughout; are moderately narrow in relation to their length; and have slightly thin, tapering legs and slightly narrow, thin backs and shoulders. The young lambs are practically devoid of fat streaking on the inside flank muscles. Their lean flesh and exterior finish must be not less than slightly soft.

The more mature lambs have traces of fat streaking on the inside flank muscles. Also, their lean flesh and external finish must not be less than slightly soft.

A carcass that has conformation equivalent to the midpoint of the Good grade and quality equivalent to the minimum for the upper one-third of the Utility grade is eligible for Good. Also, a quality that is superior to that specified as the minimum for the Good grade may compensate for a conformation that is inferior to that specified as the minimum for Good on the basis of one-half grade of superior quality for one-third grade of deficient conformation. However, in no instance may a carcass that has a conformation inferior to that specified as minimum for the upper one-third of the Utility grade be graded Good.

Utility

The Utility grade includes those lamb carcasses whose characteristics are inferior to those specified as minimum for the Good grade.

Yearling and Mutton Carcass Quality Grades

The grades are Prime, Choice, Good, and Utility for yearling carcasses and Choice, Good, Utility, and Cull for mutton carcasses.

Live Animal and Carcass Illustrations of Lamb Quality Grades

Lamb carcass quality grading is based on conformation, maturity, and evidence of quality as seen in flank streaking and flank fullness and firmness (see previous discussion and Figures 12–35 to 12–42, provided by the University of Illinois, South Dakota State University, and the National Live Stock and Meat Board). Conformation can readily be determined in live lambs, and maturity can be approximated. However, the quality traits must be estimated in the live animal by evaluation of overall finish or fatness. Sight alone is often deceiving, especially with wooled lambs, so most buyers will "wade through" the group making spot checks of fatness over the rumps, ribs, and backs on a number of the lambs.

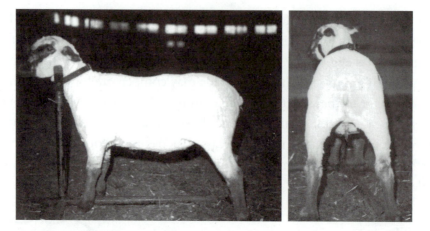

Fig. 12–35. A Prime wether. From the side, this black-faced lamb appears moderately thickly fleshed. The shoulders and hips are moderately smooth. From the rear, he is moderately wide over the back, loin, and rump. The twist is moderately deep and full, and the legs are moderately large and plump.

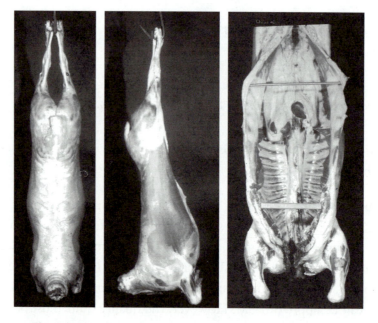

Fig. 12–36. A U.S. Prime lamb carcass. This lamb carcass has typical A maturity, as indicated by the moderately narrow, slightly flat rib bones and the slightly dark pink color of the inside flank muscles. Its plump, full legs; wide, thick back; and thick, full shoulders qualify its conformation for Average Prime. The modest streakings of fat in the inside flank muscles indicate Average Prime quality. When the conformation and quality grades are combined, this carcass qualifies for Average Prime.

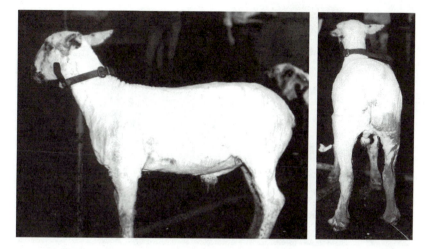

Fig. 12–37. A Choice crossbred wether. The one individual shown here is only that *one* of several sex conditions, many breeds, and types that annually qualify for this grade. The shoulders and hips show some prominence, indicating that he may be carrying a thin fat covering. From the rear, he is not as wide over the back, loin, and rump as the Prime lamb and his leg is not as fully developed.

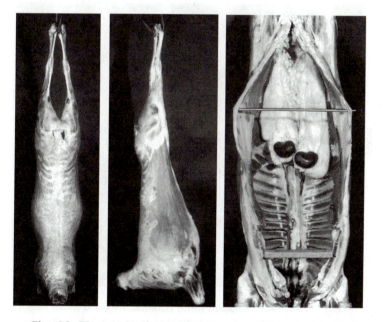

Fig. 12–38. A U.S. Choice lamb carcass. This typical A maturity lamb carcass has Average Choice conformation, as evidenced by slightly plump and full legs, a slightly wide and thick back, and slightly thick and full shoulders. The slight streakings of fat in the inside flank muscles indicate the quality requirements for Average Choice. With Average Choice conformation and quality, the final grade of this carcass is Average Choice.

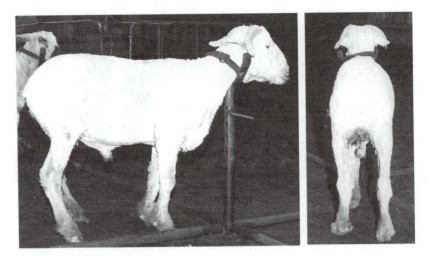

Fig. 12–39. A Good wether. (Note the *Good* grade name is retained in lamb grading.) From the side, this western lamb is moderately rangy and thinly fleshed. His hips and shoulders are moderately prominent. From the rear, he is slightly narrow over the back, loin, and rump. The twist is slightly shallow, and the legs are slightly small and thin.

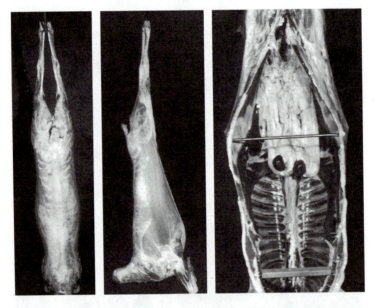

Fig. 12–40. A U.S. Good lamb carcass. This is a more mature lamb carcass (B maturity), as indicated by the slightly wide, moderately flat rib bones and the light red color of the inside flank muscles. It has Average Good conformation in that the legs tend to be slightly thin and tapering, and the back and shoulders tend to be slightly narrow and thin. Likewise, Average Good quality is indicated by the strong traces of streaking of fat in the inside flank muscles. Average Good conformation and Average Good quality combine to qualify this carcass for Average Good.

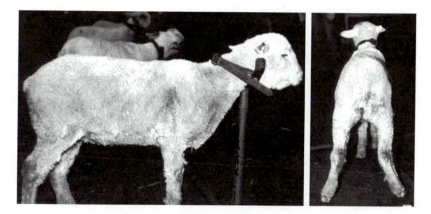

Fig. 12–41. A Utility ewe lamb. This is a rangy and angular lamb that is very thinly fleshed. She is very narrow over the back, loin, and rump and very shallow in the twist. The legs are very small and present a slightly concave appearance.

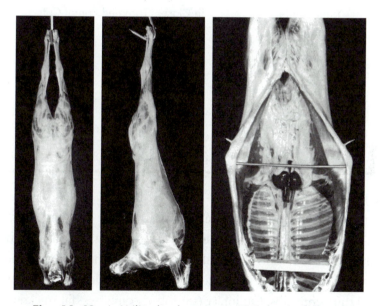

Fig. 12–42. A Utility lamb carcass. This lamb carcass has slightly wide, moderately flat rib bones and light red color of the inside flank muscles, indicating B– maturity. It has Low Utility quality because there are no fat streakings in the inside flank muscles. This carcass is slightly narrow and thin, qualifying it for Average Good conformation. Therefore, when the Average Good conformation and the Low Utility are combined, this carcass qualifies for Average Utility.

Yield Grades

Yield grades indicate the quantity of meat, that is, the amount of retail, consumer-ready, ready-to-cook, or edible meat, that a carcass contains. Whatever term you choose, the net amount available for eating is extremely important economically. Do not confuse yield grade with dressing percent, often called *yield*. Yield grade refers to the amount of edible product from the carcass, whereas dressing percent (yield) refers to the amount of carcass from a live animal.

The *lower* the number of the yield grade, the *higher* the percent of the carcass in boneless, closely trimmed retail cuts from the leg, loin, rack, and shoulder (see Table 12–8).

See Chapter 16 for a discussion of retail cuts and yields.

Table 12–8. Lamb Yield Grade Retail Yield Equivalents

Yield Grade	Percent of Warm Carcass in Closely Trimmed Primal Cuts
1	47.6–49.4
2	45.8–47.2
3	44.0–45.4
4	42.2–43.6
5	40.4–41.8

The yield grade of an ovine carcass is based on the amount of *external fat* present. The amount of external fat for carcasses with a normal distribution of fat is evaluated in terms of its actual thickness over the center of the rib eye muscle and is measured perpendicular to the outside surface between the twelfth and thirteenth ribs. On intact (unribbed) carcasses, fat thickness is measured by probing, that is, inserting a small calibrated wire or ruler to the heavy connective tissue (*fascia*) that covers the *longissimus* muscle. This measurement may be adjusted, as necessary, to reflect unusual amounts of fat on other parts of the carcass, with particular attention given to the amount of external fat on the rump, outside the shoulders, breast, flank, and cod or udder. Thus, in a carcass that is fatter over other parts than are normally associated with the actual fat thickness over the rib eye, the measurement is adjusted upward. Conversely, in a carcass that has less fat over other parts than are normally associated with the actual fat thickness over the rib eye, the measurement is adjusted downward. In many carcasses, no such adjustment is needed; however, an adjustment in the thickness of fat measurement of 0.05 inch is not uncommon. In some carcasses, a greater adjustment may be necessary.

As a guide in making adjustments, the standards for each yield grade include an additional related measurement — body wall thickness, which is measured 5 inches laterally from the middle of the backbone between the twelfth and thirteenth ribs. Measurement is made at this point because as animals fatten, this "lower rib" area fills with subcutaneous and intermuscular (seam) fat. As the amount of external fat increases, the percent of retail cuts decreases — each 0.05-inch change in adjusted fat thickness over the rib eye changes the yield grade one-half of a grade.

The yield grade of an ovine carcass or side is determined on the basis of the following equation: Yield grade = 0.4 + (10 × adjusted fat thickness, inches). The application of this equation usually results in a fractional grade. However, in normal grading operations any fractional part of a yield grade is dropped. For example, if the computation results in a yield grade of 3.9, the final yield grade is 3 — it is not rounded to 4. A yield grade can easily be determined based on adjusted fat thickness as listed in Table 12–9. The breaks between yield grades come at 0.15, 0.25, 0.35, and 0.45 inches.

The following descriptions of carcass characteristics in each yield grade provide a guide that can help in the subjective determination of yield grades.

Yield Grade 1

A carcass in Yield Grade 1 that is near the borderline with Yield Grade 2 usually has only a thin layer of external fat over the back and loin and slight deposits of fat in the flanks and the cod or the udder. There is usually a very thin layer of fat over the top of the shoulders and the outside of the legs. Muscles are usually plainly visible on most areas of the carcass.

A carcass in Yield Grade 1 with the maximum amount of fat allowed would have an adjusted fat thickness of 0.15 inch. Such a carcass with normal fat distribution and weighing 55 pounds would also have a body wall thickness of about 0.75 inch, and one weighing 75 pounds would have a body wall thickness of about 0.85 inch.

Table 12–9. Lamb Yield Grade by Fat Thickness

Fat Thickness over Rib Eye[1]	Yield Grade
(in.)	
0.00	0.4
0.05	0.9
0.10	1.4
0.15	1.9
0.20	2.4
0.25	2.9
0.30	3.4
0.35	3.9
0.40	4.4
0.45	4.9
0.50	5.4
0.55	5.9
0.60	6.4

[1]This fat thickness measurement over the rib eye muscle should be adjusted, as necessary, to reflect unusual amount of fat on other parts of the carcass.

Yield Grade 2

A carcass in Yield Grade 2 that is near the borderline with Yield Grade 3 usually has a slightly thin layer of fat over the back and loin and the muscles of the back are not visible. The top of the shoulders and the outside of the legs have a thin covering of fat, and the muscles are slightly visible. There are usually small deposits of fat in the flanks and the cod or the udder.

A carcass in Yield Grade 2 with the maximum amount of fat allowed would have an adjusted fat thickness of 0.25 inch. Such a carcass with normal fat distribution and weighing 55 pounds would also have a body wall thickness of about 0.90 inch, and one weighing 75 pounds would have a body wall thickness of about 1 inch.

Yield Grade 3

A carcass in Yield Grade 3 that is near the borderline with Yield Grade 4 usually has a slightly thick covering of fat over the back. The top of the shoulders is completely covered with fat and the legs are nearly completely covered, although the muscles on the outside of the lower legs are visible. There usually are slightly large deposits of fat in the flanks and the cod or the udder.

A carcass in Yield Grade 3 with the maximum amount of fat allowed would have an adjusted fat thickness of 0.35 inch. Such a carcass with normal fat distribution and weighing 55 pounds would also have a body wall thickness of about 1.05 inches, and one weighing 75 pounds would have a body wall thickness of about 1.15 inches.

Yield Grade 4

A carcass in Yield Grade 4 that is near the borderline with Yield Grade 5 is normally completely covered with fat. There is generally a very thick covering of fat over the back and a slightly thick covering over the shoulders and legs. There are usually large deposits of fat in the flanks and the cod or the udder.

A carcass in Yield Grade 4 with the maximum amount of fat allowed would have an adjusted fat thickness of 0.45 inch. Such a carcass with normal fat distribution and weighing 55 pounds would also have a body wall thickness of about 1.20 inches, and one weighing 75 pounds would have a body wall thickness of about 1.30 inches.

Yield Grade 5

A carcass in Yield Grade 5 has an adjusted fat thickness of more than 0.45 inch. The external fat covering on most parts of the carcass is usually greater than that described for Yield Grade 4.

National Summary of Lamb and Mutton Graded

See Table 12–10 for a summary of the total number of lamb and mutton carcasses graded, the numbers in each quality and yield grade for 1998, and the Web site for updating this information.

Table 12–10. National Summary of Lamb and Mutton Graded, 1998 (Number of Head)[1]

Quality Grade	Yield Grade					Quality Only	Total Quality Graded This Pd.	% of All Quality Graded
	#1	#2	#3	#4	#5			
Prime	6,427	86,669	129,575	33,881	6,271	0	262,823	8.2
Choice	374,846	1,281,916	1,079,461	147,791	26,313	64	2,910,391	91.7
Good	0	7	0	0	0	0	7	0.0
Utility	0	0	0	0	0	0	0	0.0
Cull	0	0	0	0	0	0	0	0.0
Yield only	0	0	0	0	0			
Total yield graded this period	381,273	1,368,592	1,209,036	181,672	32,584			
% of all yield graded	12.1	12.9	37.9	5.7	1.0			

90.7% of federally inspected lamb harvest represents total lambs graded.

[1]Information provided by U.S. Department of Agriculture, Agricultural Marketing Service, Livestock and Seed Division, Meat Grading and Certification Branch. http://www.ams.usda.gov/lsg/mgc/Reports/mn-cy98.pdf

Live Animal and Carcass Illustrations of Lamb Yield Grades

Lamb carcass yield grading is based solely on external fat thickness (see preceding discussion and Figures 12–43 to 12–52, provided by the University of Illinois, South Dakota State University, and the National Live Stock and Meat Board), each one-tenth change representing one full yield grade. Thus, the appraisal of fatness must be made as accurately as possible, meaning a "hands-on" approach. A light touch over the back, ribs, and rump with closed fingers can reveal a tremendous amount of yield grading information.

Yields and Value Differences Between Yield Grades

Value differences between yield grades naturally fluctuate from week to week and month to month because of overall meat price fluctuations and variations in cutting and trimming methods. Further discussion on yield of cuts and value difference between carcasses is found in Chapters 13 and 16.

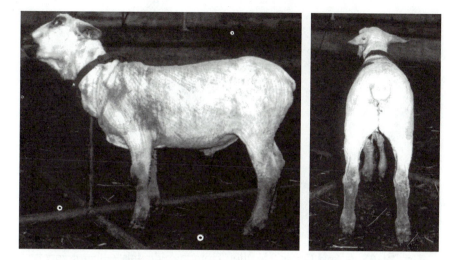

Fig. 12–43. A Yield Grade 1 wether. Only a very small percentage of market lambs qualify for this grade because it requires a high degree of trimness and muscling. This crossbred lamb is restricted just behind the shoulders, indicating that this area has not filled with fat. Also he is especially trim in the breast and flank. From the rear, leanness is apparent by the lack of fullness in the twist. Excellent muscling is indicated by his being wider and thicker through the center of his leg than he is over his back.

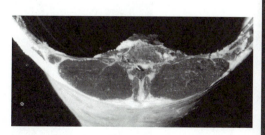

Fig. 12–44. A Yield Grade 1 carcass.

External fat thickness . . . . 0.10 inch
Yield grade 1.4

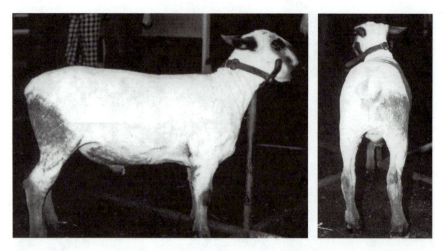

Fig. 12–45. A Yield Grade 2 wether. From the side, this wether is very trim in his breast and flank, but is uniformly smooth behind the shoulders, indicating that this area is somewhat filled with fat. This well-muscled lamb is wider through the center of his legs than up over his top, but he is carrying more finish in his twist than the Yield Grade 1 lamb.

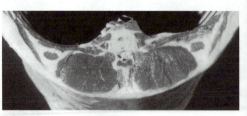

Fig. 12–46. A Yield Grade 2 lamb carcass.

External fat thickness . . . . 0.20 inch
Yield grade 2.4

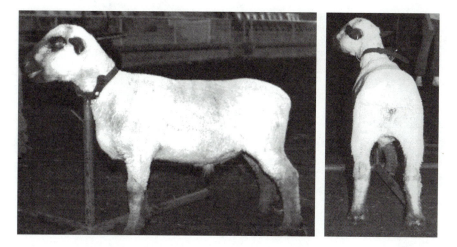

Fig. 12–47. A Yield Grade 3 wether. From the side, as well as from the rear, increased deposits of fat can be observed in the areas previously discussed.

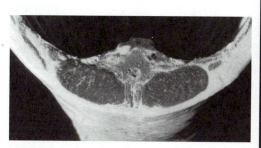

Fig. 12–48. A Yield Grade 3 lamb carcass.

External fat thickness . . . . 0.30 inch
Yield grade 3.4

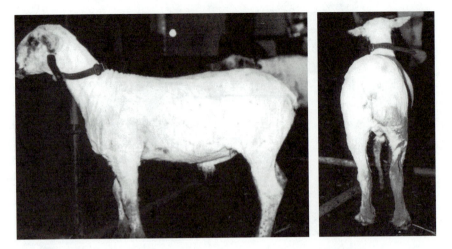

Fig. 12–49. A Yield Grade 4 wether. Although this wether is long and appears trim in the breast region, the general smooth appearance from shoulder to sirloin and rump gives evidence of his overall fatness. From the rear, he is narrow and light muscled in his legs, and his twist is filled with fat.

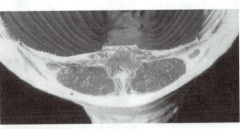

Fig. 12–50. A Yield Grade 4 lamb carcass.

External fat thickness . . . . 0.40 inch
Yield grade 4.4

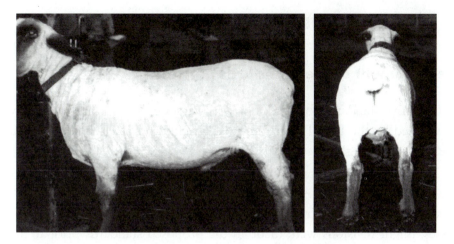

Fig. 12–51. A Yield Grade 5 wether. The heavy middle on this wether and his fullness of flank and breast portray his overall fatness. Although the lamb stands wide behind and has plump, heavy legs, the depth of twist and the squareness of dock indicate fatness.

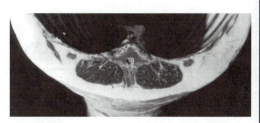

Fig. 12–52. A Yield Grade 5 lamb carcass.

External fat thickness . . . . 0.50 inch
Yield grade 5.4

PORK CARCASS GRADING

Development of the Standards[7]

Tentative standards for grades of pork carcasses and fresh pork cuts were first issued by the U.S. Department of Agriculture in 1931. These tentative standards were slightly revised in 1933.

New standards for grades of barrow and gilt carcasses were proposed by the U.S. Department of Agriculture in 1949. These standards represented the first application of objective measurements as guides to grades for pork carcasses. Slight revisions were made in the proposed standards prior to adoption as the Official United States Standards for Grades of Barrow and Gilt Carcasses, effective September 12, 1952.

The official standards were amended in July 1955. The grade designations Choice No. 1, Choice No. 2, and Choice No. 3 to were changed to *U.S. No. 1, U.S. No. 2,* and *U.S. No. 3,* respectively. In addition, the backfat thickness requirements were reduced for each grade, and the descriptive specifications were reworded slightly to reflect the reduced fat thickness requirements and to allow more uniform interpretation of the standards.

On April 1, 1968, the official standards were again revised to reflect the improvements made since 1955 in pork carcasses. The minimum backfat thickness requirement for the U.S. No. 1 grade was eliminated, and a new U.S. No. 1 grade was established to properly identify the superior pork carcasses then being produced. The former No. 1, No. 2, and No. 3 grades were renamed *No. 2, No. 3,* and *No. 4,* respectively. The former Medium and Cull grades were combined and renamed *U.S. Utility.* Also, the maximum allowable adjustment for variations from normal fat distribution and muscling was changed from one-half to one full grade to reflect the effect of these factors on yields of cuts.

In January 1985, the standards were further revised. The present grades are based on the backfat thickness of the last rib (see Figure 12–53) and muscling. Only three levels of muscling — thick, average, and thin (see Figure 12–54) — are recognized. The width of the grades remains 3 percent of four lean cuts, but the expected yield from each grade is 7.4 percent higher

---

[7]USDA–AMS. 1985. Official United States Standards for Grades of Pork Carcasses. Title 7, Chp. I, Pt. 54, Sec. 54.131–54.137, Code of Federal Regulations. (January 14).

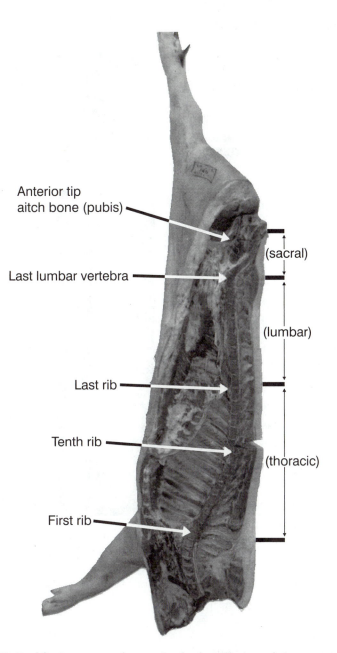

Fig. 12–53. Backfat is measured opposite the last rib. Length is measured from the front edge of the first rib where it joins the backbone to the front tip of the aitch bone on a split, unribbed carcass. Length is no longer required for a USDA grade but nevertheless is of interest to pork producers. This carcass has been ribbed between the tenth and eleventh ribs to expose the loin eye and the fat cover over it. Although not USDA required for carcass grading, exposing the loin eye and fat cover in this manner provides a very useful indicator of carcass quality and cutability that is widely used.

than previously. The lines between Grades 1 and 2, 2 and 3, and 3 and 4 are set at 1, 1.25, and 1.5 inches of last rib backfat thickness, respectively.

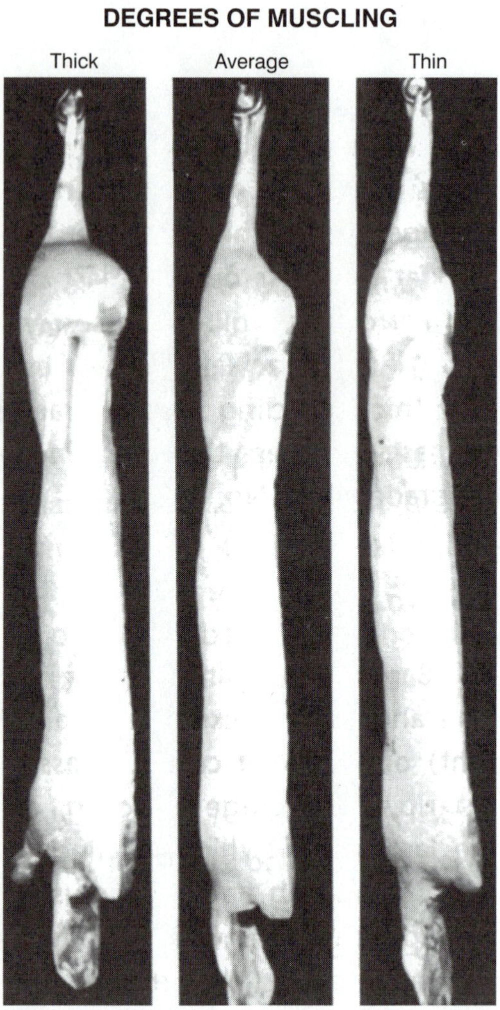

Classes of Pork Carcasses

The five classes of pork carcasses, comparable to the same five classes of hogs for harvest, are barrow, gilt, sow, stag, and boar.

"Because of the relationships between sex and/or sex condition in pork and the acceptability of the prepared meats to the consumer, separate standards have been developed for (1) barrow and gilt carcasses and (2) sow carcasses. There are no official standards for grades of stag and boar carcasses.

"The determination of sex condition is based on the following:

- Barrow carcasses are identified by a small pizzle eye [Figure 12–1 illustrates a beef pizzle eye] and the typical pocket in the split edge of the belly where the preputial sheath was removed.

DEGREES OF MUSCLING

Fig. 12–54. Degrees of muscling. Thick and Average represent the minimum for the degree. (Courtesy, Livestock and Seed Division, Agricultural Marketing Service, USDA)

- Gilt carcasses are recognized by the smooth split edge of the belly, the absence of the pizzle eye, and the lack of development of mammary tissue.

- Sow carcasses exhibit the smooth split edge of the belly characteristic of females. They differ from gilts in that mammary tissue has developed in connection with advanced pregnancy or lactation.

- Stag carcasses have the pocket in the split edge of the belly typical of males, and the pizzle eye is larger and more prominent than in barrows. In addition, other distinguishing characteristics that often may be noted are rather heavy shoulders, thick skin over the shoulders, large bones and joints, and a dark red color of lean.

- Boar carcasses have the same distinguishing characteristics as stag carcasses but to a more pronounced degree."[8]

Grades for Barrow and Gilt Carcasses

Differences in barrow and gilt carcasses due to sex condition are minor, and the grade standards are equally applicable for grading both classes.

Grades for barrow and gilt carcasses are based on two general conditions: the quality-indicating characteristics of the lean and the expected combined yields of the four lean cuts (ham, loin, picnic shoulder, and Boston shoulder).

Quality

With respect to quality, two general levels are considered: one for carcasses with characteristics that indicate the lean in the four lean cuts will have acceptable quality and one for carcasses with characteristics that indicate the lean will have unacceptable quality.

The quality of lean is best evaluated by direct observation of its characteristics in a cut surface — when a cut surface of major muscles is available, this should be used as the basis for the quality determination. The standards describe the characteristics of the loin eye muscle at the tenth rib. However, when this surface is not available, other exposed major muscle surfaces can be used for the quality determination, based on the normal development of the characteristics in relation to those described for the loin eye muscle at the tenth rib.

When a major muscle cut surface is not available, the quality of the lean should be evaluated indirectly based on quality-indicating characteristics that are evident in carcasses. These include firmness of the fat and lean, amount of feathering between the ribs, and color of the lean. The standards describe a

---

[8]National Live Stock and Meat Board. 1988.

development of each of these factors that is normally associated with the lower limit of acceptable lean quality. The degree of external fatness, as such, is not considered in the evaluation of the quality of the lean.

Carcasses that have characteristics indicating the lean in the four lean cuts will not have an acceptable quality or that have bellies too thin to be suitable for bacon production are graded *U.S. Utility*. Also graded U.S. Utility — regardless of their development of other quality-indicating characteristics — are carcasses that are soft and(or) oily. Belly thickness is determined by an overall evaluation of the belly, with primary consideration being given to the thickness along the navel edge and the thickness of the belly pocket at the flank end of the belly.

A numerical system for designating pork quality, as observed in the cut surfaces of major muscles, was published by the University of Wisconsin in 1963 in *Special Bulletin 9, Pork Quality Standards*. Although not used as such in federal descriptions, these standards served as the forerunner of the National Pork Producers Council (NPPC)[9] standards published in *Composition and Quality Assessment Procedures*.

The NPPC standards are based on three levels of color-texture-exudation in hams; six levels of color in loins, each of which is referenced and coordinated with its particular Minolta L* value; and six levels of pork loin marbling, each of which is coordinated numerically with its intramuscular lipid content. (See the color photo section.) The intermediate level in all quality traits is preferred, while the extremes at either end of the scales are cause for rejection. Two NPPC Fact Sheets describe targets[10] and critical points[11] for pork quality. An additional NPPC reference, *Variations in Pork Quality: History, Definition, Extent, Resolution*,[12] provides colored pictures of the various pork quality traits as observed in the ham-loin interface and in the center of the ham. This reference also describes the extent of pork quality variation in the industry and is a good review of past research on pork quality.

---

[9]National Pork Producers Council (NPPC). 1999. Composition and Quality Assessment Procedures. National Pork Producers Council, P.O. Box 10383, Des Moines, IA 50306. pork@nppc.org

[10]NPPC Pork Quality Solutions Team. 1998. Pork Quality Targets. National Pork Producers Council, P.O. Box 10383, Des Moines, IA 50306. pork@nppc.org

[11]Berg, E. P., and S. J. Eilert. 1998. Critical Points Affecting Fresh Pork Quality Within the Packing Plant. National Pork Producers Council, P.O. Box 10383, Des Moines, IA 50306. pork@nppc.org

[12]Kauffman, R. G., R. G. Cassens, A. Scherer, and D. L. Meeker. 1992. Variations in Pork Quality: History, Definition, Extent, Resolution. National Pork Producers Council, P.O. Box 10383, Des Moines, IA 50306. pork@nppc.org

Yield

Carcasses that have indications of acceptable lean quality and acceptable belly thickness are evaluated based on four grades — U.S. No. 1, U.S. No. 2, U.S. No. 3, and U.S. No. 4. These grades are founded on the expected carcass yields of the four lean cuts, and no consideration is given to the development of quality superior to that described as minimum for these grades. The expected yields of the four lean cuts for each of the four grades are shown in Table 12–11.

Table 12–11.   Expected Yields of the Four Lean Cuts Based on Chilled Pork Carcass Weight, by Grade[1]

Grade	Yield
U.S. No. 1	60.4% and over
U.S. No. 2	57.4% to 60.3%
U.S. No. 3	54.4% to 57.3%
U.S. No. 4	Less than 54.4%

[1]These yields will be approximately 1 percent lower if based on hot carcass weight.

These yields are based on the cutting and trimming methods used by the U.S. Department of Agriculture in developing the standards. (These standards, copies of which may be obtained from the Livestock and Seed Division, Agricultural Marketing Service, U.S. Department of Agriculture, Washington, DC 20250, generally specify 0.2-inch fat trim.) Other cutting and trimming methods may result in different yields. For example, if more fat is left on the four lean cuts than prescribed in the USDA methods, the yield for each grade will be higher than indicated. However, every method of trimming, if applied uniformly, should result in similar differences in yields between grades (see Chapter 14).

Carcasses vary in their yields of the four lean cuts because of differences in their degree of fatness and in their degree of muscling (thickness of muscling in relation to skeletal size). In the USDA grade standards, the actual thickness of backfat is a single measurement including the skin made opposite the last rib and perpendicular to the skin surface (see Figure 12–53). For carcasses that have been skinned, 0.1 inch is added to the measurement to compensate for the loss of the skin, provided no or very little fat was removed with the skin (see Chapter 5). The yield of four lean cuts from skinned carcasses will be higher than indicated in Table 12–11.

The degree of muscling specified for each of the four grades decreases progressively from the U.S. No. 1 grade through the U.S. No. 4 grade. Among carcasses of the same weight, fatter carcasses normally have a lesser degree of muscling. For purposes of these standards, three degrees of muscling — thick (superior), average, and thin (inferior) — are recognized (see Figure 12–54). These are intended to cover the entire range of muscling present among pork carcasses currently being produced.

The grade is determined by calculating a preliminary grade according to the schedule shown in Table 12–12 and adjusting up or down one grade for thick or thin muscling respectively, except for carcasses with 1.75 inches or greater backfat thickness over the last rib, which must remain in the U.S. No. 4 grade and carcasses with thin muscling, which cannot grade U.S. No. 1.

Table 12–12. Preliminary Carcass Grade Based on Backfat Thickness over the Last Rib

Preliminary Grade	Backfat Thickness Range
U.S. No. 1	Less than 1.00 inch
U.S. No. 2	1.00 to 1.24 inches
U.S. No. 3	1.25 to 1.49 inches
U.S. No. 4	1.50 inches and over[1]

[1]Carcasses with last rib backfat thickness of 1.75 inches or over cannot be graded U.S. No. 3, even with thick muscling.

Alternatively, this equation may be used: Carcass grade = (4 × backfat, inches) – (1 × muscling score). Muscling should be scored as follows: thin = 1, average = 2, thick = 3. A lower grade is more desirable, so increased muscling is designated by a larger negative number in the equation.

Specifications for Grades of Barrow and Gilt Carcasses

U.S. No. 1

Barrow and gilt carcasses in this grade have an acceptable quality of lean and belly thickness and a high expected yield (60.4 percent and over) of four lean cuts. U.S. No. 1 barrow and gilt carcasses must have less than average backfat thickness over the last rib with average muscling or average backfat thickness over the last rib coupled with thick muscling.

Barrow and gilt carcasses with average muscling may be graded U.S. No. 1 if their backfat thickness over the last rib is less than 1 inch. Carcasses with thick muscling may be graded U.S. No. 1 if their backfat thickness over the last rib is less than 1.25 inches. Carcasses with thin muscling may not be graded U.S. No. 1.

U.S. No. 2

Barrow and gilt carcasses in this grade have an acceptable quality of lean and belly thickness and an average expected yield (57.4 to 60.3 percent) of four lean cuts. Carcasses with average backfat thickness over the last rib and average muscling, or less than average backfat thickness over the last rib and thin muscling, or greater than average backfat thickness over the last rib and thick muscling will qualify for this grade.

Pork carcasses with average muscling may be graded U.S. No. 2 if their backfat thickness over the last rib is 1 to 1.24 inches. Carcasses with thick muscling may be graded U.S. No. 2 if their backfat thickness over the last rib is 1.25 to 1.49 inches. Carcasses with thin muscling must have less than 1 inch of backfat thickness over the last rib in order to be graded U.S. No. 2.

U.S. No. 3

Barrow and gilt carcasses in this grade have an acceptable quality of lean and belly thickness and a slightly low expected yield (54.4 to 57.3 percent) of four lean cuts. Carcasses with average muscling and more than average backfat thickness over the last rib, or thin muscling and average backfat thickness over the last rib, or thick muscling and much greater than average backfat thickness over the last rib will qualify for this grade.

Barrow and gilt carcasses with average muscling will be graded U.S. No. 3 if their backfat thickness over the last rib is 1.25 to 1.49 inches. Carcasses with thick muscling will be graded U.S. No. 3 if their backfat thickness over the last rib is 1.5 to 1.74 inches. Carcasses with 1.75 inches or greater backfat thickness over the last rib cannot be graded U.S. No. 3. Carcasses with thin muscling will be graded U.S. No. 3 if their backfat thickness over the last rib is 1 to 1.24 inches.

U.S. No. 4

Barrow and gilt carcasses in this grade have an acceptable quality of lean and belly thickness and low yield expected (less than 54.4 percent) of four lean cuts. Carcasses in U.S. No. 4 grade always have more than average backfat thickness over the last rib and thick, average, or thin muscling, depending on the degree to which the backfat thickness over the last rib exceeds the average.

Barrow and gilt carcasses with average muscling will be graded U.S. No. 4 if their backfat thickness over the last rib is 1.5 inches or greater. Carcasses with thick muscling will be graded U.S. No. 4 with backfat thickness over the last rib of 1.75 inches or greater, and those with thin muscling will be graded U.S. No. 4 with 1.25 inches or greater backfat thickness over the last rib.

U.S. Utility

All carcasses with unacceptable quality of lean and belly thickness will be graded U.S. Utility, regardless of their degree of muscling or backfat thickness over the last rib. Also, all carcasses that are soft and(or) oily, or are pale, soft, and exudative (PSE), will be graded U.S. Utility.

Live Animal and Carcass Illustrations of Pork Grades

Figures 12–55 to 12–63 depict live animal and carcass representatives of the various pork carcass grades. Very few, if any, pork carcasses are federally graded by the U.S. Department of Agriculture. Each packer has a grading system that is based on some of the same measurements that are described in this chapter for USDA pork carcass grades.

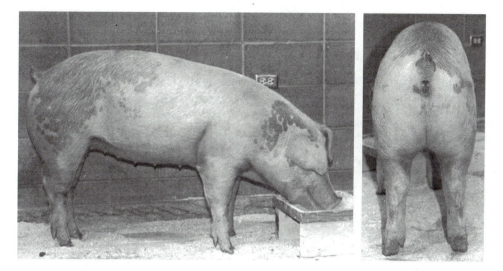

Fig. 12–55. A U.S. No. 1 grade gilt weighing 246 pounds. She is very trim in her jowls, shoulders, and middle. When she moves, her shoulder blades are clearly evident, indicating trimness. From the rear she is wider through the center of her hams than any other place on her body, indicating superior muscling. Over her top, her loins are rounded and smooth, further evidence of muscling and trimness. She is free of excess fat at the root of the tail and in the crotch. (Courtesy, South Dakota State University; photo by Kevin W. Jones)

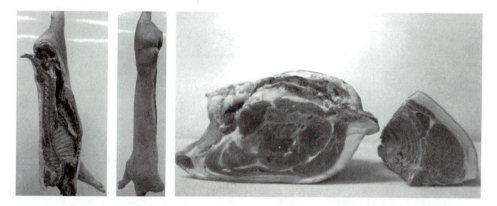

Fig. 12–56. A U.S. No. 1 grade carcass that has 0.95 inch of backfat at the last rib. Average to thick muscling is viewed over the back. The untrimmed ham and loin show a lack of excess fatness. The blade end of the loin exposed at the tenth rib shows 0.65 inch of fat cover and 5.80 square inches of loin eye area. (Courtesy, South Dakota State University; photo by Kevin W. Jones)

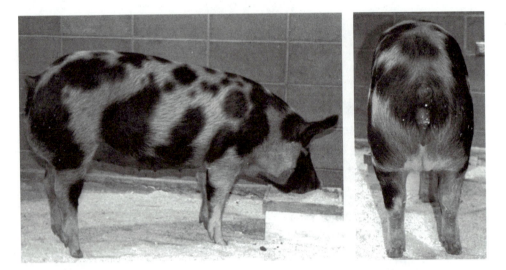

Fig. 12–57. A U.S. No. 2 grade gilt weighing 244 pounds. She shows evidence of having slightly more fat in her jowls, shoulders, and middle than the No. 1 grade gilt. From the rear some evidence of fat at the base of the tail and in the crotch is present. A significant amount of her thickness is due to fat, as indicated by the somewhat flattened turn of her top, as compared to the No. 1 hog. (Courtesy, South Dakota State University; photo by Kevin W. Jones)

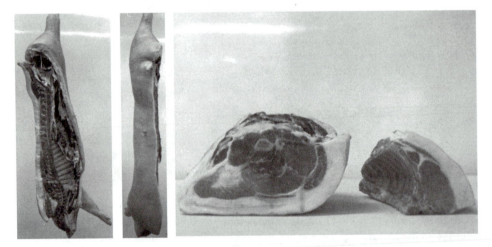

Fig. 12–58. A U.S. No. 2 grade carcass that has 1.22 inches of backfat at the last rib. Average muscling is viewed over the back. The untrimmed ham and loin show evidence of excessive fatness. The blade end of the loin exposed at the tenth rib shows 1.25 inches of fat cover and 4.50 square inches of loin eye area. (Courtesy, South Dakota State University; photo by Kevin W. Jones)

Fig. 12–59. A U.S. No. 3 grade barrow weighing 246 pounds. From the side he shows evidence of excess fat in his jowls and shoulders, over the edge of his loin, and in his belly. From the rear he shows some evidence of adequate muscling and a desirable rounded turn over his loins. His rear leg placement is somewhat narrow, indicating a lack of muscling. (Courtesy, South Dakota State University; photo by Kevin W. Jones)

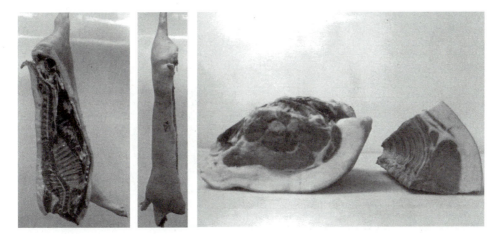

Fig. 12–60. A U.S. No. 3 grade carcass that has 1.35 inches of backfat at the last rib. Average muscling is viewed over the back. The untrimmed ham and loin show excessive fat. The blade end of the loin exposed at the tenth rib shows 1.45 inches of fat cover and 5.35 square inches of loin eye area. (Courtesy, South Dakota State University; photo by Kevin W. Jones)

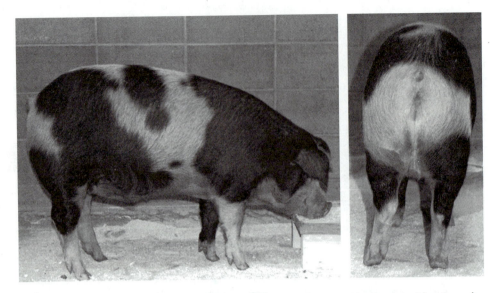

Fig. 12–61. A U.S. No. 4 grade gilt weighing 225 pounds. In the side view she clearly shows evidence of excess fatness in her jowls, shoulder, belly, and flanks. From the rear, excess fat is evident in the crotch. From the rear, her middle is wider than her hams, indicating a serious lack of muscling. (Courtesy, South Dakota State University; photo by Kevin W. Jones)

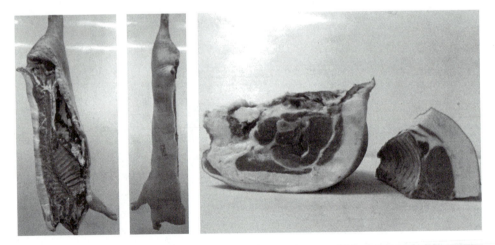

Fig. 12–62. A U.S. No. 4 grade carcass that has 1.30 inches of backfat at the last rib. Because of its thin muscling, one degree less than standard, the fat thickness is adjusted to 1.55, thus causing the carcass to grade No. 4. Noting the fatness all along the back, you can see that this carcass is leanest at the last rib. The untrimmed ham and loin clearly show excessive fat. The blade end of the loin exposed at the tenth rib shows 1.80 inches of fat cover and 4.50 square inches of loin eye area. This is definitely a No. 4 carcass. (Courtesy, South Dakota State University; photo by Kevin W. Jones)

Fig. 12–63. *(Left to right)* The blade end of loins exposed at the tenth rib from U.S. No. 1, No. 2, No. 3, and No. 4 grades shown in Figs. 12–56, 12–58, 12–60, and 12–62. Fat depth (in inches at the three-fourths loin eye distance from the chine bone) and loin eye area (in square inches), respectively: 0.65, 5.80; 1.25, 4.50; 1.45, 5.35; 1.80, 4.50. (Courtesy, South Dakota State University; photo by Kevin W. Jones)

Grade Standards for Sow Carcasses

The establishment of U.S. standards for grades of sows for harvest and of sow carcasses became effective September 1, 1956.

The grades are based on differences in yields of lean cuts and fat cuts and differences in quality of pork. In the development of the standards, sow carcass data were studied to establish relationships between measurements and carcass differences important in grading. As a result, these standards include average backfat thickness measurements as an objective guide in grading. These principles are similar to those that have received widespread acceptance in the standards for barrows and gilts, adopted in 1952.

The five grades for sows and sow carcasses are *U.S. No. 1, U.S. No. 2, U.S. No. 3, Medium,* and *Cull.* The U.S. No. 1 grade includes sows and carcasses with about the minimum finish required to produce pork cuts of acceptable palatability. The U.S. No. 2 and U.S. No. 3 grades represent overfinish, with resulting lower yields of lean and higher yields of fat. Medium and Cull grades represent underfinish, producing pork with low palatability.

The backfat measurements that qualify sow carcasses for the various grades are listed in Table 12–13.

WHAT CONSTITUTES PROGRESS?

The meat animal industry worldwide must continue to improve by openly comparing the established standards of the top-performing animals. The USDA Agricultural Marketing Service, through the Federal Meat Grading

Table 12–13. Backfat Measurement Guide to Sow Carcass Grades

Average Backfat Thickness	Grade
(in.)	
1.5 to 1.9	U.S. No. 1
1.9 to 2.3	U.S. No. 2
2.3 or more	U.S. No. 3
1.1 to 1.5	Medium
Less than 1.1	Cull

and Certification Branch of the Livestock and Seed Division, has established a set of uniform standards on which to make comparisons. These have been instrumental in fostering improvements. For the meat animal industry to continue to flourish, it must encourage the development of animals that produce high-quality, lean edible product most efficiently.

POULTRY GRADING[13]

Although grading generally involves the sorting of products according to quality and size, it also includes the determination of the class and the condition of the products. For poultry, grading may be for determining class, quality, quantity, or condition, or for any combination of these factors. Individuals, firms, or government agencies that desire to utilize the USDA poultry grading services may request them of their own volition and must pay for them. These services are performed on the basis of regulations promulgated by the Secretary of Agriculture. These regulations have been developed in cooperation with the industry, including all affected or related groups, such as health and marketing officials, producers, processors, and consumers. Ready-to-cook poultry must have been officially inspected by the USDA–FSIS (see Chapter 3) for condition and wholesomeness before it can be graded.

---

[13]USDA–AMS. 1995. Regulations Governing the Voluntary Grading of Poultry Products and Rabbit Products and U.S. Classes, Standards, and Grades. Title 7, Chp. I, Pt. 70, Sec. 70.1–70.332, Code of Federal Regulations.

USDA poultry grades fall into two categories: (1) Consumer and (2) Procurement. The U.S. Consumer grades for poultry are by far the most important, since they are the grades used at the retail level. The U.S. Consumer grades are *U.S. Grade A, U.S. Grade B,* and *U.S. Grade C.* These are applicable only to poultry that has been graded on an individual carcass basis by a grader or by a limited licensed grader working under the supervision of a grader.

The U.S. Procurement grades are designed primarily for institutional use. These grades are *U.S. Procurement Grade I* and *U.S. Procurement Grade II.* The procurement grades are based on the relative quality and the tolerances within the lot based on an examination of a prescribed sample of the lot.

Modern breeding, feeding, and management practices have resulted in improved poultry. These factors, together with more efficient marketing and procurement practices, have made it unnecessary to have wholesale poultry grades and live poultry grades. Individual quality standards for live poultry are used on a very limited basis and mostly in an academic setting.

Kinds and Classes of Poultry

The kinds of poultry, meaning the different species of poultry, such as chickens, turkeys, ducks, geese, guineas, and pigeons, and the classes of poultry, meaning the groups into which the kinds are divided, based on the same physical characteristics that are associated with age and sex, for example, fryers or hens, that are listed below are in general use in all segments of the poultry industry. The following provisions of the official U.S. standards apply to individual carcasses of ready-to-cook poultry in determining the kind of poultry and its class.

Chickens

The following are the various classes of chickens:

- *Rock Cornish game or Cornish game hen:* A Rock Cornish game hen or Cornish game hen is a young, immature chicken (usually five to six weeks of age) weighing not more than 2 pounds (ready-to-cook weight). It is usually of the Cornish breed or a Cornish-White Plymouth Rock cross.

- *Rock Cornish fryer, roaster, or hen:* The term *Rock Cornish* is often used for older classes as well. The term *Rock Cornish fryer, roaster, or hen* should apply only if the age and characteristics are appropriate according to the following descriptions.

- *Fryer:* A fryer or broiler, which is often the preferred term, is a young chicken (usually under seven weeks of age) of either sex with tender meat; soft, pliable, smooth-textured skin; and flexible breastbone cartilage. Most broilers are marketed at between six and seven weeks of age for a 3- to 4-pound ready-to-cook weight.

- *Roaster:* A roaster is a young chicken (usually three to five months of age) of either sex, with tender meat; soft, pliable, smooth-textured skin; and breastbone cartilage that may be somewhat less flexible than that of a fryer or broiler. Its dressed weight usually exceeds 4 pounds, with live weight up to 8 pounds.

- *Capon:* A capon is a surgically unsexed male chicken (usually under eight months of age), with tender meat and soft, pliable, smooth-textured skin. Only limited numbers are produced today and most of those are for specialty markets.

- *Hen:* A hen is a mature female chicken (usually more than 10 months of age), with meat less tender than that of a roaster and with a nonflexible breastbone tip. A hen is often called a fowl or a baking chicken.

- *Cock or rooster:* A cock or rooster is a mature male chicken, with coarse skin, toughened and darkened meat, and a hardened breastbone tip.

Turkeys

The following are the various classes of turkeys:

- *Fryer-roaster turkey:* A fryer-roaster turkey, often called a broiler turkey, is a young, immature turkey, usually under 14 weeks of age, of either sex, with tender meat; soft, pliable, smooth-textured skin; and flexible breastbone cartilage.

- *Young turkey:* A young turkey is a turkey, usually under eight months of age, of either sex, with tender meat; soft, pliable, smooth-textured skin; and breastbone cartilage that is somewhat less flexible than a fryer-roaster turkey. Sex designation is optional.[14]

---

[14]For labeling purposes, the designation of sex within the class name is optional, and the two classes of young turkeys may be grouped and designated as "young turkeys."

- *Yearling turkey:* A yearling turkey is a fully matured turkey, usually under 15 months of age, of either sex, with reasonably tender meat and reasonably smooth-textured skin. Sex designation is optional.

- *Mature turkey or old turkey (hen or tom):* A mature turkey or old turkey is a turkey, usually in excess of 15 months of age, of either sex, with coarse skin and toughened flesh.

Ducks

The following are the various classes of ducks:

- *Broiler or fryer duckling:* A broiler or fryer duckling is a young duck, usually under 8 weeks of age, of either sex, with tender meat, a soft bill, and a soft trachea.

- *Roaster duckling:* A roaster duckling is a young duck, usually under 16 weeks of age, of either sex, with tender meat, a bill that is not completely hardened, and a trachea that is easily dented.

- *Mature duck or old duck:* A mature duck or old duck is a duck (usually over 6 months of age) of either sex, with toughened flesh, hardened bill, and hardened trachea.

Geese

The following are the various classes of geese:

- *Young goose:* A young goose may be of either sex, with tender meat and a trachea that is easily dented.

- *Mature goose or old goose:* A mature goose or old goose may be of either sex, with toughened flesh and hardened trachea.

Guineas

The following are the various classes of guineas:

- *Young guinea:* A young guinea may be of either sex, with tender meat and flexible breastbone cartilage.

- *Mature guinea or old guinea:* A mature guinea or old guinea may be of either sex, with toughened flesh and hardened breastbone.

Pigeons

The following are the various classes of pigeons:

- *Squab:* A squab is a young, immature pigeon of either sex, with extra tender meat.

- *Pigeon:* A pigeon is a mature pigeon of either sex, with coarse skin and toughened flesh.

Standards and Grades

The differences between standards of quality and grades are sometimes misunderstood. The standards of quality enumerate the various factors that determine the grade. These quality factors, such as fat covering, fleshing, exposed flesh, and discolorations, when evaluated collectively, determine the grade of the bird.

A Quality

Conformation

The carcass or part is free of deformities that detract from its appearance or that affect the normal distribution of flesh. Slight deformities such as slightly curved or dented breastbones and slightly curved backs may be present.

Fleshing

The carcass or part has a well-developed covering of flesh. The breast is moderately long and deep and has sufficient flesh to give it a rounded appearance, with the flesh carrying well up to the crest of the breastbone along its entire length. The legs are well-fleshed and moderately long, thick, and wide at the knee and hip joint area, showing a plump appearance, with the thighs *(femurs)* moderately fleshed. The wings are also moderately fleshed.

Fat Covering

The carcass or part has a well-developed layer of fat in the skin. The fat is well-distributed so that there is a noticeable amount of fat in the skin in the areas between the heavy feather tracts. Frequently, the skin with the fat layer is removed in the preparation of boneless further processed cuts for consumption, so fat covering is of less importance with these types of products.

Defeathering

The carcass or part has a clean appearance, especially on the breast. The carcass or part is free of pinfeathers, diminutive feathers, and hair that is visible to the grader.

Exposed Flesh

The parts are free of exposed flesh stemming from cuts, tears, and missing skin (other than slight trimming on the edge). The carcass is free from these defects on the breast and legs. Elsewhere the carcass may have exposed flesh as a result of slight cuts, tears, and areas of missing skin if the aggregate of the areas of flesh exposed does not exceed the diameter as specified in the standards (reference 13).

Disjointed and Broken Bones and Missing Parts

The parts are free of broken bones. The carcass is free of broken bones and has no more than one disjointed bone. The wing tips may be removed at the joint. In the case of ducks and geese, the wing tips may be removed if both wings are so treated. The tail may be removed at the base. Cartilage separated from the breastbone is not considered as a disjointed or broken bone.

Discolorations of the Skin and Flesh

The carcass or part may have lightly shaded areas of discoloration. Any discoloration due to bruising should be free of blood clots (discernible clumps of red or dark blood cells). Any evidence of incomplete bleeding, such as more than an occasional slightly reddened feather follicle, is not acceptable. Flesh bruises and discolorations of the skin are not permitted on the breast or legs of the carcass or on those individual parts, and only lightly shaded discolorations are allowed elsewhere. The total areas affected by flesh bruises, skin bruises, and discolorations singly or in any combination cannot exceed one-half of the total aggregate area of permitted discoloration. The aggregate area of all discolorations for a carcass or a part therefrom must not exceed the diameter as specified in the standards (reference 13).

Freezing Defects

With respect to consumer-packaged poultry, the carcass, part, or specified poultry food product is practically free from defects that result from handling or that occur during freezing or storage. The following defects are permitted if they, alone or in combination, detract only very slightly from the appearance of the carcass, part, or specified poultry food product.

- Slight darkening over the back and drumsticks, provided the frozen bird or part has a generally bright appearance.

- Occasional pockmarks due to drying of the inner layer of skin (derma); however, none may exceed the area of a circle ¼ inch in diameter.

- Occasional small areas showing a thin layer of clear or pinkish-colored ice.

B Quality

Conformation

The carcass or part may have moderate deformities, such as a dented, curved, or crooked breast; or a crooked back; or misshapen legs or wings, which do not materially affect the distribution of flesh or the appearance of the carcass or part.

Fleshing

The carcass or part has a moderate covering of flesh considering the kind, class, and part of the bird. The breast has a substantial covering of flesh, with the flesh carrying up to the crest of the breastbone sufficiently to prevent a thin appearance. The legs (thighs and drumsticks) and wings all have a sufficient amount of flesh to prevent a thin appearance.

Fat Covering

The carcass or part has sufficient fat in the skin to prevent a distinct appearance of the flesh through the skin, especially on the breast and legs.

Defeathering

The carcass or part may have a few nonprotruding pinfeathers, diminutive feathers, or hairs that are scattered sufficiently so as not to appear numerous.

Exposed Flesh

A carcass may have exposed flesh provided that no part on the carcass has more than one-third of the flesh exposed. A part may have no more than one-third of the flesh normally covered by skin exposed.

Disjointed and Broken Bones and Missing Parts

Parts may be disjointed but must be free of broken bones. The carcass may have two disjointed bones or one disjointed bone and one nonprotruding broken bone. Parts of the wings beyond the second joint may be removed at a joint. The tail may be removed at the base. The back may be trimmed in an area not wider than the base of the tail to halfway to the hip joints.

Discolorations of the Skin and Flesh

Discoloration due to bruises should be free of blood clots. Any evidence of incomplete bleeding can be no more than very slight. Moderate areas of discoloration due to bruises in the skin or flesh and moderately shaded discoloration of the skin are permitted, but the total areas affected by such discolorations singly or in any combination cannot exceed one-half of the total aggregate area of permitted discoloration. The aggregate area of all discolorations for a carcass or a part therefrom must not exceed the diameter as specified in the standards (reference 13).

Freezing Defects

With respect to consumer-packaged poultry, the carcass, part, or specified poultry food product may have moderate defects that result from handling or that occur during freezing or storage. The skin and flesh should have a sound appearance but may lack brightness. The carcass or part may have a few pockmarks due to drying of the inner layer of skin (derma). However, no single area of overlapping pockmarks may exceed that of a circle ½ inch in diameter. Moderate areas showing layers of clear or pinkish- or reddish-colored ice are permitted.

C Quality

A carcass that does not meet the requirements for A or B Quality may be of C Quality. Both wings may be removed or neatly trimmed. Trimming of the breast and legs is permitted, but not to the extent that the normal meat yield is materially affected. The back may be trimmed in an area not wider than the base of the tail and to the area between the hip joints.

Standards for Quality of Specified Poultry Food Products

Poultry Roast — A Quality

This standard applies to raw poultry products labeled in accordance with the poultry inspection regulations as ready-to-cook "roasts" or with similar descriptive terminology.

- The deboned poultry meat used must be from young poultry.

- Bones, tendons, cartilage, blood clots, and discolorations must be removed.

- All pinfeathers, bruises, hair, discolorations, and blemishes must be removed from the skin, including fat where necessary.

- Skin for covering a roast may include the skin covering the crop area and the neck skin up to the whisker if the fatty blubber, spongy fat, and membranes have been removed from these areas. Fifty percent or more of the outer surface of the product must be covered with skin, whether attached to the meat or used as a wrap. Skin covering may overlap without limit in all areas provided the fatty tissue has been removed from the sternal and pectoral feather tracts. The combined weight of the skin and fat used to cover the outer surface and used as a binder shall not exceed 15 percent of the total net weight of the product.

- The product must be fabricated such that it can be sliced after being cooked so that each slice can be served with minimal separation.

- Seasoning or flavor enhancers, if used, must be uniformly distributed.

- The product must be tied or fabricated so as to retain its shape after being defrosted and cooked.

- Packaging must be neat and attractive.

- The product must be nearly free of seepage after packaging and(or) freezing; when frozen, it must have a bright, desirable color.

- The product packaged in an oven-ready container must meet all the preceding requirements except that the exposed surface of the roast need not be covered with skin. If skin is used, it may be whole or emulsified. Additionally, in such a container, comminuted meat (see Chapter 21) may be substituted for skin, but it cannot exceed 8 percent of the total weight of the product.

Boneless Poultry Breast, Thigh, and Tenderloin — A Quality

These standards apply to the raw product labeled as ready-to-cook boneless poultry breasts, thighs, or tenderloins or as ready-to-cook boneless poultry breast fillets or thigh fillets.

- Trimming is permitted around the outer edges of a whole breast, a half breast, or a thigh provided the trimming results in at least one-fourth of the breast or one-half of the thigh remaining intact and in a portion that approximates the symmetrical appearance of the original part. Trimming must result in a smooth outer surface with no angular cuts, tears, or holes in the meat portion of the product. Trimming of the inner muscle surface is permitted provided it results in a relatively smooth appearance.

- The bones must be removed in a neat manner without undue mutilation of adjacent muscle.

- For parts with skin, the skin must meet all the requirements of ready-to-cook poultry parts with respect to defeathering, exposed flesh, and discolorations and must be free of tendons and cartilage.

- Skinless parts must be free of tendons, cartilage, blood clots, and discolorations, except for minor abrasions due to preparation techniques.

U.S. Procurement Grades for Ready-to-Cook Poultry

The U.S. procurement grades for ready-to-cook poultry are applicable to carcasses of ready-to-cook poultry when they are graded as a lot by a grader who examines each carcass in the lot or each carcass in a representative sample thereof. Except for some military purchases, a very limited amount of poultry is traded in this manner.

U.S. Procurement Grade I

Any lot of ready-to-cook poultry composed of one or more carcasses of the same kind and the same class may be designated and identified as U.S. Procurement Grade I provided that 90 percent or more of the carcasses in such lot meet the requirements of A Quality, with the following exceptions:

- Fat covering and conformation may be as described for B Quality.

- Trimming of skin and flesh to remove defects is permitted to the extent that not more than one-fourth of the flesh is exposed on any part, and the meat yield of any part is not appreciably affected.

- Discoloration of skin and flesh may be as described for B Quality.

- One or both drumsticks may be removed if severed at the joint.

- The back may be trimmed in an area not wider than the base of the tail and extending to the area between the hip joints.

- The wings or parts of wings may be removed if severed at a joint, and the tail may be removed at the base.

The balance of the carcasses may be designated and identified as U.S. Procurement Grade I provided they meet the same requirements, except for having only a moderate covering of flesh.

U.S. Procurement Grade II

Any lot of ready-to-cook poultry of the same kind and the same class that fails to meet the requirements of U.S. Procurement Grade I may be designated and identified as U.S. Procurement Grade II, provided that: (1) trimming of flesh from any part does not exceed 10 percent of the meat and (2) portions of a carcass weighing not less than one-half of the whole carcass may be included if the portion approximates in percentage the meat-to-bone yield of the whole carcass.

RABBIT GRADING[15]

Classes of Ready-to-Cook Domestic Rabbit

Fryer or Young Rabbit

A fryer or young rabbit is a young rabbit carcass weighing not less than 1½ pounds and rarely more than 3½ pounds, processed from a rabbit usually less than 12 weeks of age. The flesh of a fryer or young rabbit is tender and fine-grained and is a bright pearly pink color.

Roaster or Mature Rabbit

A roaster or mature rabbit is a mature, or old, rabbit carcass of any weight, but usually over 4 pounds, processed from a rabbit usually eight months of age or older. The flesh of a roaster or mature rabbit is firmer and coarser-grained, and the muscle fibers are slightly darker in color and less tender than those of a fryer or young rabbit. The fat may be more creamy in color than that of a fryer or young rabbit.

Quality Grades

A carcass found to be unsound, unwholesome, or unfit for food cannot be included in any of the quality designations.

---

[15]USDA–AMS. 1995. Regulations Governing the Voluntary Grading of Poultry Products and Rabbit Products and U.S. Classes, Standards, and Grades. Title 7, Chp. I, Pt. 70, Sec. 70.1–70.322, Code of Federal Regulations.

A Quality

- Is short, thick, well-rounded, and full-fleshed.

- Has a broad back; broad hips; broad, deep-fleshed shoulders; and firm muscle texture.

- Has a fair quantity of interior fat in the crotch and over the inner walls of the carcass and a moderate amount of interior fat around the kidneys.

- Is free from evidence of incomplete bleeding, such as more than an occasional slight coagulation in a vein.

- Is free from any evidence of reddening of the flesh due to fluid in the connective tissues.

- Is free from all foreign material (including, but not being limited to, hair, dirt, and bone particles) and from crushed bones caused by the removal of the head or the feet.

- Is free from broken bones, flesh bruises, defects, and deformities. The ends of the leg bones may be broken due to the removal of the feet.

B Quality

The conformation and interior fat levels for B Quality are slightly less than those for A Quality.

C Quality

A carcass that does not meet the requirements of A or B Quality may be of C Quality and such carcass:

- May be long, rangy, and fairly well-fleshed.

- May have thin, narrow back and hips and soft, flabby muscle texture.

- May show very little evidence of exterior fat.

- May show very slight evidence of reddening of the flesh due to blood in the connective tissues.

- Is free from all foreign material (including but not being limited to, hair, dirt, and bone particles) and from crushed bones caused by the removal of the head or the feet.

- May have moderate bruises of the flesh, moderate defects, and moderate deformities. May have a small portion of the carcass removed because of serious bruises. May have one broken bone in addition to broken ends of leg bones that occur when the feet are removed. Discoloration due to bruising in the flesh should be free of blood clots.

USDA'S CERTIFICATION AND PRODUCT EXAMINATION SERVICE[16]

In addition to grading carcasses and cuts, the meat graders of the Livestock and Seed Division of the USDA Agricultural Marketing Service provide two more services to the meat industry. Both services are paid for by the particular segment of the meat industry using them, and both are becoming more widely used in the United States.

Meat Certification Service for Meat and Meat Products

Since 1923, the federal Meat Grading and Inspection Service has helped organizations such as government agencies, private institutions, and other purveyors of meats in their meat procurement programs. This service involves (1) assisting the purchaser in the development of specifications to assure accurate and uniform interpretation and (2) examining the product to assure its compliance with the specifications.

The Meat Certification Service is based on USDA–approved Institutional Meat Purchase Specifications, commonly called IMPS. These specifications are available for fresh beef, fresh lamb and mutton, fresh veal and calf, fresh pork, cured pork, cured beef, edible by-products, sausage products, and portion-cut meat products (see Chapter 13). Each item is numbered and belongs to a series according to the product — Series 100, for instance, is fresh beef,

---

[16]If in need of additional, detailed information, contact the USDA–AMS office in Washington, D.C., or your closest regional or area Meat Grading and Certification Service office.

and Item 104 is an oven-ready rib. IMPS can be requested from the Standardization and Review Branch of the Livestock and Seed Division.

Purchasers, be they hospitals, schools, restaurants, hotels, air lines, or steamship lines, may ask suppliers to submit bids on products based on IMPS.

When the purchaser requests delivery, the supplier asks the nearest USDA meat grading office to have a grader examine the product. The meat grader is responsible for accepting the product and certifying that it is in compliance with specifications.

The federal grader stamps each acceptable meat item, or the sealed carton in which it is contained, with a shield-shaped stamp bearing the words "USDA accepted as specified." This assures the purchaser that all products delivered met the requirements of the specifications at the time of acceptance.

This method of meat procurement assures the purchaser of a wholesome product (only meat that has passed inspection for wholesomeness will be examined for "acceptance"), of the grade, trim, weight, and other options requested. This system also encourages competitive bidding and usually results in overall lower costs, permits long-range meal planning, and eliminates controversies between the buyer and seller over compliance of product.

Product Examination Service

If a business involves meat shipments, there may be times when it is necessary to have an impartial expert officially examine a shipment's physical condition. This service is available to meat packers, wholesalers, brokers, carriers, and their insurance agents — in short, to anyone with a financial interest in the meat shipment who may want to substantiate a damage claim, protect against one, or save the time and expense of having to examine it personally. Product examinations are made on request, and a fee is charged for the service. An examination may be performed wherever the meat shipment is located — in packing plants, warehouses, trucks, or railroad cars — provided the meat is accessible to the meat grader.

During an examination, the grader impartially documents the facts about the physical condition of the meat or the meat product shipment. The meat involved may be fresh or frozen carcasses; wholesale cuts of beef, calf, lamb, pork, poultry, veal, or sausage; smoked or cured meats; etc. The grader issues an official certificate attesting to the physical condition of the shipment. Such a certificate is accepted as *prima facie* evidence by all federal and most state courts.

In a product examination, the meat grader reports only on the physical conditions that can be accurately determined and described. For example, meat or meat products are examined for extent and kind of damage, freezer burn, thawing and refreezing, cleanliness, freshness, temperature, weight ranges, and fat thickness. Also, the temperature of the conveyance and the extent of container damage are determined. Management should specify the factors to be determined by the examination and whether it is to be conducted by random sample or by 100 percent examination of the product.

NOTE: Meat that is unwholesome is not eligible for product examination. If there is any question about the meat being fit for human consumption, the meat grader will refer the problem to USDA's Food Safety and Inspection Service (FSIS) or, if appropriate, to state and local authorities having jurisdiction over the wholesomeness of meat.

THE MEAT GRADER

Government meat graders are appointed from a list of eligible people submitted by the Civil Service Commission. To qualify, a candidate must have had at least three years of suitable practical experience in wholesale meat marketing and grading or a B.S. degree in an appropriate field. College training (courses in meat and livestock judging) can be substituted for practical experience at the rate of one year of study for nine months of required experience. Before being given permanent appointments as graders, the candidates must serve a one-year probationary period, during which time they are given intensive training in the application of standards, and their work is reviewed very carefully by a grading supervisor to ascertain their ability to do the job in strict conformance with the federal standards. The government takes precautions to obtain competent people of high integrity.

Before assuming a grading or reporting assignment, cooperative training students with a civil service rank of GS–4 participate in an intensive training program for five to six months. That training is accomplished at group meetings and assignments at two different field offices. Trainees are paid travel and subsistence allowances in addition to a salary for approximately three months of the training period — while they are attending meetings and training at the first field office. They are eligible for promotion to GS–7 a year after their original employment. They are eligible for promotion to GS–9, which is the journeyperson level for most grader and reporter positions, after a year of satisfactory performance at the GS–7 grade. Higher-grade positions usually involve supervisory duties; those individuals selected to fill these openings are chosen based on their merit. Every grader's work is identified by a code in his or her grade stamp by means of a combination of letters of the alphabet (see Figure 12–21).

HOW AND WHERE MEAT GRADING IS PERFORMED[17]

Grading is only permitted in carcass form at the plant of harvest as a means of ensuring that grading is consistent. Lighting and the preparation and presentation of the carcasses are also carefully controlled.

Carcasses must be presented by two methods: stationary rails and moving chains. In stationary rail grading, the grader reviews and grades carcasses by walking from one to another as they hang from the rail. The grader is allowed a variable amount of time to study and review the grade factors. In chain grading, the grader remains at a stationary work site, and the carcasses are pushed mechanically past the grading station on a rail. Beef carcasses may move along at a rate as high as 400 per hour in a modern meat plant. The grader can grade faster because walking between carcasses is eliminated. However, most graders do not grade every carcass that comes by — those carcasses that pass by are considered later on the stationary rail. Typically, a grader may actually grade 250 to 275 carcasses per hour in a modern, high-speed plant.

COST OF THE GRADING SERVICE

The fee for grading periodically increases because of inflation. The basic rate is set for companies using graders full-time (40 hours per week), and is higher for part-time service. The rate is still higher for "premium hours" (overtime and work before 6:00 A.M. or after 6:00 P.M.) and higher yet on a federal legal holiday.

---

[17]USDA–AMS. 1988. USDA Meat Grading and Certification Service. (October).

The fees for service are based on the time required to render the service, including the time required for the preparation of certificates and the travel of the official grader in connection with the performance of the service.

THE STAMP INK

The ink used in stamping meat, whether it is the inspection stamp or the grade stamp, is made from a vegetable dye that is absolutely harmless and need not be trimmed off the meat.

The following inks are FSIS–approved for grading and inspection stamps: A.C.M.I. Violet 31, Purple Marking Ink 98P, Violet Marking Ink 2–504, Purple Meat Branding Inks GL–31 and GL–35/75, and Meat Marking Violet Inks 3–90–1 and 3–90–4.

WILL COMPUTERS REPLACE HUMAN GRADERS?

The use of computers capable of receiving video, ultrasound, and infrared images and with the capability to digitize these images into a numerical grade, based on the parameters described in this chapter, is presently being researched. Instrument grading to predict cutability has shown promise with two technologies, VIA and TOBEC. VIA stands for Video Image Analysis and consists of a computer that analyzes images from a video camera in the packing house. TOBEC is an electromagnetic scanning instrument that mea- sures TOtal Body Electro-Conductivity by passing a beef hindquarter through a cylinder surrounded by a coil that generates a magnetic field.[18] However, it will be some time, if indeed ever, that such electronic machines will replace human meat graders.

---

[18]Dr. Duane Wulf. 1999. Personal communication. South Dakota State University, Brookings, SD 57007.

13

Meat Merchandising

A bright future awaits persons choosing professional careers in the meat industry. If you are interested in the merchandising sector, you will need to master a good deal of straightforward mathematics. Also, sharpness, quickness, and accuracy are important attributes of a successful meat merchandiser.

THE SOURCE OF SUPPLY
IN THE UNITED STATES

Marketing practices have changed considerably from the days of John Pynchon, who is reputed to have been the first meat packer in the United States (Springfield, Massachusetts, 1641). The packing industry moved westward with the railroads. Cities such as Cincinnati, Chicago, and Kansas City became large packing centers. Much of the glamorized history of the Old West centers around railroads, Indians, settlers, gold, cowboys, and cattle drives, as herds of livestock were driven on foot for long distances to packing centers or to railroad loading points to be shipped to market.

Large stockyards were constructed to receive the live shipments. Commission houses and banking facilities were necessary adjuncts for the buying and selling operations. Meat packers had their buyers at the large centers, or they operated through brokers. It was to govern illegal practices in the buying and selling of livestock that the Packers and Stockyards Act was passed (see later discussion). Contract buying, by which livestock came directly to the packer holding pens, thus eliminating the intermediate handler, became popular. As the trucking of livestock took over, buyers went directly to the farmer or feeder, or to the auction, which had become popular in livestock raising communities. In recent years, the practice of dealing in futures and options has reduced some of the risk of livestock marketing.

As livestock production moved westward from eastern population centers, distribution became a problem that was solved in 1880 by the invention of the refrigerator car. Unlike the mechanically refrigerated cars of today, these cars were cooled from ice bunkers (ice and salt) built into each end of the cars. The ice bunkers had to be refilled at different points along the route. The cars sometimes derailed because of the swaying of the beef quarters suspended from the tops of them. The meat refrigerator car has largely been displaced by modern, mechanically refrigerated trucks, and the swinging carcasses have been converted to boxed, vacuum-bagged, closely trimmed primal and sub-primal cuts and to case-ready retail cuts.

Packing houses originally stood on the outskirts of towns but were soon crowded in by the growing cities. Many plants became obsolete; traffic was congested; taxes increased with land values; thus, to keep their businesses profitable, packers dismantled the old plants and erected modern ones closer to the areas of livestock production. This had an added advantage in that it cut transportation costs. Today, large harvesting and processing plants are scattered among the livestock feeding operations throughout rural areas, and thousands of refrigerated trucks travel the interstate highways delivering meat to the local retailers.

IMPORTED AND EXPORTED MEAT

The Meat Import Act of 1979 established a maximum tonnage of meat that can be imported into the United States from other countries. The act also gave the President the power to adjust that maximum allowable imported tonnage as domestic meat production rises and falls. When U.S. production is high, imports are held to a lower level; when domestic production levels are lower, imports are still restricted, but at a higher than normal level. This "counter-cyclical" law is designed to minimize economic problems for U.S. meat producers and meat consumers. However, this system is political in nature. The President sometimes needs fast-track authority to arrange trade deals, but frequently Congress does not grant it. During the period 1995–1998, meat imports and meat exports averaged out very nearly equal all years at 6.5 percent of the total red meat handled by the U.S. meat industry, according to USDA data, as reported by the American Meat Institute.[1]

---

[1]Meat and Poultry Facts 1999. American Meat Institute, P.O. Box 3556, Washington, DC 20007. See also Agricultural Statistics. USDA. http://www.meatami.org/

Ring Bologna
Knackwurst
Polish Sausage
Beerwurst
Blood and
  Tongue
Hamburger
  Loaf
Pickle and
  Pimento Loaf
Thuringer
Liverwurst
Headcheese
Fresh Pork
  Sausages

Bratwurst
Prasky
Mettwurst
Minced Ham
Bologna
Kiszka
Frankfurters
Mortadella
Minced Ham
Frankfurters
Pepper Loaf
Smoked
  Sausage Links

Sausages

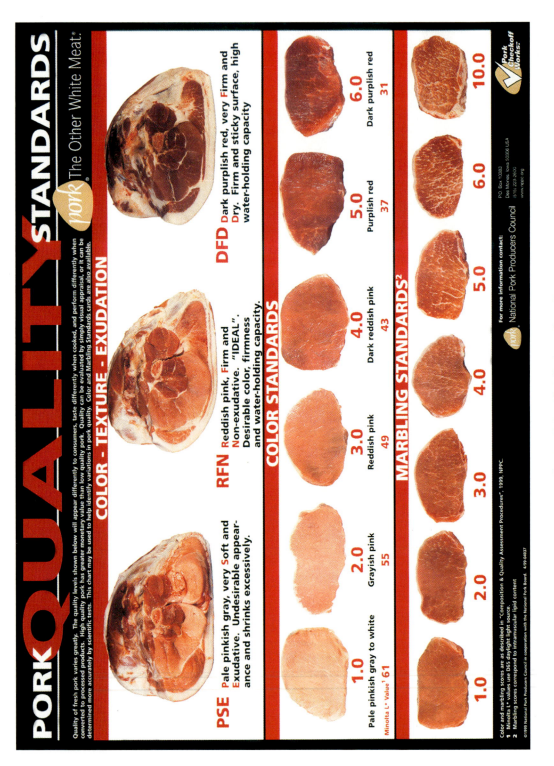

PORK QUALITY STANDARDS

pork The Other White Meat.®

Quality of fresh pork varies greatly. The quality levels shown below will appear differently when cooked, and perform differently when converted to processed products. High quality pork has greater monetary value than low quality pork. Quality can be evaluated by simply visual appraisal, or it can be determined more accurately by scientific tests. This chart may be used to help identify variations in pork quality. Color and Marbling Standards cards are also available.

COLOR - TEXTURE - EXUDATION

PSE Pale pinkish gray, very Soft and Exudative. Undesirable appearance and shrinks excessively.

RFN Reddish pink, Firm and Non-exudative. "IDEAL". Desirable color, firmness and water-holding capacity.

DFD Dark purplish red, very Firm and Dry. Firm and sticky surface, high water-holding capacity

COLOR STANDARDS

1.0	2.0	3.0	4.0	5.0	6.0
Pale pinkish gray to white	Grayish pink	Reddish pink	Dark reddish pink	Purplish red	Dark purplish red
61	55	49	43	37	31

Minolta L* Value [1]

MARBLING STANDARDS [2]

1.0	2.0	3.0	4.0	5.0	6.0	10.0

Color and marbling scores are as described in "Composition & Quality Assessment Procedures", 1999, NPPC.
1 Minolta L* values use D65 daylight light source.
2 Marbling scores correspond to intramuscular lipid content

©1999 National Pork Producers Council in cooperation with the National Pork Board.   4/99-04837

For more information contact:
National Pork Producers Council

P.O. Box 10383
Des Moines, Iowa 50306 USA
(515) 223-2600
www.nppc.org

Pork Checkoff Works.

Illustrations of Beef Marbling

Since marbling is such an important factor in judging beef, the following pictures illustrate the lower limits of six marbling degrees: Moderately Abundant, Slightly Abundant, Moderate, Modest, Small, and Slight.

It should be noted that there are nine degrees of marbling referred to in the Official United States Standards for Grades of Carcass Beef. These color photographs have been developed to assist government, industry, and academia in the proper application of official grade standards.

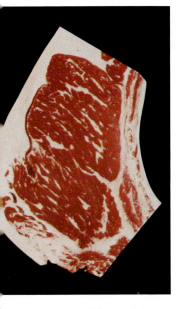

Moderately Abundant

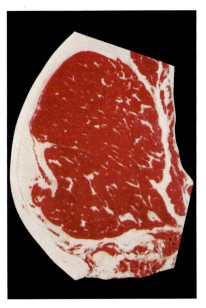

Slightly Abundant

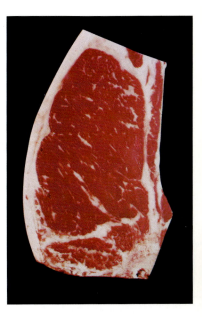

Moderate

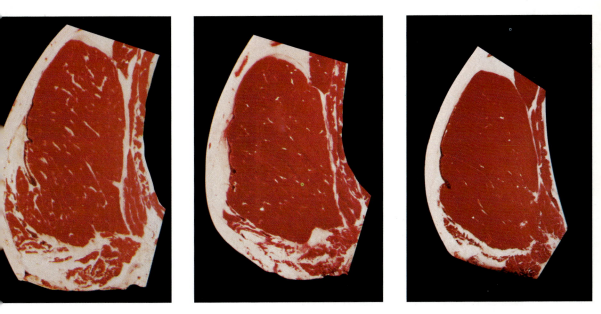

Modest   **Small**   **Slight**

The above illustrations are reduced reproductions of the Official USDA Marbling Photographs prepared by the National Live Stock and Meat Board for the U.S. Department of Agriculture. For accuracy in evaluation, it is recommended that the official set be used.

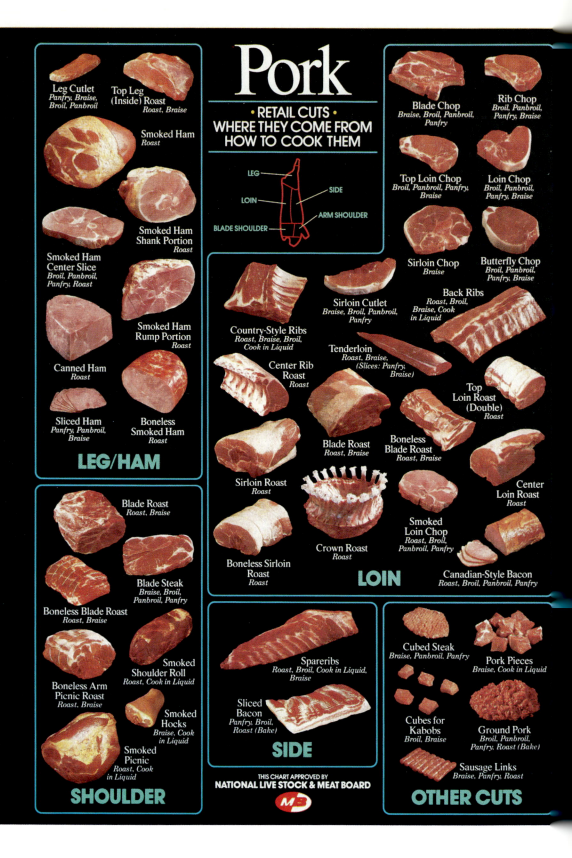

Pork

**· RETAIL CUTS ·
WHERE THEY COME FROM
HOW TO COOK THEM**

LEG
SIDE
LOIN
ARM SHOULDER
BLADE SHOULDER

LEG/HAM

Leg Cutlet
Panfry, Braise, Broil, Panbroil

Top Leg (Inside) Roast
Roast, Braise

Smoked Ham
Roast

Smoked Ham Shank Portion
Roast

Smoked Ham Center Slice
Broil, Panbroil, Panfry, Roast

Smoked Ham Rump Portion
Roast

Canned Ham
Roast

Sliced Ham
Panfry, Panbroil, Braise

Boneless Smoked Ham
Roast

SHOULDER

Blade Roast
Roast, Braise

Blade Steak
Braise, Broil, Panbroil, Panfry

Boneless Blade Roast
Roast, Braise

Boneless Arm Picnic Roast
Roast, Braise

Smoked Shoulder Roll
Roast, Cook in Liquid

Smoked Hocks
Braise, Cook in Liquid

Smoked Picnic
Roast, Cook in Liquid

LOIN

Blade Chop
Braise, Broil, Panbroil, Panfry

Rib Chop
Broil, Panbroil, Panfry, Braise

Top Loin Chop
Broil, Panbroil, Panfry, Braise

Loin Chop
Broil, Panbroil, Panfry, Braise

Sirloin Chop
Braise

Butterfly Chop
Broil, Panbroil, Panfry, Braise

Country-Style Ribs
Roast, Braise, Broil, Cook in Liquid

Sirloin Cutlet
Braise, Broil, Panbroil, Panfry

Back Ribs
Roast, Broil, Braise, Cook in Liquid

Center Rib Roast
Roast

Tenderloin
Roast, Braise, (Slices: Panfry, Braise)

Top Loin Roast (Double)
Roast

Blade Roast
Roast, Braise

Boneless Blade Roast
Roast, Braise

Sirloin Roast
Roast

Center Loin Roast
Roast

Boneless Sirloin Roast
Roast

Crown Roast
Roast

Smoked Loin Chop
Roast, Broil, Panbroil, Panfry

Canadian-Style Bacon
Roast, Broil, Panbroil, Panfry

SIDE

Spareribs
Roast, Broil, Cook in Liquid, Braise

Sliced Bacon
Panfry, Broil, Roast (Bake)

OTHER CUTS

Cubed Steak
Braise, Panbroil, Panfry

Pork Pieces
Braise, Cook in Liquid

Cubes for Kabobs
Broil, Braise

Ground Pork
Broil, Panbroil, Panfry, Roast (Bake)

Sausage Links
Braise, Panfry, Roast

THIS CHART APPROVED BY
NATIONAL LIVE STOCK & MEAT BOARD

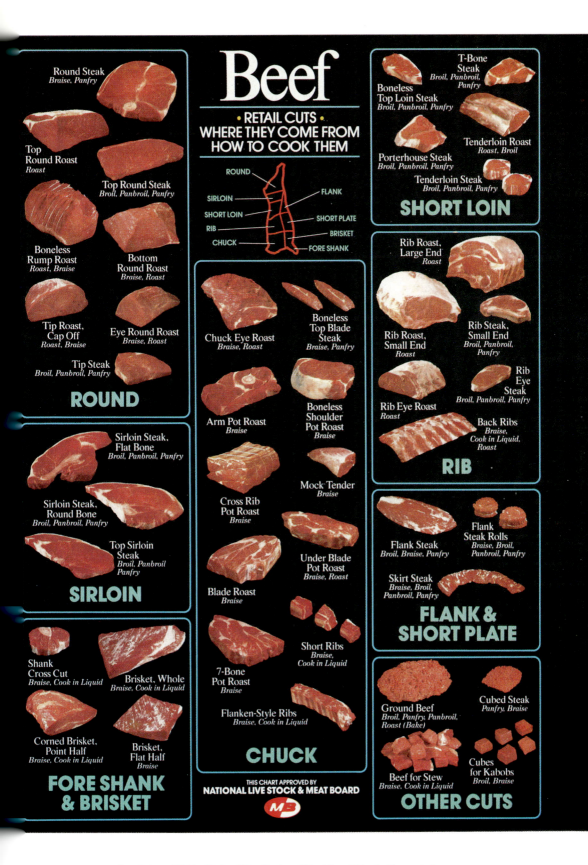

Beef

• RETAIL CUTS •
WHERE THEY COME FROM
HOW TO COOK THEM

ROUND
SIRLOIN
SHORT LOIN
RIB
CHUCK

FLANK
SHORT PLATE
BRISKET
FORE SHANK

ROUND

Round Steak
Braise, Panfry

Top Round Roast
Roast

Top Round Steak
Broil, Panbroil, Panfry

Boneless Rump Roast
Roast, Braise

Bottom Round Roast
Braise, Roast

Tip Roast, Cap Off
Roast, Braise

Eye Round Roast
Braise, Roast

Tip Steak
Broil, Panbroil, Panfry

SIRLOIN

Sirloin Steak, Flat Bone
Broil, Panbroil, Panfry

Sirloin Steak, Round Bone
Broil, Panbroil, Panfry

Top Sirloin Steak
Broil, Panbroil Panfry

FORE SHANK & BRISKET

Shank Cross Cut
Braise, Cook in Liquid

Brisket, Whole
Braise, Cook in Liquid

Corned Brisket, Point Half
Braise, Cook in Liquid

Brisket, Flat Half
Braise

CHUCK

Chuck Eye Roast
Braise, Roast

Boneless Top Blade Steak
Braise, Panfry

Arm Pot Roast
Braise

Boneless Shoulder Pot Roast
Braise

Cross Rib Pot Roast
Braise

Mock Tender
Braise

Blade Roast
Braise

Under Blade Pot Roast
Braise, Roast

7-Bone Pot Roast
Braise

Short Ribs
Braise, Cook in Liquid

Flanken-Style Ribs
Braise, Cook in Liquid

THIS CHART APPROVED BY
NATIONAL LIVE STOCK & MEAT BOARD

SHORT LOIN

T-Bone Steak
Broil, Panbroil, Panfry

Boneless Top Loin Steak
Broil, Panbroil, Panfry

Porterhouse Steak
Broil, Panbroil, Panfry

Tenderloin Roast
Roast, Broil

Tenderloin Steak
Broil, Panbroil, Panfry

RIB

Rib Roast, Large End
Roast

Rib Roast, Small End
Roast

Rib Steak, Small End
Broil, Panbroil, Panfry

Rib Eye Steak
Broil, Panbroil, Panfry

Rib Eye Roast
Roast

Back Ribs
Braise, Cook in Liquid, Roast

FLANK & SHORT PLATE

Flank Steak
Broil, Braise, Panfry

Flank Steak Rolls
Braise, Broil, Panbroil, Panfry

Skirt Steak
Braise, Broil, Panbroil, Panfry

OTHER CUTS

Ground Beef
Broil, Panfry, Panbroil, Roast (Bake)

Cubed Steak
Panfry, Braise

Beef for Stew
Braise, Cook in Liquid

Cubes for Kabobs
Broil, Braise

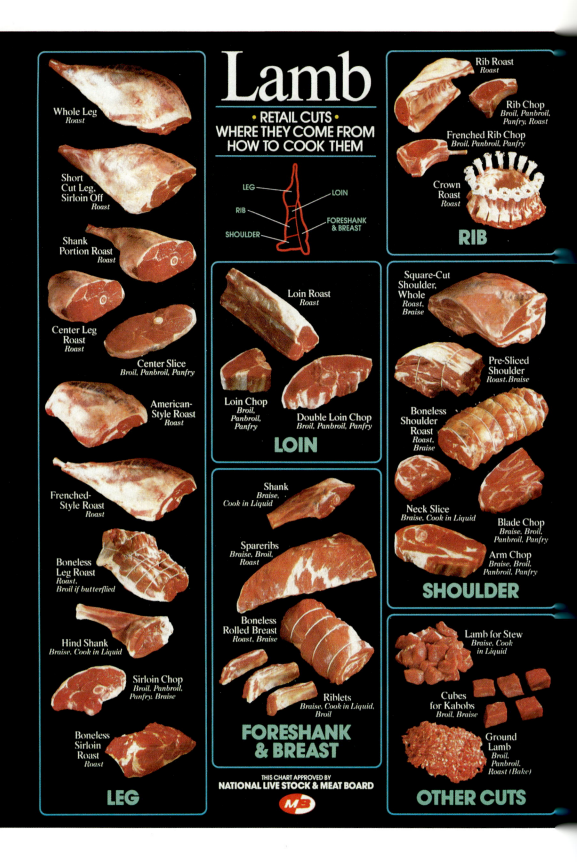

Lamb

· RETAIL CUTS ·
WHERE THEY COME FROM
HOW TO COOK THEM

LEG · LOIN · RIB · FORESHANK & BREAST · SHOULDER

LEG

Whole Leg
Roast

Short Cut Leg, Sirloin Off
Roast

Shank Portion Roast
Roast

Center Leg Roast
Roast

Center Slice
Broil, Panbroil, Panfry

American-Style Roast
Roast

Frenched-Style Roast
Roast

Boneless Leg Roast
Roast, Broil if butterflied

Hind Shank
Braise, Cook in Liquid

Sirloin Chop
Broil, Panbroil, Panfry, Braise

Boneless Sirloin Roast
Roast

LOIN

Loin Roast
Roast

Loin Chop
Broil, Panbroil, Panfry

Double Loin Chop
Broil, Panbroil, Panfry

FORESHANK & BREAST

Shank
Braise, Cook in Liquid

Spareribs
Braise, Broil, Roast

Boneless Rolled Breast
Roast, Braise

Riblets
Braise, Cook in Liquid, Broil

THIS CHART APPROVED BY
NATIONAL LIVE STOCK & MEAT BOARD

RIB

Rib Roast
Roast

Rib Chop
Broil, Panbroil, Panfry, Roast

Frenched Rib Chop
Broil, Panbroil, Panfry

Crown Roast
Roast

SHOULDER

Square-Cut Shoulder, Whole
Roast, Braise

Pre-Sliced Shoulder
Roast, Braise

Boneless Shoulder Roast
Roast, Braise

Neck Slice
Braise, Cook in Liquid

Blade Chop
Braise, Broil, Panbroil, Panfry

Arm Chop
Braise, Broil, Panbroil, Panfry

OTHER CUTS

Lamb for Stew
Braise, Cook in Liquid

Cubes for Kabobs
Broil, Braise

Ground Lamb
Broil, Panbroil, Roast (Bake)

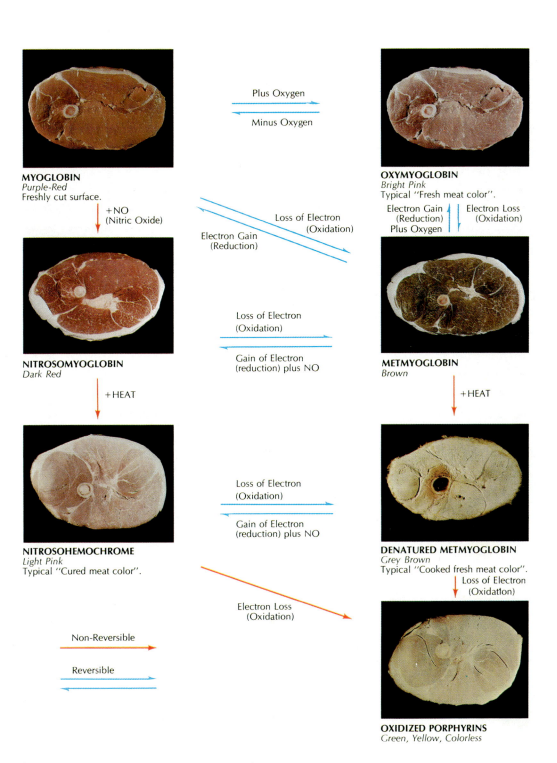

MYOGLOBIN
Purple-Red
Freshly cut surface.

Plus Oxygen

Minus Oxygen

OXYMYOGLOBIN
Bright Pink
Typical "Fresh meat color".

+NO
(Nitric Oxide)

Loss of Electron
(Oxidation)

Electron Gain
(Reduction)

Electron Gain
(Reduction)
Plus Oxygen

Electron Loss
(Oxidation)

NITROSOMYOGLOBIN
Dark Red

Loss of Electron
(Oxidation)

Gain of Electron
(reduction) plus NO

METMYOGLOBIN
Brown

+HEAT

+HEAT

NITROSOHEMOCHROME
Light Pink
Typical "Cured meat color".

Loss of Electron
(Oxidation)

Gain of Electron
(reduction) plus NO

DENATURED METMYOGLOBIN
Grey Brown
Typical "Cooked fresh meat color".

Loss of Electron
(Oxidation)

Non-Reversible

Reversible

Electron Loss
(Oxidation)

OXIDIZED PORPHYRINS
Green, Yellow, Colorless

Diagrammatic Table of Meat Color
Using Actual Photos of the Color Steps

BEEF STEAK COLOR GUIDE

Degrees of Doneness

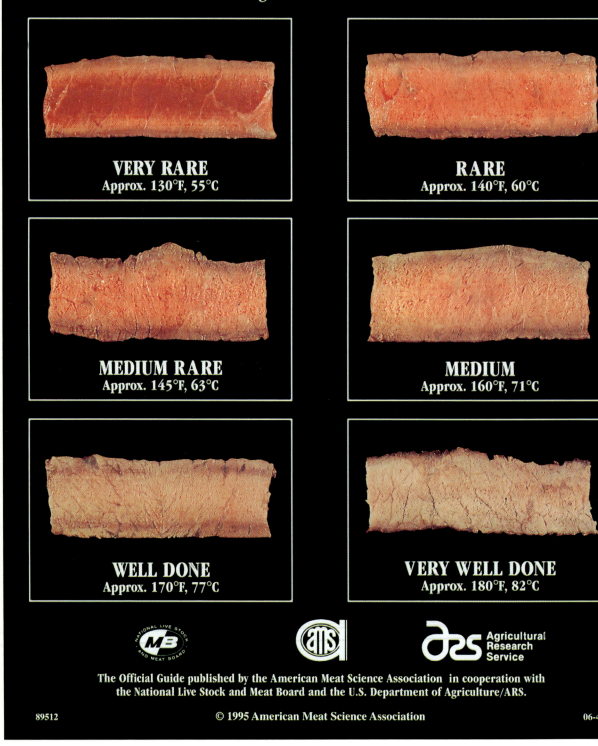

VERY RARE
Approx. 130°F, 55°C

RARE
Approx. 140°F, 60°C

MEDIUM RARE
Approx. 145°F, 63°C

MEDIUM
Approx. 160°F, 71°C

WELL DONE
Approx. 170°F, 77°C

VERY WELL DONE
Approx. 180°F, 82°C

The Official Guide published by the American Meat Science Association in cooperation with the National Live Stock and Meat Board and the U.S. Department of Agriculture/ARS.

89512
© 1995 American Meat Science Association
06-4

Most of the tonnage of U.S. meat imports is composed of fresh or frozen beef and veal from Canada, Australia, and New Zealand and processed or canned beef from Brazil, Argentina, and Canada. We in the United States export approximately twice as much pork to Japan, Mexico, Canada, and Russia as we import from Canada and the European Union. Although the tonnage is low, we import nearly 90 percent of U.S. lamb and mutton from Australia and New Zealand. We also export a great deal of chicken and turkey to Russia, Mexico, and Japan.

If you are interested in developing an export business, it would pay to visit the country where you plan to market meat, since country customs are of utmost importance to developing a market. A good source of current foreign meat marketing data and information about opportunities is a monthly publication by the Meat Export Research Center.[2] See also Chapter 11 for international trade of by-products.

All imported meats must be harvested, processed, and shipped under conditions approved by and monitored by the USDA–FSIS. Ever since the inception of meat inspection, all imported meats have had to be passed at the port of entry by FSIS inspectors before the importers can permit the products to be sold or used in the United States (see Chapter 3). For short periods, during recent years, the Canadian/U.S. border has been an "open border" for federally inspected meats passing either way. Both political events and consumer concerns determine whether meat must be inspected by the receiving country at that border at any given time.

MEAT INDUSTRY BUSINESSES

Packers

Several types of business operations make up the industry that transforms livestock to steaks, chops, roasts, and processed meat products. Conversion of the inedible portion of the livestock to useful products is also accomplished to a large extent by members of the meat industry (see Chapter 11). Delivery of both the edible and the inedible products to the ultimate consumer and user completes the function of the industry.

---

[2]Iowa State University in cooperation with the USDA, Agricultural Marketing Service, Livestock and Grain Market News, Des Moines, Iowa, and the U.S. Meat Export Federation, Denver, Colorado. http://www.ag.iastate.edu/centers/merc/news

The era of John Pynchon established the designation of *meat packer* or *packing house* for the commercial enterprise that starts with live animals and performs the necessary steps to make edible and convenient products for the consumer. Those early wholesale provisioners packed the heavily salted chunks of meat into strong, tightly sealed barrels. Modern packers buy livestock and may perform the harvesting, chilling, cutting, and processing procedures necessary to sell carcasses; primal, sub-primal, and retail cuts; processed meats (cured meats, sausages, etc.); as well as all forms of by-products.

Large, multi-plant packers may have distribution centers, branch houses, or complete packing houses throughout the United States. The larger packers sell in volume to wholesalers, large retail operators, processors, boners, breakers, and hotel and restaurant operators and suppliers. The large plants ship the products long distances from the plants in livestock production areas to the urban population centers.

Still found in some areas of the United States are those few remaining plants described as *kill-and-chill* operations, which sell only carcasses and fresh or processed by-products, and those smaller, independent packers, who buy livestock and sell products within their own regions, often specializing in products that are local favorites. The smallest packers are frequently family operations that are localized in a single community and provide custom harvesting and meat processing for livestock owners in the area. Frequently these operations sell fresh and frozen retail cuts and produce a number of processed meat items for retail sale. Because of the advent of private refrigeration, many of them no longer provide the frozen food locker rental service, but they are still often referred to as *locker* plants.

Packers and Stockyards Act[3]

The purpose of the Packers and Stockyards Act, as passed in 1921 and as continually updated, is to provide for uniform rates and practices and to prevent unfair, unjustly discriminatory, or deceptive practices. This is particularly helpful to farmers who are unable to personally supervise the sale of their livestock. Federal supervision under the act extends to trade practices, commissions, feed and yardage rate charges, weighing and scale testing, and to other services rendered at stockyards and packing plants.

---

[3]USDA, Grain Inspection, Packers and Stockyards Administration, Washington, DC. http://www.usda.gov/gipsa

Carcass weight and grade sales as well as live sales are covered under the act. When payment is made on the basis of carcass weight and grade after the animals have been harvested, factors such as place and time of harvest, carcass trim, grading and prices, condemnation terms, identification, accounting, and promptness of payment must be specified clearly prior to the sale. A final written report that accurately explains all these factors must accompany the payment to the seller. Carcass weights for payment are the hot weights prior to cooling or shrouding of the carcasses. Carcasses must be final graded before the close of the second business day following harvest. With *all* methods of sale, payment must be made for the full purchase price by the close of the next business day following the transfer of possession, or in the case of grade and yield purchases, following the determination of the purchase price.

This act also prohibits certain dual ownership relationships between custom feedlots and packing plants to prevent livestock producers from taking heavy losses if a meat packer should declare bankruptcy after the livestock have been delivered but before the payment is made. The act undergoes constant updating as the industry practices change.

Meat Packing Economics

Each year, U.S. meat packers produce approximately 50 billion pounds of carcasses from the nation's livestock. The production volume of a single pork plant may be 2.5 million pounds of carcass pork harvested and cut per day. Turnover of the product through the plant must be rapid because meat is perishable and because refrigerated storage space is limited and must be used for incoming product. The product must be sold and livestock must be purchased every day. Livestock purchase prices and wholesale meat prices change constantly.

"The four leading packers' share of steer and heifer slaughter increased from 36 percent in 1980 to 80 percent in 1997. Concentration in hog slaughter is not as high but also is on the rise, increasing from 34 percent in 1980 to 54 percent in 1997. In addition, both the slaughter and production of livestock have become more concentrated into relatively narrow geographic regions." (See Table 13–1.)

"Accordingly, GIPSA's Packers and Stockyards program is being restructured to focus on its core responsibilities under the Act — competition, trade practices, and payment protection. Eleven field offices have been consolidated into three regional offices in Denver, CO; Des Moines, IA; and Atlanta, GA,

**Table 13–1. Four-Firm Concentration in Meat Packing,
Reporting Years 1980–97[1]**

Year	Cattle[2]	Steers & Heifers	Cows & Bulls	Hogs	Sheep & Lambs
		- - - - - - - - - - - - - - (% of total commercial harvest) - - - - - - - - - - - - - - -				
1980	28	36	10	34	56
1981	31	40	10	33	53
1982	32	41	9	36	44
1983	36	47	10	29	44
1984	37	50	11	35	49
1985	39	50	17	32	51
1986	42	55	18	33	54
1987	54	67	20	37	75
1988	57	70	18	34	77
1989	57	70	18	34	74
1990	59	72	20	40	70
1991	61	75	20	44	72
1992	64	78	24	44	71
1993	67	81	25	43	73
1994	69	82	25	45	73
1995	69	81	28	46	72
1996	66	79	29	55	73
1997	68	80	28	54	70

[1]Grain Inspection, Packers and Stockyards Administration. FY 1998 Highlights. http://www.usda.gov/gipsa

[2]Includes steers, heifers, cows, and bulls.

Note: Figures for 1980–90 are based on firms' fiscal years as reported to GIPSA. Figures for 1991–97 are based on calendar year federally inspected harvest.

to allow significantly larger staffs to be located near the concentrations of beef, pork, and poultry production and slaughter. Each regional office is responsible for all trade practice and financial protection issues within its assigned region. The Denver office has access to the concentrated fed-cattle production and slaughter areas of Colorado, Kansas, Oklahoma, Nebraska, and Texas, and has nationwide responsibility for major industry and competitive issues relating to cattle and sheep. The Des Moines, IA, office, which is located in the center of the largest pork production and slaughter area, has nationwide responsibility for major industry and competitive issues relating to hogs. Atlanta's regional office is responsible for major industry and competitive issues nationwide relating to poultry.

"GIPSA established a toll-free number (1-800-998-3447) to allow producers and the public to file complaints and report market abuse. Callers can record their complaints confidentially or leave their names, addresses, and other information to discuss issues with a GIPSA representative."[4]

For years, the meat packing industry has operated on one of the lowest sales profit margins among U.S industries. Tables 13–2 and 13–3 illustrate the breakdown of the meat industry's sales income.

Table 13–2 indicates that the packing industry during 1997 returned between 65 and 75 percent of sales receipts to the suppliers of livestock. Contrary to what one might think, the top four packers paid more of their sales dollar to the livestock producer than firms in the other three categories of packer size. Also, the total cost of sales was higher for the top four packers. Table 13–2 shows that there is apparently efficiency in size, as the top four packers had lower operating expenses. Nevertheless, because of the higher cost of sales, the bottom-line operating income for the top four packers was only about one-half that of firms in the other size categories.

Table 13–3 shows some financial ratios for the various size packers. *Assets* are all properties and claims against others that may be applied directly or indirectly to cover liabilities. Equity is the residual value of a business or property beyond any mortgage or liability. The top four packers came out better than firms in the other size categories on sales per dollar of assets and equity because they also have an advantage in equity-to-asset ratio. However, the smaller packer categories excelled in most other ratios.

Some large companies have discontinued harvest operations to direct their resources toward the more profitable meat processing aspect of the business.

---

[4]Grain Inspection, Packers and Stockyards Administration. FY 1998 Highlights. http://www.usda .gov/gipsa

Table 13–2. Sales, Expenses, and Operating Income of 4, 8, 20, and 40 Largest Meat Packing Firms, 1997 Reporting Year[1]

Item	Top 4	Top 8	Top 20	Top 40
	------- (% of sales) -------			
Net sales	100.00	100.00	100.00	100.00
Cost of sales				
Livestock purchases	74.98	69.87	65.55	65.89
Total cost of sales	85.16	82.76	81.81	81.99
Gross income	14.84	17.24	18.19	18.01
Operating expenses				
Manufacturing	7.93	8.73	8.34	8.30
Advertising & selling	0.92	1.76	2.60	2.49
General & administrative	1.25	1.26	1.56	1.63
Depreciation & amortization	0.50	0.56	0.59	0.60
Interest	0.66	0.63	0.59	0.58
Other	2.49	2.52	2.43	2.36
Total operating expenses	13.76	15.46	16.11	15.95
Operating income	1.09	1.78	2.07	2.05

[1]Ranking determined by total amount spent for all livestock harvested.

Note: Reported financial figures may include information on operations other than meat packing.

Source: USDA-GIPSA. 1999. Packers and Stockyards Statistical Report, 1997 Reporting Year. GIPSA SR-99-1. (June)

Table 13–3. Selected Financial Ratios for 4, 8, 20, and 40 Largest Meat Packing Firms, 1997 Reporting Year[1]

Item	Top 4	Top 8	Top 20	Top 40
Net sales per $ of assets	4.487	3.779	3.881	3.936
Net sales per $ of equity	8.020	7.772	7.270	7.407
Gross income per $ of sales	0.148	0.172	0.182	0.180
Gross income per $ of assets	0.666	0.651	0.706	0.709
Gross income per $ of equity	1.190	1.340	1.322	1.334
Total operating expenses per $ of sales	0.138	0.155	0.161	0.160
Total operating expenses per $ of assets	0.617	0.584	0.625	0.628
Total operating expenses per $ of equity	1.103	1.202	1.172	1.182
Operating income per $ of sales	0.011	0.018	0.021	0.021
Operating income per $ of assets	0.049	0.067	0.081	0.081
Operating income per $ of equity	0.087	0.138	0.151	0.152
Equity-to-asset ratio	0.559	0.486	0.534	0.531

[1]Ranking determined by total amount spent for all livestock harvested.

Note: Reported financial figures may include information on operations other than meat packing.

Source: USDA-GIPSA. 1999. Packers and Stockyards Statistical Report, 1997 Reporting Year. GIPSA SR-99-1. (June)

Processors

Boners reduce carcasses and cuts to boneless product for further processing. Some boners purchase and harvest mature cows or other lower-grade animals, whereas other boners purchase youthful, high-quality carcasses and wholesale cuts to provide raw material for their boning lines. Boners may sell some primal and sub-primal cuts, but they principally supply boneless trim for further processing.

Although many packers are also meat processors, the term *processors* refers to those operators who purchase carcasses, primal or sub-primal cuts, trimmings, variety meat items, and nonmeat ingredients. By applying meat processing technology to the ingredients, processors are able to sell the resulting products at a profit. Some processors buy all of these raw materials and sell a wide variety of processed products. More often, processors specialize in one operation, such as sausage making, curing, canning, or frozen-food processing. There are also single-product processors, such as the Kosher corned beef processor who purchases only beef briskets. Processors may sell at all wholesale levels or through their own specialty retail stores.

The meat industry also includes *wholesalers* and *distributors* who purchase meat products in large wholesale volume. The distributors' warehouses or delivery vehicles subsequently supply the smaller retail operations with a wide variety of products from which to order on a relatively small wholesale-volume basis. Wholesalers and distributors provide access to segments of the retail market that are more difficult for packers and processors to serve.

Those operations that serve the special needs of the hotel–restaurant–institutional (HRI) meat buyers are an important and growing part of the meat industry. The HRI meat *purveyors* buy carcasses, primals, sub-primals, boneless processing meats, and variety meats in the wholesale trade. Steaks, chops, roasts, patties, cubes for stew or kabobs, and many processed items are produced in compliance with rigid specifications for grade, weight, size, quality, and uniformity. The strict weight/size specifications are achieved by skilled "portion control" personnel in the HRI purveyors' cutting rooms. Meat patties formulated within tight compositional specifications are produced in various weights and shapes by the HRI purveyors. The growth of the fast-food industry has initiated meat purveying operations that specialize in serving that segment of food retailing.

Precooking, special sauce or flavoring formulation, and other services may be provided by purveyors as well. Most purveyors recognize that the "services" they offer are very important to maintaining their customers. By purchasing from skilled and reputable HRI purveyors, the management of the food service unit of a hotel or plant cafeteria can be assured that each customer will receive a meat portion that will be appetizing to the eye and satisfying to the palate if it has been properly prepared and served.

According to the National Restaurant Association,[5] the restaurant industry, with $354 billion in annual sales, is the nation's largest private-sector employer. This sales figure includes fast-food sales. Technomic Information Services, *Chicago Tribune,*[6] reported that fast-food sales, categorized by leading menu items, were growing at the following annual percentage rates in 1998: Mexican, 6.5; sandwich other than hamburger, 6.0; hamburger, 4.2; chicken, 4.0; donut, 3.8; pizza, 3.7; and ice cream/yogurt, 2.9. Catering is popular and in demand. Airlines offer appetizing meat items pictured in their in-flight publications, with the option for free on-board order calls. All these various meat outlets are constantly monitoring food safety, and many have instituted voluntary HACCP plans to assure that no food poisoning outbreaks result from their products.

MEAT RETAILING

Meat retailers are the final link in the chain between livestock producers and meat consumers. Consumers buy meat to prepare at home from a variety of retail operations, or they buy meat cooked and ready-to-eat from an equally wide variety of food service units.

Of the total dollars consumers spend for food and beverages in grocery stores, meat, poultry, and seafood accounted for 21.1 percent in 1997. Sixty-six percent of the total 1997 consumer dollars spent in grocery stores for meat, poultry, and fish purchased fresh product. Beef made up approximately 50 percent; poultry, 28.5 percent; fresh pork, 8.7 percent; fish, 10.1 percent; lamb, 1.4 percent; and veal, 1 percent of the fresh meat sales in food markets. Of the remaining 34 percent of the total meat–poultry–fish sales dollars, nearly 9 percent was spent on sausages of various kinds, 6 percent on cured meats (bacon and ham), 7 percent on cold cuts, 4 percent on canned meats, and 8.5 percent on frozen meats.[7]

Some meat that is sold as steaks, chops, roasts, and stir-fry or kabob pieces is received at the store as "boxed" primal or sub-primal cuts. Retail meat cut-

---

[5] 1200 17th Street, N.W., Washington, DC 20036. <isal@restaurant.org>; http://www.restaurant.org
[6] http://chicagotribune.com
[7] Meat and Poultry Facts 1999.

ters trim away the fat and bone, cut smaller retail units, and then package them for sale in the retail case. Fresh (unfrozen) meat cuts are displayed in a clear film packaging material that is sealed to a rigid back board. Each package displays a label identifying the product (generally including species, wholesale cut, and retail cut); net weight; price per pound; and total value of the package. Convenience in food preparation has become more important as family and individual schedules have become extremely hectic. The number of double-income households continues to increase, as does the number of families involved in activities outside the home. The demand for meat items that are precooked, and(or) can be microwaved in the store packaging will continue to climb.

Meat cuts are coming case-ready, pre-cut and portioned out in sizes to fit the needs of today's families, and they are coming in premarinated form with added stir-fry vegetables and sauces. Case-ready meats are weighed and prepriced, depending on retailer specifications and may be large enough to meet a retailer's entire case needs. Case-ready meats have the potential to lower the costs of running a meat department while giving customers what they want when they want it. Case-ready meats can be the freshest, highest quality products available to consumers because the meat leaves the officially inspected plant in a sealed package that is not opened until the customer takes it home. This is especially true with the fresh meat counters, because the product is handled less, resulting in food safety and inventory control.

Poultry, most sausages, cured meats, and other processed meats routinely arrive at the store prepackaged, to be placed in the display area by meat department workers. Growing numbers of supermarkets are installing equipment, including smokehouses or heat processing units, needed to produce processed meat products in the store. See Chapter 3 for HACCP requirements and exceptions for such store operations.

Meat department workers may also be available to answer questions and prepare special orders upon request in self-service meat departments. Some supermarkets, and many retail meat markets, have "service" meat counters run by sales personnel skilled in meat merchandising. The focal point of a service meat department is a glass-fronted refrigerated case displaying a wide assortment of fresh and processed meats. The meat buyer may select items from the display cases or request special orders, such as extra-thick steaks. The meat cutter/sales person then removes the selected cuts from the case or prepares the special order, weighs, wraps, and labels the packaged product with the item identification and the value while the customer waits. The person working behind a service counter can assist meat shoppers with suggestions concerning type of cut, amount required, and preparation techniques for specific situations.

The food freezers in a large number of households are resupplied periodically with volume meat purchases. The freezer owner may choose to buy a whole carcass, a side, a quarter, a bundle, or several of each of a number of particular cuts. The retailer cuts, trims, wraps, labels, and freezes the purchase. The buyer may specify the steak or chop thickness, the roast size and weight, or the number of items per package. Meat retailers often provide freezer meat service as well as the self-service counter and(or) service department. Some retail meat outlets serve the freezer meat trade exclusively. Packaged, frozen meat cuts are available in retail outlets ranging from supermarkets to convenience stores, but the majority of U.S. meat shoppers continue to prefer fresh meat cuts. The marketing of fresh and frozen meat items through vending machines is a new merchandising technique in Japan and may be a major meat retailing technique in the future.

One of the most rapidly growing categories of food retailing is take-out food from supermarket delis and other food retail outlets. Deli departments report large increases in the sales of all food items. Ready-to-eat processed meats, such as cured meats, sausages, loaf items, and barbecued meats, have always been popular with deli customers. The convenience offered by these processed meats and the newer precooked meat products will continue to expand this market. Many modern meat consumers have limited time for meal preparation; they "grab a bite" whenever their schedules permit.

Online shopping for groceries is becoming commonplace because many consumers are too busy to go to a supermarket. Many markets provide Internet grocery shopping with free delivery. Estimates have been made that online grocery sales will reach $1.3 billion by 2001, and $85 billion by 2007, serving 15 to 20 million households.

Table 13–4 lists the top 10 retail supermarket chains in late 1998.

Specialty meat shops may not handle fresh meats, rather they usually specialize in various sausages (dry and semi-dry); cured, smoked meats; domestic and foreign varieties of cheese; meat supplements; meat sauces; tenderizers; condiments; and other relevant items. Most specialty shops bear names indicative of the products handled, for example, "Smoke House," "Sausage Shop," "Gourmet Cave," "Missing Link." These shops are selective in their buying and are always searching for the unusual. Because they deal in small quantities, they may experience some difficulty in getting some items. Some specialty shops furnish their own recipes to processors who make the products for them.

Table 13–4. Top 10 Supermarket Chains[1]

Rank	Supermarket Chain	No. of Stores	Annual Sales
			($ billion)
1.	The Kroger Company	2,151	40.0
2.	Albertson's	2,470	35.0
3.	Safeway Stores	2,251	26.5
4.	Wal-Mart Supercenters	436	12.0[2]
5.	Ahold USA	1,000	18.5
6.	Winn-Dixie Stores	1,217	13.2
7.	Publix Super Markets	563	11.2
8.	Great A & P Tea Co.	1,014	10.3
9.	Food Lion	1,158	10.2
10.	H. E. Burr Grocery Co.	243	6.4

[1]Food Institute, Trenton, New Jersey; reported in National Meat Association, Lean Trimmings, November 1998. http://www.nmaonline.org

[2]Grocery sales only.

Convenience stores are an increasingly important source of food for the U.S. consumer. Self-service sandwich[8] and sausage counters as well as individual packages of snack meats such as sticks and jerky serve the consumer on the go. Prepackaged deli items share the convenience store cooler with dairy products and other staples. A gas stop on the way home can net a complete take-out meal for the family.

According to the National Restaurant Association,[9] Americans are now spending more than half of their food budget on takeout and restaurant food. Within 10 years, the industry estimates that consumers will spend more than 53 percent of every food dollar on meals, snacks, and beverages away from home, compared with 25 percent in 1955. Cafeterias, fast-food counters, and all other eating establishments, therefore, retail a significant amount of meat and meat products.

The U.S. meat industry offers consumers a wide variety of purchase options. The range of options is being rapidly increased by the development of precooked, premarinated meat items; entrées combining meats and other foods, such as stir fry vegetables and sauces; processor and packer-branded, vacuum-packaged, fresh meat retail cuts; and meat products directed to consumers who have specific dietary concerns. Consumer demands have

---

[8]John Montagu, the Earl of Sandwich, is famous as the man who "invented" the sandwich when he stuck meat between two slices of bread.

[9]1200 7th St., N.W., Washington, DC 20036. <isal@rfestaurant.org>; http://www.restaurant.org.

impacted modern meat merchandising, as indicated by the development of the Meat Nutri-Facts point of purchase nutritional information program. (Nutri-Facts is based upon the sources and data discussed in Chapter 23.) Nutritional labeling of meats and other foods has been mandated by the U.S. Department of Agriculture and the Food and Drug Administration since 1993. Consumers may select meat items that provide the combination of product variety, palatability traits, nutrition, convenience, service, and economy that best serve their individual situation. The competition between the segments of the industry assures the availability of quality products and good service at a reasonable cost.

Meat Consumers

Meat merchandising is significantly influenced by changes in consumer attitudes, habits, and life styles. To be successful, the meat merchandiser must identify the consumers he or she is serving and find out what products are in demand. If other segments of the industry are involved in serving these consumers, they need to be informed and engaged as cooperators. For many years, the meat department manager's knowledge of the ethnic, economic, and age range characteristics of the customers in the trade area was said to be essential to the success of the individual meat retail operation. Correspondingly, the industry needed to react to local, regional, national, and international changes in consumer demand.

Current concerns about the consumption of meat in relation to health and physical fitness have brought about changes in the way meat is merchandised. Meat in the retail case is more closely trimmed (25 to 30 percent leaner, by many estimates) than it was in the mid–1980s. Although most consumers recognize that marbling affects the flavor and the juiciness of meat, many of them shun highly marbled cuts in their attempt to limit fat intake. They are willing to make palatability trade-offs to reduce the perceived fat hazards.

Ground product has a wide variety of uses, provides some options in composition, and can be prepared quickly and conveniently. Table 13–5 shows that, although overall beef consumption decreased from 1980 to 1999, the percentage of total beef consumed as ground product remained relatively constant during that period.

Fresh (unprocessed) meats, identified as "organic,"[10] "natural," "boutique," or "range fed" or "range produced," have increased in popularity. Some producers are raising livestock, and marketers are labeling and promot-

---

[10]Approved by USDA-FSIS on February 14, 1999.

Tabel 13–5.   Average Annual Per Capita Consumption of Beef Cuts and Ground Beef

Year	Retail Weight (Pounds)			Percent Ground
	Cuts	Ground Beef[1]	Total	
1980	42.8	33.7	76.6	44.1
1981	43.8	33.5	77.3	43.4
1982	42.1	34.9	77.0	45.4
1983	43.5	35.2	78.7	45.0
1984	42.5	35.9	78.4	45.9
1985	44.7	34.5	79.2	43.8
1986	44.0	34.8	78.8	44.4
1987	41.9	32.0	73.9	43.5
1988	41.5	31.2	72.7	43.1
1989	39.6	29.7	69.3	43.1
1990	38.1	29.7	67.8	43.3
1991	37.1	29.7	66.8	42.5
1992	36.8	29.7	66.5	42.7
1993	35.4	29.7	65.1	45.3
1994	38.0	29.0	67.0	43.3
1995	40.6	26.9	67.5	43.2
1996	37.8	29.9	67.7	44.2
1997	37.6	29.3	66.9	43.8
1998[2]	38.5	29.3	67.8	43.2
1999[3]	36.5	27.7	64.2	43.2

[1]Ground beef includes hamburger and beef used in processed sausage-type products. It is beef other than retail-type cuts, such as steaks and roasts.

[2]Consumption of both ground beef and beef cuts increased slightly during 1998. Consumption of both categories is expected to decline over the next several years as supplies decrease.

[3]Cattle-Fax estimate.

Source: Cattle and Beef Industry Statistics. March 1999. http://www.beef.org/librecon/beefstat/cbis99_03.html

ing the meat in response to consumer concerns about the safety of their food sources. Other "branded" products may be promoted as being "lean" or "low-fat" (see Chapter 3) or as having quality (palatability) traits different from non-branded product in the fresh meat retail case. The purpose of the branding or identification of these products is to promote their attributes so that these unique products can be sold at a premium.

"The principle of value-based marketing is to pay for livestock based on its performance (eating quality, leanness, optimum carcass size, etc.). Livestock, while produced in large numbers, are not manufactured like shoes; the performance traits go from selected seedstock producers through the system to the end user, and more and more the end user is selecting beef based on assurances that the product will meet center-of-the-plate satisfaction. Certification and brand identity are providing the assurances the consumer is demanding."[11] The apparent success of a limited number of "branded" fresh meat programs indicates that some consumers are willing to pay for the advertised advantages. Processors that include certain terms on the labels of their meat products must verify that their products do meet USDA requirements specified for those terms.

Meat Business Management

The management of meat operations at all levels of the industry is complex. Skillful buying, excellent and consistent quality of work, temperature control, sanitation, product features, variety, competitive pricing, friendly service, money-back guarantees, and innovative merchandising are essentials of a good meat operation. Whereas most major industries are engaged in assembling sub-units into finished products such as appliances, furniture pieces, stereos, computers, or vehicles, the meat industry starts with intact animals and disassembles them into a wide variety of consumer products with an even wider range of individual values.

Most meat businesses have multiple purchase options. Harvest operations may be selective of livestock type, weight, maturity, and grade. Processors and retailers may choose to buy carcasses, primal (wholesale), and sub-primal cuts (see meat identification section of this chapter) of various grades (see Chapter 12). By varying the degree of processing, the cutting procedure, or the packaging technique, the processor or the retailer may influence the value or the price of the unit sold. Close trimming increases the convenience and edible yield to the ultimate user, but it involves higher labor and(or) greater processing investment in a lesser weight of salable merchandise. The willingness of the buyer to pay a higher price for the added value must be adequate to compensate for the extra investment.

---

[11]National Meat Association. Lean Trimmings. July 12, 1999.

Inventory control is a constant problem in the meat industry, particularly with fresh meat. In a cutting operation, parts of the purchased unit are divided and sold to a variety of customers, some utilizing only one or two of the portions removed. Each different cut may have a different value. The demand for some cuts may be much greater than for other cuts; however, when an animal is processed by the industry, all of the parts (meat and by-products) are generated. Before another animal is processed, all of the parts of the preceding animal should be sold. How are all of the least desirable cuts sold in the same time interval as the most desirable portions? Pricing is the tool used to "control the flow" of the different-valued items. By pricing the items with a high-demand level relatively high and the low-demand items at a low level to encourage their purchase, the sellers hope to match sales with supply availability. The accumulation of any sub-unit because of the lack of a market or low demand can be economically disastrous in a meat operation.

Of course, the personnel and the meat retail displays and processing areas must be neat, clean, and sanitary to instill consumer confidence. Keeping the meat retail displays and processing areas neat, clean, and sanitary will also maximize the product's shelf life and appearance. Fresh meat in a self-service case must be sold within three to five days. Even in that short interval, the color and overall appearance of the product may deteriorate, making it unsalable.

Both the manager and the employees of a retail meat department should be aware of the income level, the ethnic background, the number of persons per living unit, and the regional preferences of their customers. Meat-portion sizes should vary to correspond to the levels of physical activity and the ages of the meat consumers who purchase from the self-service case. Each different retail cut has unique palatability traits and edible portion yields that influence its relative value.

The retail meat manager must price the product in the case to operate at a profit, to correspond to differences in product value, to account for demand preferences that are unique to the customers, and to attain a sales volume that matches the rate of production in the operation.

Meat Pricing

Accurate, dependable price reporting is essential in the wholesale meat and livestock marketing structure. There are a number of public and private sources of meat price information. The USDA Crop and Livestock Reporting

Service furnishes local market price information in the meat industry for individuals and market reporters throughout the day. Weekly reports of livestock and wholesale meat market volume, harvest weights, prices, and related data are published in *Livestock, Meat, Wool Market News,* available from the USDA Agricultural Marketing Service, *Livestock and Grain Market News,* Room 2623–S, P.O. Box 96456, Washington, DC 20090–6456 (http://www.ams.usda.gov/lsg/mncs/index.htm). Perhaps the best known of the private meat market reporting services are the *Yellow Sheet,* published by Urner Barry Publications, Inc., P.O. Box 389, Toms River, New Jersey 08754–0389, and *The Meat Sheet* (sometimes referred to by its readers as "The Pink Sheet") published by The Meat Sheet, The Total Price Report, P.O. Box 124, Westmont, Illinois 60559. Several commodity market reporting services provide current livestock, meat, and futures price information to subscribers via immediate, direct wire, monitor screen and(or) printer systems.

Retail meat prices are the culmination of a series of marketing functions. Ultimately, meat pricing is determined by supply, the availability of meat animals, and demand, the relative availability of consumer disposable income for the purchase of meat. Variables that may influence one or both of the major price determinants include livestock production costs such as feed availability and animal efficiencies, interest rates, energy costs, and labor costs. Some of the same variables influence processing charges. The state of the economy has a major influence on demand, but the price of food options, which compete for the consumers' protein dollars, also influences the demand situation. Competition between different meat sources — beef, pork, lamb, poultry, and fish — may also have a significant influence on meat pricing.

Thus, retail meat pricing is affected by a wide number of variables. Many of the costs associated with marketing individual products are identified and accounted for in a Direct Product Profit (DPP) spreadsheet model developed for retailers and other meat merchandisers. The American Meat Institute, the Food Marketing Institute, and the National Live Stock and Meat Board cooperated to produce the DPP as a tool for more informed pricing and for improving cost efficiency in meat merchandising. More recently, Texas A&M University and the National Live Stock and Meat Board have developed a software tool for retail pricing and decision making called *Computer Assisted Retail Decision Support (CARDS)*. CARDS is available for beef and pork. The simulation programs allow the manager to enter cutting and labor data that have originated in the store and to localize the results. For more information, contact: CARDS Project, Department of Animal Science, 432 Kleberg Center, Texas A&M University, College Station, Texas 77843–2471.

Meat-pricing Terms

Meat-pricing techniques utilize a number of yield factors and value adjustments. In order to follow the development of meat prices, a person must understand the terminology associated with those meat industry variables. The following terms and definitions should be helpful to anyone unfamiliar with meat-pricing principles.

- *Live weight:* Live weight, sometimes called *harvest weight* or *slaughter weight,* refers to the weight of the live animal at the time of purchase by the harvester or at the time of harvest.

- *Carcass weight:* Carcass weight is the weight of the carcass after all harvest procedures have been completed (see Chapters 5 to 8).

- *Hot carcass weight:* Hot carcass weight is the weight obtained immediately after harvest and just before the carcass enters the chilling cooler. Most beef and lamb carcasses in commercial meat packer coolers are tagged with the hot carcass weight.

- *Chilled carcass weight:* Chilled carcass weight is the weight that can be obtained at any time after the carcass has completed the postmortem chill period and before subdivision of the carcass is initiated. The chilled carcass weight is often less than the hot carcass weight due to the moisture loss during cooling.

- *Dressing percent:* Dressing percent is that proportion of the live weight that remains in the carcass or carcasses of a given animal or group of animals. Mathematically, dressing percent = (carcass weight ÷ live weight) 100. Many persons in the meat and livestock industry refer to dressing percent as *yield.*

- *Grades:* USDA and company (house) grades communicate expected value in the meat trade. Quality grades, yield grades, and carcass grades were discussed in detail in Chapter 12.

- *Cutting yield:* Cutting yield is the proportion of the weight of an original unit (carcass, side, primal cut, etc.) that is a salable product after the original unit has been trimmed and subdivided into cuts for sale.

- *By-products:* By-products are those portions of a meat animal that are not composed mainly of skeletal muscle. Fat trimmings and bone are by-products of the retail cutting operation. *Offal* items are by-products of meat animal harvest. (See Chapter 11.)

- *Offal:* Offal consists of all parts of the animal that are not a part of the carcass as it leaves the harvest area. Edible offal parts include heart, tongue, liver, head meat, and tripe (see Chapter 11). Inedible offal items include hides, hair, digestive tract contents, and other items.

- *Drop credit (hide and offal value):* Drop credit is the value of the offal portion of the harvest animal. The USDA Crop and Livestock Reporting Service reports a value based on the prices of selected beef offal items. The weekly value is quoted as a marketing guideline in dollars per 100 pounds of live weight. The actual value of the offal to a specific packer will vary widely, depending on the degree of offal processing accomplished, the volume, and the access to available markets.

- *Harvest cost (slaughter cost or kill cost):* This is the total cost of the buying process, which includes transporting the animals to the harvest facility, harvesting, chilling, transferring the harvested animals from the cooler, and handling the by-products. Harvest cost does *not* include the *purchase value* of the animal. Harvest cost is often quoted on a per head basis, but it may be based on weight in some cases.

- *Margin:* Margin is the difference in product value between buying and selling. Margin designates the portion of sales income that is utilized to operate the selling unit. Margin is used to compensate for operating costs such as rent, labor, utilities, equipment purchases and maintenance, profit, and miscellaneous expenses. It may be quoted in total dollars, in dollars per some base unit (dollars or weight) of sales (dollars per $1,000 sales), or as a percentage of sales income.

- *Markup:* Markup is the portion of the wholesale cost that is added to the wholesale cost to cover operating expenses. Markup differs from margin in that margin is based on sales value, and markup is based on purchase cost. This is an important difference when either is quoted as a percentage or as a rate per $1,000.

- *Shrinkage:* Shrinkage is the weight loss that may occur throughout the processing sequence. Cooler shrinkage = hot carcass weight – cold carcass weight. Cooler shrinkage, transportation shrinkage, and storage shrinkage are, in large part, due to moisture loss from both the fresh and the processed product. Cutting shrinkage, trimming shrinkage, and cooking shrinkage are reductions in tissue mass as well as water losses. The losses may be accounted for or identified in each case by the complementary term *yield*. Shrinkage may be identified as weight, percent, or value, depending on the situation.

- *Yield:* Yield is the portion of the original weight that remains following any processing or handling procedure in the meat merchandising sequence. *Dressing percent equals harvest yield.* Yield is usually quoted in percent and may be cited for each of the steps that were discussed as shrinkage.

- *Price and value:* The term *price* in this text refers to the cost on a per unit basis (cents per pound), whereas the term *value* refers to the total cost (worth) of a cut, of carcasses, of a load, etc. (price × weight = value).

Tests (Cutting, Yield, Price, Shrink, Labor)

The purpose of tests in meat operations is to provide actual information under working conditions that will aid management in making decisions. Because testing is an information-gathering process, a report form is necessary to record data. The data can then be evaluated and used to compare sources of supply, raw materials options, effects of purchase cost variations, processing procedures, product mix, worker efficiency, pricing procedures, and volume effects. The biological variation of animals, the wide range of cut types possible, and the constant fluctuation of prices at every level of the meat industry make testing on a regular basis essential to all meat operations. Continual updating of test data is required to monitor the effectiveness of computer-assisted meat industry management.

Examples of the forms for recording test data for three meat animal species are shown in Figures 13–1, 13–2, and 13–3. The forms contain blanks for information that relates to live animals, carcasses, wholesale cuts, and retail cuts. Margin and markup calculations are also given. A packer, processor, purveyor, or retailer may use only those portions of the forms that pertain to the data that are gathered and recorded at certain points in the marketing sequence.

BEEF CUTTING TEST

Cutting date _____ Live animal ID _____ . Carcass ID _____

Breed _____ Sex _____

Live animal weight _____ Price _____ Sales value _____

By-products: List _____ Estimated value _____

Harvest date _____ Hot carcass weight R _____ L _____ Total _____

Dressing percent _____

GRADING INFORMATION

Maturity _____ Marbling _____ Quality grade _____

Fat thickness _____ Kidney, pelvic, and heart fat weight _____ Percent _____

Rib eye area _____ Yield grade _____

I. WHOLESALE

Item	Weight	Percent of Side	Price	Value	Remarks
Side					
Forequarter					
Hindquarter				.	
Heart fat					
Kidney & pelvic fat only					
Kidney					
Round ⎤ may be combined					
Rump[a] ⎦ for pricing					
Loin					
Rib					
Chuck[b]					
Flank					
Plate					
Brisket					
Foreshank[b]					
Totals			■		
Averrage price	■		■	■	
Cutting loss or gain					

Fig. 13–1.

Beef Markup and Margin Calculations

(Based on One Side of Beef [Half Animal])

Live Animal to Carcass

Sales value of a hanging side	$ _____
Live cost × 0.5 (half animal)	$ _____
Margin per side (sales value − live cost)	$ _____
By-product value per animal × 0.5 (half animal)	$ _____
Adjusted margin/side (margin + by-product value)	$ _____
Percent markup	
(Adjusted margin ÷ live cost of half animal) 100	_____ %
Markup per pound	
(Adjusted margin ÷ pounds of half animal purchased)	$ _____
Percent margin	
(Adjusted margin ÷ sales value of side) 100	_____ %
Margin per pound	
(Adjusted margin ÷ pounds of one side sold)	$ _____

Carcass to Wholesale Cuts

Cost (value) of hanging side	$ _____
Summed value of all wholesale cuts	$ _____
Margin (wholesale cut value − side cost)	$ _____
Percent markup per side	
(Margin ÷ cost of side) 100	_____ %
Markup per pound	
(Margin ÷ side weight purchased)	$ _____
Percent margin per side	
(Margin ÷ wholesale cut value) 100	_____ %
Margin per pound	
(Margin ÷ pounds of wholesale cuts sold)	$ _____

II. RETAIL (PRIMAL CUTS)

Item	Weight	Percent of Wholesale Cut	Price	Value
Round (wholesale)				
Top				
Bottom				
Gooseneck				
Flat				
Eye				
Tip				

Item	Weight	Percent of Wholesale Cut	Price	Value
Heel				
Boneless rump roast[a]				
Lean				
Fat				
Bone				
Totals				

Fig. 13–1 (continued)

II. RETAIL (PRIMAL CUTS) (Continued)

Item	Weight	Percent of Wholesale Cut	Price	Value
Rump (wholesale)[a]				
Boneless roast				
Lean				
Fat				
Bone				
Totals				
Loin (wholesale)				
Steaks				
Wedge bone				
Round bone				
Flat bone (double)				
Pinbone				
Top-sirloin boneless				
Porterhouse				
T-bone				
Top-loin bone-in				
Top-loin boneless				
Tenderloin				
Lean				
Fat				
Bone				
Totals				
Rib (wholesale)				
Short standing rib roast				
Rib steaks				
Small end				
Large end				
Boneless				

Item	Weight	Percent of Wholesale Cut	Price	Value
Rib eye steaks				
Back ribs				
Short ribs				
Lean				
Fat				
Bone				
Totals				
Chuck, square cut or arm (wholesale)[b]				
Cross rib				
Blade roasts				
Arm roasts				
Seven-bone roasts				
Shoulder pot roasts (boneless) (clod)				
Chuck eye (roll)				
Back ribs				
Soup bones				
Cross-cuts foreshank[b]				
Stew				
Lean				
Fat				
Bone				
Totals				
Primal summary		(% of side)		
Roasts & steaks				
Other retail cuts				
Lean				
(Retail) Subtotal				
Fat				

Fig. 13–1 (continued)

II. RETAIL (PRIMAL CUTS) (Continued)

Item	Weight	Percent of Wholesale Cut	Price	Value
Bone				
Totals				
Flank (wholesale)				
Flank steak				
Lean				
Fat				
Bone				
Totals				
Brisket (wholesale)				
Boneless roast[c]				
Lean				
Fat				
Bone				
Totals				
Plate (wholesale)				
Short ribs				
Pastrami piece				
Boneless plate				
Boiling beef				
Lean				
Fat				

Item	Weight	Percent of Wholesale Cut	Price	Value
Bone				
Totals				
Foreshank (wholesale)[b]				
Cross-cuts				
Lean				
Fat				
Bone				
Totals				

III. SUMMARY (TOTALS FOR SIDE)

Item	Weight	Percent of Side	Price	Value
Roasts & steaks				
Other retail cuts				
Lean				
(Retail) Subtotal				
Fat (incl. KPH)				
Bone				
Totals				
Error	(compared to side)			

Fig. 13–1 (continued)

Beef Retail Markup and Margin Calculations

(Based on One Side of Beef [Half Animal])

Live Animal to Retail Product

Retail value $ _____

Live cost × 0.5 (half animal) $ _____

Margin per side (sales value − live cost) $ _____

By-product value per animal × 0.5 (half animal) $ _____

Adjusted margin per side (margin + by-product value) $ _____

Percent markup
(Adjusted margin ÷ live cost of half animal) 100 _____ %

Percent margin
(Adjusted margin ÷ retail value of one side) 100 _____ %

Markup per pound
(Adjusted margin ÷ pounds of half animal purchased) $ _____

Margin per pound
(Adjusted margin ÷ pounds of retail product sold from half carcass) $ _____

Hanging Side to Retail Product

Retail value $ _____

Cost of hanging side $ _____

Margin (retail value − cost of side) $ _____

Percent markup
(Margin ÷ side cost) 100 _____ %

Percent margin per side
(Margin ÷ retail sales value) 100 _____ %

Markup per pound
(Margin ÷ pounds purchased) $ _____

Margin per pound
(Margin ÷ pounds sold) $ _____

ᵃThe rump is usually part of the round but may be handled as separate wholesale cuts.

ᵇChucks are traded as arm chucks, with the foreshank on or as square-cut chucks, the foreshank being a separate wholesale cut. Square-cut chuck and foreshank may be summed to determine the wholesale price of the foreshank based on arm-chuck price, by subtracting out the value of the square-cut chuck, leaving the balance for foreshank value.

ᶜCuring and smoking charges to produce corned beef from the brisket or other cuts are:

_____ ¢/pound

Carcass fabricator _____

Data taker _____

Sources for all prices: Live _____

Wholesale _____

Retail _____

Fig. 13–1 (continued)

PORK CUTTING TEST

Cutting date _____ Live animal ID _____ Carcass ID _____

Breed _____ Sex _____

Live animal weight _____ Price _____ Sales value _____

By-products: List _____ Estimated value _____

Harvest date _____ Hot carcass weight R _____ L _____ Total _____

Dressing percent _____

GRADING INFORMATION

Backfat: Last lumbar _____ Last rib _____ First rib _____ Total _____ Average _____

Fat depth at 10th rib _____ Muscling score _____ Carcass length _____

Loin eye area _____ Carcass grade _____

I. WHOLESALE

Item	Weight	Percent of Side	Price	Value	Remarks
Side					
Ham					
Loin					
Boston shoulder					
Picnic shoulder					
Belly					
Spareribs					
Neck bones					
Jowl					
Front feet					
Hind feet					
Tail					
Fatback					
Clear plate					
Kidney fat					
Other fat					
Lean trim 50/50 (belly)					
Lean trim 90/10					
Totals					
Average price					
Cutting loss or gain					

Fig. 13–2.

Pork Markup and Margin Calculations

Live to Wholesale Cuts from One Side

Summed wholesale cuts sales value	$ _____
Live cost × 0.5 (half animal)	$ _____
Margin (wholesale value − live cost)	$ _____
By-product value × 0.5 (half animal)	$ _____
Adjusted margin (margin + by-product value)	$ _____
Percent markup (Adjusted margin ÷ live cost of half animal) 100	_____ %
Markup per pound (Adjusted margin ÷ pounds of half animal purchased)	$ _____
Percent margin (Adjusted margin ÷ sales value of wholesale cuts) 100	_____ %
Margin per pound (Adjusted margin ÷ pounds of wholesale cuts from side)	$ _____

II. RETAIL FROM LEAN CUTS

Item	Weight	Percent of Wholesale Cut	Price	Value
Ham (wholesale)				
Ham (fresh, bone-in)[a]				
Rump portion				
Shank portion				
Center slices				
Boneless				
Lean[a]				
Fat				
Skin				
Bone				
Totals				
Loin (wholesale)				
Center-cut loin chops				
Center-cut rib chops				
Blade chops				
Blade roast bone-in				

Item	Weight	Percent of Wholesale Cut	Price	Value
Sirloin chops				
Boneless sirloin roast				
Boneless blade-end roast				
Butterfly chops				
Tenderloin				
Pocket chops				
Country ribs				
Back ribs				
Lean[a]				
Fat				
Bone				
Totals				
Boston shoulder (wholesale)[a]				
Blade steaks				
Bone-in roast				

Fig. 13–2 (continued)

II. RETAIL FROM LEAN CUTS (Continued)

Item	Weight	Percent of Wholesale Cut	Price	Value
Boneless roast				
Shish kabobs				
Lean[a]				
Fat				
Bone				
Totals				
Picnic shoulder (wholesale)[a]				
Arm steaks				
Pork hocks				
Bone-in roast				
Boneless roast				
Sausage trim[a]				
Lean[a]				
Fat				
Skin				
Bone				
Totals				
Lean cut summary		(% of side)		
Roast				
Steaks				
Chops				
Other retail cuts				
Lean				
(Retail) Subtotal				
Fat				
Bone				
Totals				

III. BELLY AND MINOR CUTS

Item	Weight	Percent of Wholesale Cut	Price	Value
Belly (wholesale)[a]				
Fresh side pork				
Salt pork				
Slab bacon				
Sliced bacon				
Totals				
Spareribs				
Neck bones				
Jowl[a]				
Front feet				
Hind feet				
Tail				
Fatback[b]				
Clear plate[b]				
Salt pork				
All other fat				

IV. SUMMARY (TOTALS FOR SIDE)

Item	Weight	Percent of Side	Price	Value
Totals: Roasts, chops, &/or steaks				
Other retail cuts				
Lean[a]				
(Retail) Subtotal				
Fat				
Bone				
Grand Total				
Cutting loss or gain				

Fig. 13–2 (continued)

Pork Markup and Margin Calculations

(Based on One Side of Pork [Half Animal])

Live to Retail

Retail value (summed from one side)	$ _____
Live cost × 0.5 (half animal)	$ _____
Margin (retail value − live cost)	$ _____
By-product value × 0.5 (half animal)	$ _____
Adjusted margin (margin + by-product value)	$ _____
Percent markup (Adjusted margin ÷ cost of half animal) 100	_____ %
Percent margin (Adjusted margin ÷ retail value of one side)	_____ %
Markup per pound (Adjusted margin ÷ pounds of half animal purchased)	$ _____
Margin per pound (Adjusted margin ÷ pounds of retail product)	$ _____

Wholesale to Retail

Retail value (summed from one side)	$ _____
Cost of summed wholesale cuts (one side)	$ _____
Margin (retail value − wholesale cost)	$ _____
Percent markup (Margin ÷ cost of wholesale cuts) 100	_____ %
Percent margin (Margin ÷ retail value) 100	_____ %
Markup per pound (Margin ÷ pounds of wholesale cuts)	$ _____
Margin per pound (Margin ÷ pounds of retail product) 100	$ _____

[a]Processing from the fresh state costs: curing and smoking _____ ¢/pound, grinding and seasoning for sausage, _____ ¢/pound.

[b]Rendering pork fat into lard costs _____ ¢/pound plus the rendering shrink which approximates 25 percent.

Carcass fabricator _____

Data taker _____

Sources for all prices: Live _____   Wholesale_____   Retail_____

Fig. 13–2 (continued)

LAMB CUTTING TEST

Cutting date _____ Live animal ID _____ Carcass ID _____

Breed _____ Sex _____

Live animal weight _____ Price _____ Sales value _____

By-products: List _____ Estimated value _____

Harvest date _____ Hot carcass weight _____ Dressing percent _____

GRADING INFORMATION

Maturity _____ Flank streaking _____

Conformation _____ Quality grade _____ Leg conformation _____

Fat thickness: Top loin _____ Lower rib _____ Rib eye area _____

Kidney and pelvic fat weight _____ Percent _____ Yield grade _____

I. WHOLESALE

Item	Weight	Percent of Side or Carcass	Price	Value	Remarks
Side or carcass					
Kidney & pelvic fat only					
Kidney					
Leg					
Loin					
Rib (rack)					
Shoulder					
Flank } may be combined					
Breast shank } for pricing					
Spleen					
Totals			███		
Averrage price	███			███	
Cutting loss or gain					

Fig. 13–3.

Lamb Markup and Margin Calculations

Live to Wholesale

Sales value of a hanging side or carcass	$ _____
Live cost*	$ _____
Margin (sales value − live cost)	$ _____
By-product value*	$ _____
Adjusted margin (margin + by-product value)	$ _____
Percent markup (Adjusted margin ÷ live cost) 100*	_____ %
Markup per pound (Adjusted margin ÷ pounds of animal)*	$ _____
Percent margin (Adjusted margin ÷ sales value of carcass) 100*	_____ %
Margin per pound (Adjusted margin ÷ pounds of carcass sold)*	$ _____

_____

*If carcasses are split and sold as sides, multiply values, cost, or weights for the animal by 0.5.

II. RETAIL (PRIMAL CUTS)

Item	Weight	Percent of Wholesale Cut	Price	Value
Kidney & pelvic fat only				
Kidney				
Leg				
Frenched leg				
American leg				
Boneless leg				
Leg steaks				
Shish kabobs				
Lean				
Fat				
Bone				
Totals				

Item	Weight	Percent of Wholesale Cut	Price	Value
Loin				
Loin chops				
Lean				
Fat				
Bone				
Totals				
Rib (rack)				
Rib chops				
Rib roast (crown)				
Lean				
Fat				
Bone				
Totals				

Fig. 13–3 (continued)

II. RETAIL (PRIMAL CUTS) (Continued)

Item	Weight	Percent of Wholesale Cut	Price	Value
Shoulder				
Blade chops				
Arm chops				
Boneless shoulder				
Saratoga chops				
Shish kabobs				
Lean				
Fat				
Bone				
Totals				

Item	Weight	Percent of Wholesale Cut	Price	Value
Totals				
Foreshank				
Lamb shanks				
Lean				
Fat				
Bone				
Totals				
Neck				
Neck slices				
Lean				
Fat				
Bone				
Totals				

II. RETAIL (ROUGH CUTS)

Item	Weight	Percent of Wholesale Cut	Price	Value
Flank				
Lean				
Fat				
Bone				
Totals				
Breast				
Spareribs				
Riblets				
Boneless breast				
Lean				
Fat				
Bone				

III. SUMMARY (TOTALS FOR SIDE)

Item	Weight	Percent of Side	Price	Value
Roasts, chops, &/or steaks				
Other retail cuts				
Lean				
(Retail) Subtotal				
Fat (incl. KPH)				
Bone				
Totals				
Error (compared to side or carcass)				

Fig. 13–3 (continued)

Lamb Retail Markup and Margin Calculations

Live Animal to Retail Product

Retail value summed from carcass or side	$ _____
Live cost*	$ _____
Margin (retail value − live cost)	$ _____
By-product value*	$ _____
Adjusted margin (margin + by-product value)	$ _____
Percent markup (Adjusted margin ÷ live cost) 100*	_____ %
Markup per pound (Adjusted margin ÷ pounds of animal purchased)*	$ _____
Percent margin (Adjusted margin ÷ retail value) 100*	_____ %
Margin per pound (Adjusted margin ÷ pounds of retail product)	$ _____

Hanging Side to Retail Product

Retail value summed from all product*	$ _____
Wholesale cost of hanging carcass or side	$ _____
Margin (retail value − wholesale cost)	$ _____
Percent markup (Margin ÷ cost of carcass or side) 100	_____ %
Markup per pound (Margin ÷ pounds of carcass or side)	$ _____
Percent margin (Margin ÷ retail value, summed) 100*	_____ %
Margin per pound (Margin ÷ pounds of retail product)	$ _____

_____

*If carcasses are split and sold as sides, multiply values, cost, or weights for the animal by 0.5.

Carcass fabricator _____

Data taker _____

Sources of all prices: Live _____

Wholesale _____

Retail _____

Fig. 13–3 (continued)

Where possible, tests should report data on several of each of the whole-sale units being tested because carcasses and wholesale meat cuts do vary in composition. Tests should be designed and conducted to account for variations in the technique and ability of the employees who are, or will be, involved in the actual production. Pricing or decision errors may result from tests that are not true indicators of production conditions.

Tests are extremely important when procedural changes, equipment changes, pricing changes, or personnel changes are made. If repeat testing develops predictable patterns, the testing frequency may be reduced. Testing requires time and effort; therefore, its effectiveness should be reviewed in relation to its cost.

Not only do cutting tests provide information about operations but they also help management determine prices. For example, a retail meat department manager may want to determine the influence of an increase in whole-sale meat cost on retail prices. If the meat department operated on a 20 percent margin, by the earlier definition, 20 percent of the total retail sales income would be used to pay for in-department costs, expenses, and profits. Therefore, 80 percent of the total sales income could then be used to buy wholesale meat.

The increase in wholesale costs to the meat department in the example could be an increase in the beef rib wholesale price from $150 per 100 pounds to $165 per 100 pounds, due to a combination of seasonal change and decreased beef supplies. Since the wholesale cost represented 80 percent of the expected retail value of the rib, the manager determined that the retail value of a 30-pound beef rib must be $(30 \times \$1.65) \div 0.80 = \$61.88$.

Figure 13–4, which is an adaptation of the "Rib (wholesale)" section of Figure 13–1, demonstrates the results of a cutting test performed on one rib. The retail prices were determined by combining the relative desirability of the retail cuts, the prices in competitive stores, and the price the manager thought consumers would be willing to pay for the various retail cuts. If the retail prices necessary to achieve the required margin were too high, the manager would have to make adjustments. For instance, if competitors in the area were selling small-end steaks for $2.85 per pound, it is unlikely that large volumes of steak would have moved at $2.92 per pound. If small-end steaks had been reduced to $2.85 per pound, the retail value of the rib would have been reduced $0.42 to $61.54, subsequently decreasing the margin to 19.54 percent. The manager could have either repriced one of the other retail portions of the rib to make up the difference or increased the price of some other item

RETAIL CUTTING TEST FOR BEEF RIBS

Cut by _____ WJ _____   Date _____ 7/2/84 _____

Item _____ 103 rib _____   Source _____ Z Packing Co. _____

No. of item _____ 1 _____   Weight _____ 30 pounds _____

Wholesale price _____ $164/100 pounds _____   Wholesale cost _____ $49.50 _____

Expected retail value (20% margin) _____ $61.88 _____

	Weight	Percent of Wholesale Cut	Retail Price	Value
Rib (wholesale)	30 lb	100%	$1.65	$49.50
Short standing rib roast	4.0	13.3	2.62	10.48
Rib steaks				
Small end	6.0	20.0	2.92	17.52
Large end	10.0	33.3	2.77	27.70
Boneless				
Rib eye steaks				
Back ribs				
Short ribs	5.5	18.3	.85	4.67
Lean	1.0	3.0	1.59	1.59
Fat	1.5	5.0	–	
Bone	2.0	6.7	–	
Totals	30.0	99.6	–	61.96

Actual retail value _____ $61.96 _____

Wholesale cost _____ $49.50 _____

Margin _____ $12.46 _____

Margin as percent of sales _____ 20.11% _____

Fig. 13–4.

in the retail case to maintain the level of margin. Modification of the cutting procedure — cutting the roast into steaks or cutting a portion of the rib into boneless steaks — may have resulted in a greater retail return. Often consumers are willing to pay more for a high-yielding, "classy," boneless cut.

The manager would have reduced the display of beef rib retail cuts if beef rib wholesale prices had reached the level that made it impossible for margin requirements to be achieved at the maximum retail prices the normal consumer was willing to pay. The space normally occupied by rib cuts would have been filled with alternatives at prices the consumer would be willing to pay. The result would have been a decrease in wholesale demand for beef ribs and probably downward pressure on the wholesale rib price. At the same time, the retailer would have been attempting to maintain total sales volume by merchandising more competitively priced meat items.

The effect of margin and yield on pricing can be demonstrated with the preceding cutting test. The wholesale price increase of $15 per 100 pounds (from $150 per 100 pounds to $165 per 100 pounds) may be adjusted to account for the margin: $15 ÷ (1 − margin) = $15 ÷ 0.80 = $18.75 per 100 pounds. Assuming the total 30 pounds of wholesale rib is salable, the retail price of rib cuts would be increased an average of 18.75 cents per pound. However, in many retail situations, the fat and bone (3.5 pounds) have little or no value. The salable product, therefore, accounts for 88.3 percent of the original weight (26.5 pounds ÷ 30 pounds). Correcting the retail price change for yield: ($18.75 per 100 pounds) ÷ yield = $18.75 ÷ 0.883 = $21.23 per 100 pounds. Each pound of retail product from the rib must produce approximately 22 cents more return to the retailer if the wholesale cost of ribs increases $15 per 100 pounds (15 cents per pound). It is unlikely that the retail price of ground beef or short ribs can immediately be increased 22 cents per pound. Therefore, the higher-priced steaks and roasts must be increased more than the average 22 cents per pound to achieve the required retail value.

Animal Carcass–Retail Price Relationship

Some reasons for the differences in meat prices at different points in the meat-marketing process are given in Table 13–6. A 1,000-pound Choice, Yield Grade 3 (see Chapter 12) steer priced at $65.75 per hundred pounds cost the packer $657.50. The USDA price sources indicated that the carcass of that steer was priced at $103.75 per 100 pounds. Assuming that the steer had a dressing percent of 62 percent, the 620-pound carcass had a wholesale value of $643.25. The by-product value, quoted by the USDA source, was $5.85 ($7.67 in 1999)* per 100 pounds of live weight. The 1,000-pound steer yielded by-products valued at $58.50 ($76.70 in 1999),* which, when added to the carcass value, resulted in a total value to the packer of $701.75 ($719.95 in 1999).* The meat packer sold the carcass and by-products of the steer for $44.25* more than the cost of the live steer. The packer may have decided to sell the carcass as quarters. If so, the value of the carcass would have been $657.20. When the by-product value was added, the total value of the steer was $715.70,* as quarters. The packer's margin was increased to $58.20* because the carcass was sold as four separate units rather than as one unit.

The retailer who purchased the steer carcass would have paid $643.25 for it, providing there were no shipping costs. If the retail unit operated on a 20 percent margin, the wholesale value, $643.25, represented 80 percent of the total value of retail cuts from that carcass. Therefore, the retail value was $804.06 ($643.25 ÷ 0.80). To produce attractive, salable retail cuts, the retailer trimmed away excess bone, fat, and connective tissue. The total weight of salable retail cuts from a beef carcass will vary from approximately 60 percent to 75 percent of the carcass weight. The composition of the carcass (yield grade), degree of trimming, and proportion of boneless cuts will determine the retail yield. Therefore, the original 620-pound carcass produced from 372 to 465 pounds of retail cuts, which had a total value of $804.06.

A cutting test would have been used to establish specific retail-cut prices and to determine optimum cutting procedures for the particular retail situation. Some beef cuts, such as pot roasts, are not popular during the summer. Thus, the less desired cuts were economically priced to encourage consumers to purchase them. To compensate, the more highly demanded steaks were marked with relatively high prices in order to achieve the $804.06 retail value.

The difference between the retail value, $804.06, and the live steer value, $657.50, was used by the packers and the retailer to pay the harvest, chilling, cutting, packaging, transportation, and other costs. According to a 1999 *USDA Market News* report, harvest cost per beef animal was $38.00 and pro-

---

*Live and carcass prices of 1983=1999; but 1999 by-products (drop) are more valuable.

cessing cost was $8.75. The value increase, coupled with the weight decrease from 1,000 pounds of steer to as low as 372 pounds of meat after the nonmeat parts of the animal were removed, resulted in large price per pound increases. The average retail price of beef from the carcass is $2.16 per pound ($804.06 ÷ 372 pounds).

Table 13–6. Beef Price, Weight, and Value Relationships

Item	Price[1]	Weight	Value
	($/100 lbs.)	(lbs.)	($)
Steer, Choice, YG 3	65.75	1,000	657.50
Carcass, steer, Choice, YG 3	103.75	620.0[2]	643.25
By-product value	5.85[3]	1,000	58.50 (76.70)*
Carcass + by-product.			701.75 (719.95)*
Margin to packer as carcass			44.25
Forequarters.	91.00	322.4[4]	293.38
Hindquarters	122.25	297.6[4]	363.82
Carcass as quarters			657.20
Carcass as quarters + by-product			715.70
Margin to packer as quarters.			58.20
Carcass, retail value (20% margin).		620.0[2]	804.06[5]
Carcass, retail weight (75% yield)[6]		465.0	804.06
Carcass, retail weight (60% yield)[6]		372.0	804.06
Hindquarters, wholesale	122.25	620.0[7]	757.95
Hindquarters, retail value (20% margin)		620.0	947.44[5]
Hindquarters, retail weight (60% yield)[6]		372.0[7]	947.44

[1]Mean of four weekly averages, USDA Market News. June and July 1983 = September 1999.

[2]Assume 62 percent dressing percent.

[3]USDA quoted on basis of per hundred pounds live weight.

[4]Forequarters 52 percent, hindquarter 48 percent of carcass weight.

[5]Retail value = wholesale value ÷ (1 – margin).

[6]Approximate cutting yields: 75 percent bone-in, 60 percent boneless.

[7]Equals carcass weight to demonstrate value differences.

*Live and carcass prices of 1983=1999; but 1999 by-products (drop) are more valuable.

The retailer may have elected to purchase hindquarters only in order to increase the proportion of higher-valued steak cuts in the retail case. Purchasing an equal amount, 620 pounds, of hindquarters only would also have resulted in an equal number of pounds of retail product, 370 to 460 pounds, depending upon trimming and boning procedures. The retail value of the hindquarter cuts would have been $947.44 [(620 pounds × $122.25 per 100 pounds) ÷ 0.80]. To achieve an average price of $2.55 per pound ($947 ÷ 372 pounds) would necessitate pricing some steaks in excess of $3.50 per pound to compensate for ground beef and other lower-value cuts.

Few specific retail prices have been quoted in the preceding examples. Meat retail pricing varies widely as a result of store-to-store differences in relative demand, cutting procedures, margins, volume, by-product value, and many other factors. Meat prices are constantly changing at each level in the meat-marketing system. Successful managers at all levels must be prepared to adjust to the changes.

Livestock price increases usually result in carcass and wholesale prices moving to higher levels, and, depending on the demand situation, meat retail prices will reflect the upward trend. Similarly, downward movements in live price may gradually travel through the supply chain to the consumer. Demand reductions stemming from price increases, income loss, and(or) diet changes will result in volume reductions at the retail meat case, which will soon be detected in decreased demand and strong downward pressure on prices at the wholesale level. The harvester generally passes the effects of demand reduction on to the livestock producer as lower prices.

Fresh meat is a perishable item and must be moved through marketing channels rather swiftly. Negative demand trends fill the "pipeline" quickly and often result in rapid price decreases to salvage the product or clear out the system. The perishable food industry has an old adage: "Sell it or smell it."

Seasonal changes in price relationships may occur due to variations in the livestock and meat supply; however, consumer preferences have a significant influence on the seasonal changes in meat prices. Table 13–7 demonstrates the beef price relationship changes that occur between winter and summer. The summer prices are the prices used in Table 13–6 and the example discussed previously. The winter prices were reported by the U.S. Department of Agriculture as averages for the week ending January 16 during the same year; they were selected because the Choice, Yield Grade 3 steer carcass price on that report was the same as the summer carcass price, $103.75 per 100 pounds. This comparison also illustrates that carcass prices and livestock prices vary somewhat independently, since the live prices in this case differed $1.75 per 100 pounds while the carcass prices were identical.

Table 13-7. Beef Weight, Price, and Value Relationships in Summer and Winter

Item	Weight	Summer		Winter	
		Price[1]	Value	Price[2]	Value
	(lbs.)	($/100 lbs.)	($)	($/100 lbs.)	($)
Steer, live.............	1,000.0	65.75	657.50	64.00	640.00
Carcass..............	620.0	103.75	643.25	103.75	643.25
By-product............		5.85[3]	58.50	5.75	57.50
Carcass + by-product....			701.75		700.75
Margin to packer as carcass.............			44.25		60.75
Forequarters	322.4[4]	91.00	293.38	98.50	317.56
Hindquarters	297.6[4]	122.25	363.82	115.50	343.73
Carcass as quarters	620.0		657.20		661.29
Carcass as quarters + by-product			715.70		718.79
Margin to packer as quarters			58.20		78.79
Rounds	138.9[5]	111.20	154.46	122.50	170.15
Loins................	106.6	170.60	181.86	130.00	138.58
Flanks	32.2	53.10	17.10	56.00	18.03
Arm chucks	184.8	86.50	159.85	101.00	186.65
Ribs.................	59.5	163.75	97.43	137.50	81.81
Plates	51.5	53.25	27.42	57.50	29.61
Briskets	23.5	57.75	13.57	64.00	15.04
Fats, kidneys, etc........	22.3	10.00[6]	2.23	10.00[6]	2.23
Carcass (primals)	619.3		653.92		642.10
Margin to packer as primals[7]			54.92		59.60

[1]Mean of four weekly averages, USDA Market News. June and July 1983 = September 1999.

[2]Weekly averages, USDA Market News. January 16, 1982.

[3]USDA quoted on basis of per hundred pounds live weight.

[4]Forequarters 52 percent, hindquarters 48 percent of carcass weight.

[5]Wholesale cut weight, calculated from percents quoted by the National Live Stock and Meat Board in A Steer's Not All Steak.

[6]Estimated price.

[7]Carcass value (primals) + by-product value − live value = margin to packer.

Summer is steak season. In the winter, beef roasts are popular in many households. The $31.25 spread between forequarters and hindquarters in summer compares to a $17 spread in winter, while the carcass prices were identical. Wholesale loins from identically priced carcasses were $40 per 100 pounds more valuable, and wholesale ribs were $26 per 100 pounds more valuable during the summer steak season than during the winter. Wholesale loins and ribs together represented 42.7 percent of the carcass value during the summer, whereas in winter, the two cuts accounted for only 34.3 percent of the total carcass value. The prices increased for chucks ($15 per 100 pounds) and for rounds ($11 per 100 pounds) in the winter, the roast season.

MEAT CUTS

Wholesale and Retail Cuts

The meat industry in the United States has adopted a system of standardized *wholesale* or *primal cuts* for each species. Wholesale cuts have been defined as large subdivisions of the carcass that are traded in volume by segments of the meat industry. Relatively rigid standardization of wholesale cuts has been established for many years to provide efficient communication between high-volume buyers and sellers [see the discussion of Institutional Meat Purchase Specifications (IMPS) later in this chapter].

Guidelines long used by the meat industry in making both wholesale and retail cuts are:

- Separate tender portions of the carcass from less tender areas.

- Separate lean areas from the portions having greater amounts of fat.

- Separate thicker, more heavily muscled portions of the carcass from the thin-muscled areas.

Wholesale or primal cutting separates the legs, made up of large locomotion-muscle systems, from the back, composed of large support-muscle systems, from the thinner body-wall sections of the carcass. Subdivision of the carcass into wholesale cuts permits the purchase of only the part(s) of the carcass that may be best adapted to the buyer's purpose; for example, corned beef processors purchase only beef briskets.

Sub-primal cuts are subdivisions of the wholesale or primal cuts that are made to facilitate handling or to reduce the variability within a single cut. Sub-primal cuts are in demand by food service units, institutions, fresh meat retailers, and meat processors. Beef chucks or beef rounds are often subdivided to produce two-piece chucks or three-piece rounds. The sub-primals are easier for the packer to vacuum bag and fit into the shipping box than the primal cuts would be. Different cutting techniques applied to the portions of the large wholesale cut in the retail operation would require that the original primal cut be subdivided in the store in many cases. Thus, by purchasing the sub-primals, the retailer saves the extra cutting step and is able to leave the sub-primals in the vacuum bag until needed. One food service unit may buy only the sub-primal short loin, whereas another may purchase both the short loin and the sirloin sub-primals of the primal loin. Sub-primal cuts increase the options and offer all segments of the wholesale meat trade greater flexibility.

Retail cuts are subdivisions of wholesale cuts or carcasses that are sold to consumers in ready-to-cook or ready-to-eat forms. A large selection of different retail cuts are found in retail-meat cases due to the innovative nature of meat merchandisers. Consumers can quite easily learn to identify the cuts from various species by differences in cut size, color, and fat characteristics. Beef cuts are large and have a cherry red color and a white, firm fat. Pork cuts are intermediate in size, tend to be a grayish pink color, and have the softest fat. Lamb cuts are small, are dark pink to light red in color, and have hard, white fat.

Most of the bone-in cuts can be classified into one of seven types based on muscle/bone shape and size relationships. Familiarity with the seven primary retail-cut types — loin, rib, leg, arm, hip, blade, and belly or plate — will enable the buyer to identify the source location in the carcass, therefore making it possible to predict relative palatability as well as to identify the retail-cut name (see Figure 13–5 and the reference on page 502).

Two of the seven retail-cut types — *loin* and *rib* — have a cross section of the *longissimus* or "eye" muscle of the back as the major muscle component. *Loin* cuts often contain a section of the tenderloin muscle *(psoas major)* in addition to the *longissimus.* The *lumbar* vertebra (loin) has the typical T–bone configuration formed by the lateral process, the body, and the dorsal process of the vertebra. *Rib* cuts have no tenderloin section and often exhibit a section of rib bone.

The cross section of a round bone is typical of both *leg* and *arm* cuts. The *leg* muscle configuration of top *(semimembranosus),* bottom *(biceps femoris),*

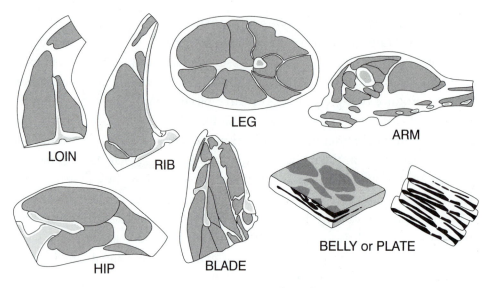

Fig. 13–5. Seven types of retail cuts.

eye *(semitendinosus),* and sirloin tip *(quadriceps)* is easily distinguished from the more diffuse arrangement of small muscles in the *arm* cuts.

Two types of cuts — *hip* and *blade* — have flat or irregularly shaped bones. The *hip* cuts are composed of a small number of relatively large, parallel muscles. The *blade* cuts display numerous small, non-parallel muscles typical of the foreleg system.

The seventh type of cut — *belly* or *plate* — is best recognized by the alternating layers of fat, lean, and rib bones that make up the body wall. *Belly* or *plate* cuts would include bacon, spareribs, and beef brisket.

If meat consumers were to apply this identification system in their purchases, they would quickly develop the ability to recognize muscle size and shape relationships that would aid them in the identification and carcass location of many boneless cuts as well. It should be noted that the seven types of retail cuts do not include all of the retail cuts that are available. (See color photos of cuts; also see Chapters 14 through 17.)

MEAT-CUT STANDARDS

Uniform Retail Meat Identity Standards

In the United States, there are estimated to be more than 1,000 names that identify the fresh cuts of beef, pork, lamb, and veal offered for sale at the retail level. Because of this situation, the Industrial Cooperative Meat Identification Standards Committee (ICMISC) was formed in 1972, composed of representatives of the entire meat industry, including meat packers and processors and all different sizes of grocery and supermarket operations. Livestock producers were represented by the National Live Stock and Meat Board.

This committee produced a most needed set of standards for the U.S. meat industry entitled *Uniform Retail Meat Identity Standards.* It includes a master list of 314 identifications for retail cuts of all species, which is designed to serve all marketing areas in the United States. This document is published and distributed at cost by:

> Order Processing Department
> National Cattlemen's Beef Association
> 5420 South Quebec Street
> Greenwood Village, CO 80111

The meat label information recommended by the ICMISC includes:

- *The kind (species) of meat* — beef, pork, lamb, or veal. It is listed first on every label.

- *The wholesale or primal cut* — loin, leg, shoulder, chuck, etc. — tells where the meat comes from on the animal.

- *The retail cut* — blade roast, spareribs, loin chops, etc. — tells what part of the wholesale cut the meat comes from.

- *Ground beef* — not less than X percent lean — tells the percentage of lean in ground beef the retailer will guarantee. ***Federal law requires that ground beef and hamburger must be at least 70 percent lean.***

If desired, the retailer may also put on the label various familiar "creative" terms describing the four preceding types of meat information. Furthermore, if cooking instructions are given, the chances for the consumer to "go wrong" are significantly lessened.

The uniform identity system is voluntary. Some states already had identification laws on the books before these standards were published. In order for the system to be truly "uniform" and thus serve *all* the people, regardless of where they move or relocate in the United States, it must be widely used by U.S. meat retailers, which it is.

The Uniform Retail Meat Identity Standards are used in this text. Also, some of the more common "fanciful" names have been added where appropriate in the following chapters. The retail cuts in the colored illustrations of beef, lamb, and pork are identified by these standards.

Institutional Meat Purchase Specifications (IMPS)

The *Institutional Meat Purchase Specifications (IMPS)* were developed by the USDA Agricultural Marketing Service to form a basis for the Meat Certification Service performed by the Federal Meat Grading Service (see Chapter 12). The IMPS stipulated three types of guidelines or requirements. The first is the "General Requirements" section, which provides specifications for refrigeration and for the packaging and handling procedures of fresh meat and processed products.

The second IMPS guideline gives detailed descriptions of and nomenclature for meat items that are routinely traded in large volume. The boxed-meat programs used by the fresh meat industry are often based on the production, vacuum packaging, and boxing of primal and sub-primal cuts. Portion-cut individual servings are also traded in volume in the food service segment of the industry. Both the meat industry and the U.S. Department of Agriculture recognized the marketing problems inherent in the virtually unlimited options for shape, composition, and degree of trim introduced by the sub-primal and boxed-meat programs. Thus, the industry has adopted and is using extensively the IMPS guidelines to standardize carcasses, primal cuts, sub-primal cuts, and retail cuts within the marketplace. Each item is assigned a number, and a detailed description of the cutting procedure and composition (bones, muscles, fat, dimensions) is outlined. Wholesale volume orders and sales are often made using only the IMPS number and grade to describe the item.

The USDA Agricultural Marketing Service has prepared and organized these standards into species or other related groupings referred to as series of Institutional Meat Purchase Specifications. The series available are listed here. Copies of the specifications may be purchased from the Superintendent of Documents.

- *Series 100* (fresh beef)

- *Series 200* (fresh lamb and mutton)

- *Series 300* (fresh veal and calf)

- *Series 400* (fresh pork)

- *Series 500* (cured, cured and smoked, and fully cooked pork products)

- *Series 600* (cured and dried and smoked beef products)

- *Series 700* (edible by-products)

- *Series 800* (sausage products)

- *Series 1,000* (portion-cut meat products)

The North American Meat Processors Association, 1920 Association Drive, Suite 400, Reston, VA 20191-1547 <namp@ix.netcom.com>, has published *The Meat Buyers Guide,* which contains excellent descriptive color pictures and accompanying discussion for the *Series 100, 200, 300, 400,* and *1,000* USDA Institutional Meat Purchase Specifications.

The third specification set of the Institutional Meat Purchase Specifications, entitled *Quality Assurance Provisions,* is a guide for sampling procedures to determine and assure product acceptability. These quality assurance provisions not only indicate how to sample lots or shipments to determine compliance with specifications but also spell out the criteria required by particular quality levels of the product.

Universal Product Code and Retail Meat Cuts

Another concept in retail meat standardization, developed by the National Cattlemen's Beef Association (NCBA) and the Food Marketing Institute in cooperation with retail food chains and equipment manufacturers, is an adaptation of the Universal Product Code (UPC) system, which is a series of lines and spaces printed on many uniform weight and labeled products. Random-weight products, such as meat cuts, presented problems in the development of scanning technology in the UPC system. The NCBA and the Food Marketing Institute encouraged scale and scanning equipment manufacturers to develop and to make available to food chains the equipment necessary to handle random-weight coding. The coded area of the package label informs the in-store computer of the species; the primal, sub-primal, and retail cuts; the packer source; the sell-by date; and other types of information the retailer may choose to code into the system. Meat-cut labels for more than 2,100 varieties of meat products are then read by the same scanner equipment used by store personnel at the checkouts, during case inventory checks, and for monitoring the movement of the meat product through the whole marketing sequence. Not only does the UPC system make the checkout procedure more accurate and efficient, but it also provides retail managers with data related to product flow, consumer preferences, and meat department efficiency.

Meat-Cut Illustrations and Recommendations

An excellent full-color brochure entitled *The Guide to Identifying Meat Cuts* (ISBN 0-88700-020-7) was released in 1998 by the American Meat Science Association (http://www.meatscience.org/), the National Cattlemen's Beef Association (http://www.beef.org/), and the National Pork Producers Council (http://www.nppc.org/), with the cooperation of the American Sheep Industry Association, the Veal Committee / NCBA, the North American Meat Processors Association, and the U.S. Meat Export Federation. In addition to identification, this handy 4" × 6" booklet covers basic labeling, nutrition labeling, wrapping, and cooking of meat.

14

Pork Identification and Fabrication

The pig is believed to be the first farm animal domesticated.[1] Pork carcasses have been fabricated into wholesale and retail cuts at packing plants since the earliest days of the meat packing business in the United States. Historically, this procedure differed from that used in beef, veal, and lamb distribution whereby whole carcasses were shipped from the packing plants to the distribution centers. Today the fabrication of carcasses of all species, as well as pork, takes place in the plants where the animals are harvested. Why then should students/consumers understand the process of fabrication if most commercially harvested pork is cut into wholesale and retail cuts before it leaves the plant? Pork has been, and will continue to be, the traditional "farmer's" meat. The pig is a simple-stomached creature that is not a fussy eater. Many hogs have been, and probably will continue to be, raised in small pens and outbuildings to provide pork for the family. The pig is small enough to make home harvest feasible with minimum facilities. Home-raised or locally purchased hogs are the primary reason for the existence of local lockers or harvest plants in many communities. This chapter will make the consumers of the home-raised pork chops and hams aware of how the pork carcass can be properly processed into chops and roasts either at home or in a federal- or state-inspected processing facility. Also emphasized in this chapter is the relationship between structure and function of the various sections of the animal and the palatability of the product. Students and consumers will also find some excellent meat identification tips.

---

[1]Archaeology Special Section. 1998. Reading the signs of ancient animal domestication. Science. 282:1448. http://www.sciencemag.org

In commercial plants, pork carcasses are chilled to an internal temperature of 40°F or less following harvest. Some plants even put hog carcasses directly into a freezer for as many as 8 hours before they move them into a cooler for tempering before cutting. This low-temperature treatment is done in an effort to improve pork quality (see color standards in the color section of this text). The chilled carcasses travel from the cooler on a moving chain to a conveyorized cutting operation. The cutting room must have a capacity equal to the harvest rate in the plant and may process more than 1,000 pork carcasses per hour. Mechanization, specialization, automation, and skill are as important on the pork disassembly lines as they are on the vehicle assembly lines. Two-thirds of the fresh pork cuts and trimmings produced by the carcass disassembly process are transferred to other areas of the plant for further processing. Fresh hams, shoulders, bellies, jowls, spareribs, and neck bones may be cured, smoked, and cooked. Some pork loins are also pumped with <0.5 percent sodium tripolyphosphate (STP) to control water (juice) retention and to reduce oxidative rancidity and sodium lactate (SL) at around 2 to 3 percent to extend shelf life. Lean trim is incorporated into many sausage items. Fat trim is rendered into lard and cracklings. Because processed pork products have been such an important part of the meat consumer's diet and because all of the processing procedures can be accomplished more efficiently and with greater uniformity on a large scale, harvest plants have cut the pork carcasses into parts since the beginning of the meat packing industry in the United States (see Chapter 13).

Although most of the pork processed today is chilled before the carcasses are cut, hot processing of pork carcasses does occur in some plants. Cutting the pork carcass before the animal heat has been removed saves the energy required to cool and reheat those portions of the carcass that are heat processed (see Chapter 19). Cured and smoked meats are cooked, lard is heated during rendering, and many sausage products are smoked and cooked. Those products that are sold fresh must be chilled immediately after the carcass has been subdivided. Although most plants are designed for the chilling period following harvest, the hot processing of pork and other carcasses may be adopted by more plants in the future.

IDENTIFICATION

The method of cutting pork is similar in all sections of the United States, even though there may be differences in regional preferences for pork cuts. (See "Meat Cuts" in Chapter 13 for information on general cut types and

PORK CHART
LOCATION, STRUCTURE
AND NAMES OF BONES

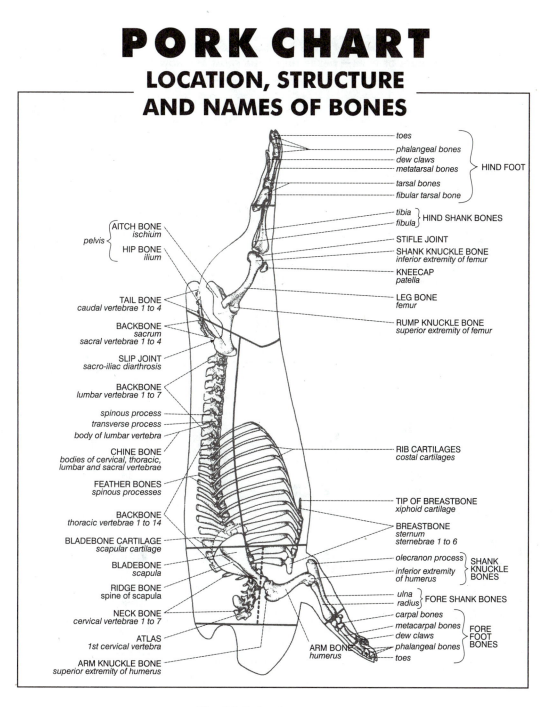

Fig. 14–1. Porcine anatomy.

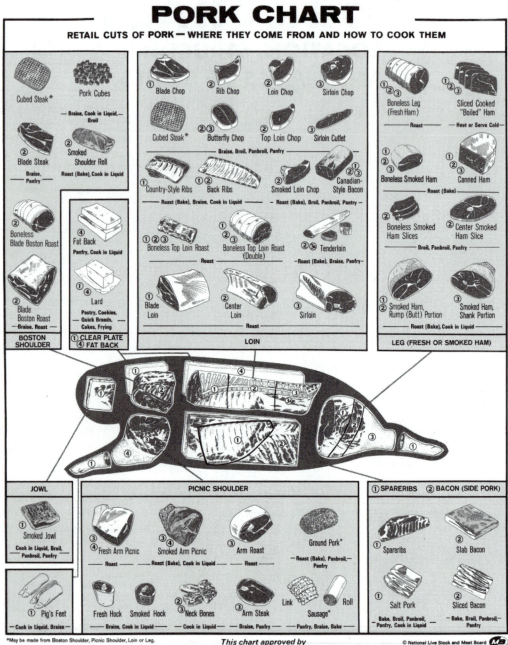

PORK CHART
RETAIL CUTS OF PORK — WHERE THEY COME FROM AND HOW TO COOK THEM

This chart approved by
National Live Stock and Meat Board

*May be made from Boston Shoulder, Picnic Shoulder, Loin or Leg.

© National Live Stock and Meat Board

Fig. 14–2.

sources of information about meat cuts.) The Uniform Retail Meat Identity Standards[2] and the Institutional Meat Purchase Specifications,[3] discussed in Chapter 13, will be used as the basis for identification in this chapter. Two valuable sources containing more information regarding muscle location and nomenclature are *Porcine Musculature Topography*[4] and *Porcine Myology*.[5] Refer also to the last paragraph in Chapter 13 for a description of *The Guide to Identifying Meat Cuts,* an excellent identification reference.

The following material will detail specific cutting procedures for pork.

FABRICATION

The skeletal structure of the porcine in relation to an outline of the wholesale cuts is shown in Figure 14–1, and a more detailed chart of wholesale and retail cuts and recommended cooking methods appears as Figure 14–2. These

Fig. 14–3. Before hog carcasses are taken from the harvest floor, they are generally split into right and left sides of approximately equal weight.

In most packing houses, hog carcasses are not completely split but are hung on single hooks or gambrels, with the two sides attached by a short area of backfat (often called *fatback*) and skin. The fresh hams may be *faced;* that is, the collar fat, a narrow strip of fat over the inside of the ham, is trimmed away on the harvest floor. Note the difference between a faced ham *(right)* and an unfaced ham *(left)*. The leaf fat, which is called kidney fat in beef, is removed on the harvest floor. The left side indicates where the leaf fat is located. Both kidneys are shown here in the left side to demonstrate their relative size and location. When the leaf fat is removed, the kidneys are also removed from the pork carcass before it is placed in the chilling cooler.

---

[2]Uniform Retail Meat Identify Standards. 1973. National Live Stock and Meat Board (now National Cattlemen's Beef Association), Denver.

[3]Institutional Meat Purchase Specifications. 1987. USDA-AMS, Washington, DC.

[4]Briskey, E. J., T. Kowalczyk, W. E. Blackmon, B. B. Breidenstein, R. W. Bray, and R. H. Grummer. 1958. Porcine Musculature Topography. Res. Bull. 206. Univ. Wisconsin, Madison.

[5]Kauffman, R. G., and L. E. St. Clair. 1965. Porcine Myology. Bull. 715. Agric. Exper. Sta., Univ. Illinois, Urbana. A new edition of Porcine Myology is now available on a CD from S. J. Jones, University of Nebraska, <ansc705@unlvm.unl.edu>.

two figures will serve as reference for the following discussion on cutting a pork carcass. The color photos of the retail cuts of pork, found near the center of the text, should also be consulted. Figures 14–3 to 14–67 appear through the courtesy of South Dakota State University. The authors and Teann Garnant, an SDSU student and meat lab employee, demonstrate the fabrication steps in the following figures.

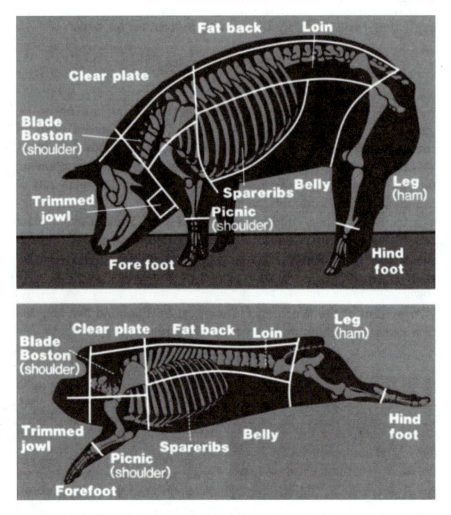

Fig. 14–4. The wholesale cuts of pork sometimes double as retail cuts. The five *primal cuts* are the pork leg (fresh ham), loin, Boston butt (Boston shoulder or blade Boston), picnic shoulder, and belly, the first four being further recognized as *lean cuts*. The pork leg (fresh ham) and the loin are the two most valuable lean primal cuts, with the Boston butt (Boston shoulder or blade Boston) and the picnic shoulder being of less value. The belly is the fifth primal cut. Other cuts are known as *minor cuts*. Of these, the clear plate and backfat are *fat cuts,* and the neck bones and spareribs are *bone cuts* of the pork carcass. The jowl is the fifth minor cut.

Generally, when hogs are finished for market they will weigh approximately 240 pounds. After harvest, the resulting carcass represents approximately 70 percent of the live weight. Therefore, a 240-pound live hog will produce a carcass weighing approximately 168 pounds.

Leg (Ham)

The terms used to identify the rear leg of the pig and the pork carcass may vary from situation to situation. For many years, *ham* was the term of preference for that portion of both the live animal and the carcass, and it continues to be widely used. However, *leg* is used today to denote the rear appendage in the fresh form, while *ham* refers to the cured form. The terms *fresh ham, ham*

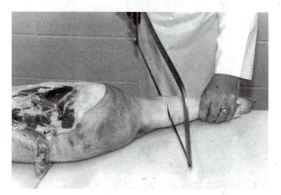

Fig. 14–5. The first step in fabricating a pork carcass is to remove the hind foot. The point of separation may vary, depending on the projected use of the ham. If a long-term, country-cured process is intended, the separation should be made through the middle of the hock joint where the solid bone will prevent bacterial contamination from invading the hollow, marrow space of the ham shank (hock) bone.

The hind foot is generally not used for human consumption because it contains an exceptionally high proportion of bone and very little edible muscle. In addition, the hind foot must be opened on the harvest floor for insertion of the gambrel or hook in the tendon to hang the carcass. This detracts from its desirability as a human food product.

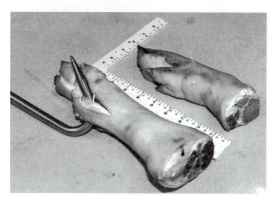

Fig. 14–6. The next step is the removal of the forefoot at the junction between the foreshank bone and the forefoot bone. The forefoot, unlike the hind foot, has a larger percentage of edible tissue composed largely of muscle and a lower percentage of tendon and bone. The forefoot is generally easier to clean and process and is therefore utilized for human consumption as pickled pigs' feet or sometimes as a raw material for sausage.

(cured), and *ham (cured and smoked)* are used in the Institutional Meat Purchase Specifications (IMPS). The Uniform Retail Meat Identity Standards label cuts as *pork leg (fresh ham)* and *smoked ham*. Since this chapter deals primarily with the fresh form, *fresh ham* or *pork leg* will be used here, and *ham* will predominate in the later chapters in which the cured product is the main subject.

The pork leg (fresh ham) is the largest single cut in the pork carcass comprising, on the average, 16 percent of the live weight and 22 percent of the carcass weight. Note its location on the live animal and the carcass (see Figure 14–4). For future reference, note the skeletal designations shown in Figure 14–1.

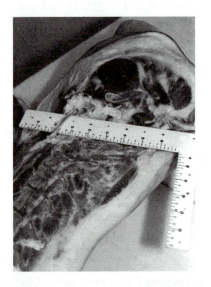

Fig. 14–7. If artery curing is anticipated (see Chapter 20), the fresh ham should be removed very carefully so that the *femoral* artery is not destroyed. Previous steps must also have been taken on the harvest floor to remove the kidney and pelvic fat with care so that the *femoral* artery was not destroyed. When the fresh ham is removed from the carcass, the *femoral* artery should not be cut too short.

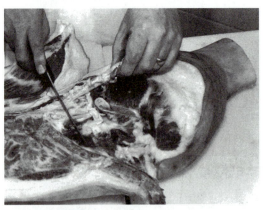

Fig. 14–8. The fresh ham is removed from the carcass at a point approximately 2½ inches cranially from (forward from) the tip of the aitch (pelvic) bone. The cut is made perpendicularly to the long axis of the pork leg and usually between the third and fourth *sacral* vertebrae.

After the bone has been severed with a saw, a knife is used to complete the removal of the fresh ham. The *lumbar* lean may serve as an indication of the overall leanness and muscling of intact carcasses.

If the flank is trimmed away from the anterior portion of the fresh ham before the separation is made, the flank is left on the belly or the side.

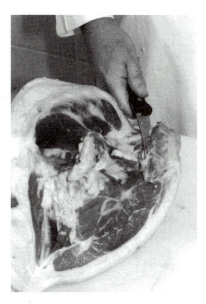

Fig. 14–9. The tail bone should be removed to further trim the fresh ham. Then, the flank side should be trimmed, and the lymph glands should be removed very carefully.

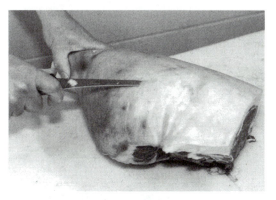

Fig. 14–10. The most common fresh ham is one that is skinned approximately three-fourths of the distance from the rump (butt) face. It is called a skinned fresh ham. The skin is removed, and the fat is beveled to a thin layer under the rump (butt) face of the fresh ham.

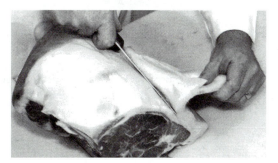

Fig. 14–11. The idea is to bevel the fat in a uniform manner so that the fat will measure approximately ¼ inch in thickness at the rump (butt) face. Different amounts of fat are trimmed, according to the market potential for the pork leg (fresh ham). All pork legs (fresh hams) are not trimmed this closely.

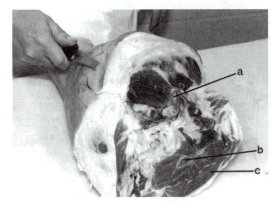

Fig. 14–12. If the fresh ham was not faced during the harvest process, the collar fat over the cushion area should be trimmed off. Note the location of the aitch, or pelvic, bone (a) as well as the shaft of the *ilium* (b), which is the round portion of the aitch bone. This round bone should not be confused with the round ham bone, or *femur,* which is located caudally to (to the rear of) the aitch bone. The large muscle in the face of every fresh ham is the *gluteus medius,* or top sirloin (c).

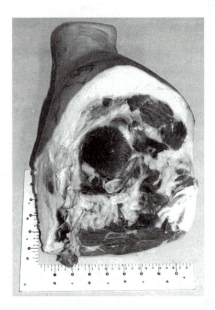

Fig. 14–13. A pork leg (fresh ham) (IMPS 402) is composed of a shank portion, a rump (butt) portion, and the center section and may be sold whole, either fresh or cured and smoked. The center section is the most valuable portion of the fresh ham; center slices are often removed for sale.

Few consumers care to purchase a whole fresh or smoked ham, except for very special occasions, because of its size and total cost. Also, the large pelvic bone present in a bone-in pork leg (fresh ham) or smoked ham may complicate carving and serving. Thus, more processors and packers are fabricating boneless legs (fresh hams), which not only prevents problems related to vacuum packaging bone-in cuts but also produces smaller, standard-weight packages.

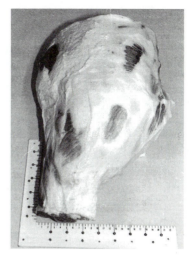

Fig. 14–14. Before a fresh ham is boned, the skin should be completely removed. The fat should be trimmed relatively close, as fat tends to be more evident in boneless meat cuts.

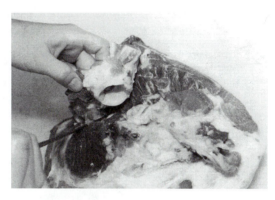

Fig. 14–15. The aitch, or pelvic, bone is removed from the rump (butt) portion of the pork leg (fresh ham). Note the socket that must be released from the *femur* ball. The *femur* is the large bone in the thigh that is often referred to as the "ham bone." When it is removed from both fresh and cured hams, it is an excellent source of flavoring for soups and sauces.

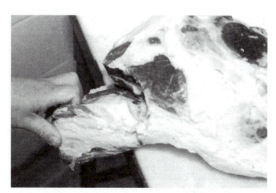

Fig. 14–16. Although the shank meat may be left on the boneless fresh ham by removing the meat from the shank before the bone is removed, a more desirable boneless fresh ham will result if the shank meat is taken off with the bone and then utilized as pork trimming in sausage. The shank could be used as a soup bone as well. Working the knife through the stifle joint between the *femur* (ham bone) and the *tibia* and *fibula* (shank bones) will remove the shank.

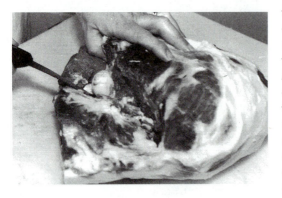

Fig. 14–17. Note the ball of the *femur* (ham bone), which has been exposed. Loosening the meat surrounding the *femur* without cutting to the outside is known as *tunnel boning.* Some processors are now completely opening up the ham at the natural seams in order to remove the fat deposits between muscles, which is necessary to produce the "90+ percent fat-free" cured and smoked boneless hams currently displayed in retail meat-cases.

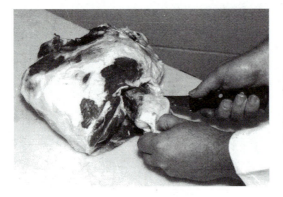

Fig. 14–18. Because the *femur* (ham bone) has been loosened from the rump (butt) end of the ham, it is a rather simple matter to remove the *femur* after the shank end has been loosened.

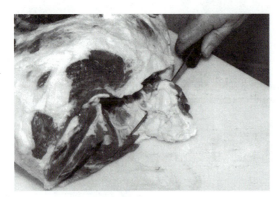

Fig. 14–19. The knee cap *(patella)* must be removed from the boneless fresh ham. This can be accomplished during removal of the *femur* or separately, as shown here.

Fig. 14–20. Removing the shank, as has been illustrated, makes it easier to locate and remove the *popliteal* lymph node (see Fig. 15–65) from the boneless fresh ham. It is located in the seam between the *semitendinosus* and *biceps femoris* muscles of the bottom or outside leg muscle system.

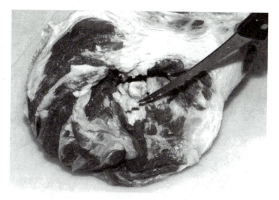

Fig. 14–21. The tunnel-boned ham has been separated from the pelvic (aitch), thigh *(femur)*, shank *(tibia and fibula)*, and knee cap *(patella)* bones as well as the *popliteal* lymph node and its surrounding fat deposit. The ham could be opened further to remove additional intermuscular (seam) fat.

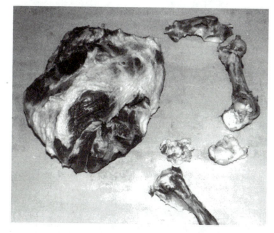

Fig. 14–22. Elastic net has made the tying of roasts with string and butcher's knots outdated. Using the netter makes tying a boneless roast a simple and quick operation. Boneless fresh hams completely separated into muscle systems for bone and fat removal can be restructured into sliceable, intact roasts by a technique known as *tumbling* or *massaging*. When muscle systems containing as much as 2 percent salt and 0.5 percent phosphate are subjected

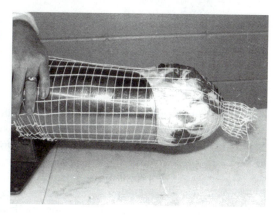

to physical impact, myosin (muscle fiber protein) is released, coating the surfaces and promoting the binding together of the different muscle systems. The coated muscles and pieces are tightly formed in a casing or netting and are bonded together during the cooking process (see Chapter 20).

Fig. 14–23. A boneless ham (IMPS 402C) may be somewhat difficult to identify; however, its main distinguishing feature is its size, for it is much larger than most other boneless pork cuts. A boneless whole shoulder may be similar in size but tends to be longer and less plump when netted. The boneless fresh ham on the right is accompanied here by a similarly prepared, cured and smoked boneless ham. The excess netting on the fresh ham will be used to hang the boneless meat in the smokehouse (see Chapter 20).

Fig. 14–24. Although most hams are boneless today, they may also be handled as bone-in roasts. Because of its large size, a bone-in ham, either cured and smoked or fresh (as shown), is often subdivided into the butt (rump) portion, center portion, and shank portion. When it is cut into two parts, the two parts are called rump (butt) half and shank half. The center portion is usually cut into slices (fresh or cured) if it is utilized separately.

The desirability of a center slice is evident, with only a small cross section of the ham bone *(femur)* present. (See also the color photos near the center of the text.) The shank portion is less desirable from a palatability standpoint, since there are more connective tissues in that lower portion of the ham. The rump (butt) portion is more palatable, but the large pelvic (aitch) bone present in this portion makes carving difficult. If the word *portion* is used in describing either the rump (butt) or the shank part, this indicates that part of the center slices have been removed. However, if the parts are designated as *shank half* or *rump (butt) half,* some of the center section must be included in the half the consumer buys.

Loin

The loin is usually the most valuable cut in the pork carcass. This lean primal cut comprises approximately 11 percent of the live weight of a hog and 16 percent of its carcass weight. Obviously, the more muscular, minimally finished hogs will have a considerably higher percent of their carcass or live weight represented by the fresh ham and the loin. The pork loin encompasses a longer area of the carcass than does the beef or lamb loin. It includes the *sacral, lumbar,* and a good portion of the *thoracic* areas of the vertebral column. Thus, because it includes the area of the ribs, a retail cut can properly be named a rib chop from the loin in pork. This is not possible in beef and lamb.

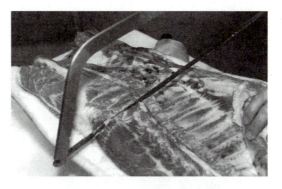

Fig. 14–25. The first step in removing the loin from the pork carcass is to separate it from the shoulder. This is often done by cutting across the third rib, or as far forward as the first rib.

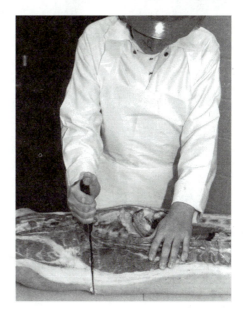

Fig. 14–26. After the bone has been cut with a saw, the final severance is made with a knife. Note that this separation is made at right angles to the long axis of the carcass.

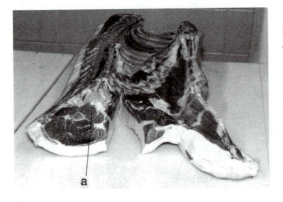

Fig. 14–27. The spareribs and belly are separated from the loin, which at this point still has the backfat remaining intact. Note in the skeletal diagrams (Figs. 14–1 and 14–4), the line that separates the loin from the spareribs. Orient yourself with the blade bone *(scapula)* on the cranial (front) end of the carcass and the sirloin end to the rear. The cut can be made with a saw or, as shown here, with a knife. The cut is made beginning from a point laterally to the tenderloin muscle (a) on the sirloin end, cranially to a point on the last rib 1 to 2 inches laterally to the tenderloin muscle. Some modification of these locations will occur with variations in cutting procedures.

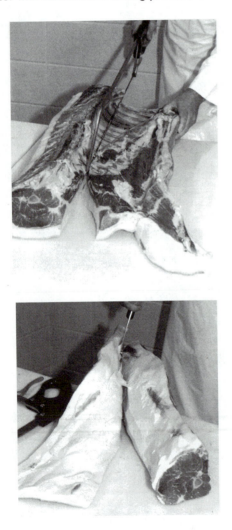

Fig. 14–28. Sawing through the ribs from the blade end necessitates placing the straight hand saw or the power saw against the third *thoracic* vertebra, with the distal (upper) end of the saw moving in the space resulting from the knife cut described in Fig. 14–27. After the ribs have been cut through, a knife is used to complete the separation of the loin and the belly. Processing plants use power rotary saws to cut through the ribs. In most plants, the loin is pulled from the belly and spareribs with a U-shaped loin knife after the saw cuts through the ribs.

Fig. 14–29. After the loin has been cut with the straight hand saw, the backfat is removed. Trimming the blade end of the loin down to the point where only ¼ inch of outside fat remains also exposes the false lean or *trapezius* muscle. The *false lean* is a small or thin muscle near the skin surface that may have larger muscles underlying it. This particular muscle must be exposed in order for proper trimming to be accomplished.

Fig. 14–30. When loins are removed under packing house conditions with a loin-pulling knife, sometimes the *longissimus* muscle is *scored;* that is, instead of removing only the fat, the knife penetrates into the loin muscle itself. This is very undesirable because it destroys a part of the most valuable edible portion of the carcass. Under the laboratory conditions shown in this fig-

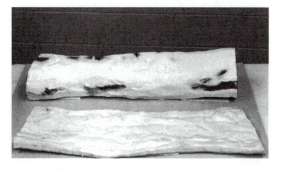

ure, the backfat is removed very carefully, without scoring the loin deeply, but exposing the muscle in a few places to demonstrate the close trim. The backfat is processed into very high-quality rendered lard. Some types of cooking utilize chunks of fresh or cured backfat, but the demand is small for such use.

Fig. 14–31. The trimmed pork loin (IMPS 410) has a higher-valued center portion and two slightly lower-valued ends. The ends include muscles of the leg (fresh ham) and shoulder systems as well as the *longissimus,* which makes up most of the musculature of the center loin.

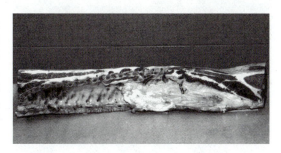

Fig. 14–32. Beginning at the cranial (front) portion of the loin, remove the *blade chops.* A blade chop is identified by the presence of the blade bone *(scapula)* and the attached cartilage.

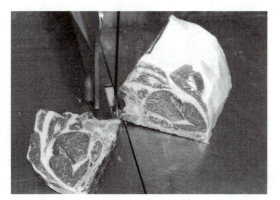

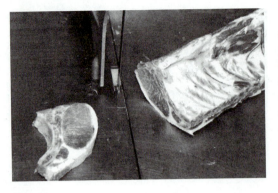

Fig. 14–33. This center-cut rib chop came from the tenth rib, but the area from the fifth rib through the last rib would be the source area for center-cut rib chops. Notice the prevalent use of the term *chop*, which originated when cleavers were used to cut bone-in meat retail cuts. Anything small enough to be chopped with a cleaver was designated a chop, while anything larger that had to be separated with a saw was called a *steak*. Thus, even though modern saws are used today to make chops, the terms *pork* and *lamb chops* but *beef steaks* remain.

Fig. 14–34. The center-cut loin chop in this example has been removed from the area of the seventh (rearmost) *lumbar* vertebra. All chops removed cranially to (forward from) this seventh *lumbar* vertebra up to the last rib are center-cut loin chops, identified by the presence of the tenderloin muscle and by the absence of a rib. All chops removed from the *lumbar* area of the vertebral column would be center-cut loin chops.

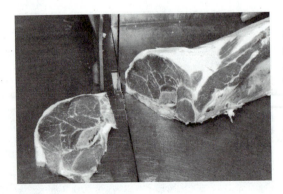

Fig. 14–35. From the caudal (rear) end of the loin, the rearmost chop removed is a *sirloin chop*. This chop is removed from the face, which was directly connected to the rump (butt) face of the ham. Sirloin chops contain the eye (*longissimus*) and the tenderloin (*psoas*), the top sirloin (*gluteus*) muscles, and the *sacral* vertebrae and portions of the hipbone (*ilium*).

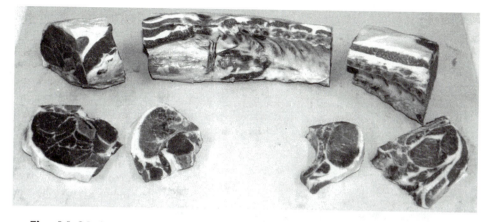

Fig. 14–36. In review, the component retail cuts from the bone-in loin are *(front row, left to right)*: a sirloin chop, a center-cut loin chop, a center-cut rib chop, and a blade chop. The pork loin chop contains a bonus, the tenderloin, which is indeed a tasty morsel! The sirloin, center, and blade roasts *(back row, left to right)* from the loin are also very desirable cuts.

Fig. 14–37. Pocket chops are an innovative modification of pork chops. Bone-in pork chops are cut approximately ¾ to 1 inch thick, but pocket chops are cut slightly thicker. Immediately adjacent to the rib, a slit is made in the loin eye muscle, which can be filled with dressing. The rib chop is most appropriate for a pocket chop, since the slit can be made under the rib and is hardly noticeable when the chop is served.

The pork loin may be boned, a merchandising technique that is gaining in popularity (see Chapter 13). Boning the main portion of the loin does not normally present a problem if the tenderloin (composed of the *psoas major* and *psoas minor*) is carefully removed first. The small *psoas minor* is detached from the *psoas major* for pork trim. From the blade end, taking out the remaining *scapular* cartilage is easy if the knife is kept flat and close to the cartilage. The backbone and the remaining portion of the ribs can be removed with the same technique. If the thoracic portion of the backbone is separated from the lumbar area and the chine (body) and feather bones (dorsal processes) are detached from the ribs with a saw, back ribs are produced. The back ribs, particularly if some meat is left on them when the loin is boned, are an excellent source for barbecued pork ribs.

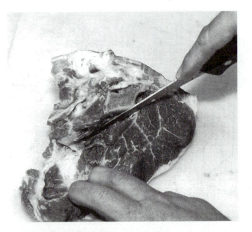

Fig. 14–38. The one somewhat difficult portion of the loin to bone is the sirloin end because it contains a portion of the aitch (pelvic) bone. Start by entering from the ham side and staying close to the hipbone. Continuing close to the bone, free it from the surrounding muscle. It can be separated from the backbone through the *sacroiliac* joint, and each can be removed separately or they can be removed together as one bone.

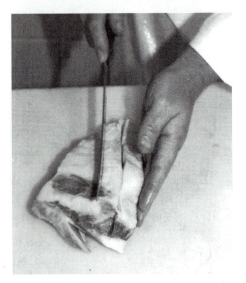

Fig. 14–39. The butterfly chop is a popular chop made from the boneless loin. This is, in reality, a double chop, since a normal (¾-inch ±) thickness cut is made but not completed all the way to the cutting table surface. A second normal thickness cut severs the double-width portion from the loin.

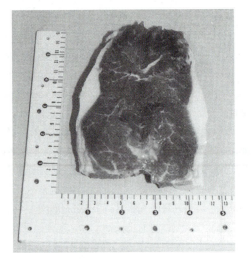

Fig. 14–40. The double chop is folded open on the first cut to become a very desirable boneless butterfly chop.

Fig. 14–41. The blade end of the bone-in pork loin can be fabricated into meaty country-style ribs by cutting through the rib bones laterally to the vertebrae, as shown, and then continuing back through the muscle but not completely through the roast.

Fig. 14–42. Cutting back under the rib ends and the backbone to form an M-shaped cut opens up the roast.

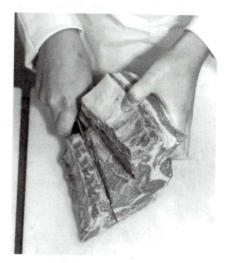

Fig. 14–43. The resulting country-style ribs produce some tender, meaty portions for barbecuing.

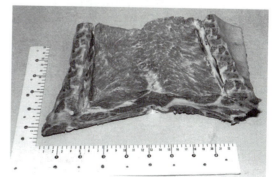

Fig. 14-44. The blade portion of the pork loin may be boned to produce a boneless roast and back ribs. The roast can be cut into butterfly chops, tied or netted with another boneless blade or sirloin roast to produce a larger double roast, or used as a single small pork roast. The *pleural* membrane, a thin sheet of connective tissue lining the thoracic cavity, should be removed from the back ribs to enhance palatability.

In summary, there are a multitude of retail cuts from the pork loin (see Figure 14–36). The most forward portion of the loin is termed *blade end*. From the blade end, the blade chops may be removed, or the blade end may be boned, and the result is a boneless blade-end roast. The roast may be cut into boneless chops, which may be single or doubled by butterflying them (see Figure 14–39). The blade end may also be utilized for meaty country-style ribs.

The most variety arises from the center section of the loin. The center itself may be utilized as a roast, but more often it is cut into center-cut loin chops, and center-cut rib chops. The rib chops may also be made into pocket chops (see Figure 14–37). The boneless center loin may be made into butterfly chops or thick-cut single, boneless chops, or it may simply be left as a boneless loin roast. The boneless loin from large packing sows, those weighing over 360 pounds alive, is oftentimes cured and utilized as Canadian bacon. When the loin is boned out, a portion of the bone that is removed (back ribs) is very desirable for barbecuing. The center section, as well as the whole pork loin, may be cured and smoked to be cut into cured and smoked pork chops, generally bone-in. A thick (1½ inches plus), smoked pork chop, properly prepared, is elegant eating at its best! *Windsor loins* or *Windsor loin chops* is the name sometimes given to cured and smoked loins and chops.

Finally, the sirloin end may be utilized intact as a bone-in roast, or it may be cut into chops. More desirably, it is boned, and two of the boneless sirloins are netted or tied together as a boneless sirloin roast.

Spareribs and Belly

When loins are pulled, with a U-shaped draw-knife, from the midsection of the pork carcasses on the packer cutting line, the backfat (sometimes called *fatback*), bellies, and spareribs remain on the moving table. Meat cutters separate the spareribs from the belly with a single pass of a straight draw-knife and cut between the backfat and the belly to complete the division of that part of the carcass. The technique may differ when a straight saw and a conventional knife are used to remove the loin (see Figures 14–27 to 14–30), but the cuts, belly, and spareribs, will be the same. The length of the ribs and the width of the belly are determined by the location of the cut separating the loin. Modern cutting methods dictate that the largest portion of the rib cage be left on the spareribs. Actually, spareribs sometimes exceed loin cuts in retail price. This may be related to supply and demand, since loins account for approximately 16 percent of the carcass weight, while spareribs account for only 3 percent of the carcass weight. The belly, one of the primal cuts, comprises 9 percent of the live weight and approximately 13 percent of the carcass weight.

Fig. 14–45. When the spareribs are removed from the belly, a small amount of meat should be left on them, thus, the name *spareribs*. All portions of the bone and cartilage, including the *sternum*, must be removed with the spareribs, since these make very unpalatable bacon.

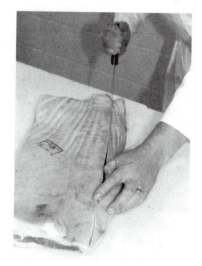

Fig. 14–46. Once the spareribs have been removed, all that remains to be done to the belly is to square it up. First of all, cutting through the center of the length of the flank pocket will square up the flank end. The flank side of the belly can be cut 1 inch longer than the loin side due to differential shrinkage of the *rectus abdominous* muscle that occurs during the smoking process. The loin side is trimmed, and then the teat line is removed so that no rudimentary mammary glands are left in the bacon slab. Because less expensive bacon is not trimmed as closely, these mammary "seeds" may be found in such bacon (hence, the term *seedy bacon*). Note the state inspection stamp on the skin surface of the belly.

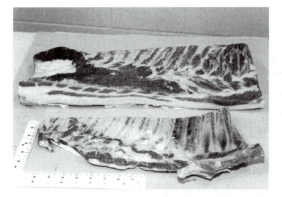

Fig. 14–47. The most common way to use the belly (skin on, IMPS 408; skinless, 409) is to cure and smoke it, which results in a cured and smoked bacon slab. The most popular form of bacon is, of course, sliced bacon. Some people enjoy fresh pork belly or fresh side pork, which also may be sliced. Others prefer the belly dry cured or salted, but unsmoked, which is called salt pork. Most bellies are skinned before being cured and smoked, but they may be processed and sold as rind on sliced bacon, with the skin still attached.

The spareribs (IMPS 416) may also be cured and smoked, but they are more often used fresh and barbecued. For best results, the spareribs are cut into individual portions, each containing two to six ribs, and cooked in liquid prior to the final heating on a grill. Spareribs are in large demand in many parts of the United States and are a favorite picnic item.

Shoulder

The shoulder area of the pork carcass, that area anterior to the third rib, is composed of two major primal cuts and three minor primal cuts. As stated earlier, the primal cuts are often retail cuts as well. Classified as primal cuts and also as lean cuts are the pork shoulder Boston butt (IMPS 406) and the

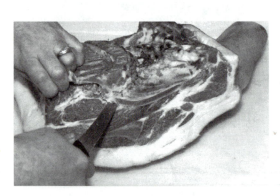

Fig. 14–48. The first step in fabricating the rough shoulder is to remove the neck bones. (By noting the anatomy drawing at the beginning of each fabrication chapter [Figure 14–1, 15–1, 16–1, or 17–1], you will see that all the species have seven neck vertebrae regardless of the length of neck. This is true of all mammals.[6]) The neck bones, which may include one, two, or three ribs, should be removed with as little lean as possible, but at the same time leaving no cartilage or bone chips in the shoulder.

---

[6]News Focus. 1999. Development shapes evolution. Science. 285:518. http://www.sciencemag.org

pork shoulder picnic (IMPS 405). The skeleton plays a very important role in identifying retail cuts from the Boston butt (Boston shoulder, blade Boston) and the picnic, as indicated in Figures 14–1 and 14–4. The three minor primal cuts originating from the anterior portion of the pork carcass are the neck bones, the jowl, and the clear plate. The pork shoulder Boston butt is nearly equal in weight to the pork shoulder picnic, and each of them represents roughly 6 percent of the live weight or 9 percent of the carcass weight of a market hog.

Fig. 14–49. Neck bones (IMPS 421) make somewhat desirable, but less expensive, "spareribs." Utilized as such, they can be very delicious barbecued. Some packers cure and smoke neck bones. They are an excellent source of stock for soups and sauces as well as for flavoring pieces in vegetables.

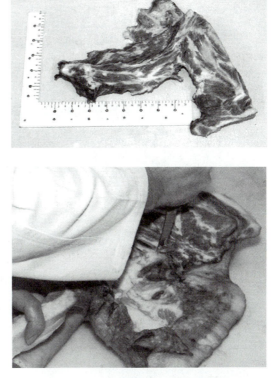

Fig. 14–50. Removing the jowl begins with a cut at the fat collar immediately above the foreshank and continues straight across the cranial portion of the shoulder parallel with the cut that separated the shoulder from the loin, belly, and spareribs. The cut should be made posterior to the "notch" that was made from the cut behind the ear when the head was removed at harvest.

Fig. 14–51. The rough-cut jowl, after removal, is trimmed into the square-cut jowl. The square-cut jowl is utilized as cured and smoked jowl bacon squares. Most jowls are utilized in sausage and loaf manufacture.

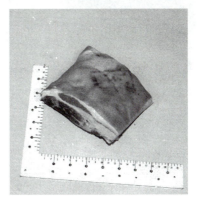

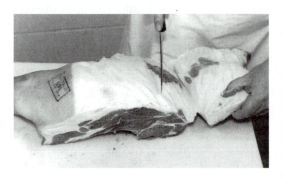

Fig. 14–52. The clear plate, a fat cut much like the backfat, is removed from the shoulder. The remaining skin should not cover more than 25 percent of the area between the elbow and the butt edge. An alternate method is to remove the clear plate from the pork shoulder Boston butt (Boston shoulder, blade Boston) after the lean cuts have been separated. The false lean of the *trapezius* muscle should be exposed (see Fig. 14–29).

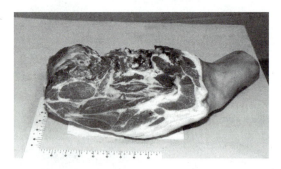

Fig. 14–53. The remaining portion, after the removal of the clear plate, neck bones, and jowl, is called the fresh pork shoulder, skinned (IMPS 404), or New York shoulder. In this form, it is a wholesale cut. Although it may be cured and smoked or used fresh as a retail roast, the whole shoulder is a larger cut than most of today's households can use efficiently. The extensive bone distribution (see Figs. 14–1 and 14–4) makes the shoulder a very difficult cut to carve.

Fig. 14–54. The shoulder is composed of two other wholesale cuts — the pork shoulder Boston butt (blade Boston or Boston shoulder) and the pork shoulder picnic. Cutting at right angles to the long axis of the shoulder at a distance approximately 1 inch below the exposed surface of the blade bone will separate the Boston butt from the picnic. (Figs. 14–1 and 14–4 show the bone structures that must be severed to make the separation.) A cut is made first with a knife, then with a saw to sever the bone.

Fig. 14–55. The pork shoulder Boston butt (Boston shoulder, blade Boston, IMPS 406) is both a wholesale cut and a retail cut. Note that the blade bone is now exposed on two adjacent surfaces of the Boston butt. The Boston butt can be utilized as a bone-in roast or cut into blade steaks. It makes an excellent boneless roast, which can be used fresh or cured and smoked.

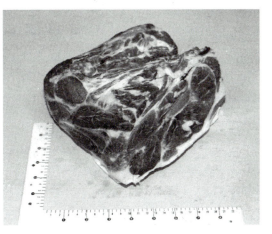

Fig. 14–56. The most popular retail cut of the Boston butt is the pork shoulder *blade steak,* which contains the blade bone *(scapula)* (a). A saw cut is made parallel to the face. Observe also the extension of the rib eye or the *longissimus* muscle (b). Since this is a larger portion than a rib or blade chop and must be removed with a saw, the term *steak* is used.

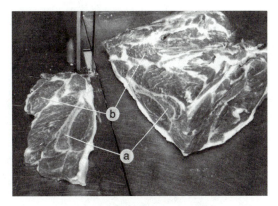

Fig. 14–57. The blade chop from the loin is very easily confused with the blade steak from the Boston butt. The key to identifying them is the presence of a rib and vertebra on the blade chop from the loin. Since the neck bones and ribs are removed in the fabrication of a pork shoulder, they cannot be present on a blade steak. Although some variation exists as to where the shoulder and loin are separated, there will usually be a large amount of cartilage in the blade chop from the loin. This results because the chop is removed farther down on the blade bone *(scapula),* where the cartilage is more pronounced. The blade bone *(scapula)* in the blade steak is more prominent, and a very small portion of cartilage is present. Also, the overall size of the blade steak is usually greater than that of the blade chop.

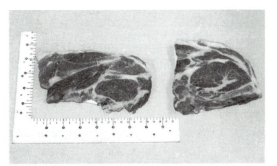

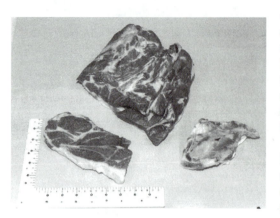

Fig. 14–58. Of all the boning operations, perhaps boning the pork shoulder Boston butt (Boston shoulder or blade Boston) is the easiest. The one bone in the Boston butt is the blade bone *(scapula)*. On the distal side is the spine of the blade bone *(scapula)* protruding from the broad, flat area of the bone.

The boneless Boston butt is highly desirable, especially when cured and smoked. In this form it often rivals the boneless cured and smoked ham and will generally be sold at a lower price.

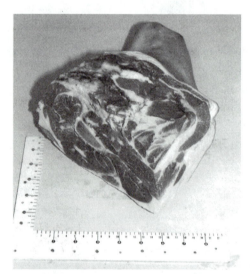

Fig. 14–59. The remaining portion of the shoulder is the other lean cut called the pork shoulder picnic (IMPS 405). In some parts of the country, it is called the *cala* because this type of cut originated in California. Most fabricators remove approximately one-third of the skin, if the clear plate was not removed earlier. The fresh picnic shoulder also doubles as a wholesale cut and a retail cut.

Fig. 14–60. The shoulder hock (IMPS 417) may be removed from the picnic shoulder. The hock is a separate retail cut that may be sold fresh, or it may be cured and smoked. Often it is boned and used in sausage production.

Fig. 14–61. The remainder of the picnic shoulder may be cut into arm roasts or arm steaks. Pork arm steaks are usually removed from the center of the picnic, with the roasts being the portions left at both ends. Many times the picnic will be sold only as a roast, either fresh or cured and smoked. The picnic shoulder is located lower on the live animal where smaller muscles, held together with large amounts of connective tissue, are required to work more as the animal moves about. The more that muscles are used in the live animal, the less palatable they are.

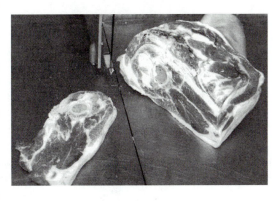

Fig. 14–62. A major portion of the bones in the shoulder are located in the picnic. The bones are the foreshank bone, which is the *radius* and *ulna* fused together, and the arm bone *(humerus)*. The shank has been removed in this illustration, since it is relatively high in connective tissue. The arm bone *(humerus)* is then removed. Note its size. Tunnel boning is not done here, but rather a cut is made from the outside to the bones. Because of the large amount of bone and connective tissue in a wholesale picnic, as contrasted with the Boston butt, the picnic has less value.

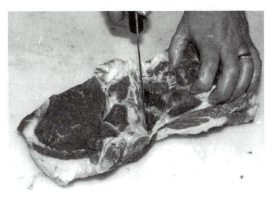

Live and carcass yields of the four lean cuts — the pork leg (fresh ham); loin; pork shoulder Boston butt (blade Boston, Boston shoulder); and pork shoulder picnic; and the primal cuts — the lean cuts plus belly; of the carcass illustrated in Figure 14–65 are summarized in Table 14–1. Adding 16 and 11 percent for the fresh ham and loin, respectively, for the live animal or 22 and 16 percent for the carcass will result in fresh ham and loin percentages of 27 and 38 percent, respectively, for the live weight and the carcass weight.

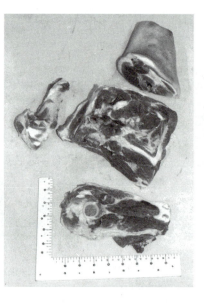

Fig. 14–63. Because of the large proportion of bone (the arm and foreshank bones) in the picnic shoulder, boning the picnic results in much more convenience for the server than boning the Boston butt, which contains only the blade bone. Often picnics are boned out to obtain high-quality sausage material.

Note the intermuscular connective tissue and fat seams in the arm steak. The picnic shoulder, in fresh or cured and smoked form, is a retail cut. When cured and smoked, the picnic gains its symmetrical shape from being hung in a stockinet as it is heated and smoked in the smokehouse. The shoulder hock may be removed from either the fresh picnic or the cured, smoked picnic. The arm steak, arm roast, and boneless picnic may also be utilized as cured and smoked product.

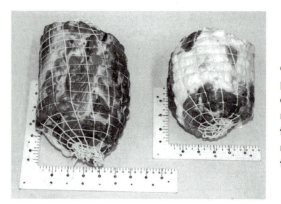

Fig. 14–64. The boneless, rolled, and netted Boston butt *(left)* and the picnic shoulder *(right)* shown here are both compact, uniformly shaped roasts that cook well in either the fresh or the cured form. They are much easier to slice and serve than the bone-in versions.

The spareribs would be foremost in value among the minor cuts. Lean trim, a by-product of carcass fabrication, as well as the jowl, neck bones, and feet are included in the minor cuts. Pork trimmings, both lean and fat, fulfill a vital role in the fabrication of processed meats, such as the multitude of sausage products, sandwich loaves, and frankfurters (see Chapter 21). The neck bones serve as a substitute for spareribs, and the jowl is utilized as an inexpensive form of slab bacon or sausage material. Fat trim is generally processed into lard.

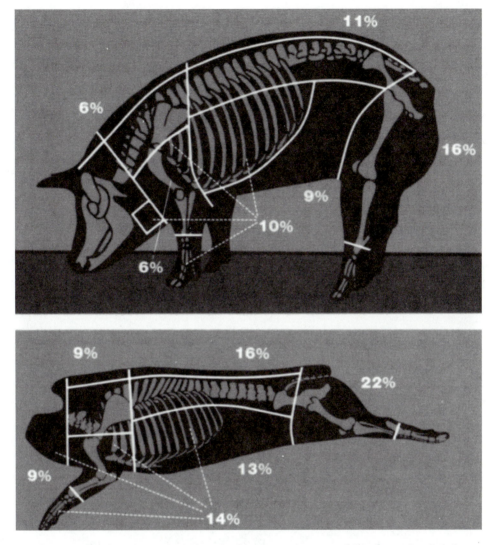

Fig. 14–65. The percentages listed adjacent to the cuts of the live animal and carcass represent the percentage of live and carcass weight, respectively, found in each trimmed wholesale cut. It must be stressed that these are average figures taken from a so-called average hog. This hog would probably have approximately 1.25 inches of fat at the last rib, be approximately 30 inches long, and have a loin eye area of approximately 4.5 square inches.

Yield data vary with differences in carcasses, cutting procedures, and other factors. Table 14–2 demonstrates the variation in pork carcass yield data as derived from only three sources. Each source has reported the information as it has acquired it. None of the data are in error, but they do differ.

Table 14–1. Percent of Live Hog and Pork Carcass Weights in Primal Cuts, Lean Cuts, Ham and Loin, Fat Trim, and Minor Cuts

Portion of Carcass	Live Weight	Carcass Weight
	- - - - - - - - - (%) - - - - - - - - -	
Five primal cuts	48	69
Four lean cuts	39	56
Ham and loin	27	38
Fat trim	11	16
Minor cuts	10	14

Table 14–2. Percent of Pork Carcass Weight in Pork Cuts

Cut	Source of Data		
	NLSMB[1]	AMI[2]	MWE[3]
	- - - - - - - - - - - - - - - (%) - - - - - - - - - - - - - - -		
Trimmed ham + trim	21.8	17.4	22
Trimmed loin	14.2	17.7	16
Bacon side + trim	13.2	16.6	13
Spareribs	3.1	4.0	3
Boston + trim	8.9	7.1	9
Picnic + trim	6.3	7.6	9
Jowls, feet, trim, etc.	11.0	10.6	11
Retail pork	79.5	81.0	83
Fat and bone waste	20.5	19.0	17

[1]National Live Stock and Meat Board. A Hog's Not All Chops.
[2]American Meat Institute. Meat Facts 1991.
[3]This text.

EVALUATION OF CARCASS MERIT AND PRODUCTION EFFICIENCY

The modern meaty hog is a result of selection and management practices that have developed over the past half-century or more. Pork carcass evaluation procedures that use the types of information listed in Table 14–1 have contributed to the production of high-quality lean pork carcasses. Pork carcass evaluation procedures utilizing percent primal or lean cuts or percent fresh ham and loin are time-consuming processes because the carcass must be cut before a comparison can be made. An evaluation procedure that results in the actual weights of salable product is the most accurate method of evaluation, if cutting procedures are standardized. However, in many instances, it is important to be able to evaluate the merits of a carcass without the expenditure of time and effort required to cut it into wholesale cuts. In commercial processing plants that cut 1,000 pork carcasses or more per hour, maintaining carcass identity on each cut and obtaining accurate cut weights from even a small number of carcasses are difficult at best and normally impossible.

In some cases, it is possible to obtain data on pork carcasses in a packing plant cooler when cut weights are not available. A popular method of pork carcass evaluation that uses carcass data is recommended and published by the National Pork Producers Council (NPPC). Sources used in formulating the NPPC procedures include research reported by Orcutt and coworkers.[7] In *Composition and Quality Assessment Procedures,*[8] both the carcass merit and the production efficiency of the animal are emphasized. Each hog and its carcass are evaluated and may be compared with other hogs on the basis of "pounds of acceptable quality lean pork gain per day of age or per day on test." The data necessary for this can be derived from the ribbed carcass in Figure 14–67.

NPPC procedures are the basis of various methods of predicting carcass composition.

---

[7]Orcutt, M. W., J. C. Forrest, M. D. Judge, A. P. Schinckel, and C. H. Kuei. 1990. Practical means for estimating pork carcass composition. J. Anim. Sci. 68:3987.

[8]National Pork Producers Council (NPPC). 1999. Composition and Quality Assessment Procedures (provided by Dr. Eric Berg, University of Missouri, and the committee).

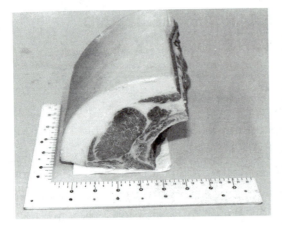

Fig. 14–66. Pork carcass evaluation often involves measurement of the size of the loin eye muscle, that is, the *longissimus*. This is done by breaking (cutting) the loin between the tenth and eleventh ribs at right angles to the backbone. If possible, this should be done before the backfat has been removed. First, some pork carcass evaluation procedures utilize a fat depth or thickness measurement (usually in tenths of inches) taken at three-fourths of the length of the eye muscle on this tenth rib eye surface. Second, if the backfat is intact, the loin eye muscle maintains its natural size and shape, while a grid is used or a tracing is made to determine the area (square inches). In some packing house situations, it is necessary to obtain the loin eye area by using grids or making tracings after the backfat has been removed from the loin. Care must be taken to prevent distortion when the supporting fat has been removed. It is important that the loin be cut perpendicularly to its axis to assure accurate measurement of the muscle area. Normally, pork loins are not broken at the tenth rib in regular packing house operations.

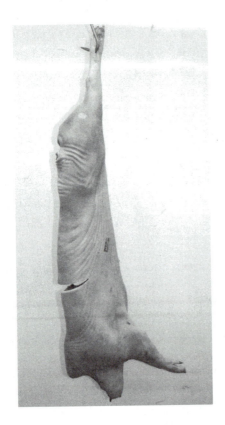

Fig. 14–67. Some pork carcass evaluators (NPPC recommended procedures) break or rib the loin while the carcass is hanging, somewhat similar to the procedure in ribbing a beef carcass (see Chapter 15). When using this technique, avoid cutting into the belly and damaging it. Maintain a cut across the loin muscle exactly perpendicular to its long axis, thus measuring its true size and assuring an accurate fat thickness measurement. This technique permits data identified in Fig. 14–66 to be obtained in the cooler before the carcass is fabricated and the backfat removed from the surface of the loin. Other data including carcass weight, backfat thickness, and carcass grade parameters (see Chapter 13) can be recorded at the same time.

Prediction Equations for Fat-Free Lean[9]

Equations to Predict Composition

The most accurate prediction equations are derived from ribbed carcasses. Ribbing of carcasses is preferred for evaluation of loin quality and overall quality. However, in a commercial setting, ribbing carcasses at the 10th rib is discouraged because of (1) safety concerns, (2) product devaluation, and (3) additional production costs associated with product handling procedures. Equations are provided for when it is not possible to rib carcasses. These equations include live animal real-time ultrasound, last rib carcass fat thickness, Fat-O-Meater, and Animal Ultrasound Services Carcass Value Technology.

1. **For live hogs using real-time ultrasound**

Equation for Lbs. of Fat-Free Lean (FFL)	Example	
1.436		1.436
+ 0.736 × sex of pig (barrow = 1 and gilt = 2)	0.763 × 1	+ 0.763
− 14.784 × 10th rib fat depth, in.	14.784 × 0.9 in.	− 13.306
+ 4.957 × 10th rib loin muscle area, in.^{2}	4.957 × 6.8 in.^{2}	+ 33.708
+ 0.331 × live wt., lbs.	0.331 × 255	+ 84.405
	Total lbs. of FFL	**= 107.01**
To convert to % fat-free lean on a live weight basis, divide by live weight and multiply by 100.	(107.01 / 255) × 100	
	% FFL on live wt. basis	**= 41.96**
To convert to % fat-free lean on a carcass weight basis, divide the % FFL (live wt. basis) by 0.74.	(41.96 / 0.74)	
	% FFL on carcass wt. basis	**= 56.70**

[9]Ibid.

2. For ribbed carcasses

Equation for Lbs. of Fat-Free Lean (FFL)		Example
11.448		11.448
− 19.926 × 10th rib fat depth, in.	19.926 × 0.9 in.	− 17.933
+ 2.593 × 10th rib loin muscle area, in.²	2.593 × 6.8	+ 17.632
+ 0.511 × adj. warm carcass wt.,* lbs.	0.511 × 188.7	+ 96.426
	Total lbs. of FFL = 107.57	
To convert to % fat-free lean, divide by adj. warm carcass weight and multiply by 100.	(107.57 / 188.7) × 100	
	% FFL = 57.01	

*Warm carcass weight is adjusted by the evaluator to account for skinning and trimming.

3. For unribbed carcasses using 10th rib loin muscle area and fat depth derived from real-time ultrasound and warm carcass weight

a. Sex of pig is known

Equation for Lbs. of Fat-Free Lean (FFL)		Example
10.548		10.548
+ 0.962 × sex of pig (barrow = 1 and gilt = 2)	0.962 × 1	+ 0.962
− 16.824 × scan 10th rib fat depth, in.	16.824 × 0.9 in.	− 15.142
+ 3.800 × scan 10th rib loin muscle area, in.²	3.800 × 6.8	+ 25.840
+ 0.451 × adj. warm carcass wt.,* lbs.	0.451 × 188.7	+ 85.104
	Total lbs. of FFL = 107.31	
To convert ot % fat-free lean, divide by adj. warm carcass weight and multiply by 100.	(107.31 / 188.7) × 100	
	% FFL = 56.87	

*Warm carcass weight is adjusted by the evaluator in account for skinning and trimming

b. *Sex of pig is not known or cannot be determined accurately (See "Classes of Pork Carcasses" in Chapter 12 and Figure 12–1.)*

Equation for Lbs. of Fat-Free Lean (FFL)	Example	
10.548		10.548
− 17.083 × scan 10th rib fat depth, in.	17.083 × 0.9 in.	− 15.375
+ 3.854 × scan 10th rib loin muscle area, in.^{2}	3.854 × 6.8	+ 26.207
+ 0.451 × adj. warm carcass wt.,* lbs.	0.451 × 188.7	+ 85.104
	Total lbs. of FFL = 106.48	
To convert to % fat-free lean, divide by adj. warm carcass weight and multiply by 100.	(106.48 / 188.7) × 100	
	% FFL = 56.43	

*Warm carcass weight is adjusted by the evaluator to account for skinning and trimming.

4. **For unribbed carcasses using last rib backfat thickness measured with a stainless steel ruler**

Equation for Lbs. of Fat-Free Lean (FFL)	Example	
24.368		24.368
− 20.049 × last rib backfat thickness, in.	20.049 × 1.10	− 22.054
+ 0.548 × adj. warm carcass wt.,* lbs.	0.548 × 188.7	+ 103.408
	Total lbs. of FFL = 105.72	
To convert to % fat-free lean, divide by adj. warm carcass weight and multiply by 100.	(105.72 / 188.7) × 100	
	% FFL = 56.03	

*Warm carcass weight is adjusted by the evaluator to account for skinning and trimming.

5. **For unribbed carcasses using the Fat-O-Meater optical grading probe to obtain fat and muscle depth between the 3rd- and the 4th-from-last rib**

Equation for Lbs. of Fat-Free Lean (FFL)	Example	
17.267		17.267
− 27.934 × fat depth, in.	27.934 × 0.91	− 25.420
+ 3.547 × loin muscle depth, in.	3.547 × 2.36	+ 8.371
+ 0.545 × adj. warm carcass wt.,* lbs.	0.545 × 188.7	+ 102.842
	Total lbs. of FFL = 103.06	
To convert to % fat-free lean, divide by adj. warm carcass weight and multiply by 100.	(103.06 / 188.7) × 100	
	% FFL = 54.62	

*Warm carcass weight is adjusted by the evaluator to account for skinning and trimming.

6. **For unribbed carcasses using the Animal Ultrasound Services Carcass Value Technology to obtain the average fat and muscle depth spanning the last rib to the 10th rib, parallel to the midline of the split carcass**

Equation for Lbs. of Fat-Free Lean (FFL)	Example	
11.839		11.839
− 15.097 × avg. fat depth, in.	15.097 × 0.91	− 13.738
+ 2.763 × avg. loin muscle depth, in.	2.763 × 2.36	+ 6.521
+ 0.514 × warm carcass wt.,* lbs.	0.514 × 188.7	+ 96.992
	Total lbs. of FFL = 101.61	
To convert to % fat-free lean, divide by adj. warm carcass weight and multiply by 100.	(101.61 / 188.7) × 100	
	% FFL = 53.85	

*Warm carcass weight is adjusted by the evaluator to account for skinning and trimming.

15

Beef Identification
and Fabrication

The beef animal is most complex. There are many types and breeds of cattle. Most countries have grading systems to categorize cattle into groups with similar attributes that consumers desire. But even with a reasonably accurate system of grading, there are large differences in the palatability of different cuts (joints) of beef carcasses within a grade. And there are many different cuts in every beef carcass. Thus, readers should learn the cutting methods that will produce cuts having the greatest potential for consumer demand and satisfaction.

IDENTIFICATION

The Uniform Retail Meat Identity Standards[1] and the Institutional Meat Purchase Specifications[2] (see Chapter 13) are used as the basis for identification in this chapter. Be sure to check the last paragraph in Chapter 13 for a description of *The Guide to Identifying Meat Cuts,* an excellent meat identification reference.

FABRICATION

The skeletal structure of the bovine in relation to an outline of the wholesale cuts is shown in Figure 15–1, and a more detailed chart of wholesale and retail cuts and recommended cooking methods appears as Figure 15–2. These

---

[1]Uniform Retail Meat Identity Standards. 1973. National Live Stock and Meat Board, Chicago (now National Cattlemen's Beef Association, Denver).

[2]Institutional Meat Purchase Specifications. 1987. USDA–AMS, Washington, DC.

BEEF CHART
LOCATION, STRUCTURE AND NAMES OF BONES

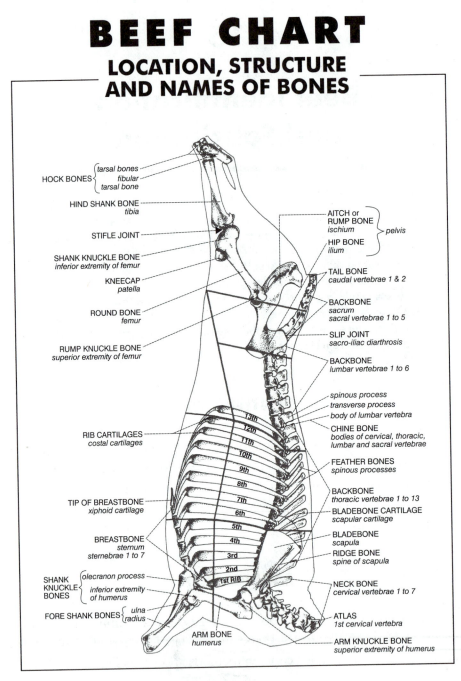

HOCK BONES { *tarsal bones*
fibular
tarsal bone }

HIND SHANK BONE
tibia

STIFLE JOINT

SHANK KNUCKLE BONE
inferior extremity of femur

KNEECAP
patella

ROUND BONE
femur

RUMP KNUCKLE BONE
superior extremity of femur

RIB CARTILAGES
costal cartilages

TIP OF BREASTBONE
xiphoid cartilage

BREASTBONE
sternum
sternebrae 1 to 7

SHANK
KNUCKLE {
BONES

olecranon process
inferior extremity
of humerus

FORE SHANK BONES { *ulna*
radius }

ARM BONE
humerus

AITCH or
RUMP BONE
ischium } *pelvis*

HIP BONE
ilium

TAIL BONE
caudal vertebrae 1 & 2

BACKBONE
sacrum
sacral vertebrae 1 to 5

SLIP JOINT
sacro-iliac diarthrosis

BACKBONE
lumbar vertebrae 1 to 6

spinous process
transverse process
body of lumbar vertebra

CHINE BONE
bodies of cervical, thoracic,
lumbar and sacral vertebrae

FEATHER BONES
spinous processes

BACKBONE
thoracic vertebrae 1 to 13

BLADEBONE CARTILAGE
scapular cartilage

BLADEBONE
scapula

RIDGE BONE
spine of scapula

NECK BONE
cervical vertebrae 1 to 7

ATLAS
1st cervical vertebra

ARM KNUCKLE BONE
superior extremity of humerus

13th
12th
11th
10th
9th
8th
7th
6th
5th
4th
3rd
2nd
1st RIB

Fig. 15–1. Bovine anatomy.

BEEF CHART

RETAIL CUTS OF BEEF — WHERE THEY COME FROM AND HOW TO COOK THEM

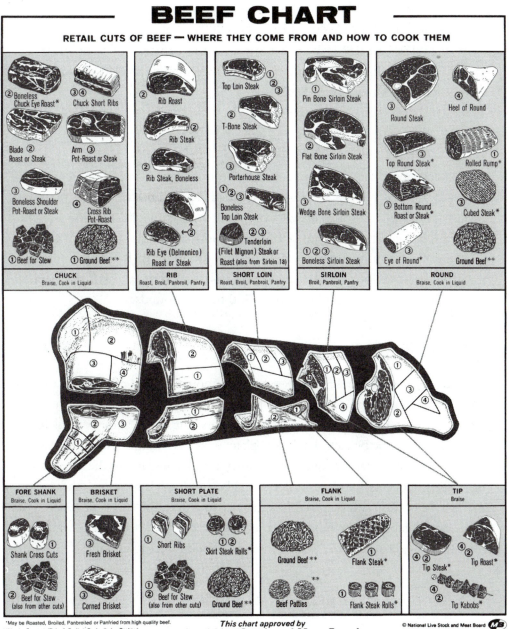

CHUCK
Braise, Cook in Liquid

RIB
Roast, Broil, Panbroil, Panfry

SHORT LOIN
Roast, Broil, Panbroil, Panfry

SIRLOIN
Broil, Panbroil, Panfry

ROUND
Braise, Cook in Liquid

FORE SHANK
Braise, Cook in Liquid

BRISKET
Braise, Cook in Liquid

SHORT PLATE
Braise, Cook in Liquid

FLANK
Braise, Cook in Liquid

TIP
Braise

*May be Roasted, Broiled, Panbroiled or Panfried from high quality beef.
**May be Roasted, (Baked), Broiled, Panbroiled or Panfried.

This chart approved by
National Live Stock and Meat Board

© National Live Stock and Meat Board

Fig. 15–2.

two figures can be referred to as the discussion moves through cutting a beef carcass. For more complete muscle identification and nomenclature, see *A Cross Sectional Muscle Nomenclature of the Beef Carcass*.[3] The center section of this text includes color photos of beef retail cuts.

Figures 15–3 to 15–105 appear through the courtesy of South Dakota State University. Figures 15–106 to 15–112 were made available by the IBP Corp., Dakota Dunes, South Dakota.

The First Steps

For demonstration purposes let us assume that generally, when beef animals are finished for market, they will weigh approximately 1,100 pounds. After harvest, the resulting carcasses represent approximately 60 to 63 percent of their live weight. Therefore, a 1,100-pound live animal will have a carcass (IMPS 100) weighing approximately 660 to 700 pounds. Before being taken from the harvest floor, beef carcasses are split into right and left sides, each side (IMPS 101) then weighing about 330 to 350 pounds.

Forequarter

The forequarter (IMPS 102) represents approximately 52 percent of the carcass weight and approximately 31 percent of the live weight. The forequarter is composed of the following wholesale cuts: the chuck and rib[4], which are primal cuts, and the brisket, shank, and plate, which are rough cuts. If the carcass or side is not all cut at the same time, it may be advisable to cut the forequarter first. The forequarter produces cuts that are best utilized as roasts or other forms that will be cooked and prepared in a manner to reduce the effects of connective tissue on palatability. The hindquarter, on the other hand, is a primary source of high-value steaks and would benefit more from additional aging time.

---

[3]Tucker, H. Q., M. M. Voegeli, and G. H. Wellington; Bratzler, L. J. (ed.). 1952. A Cross Sectional Muscle Nomenclature of the Beef Carcass. Michigan State College Press, East Lansing.

[4]Look up the following references to see why the meat industry should change its rib merchandising procedures:

> Wulf, D. M., J. R. Romans, and W. J. Costello. 1994. Composition of the beef wholesale rib. J. Anim. Sci. 72(1):94–102.

> Wulf, D. M., J. R. Romans, and W. J. Costello. 1994. Current merchandising practices and characteristics of beef wholesale rib usage in three U.S. cities. J. Anim. Sci. 72(1):87–93.

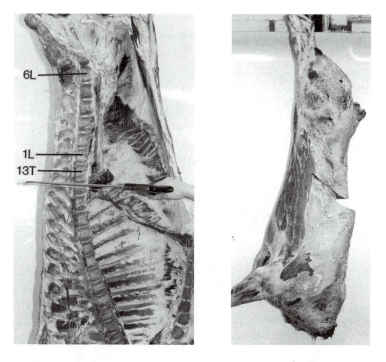

Fig. 15–3. To be properly evaluated and merchandised, beef sides must be separated into forequarters and hindquarters, a practice called *ribbing* or *quartering (left)*. Severing the half carcass between the twelfth and thirteenth ribs is the normal procedure although some local areas modify the number of ribs left on the hindquarter. In actual practice, rather than counting the ribs in a posterior direction from front to rear to find the twelfth rib, counting the exposed bodies of the *lumbar* and *thoracic* vertebrae in an anterior direction or from the rear to the front, by counting off 6 *lumbar*, Nos. 6L through 1L, and $1\frac{1}{2}$ *thoracic* vertebrae, Nos. 13T and 12T, a total of $7\frac{1}{2}$ vertebrae, reaches a point midway between the twelfth and thirteenth ribs. These exposed vertebrae are much easier to locate and to count than are the ribs. A saw is used to sever the backbone, and a knife is used to complete the cut, leaving a portion of the flank attached to the plate so that the entire half carcass may still hang from a rail *(right)*. For shipping and further processing, this plate-flank attachment is severed with a knife to complete the separation into forequarters and hindquarters. Inserting the knife at the desired spot below the rib eye muscle and making a smooth cut toward the backbone may reverse the ribbing procedure. The first method is preferred for making a nearly perpendicular cut across the rib eye muscle in order to measure its true size (see "Yield Grades" in Chapter 12).

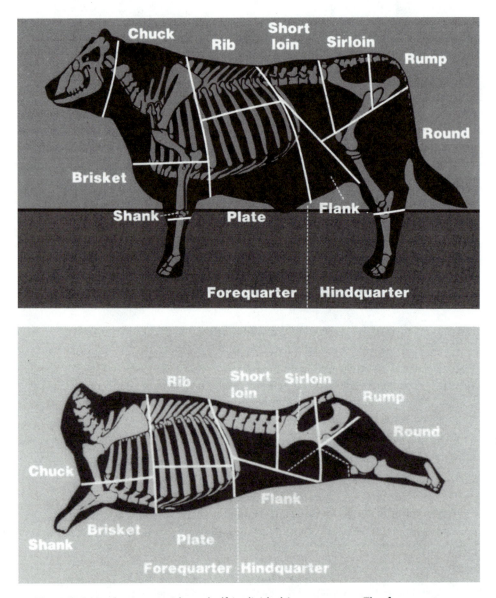

Fig. 15–4. Each carcass side or half is divided into quarters. The forequarter, composed of five wholesale cuts (the primal chuck and rib and the rough brisket, plate, and shank), is usually the heavier quarter. The hindquarter, composed of three wholesale cuts (the primal round, loin, and rough flank), is the most valuable quarter, based on retail pricing.

Fig. 15–5. The rib and plate are severed from the chuck, brisket, and shank by a cut made between the fifth and sixth ribs. The ribs are counted, and the cut is marked from the inside before the front quarter is removed from the rail. The cut can then be made from the outside, following the fifth rib closely. The back-

bone and sternum must be cut with a saw. If only the wholesale rib is removed from the forequarter, the remainder is called the beef triangle (IMPS 132). **(See Figure 15–101.)**

Fig. 15–6. The wholesale plate is separated from the wholesale primal rib (IMPS 103) by a cut made no more than 10 inches from the *longissimus* (rib eye) muscle on the large (blade) end and 6 inches from the *longissimus* at the small (loin) end. The cut illustrated here was made 10 inches from the chine bone (body of the vertebrae) on the small

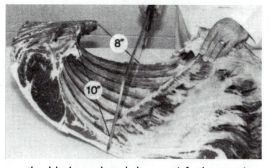

end and 8 inches from the chine bone on the blade end and does satisfy the requirements for the IMPS 103. The blade end is called thus because the blade cartilage from the blade bone *(scapula)* is exposed, while the loin end is so named because it is adjacent to the loin from the hindquarter. The completion of the separation of the rib and plate is done with a knife. The two primal ribs in wholesale form represent 5 percent of the live weight and 9 percent of the carcass weight. Steaks and roasts trimmed for retail trade from the two wholesale ribs *(right and left)* amount to approximately 3 percent of the live weight and 5 percent of the carcass weight.

Fig. 15–7. Fabrication of the rib begins when the ends of all rib bones are removed, with a saw, on a line approximately 1 inch from the lateral edge of the *longissimus* (rib eye) at the loin end. The location of this cut depends on market demand, so it may be made at a greater distance from the eye, giving more weight, but at the same time more waste, to the rib roast. USDA specifi-

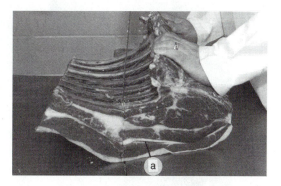

cations (IMPS 109) state that this separation should not be more than 3 inches from the eye on the loin end (small end) and not more than 4 inches from the eye on the blade end (large end). The resulting rib ends, which are removed, may be made into short ribs (see Figs. 15–43 and 15–44). In a typical blade-end view of a wholesale rib, the blade cartilage (a) should be noted. It has been separated from the blade bone itself, which remains on the chuck.

Fig. 15–8. The bodies of the vertebrae (chine bones) are removed on the saw. Removal of the blade cartilage from this cut would result in a beef rib, oven-prepared, IMPS 107. As is, it is a 107A beef rib, oven-prepared, blade bone-in.

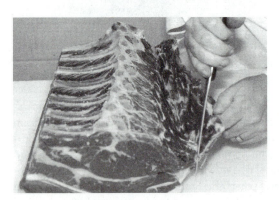

Fig. 15–9. In further processing of the rib, some of which is optional, the feather bones or *dorsal processes* of the *thoracic* vertebrae are removed. Since the main portion or the bodies of the *thoracic* vertebrae have already been removed, the only bones remaining are the ribs themselves and the blade bone and cartilage.

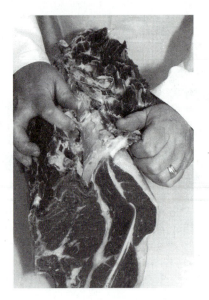

Fig. 15–10. It is essential that the yellow neck leader or backstrap *(ligamentum nuchae)* be removed, since it is composed almost entirely of elastin connective tissue, which will not be made palatable by cooking.

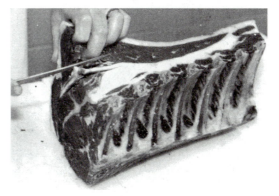

Fig. 15–11. The blade bone and(or) its cartilage is removed. From this view of the standing rib roast, note the natural rack formed by the ribs that remain. This roast would be placed in the oven inverted on this natural rack for roasting.

Fig. 15–12. Finally, the outside fat in excess of 0.25 inch, or a specified fat thickness, is removed, resulting in a finished short standing rib roast, so named for its length of ribs. The rib roast would be labeled an IMPS 109 "extra trim." The components that were removed from the wholesale rib to get this standing rib roast were the body and *dorsal processes* of the *thoracic* vertebrae (bone), the neck leader, the blade, the rib ends, and the surface fat trim. The roast on the left illustrates the blade (large) end, and the roast on the right illustrates the loin (small) end.

Fig. 15–13. Rib steaks removed from the loin (small) end are called rib steaks, small end. Note the cap muscle *(spinalis)*, which aids in rib steak identification (a).

Fig. 15–14. Rib steaks removed from the blade end are called rib steaks, large end. The considerable number of accessory muscles in the large end make it less palatable as a steak source. The large end is often most effectively utilized as a roast.

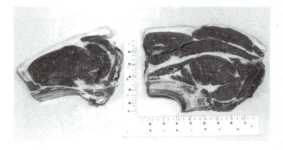

Fig. 15–15. Steaks from the rib do vary in value. A comparison between the steak from the loin (small) end and that from the blade (large) end shows that the steak from the loin end is more desirable because it has a larger rib eye muscle and less intermuscular (seam) fat. Both are called rib steaks, but the uniform labeling standards identify their value differences. (See footnote 4.)

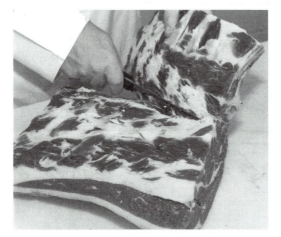

Fig. 15–16. The ribs may be removed from the trimmed rib to make an excellent boneless rib roast or very desirable boneless rib steaks.

Fig. 15–17. The removal of the cap muscles that cover the blade (cranial) end of the rib produces a rib eye roast (IMPS 112) composed of the *longissimus, spinalis, multifidus,* and *complexus.* The cap muscles, also called lifter or wedge muscles or blade meat (IMPS 109B), are composed of the *latissimus, trapezius, rhomboideus, subscapularis,* and *infraspinatus,* which provide material for cubing (see Fig. 15–67) and for other products (see Chapter 18).

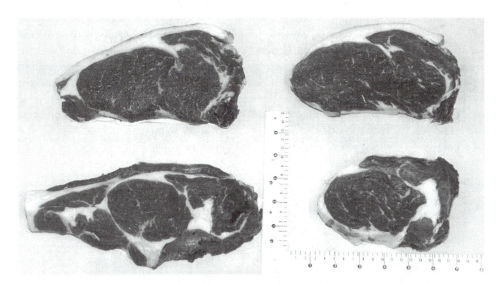

Fig. 15–18. The boneless rib steaks on the left demonstrate end-to-end differences in the rib similar to those shown in Fig. 15–15. At the top are loin-end steaks and at the bottom are blade-end steaks. The rib eye steaks on the right differ less because the cap muscles have been removed, resulting in rib eye (Delmonico) steaks (IMPS 1112). The steaks on the left (IMPS 1112A) and the roast they came from (IMPS 112A) are "lip on" because a lip composed of the *serratus dorsalis* and *longissimus costarum* remains firmly attached to the rib eye. The bone-in retail roasts and steaks from the rib equal only about 3 percent of the live weight and 5 percent of the carcass weight of a beef animal. Boneless yields would be somewhat less, depending on the fatness and muscling of the carcass. (See footnote 4.)

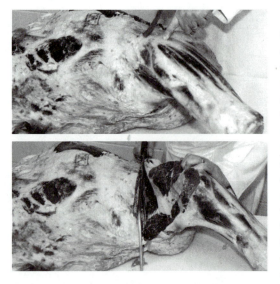

Fig. 15–19. In the portion of the forequarter that remains after the rib and the plate have been removed, the primal square-cut chuck is separated from the brisket and shank by a cut made parallel to the top side of the chuck, which severs the distal (far) end of the *humerus*. Correctly made, the cut will pass through the cartilaginous juncture of the first rib and the *sternum*. If this cut is made just slightly (½" to 1") above the point where the knuckle bone can be felt from the outside *(top)*, a butterfly–shaped cut surface on the bone will appear, signifying that the correct cut has been made. This cut must be completed with a saw *(bottom)*, as the *humerus* must be severed. If the brisket is removed and the foreshank remains attached to the chuck, it is an armbone chuck (IMPS 125). The chuck is the largest wholesale primal cut in the beef animal, and the combined right and left arm chucks represent 16 percent of the live weight and 26 percent of the carcass weight. The semi-boneless blade and arm retail roasts represent 9 percent of the live weight and 15 percent of the carcass weight. However, the boneless retail roasts secured from the chucks represent only 7 percent of the live weight and 11 percent of the carcass weight. This does not mean that the portion of the chuck not represented by roasts is all waste, since there are considerable amounts of lean trim and some miscellaneous cuts secured from the chuck in addition to these roasts. A forequarter subdivision called the arm back is the chuck and rib combined.

Fig. 15–20. This is a square-cut chuck (IMPS 113) from the animal's left side. Visualize the rib having been removed from the blade face (a) and the brisket and shank removed from the arm face (b). The chuck may be processed by several methods. Sawing parallel to the arm side 3 to 5 inches ventral to the *longissimus* at the blade end produces a square-cut chuck, divided (IMPS 113A), commonly referred to in the industry as a two-piece chuck. A retailer may cut the whole chuck, bone-in, on a band saw. The arm and blade roasts and steaks are produced by alternating cuts until the joint of the *scapula* and *humerus* is exposed. The rib bones may be trimmed from arm steaks and roasts, but they are usually left on blade cuts. A method in general use leaves the *scapula* and the *humerus* in the roasts, resulting in about a 4 percent higher roast yield on a carcass basis, raising the 11 percent quoted earlier (see Fig. 15–19) to 15 percent. An institutional cutting method removes all bone from the roasts, resulting in inside (chuck roll, IMPS 116A) and outside (shoulder clod, IMPS 114) chuck roasts.

Fig. 15–21. The square corner of the square-cut chuck where it is thinnest is called the cross-rib pot roast, formerly known as the English cut. The cross-rib may be taken from the chuck before any of the chuck fabrication methods that follow are used. A bone-in cross-rib roast is shown. The cross-rib may also be merchandised as a boneless roast. This is a low-quality roast with obvious fat seams even in this trim chuck. It should be cooked slowly with moist heat.

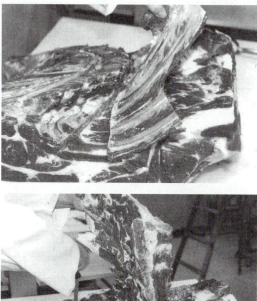

Fig. 15–22. The rib and neck bones should be trimmed out as cleanly as possible from the chuck *(top)*. Sawing across the rib bones near the backbone or sawing across the *dorsal processes* of the vertebrae can simplify this task. The leverage against the boner is reduced by making the saw cuts *(bottom)*. (See Figure 14–48 for information about the number of neck vertebrae.)

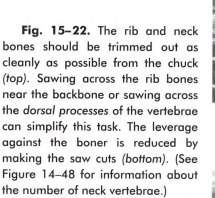

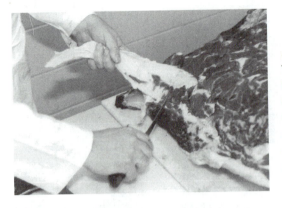

Fig. 15–23. The neck leader (*ligamentum nuchae*) is removed. This piece of connective tissue (see Fig. 15–10) is sometimes called the backstrap and is responsible for holding the animal's head erect. As discussed earlier, it is impossible to make this piece of connective tissue palatable.

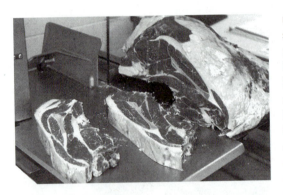

Fig. 15–24. For the most commonly used bone-in method, the square-cut chuck thus prepared is placed on a saw and two or three blade roasts are removed. The blade bone (*scapula*) appears in these roasts, thus the name *blade roast*. Note that the rib eye still remains in the chuck, although it is somewhat small in size at this point. In many retail meat cases, these roasts would contain vertebrae and ribs in addition to the *scapula*.

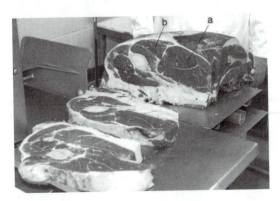

Fig. 15–25. Next, the chuck is turned 90° (right angle), and several arm roasts are cut. The arm bone (*humerus*) is very evident in these cuts, thus the name *arm roasts*. Note that there are fewer and larger muscles in the arm roasts and thus less seam fat than in the blade roasts. The main muscle systems of the arm roasts are the *triceps brachii*, long head (a), and the *triceps brachii*, lateral head (b).

Fig. 15–26. After two or three arm roasts have been removed, the remaining portion of the chuck is returned to its original position and more blade roasts are removed. As you proceed forward (cranially), the spine of the *scapula* (a) becomes evident in the blade roast. The spine of the *scapula* forms, in fact, a characteristic 7 configuration; thus, these roasts are known as seven-bone roasts. A neck pot roast may be cut

from the remainder of the chuck, with the cut being made in the same plane as the seven-bone roasts. The *pre-scapular* lymph node and the surrounding fat deposit should be removed from the neck roast, a rather low-quality roast, which would be better utilized as lean trim for ground beef.

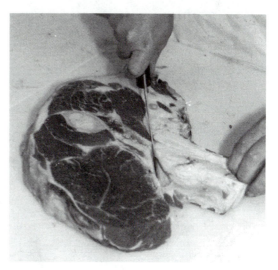

Fig. 15–27. Before the bone-in roasts from the chuck are packaged for the freezer or retail case, their fat seams may need to be trimmed.

Fig. 15–28. Because a roast should be at least 1½ inches thick to cook properly without excessive drying, the bone-in chuck roasts are larger than most consumers prefer. Therefore, the roasts are often subdivided, which permits the roasts to be cut thicker than 1½ inches.

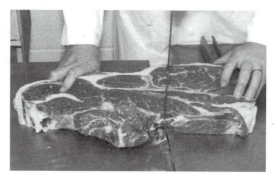

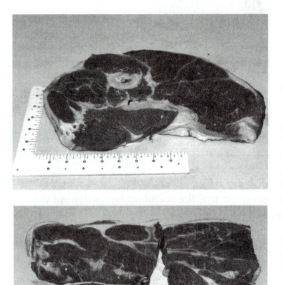

Fig. 15–29. The bone-in chuck produces arm roasts *(top)* and blade roasts *(bottom),* as illustrated. The cross-rib roast (see Fig. 15–21) and a neck roast may also result from bone-in cutting of the chuck. The remainder of the chuck is used for lean for stew, ground beef, or other products. These bone-in retail cuts equal approximately 9 percent of the live weight of the animal and 15 percent of the carcass.

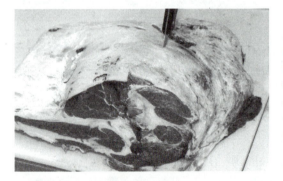

Fig. 15–30. An alternate method of fabrication is muscle boning (see Fig. 15–59), which will produce two large boneless roasts. This is sometimes referred to as the Institutional Method, since the boneless roasts are ideal for use in institutions such as hospitals and dormitories where servings of uniform size and quality are essential. One roast, the shoulder pot roast, formerly called the outside or arm chuck, lies behind the *humerus* and below the spine of the *scapula.* The term *clod* is used by the IMPS for this sub-primal, but since it is quite nondescriptive, the ICMISC recommends that *clod* not be used in retail nomenclature.

In removing the boneless shoulder (outside), enter the arm face, caudally to (behind) the *humerus* and follow this bone to its juncture with the *scapula.* Then follow the spine of the *scapula* toward the blade face, by riding the knife against the spine of the *scapula,* on the lower, or arm, side of the spine.

Fig. 15–31. As the outside chuck is trimmed away, the *scapula* is exposed. One method of removing the outside chuck is to make a cut down (toward the table) at the lower (ventral) edge of the *scapula*. That cut should expose a fat seam, which is then followed out to the arm face to remove the outside chuck roast.

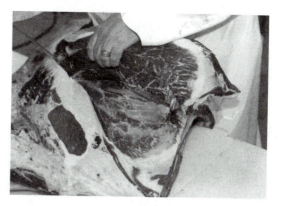

Fig. 15–32. After the boneless chuck roast has been removed, it is trimmed of excess intermuscular (seam) fat so that the lean tissue on the inner surface is exposed. The edges of the roast are squared up. The thin or upper end is trimmed considerably, resulting in a more uniform shape. This roast is what a food service manager would expect to see when a box of beef shoulder clod roasts (IMPS 114A) is opened. External fat may be trimmed to approximately 0.25 of an inch, depending on specifications.

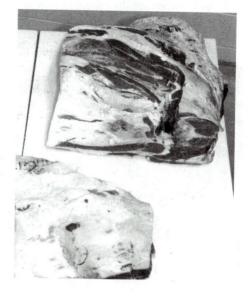

Fig. 15–33. The small muscle remaining above the spine on the upper surface of the *scapula* is the *supraspinatus* and is known as the mock, or Scotch, tender. It has a significant amount of connective tissue; therefore, it may often be tenderized mechanically to be made into minute (cubed) steaks or it may be used as lean trim (see Fig. 15–39).

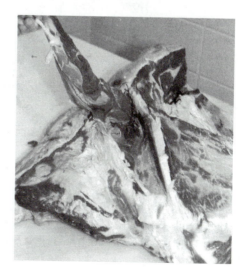

Fig. 15–34. After the removal of the thin muscle *(trapezius)* on the surface of the remaining chuck and the *humerus* and the *scapula,* the second boneless roast can be separated from the chuck. The inside chuck or chuck roll is an extension of the rib eye muscle with surrounding muscles included. With the silver dollar–shaped fat deposit in the center of the blade face as a starting place, the cut should be made parallel with the back line, to the indentation where the joint between the arm and blade was removed. The cut should be made through the boneless chuck to the table.

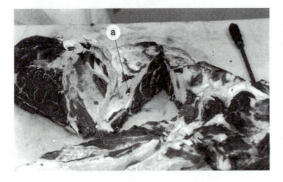

Fig. 15–35. At the joint indentation, make a 90° corner, cutting out through the back (dorsal side) of the chuck, to remove the large inside chuck roast. If the *pre-scapular* lymph node within a large intermuscular fat deposit has not been removed, this cut should bisect the lymph node as shown (a). The remainder of the chuck is used for lean trim, stew, or similar purposes.

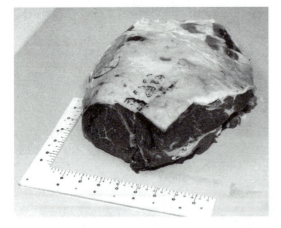

Fig. 15–36. Trimming the shoulder pot roast (outside chuck roast), also called beef chuck, shoulder clod (IMPS 114), removes intermuscular fat from the inner surface and subcutaneous fat from the outer surface. Thin muscles, particularly on the upper, or dorsal, edge, are also trimmed to square the shape of the roast and to improve its uniformity. The muscle remaining on the surface of the roast, which was over the spine of the *scapula,* is the *cutaneous omobrachialis.* This thin external muscle is often called the false lean, or rose, muscle. The fat covering this muscle is sometimes called the "frosting over the rose." If the rose and the underlying fat are trimmed to the specified fat thickness, it is known as a beef chuck shoulder clod roast (IMPS 114A).

Fig. 15–37. In this view of the finished inside chuck roll, from the caudal (rib) end, the roast is turned over so the inside surface which was removed from the first five ribs is exposed, and you can see where the ribs were once located. This serves as a useful key in identifying this inside chuck roll (IMPS 116A). The yield of roasts from this boneless method of cutting is somewhat less than that from the retail bone-in method, approximating 7 percent of the live weight and 11 percent of the carcass weight.

The roast shown could be trimmed more closely to include only the "eye" muscle and a few of the immediately adjacent muscles. The roast shown not only increases the roast yield but also provides food service units with a large unit with greater cooking and serving potential.

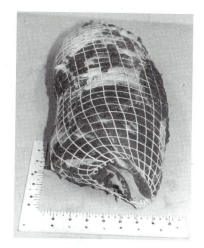

Fig. 15–38. The inside chuck roast may be tied or netted (see Fig. 14–22) to produce an even more desirable institutional roast. Both the outside chuck and the inside chuck roasts may be cut into sections to produce consumer-sized roasts.

Fig. 15–39. The *supraspinatus* may also be used as a small roast but will probably make better steaks, which are passed through a tenderizer or a cuber (see Fig. 15–67). The remainder of the boneless chuck is used for trim as ground beef, for stew beef, or as a raw material for processed meat.

Fig. 15–40. The chuck neck bones may be cut into sections for excellent soup bones to make stock or for flavoring pieces. Deboning equipment will salvage significant amounts of mechanically separated beef from these and other bones. The resulting tissue is a sausage raw material (see Chapter 21).

Rough Cuts

The shank, brisket, and plate of the forequarter and the flank from the hindquarter are called rough cuts. These four cuts combined equal approximately 14 percent of the live weight and 24 percent of the carcass weight. Various retail cuts may be fabricated from the rough cuts, although modern retailing utilizes them as ground beef or stew beef. The retail yield approximates 7 percent of live weight and 12 percent of carcass weight.

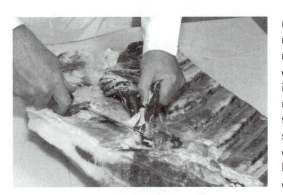

Fig. 15–41. The wholesale plate (IMPS 121) is directly ventral to the rib. It is known in the industry as the navel and(or) the short plate to differentiate it from the full plate, which includes the brisket. The diaphragm may be removed from the plate after the membrane covering the interior surface of the plate posterior to and covering the diaphragm muscle has been pulled out. This diaphragm is commonly called the outer or outside skirt in this location of the carcass.

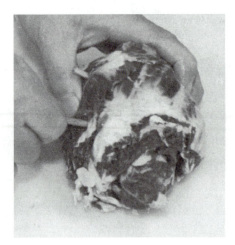

Fig. 15–42. The thin, flat skirt muscles can be rolled up and skewered. The skirts will thus produce two skirt-steak rolls (pinwheel steaks). These skirt-steak rolls look quite desirable, and when they come from high-quality beef and are properly prepared, they are. Widespread demand for skirt muscles has resulted from the increased popularity of fajitas, a traditional Mexican dish. When the muscle is thin, the diaphragm, or skirt, may best be used as lean trim.

Fig. 15–43. Short ribs (IMPS 123) can be removed from ribs 6, 7, 8, and 9 (counting from the front). The heavy muscle *(serratus ventralis)* is known as the pastrami piece (a). The piece is sometimes cured, smoked, and spiced for pastrami. These short ribs are adjacent to but ventral to those short ribs removed from the wholesale rib. Other options from the plate are boiling beef, beef spareribs, and a boneless plate. The boneless plate may be cured and sold as beef bacon, rolled and sold fresh, or used for ground beef.

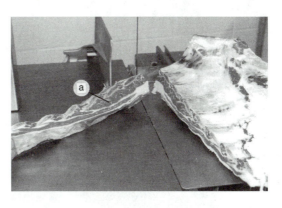

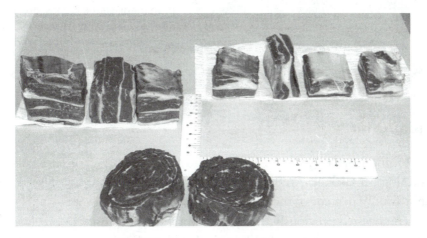

Fig. 15–44. The plate cuts are often economical cuts, which, with proper preparation, make excellent family meals. The skirt steaks, fajitas, short ribs, and plate ribs shown here add variety to the beef diet.

Fig. 15–45. The foreshank (IMPS 117) is separated from the brisket (IMPS 118) by a cut made through the natural seam that separates them.

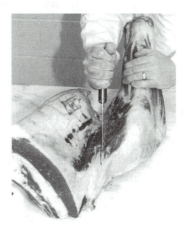

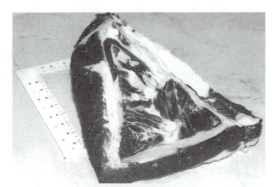

Fig. 15–46. The brisket may yield a boneless brisket roast, deckle on (hard fat and remaining intercostal muscles on the inside surface) (IMPS 119) if the *sternum* is removed. If the deckle is removed (IMPS 120), this boneless roast may be processed into corned beef.

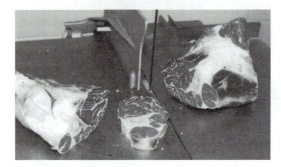

Fig. 15–47. The wholesale foreshank (IMPS 117) is easily identified. Shank cross-cuts may be obtained from the foreshank. The prominent shank bones are the *radius* and the *ulna*. The shank, being very lean, is often boned and made into lean trim for ground beef (IMPS 136). The bones (IMPS 134) are used for soup stock.

Hindquarter

The hindquarter (IMPS 155), composed of three wholesale cuts (the round, loin, and flank), represents approximately 48 percent of the carcass weight and 29 percent of the live weight.

As emphasized previously, there are several methods used to break specific wholesale units down into retail units. Local preferences, personal choices, and merchandising possibilities, as well as many other factors, may determine the best technique for a specific situation. Cutting hindquarters is an area in which a variety of options are used by meat cutters.

There are several possible methods of removing the primal round from the hindquarter. The meat industry generally acknowledges three types of beef rounds that vary according to the manner in which the sirloin tip is cut. The *sirloin tip* is that portion of the leg musculature that is anterior to the *femur* in a standing animal. It is often referred to as the knuckle in the meat industry.

The Chicago round, described in the Institutional Meat Purchase Specifications (IMPS 158), is the most common round available in the wholesale trade. (See the wholesale rounds in Figures 15–1 and 15–2, which are Chicago rounds. Note also the outlines in Figure 15–4.) The Chicago round is removed by a cut that begins at the junction of the last sacral vertebra and the first caudal vertebra and passes through a point anterior to the prominence of the aitch

Fig. 15–48. The removal of the flank is one of the earliest procedures in breaking down the beef hindquarter into wholesale cuts. The wholesale flanks represent 4 percent of the live weight of an animal and 7 percent of its carcass weight. The flank is removed by following the contour of the round and removing the cod, or udder fat, with the flank. Toward the front, the cut is marked not more than 6 inches laterally to the loin eye muscle, where a saw must be used to sever the thirteenth rib. In the illustration, the thirteenth rib was cut about 1 inch past the lateral edge of the loin eye so that more valuable loin steaks could be prepared. The same procedures illustrated here on the rail can be performed on the table.

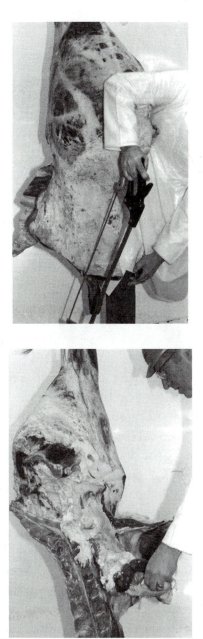

Fig. 15–49. Although the kidney and pelvic fat can be removed during harvest or before flank separation, it is sometimes easier to trim and pull it free after the flank has been trimmed away from its attachments with the round. Care should be taken not to score the tenderloin muscle that lies under the fat. The kidney fat just above the diaphragm is more obvious in the left side than in the right. This is caused by the alternate filling and emptying of the rumen in the live animal, leaving an open cavity for the kidney and surrounding fat to drop down into on the left side when the rumen is empty. For this reason, the right side is called the tight side, while the left is called the loose side in a beef carcass.

(pelvic) bone. This cut is continued on a straight line out through the flank, or ventral, portion of the hindquarter. The ball of the *femur* is exposed by the properly executed cut. However, the Chicago round cutting style can be criticized because the cut divides the sirloin tip muscle system, leaving one portion on the sirloin end of the wholesale loin and the other portion of the sirloin tip on the wholesale round. A further problem is the cutting angle because it reduces the potential value of both portions of the tip (see Figure 15–1).

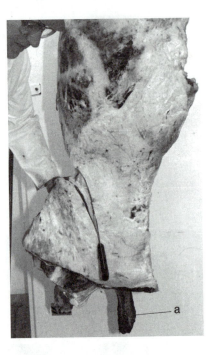

Fig. 15–50. Separation of the flank is completed with a knife. The hanging tenderloin (a) is a portion of the diaphragm muscle attached to the back in the region of the last rib, and it contracts with each breath of the live animal. Because of this, it may not be as tender as some other, less active muscles, so it is best utilized as lean trim. The hanging tenderloin may be removed when the kidney and kidney fat are removed, after the flank has been cut away, or, later, when the wholesale loin is trimmed. The true tenderloin *(psoas)* should not be cut into when the diaphragm is removed. The remaining hindquarter is referred to as being streamlined (IMPS 155B).

An alternative style results in a Diamond round (IMPS 158A), which includes the portion of the sirloin tip that is left on the loin with the Chicago style. The Diamond round is cut at the same angle through the sacral/caudal junction, exposing the ball of the *femur,* but the cut does not continue through the anterior, or flank, side of the round. The dotted line on Figure 15–4 shows the angle of the cut, which continues the angle of the aitch (pelvic) bone through the flank side of the hind and retains the complete sirloin tip on the round. Although the Diamond round results in the most useful wholesale round, it is difficult to produce without the use of a power saw to separate the loin and round on the rail.

The third option, the New York round, is a compromise — the sirloin tip muscle system is separated from the hindquarter before the round and loin are separated. The New York round is cut through the sacrum and *femur* socket at the same angle and produces the equivalent of either a Chicago round or a Diamond round with the sirloin tip removed (see the following diagram). The names *sirloin tip, round tip,* and *knuckle* (IMPS 167) all refer to the same cut.

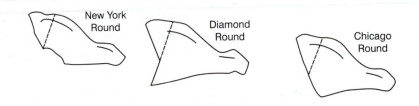

Fig. 15–51. Being one of the four wholesale rough cuts in a beef carcass, the flank is used mainly for a source of lean trim. After a strip of fat and connective tissue, 1 to 1½ inches wide, is removed from the midline edge, the membranes covering the internal surface of the flank can readily be pulled off. From each flank, only one flank steak, the *rectus abdominis* muscle, may be removed by pulling it from the interior surface.

Fig. 15–52. After its removal, the flank steak (IMPS 193) is usually scored with cuts made lightly to the surface at right angles. The remainder of the flank is trimmed of waste fat and used for ground beef. The London broil (named for London, England) may be made from a flank steak, the most popular of several cuts that might be used for this item. The London broil is marinated, broiled, and sliced thinly for serving.

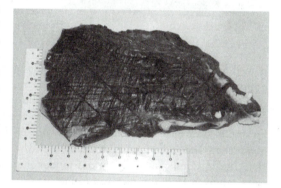

Fig. 15–53. Another method of separating the round cuts from the hindquarters without subdividing the sirloin tip is illustrated. The first cut is made caudally to (to the rear of) the *ischium* of the pelvic bone (commonly referred to as the aitch bone). By cutting parallel to and just behind the aitch bone, you must saw through the *femur* (large bone in the thigh). The rump portion of the round must be removed from the loin separately, resulting in two pieces, the equivalent of a Diamond round (see Fig. 15–69).

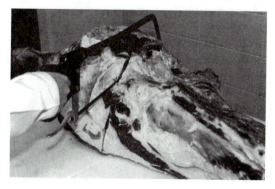

The Chicago round will represent approximately 22.5 percent of the carcass or 14 percent of the live weight. The New York round and the Diamond round, including the sirloin tip in each case, will approximate 25 percent of the carcass weight and 15 percent of the live weight. The boneless retail yield of the round in any of these forms will represent 60 to 64 percent of the wholesale round weight.

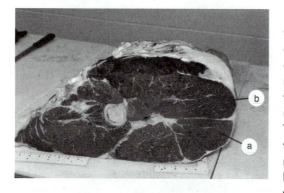

Fig. 15–54. The nomenclature of the retail cuts from the round can be explained very logically if the origin of the names is known. The outside round is the portion of the round that is toward the outside of the animal and is down on the table (a). The inside round (b) is the portion of the round that is toward the animal's midline. The outside round became known as the bottom round because when the round is cut, it is usually placed on the table or block with the outside next to the table. Therefore, it was called bottom round. The inside, being on top, was thus called top round. The eye of the round is the lower right *(semitendinosus)* muscle, and the remainder of the bottom round is the *biceps femoris* muscle. The muscle system to the left (cranial) of the *femur* is the sirloin tip (note previous discussion), made up of four muscles, collectively called the *quadriceps*. The top round is made up of the *semimembranosus* and the *adductor* muscles. The top and the tip are more tender than the bottom and the eye. If the shank were removed from the pictured round, it would be a beef round, rump and shank off (IMPS 164).

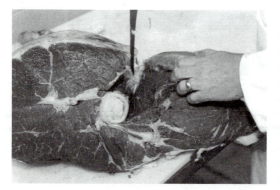

Fig. 15–55. Entering through the natural seam that separates the round tip from the inside or top round on one side and from the outside or bottom round on the other, you can remove the *quadriceps*, which compose the sirloin tip, named round tip for retail sale, and lie to the cranial (front) side of the *femur*. From this dorsal view, you can see the cross section of the *femur*.

Fig. 15–56. Entering from the stifle joint near the *patella,* at the shank knuckle bone, which is the caudal or distal end of the *femur,* you can loosen the connections sufficiently so that it is possible to grasp the tip in one hand and the knuckle of the *femur* in the other to separate the two. Final separation is made with a knife. The *patella* (a) is removed from the rough tip (beef round, knuckle) (IMPS 167).

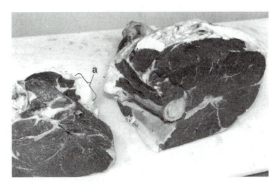

Fig. 15–57. The outside, or cap, muscle *(tensor fasciae latae)* may be removed to make the round (sirloin) tip a highly desirable retail cut called the cap off ("bald") round-tip roast (IMPS 167A) or steak. This cut is located at the junction of the sirloin and the round but is officially named *round tip,* since it is a sub-primal of the primal round. The term *sirloin tip* is well recognized by the industry but is often confusing to consumers.

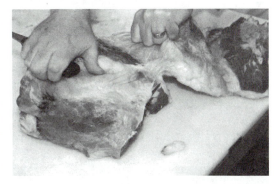

Fig. 15–58. The round (sirloin) tip makes an excellent roast and may be cut into steaks if the beef is youthful and has enough marbling to insure satisfactory flavor and juiciness. The tip is usually the most tender roast from the round, due to its location in the live animal, in front of the large round bone *(femur).* In this position, the muscle fibers are stretched when the carcass is hung in the conventional manner, thus making the roast more tender (see Chapter 22). The tip cuts can be identified by the horseshoe- or oval-shaped connective tissue line in the center of each cut. The four muscles of the *quadriceps* group are the *vastus lateralis,* largest, on the left; *rectus femoris,* middle; *vastus medialis,* smallest, on the right; and *vastus intermedius,* across the bottom of the steak, nearest the roast.

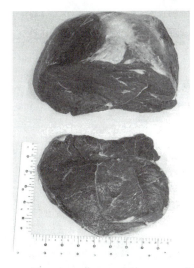

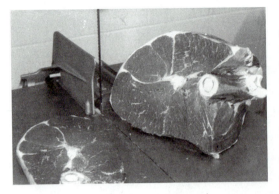

Fig. 15–59. Full round (bone-in) steaks are cut after the tip has been removed. This left round was cut with the outside round at the top in order for the bone to lead the cut through the saw. Modern meat-cutting techniques have largely eliminated full round steaks from the retail case. Tenderness differences between the top and bottom round make it difficult to effectively use the full round steak as a single unit. A technique known as *muscle boning* is used to fabricate the remainder of the round into sub-primals. The muscles and muscle groups are separated into sub-primal and retail cuts by following the natural seams as much as possible. The muscles are taken apart in much the same manner as nature put them together. Muscles grouped naturally together for function in the live animal are more uniform in palatability and thus form much more desirable retail cuts.

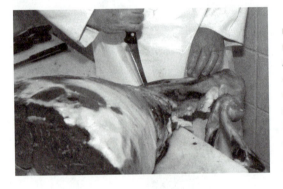

Fig. 15–60. The next step in muscle boning the round is to remove the hind shank bone *(tibia)*. Muscles of the lower round are anchored to this bone through strong connective tissues, which must be severed. The large tendon on the back of the shank is the *Achilles tendon* in which the trolley hook was inserted to suspend the intact carcass. The *tibia* is followed with a knife up into the round to the stifle joint. This is the knee joint of the animal where the *tibia* joins the large bone of the round, the *femur*. When this joint is severed with a knife, the remaining connective tissue attachments to the *tibia* are loosened, and the shank bone is removed.

Fig. 15–61. From the shank knuckle or stifle joint, the large *femur* is removed. With older methods of cutting, this large bone was left in the retail round steaks to later occupy valuable freezer space.

Fig. 15–62. The *femur* and the shank bones, along with the *patella,* represent a significant amount of the weight of the round. Bones (IMPS 134) may be sawed into lengths not to exceed 8 inches to be marketed for soup stock.

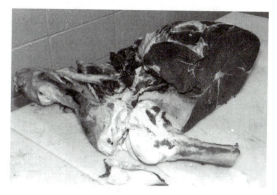

Fig. 15–63. The bottom round *(left)* can be separated from the top round by following the natural seam that separates them. The top round sub-primal is trimmed on the newly separated face as well as on the inside or top surface, where there is usually a portion of the lean *(gracilis* muscle) that has been darkened due to exposure to air for several days during marketing or aging.

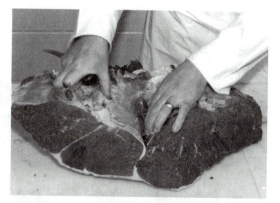

Fig. 15–64. Steaks, preferably for moist-heat cooking (braising), can be cut from the top round (IMPS 169). The top round is composed of two major muscles: the *semimembranosus* (a) and the *adductor* (b). The muscle separation is difficult to see, so top round roasts and steaks appear very solid and homogeneous. The top round and the round (sirloin) tip are generally considered to be the more tender round roasts.

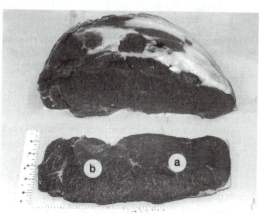

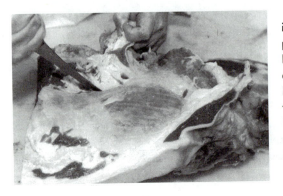

Fig. 15–65. Especially important in trimming the bottom round subprimal is to remove the *popliteal* lymph node and its surrounding fat deposit. The large fat deposit is located on the surface from which the top round was separated and between the *semitendinosus,* or eye, and the *biceps femoris* muscles. The heel of the round is separated from the lower portion of the bottom round (see Fig. 15–83).

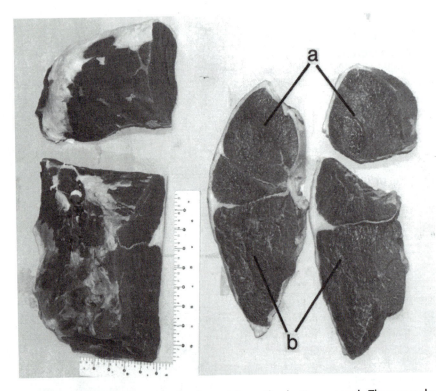

Fig. 15–66. Steaks may also be cut from the bottom round. They may be cut thicker than the top round steaks and make ideal material for swissing (see Chapter 24). Steaks and roasts from the bottom round can be easily identified by the distinct eye of the round, the *semitendinosus* muscle (a). The other large muscle in the bottom round, sometimes alone called the bottom round, is the *biceps femoris* (b). Note the muscle has two heads — thus, the name *biceps femoris* for the part near the femur. The eye and bottom round roasts may be separated to make smaller portions or to make slicing across the grain easier.

Fig. 15–67. To be used as minute (cubed) steaks, bottom round steaks may be passed through a mechanical tenderizer, a desirable method of merchandising the lower-quality bottom round steaks.

Fig. 15–68. The heel, or *Pikes Peak,* roast should be prepared with moist heat, tenderized mechanically or ground because it contains large amounts of connective tissues stemming from its location in the live animal — the lower portion of the round. The main muscle of the heel roast is the *gastrocnemius* (a). The heel of the round, along with the rest of the shank meat and trimmings, may be used as lean trim for ground beef.

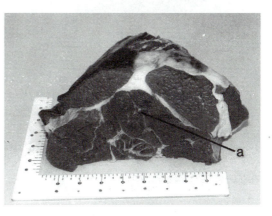

The method of removing the round from the hindquarters, illustrated in Fig. 15–53, resulted in an intact round (sirloin) tip but left the rump portion of the round still attached to the loin. However, in the industry, the rump is considered a part of the round. Thus, when rounds are quoted and traded, it is with the assumption that the rump is a part of the round. Here it is considered as a separate cut for purposes of clarity. The rump, when considered as a wholesale cut, represents 2 percent of the live weight and 4 percent of the carcass weight, while in retail form, the boneless roasts represent 1 percent of the live weight and 2 percent of the carcass weight.

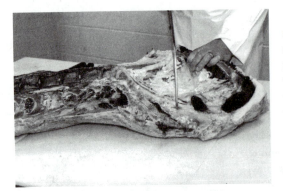

Fig. 15–69. The rump is separated from the loin by a cut made on an imaginary line connecting two points: a point on the backbone between the fifth *sacral* vertebra and the first *coccygeal* (caudal, or tail) vertebra and the anterior (front) tip of the proximal (inward) end of the *femur*. This cut is identical to the Chicago-style separation described earlier and indicated in Figs. 15–1 and 15–2, except that now the round (sirloin) tip has been removed and will not be divided by this separation. The IMPS indicate that the cut is anterior to the *femur* knuckle but exposes the *femur*. If a piece of bone from the *femur* knuckle no more than a silver dollar in size results, then an acceptable cut has been made.

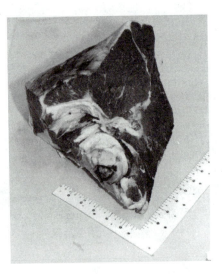

Fig. 15–70. The beef round rump roast (standing rump) normally has the knuckle from the femur removed, but it still contains the large aitch bone, which creates difficulty in carving and serving for the consumer. Note in Fig. 15–1 the large portion of the aitch bone contained in the rump area.

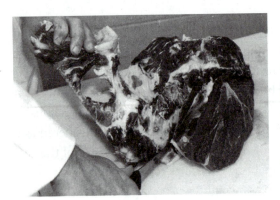

Fig. 15–71. It is more appropriate for the meat cutter to bone the rump and ultimately much more satisfying for the consumer to carve the boneless, exquisite roast for company. Cutting closely on both sides of the aitch bone with the knife will clear the bone. Note the size of the aitch bone as its last attachments are severed.

Fig. 15–72. The boneless rump *(left)* is rather diffuse, so it must be tied for ease of cooking and slicing. Any boneless roast can very easily and quickly be passed through the tube of a netter to become closely wrapped in elastic net *(right)*. The jet net is left on during roasting but should be removed before the roast is served.

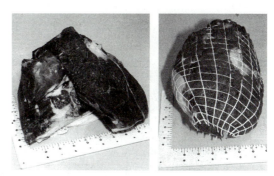

Fig. 15–73. Another approach to cutting the hindquarter on the rail begins after the flank has been removed, resulting in a streamlined hindquarter (IMPS 155B) (see Figs. 15–48 to 15–50). The round (sirloin) tip (a) is removed prior to the round/loin separation.

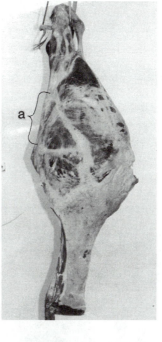

Fig. 15–74. A cut beginning in front of the aitch-bone prominence (a) is made through the round to the flank or lower side of the hindquarter. The cut is made at the same angle as the aitch-bone angle.

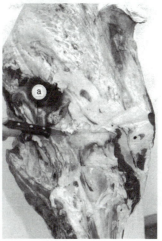

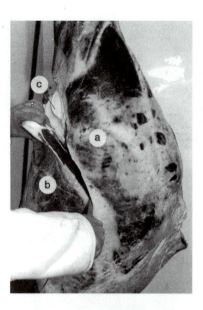

Fig. 15–75. Following the seams separating the outside round (a) (shown) and the inside round (on the opposite side) from the round (sirloin) tip (b) system, cut down from the stifle joint (c) to the cut described in Fig. 15–74.

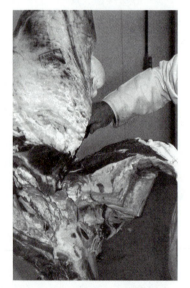

Fig. 15–76. The front of the *femur* is exposed as the round (sirloin) tip is pulled out and separated near the *femur*–pelvic joint.

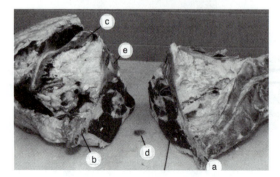

Fig. 15–77. After the round (sirloin) tip has been removed, the round with the rump on is separated from the loin on a line beginning at the fifth *sacral* vertebra (a) junction with the first *caudal* vertebra (b) to a point just anterior to the aitch-bone prominence (c). Note the small portion (d) of the *femur* ball (e) that was removed in making the cut correctly.

Fig. 15–78. The aitch bone is removed from the round. First, the fat and the small, flat, circular "oyster" muscle (*obturatorius internus*) are removed from the pelvic channel (inside) surface of the aitch bone. Some meat cutters push a steel (for aligning knife edges) through the *obturator foramen* (hole through the pelvic bone) and use it as a lever to aid in removing the bone. Note the ball of the *femur,* which must be released from its socket in the aitch

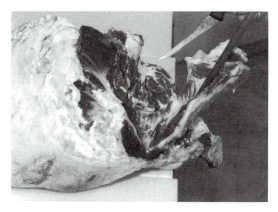

bone. After the shank bone has been removed (see Fig. 15–60), the *femur* is removed (see Fig. 15–61). Because the rump is on the round, the *femur* will be longer than in Fig. 15–62, because it will include the ball head on its proximal end. The top round and bottom round are separated on the seam between them, as in Fig. 15–63. Leaving the seam fat on the bottom round simplifies trimming. Staying on the proper seam involves making the top round as small as possible.

Fig. 15–79. The top round (IMPS 168) *(right)* can be trimmed of surface fat to no more than $\frac{1}{4}$ to $\frac{1}{8}$ inch, and the dry and knife-scored areas can be trimmed away to produce a very desirable roast, representing approximately 15 percent of the round weight. The bottom round *(left)* is known as a gooseneck round after the shank meat has been removed (IMPS 170) (see Figs. 15–65, 15–68, and 15–83). The gooseneck can be subdivided into rump, bottom round, and eye of round roasts.

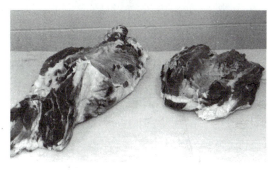

Fig. 15–80. After the fat has been trimmed from the inside surface of the outside round, and after the *popliteal* lymph node and the fat deposit surrounding it have been carefully removed (see Fig. 15–65), the rump is separated. The cut is made at the cranial end of the large boneless cut, behind the point where the aitch bone was removed and perpendicular to the long axis of the roast.

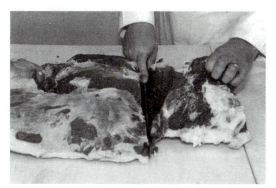

Fig. 15–81. The rump roast can be trimmed even more closely than illustrated in Fig. 15–72. The finished small roast is one major muscle *(biceps femoris)* comprising about 4 percent of the wholesale round.

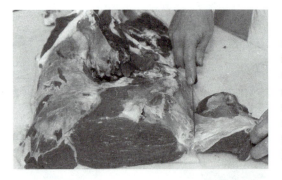

Fig. 15–82. When the round muscles are separated and trimmed, the heavy membranes of the epimysium connective tissue that surround some muscles, and especially the heavy *sacrosciatic* ligament, should be trimmed away. (Note membrane to the left of the knife blade.)

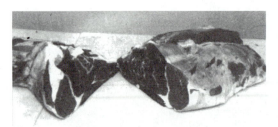

Fig. 15–83. The shank (heel of round) is separated cranially to the tapering of the heel roast. The small round muscle in the upper center of the face of the shank, the *flexor digitorium superficialis* (see white connective strands), should be trimmed from both roasts for ground beef because of its connective tissue content. The trimmed bottom round roast as shown *(right)* approximates 17 percent of the wholesale round. The heel roast is trimmed as shown in Fig. 15–68.

Fig. 15–84. The wholesale primal loin (IMPS 172) is composed of two sub-primals — the sirloin (IMPS 181) and the shortloin (IMPS 173). If full loins are fabricated into bone-in steaks, the sirloin and shortloin will not be separated, but rather the whole loin will be sawed into steaks from the sirloin end. The sirloin is more often a source of boneless steaks and roasts (see Figs. 15–99 to 15–102).

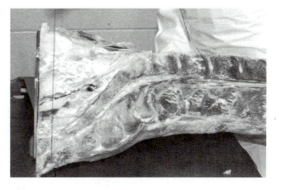

The sirloin sub-primals wholesale weight approximates 5 percent of the live weight and 8 percent of the carcass weight. The steak yield (retail yield) from the sirloins equals about 3 percent of the live weight and 5 percent of the carcass weight.

The shortloin is the sub-primal of the primal loin that remains after the removal of the sirloin. The shortloin is composed of the lumbar section of the hindquarter. As wholesale sub-primal cuts, the two shortloins from a carcass represent 4 percent of the live weight and 7 percent of the carcass weight. The steak yield from the shortloin will equal approximately 2 percent of the live weight and 4 percent of the carcass weight. Note that one rib remains on the cranial end of the shortloin. Since the steak yield is 4 percent of the carcass weight and the wholesale sub-primal yield is 7 percent of the carcass weight, the yield of the wholesale sub-primal in steaks is four-sevenths, or 57 percent. This same relationship can be used to determine the steak yield of any wholesale cut. (See Fig. 15–99 for separation of the shortloin and sirloin.)

Fig. 15–85. "Sirloin" comes from the French word *surlonge*. The *sur* means "supra" and the *longe* is "loin" — thus, "over the loin." The sirloin may be cut on a band saw into steaks that may differ considerably from one another in their value because of the amount of bone they contain. The ICMISC recommends four names for the various sirloin steaks, based on the shape of the

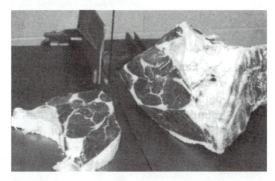

pelvic bone each contains. Beginning at the caudal (rear) end of the sirloin, this steak and one or two more cranially from (toward the front of) this steak are called wedge bone sirloin steaks, because that portion of the *ilium* (pelvic bone) has the characteristic wedge shape *(left, larger white area)*. The first steak removed is sometimes called a butt-bone sirloin steak because the pelvic bone has a depression where the *femur* engaged the pelvic bone. Note the outermost muscle, the *biceps femoris,* and its characteristic coarse texture. The next inner, larger top-sirloin, or *gluteus medius,* muscle carries throughout the whole sirloin and may be removed and sold as a boneless top-sirloin roast or cut into top-sirloin steaks. Note on the remaining portion of the sirloin, the wedge bone *(ilium)* and the *sacrum.*

Fig. 15–86. Proceeding in a cranial direction, notice the next steak, which is the round-bone sirloin steak, so named because the shaft of the *ilium* is almost round at this point. Note the larger *gluteus medius* and the smaller *biceps femoris,* as compared to Fig. 15–85.

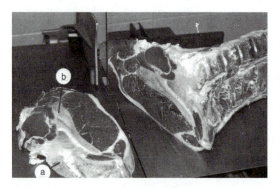

Fig. 15–87. Moving forward, observe that portion of the sirloin where the backbone or *sacrum* (a) is joined to the pelvic bone or *ilium* (b), at the *sacroiliac* joint. Since there are two bones in this particular steak, it is commonly called the double-bone sirloin steak. The *gluteus medius* and the *psoas major* (tenderloin) are the principal muscles shown here.

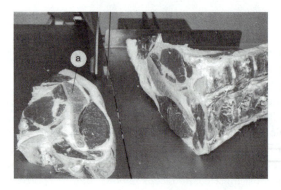

Fig. 15–88. The most cranial steak of the sirloin is the pinbone sirloin steak. The pinbone (a) or hipbone *(tuber coxae)* with its white cartilage surface comprises a large area of the steak on the saw. There is usually one pinbone sirloin steak, and the steak immediately caudal (to the rear) to it is a (double-bone) flat-boned sirloin steak. The tenderloin muscle appears quite prominent to the left of the pinbone. Note the pinbone structure remaining on the loin. The next steak will include the junction between the sirloin and the short loin, resulting in the other face of the steak having the anatomy of the porterhouse steak.

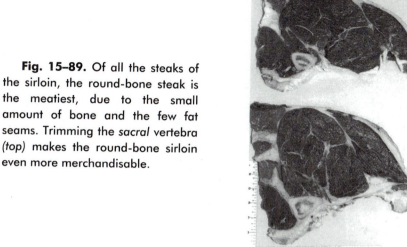

Fig. 15–89. Of all the steaks of the sirloin, the round-bone steak is the meatiest, due to the small amount of bone and the few fat seams. Trimming the *sacral* vertebra *(top)* makes the round-bone sirloin even more merchandisable.

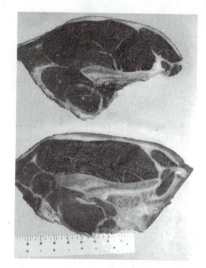

Fig. 15–90. The double-bone sirloin steak includes a significant amount of bone. To make the steak more consumer acceptable, the *sacrum* is removed, leaving the *ilium*. The steak is then called a flat-bone sirloin steak *(top)*. Note the top-sirloin, or *gluteus medius,* muscle, which persists throughout the sirloin.

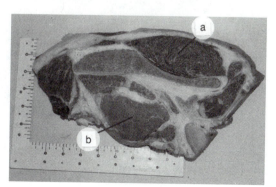

Fig. 15–91. The hipbone, or pinbone, separates the top-sirloin, *gluteus medius* (a), from the tenderloin, *psoas major* (b), in the pinbone sirloin. The pinbone sirloin is seldom found in meat cases because of the amount of bone in relation to the amount of meat. The two muscles differ in palatability and are best sold as separate boneless cuts.

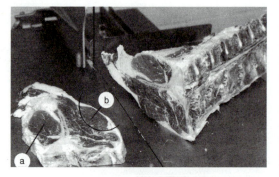

Fig. 15–92. The first steaks from the sirloin or caudal end of the shortloin (IMPS 173) are the porterhouse steaks. Porterhouse steaks may be identified by the large size of the tenderloin (a) ventral to (below) the *transverse* (T) vertebral processes. The real key to identifying the porterhouse steak is, however, the presence of an additional muscle dorsal to (above) the *longissimus*, which is actually an extension of the top-sirloin, *gluteus medius* (b). The Uniform Retail Meat Identification Standards indicate that a porterhouse steak has a tenderloin muscle that exceeds 1¼ inches in diameter.

Fig. 15–93. As the tenderloin decreases in size and the *gluteus medius* disappears in the shortloin, the steaks become T–bone steaks. The characteristic "T" is formed by the *transverse processes* of the *lumbar* vertebrae.

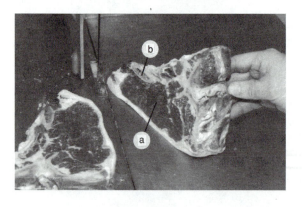

Fig. 15–94. At the anterior end of the shortloin, the tenderloin disappears, and a rib or a portion of an obliquely split rib appears in the steaks, which are identified as top-loin steaks. Previously, these steaks were called club steaks, but the term *club* was so widely used, oftentimes inappropriately, that the ICMISC decided to use the term *top-loin steak*. Note the large loin eye *(longissimus)* muscle (a) and the portion of the split rib (b). Note also that the cap muscle observed on rib steaks (see Fig. 15–13) is not present.

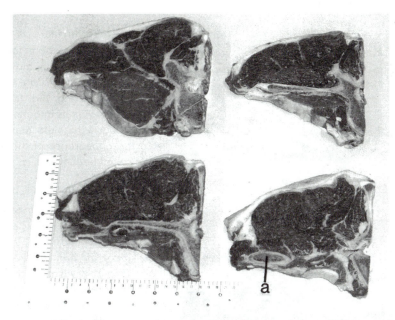

Fig. 15–95. The bone-in shortloin yields three kinds of steaks. The most caudal is a porterhouse steak *(upper left)*, since it includes a small portion of the *gluteus medius* (see Fig. 15–92). However, the upper right steak, technically a T–bone steak, can be merchandised, according to IMPS, as a porterhouse steak, because the tenderloin diameter exceeds 1¼ inches. A T–bone steak, by all definitions, is shown lower left, and the lower right steak is a top loin with the identifying bias-cut thirteenth rib exposed on the lower left area of the steak (a).

When porterhouse steaks are identified by the presence of a portion of the *gluteus medius,* and top-loin steaks by the presence of a portion of the thirteenth rib, a shortloin yields 3 porterhouse, 10 T–bone, and 3 top-loin 1-inch thick steaks. However, when the IMPS is used, the identities of the steaks from the same large, meaty loin change. Porterhouse steaks are identified as having 1¼-inch or more tenderloin diameter, and top-loin steaks are identified as having ½-inch or less tenderloin diameter, resulting in 10 porterhouse steaks, 5 T–bone steaks, and 1 top-loin steak from the same shortloin. The number of top-loin steaks decreased from three to one because two of the steaks that contained a bias-cut section of the thirteenth rib also exhibited a portion of tenderloin that exceeded ½ inch in diameter, thereby qualifying them for classification as T–bones. The word *steak* is derived from the Saxon word *steik* (pronounced "stick"), which means meat on a stick.[5]

The steak names *T–bone* and *top loin* relate to the bone configuration ("T") and location *(top)*. However, the name *porterhouse* is derived from the name given to the taverns or alehouses of the eighteenth century, which were called "porterhouses" (because the porters in London's Covent Garden market drank ale in such places). The steak itself was popularized around 1814 by a New York City porterhouse keeper named Martin Morrison. It soon became the most popular form and cut of steak in the United States.

---

[5]National Meat Association (NMA). http://www.nmaonline.org

Fig. 15–96. An alternate method of fabricating the entire loin is to remove the tenderloin muscle and process it as a fillet. Care must be taken in removing the tenderloin because scores (cuts into the muscle) would reduce the yield and detract from the appearance of these most tender of all beef steaks.

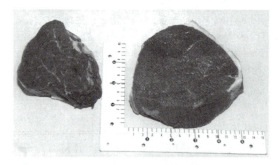

Fig. 15–97. The tenderloin is trimmed of accessory muscles (psoas minor) and membranous fat layers to produce a trimmed tenderloin (IMPS 190) from which tenderloin fillet steaks are cut. The popular *filet mignon* is a tenderloin steak wrapped with a slice of bacon. The *chateaubriand* is a large tenderloin steak from the center of the *psoas major* served with *hollandaise* sauce flavored with wine, shallots (green onions), and herbs.

Fig. 15–98. The tenderloin fillet is the most palatable muscle in the beef carcass, but it may easily be confused with the eye of the round, which is quite similar in appearance. Recall that the eye of the round is the *semitendinosus* muscle from the bottom round. During the animal's life, the *semitendinosus* muscle is used every time the animal takes a step as it moves from place to place.

Conversely, the *psoas* muscle of which the tenderloin fillet is composed, lying on the inside of the *transverse processes* of the loin, does very little work in the animal other than perhaps to help the animal maintain its posture or to aid it as it turns its body. Thus, the eye of the round is considerably less tender than the tenderloin fillet. To differentiate between the two steaks, note the coarse structure of the eye of the round as compared to the fine texture of the tenderloin fillet. The muscle bundles are much larger in the eye of the round and are enclosed in a thicker connective tissue membrane, thus causing the coarser texture. Another identifying clue is the strip of external (subcutaneous) fat, seen in the upper portion of this figure, that the eye of the round displays, since this cut is located on the surface of the leg. The tenderloin fillet, on the other hand, from the inside of the animal, has no surface exhibiting subcutaneous fat.

Fig. 15–99. As a continuation of the alternate loin fabrication procedure, the sirloin is separated from the shortloin by a cut made between the fifth and sixth *lumbar* vertebrae, which nicks the cranial or front portion of the pelvic girdle (pinbone). After the backbone has been separated with a saw, the cut is completed with a knife.

Fig. 15–100. The pelvic bone *(ilium)* and the *sacral* vertebra are removed from the sirloin. The bone should be carefully followed with the tip of the knife to avoid scoring the muscle. Note the size of the bone removed (left hand).

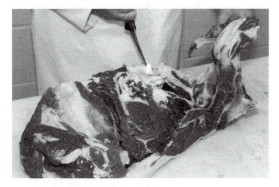

Fig. 15–101. Trimming the boneless sirloin (IMPS 182) entails detaching a 1- to 1½-inch strip along the back line to remove the connective tissue attachments to the vertebrae. Removing the ventral tail of the sirloin involves following a hard-to-find seam between the top sirloin *(gluteus medius)* (IMPS 184) and the bottom sirloin *(tensor fasciae latae)* (IMPS 185), also known as the tri-tip or triangle **(see Figure 15–5)**. The front (cranial) and rear (caudal) ends of the roast should be squared. The trimmings are largely flap meat *(abdominis internus)* (IMPS 185A) and can be used for kabobs or ground beef.

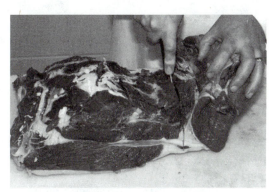

Fig. 15–102. Very desirable boneless sirloin steaks can be cut from the boneless sirloin roast (IMPS 184). The meat cutter must examine the roast carefully to assure that steaks are cut "across the grain" (perpendicular to the muscle–fiber direction). These steaks are often called top-butt steaks or boneless top-sirloin steaks.

Fig. 15–103. Further fabrication of the shortloin by this alternative method involves removing the chine bones or bodies of the *lumbar* vertebrae of the backbone, trimming the "tail" of the shortloin to approximately 1 inch lateral to the loin eye muscle (the shortloin tail is adjacent to the flank in the intact carcass), and trimming away the external fat in excess of 0.25 to 0.3 inch. This extra-short cut trimming may vary, depending on market demand and customer specifications. The resulting sub-primal is called a top loin or striploin, bone-in, or New York strip (IMPS 179 if a 3-inch tail is attached).

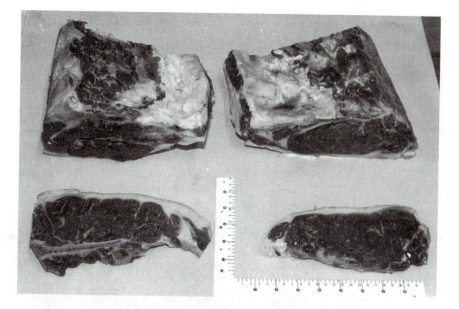

Fig. 15–104. Any steak from the top loin, bone-in, or striploin would officially be called a top-loin steak. It may be called a bone-in strip steak or a New York steak unofficially. Note the *transverse processes* of the *lumbar* vertebrae. The *dorsal processes* of these vertebrae may also remain on the steak or roast. This particular steak is actually a porterhouse steak without the bone and tenderloin muscle. If the remaining bone is removed (the *transverse process* and any remaining portion of the *dorsal process),* the resulting steak would officially be a top-loin steak, boneless (IMPS 180, 2-inch tail; IMPS 180A, 1-inch tail) or unofficially, a boneless strip steak, sometimes called a Kansas City steak.

Table 15–1. Percent of Beef Live and Carcass Weights Expected in Roasts and Steaks, Lean Trim, Fat Trim, and Bone

Component	Live Weight	Carcass Weight
	- - - - - - - - - - - (%) - - - - - - - - - - -	
Roasts and steaks	23	39
Lean trim	15	25
Total retail meat[1]	38	64
Fat trim	14	24
Bone	7	12
Total[2]	59	100

[1]Sum of roasts and steaks and lean trim.

[2]Sum of retail meat, fat trim (carcass), and bone.

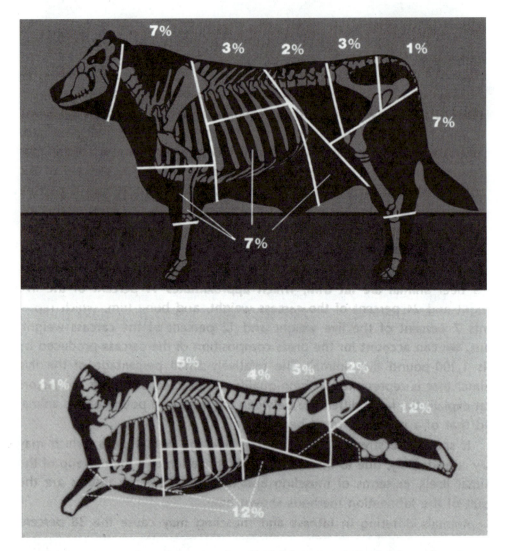

Fig. 15–105. Total retail roasts and steaks fabricated and closely trimmed as shown previously make up approximately 23 percent of the live weight of the animal (see Table 15–1). This 23 percent figure is obtained when the percentages listed in the six primal cuts — 7 percent in the round, 1 percent in the rump, 3 percent in the sirloin, 2 percent in the shortloin, 3 percent in the rib, and 7 percent in the chuck — are added together. Similar figures may be added in the carcass diagram, summing to a total of 39 percent of the carcass weight in the closely trimmed steaks and roasts (see Table 15–1). Note in the live animal diagram the figure 7 percent, which is directed toward all four rough cuts — the shank, brisket, plate, and flank. This 7 percent means that 7 percent of the live weight of the animal is represented in retail yield or, in this case, in lean trim from these four rough cuts. Likewise, the 12 percent figure in the carcass diagram has the same connotation.

RETAIL PRODUCT YIELDS

Summary of Example

Table 15–1 indicates that lean trim accounted for approximately 15 percent of the live weight. Of this, 7 percent was obtained from the rough cuts, and the remaining 8 percent resulted from trimming the primal cuts to produce desirable roasts and steaks. Likewise, in the carcass column, the 25 percent figure for total lean trim is made up of the 12 percent of the carcass weight obtained from the rough cuts, while the remaining 13 percent comes from the trimmings of the primal roasts and steaks. Total retail yield approximates 38 percent of the live weight of the beef animal or 64 percent of its carcass weight.

Note further in the table that the remaining components of the carcass of a beef animal are fat trim, which approximates 14 percent of the live weight and 24 percent of the carcass weight, and bone trim, which represents 7 percent of the live weight and 12 percent of the carcass weight. Thus, we can account for the gross composition of the carcass produced by this 1,100-pound live animal. The relatively small percentage of the live animal that is represented in high-quality edible meat is one of the factors that explain the large difference between the price of a pound of live animal and that of a pound of retail beef.

It should be stressed that these are average yield figures, which may vary considerably, due to the methods of fabrication and the makeup of the animal itself, in terms of muscling and fat. These average figures are the result of the fabrication methods shown here.

Animals differing in fatness and muscling may cause the 38 percent retail yield in the live animal to vary from 33 to 45 percent, while the retail yield from carcass weight may vary from 55 to 75 percent. The data reported here included some bone in the loin and rib steaks. The rest of the roasts and steaks and rough cuts were processed to the boneless forms indicated in the discussion.

Other Sources

Table 15–2 lists data found in Table 15–1 and in Figure 15–105 and compares this information to similar information from other sources. Although the data are similar, there are variations. That variability is one of the reasons

Table 15–2. Percent of Beef Carcass Weight in Roasts and Steaks from Each Primal Cut and in Total Roasts and Steaks, Lean Trim, and Fat and Bone, as Estimated by Three Sources

Product Type and Wholesale Cut	Source of Data		
	NLSMB[1]	AMI[2]	MWE[3]
	- - - - - - - - - - - - - - - - - - (%) - - - - - - - - - - - - - - - - - -		
Roasts and steaks			
Chuck	16.6	13.3	11.0
Rib	6.0	5.7	5.0
Loin	12.2	9.3	9.0
Round	9.6	12.7	14.0
Total	44.4	41.0	39.0
Lean trim	25.3	26.7	25.0
Fat and bone	29.8	30.5	36.0

[1]National Live Stock and Meat Board. A Steer's Not All Steak.
[2]American Meat Institute. Meat Facts 1991.
[3]This text.

why cutting tests, as discussed in Chapter 13, are so important in meat merchandising. A wide discrepancy exists between the data here and the other sources in the roast and steak yield from the chuck. The data reported in this text are based on boneless chuck roasts, whereas the other sources cut blade, arm, and cross-rib roasts, leaving the bone and some of the seam fat in the roasts. Closer trimming of the rib and loin according to the procedures described in this text results in steak cuts with less fat, bone, and "tail" on them than may be true in most cases. Correspondingly, more of the chuck, rib, and loin are found in the fat and bone portions. Much of the advantage in roast and steak yield from the round stems from the salvage of the sirloin tip as an intact roast based on procedures described earlier in this chapter. Utilizing the heel of round as a roast, rather than lean trim, contributes to the roast yield advantage in the round as well. The higher proportion of fat and bone listed may be due to a combination of more boneless cuts, closer trimming of surface and seam fat, as well as differences in carcass fat content.

The USDA equation for predicting percentage of closely trimmed retail cuts from the round, loin, rib, and chuck only is given in "Determination of a Yield Grade" in Chapter 12.

Yield Grade

Currently, the beef yield grades (see Chapter 12) can be used to predict and compare the amount of edible product that beef carcasses produce, and their use does not require the time and effort to cut, trim, weigh, and record data. The animal upon which the percentages of retail product listed in this chapter and summarized in Tables 15–1 and 15–2 might very well be our original 1,100-pound live animal with a 660- to 700-pound carcass. It might have approximately 11 to 11.6 square inches of rib eye at the twelfth rib; possess about 3.5 percent of its carcass weight in kidney, pelvic, and heart fat; and carry about 0.6 of an inch of subcutaneous fat over the rib eye at the twelfth rib. What is the USDA yield grade of this steer? (See Chapter 12 for the discussion on how to calculate yield grades.) The carcass would be stamped Yield Grade 3 by the USDA grader in a packer's cooler. Depending upon the specific carcass weight and the actual rib eye area, the yield grade to the nearest tenth would be in the range of 3.5 to 4.1.

Current Developments

The removal of fat, subcutaneous and internal, during the beef harvest process has been proposed.[6] More rapid chilling and other efficiencies may result from removing the fat as well as the feet, heads, hide, and viscera during the earliest stages of beef processing in the packing house. If adopted, the practice could result in alterations of the cutting procedures described in this chapter. Adjustments in carcass evaluation, such as yield grading procedures, would have to be made if subcutaneous and kidney, pelvic, and heart fat were removed before the carcass was chilled. Alternative methods to estimate retail yield or cutability of beef carcasses could include evaluation of fat and muscling traits on the hot carcass[7] and evaluation of intermuscular fat and other traits on the chilled carcass.[8]

---

[6]Savell, J. W., R. H. Knapp, M. F. Miller, H. A. Recio, and H. R. Cross. 1989. Removing excess subcutaneous and internal fat from beef carcasses before chilling. J. Anim. Sci. 67:881.

[7]Apple, J. K., M. E. Dikeman, L. V. Cundiff, and J. W. Wise. 1991. Determining beef carcass retail product and fat yields within one hour postmortem. J. Anim. Sci. 69:4845.

[8]Kent, K. R., G. W. Davis, C. R. Ramsey, and A. R. Schluter. 1991. Estimates of beef carcass intermuscular fat. J. Anim. Sci. 69:4836.

INDUSTRY BEEF HANDLING
AND DISTRIBUTION

Prior to 1966, most beef was distributed in carcass form; that is, it was generally trucked or shipped by rail from the packing house in forequarters and hindquarters. The quarters were eventually cut into wholesale and retail cuts in some wholesale house or in the "back room" (a meat processing room) of a retail store.

In 1976, more than two-thirds of the beef shipped to supermarkets was centrally fabricated, either at the packing house or at a meat warehouse, and the prefabricated primal and sub-primal cuts were delivered to stores in boxes instead of in the conventional carcass form. Today, a very small amount of beef is shipped in carcass form, and much of that is from one plant to another by some of the large packers.

Modern beef plants have been designed to process beef into the box. Extensive cutting, vacuum packaging, boxing, and product handling systems have been developed to process 200 to 400 beef carcasses per hour. Most of the modern operations are based on moving rails and conveyor systems, which bring the product to the meat cutters, who each perform one specific step in the preparation of a sub-primal cut. Figures 15–106 to 15–112 illustrate some aspects of carcass disassembly and the boxing operation.

Fig. 15–106. After approximately 20 hours in a chill cooler (30° to 34°F), the beef carcasses are ribbed and displayed to USDA meat graders on a moving rail. Note the excellent lighting and the plant employee applying the grade roll (see Chapter 12). (Courtesy, IBP Corp., Dakota Dunes, South Dakota)

Fig. 15–107. Inventory information, including weight and grade, is entered into the plant computers immediately after the carcass has been graded. (Courtesy, IBP Corp., Dakota Dunes, South Dakota)

Fig. 15–108. Carcasses undergo some of the initial breakdown while on the moving rail. Chuck and rib separation is an early step. (Courtesy, IBP Corp., Dakota Dunes, South Dakota)

Fig. 15–109. Still on the rail, the tip is pulled from the round and dropped on the conveyor below. (Courtesy, IBP Corp., Dakota Dunes, South Dakota)

Fig. 15–110. As the product passes down the conveyor, further trimming is accomplished. The conveyor in the foreground carries chucks, whereas ribs are sent down the one on the left. (Courtesy, IBP Corp., Dakota Dunes, South Dakota)

Fig. 15–111. Finished cuts are bagged, and the bags are evacuated and sealed under the hoods of vacuum-packaging machines. (Courtesy, IBP Corp., Dakota Dunes, South Dakota)

Fig. 15–112. The bagged product is placed in boxes. In this plant, the boxes are automatically sorted, using the Universal Product Code System (see box labels on left); palletized; and warehoused by computerized automation. The automatic robot crane can be seen at the end of the aisle moving a pallet of product about halfway up on the beam. (Courtesy, IBP Corp., Dakota Dunes, South Dakota)

16

Lamb Identification and Fabrication

Although the per capita lamb and mutton consumption in the United States continues to decline (see Chapter 1), those persons who eat lamb probably eat a generous amount of it regularly. Based on consumption data, it may be assumed that the majority of meat consumers are not aware of the many attributes of lamb and do not include it as an option when they are planning menus for the family or ordering in a restaurant. In effect, even if they so desired, many consumers cannot consider lamb as an option because it does not appear as a restaurant menu item or in the meat retail cases in some areas of this country.

Thus, the identity, relative value, and palatability of lamb wholesale and retail cuts are discussed in this chapter. Livestock producers may utilize the product knowledge in breeding animal selection and in planning lamb feeding programs to enhance consumer acceptance. Members of the meat industry may be introduced to ideas that will improve the processing, distribution, demand, and retailing in the lamb business. Educators at all levels have less access to information about lamb than some other meat species. Each reader is a food consumer, who, after reading this chapter, will perhaps find increased variety and enjoyment derived from adding lamb to his or her meat diet.

Lamb has long been noted for its flavor delicacy and for its tenderness. Starting in biblical times, and continuing through to this modern age, reference has continually been made to the desirability of lamb meat. It is featured frequently for gourmet dining at home and in hotels and restaurants. However, many people have had unpleasant experiences with lamb or mutton, due to a lack of knowledge on their part, or, perhaps, even on the part of those responsible for the processing, fabrication, merchandising, preparation, or serving of the ovine meat, be it lamb or mutton.

IDENTIFICATION

The Uniform Retail Meat Identity Standards[1] and the Institutional Meat Purchase Specifications[2] (see Chapter 13) will be used as the basis for identification in this chapter. For more complete muscle identification and nomenclature, refer to *Ovine Myology* by Kauffman, St. Clair, and Reber, University of Illinois. Be sure to check the last paragraph in Chapter 13 for a description of *The Guide to Identifying Meat Cuts,* an excellent meat identification reference.

FABRICATION

The skeletal structure of the ovine in relation to an outline of the wholesale cuts is shown in Figure 16–1, and a more detailed chart of wholesale and retail cuts and how to cook them appears as Figure 16–2. These two figures can be used for reference as the discussion moves through cutting a lamb carcass. The lamb retail cut photos in the color photo section serve as a visual aid for the many lamb cuts discussed here.

Figures 16–3 to 16–61 appear through the courtesy of South Dakota State University.

More Detailed References

In addition to those references indicated in Chapter 13, the following is especially appropriate for lamb: *How to Cut Today's New Lamb for Greater Sales and Profits,* published by the American Lamb Council, 200 Clayton Street, Denver, Colorado 80206.

The First Steps

For demonstration purposes, let us assume that generally, when lambs are finished for market, they will weigh approximately 100 pounds. Perhaps in the future, we will see lambs ready for market at 150 pounds, due to the current trend toward larger lambs. After harvest, the resulting carcass represents approximately 50 percent of the live weight. Therefore, the 100-pound live animal will have a carcass weight of approximately 50 pounds.

---

[1]National Live Stock and Meat Board, Chicago (now National Cattlemen's Beef Association, Denver).

[2]Agricultural Marketing Service, USDA.

LAMB CHART
LOCATION, STRUCTURE
AND NAMES OF BONES

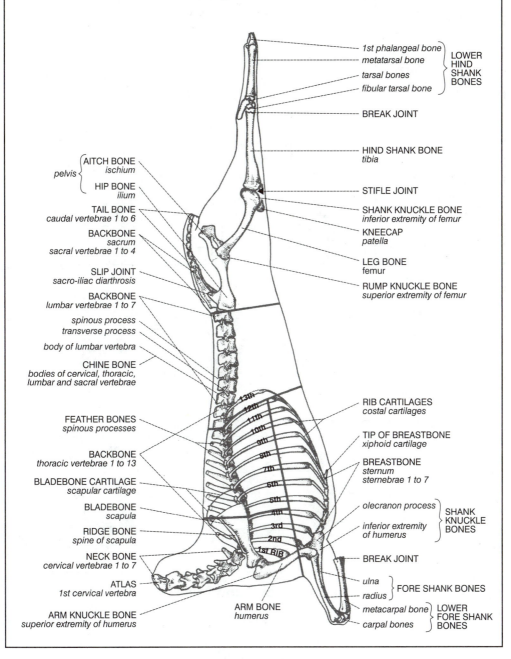

Fig. 16–1. Ovine anatomy.

LAMB CHART

RETAIL CUTS OF LAMB — WHERE THEY COME FROM AND HOW TO COOK THEM

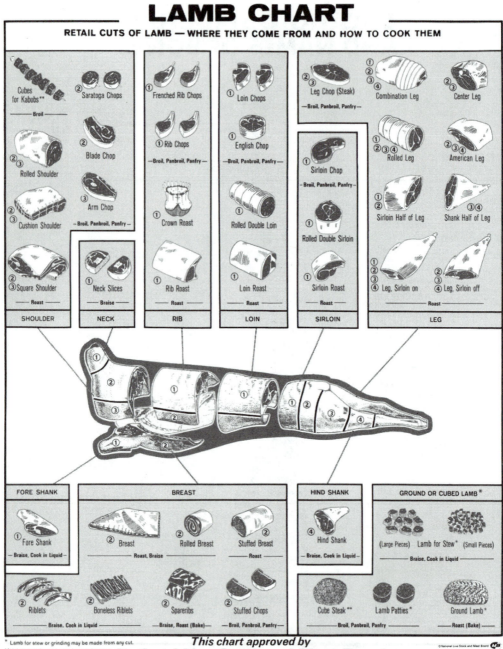

This chart approved by
National Live Stock and Meat Board

* Lamb for stew or grinding may be made from any cut.
**Kabobs or cube steaks may be made from any thick solid piece of boneless Lamb.

Fig. 16–2.

Fig. 16–3. Unlike beef or pork carcasses, lamb carcasses (IMPS 200) are not split before they are taken off the harvest floor. One reason for this is that lamb carcasses, being light in weight, can be cooled and handled as whole carcasses very readily. In addition, some cuts from the loin and rib may be double cuts if the carcasses are unsplit.

The lamb packing industry does not rib (separate foresaddle and hindsaddle) lamb carcasses under normal marketing procedures. Shipping ribbed carcasses results in shrinkage and dehydration of the exposed rib eye surface, and the lamb carcass need not be divided for ease of handling. Lamb quality and yield grading standards do not require that the carcasses be ribbed (see Chapter 12); therefore, the chilled, unribbed lamb carcasses may be transferred to the fabrication department to be cut, vacuum-bagged, and prepared for shipping as boxed lamb cuts. However, in classroom, laboratory, or carcass contest situations, in order for the subcutaneous fat thickness and rib eye area to be evaluated properly, the carcasses must be ribbed.

Ribbing a lamb carcass consists of separating the foresaddle and the hindsaddle between the twelfth and thirteenth ribs. (The terms *foresaddle* and *hindsaddle* are used because the carcass has not been split.) Because the lamb carcass is not split longitudinally, the backbone cannot be seen, so the ribber must count the ribs from the inside to find the twelfth rib. Normally, the knife is inserted toward the outside to mark between the twelfth and thirteenth ribs. A cut is then made in a manner so as to bisect the loin eye muscle *(longissimus)* at right angles to its long axis so that there is no distortion in its cross-sectional size. A skilled knife cutter may cut through the muscle and the cartilaginous connection between the *thoracic* vertebrae, thus completing the separation without using a saw. However, using a saw is very helpful in severing the backbone neatly.

After the carcass has been ribbed, the rib eye and fat cover over the rib eye can easily be measured. For fabrication, the hindsaddle is easily separated from the foresaddle when the attachment between the two, which is the flank muscle, is severed. (See Fig. 16–4.)

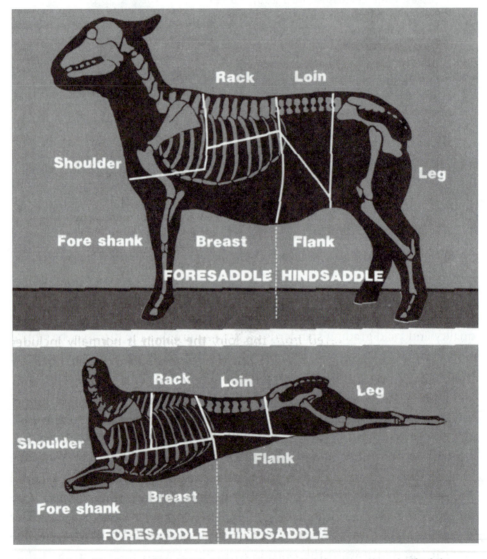

Fig. 16–4. The foresaddle (IMPS 202) comprises slightly more than half of the carcass weight, as is the case with the beef forequarter. The foresaddle, composed of the shoulder, rack, foreshank, and breast, makes up 51 percent of the carcass and 25.5 percent of the live animal, while the hindsaddle (IMPS 230), composed of the loin, leg, and flank, constitutes 49 percent of the carcass weight and 24.5 percent of the live weight. The leg, loin, rack, and shoulder are the primal cuts of lamb.

Hindsaddle

The first step in the fabrication of the hindsaddle is the removal of the flank. The wholesale flank comprises 1.5 percent of the live weight and 3 percent of the carcass weight. (See Fig. 16–5.)

When kidney chops are made, much more flank must be left on the loin, and the separation must be made farther from the back line. The resulting narrow flank can be completely removed with a knife cut that will go outside or laterally to the end of the thirteenth rib. The largest, flat, straight muscle in

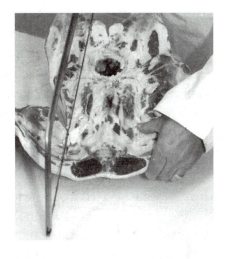

Fig. 16–5. The method in which the flank is removed may be modified in several ways. The modern process is to remove as much of the flank as possible from the loin, since the loin is much more palatable than the flank. As shown, a measurement may be made laterally or away from the end of the loin eye, a distance equal to one-half the width of the loin eye itself. In industry, a 3-inch distance from the eye is common on the trimmed loin (IMPS 232), but this leaves a rather long, undesirable "tail" on each loin chop. The closer trim, as illustrated here, means a more satisfied consumer.

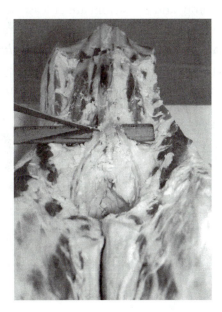

Fig. 16–6. Cutting through the muscle and connective tissue on each side of and perpendicular to the backbone at the point where the *femoral* arteries branch from the large artery *(dorsal aorta),* which lies ventrally to (under) the spinal column, will separate the leg of lamb and the loin. The cut should graze or pass immediately cranially to (in front of) the hipbones.

the flank is called the *rectus abdominis.* In beef, this muscle becomes the flank steak, but in lamb, its size, texture, and flavor dictate its use as ground lamb. Flanks are sometimes rolled and pinned to be roasted.

Kidneys are enclosed in the cranial (front) portion of the kidney fat. The kidneys and kidney fat together are sometimes called the *kidney knob.* It should be carefully trimmed, pulled away from the underlying tenderloin, and removed before the thirteenth rib is cut.

The largest wholesale and retail cut in the lamb carcass is the leg (IMPS 233). In wholesale form, it comprises 33 percent of the carcass and 16.5 percent of the live lamb. Note that these carcass and live percentage figures have a relationship of 2 to 1, since the carcass weight is one-half of the live weight. The retail leg of lamb, fabricated with the bone in (IMPS 233A), comprises about 24 percent of the carcass and 12 percent of the live weight. A boneless leg (IMPS 233B) would account for a smaller proportion of the carcass and live weight. A retail cut is ready for the consumer, while a wholesale primal cut may contain several sub-primals, or retail cuts and, as such, usually requires additional fabrication, trimming, or boning before being ready for the consumer. Relative weights help to establish a concept of live animal and carcass composition in terms of wholesale and retail cuts.

When the leg is severed from the loin, the sirloin is normally included with the leg. The resulting primal cut may be termed a *long-cut leg;* however, just *leg-o-lamb,* or *leg of lamb,* implies the whole leg with sirloin. This lamb pelvic limb fabrication differs from beef fabrication in which the round, including the rump, and the sirloin are at least two separate cuts, and pork fabrication in which the sirloin is left with the loin rather than with the leg. The separation in lamb is made at the seventh or last *lumbar* vertebra. Key skeletal landmarks in cut identification are *sacral* and *coccygeal* (tail) sections of the vertebral column, the pelvic bone, and the leg bone *(femur)* (see Figure 16–1).

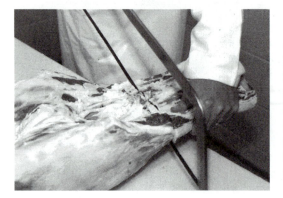

Fig. 16–7. Cutting through the vertebral column with a saw results in a flat surface on the face of the loin (IMPS 232) and the leg (IMPS 233). Complete separation can be made easily with a power saw. Using a knife and leverage at the disk between two adjacent vertebrae will separate the two cuts, but the vertebral ends will not be flat.

Fig. 16–8. The cranial (front) face of the leg is composed of muscle and bone sections, which can be identified, starting at the dorsal surface and moving ventrally on the leg face. The muscles immediately below the *subcutaneous* fat on the back are the top sirloins (a) *(gluteus medius)*. The cartilage tip of the anterior end of the pinbone (b) *(ilium)* is shown as the whiter structure in the fat on the lower right edge of the

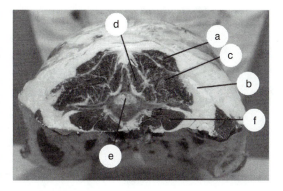

right *gluteus medius.* Ventral to the *gluteus medius* on each side are the loin eye muscles (c) *(longissimus),* which are not only the largest muscles on this cut surface but also the heaviest muscles in the whole carcass. The *longissimus* is called loin eye in the hindsaddle because it is located in the loin. It is called the rib eye when it is viewed on the rib (rack) of the foresaddle. The small muscles adjacent to the *dorsal processes* of the vertebra are the (d) *multifidus.* The large, rounded portion of the bone is the body of the seventh (e) *lumbar* vertebra with its *dorsal process* extending upward between the two *multifidus.* The *transverse processes* or T–bones extend ventrally to (under) the *longissimus.* The small muscle system ventral to the T–bones is made up of tenderloin muscles. These are the most tender muscles in the carcass. The larger muscle is the (f) *psoas major* and is associated with a much smaller *psoas minor.*

The Latin names are universal throughout the world, thus their inclusion. Common names can vary even between sections of the United States. The ISMISC standards should help eliminate this variation.

Fig. 16–9. This pair of legs must be split before the legs can be fabricated. In a young animal, the aitch bone (pubic bone) can normally be split with a knife because it is joined by cartilage. Once the aitch bone has been split, the remaining muscles can also be split with a knife, exposing the cut surfaces of the aitch bone. Then the *sacral* and *coccygeal* portions of the vertebral column must be severed with a saw to complete the separation of the pair of legs. A power saw completes the whole process in a single pass.

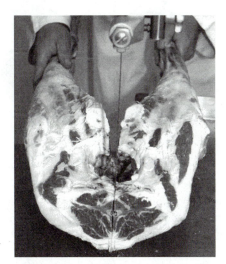

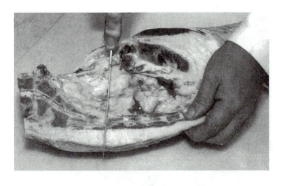

Fig. 16–10. The first fabrication step is the removal of the tail bone, composed of a number of the *coccygeal* vertebrae, the actual number depending on how long the tail was in the live lamb. Ordinarily, three tail vertebrae are left on the leg.

A lamb carcass usually has a small amount of pelvic fat in the pelvic cavity located between the aitch bone and the backbone, and it is removed when the leg of lamb is fabricated.

Fig. 16–11. Outside fat is trimmed as needed until the remaining fat cover on the leg does not exceed ¼ inch. The fell, or thin membrane, separating the pelt from the subcutaneous fat is not removed in trimming unless absolutely necessary, since this fell holds the shape of the leg and helps retain moisture and juices during cooking. When the flank side of the leg is trimmed, special care must be taken to remove the *prefemoral* lymph node.

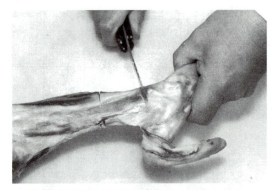

Fig. 16–12. The large tendon at the rear of the leg, the *Achilles* tendon, is severed at its origin so that the hock can be removed. There are three common types of lamb legs: Frenched, American, and boneless.

In a Frenched leg, after the tendon has been loosened, the muscle is severed about 1½ inches above the hock joint and cleared away from the bone. Slightly above the hock joint is the break joint. It is scored with a knife and may be broken across the edge of the cutting table. The hock bone, or trotter, is thus removed.

The break joint or *epiphyseal* plate is the area of growth in the long bones, so in young animals it is cartilaginous and easily broken, while in older animals whose bones have ceased growing, the *epiphyseal* plate has turned completely to bone and is tightly fused. The break joint is located above the hock joint, the joint that allows the animal to flex its legs and walk.

Fig. 16–13. The exposed shank bone is composed of the *tibia* and the *fibula,* which are fused together in the ovine. The Frenched leg is so named because the shank bone *(tibia–fibula)* is bare and exposed. *(Frenched* refers to any bone that is exposed.) Prior to Frenching the shank, this was a leg, lower shank off (IMPS 233A).

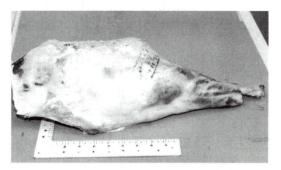

Fig. 16–14. In fabrication of the American leg, after the shank muscle and *Achilles* tendon have been loosened from the shank bone, as was done in the fabrication of the Frenched leg, the stifle joint is entered from the caudal (tail) side. This is the knee joint, which joins the shank bone with the large leg bone, the *femur* (see Fig. 16–1).

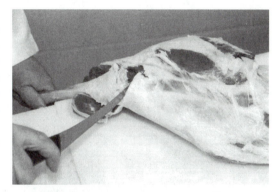

Fig. 16–15. The shank muscle (largely *gastrocnemius)* may be left attached to the American leg after the shank bone *(tibia–fibula)* has been removed. The knee cap *(patella)* is normally left in an American leg (IMPS 234A). The shank muscle can be folded over the small cavity next to the *femur,* created when the shank bone was removed and pinned in place with tooth picks or

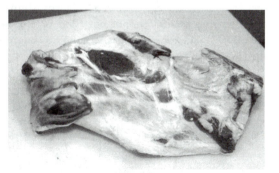

skewers. The main portion of the American leg is identical to the Frenched leg. An American leg is more valuable than a Frenched leg because more bone — the shank bone as well as the lower shank bone — has been removed.

Fig. 16–16. Note the abundant amount of connective tissue in the shank muscle, as indicated by the white or silvery streaks. Therefore, the authors prefer to cut away and separate the shank muscle from the leg and then to utilize it as ground lamb, since it does contain large amounts of connective tissue and is considerably less palatable than the leg itself. This modification of the American leg results in a more uniformly palatable bone-in leg roast.

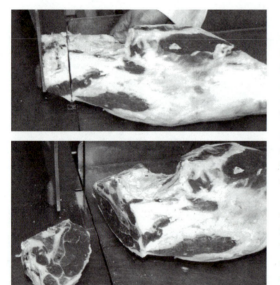

Fig. 16–17. The sirloin is normally left on the wholesale primal leg of lamb. Thus, sirloin chops can be removed from a leg of lamb. *(Top)* The location of the chop removal from the leg is indicated here. The largest muscle in a sirloin chop *(bottom)* is the top sirloin or *gluteus medius* muscle. Note the presence of the flat pelvic bone in this sirloin chop.

The whole leg may be sawed into chops and slices (steaks) starting at this cranial (front) end. The result would be four to six sirloin chops, depending on how thick they were cut. These chops are identical to beef sirloin steaks in bone and muscle structure, differing, of course, in size and color (see Figs. 15–85 to 15–91). A wedge-shaped area corresponding to the rump in beef and containing the main portion of the pelvic bone should be removed before the rest of the leg is cut into leg slices (steaks).

The rump section cannot be sliced or steaked because of the pelvic bone, but it makes excellent kabobs after the bone has been removed. The center roast and most of the shank half can then be sliced.

Fig. 16–18. The sirloin may be removed from the wholesale leg to produce a lamb leg, short-cut retail roast. The sirloin chops may be merchandised as well.

Fig. 16–19. The lamb leg may be separated into a sirloin half *(left)* and a shank half *(right)*, as shown. The sirloin half or portion, a center roast, and the shank half or portion can also be broken down. The rump section can be included with the sirloin, as shown, with the shank half, or with a center portion.

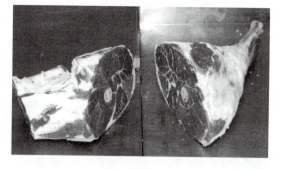

Fig. 16–20. The bone-in lamb leg offers as much versatility as any wholesale cut can. A sirloin chop and roast, a center-leg slice (steak), and shank roast are illustrated. Center-leg slices (steaks) contain the full cross section composed of the top *(left)*, bottom *(right)*, and tip *(near rule)* muscle systems, which may differ in palatability.

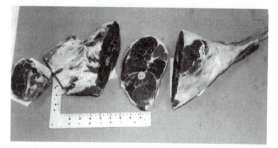

The leg can be boned out rather than steaked. The boneless leg (IMPS 234B and/or 233B) would be roasted and, as such, is almost certain to be highly palatable. There is a considerable amount of bone (the whole pelvic bone and the *femur*) left in the Frenched and American legs, which can be inconvenient to the modern consumer who serves such a roast. Boning the leg and netting or tying the resulting boneless roast before it is offered to the consumer can eliminate the problems connected with serving a leg of lamb.

Fig. 16–21. The pelvic bone is exposed between the pinbone and the aitch bone to permit the knife to enter the slip joint *(sacroiliac — see Fig. 16–1)* and to separate the pelvic bone *(ilium)* from the backbone *(sacrum)*. After separation, the *sacrum* is removed.

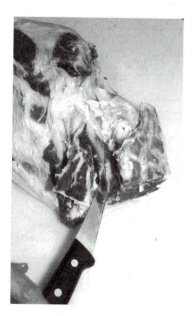

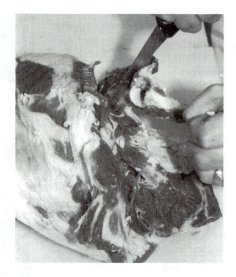

Fig. 16–22. The pelvic bone is separated from the surface of the leg and from the ball of the *femur* and then removed.

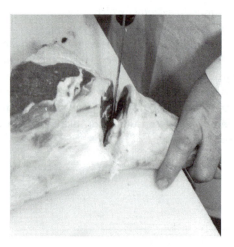

Fig. 16–23. The shank is removed at the stifle joint.

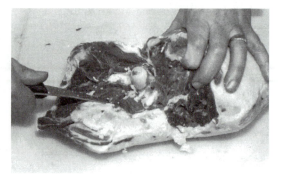

Fig. 16–24. The proximal (inward) end of the *femur* is loosened.

Fig. 16–25. The *femur* can be pulled out from the shank end without any muscles in the leg being cut through. This technique is called tunnel boning (see Figs. 13–7 to 13–18 and 20–6 to 20–9). After all bone has been removed, the sirloin end is trimmed. The *femur* may more easily be removed if the lean on the top (medial) surface of the leg is opened from one end of the *femur* to the other. The *femur* is then freed of muscle and connective tissue and can be lifted out. The knee cap *(patella)* is also removed.

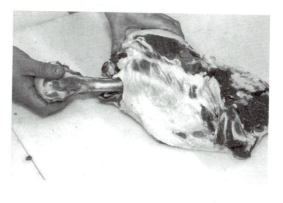

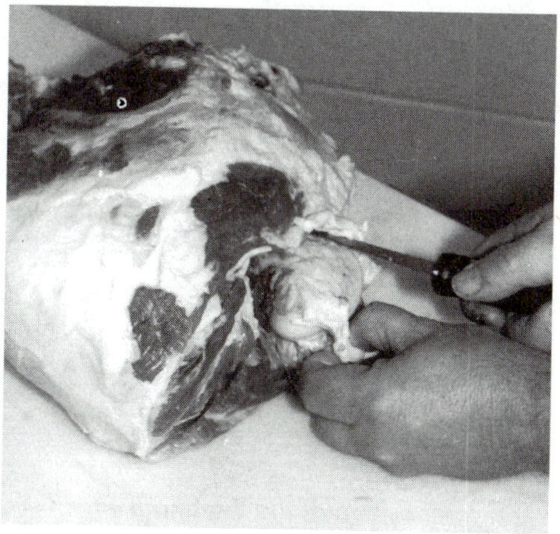

Fig. 16–26. The *popliteal* lymph node and surrounding fat should be removed from the seam between the bottom *(biceps femoris)* and eye *(semitendinosus)* muscles at the distal (rear) end of the boneless leg.

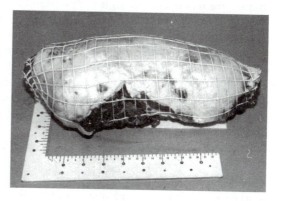

Fig. 16–27. The boneless leg (IMPS 233B), like any boneless roast, must be tied so that it can be roasted and easily sliced after it has been cooked. The jet netter and the elastic net (see Fig. 14–22) have replaced the old-fashioned string and butcher's knot. A boneless leg of lamb is identified by its size and shape. Since it is a long-cut leg containing the sirloin, the boneless leg can be recognized by its length. The keys to species identification are cut size and color. Lamb fat is harder than beef or pork fat. Also, the fell membrane may be present on lamb roasts. No such membrane is present on beef or pork cuts.

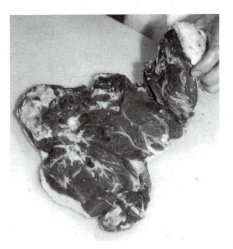

Fig. 16–28. Following the intermuscular seam between the inside *(semimembranosus)* and the outside *(biceps femoris)* muscles of a tunnel-boned leg will open up the leg to form a large portion, which, because of its thickness (1 or 2 inches), could be charcoal grilled or broiled.

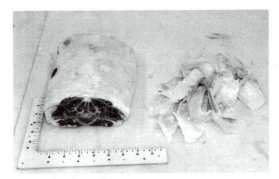

Fig. 16–29. The loin (IMPS 232) is the most valuable wholesale and retail cut in a lamb carcass, since it contains the most tender muscles of the carcass. Even so, the rack is commonly higher priced than the loin. The main reason for the high retail price of lamb rack and loin chops is that there are so few of them from each animal. Only 4 percent of the live animal is tender, juicy loin chops. Surface fat should be removed carefully, using a shaving action to prevent scoring the surface of the lean of this valuable wholesale cut.

The composition of the lamb loin varies slightly from end to end. At the seventh *lumbar* vertebra (the caudal end shown in Fig. 16–29), the *longissimus,* or loin eye, is oval-shaped; the tenderloin *(psoas)* is at its maximum size; and the *gluteus medius* (top sirloin muscle) is present (see Fig. 16–8 for muscle arrangement). From a cranial view (not shown), the loin eye is obviously larger and more symmetrically shaped. The tenderloin is not present because it originates at the last rib as a very thin muscle and gets progressively larger until it reaches its maximum size at the seventh *lumbar* vertebra. The hanging tenderloin should not be confused with the actual tenderloin (see Fig. 15–50). Since the hanging tenderloin is a part of the diaphragm muscle, it is considerably less tender than the actual tenderloin muscle because of its constant activity in the live animal; it should be removed for lean trimmings.

The fell membrane, separating the pelt from the subcutaneous fat, is removed from the loin except for the specific area containing inspection and grading stamps. Since the loin will be made into chops, which will be broiled (cooked with very high heat for a short time), the fell will shrink and distort the shape of the chops. The fell is left intact on leg roasts to hold in the juices during the slower roasting process. The subcutaneous fat on the surface of the loins should be reduced to ⅛ to ¼ inch thick when the fell is removed. The trimmed loin may be used as a roast but is most often cut into chops.

Fig. 16–30. Double loin chops may be fabricated, since the lamb carcass is not split for chilling. The advantage of cutting double chops is the increased portion size. The chop illustrated would be more appealing to customers if at least one-half of the subcutaneous fat was removed.

If the kidneys are left in the loin, they are called kidney chops (not shown), or English chops. A 2- to 3-inch chop is cut from the rib (cranial) end of the loin before the double loin is split. After the muscle has been cut with a steak knife, the separation should be completed with a saw. The resulting double or kidney chop is in demand mainly because of its uniqueness, rather than for its utility, since it does contain large amounts of subcutaneous and kidney fat. The kidneys and flank muscles are also present, neither of which compares favorably with the quality of the loin eye muscle itself.

Fig. 16–31. Removing the *lumbar* vertebra from a double loin chop produces a boneless double loin chop. The tail, or flank end, of the chop should be cut 1½ to 2 inches longer than shown in Fig. 16–30 to produce the round configuration illustrated here. The loin should be boned, rolled, and netted or tied. Individual chops (1 inch thick minimum) or slices can be cut from the larger unit to be broiled, or the boneless loin can be roasted and sliced.

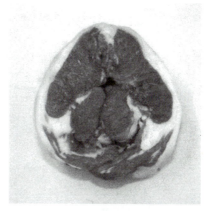

Fig. 16–32. The double loin may be split on a band saw. Each *lumbar* vertebra is split exactly in the middle, resulting in two loin roasts, each the mirror image of the other.

Fig. 16–33. The most popular method of fabricating the loin is to make individual single loin chops. The *transverse processes* of the *lumbar* vertebrae form the characteristic "T" of loin chops, just as they form the "T" in the T–bone steak of beef. The presence of the tenderloin muscle *(left)* is a major key to loin chop identification.

The term *chop* originally referred to any piece of meat fabricated (chopped off) with a cleaver. Thus, only relatively small cuts could be fabricated, while larger cuts called steaks had to be removed with a saw. This is why all lamb cuts, except leg slices (steaks), are called chops, for instance, loin chops, rack chops, arm chops, and blade chops. Today, even chops are fabricated with an electric band saw.

Foresaddle

The foresaddle (IMPS 202), which represents slightly more than one-half of the carcass weight, is composed of the primal rib (rack, IMPS 204), shoulder (IMPS 207), and the rough cuts, the foreshank (IMPS 210) and the breast (IMPS 209). Recall that the separation between the foresaddle and the hindsaddle (IMPS 230) is made between the twelfth and thirteenth ribs (see Figure 16–4).

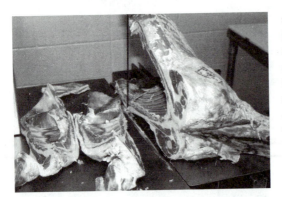

Fig. 16–34. Since breasts and shanks are still on the foresaddle (IMPS 202), they should be separated from the shoulders (IMPS 207) and the rack (IMPS 204). Saw across the arm bone *(humerus)* at a point slightly above the junction of the arm bone and the foreshank bone, which is the fused *radius* and *ulna*. The ribs and body wall are cut parallel to the backline.

Fig. 16–35. The separation of the foresaddle (IMPS 202) into the portion containing the shoulder (IMPS 206) and the portion containing the rack (IMPS 204) is usually made between the fourth and fifth ribs, leaving an eight-rib rack. Some lamb fabricators cut a seven-rib rack (as shown here by cutting between the fifth and sixth ribs). The cranial portion of the rack near the shoulder is less desirable than the caudal por-

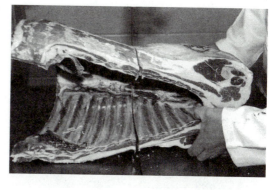

tion near the loin, since it contains more connective tissue, and the *longissimus,* or rib eye muscle, is smaller.

The largest wholesale cut in the foresaddle, and second in size in the whole lamb carcass, is the shoulder (IMPS 206). The shoulder contains a number of bones, making it difficult to carve and slice. Shoulder cuts can be priced very economically and, if fabricated and cooked correctly, can provide delightful dining. Skeletal components are especially helpful in shoulder-cut identification, so note them carefully on the carcass diagram (see Fig. 16–1).

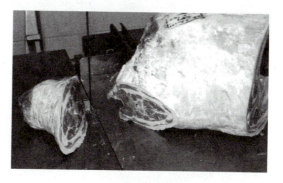

Fig. 16–36. Remove the neck from the shoulder by making a cut extending from the back. IMPS 207 indicates that the cut is perpendicular to the neck, leaving no more than 1 inch of neck on the shoulder.

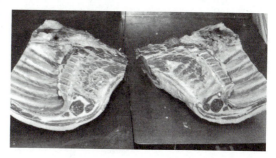

Fig. 16–37. Making a cut with a band or hand saw down the middle of the backbone will separate the pair of shoulders. The wholesale primal shoulder is often called the square-cut shoulder because it usually fits the dimensions of a square. The blade face, shown in this figure, has a portion of blade cartilage exposed, as well as the eye (*longissimus*) muscle.

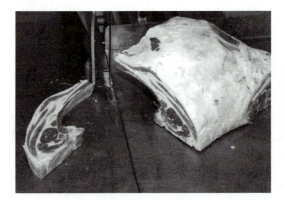

Fig. 16–38. If chops are removed from the shoulder, those from the blade face are called blade chops because of the presence of the blade bone *(scapula)*. The rib eye, or *longissimus,* extends through this area.

Fig. 16–39. At right angles to the blade face is the arm face. Thus, the chops are called arm chops. As in beef, the arm side of the shoulder is more muscular than the blade side, but less tender, since the *triceps brachii* and deep *pectoral* muscles do a considerable amount of work in providing locomotion for the live animal. Generally, the more use a muscle receives throughout the animal's life, the less palatable it will be.

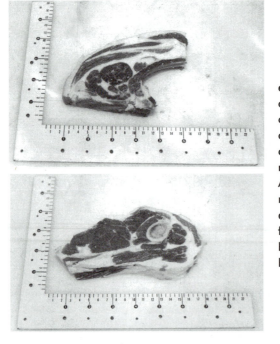

Fig. 16–40. Shoulder chops, either blade *(top)* or arm *(bottom)* chops, are less palatable than rack or loin chops, because the shoulder contains a considerable amount of connective tissue surrounding numerous small muscle systems. Arm chops from the lamb shoulder normally include sections of the ribs, as shown in Fig. 16–39, but when fat seams are large, the chops may be more closely trimmed, as shown here.

Fig. 16–41. There are several shoulder fabrication alternatives that are preferred to making chops because the resulting products are more likely to please the consumer. One alternative is the boneless blade (Saratoga) roll, which is an extension of the rib eye muscle into the shoulder. The first step in its fabrication is to take out the rib cage. Next, the eye and all muscles medially to (lying above) the blade bone are removed, following the natural seam. The eye is rolled tightly within the adjacent muscles and held in the Saratoga roll by wooden skewers.

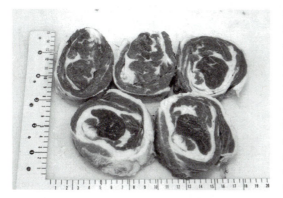

Fig. 16–42. The Saratoga roll is cut between skewers to form boneless blade (Saratoga) chops. The rib eye muscle, or *longissimus,* begins here in the shoulder and extends to the leg. It is not only the largest muscle in the lamb carcass but also one of the most palatable. Thus, boneless Saratoga chops, which are composed largely of this eye muscle, are very tender and juicy and the type of lamb cut that will encourage the consumer to repeat lamb purchases.

Fig. 16–43. The remainder of the shoulder may be diced for lamb stew (IMPS 295) or kabobs (IMPS 295A) after the blade and arm bones have been removed. Outdoor cookery is an increasingly popular leisure time activity. Lamb shish kabobs rank high on the list of desirable meats for outdoor cookery.

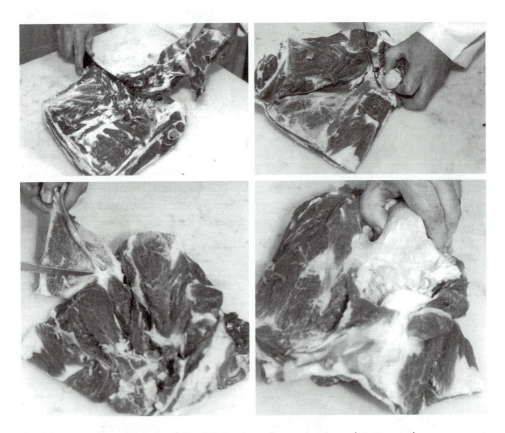

Fig. 16–44. A third shoulder fabrication alternative is to bone out the square-cut shoulder completely and then to utilize it as a boneless roast.

First remove the muscle along the inside of the neck bone (*longus colli*) and trim away any evidence of dried blood deposits remaining from harvest. The rib cage is then removed, leaving as little muscle as possible on the bones (*top left*). The yellow neck leader (*ligamentum nuchae*), which contains large amounts of elastin, must be removed because it is very unpalatable — it will not break down during cooking.

Running the knife close to the medial surface of the blade, continuing the cut through the ventral, caudal corner of the shoulder to the caudal aspect of the arm bone, splitting the lower, thin corner of the shoulder, and then folding open the medial layer of that corner will expose the large, flat blade bone (*scapula*) and the round arm bone (*humerus*). With the back and front of the shoulder used as a hinge (*top right*), the bones may be separated at the joint or removed together. In this illustration they have been separated, the arm bone removed (*top right*), and the blade bone with its protruding spine pulled free (*bottom left*).

The *prescapular* lymph node is located in the shoulder in front of the *scapula* (*bottom right*) and is surrounded by a large fat deposit. It must be carefully removed, since it would detract from the desirability of the roast. Finally, the outside fat cover is removed where it exceeds ¼ inch in thickness. The fell is left intact if possible.

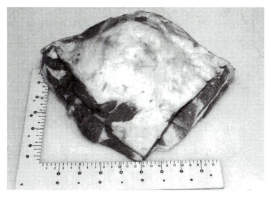

Fig. 16–45. The boneless shoulder maintains its original square shape quite well. It may be stuffed with dressing and sewn together at the blade and arm-face seams to make a desirable cushion-shoulder roast.

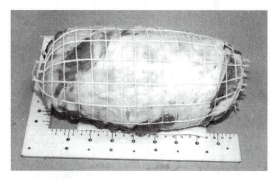

Fig. 16–46. The netted boneless shoulder (IMPS 208) roast may be the most preferred method of merchandising the shoulder. The roast is rolled around the extension of the rib eye before being placed in the netting or tied. The netting remains in place throughout roasting to the desired degree of doneness (see Chapter 24). The net is removed before the roast is served.

Fig. 16–47. When cut through the center *(left)*, even after being boned and trimmed rather carefully, the boneless shoulder still contains some seam fat and connective tissues. Yet, when properly roasted, the boneless roast will be more appealing to some consumers than broiled arm or blade chops from the bone-in shoulder might be.

The boneless leg *(right)* compared to the boneless shoulder is certainly more desirable, since it has larger muscle systems with less seam fat and fewer connective tissues, and would be preferred for special occasions. Perhaps, with a large price differential, the shoulder would provide the most desirable combination of economy and palatability.

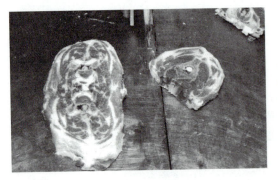

Fig. 16–48. The neck, which was removed from the paired shoulders, may be sawed into neck slices, a retail cut that lacks wide consumer demand. The lean may be salvaged for trim.

A wide variety of retail cuts originate in the lamb shoulder. The square-cut shoulder (IMPS 207) may be merchandised as a retail cut, or shoulder chops (IMPS 1207) — both arm and blade chops — may be fabricated from it. Alternately, the rib eye muscle may be utilized as Saratoga chops and the remainder for kabobs (IMPS 295A). A third method is to bone and roll the shoulder to make a highly desirable boneless shoulder roast (IMPS 208).

Fig. 16–49. The commonly accepted cutting technique for lamb (see Fig. 16–1) retains much of the rib end within the wholesale rib (IMPS 204). However, the "tails" of those ribs must be removed at a point approximately one-half the width of the rib eye away from the eye in order to make acceptable retail rib (rack) chops.

Fig. 16–50. The fell and some fat are removed from the rack roast, since like the loin, its most popular use is as rib (rack) chops, which are usually broiled. If the wholesale rack possesses more than ¼ inch of outside fat cover, it is trimmed to that thickness or less. When federally graded overfat lambs are trimmed, it may be necessary to remove the grade stamp; however, the fabricator usually tries to leave the stamp on the rib so that it can be seen on the chops in the self-service meat case.

The smallest wholesale and retail cut of lamb is the rib (rack) (IMPS 204), since on the average only 3.5 percent of the live lamb weight is actually retail rib (rack) chops. The limited supply may explain why rib (rack) chops are priced higher than other cuts of lamb. The term *rib* is used synonymously with the term *rack* and is the recommended ICMISC name. A portion of the blade or its cartilage appears in rib (rack) chops from the cranial end of the rack (see Fig. 16–1).

Fig. 16–51. Use of the unsplit, trimmed rib to produce the double chops discussed earlier would result in larger portions. However, the rack is more often split down the center of the backbone into two single lamb ribs.

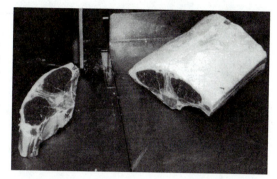

Fig. 16–52. The wholesale primal rib differs in composition from end to end. Note the blade-bone cartilage and intermuscular fat seams in the cranial (sixth rib) end rib (rack) chops *(right)*. More merchandisable rib (rack) chops may be produced if the bodies of the *thoracic* vertebrae are trimmed away with the power saw before the chops are cut (see Fig. 15–8).

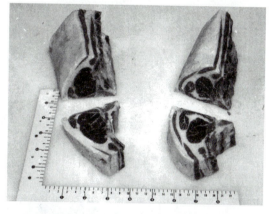

Fig. 16–53. Although a crown roast may be fabricated from a foresaddle cut, as described in Figs. 16–34 and 16–35, some modifications of cutting procedure may result in a larger, more easily shaped roast. In the example in Fig. 16–58, the breast and shank were not removed from the foresaddle, and the rib was separated from the shoulder at the third rib.

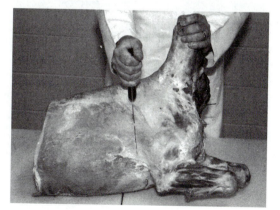

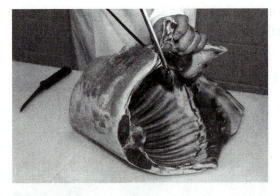

Fig. 16–54. Normally, the fabricator separates the breast from the rib (rack) by cutting laterally to the eye, a distance equal to one-half the width of the rib eye, for a close trim. However, to make a crown roast, which is a rather prestigious cut, the fabricator leaves much longer rib ends on the rack to form the crown. After the ribs have been sawed, the separation is completed with a knife.

The diaphragm is called the skirt as it appears on the foresaddle. (On the hindsaddle it was called the hanging tenderloin.) The rib eyes are large and symmetrical when viewed from this loin or caudal end.

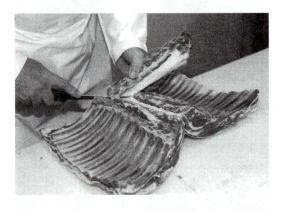

Fig. 16–55. The fabrication of a crown roast is not something that is done every day by a meat retailer; in fact, the consumer would undoubtedly have to make a special request for one. It does take valuable time to fabricate this roast, and the retail price will reflect that cost.

The chine bones or bodies of the *thoracic* vertebrae are loosened by severing the rib connections with a saw and then removed with the feather bones (*dorsal spinous processes*) attached. The blade bone and its cartilage are prominent in this cranial (shoulder) end of the rib.

Fig. 16–56. The fell and outside finish were not broken as the chine bones were removed. The rib bones were Frenched, that is, trimmed of most muscle and fat 2½ to 3 inches from the rib ends.

The portion being trimmed away, which ordinarily belongs on the breast, is composed of thin layers of muscle interspersed with layers of fat; thus, it lacks the high degree of palatability that the rib eye possesses. The muscles between the rib ends, called rib fingers or intercostal muscles, are removed so that the rib ends are completely bare, or Frenched. The blade-bone cartilage in the cranial end is removed.

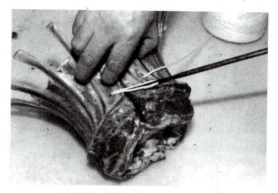

Fig. 16–57. After the tissues have been penetrated with a tying needle, the ends of the rolled rib are tied together with heavy butcher's string.

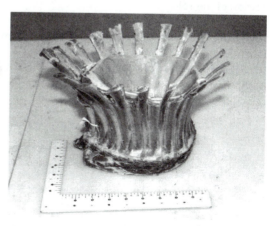

Fig. 16–58. The finished crown roast may be filled with dressing. Lean ground lamb may be used to fill the center as well. The Frenched rib ends can be decorated with small colorful paper or aluminum foil collars for serving. The properly prepared crown roast takes on a "royal" appearance.

The roast is very easy to serve, since all that needs to be done to obtain an individual serving is to cut between the ribs with a knife, giving each person one rib and the accompanying tissues (see Chapter 24).

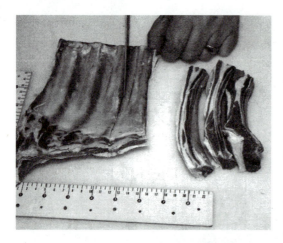

Fig. 16–59. The rough cuts of lamb are the flank of the hindsaddle and the foreshank and breast of the foresaddle. In retail form, they amount to 10 percent of the carcass weight. The excess fat should be trimmed away from the lamb flanks and the remainder should be ground for lamb patties.

The portion of the breast that corresponds to the plate in beef contains rib bones and the sternum (IMPS 209). For the several alternate fabrication methods, the diaphragm, or skirt, is removed. Cutting between each rib of the breast will yield riblets. An alternative is to remove the *sternum* and the rib cage, leaving a boneless breast composed of layers of thin muscles separated by layers of intermuscular fat. The *sternum* and rib cage are called lamb spareribs and may be prepared by barbecuing in a manner similar to the way pork spareribs are prepared. The boneless breast is usually rolled. Only thin muscles are included in the rolled breast, which may be confused with the boneless blade (Saratoga) roll. There is a vast difference in the value of a rolled breast and a boneless blade (Saratoga) roll, since the boneless blade (Saratoga) roll contains the very tender rib eye muscle. By closely examining the two, you can quickly determine the presence of the rib eye in the Saratoga roll. The rolled breast is identified by the alternate layers of thin muscle and fat in the cross section of the roll.

That portion of the breast that corresponds to the brisket in beef does contain a rather thick muscle. This may be separated from the *sternum* and utilized for cubing, much as the beef bottom-round steak is utilized.

Fabricating a breast for stuffing (Scotch roast) entails peeling away the meat portion of the breast from the bone structure, leaving a natural hinge along the full length of one side. This pocket may be stuffed with ground lamb or lamb dressing and then sewed, netted, or tied for roasting.

Fig. 16–60. Lamb foreshanks *(bottom)* (IMPS 210) may be utilized as mock duck or simply lamb shanks for braising. Lamb shanks meld with couscous and vegetables as a hearty, flavorful main course for a cold weather meal. Rear shanks removed from legs can be used in a similar manner.

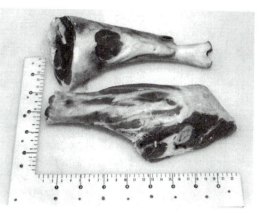

The rough cuts may be many and varied. Shish kabobs or lamb for stewing can originate from large, thick pieces of rough-cut lean (see Fig. 16–43).

Perhaps the most highly justified use for all of the rough cuts would be to process them into ground lamb (IMPS 296), which has many delightful uses. Some of the variations of rough-cut fabrication may lead to consumer dissatisfaction, due to improper cooking or simply unfulfilled expectations of good eating.

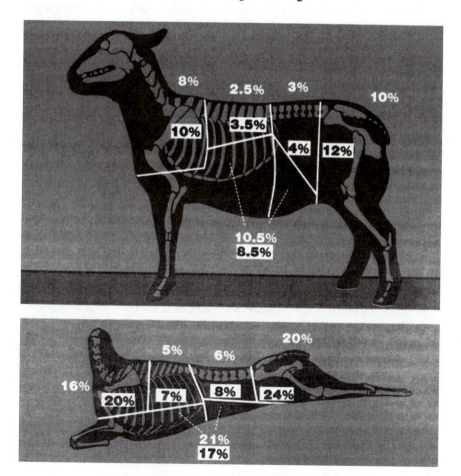

Fig. 16–61. *(Caption on following page.)*

Fig. 16–61. By studying this figure and Table 16–1, you should be able to answer the question: How much retail product does one lamb (or lamb carcass) produce?

First of all, solidify your thoughts concerning the relationship between live and carcass weights. With lambs it is simple, since a lamb carcass weighs approximately 50 percent as much as the live lamb. Thus, the bone-in retail leg represents 12 percent of the live lamb (shown in black figures in a white box on the live lamb), while this same bone-in leg represents 24 percent of the carcass (shown in black on the carcass diagram). The carcass weighs one-half as much as the live animal, so the percent of the carcass in the leg must be two times as great.

Bone-in figures (in black) and boneless figures (in white) are shown, since bone-in and boneless fabrication of the two largest primal cuts, the leg and the shoulder, have been discussed in this chapter. The fabrication of a boneless loin or rack is not shown because it is so seldom done in industry. However, the boneless yield figures mean that all cuts are boneless, including the rough cuts.

Examine the bone-in figures first. Table 16–1 indicates that roasts and chops comprise 29.5 percent of the live animal. Check these percentages on the live animal by adding 12 percent from the leg, 4 percent from the loin, 3.5 percent from the rack, and 10 percent from the shoulder, to total 29.5 percent. This same system works for carcass and live bone-in and boneless cuts. The figures placed on each cut indicate the percent which that particular retail cut represents of the carcass or live weight.

In the table, lean trim represents the trimmed boneless rough cuts as well as the lean trim produced when the wholesale primal cuts were fabricated into retail cuts. In the live lamb, lean trim from bone-in fabrication amounts to 8.5 percent of the live weight. Thus, total bone-in retail yield amounts to 38 percent of the live lamb, that is, 29.5 percent + 8.5 percent.

This same system works for bone-in and boneless retail yield based on live or carcass weights.

Check yourself:

What percent of the carcass is bone-in retail product?

> Answer: 76 percent determined by adding 59 percent for roasts and chops to 17 percent for lean trim.

What percent of live weight is boneless retail product?

> Answer: 34 percent; 23.5 percent + 10.5 percent.

What percent of the carcass weight is total boneless retail product?

> Answer: 68 percent; 2 × 34 percent, since the carcass = ½ the live weight, or also 47 percent + 21 percent from the carcass chart.

All percentage figures used here are based on averages of the detailed carcass analysis of approximately 175 lambs evaluated in the University of Illinois Live Animal and Carcass Evaluation Course during the period from 1965 through 1974. This so-called average lamb possessed the following "vital statistics":

Leg conformation	High Choice	Average fat cover	0.45 inch
Loin-eye area	2.35 sq. in.	Percent of kidney fat	3.8 percent
Fat over twelfth rib (top)	0.23 inch	USDA Yield Grade	3.6
Fat over lower twelfth rib	0.67 inch		

Since all percentage figures quoted here are averages, they can vary according to the methods of cutting and the fatness and muscling of the lambs.

Table 16–1. Percent of Live Weight and Carcass Weight Expected in Bone-in and Boneless Products of Lamb Cutting

Product	Percent of			
	Live Weight		Carcass Weight	
	Bone-in	Boneless	Bone-in	Boneless
	- - - - - - - - - - - - - - - - - - (%) - - - - - - - - - - - - - - - - - -			
Roasts and chops	29.5	23.5	59	47
Lean trim	8.5	10.5	17	21
Fat trim	6.5	7.5	13	15
Bone	5.5	8.5	11	17

A widely accepted method of evaluating lamb carcasses without cutting them is the yield grade procedure discussed in Chapter 12. Table 16–2 relates yield grades (as determined by 1982 standards[3]) to yield of sub-primal cuts from the lamb carcass. The impact of reducing both the fat thickness (trim) on surfaces of sub-primals and the length of tail left on loins and racks (see Figures 16–5, 16–31, 16–49, and 16–54) is also demonstrated by the lamb cutting data in Table 16–2.[4]

DEVELOPMENTS IN MERCHANDISING LAMB

Historically, lamb merchandising procedures have been similar to beef marketing in that lamb carcasses were cut into wholesale (primal) and retail cuts by the retailer. Some trading in foresaddles and hindsaddles and limited merchandising of untrimmed paper- or plastic-wrapped wholesale cuts occurred in the food service business. Cutting lamb carcasses and selling portions of the carcasses made the less desirable portions of the carcasses very difficult to merchandise.

---

[3]USDA. 1992. Official standards for grades of lamb, yearling mutton and mutton carcasses; slaughter lambs, yearling and sheep. Code of Federal Regulations. 7, 47:177, part 54.

[4]Garrett, R. P., J. W. Savell, H. R. Cross, and H. K. Johnson. 1990. Role of yield grade and carcass weight on the composition of lamb carcasses. J. Anim. Sci. 68:1299. See also Garrett, et al. 1992. Yield grade and carcass weight effects on the cutability of lamb carcasses fabricated into innovative style subprimals. J. Anim. Sci. 70:1829.

**Table 16–2. Percentage Yield of Carcasses by USDA Yield Grade (YG)[1]
When Fabricated into Boneless Sub-primals at Multiple Trim Levels**

Endpoint/Trim Level	YG 2	YG 3	YG 4
Boneless carcass[2]			
Untrimmed	65.08	65.12	64.94
0.25-in. (0.64-cm) trim	63.95	62.78	61.46
0.10-in. (0.25-cm) trim	61.57	59.79	58.06
0.00-in. (0.00-cm) trim	57.45	55.65	53.75
Boneless carcass[3]			
Untrimmed	62.00	61.73	61.29
0.25-in. (0.64-cm) trim	61.00	59.78	58.34
0.10-in. (0.25-cm) trim	59.06	57.29	55.54
0.00-in. (0.00-cm) trim	55.44	53.60	51.69
Boneless carcass[4]			
Untrimmed	57.84	57.11	56.04
0.25-in. (0.64-cm) trim	57.08	55.72	53.99
0.10-in. (0.25-cm) trim	55.72	53.86	51.83
0.00-in. (0.00-cm) trim	52.62	50.69	48.52

[1]Pre–1992 Yield Grades. See USDA. 1982. Official standards for grades of lamb, yearling mutton and mutton carcasses; slaughter lambs, yearling and sheep. Code of Fed. Reg. 7, 47:177, Part 54.

[2]Carcass with full-cut legs, 4-in. (10.16-cm) tails on loins and racks, and 4-rib shoulders.

[3]Carcass with full-cut legs, 2-in. (5.08-cm) tails on loins and racks, and 4-rib shoulders.

[4]Carcass with full-cut legs, no tails on loins and racks, and 4-rib shoulders.

Source: Garrett, et al. 1990. Role of yield grade and carcass weight on the composition of lamb carcasses. J. Anim. Sci. 68:1299.

The strong movement to boxed beef has been accompanied by a similar trend in lamb packer sales. Major packers who are lamb processors report that a large percent (in some cases, 100 percent) of the lamb carcasses they produce are cut into primal or sub-primal cuts (see Table 16–2), vacuum-bagged, and boxed before they leave the harvest plant. Boxes may contain three-piece carcasses (leg, back, shoulder) or one primal or sub-primal cut (leg, loin, rack, etc.). Streamlined carcasses that have been trimmed of the thin, less valuable portions of the carcass (breast, flank, shanks, kidney) are available boxed or hanging from some suppliers.

The fat and bone content of the thin (rough) cuts of the lamb carcass and the amount of labor required to obtain the usable lean product from those cuts have made them economic liabilities in the meat trade. The development of mechanical deboning equipment and the acceptance of mechanically deboned meat as a raw material have improved the utility of some portions of the ovine carcass significantly. Reducing the cost of processing those thin cuts may reduce the retail prices of the more desirable lamb cuts. Centralized cutting, boxing, and streamlining have resulted in the availability of bulk ground lamb for loaf or burger raw material to food service or other retailers.

Vacuum packaging and boxing lamb cuts is expensive because the cuts are smaller than beef cuts. The retail consumer must pay for this service, or it must be compensated for by reducing other losses. Major advantages of the vacuumized-boxed lamb are the maintenance of freshness for relatively long periods; the reduction of shipping and storage yield losses; and a more efficient utilization of rough cuts, trimmed fat, and bone at a large processing plant.

Preservation of freshness would not be a concern for the retail sales outlet in a large metropolitan area (such as Los Angeles) that has a daily, local, fresh-harvested lamb source. The added cost of bagging and boxing might only inflate the retail value of lamb in that type of situation. Contrastingly, the midwestern lamb processors with limited local markets may realize significant benefits from boxing 100 percent of the lamb shipped to population centers on the coasts.

The American Lamb Council has developed cutting methods for lamb carcasses that serve the needs of the HRI customer more adequately than the conventional cuts have in the past. The lamb processing industry recognizes that the low volume of lamb produced in the United States results in high unit costs. An increased demand in the HRI segment of the market may minimize distribution problems. A lamb carcass certification program, established by the U.S. Department of Agriculture and executed by the meat grading service, is being used by some packers and HRI buyers to reduce the fat content, increase the edible yield, and standardize the quality of lamb in the institutional trade.

17

Veal and Calf Identification and Fabrication

The definition of the word *veal* is "meat from calves of all ages and weights, from birth to 20 weeks of age." The industry recognizes four commercial classifications of veal. They are in order of increasing age and carcass weight:[1]

- *Baby veal* (bob veal) 2 to 3 days to 1 month, 20 to 60 pounds.
- *Vealers* 4 to 12 weeks, 80 to 150 pounds.
- *Calves* up to 20 weeks, 125 to 300 pounds.
- *Nature veal* or *special fed veal*, approximately 20 weeks, 180 to 240 pounds.

Veal consumption in the United States, like lamb and mutton consumption continues to decline (see Chapter 1). Like lamb consumers, persons who consume veal and calf are a small group who may eat the product rather frequently. Many people in the United States have never eaten veal. It is not available in the retail meat markets of many communities. Therefore, some potential veal consumers do not have ready access to the product. Low demand in many areas of the country combined with a reduced merchandising interval and high-moisture content influence many meat department managers to avoid handling the product.

Veal is pinkish white to grayish or light pink in color and has a mild flavor. It is often served with sauces and or spices that complement the delicate flavor of the meat. Veal is considered by some to be a "special occasion" meat.

---

[1]Kinsman, D. 1989. Veal, Meat for Modern Menus. National Live Stock and Meat Board.

Beef and veal cuts, aside from their water, fat, and ash content, differ mainly in size, color, and nomenclature. Veal is tender by nature because of its age. Calf carcasses fall between the veal and beef stage and are usually considered inferior to both. This is because the flesh of calf carcasses has developed beef characteristics without the accompanying fat covering and marbling that enhance beef flavor and juiciness qualities.

IDENTIFICATION

The Uniform Retail Meat Identity Standards[2] and the Institutional Meat Purchase Specifications[3] are used as the basis for identification in this chapter (see Chapter 13). *A Cross Sectional Muscle Nomenclature of the Beef Carcass*[4] may be helpful if more complete muscle identification and nomenclature is desired. Be sure to check the last paragraph in Chapter 13 for a description of *The Guide to Identifying Meat Cuts,* an excellent meat identification reference.

FABRICATION

The skeletal structure of the veal carcass in relation to an outline of some of the wholesale cuts is shown in Figure 17–1, and a more detailed chart of the wholesale and retail cuts and how to cook them appears as Figure 17–2. These figures can be referred to as the discussion moves through cutting a calf carcass.

Figures 17–3 to 17–18 appear through the courtesy of South Dakota State University.

More Detailed References

In addition to those references indicated in Chapter 13, the reader will find *Veal, Meat for Modern Menus* by the late Dr. Donald Kinsman, available

---

[2]National Cattlemen's Beef Association.

[3]Agricultural Marketing Service, USDA.

[4]Tucker, H. Q., M. M. Voegeli, and G. H. Wellington; Bratzler, L. J. (ed.). 1952. A Cross Sectional Muscle Nomenclature of the Beef Carcass. Michigan State Univ. Press, East Lansing.

VEAL CHART
LOCATION, STRUCTURE AND NAMES OF BONES

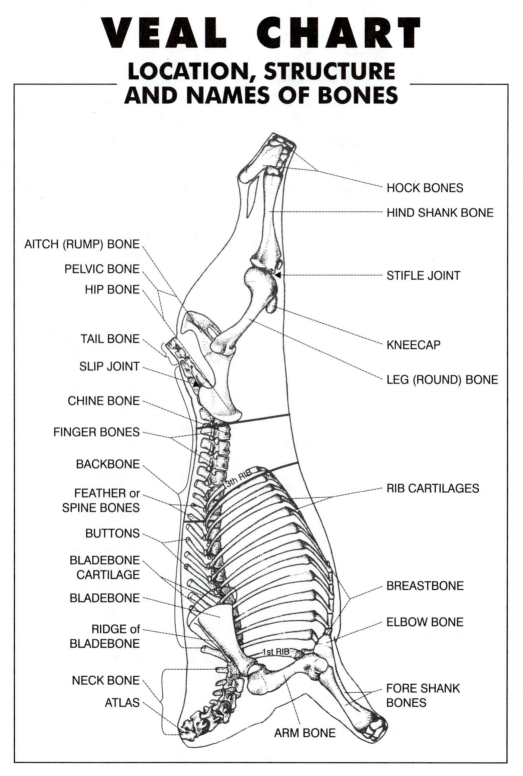

HOCK BONES

HIND SHANK BONE

AITCH (RUMP) BONE

PELVIC BONE

HIP BONE

STIFLE JOINT

TAIL BONE

KNEECAP

SLIP JOINT

LEG (ROUND) BONE

CHINE BONE

FINGER BONES

BACKBONE

13th RIB

FEATHER or SPINE BONES

RIB CARTILAGES

BUTTONS

BLADEBONE CARTILAGE

BLADEBONE

BREASTBONE

RIDGE of BLADEBONE

ELBOW BONE

1st RIB

NECK BONE

ATLAS

FORE SHANK BONES

ARM BONE

Fig. 17–1. (Courtesy, National Live Stock and Meat Board)

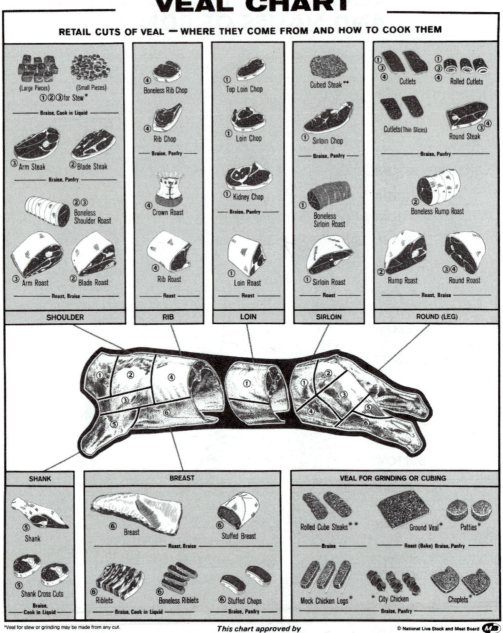

Fig. 17–2.

through the National Cattlemen's Beef Association, an excellent source of information on veal. Susan Marie Specht Overholt's MS thesis, *Veal Carcass Composition and Cutability*,[5] also contains some insightful information on veal.

The First Steps

Veal carcasses have little if any fat cover and a relatively high-moisture content. Because the small veal carcasses chill rapidly, they are not split into right and left sides during harvest. Therefore, veal carcasses (IMPS 300) are often ribbed (cut) between the eleventh and twelfth ribs into foresaddles and hindsaddles, and since cuts from those carcasses are also small, they are sold as double cuts, similar to the lamb cuts described in Chapter 16. The larger calf carcasses may be split into sides (IMPS 303) during the harvest process to facilitate chilling. Those carcasses would then be ribbed into quarters (sometimes between the twelfth and thirteenth rib, as indicated in Figure 17–1) and the cuts would more likely be smaller versions of the beef cuts described in Chapter 15.

A foresaddle is the two unsplit forequarters, anterior to the twelfth (IMPS 304) or thirteenth rib (IMPS 304A). A hindsaddle consists of the two unsplit hindquarters posterior to the eleventh (IMPS 330) or twelfth rib (IMPS 330A). Other wholesale cuts of veal are hindsaddle, long cut — two unsplit hindquarters with loin and nine (IMPS 342) or eight ribs (IMPS 342A) attached; veal chucks — split or unsplit shoulders of four (IMPS 308) or five ribs (IMPS 308A) with briskets and foreshanks attached; hotel racks — the unsplit or split ribs with seven ribs (fifth through eleventh, IMPS 305, or sixth through twelfth, IMPS 305A) on each side; loins — split or unsplit, cut in front of the hipbones forward to include the twelfth (IMPS 331) or the thirteenth rib (IMPS 331A); veal backs — single or unsplit, and cut from the fourth (IMPS 340) or the fifth rib (IMPS 340A) to the hipbone (including loin and nine or eight ribs); and legs — single or unsplit (IMPS 334), cut in front of the hipbones. Variation in wholesale cutting procedures may be greater in veal and calf than in other species (see Figures 17–1, 17–2, and 17–3).

---

[5]Specht Overholt, S. M. 1990. Veal Carcass Composition and Cutability. MS Thesis. Univ. of Connecticut, Storrs.

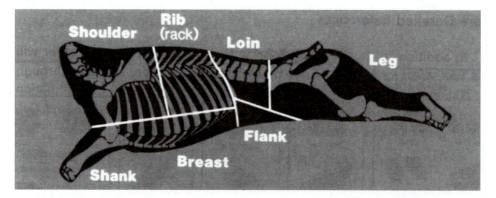

Fig. 17–3. The method of cutting veal usually follows the same pattern that is employed in the cutting of beef (Chicago style), with a few exceptions. One of these exceptions is to roll in the boned neck and brisket with the shoulder of a light veal carcass when the shoulder is rolled. Another variation is similar to cutting lamb, leaving the sirloin on the leg (see Figs. 17–1 and 17–2).

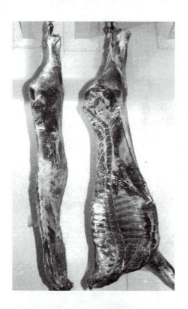

Fig. 17–4. The calf carcass used in the following discussion was split into two sides. The carcass was wrapped in film to prevent dehydration in the cooler. Note the minimal fat outside and inside the carcass.

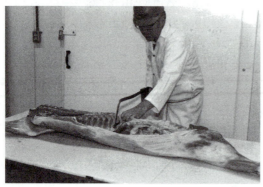

Fig. 17–5. Veal and calf carcasses are separated into forequarters and hindquarters or saddles, either between the twelfth and thirteenth ribs (IMPS 304A) or between the eleventh and twelfth ribs (IMPS 304). This separation was made between the twelfth and thirteenth ribs.

Foresaddle

Fig. 17–6. The breast (IMPS 313) and shank (IMPS 312) are separated from the square-cut chuck (IMPS 309) and hotel rack (IMPS 306) when a cut is made across the arm bone *(humerus)* just above (dorsal to) the joint between the *humerus* and the shank bones *(radius* and *ulna).* The cut shown was made parallel to the backline through the length of the forequar-

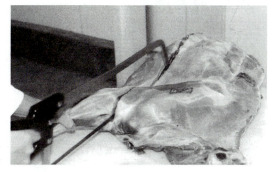

ter. The portion of this cut caudal to the fourth or fifth rib may be moved closer to the backbone to leave 4 inches or less of tail past the rib eye on the rib (IMPS specifications).

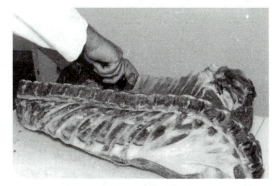

Fig. 17–7. The veal square-cut chuck and rib are separated between the fifth and sixth ribs (five-rib chuck, IMPS 309A) as in beef and as shown, or between the fourth and fifth ribs (four-rib chuck, IMPS 309).

Fig. 17–8. Although both *veal chuck* and *veal shoulder* are used, *veal chuck* is the preferred term. Small veal chucks are often boned and rolled by the same method as is followed in boning a shoulder of lamb (see Figs. 16–44 to 16–46). The chucks of calf carcasses can be made into veal shoulder clod (IMPS 310) and veal square-cut chuck, clod-out (IMPS 311) roasts, which are usually the size desired by

the trade. Although a four-rib chuck is standard, in some regions of the United States, the shoulder is cut from the carcass between the third and fourth ribs to make a three-rib chuck, while in other regions, it is cut as a five-rib chuck, shown here. A five-rib chuck is 23 to 25 percent of the weight of the carcass.

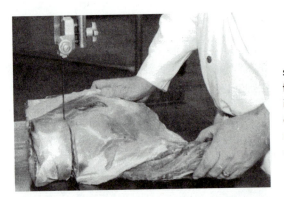

Fig. 17–9. *Veal arm steaks* are slices that are cut parallel to the arm face. An arm roast of the veal chuck is being cut here. Unscrupulous dealers have been known to misrepresent veal arm steak as veal round steak (veal cutlets).

Fig. 17–10. Turning the shoulder 90° results in the production of blade roasts and steaks of the veal chuck. Note the absence of seam fat in the blade roast as compared to other blade cuts in Chapters 14, 15, and 16. Large quantities of veal chuck are used in making veal stew and city chicken, which generally consists of 1-inch squares of veal cut ½ inch thick and placed on a 5-inch skewer with alternate layers of pork.

Veal shanks and breasts can also be boned, and the meat is diced or ground. Ground veal (IMPS 396) is commonly used in combination with pork for veal loaf (20 percent pork), and for mock chicken. When it is used for the latter, the mixture should be seasoned by the retailer before it is molded and placed on skewers. The seasoning for veal loaf is added by the consumers when it is prepared. The same combination can also be molded into patties, each patty bound by a slice of cured bacon. A mixture of 80 percent pork and 20 percent veal makes an excellent sausage.

Veal shanks (IMPS 312) are sometimes made into shank cross-cuts (see Fig. 15–47) and used for *osso bucco* (IMPS 338), a dish that originated in Italy, composed of braised veal shank cross-cuts, olive oil, white wine, tomatoes, lemon rind, garlic, anchovies, and seasoned stock. *Osso bucco* literally translates to "bone hole," referring to the bone marrow. The name of this dish has transferred to the veal cut itself.

Veal breasts (IMPS 313) can also be boned and made into breast rolls, boiled, and served as jellied cuts. Sausage breast rolls are made by rolling a layer of sausage into the boned breast or by making a pocket between the ribs and the meat (IMPS 314) and stuffing it with sausage. Merchandising briskets, short ribs, and skirt steaks increased the return from veal carcasses in a study by Susan Marie Specht Overholt (*Veal Carcass Composition and Cutability*).

The neck, shank, and breast represent 16 to 18 percent of the carcass weight.

Fig. 17–11. The veal rib, veal rack, or hotel rack (IMPS 306) may be double (unsplit) or split down the backline as shown, but the ribs cannot extend more than 4 inches past the end of the rib eye *(longissimus)*. The separation may be made 1 inch laterally to the rib eye as shown. If the rib is to be used for a crown roast (see Figs. 16–53 to 16–58), the "tails" must be cut longer. Removing

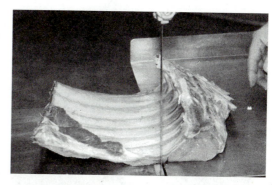

the meat from the end of the rib for a distance of 1½ to 2 inches is called *Frenching*. When ribs are to be Frenched, the "tails" must be cut accordingly. The rib ends are trimmed and may be boned out and used, as indicated for breasts, or they may be merchandised in the same manner as beef short ribs.

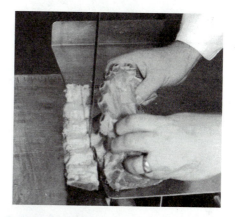

Fig. 17–12. The chine bones (central axes of the vertebrae or bodies of the vertebrae) are cut off with a cleaver or carefully passed through a band saw, as shown.

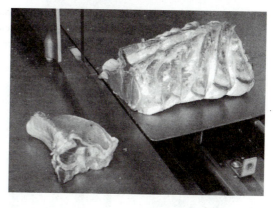

Fig. 17–13. Veal rib chops may be single, as shown, or cut as double chops if the rack was unsplit.

Veal rib chops are the ribs of veal that are cut into slices for braising. When a rib chop is cut from ¾ to 1 inch thick and a pocket is made in the eye muscle, the chop is called a *bird* or a chop for stuffing. This practice is followed for both veal and pork. If the opening (see Fig. 14–37) for the pocket is made from the inside surface of the rib, the stuffing can be inserted and will remain so without the opening being pinned. On a loin chop, the opening can be made on the flank end of the loin eye muscle. Ribs from heavy veal carcasses can be prepared for roasts either as standing ribs (bone-in) or as rolled veal ribs (boneless).

A seven-rib saddle represents 6.5 to 7.0 percent of the weight of the carcass.

Hindsaddle

Fig. 17–14. The flank is separated from the leg at the rear and cut along the lower edge of the loin. The twelfth and thirteenth ribs may be ossified sufficiently to make the use of a saw necessary, as shown. After the flank has been removed, the meat is usually diced or ground and may be utilized as indicated earlier in the discussion concerning ground veal.

Fig. 17–15. From the last rib (IMPS 331A) or the last two ribs (IMPS 331) to the hipbone is the veal loin (shortloin in beef). The front of the last (sixth) *lumbar* vertebra is the reference point for making the cut between the loin and the leg. Loin veal chops (porterhouse and T–bone steak in beef) are cut from the *lumbar* vertebrae. If the chop includes a slice of kidney imbedded in the kidney fat, the cut is a *kidney veal chop*. The veal fillet or tenderloin that lies on the underside of the vertebrae is seldom removed as a separate cut. The loin is more suitable as a roast if it is boned and rolled, preferably taking the loin saddle which includes both sides.

The loin of veal with kidney, suet, and flank represents about 17 percent of the carcass weight.

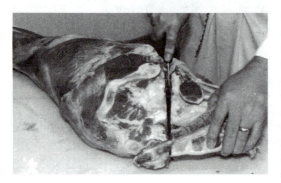

Fig. 17–16. The veal leg (IMPS 334) is the remaining portion after the loin has been removed. As in lamb, the leg includes the sirloin roast and steaks. The sirloin section can be removed by cutting on a line from the joint between the last (fifth) *sacral* vertebra and the first tail vertebra through a point exposing and just anterior to the ball of the *femur* (see Figs. 15–85 to 15–91 for a detailed discussion of the bones of the sirloin).

Fig. 17–17. Removing the rump roast entails cutting just caudally to and parallel with the aitch (pelvic) bone. The roast can be utilized as a bone-in roast or as a boneless tied roast. A larger bone-in or boneless roast results if the sirloin and rump are removed as a single unit, yielding a sirloin half of the leg.

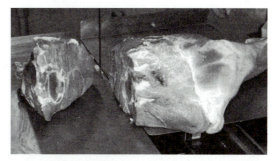

Fig. 17–18. The veal retail cut in greatest demand is the veal round steak or veal cutlet. Considering that the leg with the rump off represents 27 percent of the carcass weight and that only 50 to 60 percent of this can be cut into cutlets (round steak in beef), it must be evident that this cut is the most expensive cut in a veal carcass. Cutlets may be made in thicknesses of ½ to 1½ inches, depending upon how they are to be

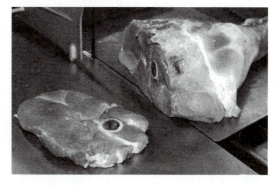

used. Unless the customer specifies otherwise, a cutlet for breading is cut ½ inch thick. The first three or four slices through the center of the leg should be priced higher than the remainder of the cuts. Slicing ceases when the stifle joint (shank knuckle bone) is reached. The meaty part on the back of the shank can be cut for a small heel of veal round pot roast. The hind shank can also be used for *osso bucco* (IMPS 338).

18

Fresh Meat Processing

Meat processing may be defined as "any mechanical, chemical, or enzymatic treatment of meat that alters the form in which it originally occurs." Another term that is often used in the industry is *further processing*, which implies that meat undergoes *comminution* (particle-size reduction by grinding and[or] chopping). The processing of meat serves a number of functions, including one or more of the following:

- Preservation and(or) shelf-life extension.
- Tenderization (by mechanical, enzymatic, chemical, or other means).
- Meat cookery.
- Manipulation and control of composition (protein, fat, and moisture content).
- Portion control (size, weight, and shape).
- Improvement in consumer convenience.

Hence, by definition, virtually all meat is processed in one way or another prior to consumption.

Preservation and shelf-life extension (improvement) are the most important processes given to meat, since it is a highly perishable commodity. The methods of meat preservation include freezing, heat pasteurization, heat sterilization, curing and smoking, dehydration, and irradiation. These important topics are covered in Chapters 19 and 20. The shelf life of meat is the length of time before the meat or meat product becomes unpalatable or unsafe for human consumption due to microbial spoilage (see Chapter 2) and(or) rancidity development (see Chapter 19). Mechanical refrigeration is the most common means of extending the shelf life of fresh meat. Without mechanical refrigeration to maintain the product temperature at 40°F or less, the shelf life of fresh meat would be limited to hours instead of days or weeks.

Since preservation is covered in Chapter 19, this chapter is devoted to the methods of and the reasons for fresh meat processing. The broad topic of fresh meat processing has gained increased importance in our society during the past 20 years, due to changes in consumer attitudes, habits, and life styles.

Consumer attitudes about meat, health, and physical fitness have brought about changes in the way meat is processed and merchandised. For example, consumers are increasingly concerned about their health and physical fitness, which have been linked to diet. This has brought about changes in eating habits and an increased demand for leaner meat products. The production of leaner meat-type animals through genetic improvement is a slow process. To compound the problem, the marketing and harvesting segments of the live-stock industry often still do not pay producers an adequate bonus for lean, heavily muscled animals. Presently, there appears to be a trade-off between meat quality (tenderness, flavor, juiciness) that is predicted by marbling (see Chapter 12) and the production of leaner meat-type animals. Since marbling levels tend to decrease with less total fat in the meat animal's carcass, two opposing forces exist between meat quality and the production of leaner meat. These opposing forces have created differences of opinion as to the direction in which livestock production should move in the future. Although numerous consumer groups, human nutritionists, meat scientists, and even many retailers are calling for the production of leaner beef, the packer often pays the top market price for Yield Grade 3 cattle that are moderately fat. As long as this marketing situation exists, there is little incentive for the producer to market leaner cattle. Similar trends are evident in the hog industry; how-ever, the factors of growth and reproductive efficiency play much greater roles in determining the type of hog marketed.

Because there is still such wide variation in the livestock harvested, compositional uniformity has become a major problem for the meat proces-sor. Compositional uniformity is important in the production of leaner prod-ucts, and extremely important in the large volume of meat products marketed through the food service industry (restaurants, cafeterias, fast-food chains, etc. — see Chapter 13).

Because of the extreme competitiveness, high volume, and low-profit mar-gins, very small variations in fat content or portion weight can account for hundreds of thousands of dollars annually in net profit differences for the larger fast-food corporations. Many specialized meat processing plants cater strictly to meeting the needs of the food service industry. Suppliers of the food service industry must meet the rigid portion size, composition, quality, and cost specifications of the retailing corporations that purchase their products; if they do not, those retailers will quickly find other suppliers.

QUALITY CONTROL

This section is very important to all phases of meat processing, not just fresh meat processing. The need for analytical programs in all meat operations is continually increasing because:

- More consumers are interested in nutritional composition, thus demanding nutritional monitoring and labeling.

- Government regulations must be strictly followed.

- The vast difference in value between fat and lean necessitates accurate control of fat–lean ratios to maintain a profitable operation.

- Consistency in quality is essential for maintaining consumer loyalty and is often required in production contracts.

Sampling

Several different analytical methods are approved to be used for determining the composition of meat products and will be discussed later. However, no analytical method is any more accurate than the sampling system used to secure the samples to test. There is no simple "cook book" sampling system that will work for all raw materials going into final meat products. Documented, approved procedures for sampling meat to determine the percent composition of fat, lean, protein, ash, or moisture were formulated by Committee F–10 of the American Society for Testing and Materials (ASTM).[1] Professor Robert E. Rust, Iowa State University, President and CEO, Rust Associates, Inc., Ames, IA <rrust@iastate.edu> was chairman of ASTM Committee F–10, when he briefly explained sampling based on his committee's recommendations.[2]

---

[1]ASTM. 1990. Standard Practice for Sampling Meat to Determine Percent in a Lot. Designation: F 427–77 (Reapproved 1990). Annual Book of ASTM Standards. (This committee no longer exists, but the sampling procedures described here are still valid.)

[2]Rust, R. E. 1976. Fat — Does that 10 gram meat sample really show the percent of fat in your 3000 lb. batch? Meat and Poultry (formerly Meat Industry). (June).

The first step in sampling is to define the *lot* or block of meat material about which composition information is needed. It may be a single 2-pound unit of ground meat for laboratory analysis, an 800-pound handtruck of ground meat, a 600-bag truck load of trimmings, or a day's production of trimmings from the cutting floor. Next, the *sampling unit* should be determined. This will probably be the smallest unit in which the product is available, such as the individual containers (bags or boxes) in which the product is packaged, the individual 1- to 3-ounce drillings taken from bags or boxes of frozen product, or the 1-pound grabs from a handtruck or mixer of ground meat. Thus, the condition the meat is found in may largely determine the sampling unit. The sampling units must be representative of the lot and approximately equal in size. Whole pieces of meat, such as chunks of boneless cow beef, must be broken through a grinder (a ¼- to ⅜-inch grind) or a chopper to be accurately sampled. The samples of similar weight must be taken randomly; i.e., a 5-pound chunk should *not* be taken from one part of the lot and a 1-pound handful from another part. The smaller the amount needed for the analytical procedure, the more critical the sampling procedure. It is usually best to use the largest sample possible.

The *number* of samples to take depends on the *variation* between sample units and the *accuracy* desired. The more variable the material sampled, and the more accurately the lot composition must be determined, the larger the number of samples that should be taken. Understanding this aspect of *quality control* entails understanding some basic *statistics*. For a more detailed discussion, please consult a statistics textbook.

Here basic statistics are applied to the determination of percent fat in a meat shipment. A measure of variation around the average (mean) of a group of samples is called the *standard deviation*. The symbol used for standard deviation is the Greek letter *sigma* (σ). Seldom is the true σ value known, but it can be estimated by an experienced person or by sampling. This estimate is indicated by *s*. Two ways to estimate *s* are (1) by calculation/computation and (2) by the range in fat content. The standard deviation is defined as *the square root of the average of the squared differences taken from the mean of the lot*. It is often referred to as the "root mean square deviation." For any group of sampling units, it can be determined as follows:

$$s = \sqrt{\frac{\sum(X_i - \overline{X})^2}{n-1}} \quad \text{or} \quad \sqrt{\frac{\sum X_i^2 - \frac{(\sum X_i)^2}{n}}{n-1}}$$

Where:

X_i = fat content of individual sample i

$\overline{X}$ = average of X from several observations

s = standard deviation

ΣX_i^2 = sum of the squares of each individual determination

$(\Sigma X_i)^2$ = sum of the individual determinations squared

n = number of determinations

n – 1 = degrees of freedom

Many microcomputer software programs include this calculation; if a microcomputer is not available, the computation can be rather conveniently carried out on a calculator.

The *range* in fat content from the leanest to the fattest sample in a lot is related to the standard deviation of the samples. A "normal" distribution is shown in Figure 18–1.

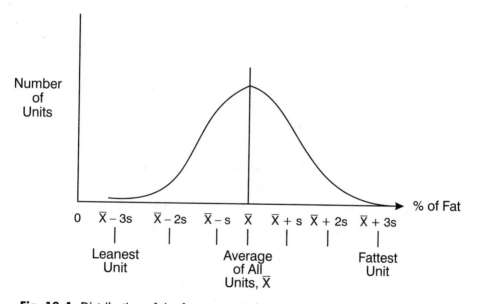

Fig. 18–1. Distribution of the fat content is found by taking a large number of units from a single lot and making an analytical determination on each one. (From ASTM Committee on Standards, 1916 Race Street, Philadelphia, Pennsylvania 19103)

A group of 24 or more units makes a reliable estimate of the standard deviation, although 10 units will give some indication. Approximately 68 percent of the samples fall within ± one *s* of the mean, while 95 percent will be within ± two *s*, and 99 percent within ± three *s*. Note that *six* standard deviations *(s)* covers nearly the full range of fat values. Therefore, if you have a good idea what the range in fat content is between samples, you can divide that by 6 to get an estimate of *s*.

To calculate the size of sample *n* needed for your required accuracy (*E* in the following formula), use this formula:

$$n = (3s \div E)^2$$

Where:

E = maximum allowable difference between the estimate of fat content of the sample and the true fat content of the lot, and

$3s$ = estimate of three standard deviations (3s) among all sampling units in the lot.

When 3s is used in the equation, the probability is only 3 in 1,000 that the sampling error will exceed E (the maximum allowable sampling error), provided the estimated standard deviation *(s)* closely approximates the true standard deviation (σ).

As examples of these relationships, consider the following:

1. You have received 500 bags of 80/20 (lean/fat) trimmings. From past experience you know that individual bags of trimmings from this supplier contained as much as 24 and as little as 16 percent fat, a range of 8 percent, which is not unusual. You need to determine the fat content of this lot to within 1 percent of the true value. How many bags are needed to make up the sample? Use the formula listed previously:

$$
\begin{aligned}
n &= (3s \div E)^2 \\
s &= \text{range} \div 6 = 8 \div 6 = 4 \div 3 \\
3s &= (3)(4 \div 3) = 4 \\
E &= 1 \\
n &= (4 \div 1)^2 = 4^2 = 16
\end{aligned}
$$

You would need to sample 16 bags.

2. You have 500 pounds of ground meat in a vat that must be proven to be within 0.5 percent of its true fat content. You have been able to

estimate that between the leanest and fattest 1-pound grab samples, the fat content varies 4 percent. How many 1-pound grab units are necessary?

$$s = 4 \div 6 = 2 \div 3$$
$$3s = (3)(2 \div 3) = 2$$
$$E = 0.5$$
$$n = (2 \div 0.5)^2 = 4^2 = 16$$

You would need sixteen 1-pound grab samples.

3. In the above example, after continued scrutiny, management discovered that occasionally 1-pound grab samples varied as much as 8 percent in fat content. How did this change affect sampling procedure?

$$s = 8 \div 6 = 4 \div 3$$
$$3s = (3)(4 \div 3) = 4$$
$$E = 0.5$$
$$n = (4 \div 0.5)^2 = 8^2 = 64$$

The two-fold increased variation required a four-fold increase in the number of samples required for equal precision. The magnitude of the change in the number of samples needed for equal precision varies as the square of the change in the range of the samples, as shown in Table 18–1.

Table 18–1. Effect of Range and Accuracy on Number of Samples Needed[1]

Range (percent)	Accuracy of Number of Samples Needed	
	1 percent	0.5 percent
2	1	4
4	4	16
6	9	36
8	16	64
10	25	100

[1]Rust, R. E. 1976. Fat—Does that 10 gram meat sample really show the percent of fat in your 3000 lb. batch? Meat and Poultry (formerly Meat Industry). (June).

In some situations it may become virtually impossible to achieve a high level of accuracy, such as 0.5 percent, unless the lot under consideration is composed of a large number of units. Even then, it would be impractical to sample the whole lot, which might be the case in example 3 above. This situation is prevented by continual testing and recording of the composition of ingoing materials.

Blending Samples to Get a Composite Sample

As Professor Rust wrote in his article "Fat — Does That 10-Gram Meat Sample Really Show the Percent of Fat in Your 3000 Lb. Batch?," "The care exercised in this final blending of sample units is critical." An Anyl–Ray® fat tester (described later) that uses 13 pounds will make the blending much easier; in fact, in this instance, each sample could be very easily analyzed. But for a 2- to 4-gram sample, the meat must be ground through a ⅜-inch plate and then mixed very thoroughly before samples can be taken. The mixing and sampling may have to be repeated to get a sample that is small enough, but still representative for the analysis. Without an accurate, representative sample, even the most sophisticated analytical test will not give a true indication of the fat level.

Analytical Procedures for Fat Determination

Two reference sources form the basis for recommended and accepted analytical procedures for fat determination. These are the American Society for Testing and Materials (ASTM)[3] and the Association of Official Analytical Chemists (AOAC).[4] As explained in ASTM Standard Practice F426,[5] the analytical requirements of the various segments of the meat industry vary in terms of accuracy, speed, equipment, operation, and application requirements. Thus, a number of standard procedures are available for the industry to use. ASTM Standard Practice F426 classifies several of the most widely

---

[3]1916 Race Street, Philadelphia, PA 19103. http://www.astm.org/

[4]Suite 400, 2200 Wilson Boulevard, Arlington, VA 22201. http://www.aoac.org/

[5]ASTM. 1990. Standard Practice for Evaluating and Classifying Methods for the Analysis of Fat in Meat and Meat Products. Designation: F 426–8 1(Reapproved 1990). Annual Book of ASTM Standards.

used procedures in terms of accuracy, precision (standard deviation of repeated measures), rate of analysis, and application.

Some analytical procedures destroy the sample, while others do not. For instance, if the raw material is simply evaluated visually, the material is still suitable for sale, as in carcass evaluation. However, it is much more difficult to visually evaluate chunked, ground, or chopped meat, and such an evaluation is not acceptable as official by regulatory agencies.

For example, consider samples from which fats are extracted by organic solvents. If not properly handled, organic solvents can be harmful to both humans and the environment. Most fat solvents can be redistilled and used repeatedly to save the environment and financial costs. Using and then disposing of fat solvents must be done with the utmost caution. It should be obvious that the dried residue remaining after fat extraction, which contains no moisture or fat, is not salable or edible.

Ether Extraction[6, 7]

Ether extraction, generally known as the AOAC method, is the most widely used method in the United States to determine the fat content of meat and meat products. Because this method is so precise, it is commonly used as a reference or standard in establishing the accuracy of other methods. Used by regulatory and food analysts, the analysis takes at least 6 hours to complete, and it may require as long as 22 hours, depending on the options chosen. Therefore, more rapid methods must be used routinely in the industry. Also the sample size is very small (3 to 4 grams = 0.10 to 0.14 ounce), which creates a sampling problem, as discussed previously.

The principle of this method is that ether is a very good fat solvent, especially of the triglyceride fraction (see Chapter 23), which is the major fat constituent of meat. Ether will not completely extract the *phospholipids,* which are important structural components of every cell wall but which compose a very small, constant portion of the total lipid content, irrespective of the triglyceride level. When ethyl alcohol undergoes dehydration (loses water), it forms diethyl ether $(C_2H_5)_2O$, the most common ether form for use as a fat solvent. Petroleum ether, a by-product of the petroleum industry, is also

---

[6]ASTM. 1990. Standard Test Method for Fat in Meat and Meat Products by Ether Extraction. Designation: F 463–76 (Reapproved 1990). Annual Book of ASTM Standards. http://www.astm.org/

[7]AOAC. 1990. Fat (Crude) or Ether Extract in Meat. Method 960.39. Official Methods of Analysis (15th Ed.): 931. Helrich, K. (Ed.); McNeal, J. E. (Assoc. Chp. Ed.). http://www.aoac.org/

acceptable as a fat solvent in this analytical procedure. Both of these solvents are very volatile and highly explosive. Thus, their use should always be confined to a fume hood. Both of them are mildly irritating to the skin and mucous membranes. Inhalation of high concentrations causes unconsciousness. Death may occur due to respiratory paralysis.

Fat is determined by the loss in weight in the dried sample after ether extraction. A 3- to 4-gram sample is weighed accurately and rapidly. Erroneous weights and results may occur if weighing intervals are extended, allowing atmospheric moisture to condense on a cold sample and(or) moisture evaporation from the sample if conditions permit. The weighed sample is dried in an oven for 6 hours at 100° to 102°C or 1.5 hours at 125°C. Fat extraction is conducted with either a Goldfisch® (see Figure 18–2) or a Soxhlet® (see Figure 18–3) extractor.

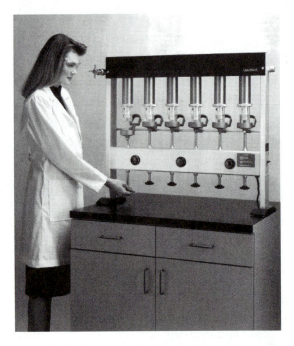

Fig. 18–2. Goldfisch® fat extractor and attendant wearing proper eye protection. (Courtesy, Labconco Corporation, 8811 Prospect, Kansas City, Missouri 64132)

The extraction period may vary from 4 hours at a condensation rate of five to six drops per second to 16 hours at two to three drops per second. After the sample has been air-dried to allow solvent evaporation and cooling, it is weighed. The percent of fat = the weight of fat ÷ the weight of the sample × 100. The weight of fat is equal to the weight of the dried sample minus the weight of the extracted sample.

Modified AOAC

The modified AOAC method[8] consists of decreasing the drying time to 30 minutes and the extraction time to 45 minutes, thus saving 4.25 hours, compared to the official method described above. The standard deviation *(s)* of this modified method was ± 0.83 percent fat, which is 1.5 times the *s* of the official method, ± 0.56 percent fat. This modified method is recommended for use as a screening method when speed is more important than precision.

Babcock

The Babcock method was originally used for fat analyses of dairy products. The principle is to use concentrated sulfuric acid, H_2SO_4, or hydrochloric acid, HCl, and heat (150°F/66°C) to break down the meat protein, allowing the encapsulated fat to separate. Both acids will cause severe skin burns on contact, so handling must be done very cautiously. The fat layer rises to the surface and up the calibrated neck of a long-necked bottle, known as a Paley bottle, to be measured. The required sample size is 9 grams (0.32 ounce), and the complete procedure takes about 30 minutes. The accuracy (±1 percent) and precision (*s* = ± 0.6) of this procedure are not equal to the accuracy and precision of the two methods described previously.

Anyl–Ray®

The most widely used fat analysis method in the patty industry is perhaps the rapid X–ray scanning

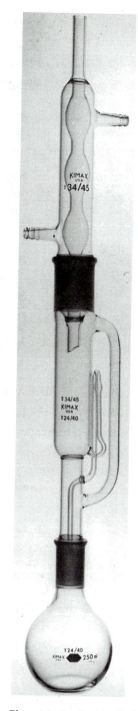

Fig. 18–3. Soxhlet® fat extraction apparatus. (Courtesy, Kimble Glass, Crystal Avenue, Vineland, New Jersey 08360)

---

[8]Pettinati, J. D., C. E. Swift, and E. H. Cohen. 1973. Collaborative study of rapid fat determination in meat and meat products by modification of the AOAC method. J. AOAC 56:1140.

Fig. 18–4. Fat content can be determined in seconds by using one of several infrared spectral analysis machines. The Anyl–Ray® machine (Kartridge Pak Co., Davenport, Iowa) shown here requires approximately 13 pounds of ground meat tightly packed in the round canister and is a non-destructive fat determination method. (Courtesy, Howard Beef, Howard, South Dakota)

technique called Anyl–Ray®,[9] which uses the principles of X–ray absorption to determine the lean–fat ratio (or percent fat) of meat products (usually fresh beef or pork) (see Figure 18–4). (See also the Appendix section on electricity to locate Anyl–Ray® X–rays in the electromagnetic radiation spectrum between gamma rays and ultraviolet rays.) Lean meat absorbs more X–rays than fat because lean contains more minerals. The X–rays that pass through the fat are measured by an electrometer, which causes an instantaneous read-out on a digital meter indicating the percent fat in the sample. The Anyl–Ray® uses a 13-pound sample, which may be fresh, frozen, cooked, cubed, or ground. The X–rays are low-level and non-destructive and thus do not damage the 13-pound sample, which is returned to production. The large sample, which makes sampling more accurate (note earlier discussion), is packed into a standard canister with plastic top and bottom and held in an exact position in the X–ray path for the reading.

A portable Anyl–Ray® (PAR) is now available for use in supermarket chains, supermarkets, and small meat and food processing establishments. The same principles apply, but a smaller sample, only 800 grams (1.75 pounds) is required. The accuracy (mean difference = ±0.3 to ±0.5 percent

---

[9]The Kartridge Pak Co., 807 West Kimberly Road, Davenport, IA 52806.

fat) and precision ($s = \pm 0.5$ to ± 1.0 percent fat) of the Anyl–Ray® compare quite favorably with the AOAC standard method.

Near-Infrared Spectroscopy (NIR)

Near-infrared light waves are just beyond the visible part of the light spectrum (see Appendix). Saturated fats reflect these light waves differently than do other food components (including other fats). The Food and Drug Administration (FDA) regulates the chemical methods of fat analysis now widely used by the food industry for controlling quality in recipes as well as for obtaining fat values published on all food labels. But, as noted above, the current methods have drawbacks, such as disposal problems posed by chemicals used in analysis. Scientists at the Agricultural Research Service, in cooperation with Foss North America, of Eden Prairie, Minnesota, are using this non-chemical alternative technology to determine fat content and type of fat in ground meat. A computer measures a sample's absorption as the light passes through the sample. NIR can measure fat levels as low as 1 percent, well within FDA requirements for precision. The approach takes less than two minutes and uses no ether or other hazardous chemicals.

Cooking

This test is based on simply heating the sample, much like a consumer would fry a hamburger, except the sample is cooked to a charred doneness. The fat and water drippings are collected in a specially calibrated test tube, and the layer (volume) of fat that rises out of the water in the test tube is measured, then converted to percent fat. This test uses a 56- to 80-gram sample, and the cook cycle takes 2 to 15 minutes. Two companies manufacture, sell, and service machines for this test.[10, 11] The accuracy (mean difference = ± 0.5 percent fat) and precision ($s = \pm 0.5$ percent fat) of this cooking method compare quite favorably with the AOAC standard method.

Rapid Microwave Moisture and Fat Analysis System

A complete explanation of the principles of microwave cooking is contained in Chapter 24. (See also the section in the Appendix on electricity to

---

[10]Hobart® FMP–1 Analyzer, Hobart Corporation, Troy, OH 45374.

[11]Univex® Fat Analyzer, Model FA–73, Univex Corporation, 3 Old Rockingham Road, Salem, NH 03079.

locate microwaves in the same region in the electromagnetic radiation spectrum as radio waves.) The application of microwave heating principles has resulted in one of the more widely used techniques for meat analysis. The validity of the method is well-documented.[12–17] The whole procedure is automated through the use of the CEM Meat Analysis System,[18] consisting of a microwave power source and a scale integrated through a microprocessor to any output system desired. A 4- to 5-gram sample is dried, and then the fat (triglycerides) is extracted with the solvent methylene chloride, also known as dichloromethane (CH_2Cl_2). This solvent, like diethyl ether, does not extract the phospholipids. In high concentrations, methylene chloride is narcotic to humans, but nevertheless, it is one of the safest fat solvents available. The CEM system allows for minimum solvent handling and automatically reclaims used solvent, resulting in a low net use of solvent and minimal environmental concerns. The whole analysis can be completed in less then 15 minutes. The accuracy and precision of this rapid microwave method compares quite favorably with the AOAC standard method.

Specific Gravity

The more fat a sample contains, the lower its specific gravity. This is the principle behind the Foss–Let system,[19, 20, 21] in which the solvent tetrachloroethylene, also known as perchoroethylene (C_2Cl_4), and a strong

[12]AOAC. 1990. Moisture in Meat and Poultry Products. Rapid Microwave Drying Method 985.14. Official Methods of Analysis (15th Ed.): 931. Helrich, K. (Ed.); McNeal, J. E. (Assoc. Chp. Ed.).

[13]AOAC. 1990. Fat (Crude) in Meat and Poultry Products. Rapid Microwave-Solvent Extraction Method 985.15. Official Methods of Analysis (15th Ed.): 931. Helrich, K. (Ed.); McNeal, J. E. (Assoc. Chp. Ed.).

[14]Committee on Foods. 1991. I: Recommendations for Official Methods. Meat, Poultry, and Meat and Poultry Products. (7) Microwave Techniques for Meat Analysis. J. Assoc. Offic. Anal. Chem. 74:182.

[15]Bostian, M. L., D. L. Fish, N. B. Webb, and J. J. Arey. 1985. Meat and Meat Products. Automated Methods for Determination of Fat and Moisture in Meat and Poultry Products: Collaborative Study. J. Assoc. Offic. Anal. Chem. 68:876.

[16]Crosland, A. R., and N. Bratchell. 1988. An evaluation and comparison of the CEM Meat Analysis System with official standard methods for the determination of moisture and fat in meat and meat products. J. Assoc. Publ. Anal. 26:89.

[17]Mann, N., A. Sinclair, M. Watson, and K. O'Dea. 1991. Evaluation of rapid fat determination in meats using the CEM automated analyser. Food Australia. 43:67.

[18]CEM Corporation, P.O. Box 200, Matthews, NC 28106.

[19]Foss Food Technology Corporation, 10355 W. 70th Street, Eden Prairie, MN 55344.

[20]ASTM. 1990. Standard Test Method for Fat in Meat and Meat Products by the Foss-Let Analyzer Technique. Designation: F 464-76 (Reapproved 1990). Annual Book of ASTM Standards.

[21]AOAC. 1990. Fat (Crude) in Meat. Rapid Specific Gravity Method 976.21. Official Methods of Analysis (15th Ed.): 932. Helrich, K. (Ed.); McNeal, J. E. (Assoc. Chp. Ed.).

mechanical shaking action are used to rapidly extract fat from a 45-gram sample. The resulting mixture is filtered, and its specific gravity is measured. The procedure takes approximately eight minutes. The accuracy (mean difference = ±0.1 percent fat) and precision (s = ±0.5 percent fat) of this specific gravity method compare quite favorably with the AOAC standard method. In high concentrations, this solvent, tetrachloroethylene, is narcotic to humans, and its defatting action on skin can lead to dermatitis. However, disposal of the used solvent is a greater concern — contamination of aquifers with chlorinated compounds such as tetrachloroethylene cannot be tolerated. This specific gravity procedure calls for use of 120 ml of solvent per sample. One solution is to reflux the used solvent and reuse it, cutting down on the amount to be disposed of. Research on using microbes to dechlorinate such compounds in the ground is underway.

Specific gravity can be used to determine the composition of whole meat cuts and carcass parts, and when used this way, it is nondestructive. The procedure requires that the product be accurately weighed both in air and in water. The difference between the weight in air and the weight in water is the weight of the water that the unit of meat displaced. Since the meat cut will displace a volume of water equal to its own volume, the weight in water and the weight in air represent identical volumes.

Example: Data to collect:

1. Weight in air
2. Weight in water

Calculations:

3. Weight of water displaced (1 − 2)
4. Density = Weight in air (1) ÷ Weight of water (3)

DATA ON PORK CUTS
WEIGHT (POUNDS)

	In Air	*In Water*	*Water*	*Specific Gravity*
Ham	19.4	1.14	18.26	1.0624
Loin	25.05	1.19	23.86	1.0499
Shld.	20.30	1.19	19.11	1.0623
Belly	15.60	0.41	15.19	1.0270

Note the belly, being the fattest primal cut, has a lower specific gravity than the ham and shoulder, while the loin is intermediate. Specific gravity is not commonly used in the meat industry.

Laboratory/Research Methods for Complete Lipid Extraction

Chloroform, also known as trichloromethane ($CHCl_3$), when combined with methyl alcohol (methanol [CH_3OH]) in the ratio of two volumes chloroform to one volume methanol, comprises a solvent that will extract all lipids, including the phospholipids in the cell walls of all tissues.[22, 23] This method has been used in research for many years, but it is time-consuming and not practical for industrial use. Also, because large amounts of solvent are needed, disposing of the used solvent is a problem. Chloroform is very volatile, and inhalation of large amounts of chloroform fumes may cause hypotension, respiratory and myocardial depression, and death. Chloroform has been banned by the Food and Drug Administration from use in drug, cosmetic, and food packaging products since 1976 and is listed as a carcinogen by the Environmental Protection Agency. Thus, anyone who works with chloroform must exercise extreme care. Another procedure that employs methylene chloride (see the preceding microwave method) as the main solvent has been developed as an alternative to using chloroform.[24] Combined with methanol in the ratio of 90 volumes methylene chloride to 10 volumes of methanol, this solvent mix completely removes all lipids from animal tissues.

Supercritical Fluid Extraction and Steam Stripping

Although the knowledge that supercritical fluids can dissolve large quantities of relatively nonvolatile materials has existed for 100 years, only in recent

---

[22]Folch, J., M. Lees, and G. H. S. Slone Stanley. 1957. A simple method for the isolation and purification of total lipids from animal tissues. J. Biol. Chem. 226:497.

[23]Bligh, E. G., and W. J. Dyer. 1959. A rapid method of total lipid extraction and purification. Can. J. Biochem. Physiol. 37:911.

[24]Maxwell, R. J., W. N. Marmer, M. P. Zubillaga, and G. A. Dalickas. 1980. Meat and Meat Products. Determination of total fat in meat and meat products by a rapid, dry column method. J. Assoc. Offic. Anal. Chem. 63:600.

years have applications been made in the food industry.[25, 26, 27] These principles have were originally applied to removing cholesterol from dairy products. However, these techniques have been applied to muscle foods, including ground beef.[28] The product remaining after fat and cholesterol extraction is suitable for use only in processed meats. At high pressures (several thousand pounds per square inch [psi]) and relatively low temperatures (100°F), a material can be brought to its supercritical state. In this state, the supercritical fluid takes on a higher density, and like a liquid, it has the ability to dissolve solids, yet like a gas, it expands to permeate everything in its vessel or container. Carbon dioxide (CO_2) is a particularly attractive solvent for use in the food industry because of its molecular weight and reaction behavior; it is nontoxic and inexpensive, and it leaves no residue. The critical point of CO_2 is at 31°C and 73 atmospheres, which are both relatively moderate. Thus, supercritical CO_2 is at a higher temperature and(or) pressure than 31°C and 73 atmospheres. The appropriate equipment necessary to apply these parameters to the solvent and the food product has been developed.

In steam stripping (distillation), the food material is superheated to 430° to 550°F in a vacuum with steam to remove cholesterol by flash evaporation. This technique will remove 92 percent of the cholesterol from lard. It is also being applied to dairy foods (see footnote 27).

Electrical Conductivity[29] and Impedance[30, 31]

. Patented electronic meat measuring equipment (EMME) has been used in the meat industry for some time. The meat sample is carried through an elec-

---

[25]Rizvi, S. S. H., A. L. Benado, J. A. Zollweg, and J. A. Daniels. 1986. Supercritical fluid extraction: Fundamental principles and modeling methods. Food Technol. 40 (6):55.

[26]Rizvi, S. S. H., J. A. Daniels, A. L. Benado, and J. A. Zollweg. 1986. Supercritical fluid extraction: Operating principles and food applications. Food Technol. 40 (7):57.

[27]Sperber, R. M. 1989. New technologies for cholesterol reduction. Food Proc. 50 (12):154.

[28]Chao, R. R., S. J. Mulvaney, M. E. Bailey, and L. N. Fernando. 1991. Supercritical CO_2 conditions affecting extraction of lipid and cholesterol from ground beef. J. Food Sci. 56:183.

[29]Boileau, R. A. 1988. Utilization of total body electrical conductivity in determining body composition. In: Designing Foods: Animal Product Options in the Market Place. National Research Council, National Academy Press, Washington, D.C., pp. 251–257.

[30]Houtkooper, L. B., T. G. Lohman, S. B. Going, and M. C. Hall. 1989. Validity of bioelectric impedance for body composition in children. J. Appl. Physiol. 66:814.

[31]Swantek, P. M., J. D. Crenshaw, M. J. Marchello, and H. C. Lukaski. 1992. Bioelectrical impedance: A nondestructive method to determine fat-free mass of live market swine and pork carcasses. J. Anim. Sci. 70:169.

tric coil on a conveyer, which causes the coil to lose energy. The energy loss is measured and then related to the lean content of the sample — lean conducts electric current about 20 times more readily than fat. Newer instruments that use the same principle have been modified to handle meat carcasses and humans. This technique is called *total body electrical conductivity* (TOBEC).

When electric current passes through a material, resistance or *impedance* develops. Instruments that have been manufactured to measure impedance are being used to estimate body composition in humans. Four electrodes are attached to the body in order to measure impedance. Research on utilizing this technology in the meat industry is ongoing.

Ultrasonics

Ultrasonic measurement is based on the principle that high-frequency sound waves (see Appendix) pass through tissue but are reflected back at the interface between two different types of tissue. Time variations for the return of the reflected signals measure distances between tissue boundaries.[32] This technique has been and continues to be used to evaluate fat thickness and loin/rib eye area in live animals and carcasses.[33, 34]

Potassium–40 Liquid Scintillation Detection

This "whole-body counting" technique is most commonly used to determine body composition in humans, but where equipment is available, it is also used for animals and carcasses.[35] Estimating the composition of a living body, or a carcass by whole-body counting, is possible because potassium (K)

[32]Topel, D. G., and R. Kauffman. 1988. Live Animal and Carcass Composition Measurements. In: Designing Foods: Animal Product Options in the Market Place. National Research Council, National Academy Press, Washington, D.C., pp. 261–262.

[33]Stouffer, J. R. 1963. Relationship of ultrasonic measurements and x-rays to body composition. Annu. N.Y. Acad. Sci. 110:31.

[34]McLaren, D. G. 1991. A study of operator effects on ultrasonic measures of fat depth and *longissimus* muscle area in cattle, sheep and pigs. J. Anim. Sci. 69:54.

[35]Siemens, A. L., R. J. Lipsey, W. M. Martin, M. G. Siemens, and H. B. Hedrick. 1991. Composition of pork carcasses by potassium–40 liquid scintillation detection: estimation and validation. J. Anim. Sci. 69:47.

has a constant relationship to lean, and thus an indirect relationship to fat.[36] Potassium is found primarily in the intercellular space, in the fat-free tissue as a constant percentage, and is not found in fat. Thus, if potassium is measured and the weight of the "whole body" is known, the fat weight can easily be determined by subtraction. [Total body weight (–) fat-free weight = fat weight].

Potassium has a normal atomic weight of ^{39}K, but a ^{40}K unstable isotope is present as a constant 0.012 percent of total potassium. As ^{40}K disintegrates, it produces gamma rays of high energy (1.46 MeV; see Appendix — electricity), which can be measured in a whole-body counter, which consists of a chamber shielded usually by 4 to 8 inches of steel against environmental or background radiation. Scintillators in the counter, which may be crystal (sodium iodide) or liquid, detect the ^{40}K disintegrations. Thus, the construction of a counter is rather expensive, and the maintenance and operation require experienced, well-trained technicians. This technique is not used in production situations in the meat industry per se, but rather as a research tool.

Final Blending of Fat and Lean Blends

In typical ground meat operations, a "leaner blend" and a "fat blend" are produced. The fat content of raw materials must be estimated as described previously in order to be combined in the right proportions to produce the correct composition in the final ground meat product. The resulting final ground product must be checked for fat content as well. There are at least two ways to compute how much of the leaner blend must be mixed with the fatter blend so that the final product will have the desired composition: the algebraic method and the method of squares (also known as the Pearson square).[37]

---

[36]Ward, G. M. 1968. Introduction to whole-body counting. In: Body Composition in Animals and Man. Proc. Symp., May 4–6, 1967, Univ. of Missouri, Columbia. National Academy Press, Washington, D.C.

[37]Bonkowski, Alexander T. 1993. Pearson Square can zero in on fat content. Meat Market & Technol. 44–45. Furnished by John Henson, formerly at South Dakota State University, presently Plant Manager, Busseto Specialty Meats, Fresno, CA.

Algebraic Method

Example:

The final product is to be 80 percent lean.

The leaner blend is 90 percent lean.

The fatter blend is 65 percent lean.

Solve two simultaneous equations:

x = pounds of 90 percent lean

y = pounds of 65 percent lean

Make 100 pounds of 80 percent lean product.

$x + y = 100$ pounds of total product

$0.90x + 0.65y = 80$ pounds of lean

$x = 100 - y$

$0.90(100 - y) + 0.65y = 80$

$90 - 0.90y + 0.65y = 80$

$-0.25y = -10$

$y = 40$ pounds of 65 percent lean

$x = 60$ pounds of 90 percent lean

Check/proof:

40 pounds × 0.65 = 26 pounds of lean

<u>60</u> pounds × 0.90 = <u>54</u> pounds of lean

100 pounds total   80 pounds of lean

Method of Squares

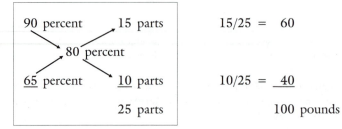

90 percent	15 parts	15/25 = 60
80 percent		
65 percent	10 parts	10/25 = <u>40</u>
	25 parts	100 pounds

Percent lean of two blends in left corners of square.

Desired lean percent in middle of square.

Subtract diagonally to get **parts** of each blend, directly to the right of the blend in question.

Add to get total parts.

Divide by total to get percent (100-pound basis).

Use the same proof as used for the algebraic method.

THE BEEF PATTY INDUSTRY

As with many meat product names, the origin of the name "hamburger" is somewhat clouded in history. Perhaps hamburger's roots can be traced to Tartary tribes from the Baltic provinces in the Middle Ages. German traders developed the Hamburg steak and brought it to America in the eighteenth century. The hamburger gained wide acceptance after exposure at the 1904 World's Fair in St. Louis. In 1921, White Castle became the first hamburger chain. The National Cattlemen's Beef Association[38] can provide information about the current ranking of hamburger chains.

The largest segment of the fresh meat processing industry is unquestionably the highly specialized beef patty industry. Ground beef products account for approximately 43 percent of the total U.S. beef consumption (see Table 13–5). Hamburgers consumed from fast-food franchises are a significant portion of this total. In fact, hamburgers and cheeseburgers comprise most (76.5 percent) of beef servings away from home (see footnote 38). The remaining 23.5 percent of beef servings away from home are composed of roast beef sandwiches, 7 percent; steak entrees, 5.2 percent; steak sandwiches, 3.2 percent; other beef entrees, 3 percent; roast beef / prime rib, 2.9 percent; and ground beef entrees, 2.2 percent.

Ground beef used in the production of patties may take on a variety of shapes and sizes from modern patty-making machines. The most common shape is round; however, square patties and oval-shaped patties are not uncommon. The most common sizes of patties are 6 per pound, 4 per pound, and 3 per pound (2.7, 4.0, and 5.3 ounces and 76, 114, and 151 grams respectively).

---

[38]National Cattlemen's Beef Association. http://www.beef.org

Sources of Raw Materials

The sources of raw materials used in the production of ground beef vary widely between processors and affect the quality of the ground beef patty produced. Typical operations will use boneless cow meat, choice beef plates, or flanks, as well as a variety of trimmings from various portions of the carcass, depending on supply and price. Some ground beef patty makers use imported frozen beef when supplies and prices are favorable (see Chapters 1 and 13). Higher cooking losses are generally associated with the use of frozen trimmings.

Low-Fat Ground Beef and Patties

Mela[39] poses the intriguing question "Why do we eat fat?" He lists flavor, texture, and learned preferences as possible reasons. Because consumers are demanding palatable, low-fat foods, the food industry is investing considerable resources to develop such products. Mela indicates that to this point, only short-term sensory acceptance has been attained, but long-term acceptance remains to be determined.

The extension of ground beef with soy protein to produce a product with nutritional quality equivalent to all-beef ground beef at a reduced cost is being widely used by the food industry. Up to 20 percent rehydrated soy concentrate in ground beef is accepted by the U.S. Department of Agriculture for school lunch and military menus. Dr. Brad Berry, U.S. Department of Agriculture, Beltsville, Maryland,[40] has been very active in researching ground beef/soy combinations that result in low-fat products. Soy concentrate is a 70 percent protein product that is available as a flour or in a coarse granular form, the latter form being more similar in texture to ground beef (see Chapter 21).

The year 1991 might be known for the "breakthrough" of the low-fat hamburger. Auburn University meat scientist Dr. Dale L. Huffman[41] is credited with the research that established the formulation for the 91 percent fat-free hamburger. The informal "standard" fat content in hamburgers to pro-

---

[39]Mela, D. J. 1990. The basis of dietary fat preferences. Trends Food Sci. Technol. 1:71.

[40]Berry, B. W., and K. F. Leddy. 1988. Effects of hot processing, patty formation before or after freezing-thawing and soy usage on various properties of low-fat ground beef. J. Food Qual. 11:159.

[41]Huffman, D. H., and W. R. Egbert. 1990. Advances in lean ground beef production. Bull. 606, Ala. Agric. Exp. Sta., Auburn Univ.

vide optimum juiciness and flavor (and verified by Dr. Huffman's work) has been 20 percent, although USDA regulations permit 30 percent fat (see "USDA–FSIS Regulations"). So 20 percent fat was the control that Dr. Huffman used to compare with his new low-fat ground beef formulation that began with a 10 percent fat ground beef product to which the following was *added:*

1. 0.375 percent salt

2. 0.188 percent hydrolyzed vegetable protein (HVP)

3. 3.0 percent water

4. 0.5 percent carrageenan

This mixture was ground through a $^3/_{16}$-inch (0.48-cm) plate.

Hydrolyzed vegetable proteins (HVPs) are formulated blends of any of the following: soy, yeast, corn, and wheat proteins that have been *hydrolyzed* (broken down into their component amino acids) to yield meat-like flavors. HVPs are available commercially in liquid and powder form and are considered a natural source of monosodium glutamate (MSG), a flavor enhancer, which, if present, should be identified for consumers.

Carrageenan is a generic term referring to a heterogeneous group of polysaccharides obtained by the aqueous extraction of certain species of red seaweed. Carrageenan is produced worldwide and is used extensively in the food industry as a thickening, gelling, stabilizing, and protein-suspending agent. It is known to improve the "mouth-feel" of a variety of food products. Molecular structures of three of the various fractions of carrageenan (alpha, beta, and iota) are clearly drawn by Nicklin and Miller.[42] A colored photo of the red seaweed source of carrageenan is shown on the front cover of *Food Technology,* March, 1991, which contains a summary by Dziezak of the use of gums in the food industry.[43]

USDA–FSIS Regulations

The ground beef segment of the meat industry is subject to specific compositional and labeling regulations by the USDA–FSIS (see also Chap-

---

[42]Nicklin, S., and K. Miller. 1989. Intestinal uptake and immunological effects of carrageenan — current concepts. Food Add. and Contam. 6:425.

[43]Dziezak, J. D. 1991. Special report — a focus on gums. Food Technol. 45 (No. 3):115.

ter 3). The Code of Federal Regulations[44] (CFR) (sec. 319.15) stipulates that:

> The terms *ground beef* and *chopped beef* are synonymous. [A product] so labeled must be made with fresh and/or frozen beef with or without seasoning and without the addition of fat as such and shall contain no more than 30 percent fat. It may not contain added water, phosphates, binders or extenders. It may contain beef cheek meat [see Chapter 6] not to exceed 25 percent. If the name is qualified by the name of a particular cut, such as *ground beef round* or *ground beef chuck,* the product must consist entirely of meat from that particular cut or part. Product labeled *ground beef, beef fat added* may have beef fat added, however the total fat may not exceed 30 percent.

Hamburger is defined in the CFR (sec. 319.15b) as:

> Chopped fresh and/or frozen beef, with or without added beef fat and/or seasonings. Shall not contain more than 30 percent fat, and shall not contain added water, phosphates, binders or extenders. Beef cheek meat may be used up to 25 percent of the meat formulation.

If ground beef or hamburger contains soy products or other extenders, it must be descriptively labeled as such. For example, *"Ground Beef and Texturized Vegetable Protein"* is the required label when a soy product or other extender is added.

If a nutritionally inferior product is used, then the product must be labeled *"Imitation Ground Beef"* or more often *"Beef Patties (Beef Patty Mix)."* The CFR (sec. 9.15c) defines *beef patties* as:

> Chopped fresh and/or frozen beef with or without the addition of beef fat as such and/or seasonings. Binders or extenders and/or partially defatted beef fatty tissue [see Chapter 11] may be used without added water or with added water only in amounts such that the product's characteristics are essentially that of a beef patty. Partially defatted beef fatty tissue is a beef byproduct derived from the low temperature rendering (not exceeding 120°F) of fresh beef

---

[44]U.S. Government Printing Office, Superintendent of Documents, Washington, DC 20402

fatty tissue. This product must have a pinkish color and a fresh odor and appearance.

Products that combine hamburger or ground beef with non-fat ingredients to make a low-fat product may be called "Low-Fat Ground Beef with an X% Solution of . . ." or *"Low-Fat Hamburger, Water, and Carrageenan Product."*

Low-Fat Meat Products in Perspective

Browner and coworkers[45] concluded that by reducing fat consumption from 37 percent of calories to 30 percent of calories, as recommended by several agencies and authorities, people might live three or four months longer at the end of their life span. Even though some fat is necessary in the human diet (see Chapter 23), meat processors in general should try to keep in step with consumer demand. Thus, many commercial firms are producing and selling the soy, carrageenan, HVP, and any number of additional products that are necessary for the production of low-fat meat products.[46]

Commercial Applications

Larger patty plants run at production capacities of 10,000 pounds per hour. They utilize a continuous system of grinding, blending, forming, freezing, and packaging such that the product flow never stops once grinding commences. In such operations, an infrared spectral analysis machine such as the Anyl–Ray® machine shown in Figure 18–4 checks the lean and fat blends almost constantly for fat content. Estimates of fat content are obtained in seconds, allowing for continual on-line adjustments of lean and fat blends to keep the product in the company's acceptable fat range.

The lean and fat blends are combined in the proper ratios to achieve the desired target fat content according to the company specifications. After a short blending period, the ground beef is typically reground through a ⅛-inch plate with a bone collection apparatus (see Figure 18–5) used to remove any bone chips or fragments from the finished product.

---

[45]Browner, W. S., J. Westenhouse, and J. A. Tice. 1991. What if Americans ate less fat? J. Am. Med. Assn. 265:3285.

[46]Hoogenkamp, H. W. 1991. Building a better burger. Meat Proc. 30 (No. 4):24.

Fig. 18–5. Bone chip collectors are used by ground beef manufacturers to eliminate cartilage and bone chips from the final product. The cartilage and bone chips do not pass through a fine grinder plate but rather come out the center tube, as shown here. A special grinder plate that has shallow grooves radiating to the center forces any bone chips to the exit tube. (Courtesy, Howard Beef, Howard, South Dakota)

In any size operation it is especially critical that all grinder knives and plates be sharp. Also the blades and grinder plate that have been used together previously must match and always be used together. The temperature of the beef to be ground should be very near freezing. Furthermore, meat should not be stuffed down into the grinder, but should be fed uniformly, allowing the screw drive to pull and drive the meat through. If these key elements are ignored, fat smearing resulting from ripping and pulling actions rather than crisp cutting will occur, which will seriously damage the appearance and texture of the ground product.

Following the final grind, the meat is transported (by bulk bin or conveyor) to a patty-forming machine (see Figure 18–6), and the subsequent patties are conveyed into a cryogenic freezing tunnel or spirulator (see Figures 18–7 and 18–8).

Rapid cryogenic freezing systems have been shown to improve patty quality by minimizing cooking losses and preserving flavor for longer periods of time. Frozen patties are boxed, generally with the aid of a patty-stacking machine (see Figure 18–9), and moved to a holding freezer for shipment.

Fig. 18–6. A Formax® patty machine (Formax, Inc., Mokena, Illinois) is shown making "four to the pound" (¼-pounder) round patties. (Courtesy, Howard Beef, Howard, South Dakota)

Fig. 18–7. Before going into the spirulating freezer, patties are perforated with serrator blades for a more rapid and uniform freeze. (Courtesy, Howard Beef, Howard, South Dakota)

Fig. 18–8. Patties are frozen to 0°F in 10 minutes inside this −80°F spirulator. The patties enter on the bottom level of the stainless steel conveyor and exit on the top level. (Courtesy, Howard Beef, Howard, South Dakota)

Fig. 18–9. Frozen patties are conveyed up to a patty-stacking machine, which reduces the boxing labor costs. Note the perforations on the patties from the serrators shown in Fig. 18–7. (Courtesy, Howard Beef, Howard, South Dakota)

Precooked Patties — Precautions and Regulations

Because meat patties are popular for school lunch programs and with the general public, and because meal preparation time is at a premium in our fast-moving society, the practice of precooking or partially precooking patties at the place of manufacture is becoming widespread. In a survey conducted by the Consumer Network of a national cross section of 5,000 shoppers, respondents said they would buy precooked ground beef if it were offered in the supermarket. They would like to see it for four reasons: preseasoning, time saving, mess saving, and lower fat content, reported *Supermarket News* on October 12, 1998. In such situations, extreme caution must be exercised so that microbial recontamination and growth do not occur somewhere in the process before the patties are consumed by humans (see Chapter 2). Since these products are uncured (see Chapter 20), water activity may be greater than 0.93 and/or a pH greater than 4.6, which means that temperature control is the only microbial barrier or hurdle.

Precooked patties are usually sold in one of three stages of doneness: (1) fully cooked, (2) partially cooked, and (3) char-marked. Each stage has a particular time–temperature requirement for safety (FSIS Directive 7370.2, 6-28-95).

Fully Cooked Patties

Patties must be held for 41 seconds at a minimum temperature of 151°F (66.1°C). This period of time is known as the *dwell time*. Patties cooked to 157°F (69.4°C) or higher need a dwell time of 10 seconds to be fully cooked. As temperatures increase between these two temperature points, the dwell times decrease concurrently. Patties must be cooled to an internal temperature of 40°F (4°C) or below within two hours after heat processing.

Partially Cooked Patties

Each patty must be heated to an internal temperature of 140°F (60°C), then cooled to an internal temperature of 40°F (4°C) or below within two hours. Cooking instructions on the label adjacent to the product name should read *"Partially Cooked: For Safety, Cook Until Well Done (Internal meat temperature 160°F)."*

Char-marked Patties

Each patty must be heated to an internal temperature of 70°F (21.1°C) but no higher, then cooled to an internal temperature of 40°F (4°C) or below. The whole process (heating and cooling) must be completed within two hours. Cooking instructions on the label adjacent to the product name should read *"Uncooked, Char-marked: For Safety, Cook Until Well Done (Internal meat temperature 160°F)."*

Many Additional Uses for Ground Beef

Ethnic foods are very popular in the United States. And the central ingredient in the very popular pizza, lasagna, macaroni, and spaghetti from Italy and the tacos, enchiladas, burritos, and tostados from Mexico is ground beef. But, probably the greatest use of ground beef is in the countless hot dishes/casseroles that have been a basic, nutritious meal for many families for many years.

TENDERIZATION — MECHANICAL AND ENZYMATIC

Another segment of the hotel–restaurant–institution (HRI) trade has specialized in offering intact muscle steaks (not ground or formed) at economy prices. Family steakhouses achieve economy by utilizing lower-quality carcasses (usually from the Standard, Commercial, and Utility grades [see Chapter 12]) and assuring uniform tenderness with mechanical and(or) enzymatic tenderization.

Mechanical tenderization is accomplished by passing the steaks (or boneless sub-primals) through a bank of needles or through a rotary steak macerator (cuber). The former method is referred to as needle tenderization (or sometimes blade tenderization) and is illustrated in Figure 18–10. The bank of needles passes through the meat, severing connective tissues and muscle fibers, making the meat more palatable. A variable speed conveyor is used to advance the meat, while the up-and-down motion of the needle bank penetrates the muscle tissue. The best results are obtained when boneless sub-primals such as boneless rib eye and boneless top sirloin are used. When sub-primals come from more mature beef (Commercial and Utility grades), they should be needle tenderized several times (two or three passes).

Fig. 18–10. *(Top)* Needle (or blade) tenderization of boneless sub-primals is one method of improving the palatability of less tender cuts. A Ross® tenderizer (Ross Industries, Midland, Virginia) is illustrated in this example. *(Bottom)* A closeup photo of the bank of needles. (Courtesy, South Dakota State University)

Fig. 18–11. *(Top)* A rotary blade tenderizer such as the Hobart® tenderizer (Hobart, Troy, Ohio) pictured here with the safety cover in place is used for making cubed steaks. *(Bottom)* The two uncovered, unprotected macerating blade assemblies rotate in opposite directions, pulling the meat through the center. (Courtesy, Hy–Vee Supermarket, Brookings, South Dakota)

The steak macerator (see Figure 18–11) can be used for tenderizing individual loin, sirloin, and round steaks; however, more often it is used to make cubed steaks from less tender portions of the carcass such as the chuck (or shoulder). Steak trimmings and end pieces are sometimes fused together in a macerator to produce high-quality cubed steaks rather than being utilized for stew or kabob meat.

Enzymatic tenderization is another means of improving the tenderness of steaks from mature cattle. Three proteolytic enzymes commonly used in the tenderization of meat are *papain, bromelin,* and *ficin.* These enzymes, derived from tropical plants (papaya, pineapple, and fig respectively), degrade muscle fibers and connective tissue to different degrees. Of the three, papain is the most widely used; however, it has the least degradative effect on collagen (the most abundant connective tissue protein). Bromelin exhibits strong degradative properties for collagen and has the least affinity for the muscle fibers and the connective tissue elastin. Ficin exhibits the greatest degradative action on both collagen and elastin; however, it also has the greatest degradative effect on the myofibrillar proteins of the muscle fibers. Care must be taken not to overtenderize and turn the meat to mush, particularly when the enzyme ficin is used. The commercial tenderizers used by many of the family steakhouses are usually a blend of the three enzymes. To be effective, tenderizers should be injected, for when they are sprayed or dipped, their penetration is limited, leaving the surface of the steaks mushy but the interior untenderized. The solution may also contain salt and phosphate.

FROZEN CONVENIENCE FOODS

Frozen convenience foods, the new generation of TV dinners, have had a substantial impact on both the retail food industry (supermarkets) and the food service industry. Frozen convenience meals emphasize health, low cost, variety, quality, and microwavable convenience with menu offerings of delicacies such as chicken cordon bleu, beef stroganoff, and veal parmesan. The aluminum trays have been replaced by attractively designed plastic dishes that allow microwave cooking. The gourmet image, more rigid quality specifications, attractive packaging graphics, and nutritional information have also contributed to the recent upsurge in sales at the retail supermarkets.

Many companies have taken the clue from calorie-conscious consumers, and have experienced phenomenal market growth. Such success stories are chronicled in many of the industry periodicals listed in the Appendix. Industry experts project that the frozen convenience meal market will be one of the hottest growth areas for the meat industry, particularly as new processing technologies for automation are developed. Frozen convenience meals have made a big hit in the airline food service market where the quality meal image is an important factor for airline companies in attracting customers.

The key to quality in processing frozen convenience meals is in carefully monitored sanitation programs and rapid freezing processes. Meat, pasta, and vegetable items are precooked (vegetables are sometimes only blanched), loaded onto plastic dishes moving down a conveyor line, frozen in a cryogenic tunnel, and sealed with a thin plastic film over the top before being packaged. The end result is a complete gourmet-style hot meal that can go from freezer to plate in less than 10 minutes — that's convenience!

RESTRUCTURED MEAT PRODUCTS

Restructured meat products are products that have been ground, flaked, or chopped and formed into steak/chop- or roast-like products with a texture that is closer to that of an intact steak than that of ground meat. Technology in this area is continually advancing, and with the development of products composed of whole muscles formed together, the term *structured* has evolved.

Researchers Dr. Dale L. Huffman (Auburn University), Dr. Roger W. Mandigo (University of Nebraska), and Dr. Glenn R. Schmidt (Colorado State University) and their coworkers are widely recognized as early "pioneers" and current leading experts in the field of structured meat products. Dr. Roger C. Johnson *(Triumph Pork Group, LLC)* initiated the new work in whole muscle structuring. The National Live Stock and Meat Board, during the period when Dr. B. C. Breidenstein was Director of Research, published several excellent sources of information on manufacturing practices for restructured products based largely on the work of the previously named researchers.[47, 48, 49] (Anyone interested in the current status of structured meat research and development can search the literature for publications by these

[47]Breidenstein, B. C. 1982. Intermediate value beef products (restructured beef products). National Live Stock and Meat Board, Chicago (now National Cattlemen's Beef Association, Denver).

[48]National Live Stock and Meat Board. 1983. Manufacturing guidelines for processed beef products. Chicago. (c/o National Cattlemen's Beef Association, Denver.)

[49]National Live Stock and Meat Board. 1985. Manufacturing guidelines for processed non-cured pork products. Chicago. (c/o National Cattlemen's Beef Association, Denver.)

researchers.) The National Live Stock and Meat Board provided much of the information that is given on structured meat products in this text.

Depending on the method and the extent of particle-size reduction, structured products are also referred to as *flaked and formed, chunked and formed, sectioned and formed, chopped and formed,* or *whole muscle.* Unlike ground beef, pork, chicken, turkey, or lamb, the meat is mixed after comminution (particle-size reduction) with some agent, usually salt, phosphate, and(or) protein or carbohydrate binders, that will either directly or indirectly bind the particles back together. The meat is then formed into the desired shape and, after being cooked, will maintain that approximate shape.

A number of processing considerations important in the production of structured products are raw materials source and quality, comminution method and particle size, nonmeat ingredients, and processing procedures, such as mixing time, processing temperature, forming method, processing equipment, and packaging system(s).

Raw Materials Source and Quality

A wide range of muscle tissue and meat by-products can be used in the formulation of restructured products, depending on the desired cost and quality of the finished product. The selection of raw materials should be largely influenced by the market segment the processor intends to enter and whether the product as formulated will offer a competitive value alternative to the prospective consumer. For example, if the intended consumer is within an institutional market, minimizing the raw materials costs while meeting the quality specifications for that market is an important consideration.

The raw materials that have the greatest economic value for restructuring processes are those that have a lower wholesale value or that are lower-valued by-products of the processing industry. Such materials allow greater room for the higher production costs associated with restructuring (which is necessary to upgrade the ultimate quality of the finished product) and yet still maintain an economically priced item. Primal cuts, boneless primal cuts, selected trimmings, non-selected trimmings, mechanically deboned tissue, and(or) partially defatted tissue can effectively be used as raw materials. Primal cuts that may be used are generally limited to the chuck (or shoulder) and round (or leg). The loin meat is generally too costly to be used as a raw material. Furthermore, restructuring would not generally result in quality improvement of the product, since the loin is more tender and contains less connective tissue than a lower-valued cut such as the beef chuck. The breast meat from spent fowl, which is usually very tough, has been made very acceptable by this technique.[50]

Raw materials should be as fresh as possible to prevent problems caused by bacterial, enzymatic, and oxidative degradation, which results in off-flavors, off-odors, and a more rapid deterioration of desirable color. Pork and poultry fat is more unsaturated than lamb and beef fat and thus is more subject to the development of off-flavors (oxidative rancidity) (see Chapters 19, 21, and 23). Ideally, meat from pork and poultry carcasses should be boned within 24 hours postmortem and further processed within 48 hours postmortem. Beef and lamb should be further processed within seven days postmortem.

Comminution Method and Particle Size

The method and degree of comminution is largely determined by the available equipment and the connective tissue content of the raw materials being used. Generally, meats containing moderate to high levels of connective tissue, such as certain muscle systems of the shoulder and shank regions, should be reduced to a fine particle size to insure uniform palatability. Raw materials that have lower concentrations of connective tissue are palatable in larger particle sizes and contribute a more desirable texture to the finished product. It is important to remove the dense connective tissue membranes from larger muscle systems if a large particle size is used in the restructured product. Removal of these connective tissue membranes, called *denuding*, results in extremely high-quality raw material. Considerable differences in connective tissue content and tenderness exist between different muscles in the beef forequarter.[51]

Particle-size reduction may be accomplished by sectioning, chunking, slicing, flaking, grinding, or chopping. Each method contributes different textural properties to the final restructured product.

Sectioning

Sectioning involves the separation of entire muscles (or muscle systems) by seaming between the muscles with a knife. Sectioning is a widely used

---

[50]Seideman, S. C., P. R. Durland, N. M. Quenzer, and C. W. Carlson. 1982. Utilization of spent fowl muscle in the manufacture of restructured steaks. Poult. Sci. 61:1087.

[51]Johnson, R. C., C. M. Chen, T. S. Muller, W. J. Costello, J. R. Romans, and K. W. Jones. 1988. Characterization of the muscles within the beef forequarter. J. Food Sci. 53:1247.

practice in the commercial ham industry for producing *sectioned and formed hams*. This technique has been used to remove the *serratus ventralis* muscle from beef forequarters to make structured steaks.[52] It is desirable to denude the sectioned muscles that have heavy connective tissue membranes. The tenderness of the individual muscle and the amount of internal connective tissue and fat will affect the usefulness of the sectioned portions.

Chunking

Chunking is the process of making coarse particle sizes (but smaller than sectioned muscles). Meat can be made into chunks with a very coarse grinder plate (see "Grinding"), a meat dicer, a bowl chopper, or an ordinary knife. Coarse grinding and dicing are the two most common methods of producing desirable-sized chunks.

Slicing

Thin slicing frozen and tempered (partially thawed) meat on a high-speed slicer or cleaver is particularly useful for tissues such as the beef plate and Boston butt, which are often relatively high in fat. Thin slicing these cuts and formulating with a lean fraction results in a restructured product that has an appearance very similar to natural marbling.

Flaking

Flaking is the process of reducing particle size with the Urschel Comitrol® (Valparaiso, Indiana) or similar equipment. An example of a laboratory-sized Comitrol® and a flaking head used in particle-size reduction are shown in Figure 18–12. Meat should be frozen and tempered (24° to 26°F) prior to flaking in a Comitrol® to insure uniform flake sizes. The flaking temperature of lean and fat fractions is also critical to insure good protein extraction necessary for product bind as well as dispersion of the fat throughout the product. A wide variety of flaking heads that result in flake sizes ranging from very coarse to very fine are available.

---

[52]Johnson, R. C., J. R. Romans, T. S. Muller, W. J. Costello, and K. W. Jones. 1990. Physical, chemical and sensory characteristics of four types of beef steaks. J. Food Sci. 55:1264.

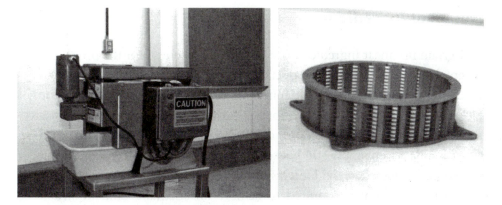

Fig. 18–12. *(Left)* A laboratory-sized Comitrol® (Urschel Laboratories, Inc., Valparaiso, Indiana) machine used for flaking meat in restructured products. *(Right)* An example of a flaking head used by this machine. A rotating impeller (not shown) forces frozen meat through the openings on the flaking heads, which are available in many sizes. (Courtesy, South Dakota State University)

Grinding

Grinding is a more restrictive term than *chunking*. It is accomplished by passing meat through a grinder plate and cutting the meat with a grinder knife that fits snugly against the plate. Grinder plates are available with a large variety of hole sizes ranging from $\frac{1}{16}$ inch to large kidney plates that have dimensions of $1\frac{1}{2}$ by 3 inches per hole or larger. As with most comminution equipment, the importance of sharp knives cannot be overstressed. Sharp grinder knives are critical for efficient operation, particle definition, and the prevention of fat smearing.

Chopping

Chopping is done in bowl choppers or silent cutters. These machines are available with vacuum domes, and some models have the capabilities for simultaneous chopping and cooking. A large bowl chopper (see Figure 21–10) consists of a rotating bowl that forces the meat to pass through a series of rotating knives that are closely adjusted to the surface of the bowl. Chunking meat in a bowl chopper necessitates exercising extreme caution so that batch-to-batch variation in chunk size is minimized. Nonetheless, the bowl chopper has the greatest versatility for creating different particle sizes (large to small) without requiring a change of parts (that is, plates, cutting head, dicing head, etc.).

Nonmeat Ingredients

Certain nonmeat ingredients are considered essential for the production of acceptable structured meat products. Of these, salt (NaCl) is the most widely used nonmeat ingredient. Salt functions to solubilize proteins that form a tacky exudate (the muscle myofibrillar protein myosin) that, when heated, acts to bind the meat particles together in a continuous mass. Salt also functions as a flavor enhancer at the levels commonly used in structured products. Research conducted at the University of Nebraska showed that a salt level of 0.5 to 0.75 percent is required to achieve an adequate bind in structured products.[53] Since salt acts as a pro-oxidant leading to the development of oxidative rancidity (see Chapters 19, 21, and 23), it is extremely important that high-purity–grade salt be used in manufacturing structured products. Common table salt typically contains small amounts of copper, iron, and chromium impurities, which can act as catalysts in the development of oxidative rancidity.

Polyphosphates are another group of compounds commonly added to structured products at a level of 0.15 to 0.40 percent. Polyphosphates function to improve the protein extraction of the myofibrillar proteins necessary for good product bind and to decrease the shrinkage that occurs during cooking. Phosphates are also reported to improve the fresh color of structured products (which is often a problem) and to inhibit oxidative rancidity. (For a more detailed discussion of phosphates, see discussion at the end of this chapter and Chapter 20.)

A binding system utilizing the hydrocolloid alginate (a gum extracted from brown seaweed) has been patented[54] and approved for commercial production.[55] This system utilizes the ability of alginate to form instantaneous gels by reacting with calcium salts to bind comminuted structured products in both the raw, refrigerated form and the cooked state. The adhesive properties of hydrocolloids are due to their ability to interact with proteins and lipids commonly found in food and because of their high affinity for water. Research conducted at South Dakota State University indicated that an

---

[53]Mandigo, R. W. 1982. Restructured Meats. Proc. of the Fourth Annual Sausage and Processed Meats Short Course, Iowa State Univ.

[54]Schmidt, G. R., and Means, W. J. 1986. Process for preparing algin/calcium gel structured meat products. U.S. Patent 4,603,054.

[55]USDA–FSIS. 1986. Binder consisting of sodium alginate, calcium carbonate, lactic acid, and calcium lactate. Fed. Reg. 51 (159):29456.

alginate/calcium/adipic acid gel could be used as an adhesive binder in the production of structured beef steaks composed of whole muscles or large muscle pieces.[56] The pieces remained bound not only in the cooked product but also in the raw, refrigerated state.

Other ingredients that may be used to add variety to restructured products or to reduce the cost include seasonings, binders, extenders, and nitrites. These nonmeat ingredients are discussed in Chapter 21 and serve the same functions as those outlined for sausage ingredients.

Processing Procedures for Structured Products

Following the selection of raw materials, particle-size modification, and selection of nonmeat ingredients to be used, the processor is ready to begin manufacturing structured products. The next processing steps depend on whether the product is to be produced from comminuted or sectioned raw material. Temperature considerations are extremely important in maintaining structured product integrity.

Comminuted Products (Chunked, Sliced, Flaked, Ground, and Chopped)

The raw materials are combined with salt, phosphates, etc., in a mixer (blender, massager, or tumbler). Fresh structured products are typically blended for 5 to 15 minutes, depending on meat temperature, particle size, and blending action to get the necessary protein extraction required for binding. Low temperatures (approximately 30° to 32°F) are critical to preserving a desirable oxygenated fresh meat color. After blending, the products are typically stuffed in polyethylene bags or other low-cost casings and tempered to a temperature of approximately 26°F. Hydraulic meat presses can then form the restructured products to virtually any shape desired. A uniform temperature of the products is extremely critical to minimize pressing losses (if too warm) or to insure that the products do not break apart later when cleaving into individual portion sizes (if too cold). The products must remain frozen

---

[56]Johnson, R. C., T. S. Muller, J. R. Romans, W. J. Costello, and K. W. Jones. 1990. Effects of algin/calcium and adipic acid concentration on muscle-juncture formation. J. Food Sci. 55:906.

until they are cooked. If they are thawed prior to being cooked, they will not retain their shape and may fall apart.

The inclusion of calcium alginate in the formulation eliminates the need for freezing and the critical tempering times required for hydraulic pressing as a forming method. The products can be formed into individual portion sizes much like hamburger patties are made. Using such a modified patty-forming machine eliminates not only the lengthy and critical tempering times (to 26°F) but also the need for cleaving into individual portions. This system is less labor- and energy-intensive.

Sectioned and Whole Muscle Products

Two whole beef *serratus ventralis* (SRV) muscles were trimmed of all outside fat and the surface of one was coated with an alginate/calcium/adipic acid gel solution. A second SRV muscle was stacked on the first, and the new muscle mass was vacuum-packaged and held at 41°F (5°C) for 15 hours before being cut into steaks. These steaks were handled as any intact muscle steaks (stored in a freezer and thawed prior to being cooked), and they kept their shape and integrity throughout until they were eaten. (See Chapter 20 for a discussion of sectioned and formed hams.)

EXTRUSION

Another forming method that is becoming more widely used in the food industry is the extrusion process. The extrusion process involves forcing a material to flow under a variety of controlled conditions and then to pass through a shaped hole or slot at a predetermined rate. The first applications of extrusion to the food industry date back to the mid- to late 1800s in the production of sausages and processed meats. These extrusion machines were the forerunners of the present-day grinders and stuffers (see Chapter 21). Extrusion is now used to produce pasta, ready-to-eat cereals, confectionery items, chewing gum, and many other popular nonmeat foods. A very complete review including pictures and diagrams explaining and illustrating extruders and the products produced with them was compiled by Dziezak.[57]

---

[57]Dziezak, J. D. 1989. Single- and twin-screw extruders in food processing. Food Technol. 43 (No. 4):115.

Extrusion is an energy-efficient food processing method because it combines several steps, such as mixing, heating, shearing, texturizing, and drying in a single operation. Aside from the very early and continued application of the principles of extrusion to sausage manufacture, present-day use in the meat industry is in the developmental stages. Theoretically, structured product could be extruded from specially designed stuffing horns onto a conveyor, subsequently frozen or tempered and sliced or cleaved into individual portions, such as kabobs. Michigan State University researchers used a twin-screw extruder to experimentally produce a variety of products containing mechanically deboned chicken in combination with three nonmeat binders: corn starch, soy protein isolate, and wheat gluten.[58] The mechanically deboned chicken could not be extruded without nonmeat binders because the extruder barrel clogged. The free water and fat released in the extruder disrupted flow patterns and led to die plugging. In this study, corn starch was the best binder. Research on the application of extrusion technology to red meats is continuing.

MECHANICALLY SEPARATED MEAT, POULTRY, AND FISH[59]

Defined briefly, mechanically separated meat (MSM) (beef, veal, lamb, and pork), mechanically separated poultry (MSP) (chicken and turkey), and minced fish result when meat is removed from the bones by grinding the meaty bones into a thoroughly comminuted product and forcing this against a sieve, which catches bone particles but allows the meat to pass through. Machines powerful enough to accomplish these separations are large and costly. Usually, bones of the vertebrae, especially neck bones, are used for MSM because it is a difficult and slow job to hand-bone them, and even then, a good deal of nutritious muscle is left on the bones. Chicken and turkey backs and necks are good material for MSP. Meaty bones, except fish bones, are ground through a 1.3- to 3.0-cm plate before being put into the pressure deboner. The small bones in fish do not need to be preground.

---

[58]Alvarez, V. B., D. M. Smith, R. G. Morgan, and A. M. Booren. 1990. Restructuring of mechanically deboned chicken and nonmeat binders in a twin-screw extruder. J. Food Sci. 55:942.

[59]Dr. R. A. Field, University of Wyoming, is primarily responsible for the research in this area. See his chapter, "Mechanically Separated Meat, Poultry and Fish," in Pearson, A. M., and T. R. Dutson (Ed.). Edible Meat By-Products, Advances in Meat Research. 1988. Vol. 5. Ch. 4. Elsevier Science Publishing Company, New York.

In producing MSM, MSP, and minced fish, heat is generated by the friction of the grinding and forcing action as bone is forced against the sieve. Heat promotes microbial growth so the bones must be kept clean and cold before they are processed. Even when these precautions are taken, MSM, MSP, and minced fish deteriorate rapidly. A short holding time and immediate, rapid mixing retard oxidation and microbial degradation.

MSM, MSP, and minced fish contribute significantly to the world's meat supply. These products are nutritious, economical, and palatable. They are especially useful in processed meats (see Chapter 21) because of their emulsion and water-holding capabilities, which often exceed hand-boned meats. MSM, MSP, and minced fish contain a higher level of calcium than hand-boned meat because of the extremely tiny bone chips remaining in the meat. This calcium is readily absorbed by humans, and the tiny bone chips do not reduce the palatability of the final product.

Governmental regulations concerning the use of these three products are not consistent. The U.S. Department of Agriculture regulates the use of MSM and MSP, while the use of minced fish is regulated by the Food and Drug Administration with input from the National Marine Fisheries Service of the Department of Commerce. At the present time, federal law does not specifically limit or regulate the amount of minced fish used in the various fish products. Since the seafood industry now has a HACCP plan[60] (also see Chapters 3 and 10), eventually regulations more closely monitoring the use of miched fish will be forthcoming. The use of minced fish in the production of surimi (see "Surimi") has greatly expanded in recent years.

These USDA regulations[61] are in effect for the use of MSM:

- At least 98 percent of the bone particles present can be no larger than 0.5 mm (0.02 inch) in the greatest dimension, and no particle can be larger than 0.85 mm (0.03 inch).

- The calcium content cannot exceed 0.75 percent, equal to not more than 3 percent bone.

- The protein content cannot be less than 14 percent, and fat no more than 30 percent. If MSM is to be used in processed products which have final protein and fat requirements, the MSM is not required to meet these requirements but must be labeled *"Mechanically Separated*

---

[60]Title 21 of the Code of Federal Regulations Part 123 — Fish & Fishery Products, Appendix 7: Seafood HACCP Regulations.

[61]Code of Federal Regulations. January 1999. 9CFR. 319.5.

(Species) for Processing." The fat and protein content of MSM will be evaluated as these components contribute to the fat and protein content of the final product.

- The minimum protein biological value must be equal to intact muscle or 33 percent of the total amino acids (see Chapter 23).

- MSM can constitute up to 20 percent of the meat portion of the product, but it *cannot be used in baby food, hamburger, ground beef, fabricated steaks, barbecued meats, roast or corned beef, cured pork products, beef and gravy, or meat pies.*

- MSM can be used in beef patties, in pressed and chopped ham, and in many sausage items, stews, and spreads, etc.

- When MSM is used as an ingredient, it must be identified on the label as *"Mechanically Separated (Species)."*

The USDA regulations for mechanically separated poultry (MSP) (chicken or turkey) are only slightly different from the regulations for mechanically separated meat (MSM) noted above. MSP is defined as "a product resulting from the mechanical separation and removal of most of the bone from attached muscle and other tissue of poultry carcasses and parts of carcasses that has a paste-like form and consistency, that may or may not contain skin with attached fat."[62] At least 98 percent of the bone particles present can be no larger than 1.5 mm (0.06 inch) in the greatest dimension, and no particle can be larger than 2 mm (0.08 inch). The calcium content cannot exceed 0.235 percent for mature poultry and 0.175 percent for other poultry. Other regulations and uses are as listed for MSM above. MSP can be used in red meat franks and bologna, but it is limited to 20 percent (see Chapter 21).

Although regulations may vary around the world, the use of MSM and MSP is encouraged by many countries. There is also a market in pet food.

ADVANCED MEAT RECOVERY SYSTEMS[63]

Advanced meat recovery (AMR) machines do not grind, crush, or pulverize bones to separate muscle tissue, and the bones and the interconnecting soft tissues that link bones emerge from the process in a manner consistent

---

[62]Code of Federal Regulations. January 1999. 9CFR 381.173.

[63]Federal Register, April 13, 1998, Volume 63, Number 70.

with hand-deboning operations that use knives. The advanced meat recovery systems produce distinct whole pieces of skeletal muscle tissue with a well-defined particulate size similar in consistency to any species trimmings derived by hand deboning and used to formulate processed meat products. The color is similar to that of regular trimmings. The meat derived through use of these machines has the functional and chemical characteristics of meat; there are no powdered bone or constituents of bone, e.g., bone marrow, that are not in conformance with the definition and expectation of meat or that would render the product adulterated or misbranded. The meat/bone separators and recovery systems are fundamentally different from the machines used to manufacture mechanically separated meat and mechanically separated poultry.

However, possibilities remain that incorporation of spinal cord and bone marrow in products resulting from advanced meat/bone separation machinery might occur. The meat industry recognizes concerns about the potential health risk from consumption of spinal cord due to the possible link between bovine spongiform encephalopathy (BSE) and the human variant Creutzfeldt-Jacob disease (vCJD). Yet, government scientists and public health experts agree there is no evidence of BSE or vCJD in humans in this country. For an update on these diseases, consult this Web site: http://w3.aces.uiuc.edu/AnSci/BSE/

Much more care is being exercised on harvest floors to remove spinal cord.[64] This is especially true when the related vertebrae are going into advanced meat recovery (AMR) systems, because there is an organoleptic (sight, smell, etc.) inspection. One interesting development is the flushing of the spinal cavity with a steam nozzle on the harvest floor to loosen the spinal column and make it easier to remove. This is followed by another flush of the column. This technique is being used in this country and in Britain. Regulations regarding the use of AMR meat are pending. Readers may contact Dr. Daniel L. Engeljohn, FSIS, Washington, D.C., (202) 720-5627, for updated information.

SURIMI

Surimi is a term that originated in Japan many years ago for minced fish (see previous section) that has been washed with large amounts of fresh

---

[64]Rosemary Mucklow, Executive Director, National Meat Association, <rosemary@nmaonline.org>.

water, perhaps five volumes, to clean it up. The resulting product is a wet, concentrated source of fish myofibrillar protein. Surimi is very functional in processed meat products and may compete with soy proteins, whey, non-fat dry milk, etc., for this purpose (see Chapter 21). The use of surimi is becoming more prevalent in the United States, and the USDA–FSIS has approved a number of labels for its use in meat products. Publications by Dr. Tyre C. Lanier, researcher at North Carolina State University[65, 66] detail the procedures for producing surimi and discuss how it is being used in the meat industry.

Researchers at the University of Illinois[67] are producing surimi-like material from lean beef and pork and beef by-products (hearts, weasand meat, head meat, and tongues). They found that surimi-like products from beef and pork had functional properties similar to or better than fish surimi. Of the by-products tested, beef hearts produced the most acceptable surimi-like material.

SOUS VIDE[68, 69, 70]

This food processing method, developed in France, is becoming popular in the United States. *Sous vide* translated literally from French means "under vacuum." The food (including meat) is vacuum-packaged in water- and air-impermeable "plastic/vinyl" bags and is cooked slowly, perhaps for four hours, usually in circulating water. This procedure allows the meat to cook in its own juices, which improves the flavor and breaks down connective tissues, making the meat more tender. The internal temperature of the meat must reach 160°F (71°C) according to USDA–FSIS regulations. Immediately after being cooked, the product must be cooled to a temperature just above freezing and then stored and distributed at that same temperature. (Note the discussion earlier in this chapter on the safety of precooked beef patties and apply that information to this *sous vide* procedure.) If the cooking/chilling

---

[65]Lanier, T. C. 1986. Functional properties of surimi. Food Technol. 40 (No. 3):107.

[66]Holmes, K. 1991. Surimi and meat — a future together? Meat and Poultry. 37 (No. 3):22.

[67]McKeith, F. K., P. J. Bechtel, J. Novakofski, S. Park, and J. S. Arnold. 1988. Characteristics of surimi-like material from beef, pork and beef by-products. Proc. 34th International Congress Meat Sci. and Technol., p. 325.

[68]Baird, B. 1990. *Sous vide:* What's all the excitement about? Food Technol. 44 (No. 11):92.

[69]Adams, C. E. 1991. Applying HACCP to *sous vide* products. Food Technol. 45 (No. 4):148.

[70]Beauchemin, M. 1990. *Sous-Vide* Technology. Proc. 43rd Reciprocal Meat Conf. 43:103.

procedures are strictly followed, any microorganisms that could potentially cause food poisoning problems will be controlled (see Chapter 2). The vacuum package provides anaerobic conditions that would promote the growth and deadly toxin production by *Clostridium botulinum* if present and if other conditions permitted growth. Also, because the meat is uncured, there is no sodium nitrite present to protect against *Clostridium botulinum*. Furthermore, the water activity and pH are both high under these uncured conditions, so there are no secondary protective barriers/hurdles against microbial growth other than the temperature control.

Since *sous vide* products are fully cooked, they can quickly be warmed up in a microwave oven for serving. It is for this reason that this food- (meat-) handling procedure is becoming popular in some restaurants. The variety and ease of preparation allows the owner to decrease the number of highly qualified help and yet provide a wide variety of high-quality menu items. *Sous vide* is also popular with the airlines and in more and more homes.

SODIUM TRIPOLYPHOSPHATE PUMPING[71]

The pork and poultry industries have implemented the process of pumping fresh pork loins and chicken breasts with a sodium tripolyphosphate (STP) solution. STP is an alkaline phosphate that is commonly used because its solubility and properties present a desirable combination for curing. A 1 percent solution of STP has a pH of 9.9 (Price and Schweigert).[72] Phosphate level is limited by regulations to 5 percent in the brine (10 percent pump) or 0.5 percent in the product.[73] Phosphate pumping of fresh pork and poultry has yield and sensory advantages.

Addition of alkaline phosphate to meat products results in an increase in the water-holding capacity (WHC) by increasing pH and protein solubility.[74] Sutton et al.[75] observed that pump yield was increased 2.76 and 6.41 percent for fresh pork loins pumped to a final STP concentration of 0.2 and 0.4

[71]Bidner, Brian Scott. 1999. The Effects of *rn* Genotype, Feed Withdrawal Prior to Slaughter, Lysine-Deficient Diet, and Sodium Tripolyphosphate Pumping on Pork Quality and Sensory Characteristics. MS Thesis. Department of Animal Sciences, University of Illinois, Urbana.

[72]Price J. F., and B. S. Schweigert. 1987. The Science of Meat and Meat Products, 3rd ed. Food and Nutrition Press, Inc. Westport, CN.

[73]Ibid.

[74]Ibid.

[75]Sutton, D. S., M. S. Brewer, and F. K. McKeith. 1997. Effects of sodium lactate and sodium phosphate on the physical and sensory characteristics of pumped pork loins. J. Muscle Foods. 8:111.

percent. STP (0.2 and 0.4 percent) reduced purge loss (1.27 and 2.63 percent) but did not impact drip loss of a 1.3-cm fresh chop or moisture loss from a cooked chop.[76] The pH of fresh pork pumped with STP (0.2 and 0.4 percent) was increased (0.11 and 0.20 points) compared to that of water-pumped control pork loins.

Palatability attributes of cooked or reheated pork can be substantially improved by pumping with STP. As a result of increased moisture retention, juiciness is improved when pork is served directly following cooking or after reheating (Smith et al.).[77] Similar results were found when beef roasts were pumped with STP. Brewer et al.[78] also observed juiciness advantages when PSE, normal, and DFD loins were pumped with STP. Sutton et al. (see footnote 75) found that juiciness was improved when loins were pumped to a final concentration of 0.4 percent STP. No differences were observed with loins pumped to 0.2 percent STP.

Tenderness is also improved as a result of disruption of muscle proteins. Smith et al. (see footnote 77) reported that tenderness was improved for pork loins that were served directly or reheated. Sutton et al. (see footnote 75) and Brewer et al. (see footnote 78) also reported that STP improved tenderness of pork.

Phosphates can have both positive and negative effects on flavor. Smith et al. (see footnote 77) found that STP reduced warmed–over flavor in reheated pork loins. STP pumping did not impact flavor and off-flavor intensity. Smith et al. (see footnote 77), however, explained that panelists observed some soapy or metallic flavors associated with STP-pumped roasts. Sutton et al. (see footnote 75) found STP did not consistently impact pork flavor, salt intensity, or alkalinity. Brewer et al. (see footnote 78) also reported that STP did not impact off-flavor intensity.

Regulations governing the amount of weight gain permitted by pumping STP solutions into pork and poultry are somewhat complicated and involved. Fortunately, products must be labeled in order for consumers to know exactly what they are getting. Readers may contact authors of references 71, 75, 77, and 78 for details.

---

[76]Ibid.

[77]Smith, L. A., S. L. Simmons, F. K. McKeith, P. J. Betchel, and P. L. Brady. 1984. Effect of sodium tripolyphosphate on physical and sensory properties of beef and pork roasts. J. Food Sci. 49:1636.

[78]Brewer, M. S., M. Gusse, and F. K. McKeith. 1999. Tripolyphosphate effects on pork characteristics from PSE, normal and DFD carcasses. J. Muscle Foods. In press.

19

Preservation and Storage of Meat

Preservation of red meat, fish, and poultry products is accomplished by creating an unfavorable environment for the growth of spoilage organisms (bacteria, yeasts, molds, and parasites), controlling the action of certain enzymes within the tissues, and preventing the chemical oxidation of lipids, which leads to rancidity. The various methods of preserving meat that have evolved include drying, smoking, salting, freezing, canning, freeze-drying, atmospheric modification and irradiation, and combinations thereof. Throughout recorded history, curing or salting has played the most significant role in keeping meat safe for later consumption; curing is covered in Chapter 20, and the use of salt is covered in Chapter 21.

Most of the processed meats available today originated from techniques that were developed to extend the length of time between the harvest of the animal and the consumption of the meat derived from that animal. Early peoples (and primitive societies today) recognized that salt, drying, cold temperatures, and cooking would extend the postmortem period of meat palatability and safety. Although modern processed meats are produced to enhance or to alter palatability, to increase convenience, or to expand the use and demand for less valuable meat items, shelf life is almost always increased by the processing procedures.

WHY PRESERVATION?

Meat Spoilage and Safety

Meat is an unstable product largely because microorganisms thrive on its rich supply of nutrients, just as humans thrive on them. (Chapter 2 lists and discusses many of the microorganisms that may impact meat palatability and safety.) Meat spoilage results when undesirable odors and flavors are produced by changes in meat. Meat preservation techniques were devised by early peoples to prevent or, more accurately, to slow meat spoilage. When not adequately preserved, meat often spoils before potentially harmful food-poisoning organisms or their products are present in toxic amounts. Most humans view spoiled meat as inedible; therefore, potentially hazardous meats are often not consumed. Spoilage serves as a protective or warning system. However, it is possible for meat spoilage to be controlled in food that continues to support the growth and toxin production of food poisoning organisms. In such cases, inadequate preservation can contribute to serious human health problems.

Enzymes, compounds that catalyze chemical reactions within the meat, are able to produce chemical and physical changes that alter the physical and palatability characteristics of the meat. These alterations may produce meat spoilage. Many meat enzymes function better and change the product most rapidly at or near normal body temperature and neutral pH. Low temperature, the presence of ions such as sodium and chloride (salt), pH changes, and reduced moisture slow enzyme activity. Cooking and other treatments may inactivate enzymes permanently.

Some enzymes, as well as other factors, may enhance the oxidation of fatty acids in meats. Consumers detect the presence of oxidized fatty acids as rancidity or off-flavors. The prevention of fatty acid oxidation in stored or processed meat continues to challenge both the meat processor and the meat scientist.

General Meat Preservation Concepts

A meat preservation process should be practical and usable, should not make the product unpalatable or worsen its appearance, and must not in any way harm those who consume the meat. Since meat is "preserved" primarily from microorganisms, sanitation is the first big step in successful meat preservation. Minimizing the number of microorganisms that contact and contaminate the product will lengthen the shelf life of any food item. Thus, meat inspection standards (see Chapter 3) and industry sanitation practices (see Chapter 2) are designed to extend the safe shelf life of meat by reducing the possibility of contamination.

Meat processors can control the factors that affect microbial growth and thus give meat a longer shelf life, meaning that the meat stays in wholesome condition longer. For example, beef carcasses can hang in a 33° to 35°F cooler for perhaps three weeks to age (not a standard practice), and, depending on the relative humidity of the cooler, perhaps only a few molds and some harmless bacteria such as *Pseudomonas* would grow on limited areas of the carcasses. However, grinding meat, which distributes bacteria that are present over many surfaces, and holding it at 40°F will result in serious spoilage within a few days.

REFRIGERATION AND FREEZING EQUIPMENT

On April 9, 1626, English writer and statesman, Sir Francis Bacon, died of a cold he caught while conducting experiments on his theory that keeping meat frozen would keep it fresh.

Mechanical Refrigeration

Low temperature is an important factor in meat preservation. Ice was the major source of refrigeration until about 1930. The development of "artificial" or mechanically produced refrigeration impacted meat preservation more than any other recent technological advancement.

Mechanical refrigeration systems are based upon the heat gained and lost in a material when it changes state between the gaseous and liquid forms. Electrical or mechanical energy is used to pressurize a gas until it becomes a liquid (done by a motorized compressor). The heat in the liquefied gas, generated during the change of state, is removed by transferring it to water or air moving around the pipes containing the gas (in a condenser). That heat is transferred into the environment or drained away with the waste water. The cooled, liquefied gas is allowed to expand through a valve into a low-pressure coil, where it returns to the gaseous state. The change of state from liquid to gas requires that heat be absorbed by the expanding material. The expansion to a gas occurs in a device called an evaporator, which is located in the chamber to be refrigerated. Fans move the air in the chamber past the gas-containing coils of the evaporator. The expanding gas absorbs the required heat from the moving air, which is redistributed throughout the area, thus lowering the temperature of the chamber. Brine passed through an evaporator coil will transfer its heat to the expanding refrigerant gas and thus be chilled. The gas leaving the evaporator and the refrigerated area is pulled away by suction to the compressor, and the process is repeated.

Large refrigeration systems use the more efficient ammonia (NH_3) as the gaseous medium in the system. Most systems, particularly home refrigeration units, refrigerators, freezers, and air conditioners, use freon gas refrigerants because the freon systems have lower maintenance requirements, and much less physical distress results from freon gas leaks than from ammonia leaks.

Meat Storage Units

Home Storage Units

The refrigerated storage of meats, both chilled and frozen, should be in well-designed units that maintain constant temperatures over long periods of time. Meat refrigerator (cooler) temperatures should be maintained at 27° to 32°F, and freezer temperatures should be 0°F or less to maximize the storage life of the product. The storage units should be designed to achieve uniform temperatures throughout the volume of the unit and insulated to maintain the optimum operating temperature by reducing the heat uptake from the environment. Units designed for frozen food storage in the home may be capable of achieving temperatures lower than 0°F, but they may operate inefficiently at lower temperatures.

Although home freezers are designed to store all types of frozen foods, including meat, most home units are not well-designed to freeze large volumes of product rapidly. The most common home frozen food storage unit is part of a refrigerator–freezer combination unit. One section is designed to store foods at temperatures above freezing and a second section is maintained at sub-freezing temperatures. Wide ranges of size options and freezer locations are offered in the combination refrigerator–freezers with correspondingly wide price ranges. Combination units with larger freezer compartments will have freezer sections designed much like the upright or side-door home freezers.

The upright freezer looks very similar to a refrigerator, but it has a single chamber that operates only at sub-freezing temperatures. The advantages of upright freezers and combination units are high visibility of products in the freezer and relatively easy product accessibility for a standing person (the unit has shelves that allow for some segregation and organization of the contents). Most upright freezers have freezing coils installed in some of the shelves, which do provide direct contact and more rapid freezing rates for unfrozen product that is placed on those shelves. Upright or side-door freezers are more likely to spill cold when the doors are opened than horizontal freezers and, unless properly adjusted and sealed, may leak more cold around the doors when they are closed.

Horizontal or chest freezers are top opening units. Dividers are an option that permits some organization of the storage space. In order for the product to be placed into and removed from a horizontal freezer, a person must stoop over and reach down into the unit. However, since cold air tends to drop, the top door is much less subject to cold leakage, and there is little spillage of cold when the door is opened.

Walk-in Storage Units

Meat retail and food service operations require large volumes of refrigerated and frozen storage. Some homeowners may also be able to utilize the additional volume provided by walk-in coolers and freezers.

Manufactured units of this type usually are prefabricated at the plant and assembled on the owner's premises. Many businesses buy the refrigerating units and construct the insulated storage areas. This permits the owner a wider choice in capacity, and generally it is cheaper. Coolers and especially freezers require special construction details and should be designed and built by persons having knowledge and experience with refrigerated storage units. The most popular insulation consists of 4 or 6 inches of styrofoam, sandwiched between laminated steel or plastic panels. The cooler should be capable of operating efficiently at 35°F. A 0° to –10°F walk-in freezer may be built within the cooler to economize on insulation materials, or it may be constructed as a separate unit.

Single compressor and condenser systems may serve evaporator units in more than one refrigerated area. It is important to have the refrigeration system sized to correspond to the demand. Overdesigning results in unnecessary capital investment and less efficient operation. Having a separate refrigeration unit for each cooler and(or) freezer may be advisable. This will save closing down the plant in case one unit goes bad. Freon–12 gas has been the most satisfactory refrigerant to use, and forced-air cooling units that are self-defrosting are in equal favor with gravity units.

Cryogenic Freezing

Cryogenic freezing utilizes (very) low temperatures to freeze product rapidly. Various systems of producing cold are utilized. One of these freezing systems, which most closely resembles a conventional refrigerator or freezer, employs a brine ($CaCl_2$ solution) at –40°F. The brine circulates from a tank through plate fin coils. These coils are above a stainless steel belt that carries the meat product through a chamber. The moist product (normally steaks, chops, or patties) freezes instantly to the belt as it makes contact and later pops off at the end where the belt turns under to begin to return. Freezing capacities of up to 3,600 pounds of 0°F product per hour are available.

Liquid nitrogen is more widely used as the refrigerant in similar conveyor–tunnel–chamber arrangements. Liquid nitrogen itself has a temperature of –320°F, but very seldom is meat immersed directly into the liquid; rather, the liquid is placed under pressure of 15 to 22 psi (pounds per square inch) and sprayed through nozzles over the product. The closer the product is to the nozzle, the nearer to –320°F is the temperature. Normally, freezing temperatures in liquid nitrogen systems range from –100° to –200°F.

Liquid carbon dioxide (CO_2) is also used as a refrigerant in cryogenic systems, with the same principles that were described for liquid nitrogen applied. Liquid CO_2 has a temperature of –180°F and flashes to CO_2 snow at –109°F, with the average temperature inside the machine being –80°F.[1]

Cryogenic spirulators have gained widespread acceptance in recent years as a rapid-freezing technique in the beef patty industry. Spirulators are particularly beneficial when large freezing volumes are needed but when only limited floor space is available. Inside the cube-shaped chambers are conveyors designed like large corkscrews. Spirulators can be purchased as self-contained units as small as 8- by 8-feet square, or as larger models with freezing capacities of over 12,000 pounds per hour. A diagrammatic illustration of an Airco® spirulator (Murray Hill, New Jersey) is shown in Figure 19–1.

THE FREEZING OF MEAT

In developed countries, freezing is the most common method of preserving fresh (uncured) meats for extended periods of time. Even though the bulk of beef, pork, and lamb is purchased unfrozen in retail meat counters, it is often subsequently frozen once consumers take it home. Consumers are still suspicious of frozen meat at the retail level. Unfortunately, they prefer to buy fresh retail cuts and to freeze them at home, often in inadequate freezers and

---

[1]Cryogenic catalyst. 1988. Meat Proc. 10:66 and 67.

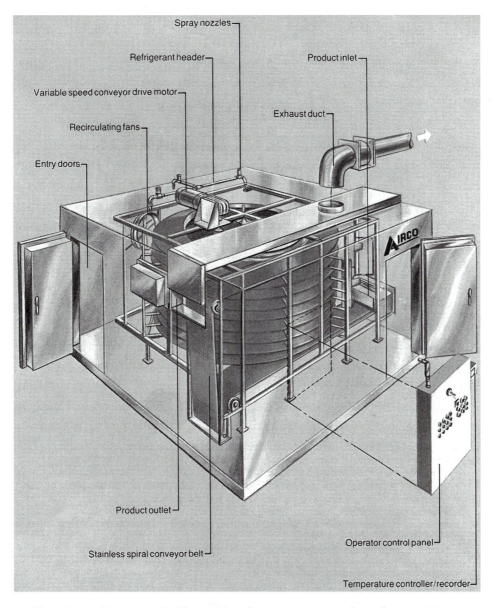

Fig. 19-1. Diagrammatic illustration of a cryogenic spirulator freezer. (Courtesy, Airco Kryofoods, Murray Hill, New Jersey)

in inadequate packaging. A rise in frozen retail meat sales has been predicted since the 1950s, but it has been slow to come. However, as consumer habits and buying patterns change, there will undoubtedly be a gradual trend toward the purchase of frozen meat. The advantages offered by frozen fresh meats to both the industry and the consumer should stimulate continued research and development of this marketing concept.

Frozen meat has achieved acceptance in the food service industry. Many portion-controlled (see Chapter 13) meat entrées go directly from the freezer to the chef's grill. Since the cooked appearance of frozen cuts does not differ from fresh cuts, the consumer is not able to differentiate between the two. The shelf-life extension and the purchasing and inventory flexibility offered by frozen meat items are valuable assets in the food service industry.

In an extensive study conducted by Kansas State University,[2] the distribution costs, acceptance, cooking, and eating qualities of frozen meat were investigated. The processing steps that met with the most success included fabricating retail cuts at least 48 hours postmortem in a room no warmer than 55°F; allowing a 30-minute period for bloom to develop before freezing or packaging; freezing initially at –70°F; and storing in the dark at –15°F or lower. Retail displaying at case temperatures of –20°F or lower; packaging in a film with moderate- to high-oxygen permeability to maintain color; and using a 5- to 6-second water dip at 90° to 95°F to remove color bleaching completed the most successful merchandising sequence.

The results of comparative cooking tests of frozen versus unfrozen steaks and chops showed that freezing did not impair the palatability of beef steaks, but that fresh, unfrozen pork chops were preferred to frozen chops, even though the frozen chops were acceptable. Frozen chops and beef steaks generally had higher cooking losses than their fresh counterparts. In the retail market, most of the customers who purchased the frozen meat liked its palatability; 23 percent were dissatisfied with the tenderness, juiciness, or texture. More than 85 percent of the nonpurchasers indicated that the frozen meat was unappealing and unappetizing or appeared artificial.

Physical Effects of Freezing

Freezing acts as a preservation method by almost completely inactivating the meat enzymes and inhibiting the growth of spoilage organisms. The lower the temperature, the greater the inhibitory action and the longer the period of satisfactory storage. Low temperatures do not destroy vitamins. Most of the vitamin loss is caused by heat or light, or vitamins are lost in the juices that escape (see Chapter 24). To secure and maintain very low temperatures requires expensive construction and entails high operating costs. The industry has been utilizing temperatures ranging from 0° to –32°F.

---

[2]Frozen meat. 1973. Res. Publ. 166. Agric. Exper. Sta., Kansas State Univ. (September).

Lean meat averages from 65 to 75 percent water, which expands at both high and low temperatures. The actual point at which meat juices will freeze solid is not 32°F but 28° to 29°F. The rate of crystallization and the size of the crystals formed depend upon the temperature. Slow freezing causes the water to separate from the tissue into pools that form large crystals. These stretch and rupture some of the surrounding tissue. Rapid freezing results in very little water separation; therefore, the crystals are small and less expansive. Portions (patties, steaks, chops) frozen in the cryogenic systems discussed earlier are examples of rapidly frozen products and may be referred to as IQF (individually quick frozen) products. Because there is practically no pool crystallization in very low temperature freezing, the drip is considerably less than from meats frozen at higher temperatures. Drip loss from thawing meat includes protein, vitamins, and other nutrients, in addition to moisture, and results in decreased cooked yields and juiciness.

Meat temperatures must be brought down to 40°F within a 16-hour period to prevent the growth of spoilage-producing bacteria deep in carcass tissues or in the center of containers of warm meat. If hot meat goes directly to the freezer, it must reach 0°F within 72 hours to prevent the growth of putrefactive bacteria. For large packs of hot meat, a freezer temperature of –5°F with air velocities of 500 to 1,000 feet per minute (–21°F wind chill effect) is recommended. For some freezers, the amount of meat frozen at one time should not exceed 2 pounds per cubic foot of freezer space. A cubic foot of meat weighs 35 to 45 pounds, so 2 pounds would occupy less than 10 percent of the available space. More than this amount raises the freezer temperature and slows down the freezing process. The key rule is "fast to 0°F or below."

Clostridium perfringens does not grow in cured meat or poultry products during an extended cooling period because nitrate has a demonstrated inhibitory effect on the pathogen. The National Food Processors Association (NFPA) requested the following language be adopted by FSIS: "Cured meat and poultry products may be cooled such that the products' maximum internal temperature is reduced from 130°F to 80°F in 5 hours and from 80°F to 45°F in 10 hours (15 hours total cooling time)." Dan Engeljohn, Director of the USDA Regulations Development and Analysis Division, responded that the preceding cooling procedure can now be referenced as a "safe harbor" and will be included in the compliance guidelines for 9 CFR 318.17 and 381.150.

Mechanically refrigerated chambers specifically designed to freeze food products rapidly are called "quick," "sharp," "fast," or "blast" freezers. A combination of 28 miles per hour air velocity with –20°F, producing a minimum windchill effect of –77°F, is necessary in blast freezers. (See Appendix for calculating windchill effect.) Small packages, lean meat, low temperatures, and high air velocity all enhance the freezing rate of meat.

High-quality frozen meat products result when:

- The animal is physically sound (in good health).

- The animal is properly harvested in sanitary facilities.

- The animal is properly chilled (chill room, 29° to 36°F).

- The aging period is restricted.

- Cutting and processing are done quickly in sanitary facilities.

- The wrapping material and technique are of good quality.

- The holding temperature is 0°F or lower.

Cutting Method for
Frozen Storage Meats

Tests conducted at Kansas and Michigan experiment stations show that boning meat has no effect on the flavor or juiciness of the cooked meat and that packaging boneless meat is easier, causes less damage to wrappers, and saves up to 35 percent of the frozen storage space. The expense of boning adds to the labor charge over the bone-in method, which is absorbed in part because less space is required, a smaller amount of paper is needed, and cooking and carving can be done with greater ease and satisfaction.

Methods for breaking down a beef, pork, lamb, or veal carcass for subsequent boning are explained in Chapters 14, 15, 16, and 17. Regardless of the method used, the cuts should be made ready for cooking and in such sizes as will best meet the needs of the family. Some cuts other than the standard ones, such as the top round muscle sliced into chipped steak (about ¼ inch on the slicing machine), make for variety and aid in menu planning. Giving the top round a slight freeze (not solid) will make it slice evenly.

Oxidative Rancidity

Since the growth of spoilage microorganisms is inhibited at freezer temperatures, oxidative rancidity is the principal factor limiting the storage life of frozen meats (see Table 19–1 for the maximum recommended storage times for various frozen meats).

The development of rancidity in animal fats depends upon their ability to absorb oxygen from the air. This weakness for oxygen varies with the basic chemical structure of the fat involved. Any fatty acid that has one or more double bonds in the carbon chain will be vulnerable to a cleavage caused by the oxygen taking the place of the double bond and forming aldehydes (the reactive group CHO–) and shorter fatty acids (see Chapter 23). These products, so formed, generally are no longer pleasing in taste or odor. As a result, they affect the palatability of the fat and the adjoining lean.

Since pork fat is fairly high in unsaturated fatty acids (e.g., oleic acid — one double bond; linoleic acid — two double bonds; linolenic acid — three double bonds), which have the ability to absorb oxygen, it follows that its storage life is lessened considerably. Beef and lamb, on the other hand, have a higher proportion of saturated fatty acids with no double bonds in the carbon chain and therefore are less susceptible to oxygen absorption and oxidative rancidity, with a subsequently longer storage life.

The obvious ways to combat oxidative rancidity are to eliminate the air and to use antioxidants. The elimination of air can be done in several ways, the most practical of which is to use a wrapping material that is airtight, moistureproof, and properly applied. The loss of moisture from meat or any other food is usually termed *shrink* or *dehydration*. The loss of moisture from the frozen surface of meat has been dubbed *freezer burn*. A good wrapping material will reduce both oxidative rancidity and freezer burn.

Australian workers have shown that freezer burn occurs less when meat is frozen at –4°F as compared to 14°F. This is of practical use for the storage of large cuts and carcasses, which are too unwieldy to wrap properly.

Factors That Stimulate Oxygen Absorption

Increased temperatures accelerate oxygen absorption, as mentioned previously. The ultraviolet light used in the sterile lamps that are part of the equipment of some coolers accelerates oxidation. The minerals copper, iron, manganese, cobalt, and lead also enhance oxidation. Salt (NaCl) increases the susceptibility of fats to oxidation.

Aging

Aged meat has shown higher peroxide values and a shorter storage life than 48-hour chilled meat. Although aged meat was slightly more tender during the first month of frozen storage, in subsequent months, the fresh and aged meats were equal in tenderness. Thus, aging meat for the development of flavor, aside from its tenderizing effect, becomes a questionable practice for meat that is to be held in zero storage for more than six months. Experiments show that the length of the holding (aging or ripening) period has a direct bearing on the storage life because it permits oxygen absorption by the exposed fat.

Trimming Fat

What about the fat on meat? Again, the results indicate that it is advisable to trim closely before freezing. The fat will probably not be eaten, even if it is palatable; it will taint the lean if it oxidizes, and it will take up freezer storage space. In the case of pork, the nature of the fat makes it more vulnerable to oxidation, shortening its storage life below that of beef, veal, and lamb. It is very important, therefore, to trim closely or to freeze only those cuts that are quite lean. For example, tests on sausages of different degrees of fatness showed that the lean sausages had a longer storage life than those containing more fat. It has also been demonstrated that pork that was frozen after 48 hours of chill had a longer storage life than pork that was chilled for 7 to 14 days before it was frozen. The same was true of the sausage made from such pork.

Freezer Storage of Seasoned Meats

Salt affects the rate of fat oxidation, causing cured meats or meat products seasoned with salt to acquire a flat, rancid taste in a shorter time than the unseasoned product. Whole cured hams, picnics, or butts, properly wrapped in a good grade of locker paper, will maintain the original flavor for two months or possibly more; however, half-hams or sliced bacon will lose flavor within the month.

The freezer storage life of fresh sausage can be lengthened by omitting the seasoning and then adding the seasoning after the sausage has been thawed. An anti-oxidant such as BHA, BHT, or propyl gallate (see Chapter 21) may be added to the sausage at the second grinding. Pork trimmings can be frozen for future sausage making, but the holding period should not exceed one month. Smoked sausage has a longer storage life than fresh sausage.

Packaging Meat for the Freezer

Wrapping materials used for frozen meats must be excellent barriers to prevent moisture loss or dehydration of the product and to prevent oxygen and other volatile flavor modifiers from entering the package (see "Packaging Materials") during storage. The wrapper should be pressed tightly to the meat to exclude all the air possible, making the package practically airtight. Effective packaging is easier if the cuts of meat are compact and square or rectangular in shape. There should be no sharp edges of bone protruding to puncture the wrapping material. A double layer of waxed paper or other interleaf material should be placed between cuts, to make it possible to separate the frozen cuts, if several are wrapped in the same package.

The most desirable paper wrapping method from the standpoint of maximum air seal is the apothecary or drugstore wrap, although work at Kansas showed as good results with the use of the butcher's wrap, a quicker and more rugged method, which gives the package a double thickness of paper. With a little practice, most people become quite proficient and speedy with the drugstore wrap. It does not pay to economize on paper, either in quality or in quantity. An adhesive tape made especially for low temperatures is used for securing the package. The tape may also have the best surface upon which to write the identification of the wrapped product.

Vacuum packaging in an abrasion-resistant, moisture-impermeable film provides the best packaging system for shelf-life longevity; however, it requires the use of specialized equipment that is generally not available to most consumers.

Temperature, Length of Storage, and Thawing

The lower the temperature, the longer the period of successful storage. Recommendations are definitely for zero or lower, the limiting factor being the cost of the equipment and the cost of maintaining the lower temperatures.

The length of the storage period should not be over 12 months, for economic reasons if for no other. With a proper wrap in good-quality paper and a zero temperature, lean meat will keep well for six to eight months, with some exceptions. These exceptions have to do with products that contain salt, such as seasoned sausage, liver pudding, scrapple, sliced ham, and bacon slices. Table 19–1 summarizes the recommended storage times for red meat.

Table 19–1. Storage Time Chart[1] (Maximum Refrigerator–Freezer Recommendations for Fresh, Cooked, and Processed Meats[2])

Meat	Refrigerator (38° to 40°F)	Freezer (at 0°F or lower)
Beef (fresh)	3 to 4 days	6 to 12 months
Veal (fresh)	1 to 2 days	6 to 9 months
Pork (fresh)	2 to 3 days	6 months
Lamb (fresh)	3 to 5 days	6 to 9 months
Poultry (fresh)	2 to 3 days	3 to 6 months
Ground beef, veal, or lamb	1 to 2 days	3 to 4 months
Ground pork	1 to 2 days	1 to 3 months
Variety meats	1 to 2 days	3 to 4 months
Luncheon meats	3 to 5 days	1 to 2 months
Sausage, fresh pork	2 to 3 days	1 to 2 months
Sausage, smoked	7 days	1 to 2 months
Frankfurters	3 to 5 days	1 to 2 months
Bacon	7 days	1 month
Smoked ham, whole	7 days	1 to 2 months
Smoked ham, slices	3 to 4 days	1 month
Beef, corned	7 days	2 weeks
Leftover cooked meat	3 to 4 days	2 to 3 months
Frozen combination foods		
Meat pies (cooked)	—	2 to 3 months
Swiss steak (cooked)	—	3 months
Stews (cooked)	—	3 to 4 months
Prepared meat dinners	—	2 to 3 months

[1]Lessons on Meat. 1991. National Live Stock and Meat Board, Chicago.

[2]The range in time reflects recommendations for maximum storage time from several authorities. For top quality, fresh meats should be used in two or three days; ground meat and variety meats should be used in 24 hours.

W. L. Sulzbacher, former bacteriologist for the U.S. Department of Agriculture, reported: "There is no indication that frozen meat becomes more perishable after thawing than fresh meat."

There is no reason to hesitate to refreeze meat when occasion demands, but it should be done within the day. If it is to be used the following day, it should be placed in the rear of the refrigerator rather than refrozen for that short period. Thaw in refrigerator or microwave, as noted in "Microwave Cookery" and "Handling Frozen Meat in the Home and the Institution," in Chapter 24.

Preparing and Freezing Poultry

Broilers and fryers are usually cut into halves or quarters; however, birds can be left whole for roasting, cut into stewing joints, or boned. Cut-up birds are popular because second joints, drumsticks, and white meat can be packed separately from the less desirable wings, backs, and necks. An added advantage of cut-up birds is compactness, eliminating the large body cavity that traps considerable air, thus requiring less storage space. Roasters frozen whole should have the excess internal fat removed because it will oxidize and become rancid much more rapidly than the rest of the bird. The giblets are wrapped in absorbing parchment or foil and placed inside the bird.

In large-scale operations, whole birds — poultry, turkeys, ducks, geese, and other types — are usually vacuum-packed. The birds are placed in Cryovac® or other impervious bags, and the air is exhausted with a vacuum pump. The end of each bag is then made airtight by sealing with heat or placing a metal clamp around the opening. It is then immersed in hot water to give it a skintight shrink. Storage at 0°F will keep properly wrapped fowl edible for four to six months.

FSIS regulations now read that all raw poultry that has not been chilled to an internal temperature below 26°F will be permitted to be labeled "fresh." All other poultry must be labeled "frozen" or "previously frozen." Chicken and turkey tenders, wings, and giblets are exempt from this labeling regulation.

Freezing Fish

Rapid freezing of fish and shellfish is desirable to prevent the formation of large ice crystals that can damage cells and allow loss of moisture ("drip loss") and subsequent loss of texture upon thawing.

Fish, including shellfish, can be divided into two main groups, based on the oil content of the flesh. The non-oily fish (less than 3 percent oil) — cod, haddock, halibut, and swordfish, to mention a few — store their oil in the liver rather than in the flesh. The oily fish (over 3 percent oil) — herring, mackerel, and salmon, for example — have the oil distributed throughout the flesh.

The chief type of spoilage in frozen fish, as in warm-blooded animals, is caused by the oxidation of the fats, resulting in rancidity. The action of bacteria and enzymes is inhibited by low temperatures, but air must be excluded if oxidation is to be held to a minimum. Therefore, eviscerated fish should be wrapped either in the round (unsplit) (see Chapter 10) or as fillets, in plastic film, with as much air as possible excluded, and then frozen at 0°F or below. Fish that are too large to be wrapped should be quick-frozen and dipped in cold water several times to cover them with a glaze of ice. The ice glaze will evaporate within several months unless the humidity of the holding room is very high. Reglazing or wrapping in moisture–vapor-proof material is then necessary. Frozen fish that have been well-wrapped can be stored with other foods without imparting or transferring any odor or flavor to them. Table 19–2 lists the maximum recommended frozen storage periods for some common species of fish.

Table 19–2. Maximum Storage Periods for Frozen Fish

Species	Round or Headed and Gutted	Wrapped, Packaged
Croaker	6 to 8 months	8 to 10 months
Grouper	6 to 8 months	8 to 10 months
Lake herring	6 to 8 months	8 to 10 months
Ling cod	6 to 8 months	8 to 10 months
Mackerel (Spanish and Boston)	6 to 8 months	8 to 10 months
Mullet	6 to 8 months	8 to 10 months
Red snapper	6 to 8 months	8 to 10 months
Rockfish	6 to 8 months	8 to 10 months
Rosefish (ocean perch)	6 to 8 months	8 to 10 months
Sablefish	6 to 8 months	8 to 10 months
Salmon	6 to 8 months	8 to 10 months
Sea trout	6 to 8 months	8 to 10 months
Shrimp	6 to 8 months	8 to 10 months
Cod	8 to 10 months	10 to 12 months

(Continued)

Table 19–2 (Continued)

Species	Round or Headed and Gutted	Wrapped, Packaged
Flounder (sole)	8 to 10 months	10 to 12 months
Haddock	8 to 10 months	10 to 12 months
Halibut	8 to 10 months	10 to 12 months
Pike (all species)	8 to 10 months	10 to 12 months
Pollock	8 to 10 months	10 to 12 months
Porgie (scup)	8 to 10 months	10 to 12 months
Sole	8 to 10 months	10 to 12 months
Whiting	8 to 10 months	10 to 12 months
Smelt	8 to 10 months	8 to 10 months
Whitefish	8 to 10 months	8 to 10 months

CHILLED MEAT PRESERVATION

The Japanese market for U.S.–produced beef and pork has expanded rapidly following the liberalization of import regulations. Prior to the liberalization, most of the meat shipped to Japan from the United States was frozen, but a large Japanese demand for high-quality, chilled, unfrozen product can now be served by U.S. processors. The time required for shipment throughout the world and for movement through the marketing system in the receiving country is significantly longer than the time in which the fresh carcasses and wholesale cuts formerly marketed by U.S. packers would remain palatable. Thus, the use of vacuum packaging allows carcasses to be cut, at the plant of origin, into primal, sub-primal, or even retail cuts, after which each cut is placed in an airtight envelope. Vacuum-wrapped cuts, at the proper temperatures, may be held for extended time periods during marketing without the occurrence of damaging microbial growth. Many of the organisms that influence meat spoilage require the presence of air (oxygen) to grow. The vacuum-packed product is also "preserved" from weight loss (shrinkage) and discoloration that results from evaporation of moisture during storage.

The shelf life of chilled cuts is maximized by aggressive sanitation programs throughout the harvest, chilling, cutting, and packaging operations. Reducing initial microbial contamination of the product significantly increases the time interval before microbial growth causes changes in odor, appearance, and flavor (see Chapter 2). Carcasses in post-harvest chill coolers may be sprayed with organic acids or other anti-microbial sprays to reduce surface counts.

Modern chilling coolers operate at lower relative humidity than in the past to reduce condensation, a source of contamination, from overhead surfaces. The lower relative humidity enhances evaporation of water, which is periodically sprayed on the carcass surface. The evaporation of the water requires the absorption of heat from the carcass surface, speeding up the post-harvest chilling process and minimizing microbial growth. The recent refinement of refrigeration technology that will maintain the product temperature to an accuracy of 1°F within the "superchill zone," between 32°F and the temperature at which ice crystals form in the product (29°F),[3] will further extend the shelf life of chilled fresh meats.

Minimizing microbial loads and chilling rapidly, followed by effective vacuum packaging and the maintenance of storage and shipping temperatures that are consistently close to freezing, may result in a shelf life of 100 days or more for fresh, chilled beef.

"Boneless chilled pork destined for the retail sector of Pacific Rim export markets requires a storage life of at least 10 weeks.[4] Five keys to maximizing storage life of chilled pork, in order of their importance are: (1) Keep in cold. (2) Keep it clean. (3) Remove essentially all of the oxygen from the package. (4) Prevent oxygen from reentering the package. (5) Provide an environment containing at least 25% carbon dioxide within the package." These rules apply to the processing of all meats, whether for domestic or foreign consumers.

More "chilled meat" merchandising concepts are *sous vide* and *modified atmosphere packaging*. (See Chapter 18 for more detailed information concerning *sous vide,* precooked entrée items in vacuum packages that are distributed unfrozen.) Modified atmosphere packaging (MAP) is a technique that involves removing the air from the food package by vacuum and introducing another atmosphere into the package. The introduced gas may be nitrogen, carbon dioxide, or other gases or mixtures. The purpose is to minimize microbial growth, while reducing physical changes such as color, shape, and purge (separation of juice from the meat cut and accumulation in the package) that may occur in vacuum packaging. The package may appear to be puffed or pillow-shaped in some cases. Some *sous vide* products may be packaged in modified atmosphere packaging.

---

[3]Morris, C. E. 1991. Forging new links in the cold chain. Food Eng. 7:61.

[4]National Pork Producers Council. 1997. Extension of Chilled Pork Storage Life. Facts #04282.

THE CANNING OF MEAT

The sterilization of food products by canning is an efficient method of food preservation. The Napoleonic Wars created a need to preserve foods for sea service and military stores. The French government offered a prize for the most practical method of keeping foods in good condition. The result, in 1809, was hermetic sealing, attributed to Frenchman Nicolas Appert, who used bottles and jars.

A year later, Peter Durand, an Englishman, initiated the use of tin-coated iron cans. This led to the first commercial canning operations. With the invention of the can came the term *canning,* which was broadened in later years to include preservation in other forms of containers.

The canning of meat represents the second most common preservation method of meat products for extended periods of time (freezing is the most common). Canned-meat products may be conveniently grouped into two categories: sterilized products and pasteurized products. Sterilized products are shelf-stable (need no refrigeration), while pasteurized products require refrigeration to inhibit spoilage.

Canned foods are preserved by hermetically sealing (preventing the escape or entry of air) the product in a container and destroying, through the application of heat, those microorganisms capable of producing spoilage. Spoilage microorganisms, such as bacteria, yeasts, and molds, are universally present throughout the environment. The organisms of greatest concern in the canning industry are certain anaerobic bacteria that are capable of forming protective endospores (spores within the bacterial cell). Some genera of bacteria form spores that are capable of resisting adverse conditions such as high temperatures, low moisture, and antiseptics. In fact, considerable time at 240°F is required to kill some spores whose vegetative cells can cause food poisoning in humans. When sufficiently adverse conditions are encountered, the vegetative bacterial forms are killed, leaving behind the spore forms. When spores again meet favorable conditions, they may slowly return to the actively growing (vegetative) form and then multiply very rapidly (see Chapter 2).

Sterilization

A food substance becomes absolutely sterile only when no vegetative microorganisms or spores are present, either because there were none originally or because they have all been killed. Absolute sterility seldom exists in commercially canned products. Commercial sterility refers to the destruction of spoilage organisms and their spores such that a product will not undergo spoilage, even under indefinite storage periods. The principal method of assuring commercial sterility is by heat treatment in a sealed container. Sanitation is extremely important in meat-canning plants, since the fewer the number of organisms originally present, the more effective is the sterilization process.

A time–temperature relationship is required for the destruction of most microorganisms. That is to say that simply reaching a specified internal temperature is often not sufficient to assure the destruction of certain spores. The product must be held at this minimum temperature for a specified period of time to achieve the destruction of spores of concern. This concept is known as the F_o *value*, which is defined as the equivalent time in minutes at 250°F required to destroy the vegetative and spore forms of a given organism. See also the discussion in the section covering temperature and time in Chapter 2 and use of the indicator letter "D." The usual practice is to calculate this in terms of the destruction of *Clostridium botulinum* for an equivalent number of minutes at 250°F. While the F_o value is a somewhat arbitrary way of expressing the effectiveness of a process, it does provide a useful tool for the determination of safe, commercial sterilization levels. A "safe cook" is considered to have an F_o value of 2.78 or more. This means a minimum of 2.78 minutes at 250°F. Three minutes is often used as the minimum time at 250°F for a safe cook.

Canning at the sterilizing temperature (250°F) necessitates processing the cans in a *retort cooker* in order to achieve the needed temperature. A retort cooker is simply a giant pressure cooker that operates under pressures of 12 to 15 pounds per square inch (psi). This enables the cooking–water temperature to rise above the normal boiling point of 212°F (at sea level). *Retort processing* is basically cooking under pressure.

The advantages of canning meat products using sterilization temperatures are numerous. Perhaps the greatest advantage is the indefinite storage life over a wide temperature range. Canned-meat products, opened after many years of storage, have still maintained an edible quality (although flavor deterioration had occurred). Canning provides an effective preservation method for high-quality food in times of catastrophes (natural or human-made), which result in the disruption of power and thus mechanical refrigeration. Other advantages include:

- Energy savings (because no refrigeration is required).

- Less critical merchandising speed.

- Minimal product spoilage during merchandising (some "leakers" do occur).

- A stabilizing effect on market prices when raw material supplies become low.

- Convenience, since no cooking is required after the can is opened.

The principal disadvantage of utilizing sterilization temperatures in the canning of meat and meat products is the deterioration in food quality that occurs. Sterilization of solid meat portions exceeding 1 pound is limited. The heat treatment required for sterilization of larger units will liquefy or cause excessive palatability alterations of the product adjacent to the can surface. Products that will "flow" inside the cans, such as stews and hashes, can be sterilized with less damage because heat transfer inside the cans occurs by convection in addition to conduction (see Chapter 24).

Examples of popular canned-meat products that have undergone commercial sterilization are the potted-meat products, Vienna-style wieners, corned beef hash, roast beef, corned beef, beef stews, and pickled pigs' feet.

Pasteurization

Pasteurized meat products undergo a less severe heat treatment (generally 155° to 165°F internal temperature), and the thermophilic spore-forming microorganisms are not destroyed. However, under good sanitation practices and proper refrigerated storage conditions (near 32°F), these canned products will typically have a shelf life of six months. This relatively long shelf life (compared to other cooked products) is possible because no recontamination of the product can occur while it is in a sealed container. Since it was cooked in this container to a temperature of at least 155°F, virtually all vegetative forms of microorganisms have been destroyed. A refrigeration temperature of near 32°F effectively inhibits the outgrowth of bacterial spores that survived the heat treatment.

Because of the concern for *Listeria monocytogenes* (see Chapter 2) many plants employ a post-cook, post-packaging pasteurization process for sausage products, including frankfurters.

Most canned hams are pasteurized products. Because cured meats, such as hams, often pick up an undesirable metallic flavor when cooked to sterilization temperatures, they are generally pasteurized. The consumer should be very cautious about purchasing a canned ham that is not refrigerated. Pasteurized products, by law, must display in bold type on the principal display panel of the can the words **"Keep Refrigerated."**

Aseptic Canned Foods

As indicated earlier, the rather severe heat treatment required to produce canned products causes physical and chemical changes in some foods that make them unpalatable or even inedible. Some heat-sensitive foods will tolerate short intervals of heat treatment sufficient to eliminate vegetative microorganism forms. The product may be passed through heat exchangers that rapidly increase the temperature of the product sufficient to pasteurize or even sterilize but cause minimal heat damage to the product. Within a closed system, to avoid recontamination, the heat processed food may be filled into sterile containers (cans), which are closed and sealed under sterile (aseptic) conditions. The technique of "hot pack" home canning of foods that was practiced in the first half of this century was a variation of aseptic food processing.

Retortable Pouches

Foil-laminated and film pouches (see "Packaging Materials") have become a popular and more economical means of canning in recent years. Although nonmeat foods are the most common contents of the retortable pouches, some meat products are being preserved in this manner. Foil-laminated retortable pouches are in widespread use in the food industry for producing shelf-stable, sterilized products. The advantages of the retortable pouches are lower container costs and generally faster processing (retorting) times due to the flatter configuration of a pouch compared to a can. Figure 19–2 illustrates several meat items offered in the U.S. military's MRE (Meal, Ready to Eat) ration. The lighter weight of the retortable pouch was a major factor in the military's decision to switch from the can to this type of package. The military was satisfied with the performance of the system during the 1990–1991 war with Iraq and continues to make modifications involving menus and serving sizes within the same basic concept.[5] Figure 19–3 illustrates how flexible retortable pouches are used in the entire MRE ration.

---

[5]Baird, B. 1991. Hot meals in a hot spot. Food Technol. 2:52–56.

Fig. 19–2. Examples of flexible retort packaging of meat products used in the U.S. military's MRE (Meal, Ready to Eat) rations. The meat products pictured here are shelf-stable and heat-sterilized. (Courtesy, U.S. Army Natick Research Center, Natick, Massachusetts)

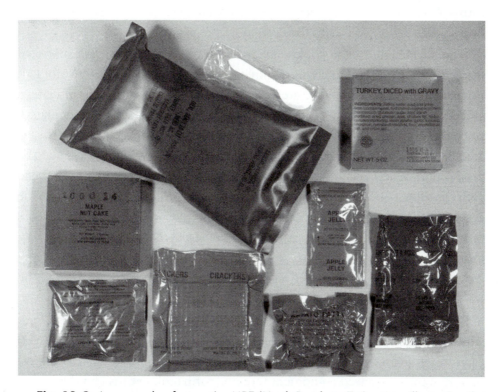

Fig. 19–3. An example of an entire MRE (Meal, Ready to Eat) ration, illustrating the extensive use of flexible packaging. The elimination of the tin can in military rations has decreased both the cost and the weight of the ration. (Courtesy, U.S. Army Natick Research Center, Natick, Massachusetts)

Cook-in Processing

Cook-in processing has gained widespread acceptance among larger processors in the meat industry. The cook-in process refers to the cooking of products (usually boneless hams and boneless turkey breasts) in a plastic, formable film and then using this same film as the merchandising container. Actually two types of film — a forming film and a non-forming film for the lid — are used. Both plastic films are composed of multiple layers of materials designed to meet their specific functions (see "Packaging Materials"). High-speed complex packaging machines, fed by rolls of the two films, enclose the product, remove the air, and seal the packages.

The forming film is fed through a heated chamber, which makes a cavity in the film that is filled with the product. The non-forming film is then conveyed over the top of the product-filled cavities, a vacuum is drawn, and the two films are sealed together to form an airtight container ready for water or steam cooking. Cook-in films are available in both transparent and opaque forms. Cook-in hams are generally placed in stainless steel molds prior to and during cooking to give them the desired shapes.

The advantages of the cook-in ham processing system over conventional water cooking systems include elimination of cooking losses, a twofold to threefold increase in shelf life, greater efficiency in packaging costs and labor, and a reduction of sewage-treatment costs. Cooking losses are eliminated, since the product is merchandised in the cooking film. The skintight evacuated package allows little or no moisture purge between the products and the plastic film. The shelf life is increased because after the product has been cooked, there is no recontamination by packaging lines and handling of the product. There is also a labor savings, since there is no repackaging after the product has been cooked. Furthermore, the task of cleaning the cooking molds is greatly simplified when a sealed plastic film is used. Finally, sewage-treatment costs are significantly reduced because the rendering of fat and the leaching of meat proteins into the cooking water are eliminated.

When compared to conventionally canned hams, the cook-in process saves considerably in cost. Obviously, the container costs are substantially lower for the cook-in process. Equipment costs are also greatly reduced with the elimination of the specialized equipment required for handling cans and the retorting chamber; however, the canned ham will have a longer shelf life with its sturdy container that is less apt to become punctured or damaged during merchandising.

Microwave Reheatables

Meat-based and meat-containing, shelf-stable products are packed in non-metallic, single-serving sized containers with pull-off metal tops. Not only are the products stable without refrigeration and easier to open than convention-ally canned stews, hashes, and related products, but the containers are microwavable (after the metal top is removed) and are serving dishes as well. The small size of the containers reduces the need for extended heat cycles to achieve sterilization of the product. The products most often merchandised in such containers will flow at the cooking temperatures required, reducing the processing time due to heat transfer by both conduction and convection. The negative aspects of individual merchandising/serving containerization are higher cost, greater use of nonrenewable resources per pound of food pro-cessed, and detrimental impact on the environment because the package is discarded.

PRESERVATION OF MEAT
BY DRYING

Drying in the sun and above the cooking fires has been a meat preserva-tion technique since the days of cave dwellers. The removal of moisture from meat is an effective means of preservation because moisture is a critical ele-ment required by all microorganisms for growth. This method of preserving meats lowers the moisture content of the product to a point at which the activity of food-spoilage and food-poisoning microorganisms is inhibited. The sun and the cooking fires accomplished the task slowly and effectively. Thermostatically controlled electric ovens that maintain a constant, but rela-tively low temperature are available to achieve similar results today.

Moisture requirements for microbial growth should be defined in terms of water activity (a_w) in the environment (meat) of the microorganisms. Water activity is the ratio of the water vapor pressure of the food substrate (P) over the vapor pressure of pure water (P_o) at the same temperature (e.g., $a_w = P/P_o$). The a_w of fresh meat is above 0.99 (see Chapter 2). The minimum a_w values required for the growth of various classes or genera of microorganisms are reported in Table 19–3.

Foods preserved by dehydration are conveniently divided into two groups, based on a_w levels: low-moisture foods and intermediate-moisture foods. Low-moisture foods are defined as those having an a_w of less than 0.60 and

Table 19–3. Approximate Minimum a_w Levels Required for Growth of Microorganisms and Some Benchmark a_w Values

Microorganisms	Minimum a_w
Groups[1]	
Most bacteria	0.90
Most yeasts	0.88
Most molds	0.80
Halophilic[2] bacteria	0.75
Xerophilic[3] molds	0.60
Osmophilic[4] yeasts	0.60
Specific organisms	
Acinetobacter[5]	0.96
Enterobacter aerogenes[5]	0.95
Bacillus subtillis[5]	0.95
Clostridium botulinum[1]	0.94
Escherichia coli[5]	0.96
Pseudomonas[5]	0.97
Staphylococcus aureus[1]	0.83
Saccharomyces rouxii[5]	0.62
Benchmark values[1]	
Ice at −15°C	0.86
Intermediate-moisture foods—upper limit	0.85
Saturated solution of NaCl	0.75
DNA instability	0.55

[1]Montville, T. J. (ed.). 1987. Food Microbiology. CRC Press, Inc., Boca Raton, Florida.
[2]Requires or prefers presence of salt.
[3]Requires or prefers dry environment.
[4]Requires or prefers high osmotic pressure.
[5]Jay, J. M. 1978. Modern Food Microbiology. Van Nostrand Reinhold Co., New York.

containing less than 25 percent moisture. Intermediate-moisture foods have an a_w between 0.60 and 0.85 and contain less than 50 percent moisture.

Moisture removal may be accomplished by low-temperature drying (less than 120°F), high-temperature drying (greater than 200°F), freeze-drying, and(or) salting. Low-temperature drying systems are generally less costly to operate, but they require more time. However, liquid or particulates up to ¾ inch, may be dried in large surface area, low-temperature dryers that flow the product to be dried down through moving balls heated by hot air (up to 160°F) entering the units from the bottom. High-quality dried products may be produced at half the cost and in half the floor space required by freeze-drying equipment.[6]

Low-moisture meat products include freeze-dried foods produced by specialty companies for camping and backpacking and for the military. Freeze-dried products maintain dimensions and rehydrate more readily than products dried by other methods. Freeze-drying is costly since the products must first be frozen and then dried under vacuum, both of which are high energy–consuming processes. The moisture in the frozen product is sublimated, proceeding directly from the solid to the vapor state.

Meat powders for soup stock and bouillon are also low-moisture foods but are generally produced by a high-temperature drying procedure such as spray or drum drying. Beef jerky (see Chapter 20) sometimes falls into the upper end of the low-moisture category with a_w values typically ranging from 0.55 to 0.70.

The U.S. military has developed low-moisture compressed food bars designed for long-range patrol operations. "Food Packet, Assault" rations, as pictured in Figure 19–4, are designed to be light weight, high-density meals with a relatively long shelf life, suitable for both cold and hot climates. Foods in the assault packet are all low-moisture or intermediate-moisture and can be consumed dry or rehydrated.

The true dry sausages fall into the intermediate-moisture category with a_w values typically ranging from 0.65 to 0.80. As such they require no refrigeration; however, they are often subject to external mold growth unless treated with a mold inhibitor such as potassium sorbate (see Chapter 21).

The a_w of meat products may be lowered by the addition of salt; however, the addition of salt alone requires extremely high levels to reach the a_w level required for the preservation of intermediate-moisture foods. Table 19–4 gives the effects of salt concentrations on a_w. For this reason, salting is generally accompanied by drying to produce intermediate-moisture foods. As noted earlier in the chapter, meat preservation processes often combine two or more basic techniques.

---

[6]Swientek, R. J. 1988. Low-temperature drying saves energy. Food Proc. 1:45 and 46.

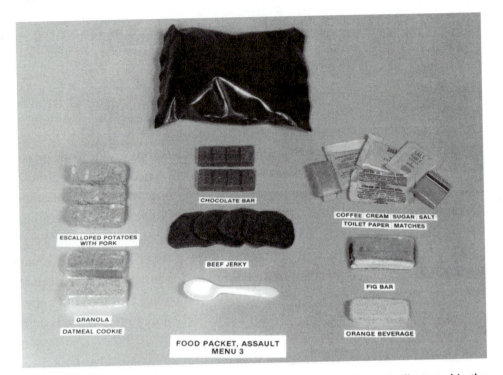

Fig. 19–4. An example of low- and intermediate-moisture foods is illustrated in the U.S. military's Food Packet, Assault Menu. (Courtesy, U.S. Army Natick Research Center, Natick, Massachusetts)

Table 19–4. Relationship Between a_w and Concentration of Salt Solutions[1]

Water Activity (a_w)	Sodium Chloride Concentration	
	Molal[2]	Percent (weight/volume)[3]
0.995	0.15	0.9
0.99	0.30	1.7
0.98	0.61	3.5
0.94	1.77	10.0
0.90	2.83	16.0
0.86	3.81	22.0

[1]Price, J. F. and B. Schweigert. 1987. The Science of Meat and Meat Products. Food and Nutrition Press, Inc., Westport, Connecticut.

[2]One gram-mole (58.5 grams of NaCl) dissolved in 1 liter (1,000 ml) of water equals 1 molal (e.g., 0.61 molal = 35.7 grams of NaCl in 1 liter of water).

[3]Weight/volume [e.g., 3.5% = 3.5 pounds NaCl per 100 pounds solution or 0.3 pound (4.8 oz.) per gallon (8.3 pounds) of solution].

PRESERVATION OF MEAT
BY IRRADIATION

Background

The preservation of meat or other food products by exposure to ionizing radiation dates back to a 1909 French patent by O. Wust. However, it was not until the early 1950s that significant progress was made utilizing irradiation as a practical method of food preservation. Since the late 1950s, the U.S. government has sponsored much of the research in this country related to food preservation by irradiation. The U.S. Army has probably played the most significant role in the development and testing of the process. Governmental support for irradiation research was cut significantly after 1968, when the FDA refused approval of a petition to produce irradiated ham for public consumption. Nonetheless, international interests in the irradiation process have continued to be strong.

A recommendation by the Joint Expert Committee on the Wholesomeness of Irradiated Foods, convened by the World Health Organization (WHO), sparked new interest in the irradiation process.[7] After reviewing approximately 30 years of safety and toxicology testing of irradiated foods, the committee concluded that any food irradiated to a dose of 1.0 Mrad (10 kiloGray [kGy])[8] or less is a wholesome product for human consumption and therefore should be approved without further testing.

At that time, the WHO Joint Expert Committee deferred general recommendations for foods irradiated at levels higher than 1 Mrad (10 kGy), until data were available from numerous long-term ongoing studies. In 1984, the international authority on food regulations, *The Codex Alimentarius,*[9] published the "Codex Standard for Irradiated Foods" (Worldwide Standard), which stated that the overall average dose absorbed by a food should not exceed 10 kGy.

---

[7]WHO. 1981. Wholesomeness of Irradiated Food. Report of a Joint FAE/IAEA/WHO Expert Committee, October 27 to November 3, 1980, World Health Organization Tech. Rep. Series, No. 659, Geneva, Switzerland.

[8]Gray (Gy) is an internationally accepted unit of absorbed radiation dose. It is defined as the mean energy imparted by ionizing radiation to matter per unit mass. One Gy is equal to 1 joule per kilogram. The older unit, the rad, equals 0.01 Gy. From Food Irradiation, a Technique for Preserving and Improving the Safety of Food. 1988. World Health Organization and the Food and Agriculture Organization of the United Nations.

[9]From Food Irradiation, a Technique for Preserving and Improving the Safety of Food. 1988. World Health Organization and the Food and Agriculture Organization of the United Nations.

In 1986, the Food and Drug Administration issued rulings permitting irradiation treatments of food at a maximum of 1 kGy, of spices and vegetable seasonings at a maximum of 30 kGy, and of dry enzyme preparations at a maximum of 10 kGy. International labeling standards were outlined by *The Codex Alimentarius* Committee on Food Labelling in 1989. Those standards proposed the international radura logo/symbol (see Figure 3–2), which is currently required by the Food and Drug Administration on foods that have been subjected to radiation processing.[10] (See "Hide Preservation" in Chapter 11.)

Irradiated strawberries were first available in U.S. stores during the 1992 season and were apparently well-accepted by consumers. Irradiation extended the shelf life of the strawberries, which were processed by a commercial food irradiation facility located in Florida. The Florida Department of Agriculture is also developing an irradiation unit to process fruit. Both the Food and Drug Administration and the U.S. Department of Agriculture approved pork and poultry as the first meats to be irradiated. In 1997 the FDA approved beef irradation. In 1999 the USDA also approved beef irradiation but waived its authority over ready-to-eat (RTE) products in order to expedite the FDA approval process for them. A facility specifically designed to irradiate meat is located at Iowa State University. The Iowa facility is powered by electricity and does not require radioactive elements.

Pasteurization by Irradiation (Radurization)

Meat pasteurization (elimination of non-spore forms of spoilage and pathogenic microorganisms) can be accomplished with medium radiation doses of 1.0 to 7.0 kGy, a process referred to as *radurization*. Like heat pasteurization (e.g., canning hams), radurization requires post-processing refrigerated storage of the product to prevent spoilage initiated by the more resistant forms (spores) of microorganisms. Radurization is designed to kill or inactivate food-spoilage organisms without the use of heat and in a wide variety of packaging. This gives the process tremendous potential applications in merchandising fresh meat, poultry, and fish products. For example, fresh (unfrozen) meats could be effectively exported to any country in the world

---

[10]Council for Agricultural Science and Technology (CAST). 1996. Radiation Pasteurization of Food. Issue Paper Number 7. Ames, IA.

and still maintain a fresh, acceptably long shelf life upon arrival. Poultry products often create merchandising problems for retailers because of their short display period before bacterial spoilage occurs. Fresh seafoods from coastal regions could be shipped to the interior portions of larger countries without the expense of rapid air freight, the method by which shipments are now made. The shelf life of other products, such as fresh pork sausage, bacon, corned beef, frankfurters, and luncheon meats, could be extended from 3 to 10 times their present safe-storage period. Savings from product spoilage would be passed on to the consumer, since the retailers must include in their operating margin an allowance for perishable products that spoil before being sold.

Radurization of poultry meat provides a very effective means of controlling contamination with salmonellae. Salmonella contamination can be serious enough that some health authorities have suggested that radurization should be required for poultry products to bring the problem under control.

Radurization effectively inactivates parasites that may be transferred to humans and other animals through the food chain. *Trichinella spiralis,* the organism, which is responsible for the debilitating and sometimes fatal trichinosis, can be inactivated in pork by a 0.15 kGy dose of irradiation.[11] Other parasites of meat animals and fish are also rendered harmless by low-dose treatments.

Sterilization by Irradiation (Radappertization)

Sterilization by irradiation is commonly referred to as *radappertization.* Depending on the food, spoilage and disease organisms (vegetative and resistant forms) are killed at radiation doses of 10 to 50 Gy. Similar to their heat-sterilized (canned) counterparts, radappertized meat products may be stored for years without refrigeration (providing the sealed containers are not punctured or corroded).

Before radappertization processing, meat must be cooked to 140° to 150°F in order to inactivate any autolytic (tissue degrading) muscle enzymes that could produce off-flavors during prolonged storage at room tempera-

---

[11]Food Irradiation, a Technique for Preserving and Improving the Safety of Food. 1988. World Health Organization and the Food and Agriculture Organization of the United Nations.

ture. After being cooked, the meat is frozen to –40°F and irradiated while frozen, which further minimizes flavor changes that would otherwise occur at the higher radiation doses (10 to 50 kGy). In contrast, cooking is not necessary for radurization processing because the meat must be stored under refrigerated conditions, and the shelf life is still limited.

Radappertized ham was consumed by the *Apollo* astronauts who landed on the moon and by the U.S. astronauts and the Soviet cosmonauts who participated in the joint U.S.–Soviet *Apollo–Soyuz* space flight in 1975. During the *Columbia* space shuttle flights, U.S. astronauts dined on radappertized beef, pork, smoked turkey, and corned beef. Radiation-sterilized food products are approved for use in special hospital meals needed by patients who must have sterile foods because their immunity to disease has been altered by illness such as AIDS or by medical treatment such as chemotherapy. These special diets have shown no effect on the rate of disease progression. Some examples of food preservation applications by radappertization are presented in Table 19–5.

Methods of Irradiation Processing

Two types of ionizing irradiation processes, gamma radiation and electron radiation, are in commercial use today. The gamma radiation facilities utilize the isotopes cobalt–60 and cesium–137. Gamma radiation has a very short wavelength (see Appendix — electricity) and a high penetrating ability of the same nature as short X–rays while at the same time imparting more energy into the penetrated mass than X-rays. Electron radiation is created by large electron accelerators that range in power from 4 to 10 Mev (million electron volts) and is limited in penetrating ability. Both processing methods cause the material they strike to produce electrically charged particles, called ions. The food preservative effect may be the result of the alteration of essential physical structures or enzymes within the pathogenic and spoilage organisms. Neither gamma nor electron irradiation preservation systems impart radioactivity to the food product. Irradiation preservation methods leave no more radiation in a product than what was originally there by nature (from the sun and the environment).

Substantial energy savings are incurred for radappertized products during processing, merchandising, and storing when compared to other methods. Table 19–6 compares several preservation methods and the amount of energy required to maintain the wholesomeness of the product. At the time the table

Table 19–5. Dose Requirements in Various Applications of Food Irradiation[1]

Purpose	Dose (kGy)[2]	Products
Low dose (up to 1 kGy)		
Inhibition of sprouting	0.05–0.15	Potatoes, onions, garlic, gingerroot, etc.
Insect disinfestation and parasite disinfection	0.15–0.50	Cereals and pulses, fresh and dried fruits, dried fish and meat, fresh pork, etc.
Delay of physiological process (e.g., ripening)	0.50–1.0	Fresh fruits and vegetables.
Medium dose (1–10 kGy)		
Extension of shelf life	1.0–3.0	Fresh fruits and strawberries, etc.
Elimination of spoilage and pathogenic microorganisms	1.0–7.0	Fresh and frozen seafood, raw or frozen poultry and meat, etc.
Improvement of technological properties of food	2.0–7.0	Grapes (increasing juice yield), dehydrated vegetables (reduced cooking time), etc.
High dose (10–50 kGy)		
Industrial sterilization (in combination with mild heat)	30–50	Meat, poultry, seafood, prepared foods, sterilized foods for hospital diets.
Decontamination of certain food additives and ingredients	10–50	Spices, enzyme preparations, natural gum, etc.

[1]Source: Food Irradiation, A Technique for Preserving and Improving the Safety of Food. 1988. Published by World Health Organization and Food and Agriculture Organization of the United Nations, Geneva, Switzerland.

[2]Gy: Gray—unit used to measure absorbed dose (see text footnote 8).

was developed, heat sterilization (canning) required over five times as much energy as radappertization to achieve sterility. Storing frozen meat at 13°F for 25 days required over 30 times the energy needed to radappertize a product.

As one would expect from these energy-usage values, the preservation cost per pound of irradiated products is economical. Although large initial capital investments are required for irradiation facilities ($2 to $10 million), the volume of product that can be processed makes the investment economical on a

Table 19–6. Typical Energy Values Used in the Processing of Food[1]

Preservation Method	kj/kg[2]
Radurization (with 0.25 Mrad)[3] (2.5 kGy)[4]	21
Radappertization (with 3 Mrad) (30 kGy)	157
Heat sterilization.	918
Blast freezing chicken meat from 40° to –10°F (4.4° to –23.2°C)	7,552
Storing product at –13°F (–25°C) for 25 days	5,149
Refrigeration storage for 5.5 days at 32°F (0°C)	318
Refrigeration storage for 10.5 days at 32°F (0°C)	396

[1]Brynjolfsson, A. 1980. Food Irradiation in the United States. Proceedings of the 26th European Meeting of Meat Research Workers, Vol. I, Colorado Springs, American Meat Science Assn., p. 172.

[2]Kilojoules per kilogram. Standard International (SI) units of measure for energy and mass (weight).

[3]Older measure of energy absorbed, 1 rad = 0.01 Gray (1 Megarad = 1,000,000 rads).

[4]Modern measure of energy absorbed by an irradiated mass, one Gray = one joule per kilogram of mass (1 kGy = 1,000 Grays).

per pound basis. At a cost of $10 million, such a facility would have a capacity to handle 100 million pounds of product per year.[12]

MEAT PRESERVATION BY OZONE

Ozone, consisting of three oxygen atoms, instead of two, is highly reactive. Exposing regular oxygen to electrical energy creates ozone. One can see the everyday effects of ozone in the rusting of metal. Ozone can be infused in water and used to preserve meat; however, at this writing, FSIS approval has not been granted to ozone meat preservation.

---

[12]Brynjolfsson, A. 1980. Food Irradiation in the United States. Proc. of the 26th European Meeting of Meat Research Workers, Vol. I, Colorado Springs, American Meat Science Assn., p. 172.

PACKAGING

The ultimate success of many meat preservation techniques requires that the product be properly packaged post-processing to maintain palatability and other economically important product characteristics. Canning, pouch retorting, cook-in processing, and radiation techniques include the package as an essential factor in the preservation process. Other preservation techniques involve packaging prior to, during, or following the application of other phases of the preservative system.

Many processed meat products are wrapped to provide a barrier between the product and its immediate environment. The following general groups of wrapping materials are suitable for meats.

- Wax- or paraffin-treated kraft papers.

- Aluminum foil.

- Laminated foils.

- Films (polyethylene and pure or mixed polymers or copolymers of vinyl chloride, vinyl acetate, or vinylidine chloride).

- Food-grade "plastic" dips (Dermatex®).

The characteristics of these wrapping materials differ, and their suitability for packaging meats, vegetables, and fruits depends upon how closely they come to meeting the following criteria:

- Have low moisture-vapor transmission.

- Have differing oxygen permeability, depending on use.

- Have good tensile strength and are puncture-resistant.

- Are pliable.

- Will maintain pliability and tensile strength at sub-zero temperatures.

- Are non-toxic.

- Are odorless.

- Are easy to mark for identification.

- Have good stripping qualities (will peel from meat when frozen).

- Are greaseproof and stainproof.

- Provide good sealing properties.

The moisture loss, or shrinkage, of any food during freezer storage must be held to a minimum. An excess of 8 percent shrink in meats and 3 percent in poultry is considered to make them unacceptable as fresh meat. This loss in weight is easily measured, and the changes in color, aroma, flavor, and texture are in about the same proportion as the loss in weight. To hold dehydration to a minimum necessitates the use of a material that has a low moisture-vapor transmission at low temperatures.

Wax- or Paraffin-treated Kraft Papers

These packaging materials are basically wood pulp papers. Kraft (German word meaning *strong)* is probably the wrapping material most often used in the home and by local meat processors for freezing meat and for wrapping other processed meats for refrigerated or dry storage. The effectiveness of packaging meat in paper wrappers is influenced by the care and technique used in wrapping the product (see "Packaging Meat for the Freezer.") These materials come in many forms having different qualities that give protection against oil, grease, chemicals, molds, moisture-vapor and oxygen transmission, and water. Paper is not a good heat conductor (see Chapter 24) and therefore may slow meat freezing rates.

Vegetable parchment is another wood-pulp paper that has many uses as a meat wrapper, depending upon its treatment. Some grades are impervious to oxygen, carbon dioxide, and nitrogen. The coated parchment is used for freezer-wrapped meats. Silicone-treated vegetable parchment has anti-sticking properties that make it particularly useful as dividers for frozen meat cuts and for hamburger, sausage, and lamb patties.

Aluminum Foil

Aluminum foil has been used primarily by consumers for home freezing and other meat storage because it is readily available, is easy to work with, is an excellent heat conductor, and has the additional advantage in that it can be used as a cooking wrap as well as a storage wrap. The primary disadvantages of aluminum foil are that it is easily torn, has a low resistance to punctures from bones or other sharp objects, and cannot be used in a microwave oven. The increased availability of a wide variety of films and bags has also reduced the popularity of aluminum foil as a household food wrap.

Films

The greatest technological advancements in the packaging industry have occurred with the copolymer films. General-purpose films are used to wrap meat and other food for home storage. Bags and pouches made from films are used by the consumer for freezing and storing various foods, including meats. Meat-packaging films are generally multi-layer structures that use a variety of polymer resins (long-chain structural chemical compounds) and coatings to meet the specific needs of various products in the industry. They usually contain three or more coextruded layers, with each layer contributing a specific property to the structure (see Figure 19–5).

Most processed meat is vacuum-packaged and sometimes flushed with an inert gas such as nitrogen or carbon dioxide to reduce the speed of lipid oxidation, which causes oxidative rancidity. The packaging film must keep atmospheric oxygen from getting back into the package. The ability of different

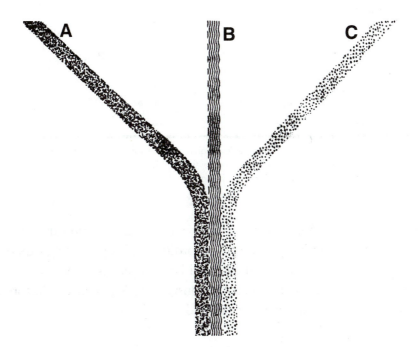

Fig. 19–5. Composite polymer film.

Layer A: The outside layer must be scuff- and abrasion-resistant.

Layer B: The middle layer generally provides the barrier to oxygen.

Layer C: The sealant layer must be capable of being melted and welded under pressure to the sealant layer of the other film to make a gastight seal.

films to slow down the transmission of oxygen depends on a number of factors, such as the type of polymer, thickness, ambient temperature, and humidity. Some films are designed as oxygen barriers (frozen meat storage wrap), while other films are designed for high oxygen transmission to give fresh meats a desirable, oxygenated surface color (self-service retail meatcase wrap). Oxygen transmission values are available for most films and are generally reported as milliliters (cubic centimeters) of oxygen transmitted through 100 square inches (ml O_2 per 100 square inches) of film in 24 hours under standardized conditions. Typical oxygen transmission values for some commonly used films are reported in Table 19–7.

Table 19–7. Typical Oxygen Transmission Values[1]

	ml O_2/100 sq. in./ 24 hrs., 73°F, 0% RH
Ethylene vinyl alcohol (EVOH)	0.1
Saran (PVDC) coatings	1.5–1.0
1-ml nylon	2–3
0.5-mil polyester	9
0.75-mil oriented polypropylene	200
2-mil sealant (EVA or Surlyn)	300

[1]Gehrke, W. 1983. Film Properties Required for Thermo-formed and Thermal-processed Meat Packages. In Proceedings of the 36th Reciprocal Meats Conference, National Live Stock and Meat Board, Chicago, p. 55.

Willard Gehrke, an expert in the meat packaging film industry, provided most of the following information.[13] According to this packaging expert, the oxygen transmission of the finished multi-layered film is a complicated subject, and a number of factors must be considered. A few of the factors that influence the oxygen barrier of the finished package are:

- Films for packaging processed meat are generally a combination of various polymer films. Each layer in the finished film will contribute something to the barrier properties of the finished structure.

---

[13]Gehrke, W. 1983. Film Properties Required for Thermo-formed and Thermal-processed Meat Packages. In: Proc. of the 36th Recip. Meats Conf., National Live Stock and Meat Board, Chicago, p. 55.

- Some of the polymer layers in the film structure will vary in initial thickness.

- The forming film will be considerably thinner after forming than it was originally. This thinning is not uniform and varies with the size and shape of the cavity, as well as the technique used to do the thermoforming. Thin films permit more oxygen to pass through and enter the package. Each layer in the packaging film will be thinner after forming.

- The entire package will be handled in packaging, boxing, shipping, and merchandising. Any abuse will tend to harm the barrier properties.

- The amount of moisture in the air can change the rate of oxygen transmission for certain packaging films.

- Processed meats are generally stored at refrigerated temperatures. Less oxygen is transmitted through all polymer films as the temperature is reduced.

Films are manufactured for many different purposes that depend upon the product to be packaged and the equipment on which the packaging is accomplished. A good seal is paramount in the production of a vacuum-packaged product. The sealant is usually the thickest layer of packaging film and is generally composed of one of several compounds, each having slightly different properties to meet different packaging needs.

The four sealants most commonly used in meat films are:

1. Low-density polyethylene (poly)

2. Ethylene–vinyl acetate (EVA)

3. Ionomer (Surlyn)

4. Linear low-density poly (LLDPE)

Film-packaged processed meats such as sliced sandwich meats and sausages have been sealed well until the consumer breaks the seal and opens the package the first time. Most meat packages have not provided adequate protection for the product that is to be used at some later time because the original seal could not be re-established. A secondary seal that is based on the zipper technology (e.g., on food and freezer storage bags) is being added to some film packages used for meat products that are not completely consumed at one time. Many processed meat products are now packaged on machines that require two different types of laminated films — a non-forming film and a forming film.

The packaging system generally employs a machine that performs most of the following steps:

1. Heats the forming film.

2. Forms it into a cavity of the desired size.

3. Provides a loading area — so the product can be placed in the formed cavity.

4. Brings the non-forming film in place over the forming film containing the product.

5. Pulls a vacuum to remove air inside the package.

6. Introduces an inert gas, such as nitrogen, into the package, if desired.

7. Seals the package. This is done with a heat-seal bar that melts the two sealant layers and welds them together under pressure.

8. Cuts or trims the films to provide individual packages.

The most widely used non-forming film is a structure consisting of oriented polyester/saran/sealant. The polyester layer is usually ½ mil (0.0005 inch) in thickness. The sealant layer is usually 2 to 3 mils (0.002 to 0.003 inch) in thickness. The non-forming film is frequently printed. Any printing is generally buried in the structure — that is, placed between the saran and the sealant. Each layer in the structure has a specific purpose.

- The polyester is resistant to the temperature needed to melt the sealant and is scuff- and abrasion-resistant.

- The saran layer provides the barrier to oxygen.

- The sealant must be capable of being melted at a relatively low temperature to make a hermetic bond with the sealant layer of the forming film.

The forming film structure used for the largest number of processed meat packaging applications is nylon/saran/sealant. The gauge of the nylon and the gauge of the sealant are adjusted, depending on the size and shape of the cavity the film must be thermo-formed into to accommodate the product. The total thickness will vary from approximately 2.75 mils for a product such as bacon to 11 mils or more for a product such as a large ham.

Semi-rigid films are another popular class of forming films in the industry. Semi-rigid films were developed for use on finished packages of processed meat that are hung on pegs for display in stores. A stiffer structure is used — generally consisting of a 7.5- to 12-mil semi-rigid polyester sheet combined with a barrier layer and a sealant.

In order to allow for the thinning that takes place in thermo-forming, the starting film must have sufficient material to insure that the side walls and corners of the shaped package have adequate thickness to provide the desired protection. Each layer in the formed web structure has its specific purpose.

- The nylon or semi-rigid film must be capable of being softened by heat and formed into a cavity of the desired size and shape, yet be resistant to temperatures needed to melt the sealant. It must also be scuff- and abrasion-resistant.

- The barrier layer provides the main resistance to oxygen transmission.

- The sealant must be capable of being melted at a relatively low temperature to make a hermetic bond with the sealant layer of the non-forming film.

The process of heating and forming a film into a cavity produces a shaped film that is considerably thinner than the original flat film. The degree of thinning and uniformity of the shaped film is a complicated subject and will depend on a number of factors, including:

- The type and original thickness of the forming film.
- The size of the area being formed.
- The depth and contour of the cavity.
- The procedure used to heat and draw or push the heated film into the cavity.

Forming Films for Thermal Processing

Perhaps the most demanding requirement for forming films is for those applications in which the thermal processing (or cooking) is done in the same package used for shipping and merchandising. Cooked hams and poultry items are the most common meat items for which the cook-in technology is being used.

These applications are some of the most demanding for forming films, because the film must shrink with the product as the product is cooked and cooled. In addition, the film must have a good appearance and be tough enough to stand abuse in shipping and handling.

The most important characteristic of forming films is their ability to shrink. Films formed on a conventional processed meat-packaging machine by vacuum alone do not shrink to an acceptable degree. To have shrink, the films must be formed at lower temperatures. To do this forming at the critical temperature, the packaging machines are designed with plugs to assist in getting the films properly in the cavities.

Laminated Foils

Laminated foils combine the multi-layer concept of packaging films with the metallic foil wrapping material. The foil is "laminated" with film(s) to form a multi-layered structure, as illustrated in Figure 19–5. Laminated foils have become popular packaging material for intermediate-moisture foods such as beef sticks, jerky, and meat products that are heat sterilized in laminated foil retortable pouches. Since the pouches are subjected to the high-temperature/pressure combination of the canning retort, the lamination and sealing bonds must be exceptional. The laminated foils offer high resistance to puncture and abrasion, making them an ideal economic means of offering sterilized, shelf-stable products.

Food-Grade Dermatex®
(Acetylated Monoglycerides)

Another approach to packaging cooked, processed products consists of dipping the products in food-grade monoglycerides (trade name, Dermatex®). This compound is made from highly distilled monoglycerides from animal fats. Dermatex® is marketed in wax-like blocks, which are subsequently submerged into a dipping tank and melted. Although Dermatex® is available in a number of colors, transparent Dermatex® seems to be the most popular. Dermatex® is approved by the U.S. Department of Agriculture for use as a packaging film in cooked sausage products such as summer sausage, liver sausage, and other "chub" items. The principal advantage of this innovation is that it provides a more economical alternative to vacuum packaging with shelf-life properties similar to vacuum-packaged products.

20

Meat Curing and Smoking

This chapter will discuss the history of meat curing and smoking, the functions of these processes in the modern meat industry, the ingredients used in curing, the nitrite controversy, meat-curing chemistry, the basic methods of meat curing, commercial curing procedures, smoking and cooking, and cured and smoked products.

HISTORICAL ORIGINS

The salting, or curing, of meat is the oldest known form of preservation for this perishable commodity. The ancient Sumerian culture that flourished around 3,000 B.C. is believed to have been the first to make use of salt for meat and fish preservation. Between the years 3,000 B.C. and 1,200 B.C., historical evidence indicates that the Jewish people gathered salt from the Dead Sea to use in the preservation of various foods. The Chinese and Greeks are also reported to have used salted fish in their diets and are credited with passing this practice on to the Romans, who included pickled meats in their diets.[1]

The following was written by the Roman scholar Cato,[2] who lived in the third century B.C.

Salting of hams and of the small pieces such as are put up at Puteoli

Hams should be salted in the large storage jars or in the smaller jars in this way: (1) When you buy the hams cut off the feet; (take)

---

[1]Jay, J. 1978. Modern Food Microbiology. D. Van Nostrand Reinhold Co., New York.
[2]Brehaut, E. 1933. Cato, The Censor on Farming. Columbia University Press, pp. 145–146.

for each ham a half-modius[3] of Roman salt ground in a mill, sprinkle the bottom of the large jar or of the smaller one with the salt, then put in a ham, skin side down, and cover completely with salt. (2) Then put a second on top, cover the same way, take care that meat does not touch meat. Cover them all the same way. When you have placed them all, cover above with salt so that the meat will not show. Make the salt level. When they have been five days in the salt, take them all out, salt and all. Place on the bottom those that were on top, and cover and arrange in the same way. (3) After twelve days in all, take the hams out, wipe off all salt and hang them in the wind for two days. On the third day wipe them off well with a sponge, rub thoroughly with olive oil and vinegar mixed, and hang them up on the meat rack. Neither moths nor worms will touch them.

Just how dry salting originated is not definitely known. It is quite likely that the use of salt for preserving meat was entirely accidental. Since saltpeter was probably an impurity in the salt that was used, it remained for the chemist to identify the agent responsible for the characteristic cured meat color. The salted meats of the ancients were very unevenly cured and were objectionably dry and salty. The latter part of the eighteenth century marked the beginning of salt curing of meat on a scientific basis.

Historical information about the smoking of meat is more scanty; however, it is quite likely that the process evolved as a result of meat being cooked on open fires. On the North American continent, the Indians often hung strips of meat in the tops of their tepees out of the reach of dogs. As the strips dried, the meat took on a smoky flavor from the campfire smoke inside the tepees. The dried strips, which were very hard and inflexible, were beaten with stones or wooden mallets into powder, which was then mixed with dried fruits and vegetables to make *pemmican*. In this form, the dried meat was transported in skin sacks or bladders and was the principal food whenever the tribes were migrating.

Modern meat-curing and smoking practices are used primarily to create flavor and appearance variations in the food supply and to inhibit the outgrowth of the *C. botulinum* spores that cause the lethal food poisoning botulism (see Chapter 2). Preservation is no longer the principal reason for curing

---

[3]A unit of volume of Cato's time, approximately equal to ½ U.S. gallon.

and smoking meat in modern societies with readily available mechanical refrigeration. As a result, the hams processed today are considerably milder (less salty) than those produced years ago.

Production of country-cured meats by dry-curing methods is a major processing industry, concentrated in the southern regions of the United States. Preservation by salting and dehydration of country-cured ham and bacon is of major importance for these products, since they are often unrefrigerated during storage. Preservation by salting requires much higher salt levels than are present in commercial hams and other cured products that are not country-cured. Many people outside the southern regions of the United States consider the country-cured hams to be too salty and unpalatable unless they are boiled for long periods to remove the high salt content from the meat.

FUNCTIONS OF CURING AND SMOKING

The functions of modern commercial meat curing and(or) smoking processes are the following:

- Food safety

- Refrigerated shelf-life extension

- Flavor development

- Color development (internal and external)

The microbiological safety that is assured from the addition of nitrite in cured meats is perhaps the most important function of the curing process. Nitrite is the most effective antibotulinal agent known in the prevention of the sometimes fatal botulism food poisoning (see Chapter 2).

The extension of product shelf life is another function of the curing process that is related to the *bacteriostatic* (inhibiting growth of bacteria but not destroying or killing them) properties of salt. Since most cured-meat products are fully cooked prior to being sold (except country-cured hams, bacon, and dry sausage), the thermal processing extends the product shelf life before spoilage occurs.

Cured meats offer the consumer a convenient, versatile product that features a salty flavor, which many people enjoy. There is nothing quite like a juicy bacon, lettuce, and tomato sandwich on a hot summer day!

The characteristic pink cured-meat color associated with the lean of cured products provides important aesthetic qualities. (Nitrite, the preservative chemical, and the myoglobin/nitrite pigment chemistry, responsible for the development of the pink color in cooked cured meats, are discussed in detail later in the chapter.) The external color of smoked cured-meat products is primarily determined by the amount of smoke deposition (natural or liquid) on the product surface during thermal processing. A moderately intense chestnut color is usually considered most desirable. Meat curing is considered a bacteriostatic process, which creates an unfavorable environment for microbial growth. Most bacteria, yeasts, and molds that cause food spoilage have a relatively low tolerance to salt (NaCl). However, there are numerous spoilage and *pathogenic* (disease-causing) bacteria that can tolerate the salt levels present in most cured products (2 to 4 percent). These organisms are collectively referred to as *halophilic* (salt-loving) organisms.

MEAT-CURING INGREDIENTS

Salt

Salt has been very important to world history, as well as an essential ingredient in food.[4,5] Roman soldiers were given special salt rations called *salarium argentum*, the forerunner of our word "salary."[6]

The most important curing ingredient is salt (NaCl). It makes up the bulk of the curing mixture because it is a good preservative and because it provides the most desirable flavor. Salt acts as a dehydrating agent, osmotically lowering the water content of the product and of bacterial organisms, thus limiting their ability to thrive and reproduce. (See the discussions of the impact of salt on water activity, a_w, in Chapters 2, 19, and 21, specifically Table 19–4.) After salt has been applied to the surface, it moves into the meat by the process of diffusion and causes moisture movement to the surface. Both actions attempt to equalize the salt concentration in and around the meat cut. When salt is added to water as a solvent and a carrier, the solution is called a pickle or brine (see "Salt" in Chapter 21).

---

[4]Pszczola, D. E. 1997. Salty developments in food. Food Technol. 51:79–83.

[5]Eddy, K. 1997. It reigns — Bow down before a substance that's essential in the kitchen. Chicago Tribune (October 1). Tempo 4–5.

[6]Salt Institute. www.saltinstitute.org

Sugar

Sugar, a secondary ingredient in the curing formula, counteracts the astringent quality (harshness) of the salt, enhances the flavor of the product, and provides the substrate for the formation of acid, thus lowering the pH of the cure. Its role in color development and color stability under present commercial nitrite-based curing practices has been found to be negligible. When nitrates are used as curing agents, sugar enhances the growth of microorganisms that reduce the nitrate to nitrite, the first step in the curing chemistry sequence. The sugars most frequently used in meat curing are sucrose and dextrose.

Nitrate and Nitrite

Sodium nitrate is a naturally occurring substance in vegetables, water, and soil. Originally discovered as an impurity in salt, it has been used in small amounts for thousands of years to cure meats. When nitrate is used to cure meats, it is readily converted to nitrite by microbes or by a reducing agent such as ascorbate. *Nitrite* is the active meat-curing agent. Sodium nitrite is used more commonly today to cure processed meats. Both nitrate usage and nitrite usage are allowed under the Meat Inspection Act, but nitrate usage is permitted only in the production of country-cured hams and in some sausages that are cured for several weeks[7] (see Chapter 3).

Nitrite is considered essential in cured meats because it performs several vital functions. They are:

- It prevents botulism and has other bacteriostatic properties to safeguard against the mishandling of meat by manufacturers, distributors, retailers, and(or) consumers (failure to refrigerate properly because of mechanical malfunction, negligence, or ignorance).

- It retards lipid oxidation, which otherwise causes an undesirable (warmed over) flavor.

- It gives cured meats their characteristic cured flavor.

- It is responsible for the color fixation of the meat pigments, which results in the characteristic cured pink color.

---

[7]Federal Register. 1989. Vol. 54, No. 202, p. 3042 (October 20).

Nitrite is the only substance that will do all these things. No substitute has been found, even though more than 700 substances have been tested as possible replacements. Without the addition of nitrite (or nitrate), cured meats as we know them would not exist. Rather, ham and other products would have an unattractive grayish green color and a flat, salty flavor.

Generations of consumers have become accustomed to the bright pink cured-meat color. The addition of nitrite to meat indirectly induces the development of that characteristic cured color. This color change from the purple-red of fresh meat to the pink of cured meats involves the meat protein *myoglobin,* which is the vehicle for oxygen storage in the muscle. In meat curing, nitrite (NO_2^-) is broken down to nitric oxide (NO),[8] which then binds to the myoglobin molecules to induce the color change. Nitric oxide cannot be added directly as an ingredient: it must be generated within the meat from nitrite. The color chemistry of nitrite and myoglobin is discussed later in this chapter. (See also the variations of cured-meat color in the color photo section.)

FSIS regulations (CFR secs. 318.7b, 318.7c) permit the use of sodium or potassium nitrate only in dry-cured country ham and in dry sausages. Approved usage levels for nitrate in dry-cured country ham are 3½ ounces for each 100 pounds of meat in dry salt or dry cure and 2¾ ounces in 100 pounds of chopped meat and(or) meat by-product.

Current FSIS regulations permit the use of sodium or potassium nitrite in all products except pumped bacon at the following levels: 2 pounds in 100 gallons of pickle at 10 percent pump level; 1 ounce for each 100 pounds of meat in dry salt or dry cure; and ¼ ounce in 100 pounds of chopped meat and(or) meat by-product.

The use of nitrites or nitrates, or combinations thereof, is not permitted to result in more than 200 parts per million (ppm) of nitrite, calculated as sodium nitrite in the finished products other than in bacon. In pumped bacon, in-going nitrite levels may not exceed 120 ppm and must be accompanied by at least 550 ppm of sodium erythorbate or ascorbate as a cure accelerator. The immersion-cured bacon and dry-cured bacon in-going nitrite level limits are 120 ppm and 200 ppm, respectively. Residual nitrite levels in the finished pumped bacon cannot exceed 40 ppm.

The nonmeat ingredients for meat-curing operations must be formulated with great caution. ***Both nitrate and nitrite are toxic to humans when con-***

---

[8]Named molecule of the year (1992) by Science. Science. 1992. 258:1862 and 1898.

sumed at levels higher than allowed in cured-meat products. Weighing the minute amounts of nitrite and nitrate must be done carefully on accurate scales. Equally important, the nitrite and(or) nitrate must be uniformly mixed with other cure ingredients to avoid the possibility of a toxic level occurring in a single meat serving or in one lot of product. Because such small amounts must be accurately weighed and mixed, using well-known sources of commercially produced nitrite- and(or) nitrate-containing meat-curing mixes is recommended.

Sodium Ascorbate and Erythorbate

The time required for color development and retention has caused some problems in cured meats and particularly in emulsion-type products that are heat processed immediately. It has been found that ascorbic acid and erythorbic acid and their salts hasten color changes in the curing process. Chemical reactions, in the presence of ascorbate or erythorbate, may involve either the nitrite by producing more nitric oxide, or the meat pigment, enhancing the reduction of metmyoglobin to myoglobin.

Federal regulations permit the addition of 75 ounces of ascorbic acid or 87½ ounces of sodium ascorbate or erythorbate to 100 gallons of pickle and ¾ ounce of ascorbic acid or ⅞ ounce of sodium ascorbate to each 100 pounds of chopped meat (550 ppm).

Spraying the surface of cured cuts prior to packaging with 5 to 10 percent ascorbic acid or ascorbate solution, the use of which will not result in a significant addition of water to the product, has been found to deter color fading caused by light. The ability of ascorbic-treated cured meats to resist fading is ascribed to residual ascorbic acid maintaining reducing conditions on the exposed cured-meat surface. In addition to its beneficial effects on color, ascorbate, commonly known as vitamin C, has been shown to inhibit the formation of *nitrosamines* in cured meats. Nitrosamines are known *carcinogens* (cancer-causing agents) and are discussed more thoroughly later in this chapter (see "The Nitrite Controversy").

Alkaline Phosphates

The primary purpose of alkaline phosphates is to decrease the shrinkage in smoked meat and meat products and the "cook-out" in canned-meat products. The ability to boost the water-binding quality of meat results in increased yields of up to 10 percent. Greater yield increases are possible when phosphates and other added substances are combined in meat curing.

The phosphates approved for use are disodium phosphate, monosodium phosphate, sodium metaphosphate, sodium polyphosphate, sodium tripoly-phosphate, sodium pyrophosphate, sodium acid pyrophosphate, dipotassium phosphate, monopotassium phosphate, potassium tripolyphosphate, and potassium pyrophosphate. These phosphates may be added to the pickle for hams, bacon, pork shoulders, picnics, Boston butts, boneless butts, and pork loins. The pickle shall not contain more than 5 percent of an approved phosphate, and the finished product shall not contain more than 0.5 percent phosphate. Phosphates do not dissolve easily in water, so phosphate-containing pickles must be formulated carefully. Curing-ingredient suppliers have specially designed meat-curing phosphate products that go into water solutions more readily.

Spices and Flavorings

A number of spices and flavorings are often used to season ham and other cured-meat products to give them unique flavor characteristics. Soluble spices, usually on a dextrose or salt carrier, may be dissolved in the pumping pickle to be incorporated into the cured product. Alternatively, some hams such as Prosciutto hams (see "Prosciutto (Parma) Hams" later in this chapter) are rubbed with spices on the surface. Spices commonly used in seasoning various hams are pepper, cinnamon, clove, nutmeg, and allspice (see Table 21–4).

Water

Water is an often overlooked curing ingredient used in pumping- or immersion-curing techniques. The primary functions of water in modern meat-curing practices are:

- To act as a dispersing medium for salt, nitrite, sugar, phosphates, and other curing ingredients.
- To assist in maintaining a moist, juicy end product.
- To compensate for moisture loss during thermal processing.
- To reduce product cost by merchandising products that have water levels greater than were naturally present in the raw meat.

The quantity of water contained in processed-meat products (including poultry) that is in excess of the normal water content in the unprocessed form

Table 20–1. Protein Fat-free (PFF) Values for Determination of Added Water in Cured Pork Products and Labeling Implications[1]

Type of Cured Pork Product	Minimum Meat PFF Percentage[2]	Product Name and Qualifying Statements
Cooked ham, loin[3]	20.5	(Common and usual, i.e., ham, loin)
	18.5	(Common and usual) with natural juices
	17.0	(Common and usual) water added
	<17.0	(Common and usual) and water product; –X% of weight is added ingredients[4]
Cooked shoulder, butt, picnic[3]	20.0	(Common and usual)
	18.0	(Common and usual) with natural juices
	16.5	(Common and usual) water added
	<16.5	(Common and usual) and water product; –X% of weight is added ingredients[4]
Uncooked cured ham, loin	18.0	Uncooked (common and usual)
	<18.0	Uncooked (common and usual) and water product; –X% of weight is added ingredients[4]
Uncooked cured shoulder, butt, picnic	17.5	Uncooked (common and usual)
	<17.5	Uncooked (common and usual) and water product; –X% of weight is added ingredients[4]
Ham patties, chopped ham, pressed ham, and spiced ham	19.5	(Common and usual)

[1]Federal Register. Vol. 49, No. 73, Friday, April 13, 1984. "Rules and Regulations." Also Code of Federal Regulations 9CFR 319.104. 1999.

[2]The minimum meat PFF percentage should be the minimum meat protein that is indigenous to the raw, unprocessed pork expressed as a percent of the non-fat portion of the finished product; compliance is determined under sec. 318.19 of the Code of Federal Regulations.

[3]The term *cooked* is not appropriate for use on labels of cured pork products heated only for the purpose of destroying possible live trichinae. See Chapter 3 and cooking section in this chapter.

[4]Processors may immediately follow this qualifying statement with a list of the ingredients in descending order of predominance rather than having the traditional ingredients statement. In any case, the maximum percent of added substance in the finished product on a total weight percentage basis would be inserted as the X value; e.g., *Ham and Water Product—20% of Weight Is Added Ingredients*. A prerequisite for label approval of these products is a quality control program approved by the USDA under sec. 318.4 of the Code of Federal Regulations. See page 739 for the formula for determining PFF values.

is called *added water (AW)*. The retention of a portion of the water added during the formulation of cooked sausage and loaf items is possible because gels or emulsions are formed (see Chapter 21) or because binder ingredients are used to hold the water.

The retention of a portion of the water added to cured meats by pickle (brine) pumping is possible, following the cooking process, only if phosphate is an ingredient in the curing pickle. Added water in cured meats may increase the yield of finished products to exceed the original weight of the raw material, even after cooking. A USDA labeling regulation, which took effect April 15, 1985, is designed to inform consumers of the added water and added weight in cured and cooked pork and poultry products.

The current cured-meat composition regulation, which was published in the *Federal Register* on April 13, 1984, identifies the presence of added water and increased yield based on the protein fat-free (PFF) method and replaces the previous standards, which also limited added water and other substances. PFF regulations describe the procedures for measuring and set standards for meat protein content in cured pork products on a fat-free basis. The regulations identify product label standards based on PFF values, outlined in Table 20–1, and list the procedures for determining compliance. Cured pork products that do not measure up to the prescribed PFF levels shall be labeled to reflect this and are called "nontraditional" products. The formula for determining PFF values is PFF value = [percent of meat protein ÷ (100 − percent of fat)] × 100.

THE NITRITE CONTROVERSY

Both the food industry and various government regulatory agencies make every effort to insure that the risks associated with the food supply are kept as low as possible. However, attempts to simplify scientific concerns in effective legislation have led to some dilemmas. One such case occurred in 1959 when the U.S. Congress enacted the Food Additives Amendment to the federal Food, Drug, and Cosmetic Act. This amendment, commonly referred to as the Delaney Clause, provided that *no cancer-causing substance or carcinogen could be added in any concentration to the food supply.*

Most scientists would probably agree that the passage of the Delaney Clause was beneficial in 1959, considering the state of the development of the science of toxicology at that time. However, since the enactment of this legislation, technological capabilities to detect minute traces of substances have greatly improved. This has brought about controversies regarding the interpretation of the Delaney Clause because *no* was intended in the legislation to mean *zero,* a finite entity. However, nature does not follow finite rules. If a proven carcinogen is formed from a food additive at "parts per billion" or "parts per trillion" levels (which are now detectable), does it pose a threat to the food supply? Should it be banned outright as a food additive, or should the benefits the compound provides be weighed against the potential risks it may pose? Such has been the case with sodium nitrite, which was very nearly banned in 1980.

Events leading to the nitrite controversy began in the early 1970s when several parts per billion (ppb) of N–nitrosopyrrolidine (NPYR), a known carcinogen in laboratory animals, were detected in some samples of bacon fried at high temperatures (350° to 400°F). Subsequent research found traces (less than 5 ppb) of two other nitrosamines, dimethylnitrosamine (DMNA) and N–nitrosomorpholine (NMOR), present in fried bacon. Nitrosamine formation occurs as a reaction between nitrite (NO_2^-) and secondary amines. Both nitrite and secondary amines occur naturally at varying levels in most food substances. These early findings brought a flurry of intensive research on nitrite and cured-meat products and their subsequent safety as foods.

The Risk–Benefit Concept

The problem of defining the carcinogenic risk of a very low level of a chemical in foods is a dilemma for regulatory officials. The present "yes" or "no" legal framework for accepting or rejecting makes the decision process relatively easy. However, many scientists wonder if it is really in the public's best interest, since there is no room for scientific judgment as to the degree of risk versus the overall potential benefit.

Scientific and historical evidence have effectively proven that the minute, unknown risks of cancer posed by the inclusion of nitrite from meat products in the diet are far outweighed by the important benefits this preservative provides to our society. In addition, nitrites have a proven safety record of thousands of years of use in cured-meat products. Most important, nitrite is the only known chemical substance that can be added to meat that prevents the highly lethal food poisoning botulism (see Chapter 2). Since *C. botulinum* produces a heat-stable spore, it survives the cooking temperatures that are used for most meat products. The deadly toxin is produced only by vegetative (growing) forms of the cell. Nitrite prevents the outgrowth of the microorganism's spores and thus the subsequent possibility of toxin production.

The importance of this inhibitory action of nitrite on *C. botulinum* is often not recognized by the consuming public. Without the use of nitrite, there would be an increase in the number of deaths caused by botulism. Meat products are constantly being temperature-abused (held unrefrigerated for periods of time) either by consumers or during the distribution of meat through the merchandising channels. The brown bag lunches consumed in public schools and in many places of employment often sit for hours without refrigeration. These could provide prime incubation periods for *C. botulinum* growth. Pasteurized canned hams and other meats are of even greater concern. One never knows if these canned products have been temperature-abused during distribution, which would allow for the growth of *C. botulinum* if no nitrite were present. Although "zero risk" or "absolute safety" of food is an unattainable goal, it is nonetheless a worthy goal, and some industries have approached it.

"For the purpose of surveillance, . . . [the Centers for Disease Control and Prevention (CDC)] defined a foodborne outbreak as one or more cases of botulism in which a contaminated food source was implicated. In the period 1899–1949, 477 foodborne botulism outbreaks were recorded in the United States, and in the period from 1950 through 1996, an additional 444 outbreaks were reported to CDC, for a total of 921. The average number of outbreaks per year has changed little, 9.7 per year for the earlier time period and 9.4 per year since 1950.

"For the period 1899–1949, 1281 cases of botulism were reported, and in the time from 1950 through 1996, an additional 1087 cases were reported, bringing the total to 2368 cases. The average number of cases per outbreak has remained constant: 2.6 cases/outbreak in the first half of this century and 2.5 cases/outbreak so far in the second half."[9]

"From 1950 through 1996, 289 (65.1%) botulism outbreaks have been traced to home-processed foods and 31 (7%) to commercially processed foods, including foods served in restaurants. The type of food processing responsible for 124 (27.9%) outbreaks is unknown.

"Vegetables were the most important vehicle for the botulism toxin in the United States from 1950 through 1996. Fish and marine mammals were also responsible for a large number of the botulism outbreaks during this period. . . . Beef, milk products, pork, poultry, and other vehicles caused fewer outbreaks."[10]

---

[9]CDC. 1998. Botulism in the United States, 1899–1996: Handbook for Epidemiologists, Clinicians, and Laboratory Workers. Centers for Disease Control and Prevention, Atlanta, GA. http://www.cdc.gov/ncidod/dbmd/diseaseinfo/botulism.pdf

[10]Ibid.

On July 23 and 24, 1996, the U.S. House and Senate, respectively, passed, and on August 3, President Clinton signed, the Food Quality Protection Act of 1996, which replaced the Delaney Clause.

Other Sources of Nitrite

Fred Deatherage, a renowned biochemist at The Ohio State University, now deceased, reported that the total dietary intake of nitrite or nitrate from cured-meat products represents only 2 to 4 percent of the nitrite getting into the human body.[11] He indicated that certain vegetables have nitrate levels many times that found in cured meats. Celery has 1,600 to 2,600 ppm, radishes 2,400 to 3,000 ppm, lettuce 100 to 1,400 ppm, zucchini squash 600 ppm, and potatoes 120 ppm. Deatherage also determined that the salivary glands and intestines are significant production sites (from bacterial nitrification processes) of nitrate and nitrite in the human body. The average person ingests 8 milligrams of nitrite daily from salivation — more than twice the amount consumed from cured meats! Intestinal nitrite production has been estimated to be approximately 80 to 130 milligrams per day, which has lead some experts to believe that *nitrite might represent a natural chemical defense system protecting the human body from botulism.* For some reason, *C. botulinum* spores do not germinate to active vegetative cells in the gut. Deatherage further stated that "evidence suggests the nitrite produced by the gut (and that ingested from foods) inhibits germination of *C. botulinum* spores in the gut just as it does in cured meats."

One explanation for sudden infant death syndrome, whereby babies die in their cribs for no apparent reason, is that the outgrowth of *C. botulinum* spores to active vegetative bacteria, in the infant's intestine, produces enough toxin to kill the baby. One theory, which needs further study, is that normal bacterial nitrification processes producing nitrite have not been sufficiently developed in the first months of life to protect some babies.[12] The Sioux Honey Association has issued a warning that honey should not be fed to infants less than one year old. Honey has been shown to contain *C. botulinum* spores (see footnote 11).

In response to the finding of nitrosamines in fried bacon, the meat-processing industry immediately set out to investigate corrective-processing methods. A task force was set up to examine bacon-processing methods and

---

[11]Deatherage, F. 1978. Man and His Food and Nitrite. National Live Stock and Meat Board, Chicago.

[12]Ibid. Also M. P. Doyle (Ed.). 1989. Foodborne Bacterial Pathogens.

to determine how these methods could be altered to eliminate nitrosamine formation after the product had been cooked. The task force concluded that reducing in-going nitrite levels in the bacon to 120 ppm and achieving residual levels (after processing) of less than 40 ppm would reduce nitrosamine formation to non-detectible levels or at least trace levels (less than 5 ppb). It was found that the 40 ppm residual levels could be attained by adding the maximum amount (550 ppm) of ascorbate, a vitamin C derivative, which acts as a reducing agent. Also, the residual nitrite level could be further reduced by lengthening the cooking and processing times. These levels (120 ppm in-going) were still sufficient for controlling the outgrowth of *C. botulinum* spores. Most of the bacon-processing industry was using the lower nitrite levels long before they were enacted into law.

Readers can rest assured that cured meats are safe.[13]

MEAT-CURING CHEMISTRY

Fresh Meat Color — Myoglobin

Myoglobin is the meat protein principally responsible for the color of all meat. The myoglobin content of muscle tissue varies with the species, sex, and chronological age of the animal and the particular muscle. The oxygen needs of a muscle largely determine its myoglobin content. The myoglobin content of lean muscle tissue in veal and pork ranges from 0.5 to 2 milligrams per gram of wet tissue. Young beef contains 2 to 4 milligrams of myoglobin per gram of wet tissue, while mature beef contains 4 to 8 milligrams per gram. The myoglobin content of poultry white meat is generally less than 0.5 milligrams per gram, while that of the dark meat ranges from 2 to 4 milligrams per gram. Lean muscle tissue from lamb and mutton contains 4 to 8 milligrams of myoglobin per gram of wet tissue.

Chemical Structure of Myoglobin

Myoglobin is a very dynamic protein that can readily undergo changes in color, depending on its immediate environment. An explanation of how and why these changes occur is difficult without an understanding of the structure of myoglobin, illustrated in Figures 20–1 and 20–2. Myoglobin is a complex protein structure that includes both a protein portion and a non-protein por-

---

[13]National Pork Producers Council. 1997. Safety of Cured Pork Products. FACTS. Des Moines, IA.

tion. Evidence exists that protein conformation changes result in "motion" within the molecule. Myoglobin protein molecular motions can be characterized and classified.[14] The non-protein portion is called a "heme group" and is composed of two parts, an iron atom and a porphyrin ring (see Figure 20–1). For this reason, myoglobin is often called a "heme protein" and is the main source of dietary iron in meat.

Examination of Figure 20–1 reveals that there are six binding sites for the iron (Fe) atom in the heme group. Four of these sites are used in stabilizing the porphyrin ring, one binding site is used to link the protein portion (the globin) to the heme group, and the sixth site is free to interact with a number of chemical elements.

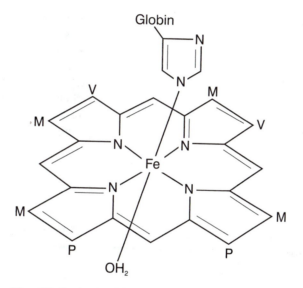

Fig. 20–1. A simplified schematic illustration of the heme group moiety of the myoglobin molecule. *M, V,* and *P* represent methyl, vinyl, and propyl groups respectively.

The color state of meat is largely determined by the chemical compound present at this sixth site and redox state of the iron atom (Fe^{+2} or Fe^{+3}). The physical state of the globin also influences the color complex that is formed. Heat denaturation (from cooking) of the globin results in a brown to gray color of the meat, unless nitric oxide is present at the sixth binding site. Table 20–2 lists the most common pigments (colors) of the myoglobin molecule in meat, the redox state of the iron atom, and the complexed element at the sixth binding site. (Also see the myoglobin figure in the color photo section.)

---

[14]Frauenfelder, et al. 1991. Science. 254:1598.

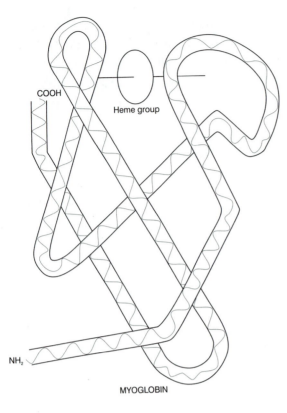

COOH

Heme group

NH₂

MYOGLOBIN

Fig. 20–2. A three-dimensional representation of the complex myoglobin molecule and orientation of the heme group within the molecule. Note the small size (proportionally) of the heme group which contributes the important color properties to meat. The actual myoglobin molecule is much more tightly compacted, and the heme group is held in a stationary position by hydrophobic interactions. The dotted line represents the approximate configuration of the polypeptide chain (chain of amino acid residues composing the protein).

Table 20–2.   Heme Pigments of Muscle

Pigment Forms	Fe Redox State	Element at 6th Site	State of Globin	State of Heme Group	Color
Myoglobin	Fe^{+2}	H_2O	Native	Intact	Purplish red
Oxymyoglobin	Fe^{+2}	O_2	Native	Intact	Bright red
Metmyoglobin	Fe^{+3}	H_2O	Native	Intact	Brown
Denatured myoglobin	Fe^{+2}	H_2O	Denatured	Intact	Brown
Nitrosomyoglobin	Fe^{+2}	NO	Native	Intact	Bright red
Nitrosohemochrome (see text)	Fe^{+2}	NO	Denatured	Intact	Bright red (or pink)
Sulfmyoglobin	Fe^{+3}	H_2S	Denatured	Intact but reduced	Green
Cholemyoglobin	Fe^{+2} or Fe^{+3}	H_2O_2	Denatured	Intact but reduced	Green
Free and oxidized porphyrins	Fe	Absent	Absent	Ring structure destroyed—open chains	Yellow Colorless

Chemistry of Cured-Meat Color

The color of ham and other cured-meat products is one of the principal factors that affect their selection from the retail meat display case in the super-market. A properly cured ham should have a uniform bright pink color that is free of uncured gray spots or other color defects, and it should, under proper packaging methods, maintain this desirable color.

Numerous factors affect the ultimate color of cured-meat products and their ability to retain this color. These factors include the myoglobin content of the muscle, pH, amount and uniformity of nitrite dispersion in the muscle, microbiological condition of the meat, exposure to light and oxygen after processing, and type of packaging employed. To properly understand how these factors affect the color of cured meat, one must understand the chemical reactions that are involved in curing processes.

When nitrite (NO_2^-) is combined with meat, it will eventually be reduced to nitric oxide (NO). Reduction to nitric oxide is often accelerated by the use of ascorbate or erythorbate, which acts as a strong reducing agent. Nitric oxide has a greater affinity for the sixth bind site of the heme group than water does. Hence, the myoglobin pigment is readily converted to *nitrosomyoglobin,* a relatively unstable intermediate in the curing process. Cooking the meat (heat) denatures the globin (protein portion of myoglobin), which in turn, has a stabilizing effect on the bond between nitric oxide and the heme group. The subsequent pigment, called *nitrosohemochrome,* results in the typical bright pink color associated with cured meats. The general reaction responsible for cured-meat color is given here:

$$NO_2^-$$
$$\searrow$$
$$\text{Myoglobin} + \text{NO} \rightarrow \text{Nitrosomyoglobin} + \text{Heat} \rightarrow \text{Nitrosohemochrome}$$
$$\searrow$$
$$H_2O$$

The stability of the nitrosohemochrome pigment is influenced by oxygen tension, light, and other factors. Fading of the desirable bright pink color of cured meats will occur when they are exposed to light and oxygen. For this reason, sliced ham, bacon, and other cured-meat products are generally vac-uum-packaged to exclude exposure to atmospheric oxygen during storage and merchandising. Center-ham slices are sometimes merchandised in an oxygen-permeable film and are turned face down in the display counter to minimize exposure to light and subsequent color fading. However, the recommended practice is to vacuum package cured-meat products when surface color is critical to merchandising. Vacuum packaging is less critical in whole hams and bellies (bacon), since the interior of the product is not exposed to light and is exposed to only minimal oxygen.

Myoglobin Oxidation State and Doneness Indications in Ground Beef Patties

To document that proper endpoint temperatures had been reached when cooking ground beef patties, Kansas State researchers[15] conducted this experiment: Ground beef patties known to develop a normal or premature brown color when cooked to 131°F (55°C) received no treatment or were reduced or oxidized before cooking (see the column headed "Fe Redox State" in Table 20–2). Internal raw appearances were red/purplish-red for normal patties, purplish-red/red for patties with reduced myoglobin, and brown for premature brown patties and oxidized groups. Cooked colors of normal–no treatment and reduced and of premature brown–reduced were more red than those of the normal-oxidized and premature brown–no treatment and oxidized patties. Cooked colors of premature brown–reduced and normal–no treatment were similar, and those of normal-oxidized and premature brown–no treatment were similar. Oxidized pigment produced a premature brown color, whereas reducing conditions produced normal cooked color. Thus, to be assured that beef patties have reached the proper temperature (160°F) before serving, one must **not** depend on the internal color of the patties but on an accurate reading from a thermometer inserted into one or more representative patties.

BASIC METHODS OF MEAT CURING

The basic principles of curing do not differ whether they are accomplished in a processing plant, curing thousands of pounds per day, or in a home, curing the cuts from one or two hogs at a time. Discussion of methods that follow will be directed to a small-scale curing project. Prolonged exposure of

---

[15]Warren, K. E., M. C. Hunt, and D. H. Kropf. 1996. Myoglobin oxidative state affects internal cooked color development in ground beef patties. J. Food Science. 61:513.

meat to salt action results in excessive shrinkage and a high salt content. Therefore, it is important in using any of these methods to follow quantitative measurements and time schedules.

Definition of Curing Terms

Artery curing. Injecting cure into the *femoral* artery of ham and *brachial* artery of the shoulder (see Figures 20–4 and 20–5).

Brine. A combination of salt and water. Also called a pickle.

Cover pickle. The pickle (see definition under "Pickle") in which curing meats are submerged.

Immersion curing. Curing in a cover pickle.

Injection curing. Introducing the pickle into the meat with penetrating needles (see Figure 20–3).

Overhaul. The rehandling or repacking of meat during the pickling period to permit a more uniform distribution of pickle.

Pickle. A combination of salt and water. Also called a brine.

Pumping. The forcible introduction of pickle into a ham by means of a ham pump and needle.

Quick cure. A pickle containing sodium nitrite.

Salimeter (salimeter or salinometer). A ballasted glass vacuum tube graduated in degrees and used for testing the strength (density) or salinity of pickle.

Stitch or gun. A single insertion of the needle in the pump method. The number of stitches or guns given each ham varies with the size of the ham and the strength of the pump pickle. It is more desirable to stitch small amounts in numerous locations than large amounts in a few locations.

Sweet pickle. The addition of sugar to a sodium chloride brine.

Dry Salt Cure

The dry salt cure was the original method employed by our ancestors who practically had to pick the salt out of their teeth. It involved the rubbing and packing of meat in salt for considerable periods of time. This method may be used today to produce salt pork from fatbacks, heavy jowls, and occasionally, heavy sow bellies that are packed or rubbed with dry salt. Salt pork finds favor as a "seasoning meat" with greens or other dishes. Adding 10 ounces of sodium or potassium nitrite to each 100 pounds of salt and using 10 pounds of cure per 100 pounds of pork as it is layered gives the salt pork some color and improves the flavor. (Note the weighing and mixing precautions in the "Nitrate and Nitrite" section of this chapter.) The salt pork should cure for two weeks or more.

Dry Sugar Cure (Dry Country Cure)

The dry sugar cure has proven to be the safest method for individuals and processors who do not have refrigerated curing rooms or the equipment for injection curing. Its chief advantages are that:

- The rate of cure is faster than immersion curing because the curing ingredients are applied directly to the meat surface in their full concentration.

- The curing can be conducted safely at higher and wider temperature variations than is possible with immersion curing.

- The time schedule is not exacting.

- Less spoilage occurs in the hands of the novice or under unfavorable curing conditions.

A simple and time-tested formula consists of mixing 8 pounds of table or curing salt, 3 pounds of cane sugar, 2 ounces of nitrate (saltpeter), and 1 ounce of sodium or potassium nitrite. This is commonly referred to as the *8–3–2–1 formula,* which can only be used for hams. (Precautions regarding the use of nitrate and nitrite are discussed in the "Nitrate and Nitrite" section of this chapter.) A 1989 ruling removed nitrates from curing mixes for bacon, and earlier regulations required that they be omitted from other products. The authors suggest that the nitrate level be reduced by half to prevent the presence of greater than necessary amounts of nitrate and nitrite in the finished product. For each pound of pork, 1 ounce of cure should be used, which will require three separate rubbings for hams at three- to five-day intervals; two rubbings for picnics and butts; and one thorough rubbing for bacon, with a light sprinkling over the flesh side of each bacon after it has been rubbed. For heavy hams (over 20 pounds, 1½ ounces of cure per pound of ham or four rubbings should be used. The rubbed meats should be placed in boxes, on shelves, or on non-metal tables to cure — not in tight boxes or barrels where they will rest in their own brine.

The length of the curing period is seven days per inch of thickness. Since most hams weighing 12 to 15 pounds measure 5 inches through the cushion (see Figure 25–8), they will cure in 35 days; whereas bacon 2 inches thick will cure in 14 days. If the cured cuts remain in the cure for a longer period of time, they will not become any saltier; thus, it is possible to smoke all of them at the same time.

If some salt if forced into the aitch bone joint to guard against bone souring, the curing can be done at a higher temperature (50°F) and in a shorter period of time. This is because salt absorption is more rapid at the higher temperatures.

Box Curing (Pressure)

The pressure method of box curing was a popular method before curing became so highly mechanized. Applicable only to bacon, it was practiced by packers who produced a mild-cured product. The amount of cure used was about ¾ ounce to each pound of bacon. Tight boxes and in many cases ordinary curing vats were used. The size of the box was such that the bacon fit in snugly without overlapping, regardless of the number of rows. A lid fit loosely inside the box or vat, upon which considerable pressure was exerted either by a dead weight or by a screwjack on a crossbar attached to the sides of the container.

The pressure on the bacon caused the brine that was formed to rise to the top and cover the meat, thereby sealing it from the air. With just the right amount of curing salt having been added, the length of time to cure the bacon was of minor importance because it could not become too salty. Packers used this method of curing bacon as a means of storage, allowing it to remain in cure for as long as 90 days.

Hot Salt Cure

A practice followed in some communities and tested at the Pennsylvania Agricultural Experiment Station with success is as follows: Rub the cushion side and butt of the ham with 1 ounce of saltpeter (usually potassium nitrate, sometimes sodium nitrate). Follow immediately with a rubbing of granulated or brown sugar over the entire ham. Allow the ham to absorb these ingredients for several hours. Then, heat sufficient salt so that it is uncomfortable to the hands (wear cotton gloves). Place the ham in the hot salt and cover for five minutes to get the ham in a soft condition. Using a clean, round, pencil-size stick, force the hot salt into the aitch bone joint. Thoroughly rub the ham with the hot salt. An accurate measure would be an increase in weight ¾ ounce per pound of ham. Allow the ham to absorb the cure for five to seven days, rub with black pepper, and then smoke.

Sweet Pickle Cure

Sweet pickle with a salimeter reading of 75° to 85° is recommended for home curing. (A salimeter is a necessary piece of equipment for making up curing solutions of different salinities.) Table 20–3 gives the amounts of the different ingredients and the water necessary to make such pickles. Modern curing mixes will readily dissolve in cold water. Brine should be mixed and stored in clean containers that are resistant to the corrosive effects of the salt/nitrite/phosphate solution. Water heated to the boiling point will dissolve the cure ingredients more readily and may minimize the microbial population in the pickle. However, because the hot curing pickle must be chilled before it is poured over the meat, it is vulnerable to contamination during the chilling period.

Table 20–3. Sweet Pickle Formulations

Salt	Sugar	Sodium Nitrite	Cold Water		Degree of Pickle by Salimeter 40°F
(lbs.)	(lbs.)	(oz.)	(gal.)	(lbs.)	
10	3	¼	4	33⅓	95
9	3	¼	4	33⅓	90
10	3	¼	5	41⅔	85
8	3	¼	4	33⅓	85
8	3	¼	5	41⅔	75
6	3	¼	4	33⅓	70
7	3	¼	5	41⅔	65
6	3	¼	5	41⅔	60

Cold water weighs 8.33 pounds per gallon; hot water 8 pounds per gallon.
Seven pints of salt weigh 8 pounds.
One quart of syrup weighs 3 pounds.
If salimeter is calibrated for reading at 60°F, subtract 0.116 per degree below 60°F.

Containers

Plastic or fiberglass barrels, large stone crocks, and stainless steel containers make suitable curing receptacles. ***Containers that have held any sort of spray material or other agricultural or industrial chemicals must never be used.*** Using metal containers that will corrode should be avoided. Lids and weights must be clean.

Temperature

The best temperature for the curing room is from 35° to 40°F. Unless curing vessels have been previously contaminated with ham-souring bacteria, very little spoilage is experienced at these temperatures. Successful sweet pickle curing can be done at temperatures ranging from 40° to 50°F. Temperatures of 50°F and over are too high for safe pickle curing. The brine will sour and become ropy, indicating bacterial growth, and the hams will develop an off-flavor or will sour around the bone.

Length of Cure

It is necessary to follow rather closely the length of cure prescribed for the different strengths of pickle indicated here: 85° pickle cure: 9 days per inch; 75° pickle cure: 11 days per inch; 60° pickle cure: 13 days per inch.

Hams should be measured through the cushion back of the aitch bone to determine their thickness (see Figure 25–8). Hams measuring the greatest thickness should be placed in the bottom of the curing barrel or vat, and the lighter hams placed on top. When fresh pork from a harvest session is ready to be cured, a system of filling the barrel or vat should be followed. The chilled hams, shoulders, and bacon should be packed into the barrel or vat in the order named, and sufficient cold pickle should be poured over the pack so that it will be covered when the lid is weighted down. Four gallons of pickle will cover 100 pounds of closely packed meat, but 4½ to 5 gallons are necessary if the meat is loosely packed. The thickness of each layer should be recorded on a card or in a book so that the date when the layers are to be taken out can be determined. Overhauling (moving and turning) meat once or twice during the curing period is desirable to permit the pickle to reach all parts of the meat. As an example, a barrel of pork consists of hams 5½ inches thick, on top of which are shoulders 3½ inches thick, over which are bacon 2 inches thick, curing in an 85° pickle (see Table 20–3). According to the schedule of 9 days per inch in cure, the bacon must come out in 18 days, the shoulders in 30 days, and the hams in 50 days. If 5 gallons of water is used to dissolve the 8–3–2–1 formula, making a 75° pickle, it will be necessary to cure at the rate of 11 days per inch. A 75° pickle is preferable, if the curing room temperature does not rise above 45°F.

A large part of the pork spoilage occurring in home-curing projects could be eliminated if 4 to 8 ounces of pickle were pumped into the center of the ham and around the hip joint soon after harvest. A good syringe-type pickle pump costs approximately $50.

Pump Pickling (Stitch or Spray Pumping)

To hasten the introduction of the cure to the center of the ham, processors use the practice known as pumping. This consists of forcing the curing pickle into the center of the ham through a needle attached to a plunger-type pump or syringe or with a mechanically operated pump (see Figures 20–3 and 20–5). The curing of meat cuts is hastened considerable because the curing ingredients diffuse from inside the cut as well as from outside. The pickle is most often introduced by stitch pumping, which penetrates the cut repeatedly and injects small pockets of cure throughout the mass of the cut (see also Figure 20–7). Another technique, artery pumping, will be described later.

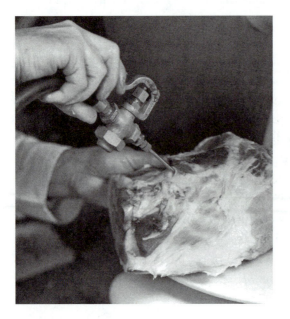

Fig. 20–3. Giving the ham a stitch at the hip joint (pump pickling). Multiple stitches should be given in the cushion, stifle joint, hock end, and butt end.

The strength of the pickle varies, depending on the amount of pickle to be pumped into the meat, the intensity of salt flavor desired, and the environmental storage conditions. Hams should not be stitch pumped with more than 8 percent of their weight of 85° pickle or they will be salty. That would mean that 1½ pounds of 85°pickle is pumped into a 20-pound ham. For a 60° to 75° pickle, the cut may be stitch pumped with 10 percent or 2 pounds of pickle in the 20-pound ham.

Following injection of the curing pickle into the product, time is required for the concentration of the cure ingredients to equalize throughout the product by diffusion. Pumped hams or other cuts may be immersed in a cover pickle of equal or greater strength for 7 to 14 days. Combining pumping and dry curing will cure hams very successfully. The procedure is to use an 85° pickle at the rate of 8 percent of the weight of the ham and then to rub with one-half its usual application or ½ ounce of the recommended formula per pound of ham and cure it 2½ days per pound.

The advantage of pump pickling are:

- The salt is introduced to the center of the ham before spoilage has a chance to take place.

- The curing period is shortened by almost one-third.

- A quicker turnover is effected.

Artery Curing

Artery curing consists of forcing a pickle into the *femoral* artery on the inside butt end of the ham by means of small needle attached to a hose and connected to a pump that exerts a pressure of 40 to 50 pounds (see Figures 20–4 and 20–5). In order for artery pumping to be successful, when the hog is harvested and the carcass is cut, the *femoral* artery must be preserved as it enters the ham, and curing must be done before the vascular system dries out. The artery-pumped hams are either rubbed with the dry mix or placed in a pickle of similar strength for five to seven days to complete the curing. The advantages of artery curing compared to hand stitch pumping are speed and uniform flavor. Artery curing is particularly adapted to small pork processors and those processors doing custom curing because the capital investment in equipment is minimal.

The salinity of the pickle to use will depend upon how long and under what conditions the hams are to be stored. A pickle strength of 60° to 65° on

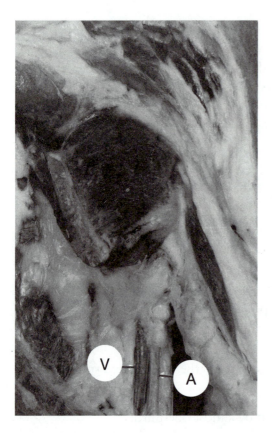

Fig. 20–4. Prior to removing the ham from the pork carcass, carefully loosen the leaf fat over the inside of the ham butt and separate the *femoral* artery A from the fat. Cut the artery long. It differs from the vein V in that it is strong and elastic.

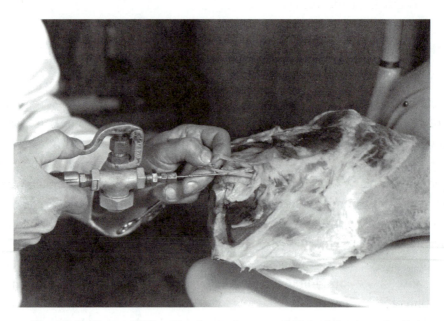

Fig. 20–5. Pump pickling the *femoral* artery ahead of the branch. If the artery is too short, each branch should be pumped separately.

the salimeter is desirable when hams are to be held under refrigeration (less than 40°F). Hams pumped with a 75° pickle, according to the formula recommended in this text, have a longer storage life and a more intense flavor than those hams cured with a 60° or 65° pickle. The artery-pumped hams are given a light rubbing of the dry cure mixture and then shelf-cured for several days before being smoked. The use of an 85° pickle (10 percent by weight) and one thorough rubbing of the dry cure produces hams with more salt flavor and good holding quality.

The Combination Cure

Cured-meat products can be produced from a combination of pumping and immersion in pickle or under a dry rub for a period of time. Another successful method of curing hams or other meat cuts is the combined use of the dry cure and the sweet pickle cure. The quicker the salt gets to the center of the ham, the less danger there is of loss from spoilage. Experiments show that when 1½ to 2 pounds of salt per 100 pounds of pork is rubbed into hams 24 to 48 hours before they are placed in a 75° pickle, they will cure in nine days per inch. This initial rubbing of salt is all absorbed overnight and has more rapid penetrating qualities because it is neither mixed with other ingredients nor dissolved in water.

Presmoke Soaking

All dry or sweet pickle–cured pork should be rinsed or soaked in cold water before it is placed in smoke. The soaking removes the excess salt on the outside and eliminates the formation of salt streaks on the meat when it is exposed to the heat of the smokehouse. Soaking pickled bacon and shoulders one hour and hams about two hours is sufficient. Dry sugar–cured meats cured with 1 ounce of the cure per pound of meat need not be soaked, but they should be rinsed of surface salt. Rinsing is sufficient for most pumped cuts also.

Cured-Meat Shrinkage

There is very little difference in the shrinkage of pork cured by the dry and sweet pickle methods at the end of a 60-day aging period. Sweet pickle–cured hams will gain an average of 5 percent in weight as compared to a loss of 5 to

7 percent for hams dry-cured. Sweet pickle–cured hams lose about 5 percent during smoking as compared to 2 percent for dry-cured hams. Bacon will shrink about 5 percent more than hams because of the large surface area.

Pumped products may hold most of the injected pickle through the presmoking period but lose the added moisture in the smokehouse unless phosphates have been included in the pickle. To be labeled as *"hams"* under federal requirements, the meat, while in the smokehouse, must be shrunk back to the original fresh weight. If they do not come down to their fresh weight, the hams are returned for further heating and shrinking, or they must be labeled *"Ham, Water Added"* or *"Ham and X% Water Added"* (see Table 20–1).

COMMERCIAL CURING PROCEDURES

Mechanized Pumping

Both artery pumping and stitch or spray pumping have been mechanized to a high degree in the meat industry. Figure 20–6 shows the Comcure®,[16] a machine designed to eliminate problems of control, yield, and production in artery pumping hams. Each of the four stations is independent and may be turned off without affecting any of the others. The desired percentage is dialed on the front of each station and reads to tenths of a percent. The operator has no decisions to make regarding the correct amount of pickle. He or she merely places the ham on the platform, clamps the gun onto the artery, pushes the programming start button, and proceeds to the next station. A production figure of 240 per hour is a conservative one to allow for situations involving heavy hams at high percentages. Pumping hams in the 12- to 16-pound range at percentages of 10 to 15 percent will exceed the 240 per hour figure easily. Few ham processors utilize the artery pumping procedures today because most commercially produced hams are boneless.

Most cured-meat products consumed in the United State are pumped with units similar to the one shown in Figure 20–7, which is an InjectoMat®,[17] a machine that stitch or spray pumps through needles that introduce the brine at hundreds of points in a large cut. The pressure of the brine forces relatively uniform brine distribution throughout the meat. Machines like the InjectoMat® are fully automatic. Large (up to 500 pounds or more) combo

---

[16]Vogt, Inc., Clawson, Michigan.

[17]Koch Supplies, Inc., Kansas City, Missouri.

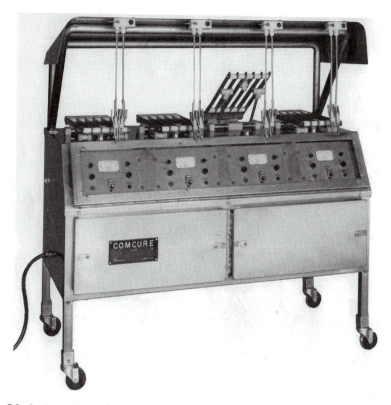

Fig. 20–6. An automatic electronically controlled artery pumping machine. (Courtesy, Vogt, Inc.)

Fig. 20–7. An automatic bone-in pickle injector. *(Left)* An overall view showing volume and stroke adjustment controls. *(Right)* A closeup of InjectoMat® injecting a ham.

bins of fresh — bone-in, semi-boned, or boneless — product are dumped onto conveyors that feed the automated stitch-pumping units.

Needles operate in unison until they strike an obstruction; spring-loading protects each needle when it strikes a bone. A control bar aligns each needle and holds the product in position during the instant of the injection stroke. The quantity of pickle is adjusted by belt speed, volume, and strokes per minutes. Hams are pumped at the rate of 6 to 12 per minute, up to 6,000 pounds per hour. Smaller product units, like beef briskets, may be loaded onto the conveyor belt 2 wide for more than 12 per minute production speed. After injection, the pumped product is conveyed away to a vat or a truck or to a station where the cured meat is loaded onto smoking/cooking racks.

Boneless Hams

Most commercial hams produced today are boneless. Boneless hams are produced in two forms: whole boneless hams and sectioned and formed hams. Traditional whole boneless hams are made from one ham only and are often tunnel-boned. As discussed in previous chapters, tunnel boning is the process of removing the *femur* by tunneling around it with a knife and leaving the ham in one piece, with all muscles attached together. This boning procedure is illustrated in Figures 20–8 to 20–11. The ham may be boned before or after it has been cured, using previously outline methods. If it is boned afterwards, it is important to have a market or method of utilizing the miscellaneous cured trimmings that are created but not used in the ham (e.g., the shank meat).

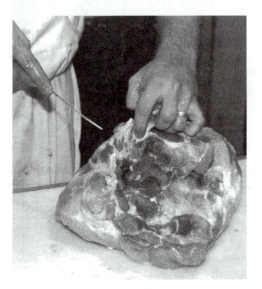

Fig. 20–8. The first step in tunnel boning a ham is to remove the aitch bone (shown here) and the ham's shank through the stifle joint (not shown).

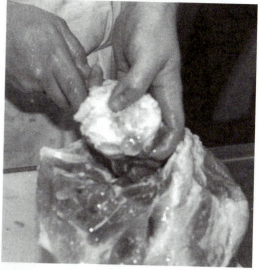

Fig. 20–9. Closely encircle the *femur* with a boning knife to remove it.

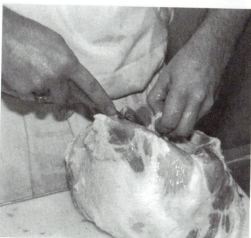

Fig. 20–10. Tunnel boning will remove the internal seam fat to produce a high-quality ham. Note the boneless ham is in one piece and has not been cut open.

Fig. 20–11. Boneless hams are often stuffed into fibrous casings, clipped tightly, and placed in spring-loaded molds to flatten the hams (not shown). This prevents the occurrence of air pockets in the center region of the hams where the *femur* was removed.

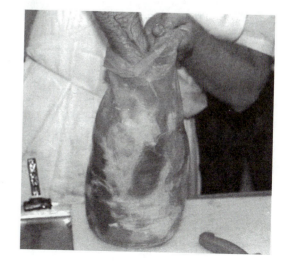

A high percentage of the commercial boneless hams that are produced are the sectioned and formed variety, which includes the majority of the hams that are labeled *"Water Added"* or *"With Natural Juices"* (see Table 20–1). Hams manufactured in this manner are generally boned by seaming out the three major muscle groups of the pork leg (fresh ham) — the knuckle, the inside and outside muscle groups that correspond to the round (sirloin) tip, and the top and bottom round sub-primals of beef illustrated in Chapter 15. Grouping like muscles produces a uniformly colored sectioned and formed ham. The boneless muscle chunks are typically stitch pumped in a high-speed pumping machine similar to the one shown in Figure 20–7.

After the pumping process, the muscles are *tumbled* or *massaged* to extract meat proteins that coat the outer surfaces and function to bind the muscles back together. Both tumbling and massaging accomplish the same function, extracting protein to the surface of boneless hams, but they utilize slightly different principles. Figure 20–12 shows a small laboratory vacuum tumbler. Tumbling relies on gravitational impact (from meat pieces dropping from the top to the bottom of the drum as it rotates) and abrasion against other meat pieces to extract the myofibrillar proteins necessary for binding meat sections together. Massaging relies mainly on abrasion of meat against meat or meat against stainless steel rotating paddles. Most massagers have a heavy, blunt paddle positioned vertically in a square vat. The paddle rotates slowly to gently massage the meat while minimizing tearing of muscle tissue and particle-size reduction. Both tumbling and massaging have a tenderizing effect on the ham muscle and result in greater pickup and retention of moisture by the meat.

Fig. 20–12. This small Vortron® vacuum tumbler (Vortron, Inc., Beloit, Wisconsin) can handle up to 250 pounds of product. The presence of vacuum is reported to reduce tumbling time required for sufficient protein extraction. (Courtesy, South Dakota State University)

SMOKING AND COOKING

Equipment

Cured and smoked products in the retail meatcase are virtually all cooked and ready to eat. The smoking/cooking process is often accomplished in a *heat processing unit,* the modern meat industry terminology for what was once called a *smokehouse.* Using gas, steam, or electric heat, the commercial stainless steel units are often controlled by microcomputers, processing as much as 20,000 pounds of product per batch. The temperature, air speed, relative humidity, smoke density, and length of each stage of the process can be accurately controlled by computer, which activates a variety of pneumatic valves, switches, dampers, thermostats, and smoke generators in response to signals from timers, temperature probes, and humidity sensors.

Most commercial units are connected to smoke generators that discharge into the product chamber. Sawdust or wood chips, metered and distributed on a surface heated by electricity or gas, smolder and smoke. Other units may actually burn wood logs, chips, or sawdust to produce the smoke. Using dampened wood and controlling the oxygen supply may be needed to maximize smoke production, control burning temperature, and minimize flaring. The location of smoke inlets and the air movement during the smoking cycle must produce uniform smoke distribution and environmental conditions throughout the house to assure finished product uniformity. Many smoking units have atomizing nozzles in the chambers to introduce *liquid smoke* (discussed later under "Smoking"), which is absorbed by the meat surface in the same manner as the components of the combustion-produced smoke. The large commercial smoking chambers must be engineered to limit atmospheric emissions to levels designated by the Environmental Protection Agency or by local agencies. Heat processing units may be required to use liquid smoke in order to meet pollution standards.

Heat processing units offering wide range of humidity conditions must be equipped with steam or water atomizers or sprays to introduce moisture into the chambers during the various stages of processing. Many units are supplied with refrigeration coils to prevent excessive temperature rise when cool-smoking the uncooked product or to provide rapid cool-down of the finished product. In large processing plants, heat processing units may be teamed with equal-sized, high-capacity cooling units into which the hot, cooked product is directly transferred for quick chill-down. Small processors and larger processors of specialty products continue to use brick or tile smoking/cooking units, sometimes referred to as chimney smokehouses. However, many country-cured products today are processed in stainless steel units.

Smoking/cooking units have been constructed by private citizens for "home processing" meat products since the days of the early Egyptians. Relatively large volumes of meat for family use can be smoked, aged, and stored (country-cured products that do not require refrigeration) in "houses" with 40 to 50 square feet of floor space and 9 feet of vertical clearance between the floor and the rods or beams, upon which the product is hung. Fireproof brick or tile construction provides insulation benefits, but tight wood construction is acceptable if adequate fire prevention measures are taken. A concrete floor and tightly screened ventilation openings will prevent rodents and insects from contaminating the products. Ventilation is essential, especially if the unit is to be used for drying, aging, and storage. If cooking is to be done in the unit and the smoking fire is to be the heat source, a grating, firebox, or burner is necessary. The meat should not hang directly over the fire or other heat source; therefore, the fire should be in an offset or beneath a heavy sheet metal baffle plate. Cool smoking can be accomplished by "piping" smoke into the house from an external smoke generator, firebox, or grating. Smaller units may be made from wooden or steel barrels or drums into which smoke is piped. Some arrangement for hanging products (two hams or bellies at a time) and a lid with a humanpower-operated vent are necessary. Old refrigerators may also be used in a similar fashion and have the added advantage of insulation. Before the product is smoked in a modified refrigerator or a similar chamber, all components should be determined to be physically and chemically stable to the heat and smoke. The refrigerator or similar box must also be vented. A firepan inside the unit or an external source of heat and smoke may be used with the refrigerator or a similar chamber. *When not in use, the refrigerator-smoker doors should be locked open or closed. Removing the latch or the door will also prevent both humans and animals from being accidentally trapped in the chamber.*

Smoking and Cooking Procedures

The sequence and timing of the various stages of smoking and cooking meat products as well as the product, the equipment, and the processor will determine the specific procedures utilized. Cured meats that are smoked and cooked may be subjected to surface drying, smoking, cooking, tenderizing, showering, and cooling stages of a single processing cycle in a modern heat processing unit.

Drying

It is critical that the surface of the product be slightly moist before the product is smoked. If the surface is too wet, the smoked product will have a streaked appearance. If the surface is too dry, the product will be pale in color because the smoke volatiles will adhere poorly to the product. Therefore, the product should generally be exposed to a short drying period prior to being smoked. Moderate to low chamber temperatures, 100°F ± 20° (37°C ± 11°); low relative humidity, ≤45 percent; and rapid air movement in a vented chamber for a short time, 30 minutes or less, will remove excess surface moisture from the product. Smoke will not adhere to dry surfaces, so drying cycles must be adjusted to remove the excess moisture but to leave the surface moist enough to absorb smoke. Experience with each product in the individual heat processing unit will establish the optimum combination of drying conditions.

Smoking

The purpose of smoking most meat products today is to impart a unique smoke flavor and aroma and to give the surface a uniform smoked color. Country-cured hams and similar cured and smoked meats that do not require refrigerated storage owe their stability to the combination of low moisture and high concentrations of curing agents and heavily smoked surfaces. The length of time that the product is to be subjected to smoking will be determined by the type of product, the density of natural or liquid smoke generated in the unit, and the affinity of the meat surface for the smoke.

Natural wood smoke is generally produced from hardwood sawdust, wood chips, or logs. Although the wood of the hickory tree is the most popular source of smoke, other hardwoods, such as oak, maple, ash, mesquite, apple, cherry, and other fruitwoods, are commonly used. Pine and other coniferous trees should be avoided as a source of smoke because of their high tar content and bitter flavor. The smoke may be produced from an elaborate electronically controlled smoke generator or from a variety of much simpler versions, ranging from log burning to humanpower-controlled smoke generators.

Natural wood smoke is composed of three principal phases: (1) *solids* — fly ash and tar, (2) *noncondensibles* — air and combustion gases, and (3) *condensibles* — acids, carbonyls, phenolics, and polycyclic hydrocarbons. The solid and noncondensible phases do not contribute significantly to the flavor, aroma, or preservative properties of smoked products. Dr. Hugo Wistreich, former president, B. Heller and Co., Chicago, identified the *phenolic fraction* as the primary source of the smoky aroma and flavor.[18] The phenolic fraction is also thought to be responsible for the preservative properties of wood smoke. The *carbonyl fraction* has been demonstrated to be the source of the desirable amber-brown color generated during the smoking process. German research suggests that "the aroma of smoked meat is due to a whole spectrum of compounds found in wood smoke" and does not result from the presence of the phenols only.[19] Wood burned at 1,100° to 1,400°F (600° to 750°C) optimizes the production of smoked meat aroma compounds. The amount of smoke deposited on the product surface, under given environmental conditions, will be determined largely by the length of time the unit is filled with smoke. Variation in smoke density within the chamber will also affect the degree to which the smoke components adhere to the product surface. At the end of the smoke cycle, usually determined by time or by observation of the product, smoke generation ceases, and the vents are opened to purge the unit of smoke.

Smoke may be applied to the surface of products in one of two forms: natural and liquid. Liquid smoke (sometimes inappropriately called "artificial smoke") is manufactured from the condensible fraction of natural wood smoke through a water permeation–distillation process. Through further processing, the essential volatiles are refined and marketed in various forms (e.g., oil-based, water-soluble, or dry on a dextrose or salt carrier). The application of liquid smoke in meat processing is a well-established practice. The principal methods of application are (1) *drenching* or *dipping*, (2) *atomizing*, and (3) *regenerating*. High-volume continuous frankfurter lines often dip the franks into an acidic bath containing liquid smoke. In addition to imparting flavor, the acidic liquid smoke bath acts to "set the skin" prior to thermal processing, thus greatly aiding the peelability properties of the finished product (see Chapter 21). Atomizing the liquid smoke into very fine particles in a modern smokehouse is also a common method of application. Smoke is regenerated by superheating the atomized liquid smoke prior to spraying it in the modern smokehouse.

Liquid smoke or granular dry smoke on a dextrose carrier may be added directly to the formulation of processed meat products to impart a uniform smoky flavor. When liquid smoke is added, not more than ⅛ ounce liquid smoke per 100 pounds of meat should be used. Liquid smoke is extremely purified and potent (particularly the oil-based type) and as such will impart an unacceptable bitter flavor at high levels.

---

[18]Wistreich, H. G. 1977. Smoking of Meats. Proc. of Meat Industry Res. Conf., Chicago, p. 37.

[19]As reported by A. M. Pearson. 1989. The Natl. Provis. (January 14).

Cooking

Since heat enhances product surface drying, which in turn reduces smoke uptake, only a moderate temperature rise, if any, should occur during the smoke cycle. Following the smoking phase of the processing sequence, the temperature within the heat processing unit should gradually be increased during the cooking phase. A rapid increase in environmental temperature may result in drying and overcooking of the product surface, termed *case hardening,* before the desired internal temperature is reached. As the chamber temperature increases in increments, heat is conducted through the product to minimize the difference between surface and internal temperature. Therefore, the maximum temperature attained inside the heat processing unit is lower ($\pm200°F$) than the temperature used in most cooking operations. Tenderization is maximized by the slow cooking, particularly if the relative humidity is high, resulting in long, slow cook cycles for "tenderized" cooked products.

The product will determine the environmental conditions in a heat processing unit during the cooking cycle. Moist, high-yielding finished products will be cooked in high-humidity, minimum air velocity closed units; whereas drier products will be cooked at lower humidities, with higher air velocities and open vents to remove moisture from the system. The drying rate must be controlled to avoid excessive surface drying, which may reduce the unit's ability to remove inner moisture. Injecting steam into the house or introducing water to humidify the warm air in the heat processing unit can reduce surface drying as well as cooking shrink. Fans move and mix the air throughout the chamber to maintain uniform temperature and humidity. Closing or regulating the vents to prevent or control the loss of heat and humidity to the environment will minimize or control shrinkage.

Uniformly shaped products may be cooked according to a series of timed intervals at specified temperature, humidity, or air movement conditions in the chamber. Experience in cooking each product in the individual units will help cooking operations to establish an optimum timed cooking schedule. To accommodate non-uniform products that may vary in composition, initial temperature and shape, most cooking operations utilize probes in products to determine when target internal temperatures have been reached and when the cooking cycle should be terminated. Cooking cycles may be programmed to change conditions, such as temperature and humidity in the cooking chamber, based on probe temperature as well.

Temperatures Required to Achieve Certain Finished Product Characteristics[20]

USDA-FSIS allows terms such as "Ready-to-Eat," "Cooked," "Fully Cooked," "Thoroughly Cooked," and "Ready to Serve" to be marked on heated and smoked products provided the products show cooked characteristics — i.e., partial meat separation from bone, easy tissue separation, and cooked color, texture, and flavor. This usually requires a minimum internal temperature of 148°F.

Specifically:

- **Completely Cooked** — This term has been approved for use on labels identifying ham commodities reaching an internal temperature of 158° to 162°F during processing. This heat results in a product that is relatively dry and quite pliable, with the tissues separating readily and with the fat rendered to a marked extent.

- **Ham, Cooked** — The product is cooked in conventional manner. If labeled "Completely Cooked," the ham should reach an internal temperature of 158°F.

- **Poultry, Fully Cooked, Ready-to-Eat, Baked, Roasted** — If uncured, the meat must be cooked to 160°F; if cured and smoked, it must be cooked to 155°F.

Showering and Chilling

Smoked and cooked products are normally showered or washed immediately following heat processing to remove salt and fat streaks from the surface. Cold water also initiates the cooling process and moistens the surface. Showering should be sufficient to produce a uniform surface appearance but should not bleach or remove the smoked color.

Cooked products should be quickly cooled to 40°F (5°C) or less. Cooked-meat chilling units must have high heat-removing capacity, while at the same time minimizing shrinkage. Maintaining sanitary conditions and avoiding contact with uncooked meat or surfaces that have been in contact with uncooked product will help lessen recontamination of the surface of cooked product with spoilage organisms.

---

[20]USDA-FSIS Directive 7370.2; provided by James Anderson, USDA-FSIS Inspector, University of Illinois, Urbana.

CURED AND SMOKED PRODUCTS

Bacon

The pork belly is trimmed into a rectangle (described in Chapter 14), cured according to the methods described previously, and partially cooked and smoked. The belly may be processed with the skin (rind) on, but it is most often skinned before it is cured.

Following the heating and smoking process, the bacon slab is routinely sliced into narrow, thin slices, cut perpendicular to the long axis of the slab. Making consumer packages to be displayed in the retail meatcase entails laying each slice flat on a paper "board," with adjacent slices partially overlapped in a *"shingled"* pattern. Most consumer packages contain 12 or 16 ounces of product that is usually partially visible through the vacuum-packaging material. Larger "family packs" as well as more compactly boxed and vacuum-overwrapped packages may also be available. Institutional (HRI) bacon slices may be placed as single slices on coated paper layered in boxes containing as much as 10 pounds each. Bacon may also be merchandised as unseparated sliced in bulk boxes.

Bacon must be cooked — fried, grilled, cooked on a broiling rack, or microwaved — before it can be consumed. Because overcooking, "charring," bacon may increase the possibility of nitrosamine formation, it should be avoided (see "Nitrite Controversy").

"Slab bacon" (unsliced) is available for consumers or food service users who may prefer to slice the product themselves. Users who add it as a flavoring to other foods may prefer to portion it in a form other than slices. The slicing operation produces "bacon ends," which are non-uniform slices or chunks produced from the ends of slabs. The bacon ends are lower in cost and are useful as a flavoring agent.

Tightly rolling the belly into a cylinder and forcing it into an elastic netting tube or casing or tying it before it is smoked and cooked will yield round bacon. The heating process "sets" the shape, and the resulting slices will be round and will fit on a bun.

Salt pork may be produced from fresh bellies or other pork cuts by one of the curing methods discussed earlier. This product is not smoked or cooked. It is cooked at the time of use and may be fried or added as a flavoring in the preparation of other foods, such as vegetables.

Beef Bacon

The meaty portions of lean, boneless beef plates may be cured, using the methods described earlier to produce beef bacon. Compared to pork bacon, beef bacon has greater lean content, minor flavor variations, slightly coarser lean texture, and increased chewiness.

Lean or Formed Bacon

Bacon-shaped slices of cured pork, beef, or turkey that contain significantly less (50 percent less, according to the labels) fat than conventional bacon products are available in the retail meatcase. Lean and fat are subjected to particle-size reduction (grinding, flaking, cubing, chunking), mixed with curing ingredients, and formed into slab-like shapes (see "Restructured Meat Products" in Chapter 18). The slabs are cooked and smoked (either in a heat processing unit or by the addition of liquid smoke), and then sliced, and the slices are vacuum-packaged into bacon-style retail units. The fat and lean composition of the product can be formulated closely, avoiding the relatively wide biological variation between animals and between locations within the belly or plate that are encountered in conventional bacon.

Canadian Bacon

The boneless loin eye muscle *(longissimus)* and(or) sirloin muscle *(gluteus medius)* from heavy pork carcasses (sows) are mildly cured, using one of the methods described earlier. Most commercial processors brine pump the loins, follow with a tumble or massage cycle, stuff into permeable casings (see Chapter 21), and then fully cook and lightly smoke them. A long, slow cook cycle, complemented by the brine, tenderizes the more mature loin muscle.

Moderately thin slices of Canadian bacon are fried or broiled and served with eggs and(or) pancakes or French toast as part of a breakfast menu. Thinly sliced Canadian bacon is a popular high-quality meat topping for pizza. The tender, delicately flavored, cured pork product may also serve as an entreé in a sliced form similar to ham.

Corned Beef

Historically, beef preserved by the additional of salt was called corned beef because grains or "corns" of salt (sixteenth century terminology) were added to the beef. Today, corned beef generally refers to the boneless brisket muscle that has been cured by one of the methods described earlier in this chapter. Other muscles may serve as raw material for corning. The addition of garlic, allspice, bay leaves, and whole peppers or other spices is also common in the production of corned beef. Corned beef is usually not cooked when sold and will have to be cooked by the user. In the retail meatcase, corned beef is often packaged in strong, clear bags that enclose the product and a small amount of the brine and whole spices used in processing the product. Slow, moist-heat cooking will enhance tenderness and flavor of this hearty product. Corned beef as a main dish item is relished by many consumers. When sliced thin, across the grain, it is an excellent sandwich meat. It is the basis of the popular *Reuben* sandwich.

Cured Beef

Beef muscles [including the closely trimmed top and bottom round, the round (sirloin) tip, and the larger muscles of the chuck (see Chapter 15)] may be pumped with a curing brine (pickle) and tumbled or allowed to stabilize in brine. Cured beef is fully cooked in a heat processing unit with or without smoking. The color is darker red, the flavor is richer, and the texture is chewier than that of ham. When thin sliced, cured beef can be used alone or with other ingredients for sandwiches and snacks. It adds color and variety to meat trays and buffet presentations.

Dried Beef

Dried beef raw material and processing procedures are the same as those described for cured beef. As the name implies, dried beef has a lower moisture content than cured beef. The moisture–protein ratio in dried beef must be no higher than 2.04:1; whereas, a moisture–protein ratio of approximately 4:1 is required in cured pork products. Reducing the humidity and cooking temperatures during the cooking process will extend the cooking interval and increase the moisture loss. Dried beef is smoked to varying degrees, ranging from a hint of smoke aroma and flavor to a well-blackened surface. Drying periods in temperature- and humidity-controlled chambers following the smoking and cooking processes will slowly reduce the moisture content. Although relatively dry, modern dried beef products should be held in refrigerated storage. Dried beef is fully cooked and ready to eat. *Buendnerfleisch,* which originated in Germany, is a pressed and dried beef round that is cured with wine.

Beef Jerky

The term *jerked beef* is derived from the Spanish word *charqui*, which describes dried meat strips. The product name eventually evolved to *jerky,* which today may be produced using a combination of curing, smoking, and drying procedures. The raw material is lean beef cut parallel with the fibers into thin strips, approximately 1 inch wide, ⅛ inch or less thick, and of various lengths. Dry or immersion curing (marinating) overnight in a refrigerated chamber will penetrate the narrow strips. The salt–nitrite cure mix is often supplemented with other flavorings and spices in the production of jerky. The home meat processor may find using prepared meat sauces or dressings and small amounts of liquid smoke more convenient as sources of preferred flavors. The cured strips should be rinsed of excess curing ingredients, spread on a screen, and then placed in the smokehouse to be smoked, cooked, and dried. The relatively large surface area reduces the length of time to smoke and cook jerky as compared to other products. The drying and cooking should be done slowly to prevent excessive surface crying, called case hardening, which could seal in deeper moisture, thus slowing the drying and toughening the final product. The home-processed product may be cooked and dried on racks or a screen in a cool oven (200°F) with the door slightly ajar. Properly dried jerky is shelf-stable (does not need refrigeration) if packaged to prevent rehydration and contamination from the environment during storage. Jerky processing techniques may be used for strips of game, lamb, and mutton meats as well as for beef.

Jerky strips are also know as "sticks." Different type of sticks can be made using the same formulae and procedures discussed in Chapter 21 for dry and semi-dry sausages. At the University of Illinois Meat Lab, manager Chuck Stites uses the semi-dry summer sausage recipe and procedure, but forces the mixture through a flattened stuffing horn so that flat strips come out onto a stainless steel screen for cooking and drying in the smokehouse.

Hams

The commercial production of hams — bone-in, boneless, chunked, and formed — was discussed in detail earlier in the chapter. Table 20–1 and the related text outline the various label options for hams and other commercial cured pork products permitted by the meat inspection regulations. Boneless pork and poultry (turkey principally) are cured and merchandised within the various ham compositional categories as round or oval-shaped roasts.

A number of imported and domestic speciality hams are also available to cured-meat consumers. Most of the specialty hams are dry-cured products, but each has unique traits or a background that may be of interest to the reader.

Canned Hams

Canned hams are stitch-pumped, boneless ham portions that are sealed, under vacuum, and cooked in the cans. The product weight is adjusted, usually by hand, to the weight specified on the label. Canned hams will often be labeled *"Ham, with Natural Juices"* because the sealed system inside the can prohibits the loss of moisture by evaporation during cooking (see the discussion on canning in Chapter 19).

Country Hams[21]

The curing procedure for country hams was discussed earlier as the dry sugar cure. Because the hams are not cooked, the fresh pork legs (hams) used in country ham production must be verified as being "trichina-free" (see Chapter 3), as defined by the USDA–FSIS meat inspection regulations. Country hams must also be dry-cured, must contain at least 4 percent salt, and must shrink at least 18 percent of their fresh weight during processing.

Country hams are cured for 30 to 45 days under refrigeration. Country ham processors may cure hams in stacks by layering the hams 4 feet or higher in the curing coolers. The hams are overhauled at least twice during the curing period. If country hams are smoked, the smoking stage may vary in length from two to more than seven days, due to differences in the intensity of the smoke, in the number of hours of smoking per day, and in customer preference. The properly smoked country ham has an amber or mahogany surface color. The temperature of the house is maintained below 100°F (38°C), preferably 70°F to 90°F, to prevent ham spoilage during the smoking cycle.

---

[21]Dr. Dwain H. Pilkington, Red Meat Extension Specialist, North Carolina Cooperative Extension Service, contributed to this discussion.

A three- to four-month aging period completes the country ham processing cycle. Physical and chemical changes occurring during the aging period result in the flavor, texture, and appearance typical of country hams and other dry-cured meats that undergo a long post-processing aging period. Studies suggest that the presence of curing salts stabilizes *cathepsin enzymes* in the muscle through the curing process.[22] These enzymes are then present to act upon muscle proteins in the aging period and may have a significant role in the development of the characteristic flavor of these unique products.

Some country hams are rubbed with finely ground black or red pepper at a point in the process, which may vary from before the hams are smoked to after the hams are aged and are ready to sale. The finished hams are dry, very firm, with dark red lean and yellow fat. Whereas commercial ham processors have modified the salt content and process to result in less intense flavor, country hams maintain the rich, distinctive flavor that makes them unique. Before being eaten, country hams should be *parboiled* (cooked in water, preferably with several changes of water) to remove some salt and to soften the hams. The hams may ben be baked, sliced and fried, or cooked as any other hams before they are served.

"Virginia Ham" and "Smithfield Ham" are examples of products that may be somewhat unique due to the raw material and processing techniques employed. These products must be produced in Virginia or in the community of Smithfield, as specified by a Virginia law passed in 1925. The hogs, from which the early American hams made in Smithfield, Virginia, originated, were produced in the peanut belt of Virginia and North Carolina. In the early days of Virginia ham production, curing and smoking these classic hams was considered to be an *art*. According to folklore, in those days the hogs were harvested in November at a time when the points of the moon were up, which would assure success. At other stages of the moon, the ham was likely to lose weight, wither, and result in a product unfit to eat.

Imported European Canned Hams

Denmark is a major exporter of pork products, particularly canned hams. Depending upon the internal economy, Poland has periodically been a major source of imported canned and cured pork products as well. "Krakus Ham" is an example of the Polish imported canned and cured meats. Cured products entering the U.S. market must comply with the same compositional regula-

---

[22]Toldra and Etherington. 1988. Meat Res. 23:1.

tions (see Table 20–1) as domestically produced products and must be produced in plants that comply with the sanitary, structural, and production restrictions imposed by the FSIS (see Chapter 3). The various imported canned hams may have distinctive flavors or other characteristics related to production techniques that are unique to the countries of origin. They offer the cured-meat consumer increased variety.

Prociutto (Parma) Ham

Prosciutto hams are dry-cured, uncooked, bone-in, aged hams that have been produced in Italy for centuries. The aitch (pelvic) bone is removed, and the hams are kneaded and flattened to approximately 2 inches thick and then shaped like pears or chicken drumsticks. After being seasoned with pepper, nutmeg, and other spices, the hams undergo a year-long processing and aging sequence. Companies producing Prosciutto hams in the United State use the traditional practices. One Italian company selected the location for a plant in the United States because the environmental conditions (temperature and humidity) were similar to those in the Prosciutto processing area of Italy.[23] Prosciutto hams from the Parma region of Italy are said to be the ultimate in quality because of the area's environment and the effect it has on the aging of the hams. Italian laws define the boundaries of the area in which hams identified as Parma hams can be produced, and a consortium of producers are chartered by the government to use the Parma name and label.

Italian hams were banned from the United States from 1968 to 1988 because of the threat of importing African swine fever with pork products. In 1988, U.S. animal health regulators determined that the production procedures required by the Parma consortium were strict enough to insure that exotic animal disease would not be introduced with the product. Since that time, only Parma hams that have undergone a processing and aging period of at least 400 days may be imported from Italy.[24]

Westphalian (German) Hams

The unique flavor of hams produced in the Westphalian area of northwest Germany is the result of smoking the hams with juniper twigs and berries

---

[23]Pietraszek, G. 1988. Virginia town vies with Italy for production of Proscuitto. The Natl. Provis. (October 8).

[24]Thompson, K. 1988. Parma ham: On its way. Meat and Poultry (formerly Meat Industry). (October). p. 20.

over a beechwood fire. Following a combination of dry and immersion curing, the hams are aged or ripened for a month under refrigeration. Then the hams are smoked for a period of seven or eight days.

Loins (Pork)

Boneless and bone-in loins are lightly cured, cooked in a heat processing unit (smokehouse), and often smoked. The whole loins may be vacuum-packaged and wholesaled to food service establishments and retailers. Chops are cut from the loins, to be reheated before they are served as an elegant meat main dish. Processors also cut the chops at the plant and vacuum package them for retail and food service markets. Menus often identify these cured and smoked pork chops as *Windsor Chops*.

Shoulders (Pork)

Although most cured and smoked pork shoulders produced today are boneless, they may be cured with the bone in. Bone-in cured and smoked shoulders are processed in the same manner as hams and may be utilized as roasts and steaks similar to the way ham is served. The cured shoulder is a less desirable cut than ham because it has a lower meat-to-bone ratio and more intermuscular seams. Removing the shank or hock portion and boning will improve the cured shoulder. After the boned shoulder has been pumped with cure, it is tumbled, rolled, and stuffed into elastic netting or fibrous casings to be smoked and cooked. The ham-curing procedures described earlier in this chapter are all appropriate for boneless shoulders. The cured and smoked, boneless, rolled shoulder may be used in the same manner as a boneless ham would be used. The shoulder includes a larger number of small muscles that have diffuse orientation. Connective tissue and fat seams are more numerous in the boneless shoulder than in the ham, thus resulting in a less tender product.

Boston Butt (Blade Boston, Boston Shoulder)

The pork shoulder may be divided into the wholesale Boston butt and the wholesale picnic shoulder (see Chapter 14). The Boston butt is the dorsal portion of the shoulder and contains a large portion of the *scapula* (blade bone). The Boston butt may be cured bone-in, or it may be boned and rolled before it is cured in a manner similar to that for the whole shoulder.

Cappicola

Cappicola are boneless cured shoulder butts (Boston butts) that are lightly rubbed with pepper and other seasonings. Smoking followed by a drying period gives this product a unique aged flavor.

Picnic Shoulder

The ventral or shank portion of the shoulder (also known as the *cala* — see Figure 14–59) has a lower ratio of muscle to bone and a more diffuse muscle directional orientation than the blade portion. In addition to being an economical cured pork cut, the cured bone-in picnic shoulder can serve as a source of cured-meat flavoring for vegetable dishes and as a source of "ham bone" flavor for soups and stocks that are used for gravies and sauces. If a fresh picnic is to be served as cured and smoked roast or slices, the shank or hock should be removed, and the rest of the picnic boned before it is cured. The cured boneless picnic should be rolled and netted or stuffed into a casing before it is smoked and cooked.

Specialty Cured Pork Products

Spare Ribs and Neck Bones

Immersion curing is employed for cuts with high bone content. The curing time is brief because of the limited thickness of lean that the salt must permeate. The cuts are then placed on racks to be smoked and cooked. In some cases, these cuts may be cooked without being smoked. Although cured and smoked spare ribs can be eaten as they are, additional cooking under moist-heat conditions will further tenderize the lean and increase the ease of separating it from the bones. Neck bones may be served like spare ribs, but they yield less edible meat. They are also excellent sources of cured "ham bone" flavor in the cooking of other foods. The lean and bones will separate; thus, the bones can be removed before the food is served.

Hocks (Ham Hocks)

Hocks, skin on, may be removed from hams or shoulders after being pumped with cure, or they may be immersion-cured after they have been removed from fresh hams. They may be smoked and cooked on racks. Ham hocks are used to complement the flavor of other foods, such as beans.

Jowls

The same techniques that are used to cure bacon are used to cure jowls. Jowls may be utilized as *salt pork* without being smoked and cooked, or they may be smoked. Cured and smoked jowls are a low-cost bacon-like product most often used as a flavoring agent.

Poultry

Whole birds may be cured by immersion or pumping. Milder cures with lower salt levels may be preferred in the milder-flavored turkey, capon, duck, and goose meats. The whole birds are bound in netting, which will hold the wings and legs against the carcass and will facilitate hanging the birds in the smokehouse. Controlling the smoke intensity and the duration of the smoke cycle will affect the smoke flavor. Hollow birds will cook more rapidly than solid hams and other cuts. Boneless breasts and bone-in or boneless legs (thighs and drumsticks) may be cured, with smoking and cooking done in netting or on racks, as indicated for neck bones. Differences in salt or smoke flavors may be desired in the light and dark meats. Fabrication of the carcasses into separate cuts and then trimming the fat may be preferable for birds with significant amounts of carcass fat (ducks and geese). The low melting point of the fat may result in excessive grease drip and streaking of the smoked surface during the cooking phase when the birds are processed as whole carcasses.

Game

Most game animal and fowl meats may be cured and smoked to enhance their keeping qualities or to provide consumers with a variety of eating alternatives. The level of salt and other curing ingredients may be adjusted to provide the effect most desired by the individual. As indicated earlier, adding spices to the dry-curing mixes or to the curing brines will render other flavors. The added ingredients should be soluble in water or in dilute salt solutions if immersion- or injection-curing methods are to be used. Seasonings, such as spices, may be rubbed onto the surface of cured meats before or after heat processing to impart unique flavors to the product.

21

Sausages

This chapter will present the history of sausage and relate the role of sausage in our modern lives. Then, it will discuss sausage identification, terminology, ingredients, meat emulsions, sausage casings, and the principles of sausage production. Other topics will include product defects, least cost sausage formulations, and preblending. The chapter will close with a description of a modern sausage kitchen.

Sausage is one of the oldest known forms of processed food and has been highly prized throughout various stages of history. The word *sausage* is derived from the Latin word *salsus* meaning *preserved,* or literally, *salted.*

The earliest known recorded reference to sausage was in Homer's *Odyssey* written in the eighth century, B.C. There is definite historical evidence that the ancient Babylonians produced and consumed sausages almost 3,500 years ago. Sausage became so popular during the wild festivals in Rome that as the Christian Era began, it was banned. Strong public protest eventually forced a repeal of the ban.

By the Middle Ages, many varieties of sausages were being produced throughout Europe. Each variety became unique, dependent on the climate of the region and the availability of various spices. The warmer climates of Italy, southern Spain, and southern France led to the development of dry and semi-dry sausages. The cooler climates of Germany, Austria, and Denmark produced fresh and cooked sausages because preservation was less of a problem.

It was during this period (the Middle Ages) that spices became very important and valuable commodities. Black pepper was in such demand by sausage makers that it was traded as a form of currency. The discovery of the Western World was indirectly a result of the popularity of sausage throughout Europe and the essentiality of spices for its production. When Columbus made his eventful voyage, he was searching for a shorter route to the Spice Islands of the Far East to obtain spices used in the production of sausage. It is remarkable and ironic that he should land in the area of what is now known as Jamaica, the one place in this hemisphere where spices flourish.

Partly because of the increasing availability of different spices from the Far East, sausage makers, or *wurstmachers,* became very skilled at creating distinctive types of sausages throughout Europe. Many sausages became known by the name of the city or town from which they originated. Specific examples include Genoa and Milano salamis from Genoa and Milan, Italy; bologna from Bologna, Italy; frankfurters from Frankfurt, Germany; and braunschweiger from Brunswick (Braunschweig), Germany.

The early Native Americans in the United States produced a form of sausage known as pemmican by combining dried berries and meat into a cake and then smoking or sun drying them. However, the U.S. sausage industry did not really develop until during and after the Civil War, when the Industrial Revolution began. With the Industrial Revolution came many jobs and many European immigrants. During this time period, a great influx of German, Italian, Polish, Dutch, Danish, and numerous other peoples brought with them Old World sausage recipes and skills to satisfy their ethnic tastes.

In 1901, while attending a baseball game in New York, cartoonist Tad Dorgan "heard vendors hawking red hot dachshund sausages (developed first in Germany as 'dachshund' or 'little dog' sausages). Not knowing how to spell 'dachshund,' Dorgan simply wrote 'hot dog' on his cartoon of barking sausages nestled in warm rolls. The cartoon was a sensation, and the term 'hot dog' was born."[1]

The contemporary role of sausages fits conveniently into modern life styles as an elegant entree for entertaining as well as the main course in "quick-and-easy" meals. The modern sausage industry is extremely diversified in the size and types of plants in operation and represents a unique blend of art and science. Most of the major sausage processing plants are highly mechanized and automated to handle large volumes of products with great efficiency. Most meat industry periodical publications (see Appendix) routinely contain articles on sausage processing and processors and feature an annual "Sausage Issue" in the spring.

---

[1]National Meat Association. Lean Trimmings, June 12, 1995.

The art of making sausage developed slowly throughout the centuries, with sausage makers having little or no understanding of the scientific principles underlying the production of sausages. Today, science plays an integral role in sausage formulation by minimizing the batch-to-batch variation that once occurred and virtually eliminating the unsuccessful batches that had to be thrown away. During the past century, a phenomenal amount of scientific data and literature has been compiled on the topic of sausage production. The modern meat processor must have a basic understanding of the scientific principles responsible for the desired textural, water binding, flavor, color, safety, and shelf-life characteristics of the finished product.

The production of sausage constitutes an important segment of the U.S. meat industry. Through sausage production, virtually unlimited variety is created in the meat supply.

By contacting the National Hot Dog and Sausage Council in Washington, D.C.,[2] you can find the full, current status of the industry and current consumption amounts of the various sausage products.

SAUSAGE IDENTIFICATION

Sausages and ready-to-serve meats are generally grouped according to the processing method used. The following classifications should be helpful in the identification, selection, and care of these products.

Fresh Sausages

Fresh sausages are made from selected cuts of fresh meat (not cooked or cured) and must be stored in a refrigerated (or frozen) state prior to being consumed. They should not be held in a refrigerator for more than three days before being eaten, and they should be cooked thoroughly before they are served. Examples of these products are fresh pork sausage (bulk, patty, or link); country-style pork sausage; fresh kielbasa (Polish); Korr (Swedish); Italian sausage; bratwurst; bockwurst; chorizo (fresh); and thuringer (fresh).

Uncooked Smoked Sausages

Uncooked smoked sausages are smoked but not cooked prior to being sold. Sausages in this class may be either cured or fresh. They must be held

---

[2]http://www.hot-dog.org/

under refrigerated conditions and for no longer than seven days. Uncooked smoked sausages should be cooked thoroughly before they are served. Examples are smoked pork sausage, kielbasa, mettwurst, and smoked country-style sausage.

Cooked Sausages

Cooked sausages are usually made from fresh meats that are cured during processing, fully cooked, and smoked. Cooked sausages constitute the greatest tonnage of sausage produced in the United States. Cooked sausages should be refrigerated until they are consumed. Since many are vacuum-packaged, the freshness date on the packages should be checked for storage life. Generally, cooked sausages will keep seven days, under refrigerated conditions, after a package has been opened. Since they are fully cooked, these sausages are ready to eat (although some are generally served hot). Examples of cooked sausages are frankfurters (wieners), bologna, beerwurst (beef salami), New England sausage (berliner), mettwurst, cotto salami, German-style mortadella, knackwurst, smoked thuringer links, teawurst, and Vienna sausage.

Dry and Semi-dry Sausages

The dry and semi-dry sausages are made from fresh meats that are cured during processing and may or may not be smoked. A carefully controlled bacterial fermentation produces a lower pH (4.7 to 5.3), which aids in the preservation and produces the tangy flavors associated with this class of products. The dry sausages, which include the Italian salamis and pepperoni, are generally not cooked. Dry sausages require long drying periods (generally from 21 to 90 days, depending on product diameter), whereas semi-dry sausages are often fermented and cooked in a smokehouse. Both types are ready to eat. Cool storage is recommended for dry sausages, while semi-dry sausages should be refrigerated. Examples of dry and semi-dry sausages are summer sausage; cervelat (many variations); thuringer; the salamis Genoa, Milano, Sicilian, B. C., D'Arles, and hundreds more; chorizos; frizzes; Lebanon bologna; pepperoni; mortadella; Lyons; landjaegar; and sopressata.

Luncheon Meats/Loaves

Loaf products are made from fresh meats that are cured during processing, are fully cooked, and may be smoked (although they usually are not). Luncheon meats may be cooked in loaf pans or casings or water-cooked in stainless steel molds. Luncheon meats are often sliced and vacuum-packaged at the processing plant. Loaf products are ready to eat and should be refrigerated; they may be kept approximately seven days after the package has been opened. There is usually a freshness date on the package. Examples of luncheon meats are loaves, such as Dutch, ham and cheese, honey, jellied tongue, old-fashioned, olive, pepper, pickle and pimento, headcheese, and chopped ham; scrapple; and Vienna sausage. Ham, turkey (often smoked), chicken breast, and roast beef are whole muscle products that are often sliced and sold along side luncheon meats in the meatcase.

SAUSAGE TERMINOLOGY

The matter of sausage terminology is a tangle of Gordian complexity, a strange and untidy knotting of history, tradition, secrecy, myth, religion, politics, and stubborn pride. Considerable confusion has arisen from the meshing of Old World and New World traditions as to the naming of sausage products. The following glossary is a descriptive list of selected sausage types and styles. This list is by no means complete, but rather a representation of some of the more common products and styles of products on the market in this country. An anonymous author in *Meat Industry* (now called *Meat and Poultry*) magazine[3] summed up the tumultuous task of compiling a glossary:

> Even a brief glossary such as this reveals the new world translations of old world products that plague accurate sausage identification. Take a word like "cervelat," for instance: "Cervelat" properly refers to a family of semi-dry sausages, to which summer sausage and thuringer belong. But some U.S. processors make a summer sausage and call it thuringer, others make a thuringer and call it summer sausage, and still others make both and call it cervelat.
>
> Sorting out salamis would take a team of Italian geographers, and even they wouldn't be able to solve the problem of how to spell the word. "Salami" or "Salame"? San Francisco Bay Area processors use the latter; everyone else uses the former. In Italy, it's spelled both ways.

---

[3]Anonymous. 1983. A glossary of old world sausage. Meat Industry. 29 (6):46.

Then there's kielbasa. It would take pages to talk about kielbasa. Suffice it to say that in Poland, frankfurters and salamis are kielbasa, but here kielbasa is Polish sausage. But not always — sometimes Polish sausage is a little less spicy than kielbasa. Sometimes Polish sausage is called kielbasa anyway. But salami is never kielbasa — except in Poland.

In 1938, Paul Aldrich[4] stated:

Some standardization in types and uniformity in the use of names appears to be necessary if dry sausage is to achieve real popularity in the United States. The word *frankfurter* carries with it a fairly definite idea of shape, color, flavor and texture of the product named; *salami* [and] *cervelat* mean little to the housewife [consumer] because these terms are so loosely used and so indefinite.

Unfortunately, this observation still holds true today.

Sausage Glossary

Alpina Salami. A spicy Italian-style salami of U.S. origin.

B. C. Salami. A dry Italian-style salami stuffed into a natural beef casing.

Balleron. A mildly seasoned bologna-type product made of finely chopped veal and pork with bits of smoked beef tongue and pistachio nuts added for flavor. A sandwich meat, stuffed into a dark casing.

Bangers. English breakfast sausage, though technically not allowed to be called *sausage* by USDA standards because the formula calls for bread crumbs to be mixed with the pork emulsion. A plump product — usually about 4 inches long and $1\frac{1}{2}$ inches thick.

Beerwurst.* One of the big sausages — usually $2\frac{1}{2}$ to $3\frac{1}{2}$ inches in diameter — stuffed into distinctive veined natural casings or vein-decorated artificial casings. It is basically a coarse-ground German-style salami, made of beef and pork, spiced with garlic.

---

[4]Aldrich, P. I. 1938. Sausage and Meat Specialties. The Packers Encyclopedia, Part III. The Natl. Provis., Chicago.

*These sausages are pictured in color in the textbook center section.

Blood Sausage. Known as blutwurst in Germany. Primary meat constituent is diced pork or pork fat, mixed with beef blood and mild spices. Very dark in appearance and often available in both chubs and rings.

Blood-and-Tongue Sausage.* Distinguished from blood sausage by its main ingredient — chunks of cured, cooked tongue in various sizes and shapes. This sausage is also available in a variety of conformations: ring, stick, and loaf are typical. Sometimes it is spiced with a small amount of ground clove.

Bockwurst. The traditional way to sell bockwurst is when it is fresh, though several cooked bockwurst products are now available at retail. Fresh bockwurst, like the cooked varieties, is a veal and pork sausage with a taste and bite similar to that of frankfurters. Styles include the coarse-ground farm-style bockwurst.

Bockwurst (cooked). Until recently, most bockwurst was sold fresh. However, increased shelf life has brought about the retailing of cooked bockwurst. White bockwurst are similar to frankfurters in flavor, white in color, and generally thicker in diameter.

Veal is the primary meat constituent, supplemented by pork. Other ingredients are milk, eggs, and, often, chopped parsley. The casings tend to be very thin, and cooked bockwurst is usually boiled at the meat plant, not smoked. Regional bockwurst products in the United States are farm-style (similar to bratwurst), Swiss-style (very mildly seasoned), and thuringer bockwurst (coarse texture with a speckled appearance).

Bologna.* The second most popular sausage in the United States, bologna accounts for nearly 20 percent of all the sausage sold in the United States, second only to frankfurters. In generic terms, bologna is a fully cooked, mildly seasoned sausage made from beef and(or) pork. But several styles and shapes of bologna are produced in volume, among them:

- **Berliner or New England Sausage.** A product made of coarse-ground pork with pieces of ham or chopped beef interspersed in the emulsion. When Berliner is in small chub form, it is sometimes called Leona bologna; when in stick form, it is known as jagdwurst. Generally it is stuffed into large casings, similar to beerwurst.

- **German.** A light-colored bologna with medium-coarse texture. The meat ingredients are beef and pork, and the spices are dominated by garlic. Also stuffed into large-diameter casings.

- **Ham-Style.** According to regulation, the primary ingredient in ham-style bologna must be cured ham pieces. These pieces are suspended in a fine emulsion, and the product is both smoked and cooked.

- **Vienna.** Sometimes called garlic. Another large-diameter sausage with a creamy meat texture. Pistachio nuts may be used for seasoning in place of garlic. Vienna sausage is produced in both rings and chubs.

Bratwurst or White Hots.* Produced both fresh and cooked. In cooked form these sausages are known regionally as white hots and are made in the same way as frankfurters, except that the nitrite is omitted (hence, the white color). Main meat ingredients are pork and(or) veal. Fresh bratwurst (uncooked) is prevalent in much of the upper Midwest with German-style, Wisconsin-style, and Swiss-style the predominating names. Each is coarse-ground, combined with various seasoning combinations, and stuffed into hog casings.

Braunschweiger. Basically, a creamy-textured liver sausage that has often been smoked. See Liverwurst.

Caserta Pepperoni. An Italian product consisting of 75 percent pork and 25 percent beef, stuffed into hog casings and linked in pairs (12 ounces to each piece). Generically, pepperoni means red pepper pods, and Caserta is a city in southern Italy.

Cervela. See Garlic Sausage.

Cervelat. *Cervelat,* like *salami,* is one of those sausage terms encompassing a whole family of products. It is used in the United States interchangeably with *summer sausage* and *thuringer,* making accurate identification confusing at best. In general, cervelat is a mildly flavored semi-dry sausage, somewhat softer in consistency than the Italian salamis. For the purposes of this glossary, thuringer, in its semi-dry form, is considered to be a member of the cervelat family; summer sausage falls into the farm-style group of the cervelat family.

There are many sausage styles and types of cervelat. These include (in alphabetical order):

- **Farm-Style.** Made of equal parts of beef and pork, rough chopped and mildly seasoned, and stuffed into 1½-inch diameter casings of varying lengths.

- **Goettinger.** A distinctively flavored hard cervelat.

- **Göteborg.** A salty and heavily smoked, coarse-chopped cervelat.

- **Gothaer.** Made of fine-chopped lean pork.

- **Holsteiner.** Similar to farm-style except that it is packed in rings.

- **Landjaegar.** A flattened and smoked cervelat about the size of a frankfurter.

- **Thuringer.** A tangy semi-dry sausage, mildly seasoned and characterized by a relatively soft consistency.

Chorizo. A Spanish dry sausage made from coarse-cut pork. About the size of a large frank, chorizo is a highly spiced, hot sausage. The true Hispanic product contains pork salivary glands; the Mexican variety of chorizo may contain a combination of beef, pork, and fillers.

Chourico. A Portuguese dry sausage similar to chorizo, except it is an all-pork product with no fillers that is spicy, rather than hot like chorizo.

Coney Island Red Hots. High-quality frankfurters, which were originally formulated to be roasted (not boiled) on hot plates by sandwich stand vendors. The red hots originated in New York City, were dyed red, and are considered to be the namesake of the hot dog.

Cooked Salami. The various types of cooked salami include beer, cooked, and Kosher sausages. Each is produced from spice-enhanced cured meats or from fresh meats to which cure and spices are added. Generally, these are beef-dominated salamis (as opposed to the dry salamis, which are usually pork-dominant), stuffed into natural or fibrous casings, and smoked to a deep red.

D'Arles Salami. A very coarse-textured dry French salami that is stuffed into 18-inch hog bungs and wrapped with cord designed to show a diamond-shaped pattern.

Farmer Sausage. A mildly seasoned dry or semi-dry sausage that originated with the farmers of northern Europe. It is made of 65 percent beef and 35 percent pork and is chopped medium fine, seasoned, stuffed into beef middles, and heavily smoked. Each piece weighs from 1 to 2 pounds.

Frankfurter.* In its various forms, the frankfurter (or frank) accounts for 25 percent of all sausage sold in this country, easily making it the most popular sausage on this side of the Atlantic. Nearly 95 percent of frankfurters are sold skinless, and the most popular size is 1.6 ounces, or 10 to a pound. Most frankfurters in the United States are a blend of beef, pork, and(or) poultry meat; mildly seasoned with paprika and other spices; and smoked. In recent years, there has been a substantial increase in the volume of poultry frankfurters entering the marketplace. Poultry meat is often combined with red meats to produce frankfurters at a lower cost. German-style frankfurters are made of veal, and "old-fashioned" or "old-style" franks are almost always stuffed into natural casings. Cocktail franks are the same as regular franks except that they are much smaller and are used as appetizers. Vienna-style sausages (frankfurters) are cooked and canned.

Frizzes. A highly seasoned dry sausage, made from lean, cured pork that has been rough chopped and stuffed into hog middles.

Galician Sausage. A smoked beef and pork sausage that is seasoned with garlic and stuffed into beef rounds 17 to 20 inches long. The casing is broken in the middle, and the ends are tied together. During the Easter season, Galician is favored by the Polish people and by the people of the western regions of the former USSR.

Garlic Sausage. Also known as cervela and knoblauch. It is similar to frankfurters except that garlic is the prime spice ingredient. The shape is also distinctive: very short and squat, typically about 5 inches long and 2 inches thick. Usually stuffed into natural casings.

Gelbwurst. A luncheon meat that is similar to the mildly flavored Swiss-style bockwurst. Made from veal, pork, milk, and eggs. A typical gelbwurst chub is 24 inches long and about $2\frac{1}{2}$ inches thick.

Genoa Salami. A moderately spiced dry salami of Italian origin. Traditionally, this salami is stuffed into hog bungs, 20 to 24 inches long, and wrapped vertically and horizontally with flax twine, starting at the small end, with loops 2 inches apart joined with slip-hitch knots (see Figure 21–6).

German Salami. A mildly spiced dry salami that is heavily smoked. Traditionally this salami is stuffed into calf bladders or beef middles and tied with flax twine.

Göteborg, or Swedish, Sausage. A salty, heavily smoked cervelat that is somewhat coarsely chopped. Göteborg is traditionally soaked in a salt brine before it is smoked. This sausage gets its name from the Swedish city of Göteborg.

Gothaer. A summer sausage reported to be impossible to manufacture in any but the winter months (believe it or not).

Haggis. A Scottish food made from the heart, lungs, and livers of sheep or calves, highly seasoned, mixed with oatmeal, stuffed into a sheep's or pig's stomach, and then boiled.

Hamburger Loaf.* Composed of fresh beef trimmings, flour, onions, and seasonings.

Headcheese.* Available in the United States in two styles: French and German. French headcheese consists of cured pork pieces suspended in vinegar-flavored gelatin, with bits of pickles and pimentos added for flavor. There is no casing. German headcheese, or Suelze, is also composed of cured pork pieces, plus some beef, but the gelatin is produced by heating to resemble more of an emulsion. German headcheese requires a casing; normally natural casings are used.

Holsteiner. Similar to farmer sausage, except that the ends are tied together (sometimes called horseshoe sausage). Dried and smoked, it appears on the market in pieces weighing about 1 pound each.

Hungarian Salami. Similar to German salami. Mild in flavor but heavily smoked.

Italian Salami. Characterized by a unique "tang" in the flavor. Genoa, Milano, Sicilian, and southern Italian, plus hundreds more, are regional specialties. In this country, Italian-style salami produced in the San Francisco Bay area, where it is spelled *salame,* is allowed to develop an outside coating of white mold; in other parts of the United States, the mold is brushed off the casing. The basic meat ingredients of Italian salami are coarse-chopped cured pork and fine-chopped lean beef. The drying time is typically three months.

Italian Sausage. A fresh sausage characterized by the spices pepper, fennel, and(or) anise. Other spices may also be included. It is typically made from pork, but it may contain beef and(or) veal if the sausage is properly labeled as containing these ingredients.

Jagdwurst. Berliner bologna in stick form. See Bologna.

Jaternice. A ring liver pudding of Bohemian origin. Made of liver, head meat, bread crumbs, and cornmeal or other fillers. Traditionally contains too much filler to be classified as a sausage in the United States.

Kielbasa. Some sausage producers call kielbasa "Polish sausage"; others who make both products give kielbasa a little spicier taste. It is a coarse-ground beef and pork sausage, usually smoked to a medium red. See Polish Sausage.

Kiszka.* A smoked beef and pork sausage, originating in Hungary, that also contains rice and pork fat pieces. The primary spice is garlic.

Knackwurst.* Also known in Switzerland as *schueblig.* A coarse-ground beef, pork, and veal sausage smoked to a reddish brown. Garlic or onion dominates the several spices used in knackwurst. The product is typically stuffed into a natural casing.

Korv. A Swedish sausage made of fresh pork and raw potatoes and heavily spiced. Stuffed into hog casings.

Knoblauch. See Garlic Sausage.

Landjaegar. A Danish "hunter's sausage" made of beef and pork. This unique dry to semi-dry sausage is stuffed into hog casings, cut into double links, and pressed flat between weighted boards to give it a square appearance. Garlic and caraway seeds give it a distinctive flavor.

Lebanon Bologna. Not actually an Old World sausage but a development of the Pennsylvania Dutch culture, this semi-dry bologna is an all-beef product, characterized by a very heavy smoke and a very acid flavor.

Linguisa. In cooked form, this Portuguese sausage is made of coarse-ground, highly spiced pork. Typically smoked to a medium reddish orange.

Liverwurst.* A very fine–ground cooked sausage seasoned with onions, spices, and sometimes pistachios. The primary meat ingredients are pork and liver. Regional varieties are French-style, which is a coarser mixture that is slightly darker in color, and country-style, which is flavored with bacon bits. By regulation, liverwurst must contain at least 30 percent liver, and if a pink color is desired, the meats must be cured.

Loaves. See Luncheon Meats.

Lombardia Salami. This coarse-cut Italian salami has a moderately high fat content and contains brandy. Like most of the Italian salamis, Lombardia has its own characteristic twine wrap, which consists of a 1-inch square pattern.

Luncheon Meats. This large group of loaf products is primarily of U.S. origin and does not require a casing for cooking. Various condiments are added to fine-ground or comminuted meat batters to create infinite variety. Some examples are pickle and pimento loaf, olive loaf, and ham and cheese loaf.

Lyons Sausage. A French all-pork sausage, seasoned with garlic and other spices and stuffed into large-diameter casings. Traditionally, the pork is portioned in parts as follows: four parts of fine-chopped lean pork and one or two parts of diced pork fat.

Mettwurst.* A pork, veal, and garlic spreadable sausage produced in a ring. If the same product is produced as chubs, it is called *teawurst.* Both mettwurst and teawurst are of German origin and are designed for sandwich spreading.

Milano Salami. A beef and pork dry sausage of Italian origin. Milano is generally stuffed into large hog bungs and wrapped with twine, three coils per inch.

Minced Ham.* Cured pork trimmings that are seasoned, ground (chopped), and stuffed into beef bladders.

Mortadella.* A sausage that looks very similar to common bologna, except that cubes of pork fat are suspended in the emulsion. Originating in Bologna, Italy, mortadella is a delicately spiced sausage made from both pork and beef, very finely chopped, and usually smoked briefly at a high temperature prior to being air dried. In the earliest recipes, mortadella (from the Latin word for myrtle) was flavored with myrtle berries. It is nearly always stuffed into large-diameter beef bladders.

New England. A ham-like sausage that is composed of 85 to 90 percent cured lean pork and 10 to 15 percent cured lean beef. This sausage is fully cooked and smoked.

Old-fashioned Loaf. A coarse-ground combination of pork and beef, mildly seasoned with onion.

Pepper Loaf.* A combination of cured pork and beef and seasonings, with a coating of fine-ground black pepper over the top of the loaf.

Pepperoni. A hotly spiced, coarse-ground pork and beef sausage with a very distinctive flavor, making it ideal as a pizza topping (the overwhelming dominant use in the United States). Ground red pepper is the major flavor ingredient. In some regions of Europe, pepperoni is known as *kabanossi*.

Pickle and Pimento Loaf.* A sausage that is composed of cured pork and beef, seasonings, sweet pickles, and diced pimentos.

Polish Sausage.* A Polish-style coarse-ground sausage that is similar to kielbasa. Pork is the main meat ingredient, and the flavor is accented toward garlic. The sausages — usually 4 to 5 inches long, $1\frac{1}{2}$ inches in diameter — are smoked thoroughly. In Poland, *kielbasa* is a generic word meaning *sausage*. In the United States, *kielbasa* is used exclusively to describe Polish or Polish-style sausages.

Polka Sausage. A sausage with Old World predecessors, but in the United States strictly a West Coast specialty. Made of pork, beef, and onions, blended to make the product an ideal complement to beer. Generally stuffed in a natural casing and smoked to a dark red.

Pork and Potato Ring, or Korv. A sausage of Swedish origin, still popular there at Christmas. The basic meat is pork, to which potatoes, onions, and spices are added. It is usually uncooked and sold frozen; contains no nitrite.

Pork Sausage.* A fresh, uncooked sausage made entirely from pork and seasoned with salt, pepper, and sage. Pork sausage is sold in bulk 1- or 2-pound chubs or in link or patty form.

Prasky.* Sausage of Polish or Czechoslovakian origin. Made from pork trimmings cured with salt and seasoned with sugar, pepper, and garlic. Prasky is traditionally stuffed into beef weasand (esophagus) casings.

Salami. Like cervelat, salami refers to a family of sausage products, mostly of Italian origin but also from other parts of Europe as well. Most dry salamis are made from pork and have a coarse-ground texture; garlic is the main spice, enhanced with others. The mild Italian winters are perfect for the extended drying periods required for high-quality dry salamis, and hundreds of regional salami specialties have developed in cities from Brindisi to Venice. While the production of most of these salamis goes back several hundred years, it was not known until after World War II that the operative factor in the development of dry salami flavor and quality was bacteria (specifically, lactobacilli). Traditional salamis are characterized with different variations of twining or cording to provide a distinctive decorative effect and to give support to the hanging sausage during fermentation and drying. B. C. salami, for example, has only a few vertical and horizontal cordings, whereas Genoa salami has many wrappings of twine, both vertical and horizontal, in a basket-weave effect (see Figure 21-6). Some of the salami sausages on the market are as follows: Alessandria, Ancona, B. C. salami, Bobbio, Capri, Catania, cooked salami, Corti, D'Arles, H. C. salami, Genoa, Kosher salami, Lazio, Liguria salami, Lombardia, Lola, Milano, Nola, Novaro, Savonna Genoa, Sicilian or Sicani, Sorrento, and many others.

Schinkenwurst. An Old World sausage closely related to berliner bologna. Ham pieces and diced pistachios are interspersed in a fine emulsion of pork and veal. Schinkenwurst is distinguished from berliner by the fineness of the emulsion and a non-garlic flavor.

Schueblig. See Knackwurst.

Scrapple. A product of Pennsylvania Dutch origin that is not a sausage, but rather a cornmeal-based suspension of head meat and(or) pork sausage. Scrapple is sliced cold and fried before it is served — usually for breakfast.

Sopressata. In Italy, any sausage stuffed into the wrinkled casings made from hog middles. In the United States, a sausage made from coarse-chopped pork and seasoned, in part, with whole black pepper.

Souse. A loaf product similar to headcheese that is bound together with gelatin. Souse typically will contain cured and cooked head meat, tongues, hog lips, and hog snouts, and pickles and pimentos for an attractive color. In the holiday seasons, souse often contains a small amount of Scotch or Bourbon.

Smoked Sausage.* Typically made from chopped/ground cured pork, stuffed into hog casings and smoked.

Suelze. See Headcheese.

Summer Sausage. See Cervelat and see Thuringer.

Teawurst. See Mettwurst.

Thuringer.* Some thuringers are fermented semi-dry sausages, while others are not fermented, but sold fresh. In fact, considerable confusion surrounds the definition of *thuringer,* because in certain parts of the United States, the name is used interchangeably with *summer sausage.* Other sausage manufacturers produce both a thuringer and a summer sausage. As identified here, cooked thuringer sausage is made mostly from pork (sometimes beef and veal are added in small quantities), with similar flavorings used as in fresh pork breakfast links, though sage is usually not used to season thuringer. In years past, this sausage was sold fresh, but — as in the case of bockwurst — many processors are offering thuringer in cooked form for better shelf life.

Vienna Sausages (Wieners). Vienna wieners are a popular canned sausage, similar to frankfurters, although usually more bland in flavor and smaller in diameter. Vienna wieners are generally not linked, but are cut into uniform lengths prior to being canned.

Zampone. An Italian sausage product made from stuffed pork trotters (feet) that is sliced and sold as a ready-to-eat item.

Information for this glossary was obtained from the following sources:

Encyclopedia of Labeling Meat and Poultry Products. 1989. Meat Plant Magazine, St. Louis.

Heller's Secrets of Meat Curing and Sausage Making. 1929. B. Heller & Co., Chicago.

Meat Industry. 1983. 29:6.

Sausage and Meat Specialties. 1938. Part III, The Packers Encyclopedia. The Natl. Provis., Chicago.

Standards and Labeling Policy Book. 1986 (periodically updated). USDA–FSIS, MPS Tech. Ser., Standards and Labeling Division.

SAUSAGE INGREDIENTS

A basic understanding of the properties and(or) function(s) of raw meat materials and nonmeat ingredients and of the interaction of the two is essential in order for the reader to comprehend why manufacturing practices are different from one product to the next.

Raw Meat Materials

Large numbers of different raw meat materials are used in the production of all types of sausage, with each contributing particular properties to the finished product. Raw materials vary considerably in their proximate composition (percentage protein, moisture, fat, and ash), color, connective tissue (collagen) content, and binding ability. Table 21–1 lists 27 of the more common raw materials for sausage production. The values represented in this table serve only as a guide for average values. Actual values may vary considerably from lot to lot of the same raw material. Other raw meat materials that are not listed in this table but that are in widespread use are mechanically separated meat (MSM) (beef, veal, lamb, and pork) and mechanically separated poultry (MSP) (chicken and turkey), both of which were discussed in Chapter 18, and mechanically desinewed meat.

Larger processors who utilize preblending practices and computer formulations (see discussion later in chapter) rely heavily on values for protein, moisture, fat, collagen, color, and bind.

In recent years the use of poultry meat has had a considerable impact on the production of sausage. The use of turkey meat to produce turkey frankfurters, bologna, and breakfast sausage has grown tremendously. This growth can be attributed to two primary factors. First, processed poultry products are generally cheaper than their red meat counterparts (see page 241). Second, many consumers perceive processed poultry products to be leaner and more nutritious than the traditional sausage counterparts. The nutritional composition of all meats is discussed in Chapter 23.

The variety meats most commonly used in sausage production are hearts, tongues, livers, kidneys, tripe (beef stomachs), and pork stomachs. The amount and type of variety meat that may be used is dependent on the product being produced and on the quality of the product. Certain products such as braunschweiger and liverwurst are composed of predominantly liver (a variety meat). Other products such as frankfurters and bologna may contain no variety meats or as much as 30 to 40 percent of variety meats, although the law permits up to 85 percent of the meat block to be variety meats if the final product is properly labeled. If variety meats are included in a *cooked sausage* product, such as frankfurters, the words *"Contains Variety Meat"* must be prominently identified on the label. In addition, the particular variety meat, e.g., pork hearts, must be listed in the ingredient list in order according to predominance. For dry and semi-dry sausages, the particular variety meat used need only be listed in the ingredient statement. When low-binding meats such as tripe and pork stomachs are used in formulations, they should be limited to 15 percent or less of the formulation, or product instability may occur. Hearts provide a relatively good raw material source for those products

Table 21–1. Percent of Protein, Moisture, Fat, and Collagen and the Color and Bind Values of Meats Most Often Used in Commercial Sausage Production, with All Bones and Bone Fragments, Heavy Sinew, and Gristle Removed[1]

Meat	Pro-tein	Mois-ture	Fat	Colla-gen[2]	Color[3]	Bind[4]
	------------- (%) -------------				---- (index) ----	
Bull meat, full carcass	20	68	11	20	100	100
Cow meat, full carcass	19	70	10	21	95	100
Beef shank meat	19	73	7	66	90	80
Beef chucks	18	61	20	30	85	85
Beef trimmings, 90% lean	17	72	10	30	90	85
Beef trimmings, 75% lean	15	59	25	38	85	80
Beef plates	15	34	50	—	—	—
Beef flanks	13	43	42	—	55	50
Beef head meat	17	68	14	73	60	85
Beef cheeks, trimmed	17	68	14	59	10	85
Beef tissue (partially defatted)	20	59	20	—	30	25
Veal trimmings, 90% lean	18	70	10	—	70	80
Mutton	19	65	15	—	85	85
Poultry meat (dark)	19	67	12	—	80	90
Pork trimmings, 80% lean	16	63	20	24	57	58
Pork trimmings, 50% lean	10	39	50	34	35	55
Pork blade, 95% lean	19	75	5	23	80	95
Picnic trimmings, 85% lean	17	67	15	24	60	85
Pork jowls	6	22	72	43	20	35
Pork cheeks, trimmed	17	67	15	72	65	75
Pork tissue (partially defatted)	14	50	35	—	15	20
Pork hearts	16	69	14	27	85	30
Pork tripe	10	74	15	—	20	5
Beef hearts	15	64	20	27	90	30
Beef tripe	12	75	12	—	5	10
Beef lips	15	60	24	—	5	20
Beef weasand meat	14	75	11	—	75	80

— = No data available.

[1]Hickey, J. C., and A. W. Brant. 1974. Sausage Makers Handbook. Professional Food Flavors, 3970 Commerce Drive, West Sacramento, CA 95691.

[2]Values represent the percentage of the protein content that is collagen.

[3]Color values are expressed as an index in which 100 represents a raw material with the most intense red color, and 0 (zero) represents a raw material with virtually no contribution of red color.

[4]Bind values are expressed as an index in which 100 represents a raw material which contributes the maximum bind and 0 (zero) represents a raw material which contributes little or no bind.

in which a dark red color is desired, e.g., summer sausage. Hearts are also commonly used in emulsion-type products such as bologna. Tongues that have been cured are the only meat source in jellied-tongue loaf and are a major constituent of blood-and-tongue loaf. Tongues are also used in numerous other sausage products, depending upon the price and the supply. Kidneys are the least used of the variety meats in sausage production. The vast majority of them are either exported or used in pet food.

The proper selection of meat ingredients is essential for the production of sausage of uniform quality. This does not mean that only the more costly meat ingredients should be used in all situations, but rather that the meat ingredients should be combined to meet predefined standards of fat content, color, binding properties, and other characteristics. The raw materials should be as fresh as possible with very low microbial growth. Inventory control of raw meat materials should be set up on the FIFO (first in, first out) principle to insure a constant and uniform turnover of raw materials from either the holding cooler or the freezer. Experienced meat processors will usually avoid "bargain" raw materials, since there is usually a reason for the bargain, such as temperature abuse that results in a high microbial load. When it comes to the use of meat ingredients, the end product can only be as good as the quality of the raw materials going into its production, i.e., "Garbage in, garbage out."

Meat ingredients will vary considerably in their moisture-to-protein ratio, in their lean-to-fat ratio, in their water-binding properties, and in their relative amounts of pigment (red color). Binding consists of (1) the ability of the meat ingredients to hold and entrap fat and water to produce a stable emulsion and (2) the ability of lean meat particles to be held or glued together. The specific factors responsible for variations in the binding properties of meat ingredients used for sausage are very complex and are discussed in the "Meat Emulsions" section.

Raw meat materials are classified according to their binding ability as *binder* or *filler* meats. Binder meats are further subdivided into high, medium, and low, based on their ability to bind the product. Skeletal meat from the beef animal — bull meat, cow meat, shank meat, and boneless shank, for example — is considered to have the highest binding properties. Meat ingredients having an intermediate binding ability (medium) include head meat, cheek meat, and lean pork trimmings. Low-binding meats usually contain a large proportion of fat or they are non-skeletal meats. Regular pork trimmings, such as jowls, ham fat, beef briskets, hearts, hanging tenders (diaphragm), and tongue trimmings, are low-binding meats. Filler meats are meats with little or no binding ability. They include tripe, pork stomachs, lips, snouts, skin, and partially defatted pork and beef tissue. While these meat ingredients are nutritionally acceptable, their use in sausage products must be severely limited, if a high-quality product is to be produced. Perhaps the most desirable means of using filler meats is in the jellied-loaf products (for example, souse and headcheese) because the binding ability of the meat is not important in these products.

The moisture-to-protein ratios of the various raw meat materials are important to the processor in predicting the composition of the final product. When using meat with a low–moisture-to-protein ratio (such as bull meat), processors can add more water to the meat during mincing than with meat containing a high–moisture-to-protein ratio (such as beef hearts). However, economics dictates the formulation of ingredients to a large part, and meats having a low–moisture-to-protein ratio are generally more expensive — even though slightly more water may be added. Table 21–2 lists the moisture–protein ratios of selected meats used in sausage formulations. These values reflect averages of the various ingredients listed. Actual moisture–protein ratios will change somewhat from sample to sample, depending on the condition of the meat and the fat content.

Table 21–2. Percentage of Approximate Moisture and Protein Content and M:P Ratios of Some Meats Used in Sausage Making

Sample	Moisture	Protein	M:P Ratio
Pork trimmings, 50% lean	39	10	3.9
Pork cheeks, trimmed	67	17	3.9
Pork jowls	22	6	3.7
Pork fat	6	1.3	4.6
Beef chucks	61	18	3.4
Bull meat	68	20	3.4
Beef tripe	75	12	6.3
Beef trimmings, 90% lean	72	17	4.2
Beef flanks	43	13	3.3
Beef hearts	64	15	4.3

Nonmeat Ingredients

Various nonmeat ingredients are used in the production of sausage to provide certain functional properties, to create variations in flavor and appearance, to prolong the shelf life of the product, and to insure the microbiological safety of the final product. Some of these ingredients, such as salt, cannot be omitted. Others, such as cereal grain fillers, are adjuncts whose primary function is to reduce the cost of the products.

The following ingredients are the ones most commonly used in sausage production. The regulatory limits of certain ingredients are subject to change.

Water

Water is often an overlooked ingredient in sausage formulations; nonetheless, it is an extremely important nonmeat ingredient. FSIS meat inspection regulations permit added water (AW) of 3 percent for fresh sausages and luncheon meats and meat loaves and formerly dictated that the moisture content of the final cooked sausage product (frankfurters, etc.) could not exceed four times the protein content plus 10 percent (4 P + 10), i.e., 10 percent AW. However, because of the emphasis on leanness in all meat products, the regulations were changed, effective April 14, 1988, to allow added water to be substituted for fat. Maximum fat content remains at 30 percent, but added water is no longer limited to 10 percent. Rather, the sum of fat and added water in the final product cannot exceed 40 percent. Added water is still determined by computing the difference between four times the protein content and the actual moisture content.

All-meat franks (see the frank examples in Table 21–3) usually contain up to 3 percent salt and 1.2 percent other minerals (ash), non-protein nitrogenous substances, and carbohydrates, totaling 4.2 percent. Therefore, protein, fat, and water sum to 95.8 percent for the franks in Table 21–3 (95.8 + 4.2 = 100). Frank #1 contained 12.5 percent protein, so 4 × 12.5 = 50 percent is the estimate of moisture naturally occurring in the product, and the starting percentage to compute added water. Prior to 1988, the maximum water allowed would have been 60 percent (50 + 10). Frank #1 then could legally have this percent composition: water 60, protein 12.5, fat 23.3 (100 – 60 – 12.5 – 4.2 = 23.3). However, as can be seen in Table 21–3, the manufacturer chose to reduce the added water to 6.3 percent and to keep the fat at 27 percent, still meeting regulations.

Table 21–3. Standard Formula Analysis for Frank Samples

Attribute	Regular Frank Samples				
	#1	#2	#3	#4	#5
	- - - - - - - - - - - - - - - (%) - - - - - - - - - - - - - - -				
Fat	27.0	10.0	23.5	30.8	29.0
Protein	12.5	12.5	12.4	12.1	11.6
Moisture	56.3	73.3	59.9	52.9	55.2
Protein, × 4	50.0	50.0	49.6	48.4	46.4
USDA–AW[1]	6.3	23.3	10.3	4.5	8.6
Fat + AW[1]	33.3	33.3	33.5	35.3	37.8

[1]AW (Added water). The sum of fat and added water in the final product cannot exceed 40% according to USDA regulations. Water naturally part of the meat is estimated to be 4 × protein content. The difference between total moisture and protein × 4 = USDA–AW values. Calculations from moisture, fat, and protein analyses indicate that batch #4 is out of compliance by 0.8 percent for fat and that batch #2 can be labeled low-fat and(or) 90% fat-free. Batch #s 1, 3, and 5 could have had additional water and(or) fat added to a combined level not exceeding 40%. See discussion in text.

In order for any product to be labeled low-fat, except ground beef/hamburger, the product cannot contain more than 10 percent fat by weight (see Chapter 3). A low-fat product is more clearly and accurately labeled 90 percent fat-free. The fat must be lowered to 10 percent and water must be substituted for the fat that was cut in order to make a new low-fat, 90 percent fat-free frank (#2). The new low-fat frank now has the following percentage composition: water 73.3, + protein 12.5, + fat 10 + salt, etc., 4.2 = 100. The added water is 23.3 percent (4 × 12.5 = 50; 73.3 − 50 = 23.3) and the fat + added water = 33.3 percent, within the 40 percent limit regulation. Most manufacturers who produce 90 percent fat-free franks will use additional nonmeat binders and extenders.

Protein must be defined in order for this regulation to be enforced. The protein normally occurring in the meat ingredients is known in the regulations as "Group 1 Protein." But there are other meat proteins that may have been further processed by hydrolysis, extraction, concentration, or drying plus protein from numerous nonmeat ingredients. These are known as "Group 2 Proteins." The FSIS regulation states that a maximum 1 percent allowance will be made for all Group 2 proteins. So, in the preceding example, the 12.5 percent protein would be composed of 11.5 percent Group 1 and 1 percent Group 2 proteins.

Emulsion products would be very dry and unpalatable if they contained only the moisture inherent in the meat ingredients. Water is often added in the form of ice during the chopping or mincing process to hold the temperature of the meat batter down, thus insuring stability of the batter. Moisture makes the final product juicier and works with salt in helping to solubilize the meat proteins (see later discussion).

Salt

Review the material on salt in Chapter 20. Also, see Lowell Dearinger's interesting article on salt.[5] Other than meat, salt is the most critical ingredient in sausage manufacturing. Without salt, sausage, as it is known today, could not be made. Salt has three primary functions: preservation, flavor enhancement, and protein extraction to create product bind. In modern sausage production, the preservative effect of salt is not as important (except in dry sausages) as it was prior to the advent of refrigeration. To be a totally effective, single preservation method, brine concentrations in the product must be around 17 percent. The percentage of brine concentration in the product may be calculated as follows:

$$\frac{\% \text{ salt added meat}}{\% \text{ water in meat} + \% \text{ salt}} \times 100 = \% \text{ brine concentration}$$

Example: Our franks have 2.5 percent salt and 60 percent water.

$$\frac{2.5}{60 + 2.5} = \frac{2.5}{62.5} = 0.04 = 4 \text{ percent brine concentration}$$

Most sausages will have a brine concentration of 4 to 6 percent, which corresponds to 2 to 3 percent added salt.

The two most important functions of salt are to impart flavor and to solubilize the meat proteins actin and myosin (see Chapter 22), which are responsible for the product texture and for water binding. The solubilized proteins also stabilize the sausage batter so that the fat does not coalesce into large fat pockets or migrate to the surface to form fat caps during cooking.

---

[5]Dearinger, L. 1977. When salt was an industry. Outdoor Illinois. (January.)

Sodium chloride (NaCl) is the most common salt used in sausage manufacturing because it has the most desirable flavor, is readily available, and has the greatest protein solubilization properties. However, sodium has been linked to hypertension, which can lead to coronary heart disease and stroke in some people. For this reason, patients suffering from high blood pressure have often been placed on reduced sodium diets. In recent years, the major meat processors have responded with low-sodium luncheon meats and other sausages by replacing a portion of the sodium chloride with potassium chloride (KCl), calcium chloride ($CaCl_2$), or other salts. However, these salts are generally more bitter than sodium chloride. Studies have shown that when a 50/50 combination of NaCl and KCl is used, the objectional taste of KCl is almost totally suppressed by equal concentrations of NaCl.

Nitrite and Nitrate

Historically, nitrate was present as a naturally occurring contaminant in the salt used for sausage production. Early sausage makers found that salts from certain areas of Europe produced superior sausages; however, they did not recognize that this was due to a contaminant. Chemists later identified nitrate (NO_3^-), and it was then added intentionally in the form of saltpeter (potassium nitrate [KNO_3]). Eventually, chemists recognized that *nitrite* (NO_2^-), not nitrate (NO_3^-), was responsible for the beneficial color and flavor-enhancing properties of sausage products. Some of the nitrate is converted to nitrite by bacteria during long curing processes. With today's rapid-processing techniques, nitrate is seldom used except in dry-sausage production. Nitrite is added directly to the sausage batter — usually in the form of sodium nitrite ($NaNO_2$). Potassium nitrite (KNO_2) can also be used.

Nitrite has four primary functions in cured meat and sausage products. They are: (1) to provide bacteriostatic properties, (2) to serve as a powerful antioxidant, (3) to improve the flavor, and (4) to develop the characteristic pink cured-meat color. The bacteriostatic properties of nitrite are extremely important in the thermally processed, vacuum-packaged products such as frankfurters and luncheon meats. Without nitrite, the safety of these products would be jeopardized by the organism *Clostridium botulinum,* the bacterium that causes botulism, the lethal food poisoning (see Chapter 2). At normal meat processing temperatures, clostridia do not die; instead, they form protective spores. Nitrite prevents the outgrowth of *C. botulinum* spores and the subsequent production of one of the world's most deadly toxins. Nitrite also inhibits the oxidation of the lipids (fats) in meat, which would otherwise lead to the development of oxidative rancidity. Without nitrite, a stale, warmed-over flavor would exist in most products. Oxidative rancidity is often a problem in fresh sausages because they do not contain nitrite. Development of the cured-meat color was discussed in Chapter 20 and is illustrated in the color section near the center of this text. About 40 parts per million (ppm) of nitrite in the finished product is considered necessary for the formation of the cured color.

Nitrite and nitrate are the most regulated and most controversial of all the sausage ingredients. Sodium or potassium nitrite may be used at a level not to exceed $\frac{1}{4}$ ounce per 100 pounds of meat (156 ppm) in sausage products. Sodium or potassium nitrate may be used at a level not to exceed $2\frac{3}{4}$ ounce per 100 pounds of meat. Extreme caution must be exercised in adding nitrite or nitrate to meat, since too much of either ingredient can be toxic to humans. Since such small quantities are difficult to weigh on most available scales, *it is strongly recommended that a commercial premixed cure be used when nitrite and(or) nitrate is called for in the formulation.*

Ascorbates and Erythorbates

Ascorbates and erythorbates (the optical isomer of ascorbate) are strong reducing agents that accelerate the conversion of metmyoglobin and nitrite to myoglobin and nitric oxide (see Chapter 20). These vitamin C (ascorbic acid [$C_6H_8O_6$]) derivatives are also known as cure accelerators, since they act to speed the curing reaction. Residual amounts of these compounds present in the finished product also add stability to the cured color by reducing the deterioration rate of the nitrosohemochrome pigment. A further beneficial function appears to be that ascorbates and erythorbates seem to inhibit the formation of nitrosamines (see Chapter 20). Examples of specific compounds approved for use as cure accelerators in sausages and the maximum amounts permitted per 100 pounds of meat by the USDA–FSIS include:

ascorbic acid: $\frac{3}{4}$ oz.

erythorbic acid: $\frac{3}{4}$ oz.

sodium erythorbate: $\frac{7}{8}$ oz.

sodium ascorbate: $\frac{7}{8}$ oz.

citric acid: May replace up to 50 percent of those listed above.

sodium citrate: May replace up to 50 percent of those listed above.

sodium acid pyrophosphate: Alone or in combination with any of the above may not exceed 8 oz. (0.5 percent in the finished product).

glucono delta lactone (GDL): 8 oz.

In Genoa salami: 16 oz.

Many supplier companies are in business to furnish these many food additives essential to the meat industry.

Sugars

A variety of sugars are commonly used in different sausage products, ranging from sucrose (cane or beet sugar [$C_{12}H_{22}O_{11}$]) to dextrose (corn sugar [$C_6H_{12}O_6$]). Included in this latter group are corn syrup, corn syrup solids, and sorbitol. Sugars are used primarily for flavoring to counteract the salt-flavor intensity and to provide food for microbial fermentation in the fermented sausages. Most sugars (except sorbitol) increase the browning of meat during cooking, which may or may not be desirable, depending on the product. Dextrose is essential in fermented sausages because fermentation bacteria require a simple sugar to produce lactic acid. Dextrose is usually added at the 0.5 to 1 percent level (of the meat weight) in the formulation for fermented sausages. Sorbitol has been credited with reducing the charring of frankfurters on the grill.

Antioxidants

Several compounds may be added to fresh and dry sausage to retard the development of oxidative rancidity. As discussed in Chapters 19 and 23, oxidative rancidity develops at the unsaturated (double) carbon-carbon bonds in fatty acids present in meat. Salt, light, heating and freezing, and traces of certain metals all increase rancidity development. Grinding and chopping meat for sausage production exposes more of the membrane fatty acids to oxidation. Antioxidants react at the double-bond sites, protecting them from oxidation. The most common of the antioxidant compounds are BHA (butylated hydroxyanisole), BHT (butylated hydroxytoluene), and propyl gallate. These three compounds, all of which have the aromatic hydrocarbon benzene ring as their basic structure, have been permitted in foods for many years to prevent deterioration caused by fat oxidation. In fresh sausage, the allowable level that can be used is 0.01 percent of the fat content for any one of the above or 0.02 percent for any two or more used in combination. In dry sausage, the allowable level is 0.003 percent of the meat block weight for any one or 0.006 percent for two or more used in combination. Regulations of allowable levels are based upon the fat content of fresh sausages and the total meat block weight for dry sausages. There are many commercial offerings of antioxidant and antioxidant/flavoring combinations available in the market-place.

Phosphates

Review the material on phosphates in Chapter 18. Phosphates improve the water-binding capacity of the meat, solubilize proteins, act as antioxidants, and help protect and stabilize the flavor and color of the finished product. Through the use of phosphates, processors can attain a longer product shelf life and improve the smokehouse yield.

Phosphates are approved at a level not to exceed 0.5 percent in the finished product. There is about 0.1 percent naturally occurring phosphate in muscle tissue that must be considered in the analysis when phosphates are added. The following phosphates are approved for use: disodium phosphate, mono-sodium phosphate, sodium meta-phosphate, sodium polyphosphate, sodium tripolyphosphate, sodium pyrophosphate, sodium acid pyrophosphate, dipotassium phosphate, monopotassium phosphate, potassium tripolyphos-phate, and potassium pyrophosphate.

Mold Inhibitors

Mold growth is a common problem in the production of dry sausage. To inhibit mold growth, sausages may be dipped in a 2.5 percent solution of potassium sorbate ($CH_3CH:CHCH:CHCOOK$) or a 3.5 percent solution of propylparaben (propyl–p–hydroxy benzoate [$C_{10}H_{12}O_3$]).

Glucono Delta Lactone (GDL)

Glucono delta lactone (GDL), which is a derivative of gluconic acid, has the following formula:

$$
\begin{array}{c}
\overline{\, \text{O} \,} \\
CH_2OH \bullet CH \bullet (CHOH)_3 \bullet CO
\end{array}
$$

This compound serves as a cure accelerator and also produces an acid tang similar to that produced by natural fermentation. It is used at the 0.5 percent level in some fermented sausages, with the exception of Genoa salami, for which its use is permitted at the 1 percent level.

Monosodium Glutamate (MSG)

Monosodium glutamate (MSG) is the sodium salt of glutamic acid, which is one of the common, naturally occurring, nonessential amino acids found in protein. This compound has the following formula: $COOH(CH_2)_2CH(NH_2)COONa$. MSG blends out food tastes without contributing any noticeable odor or taste. It may be used in amounts sufficient for the purpose but must be listed separately in the ingredients statement on the label. Since MSG is a potential protein component, it must be considered in the Group 2 proteins when added water is computed. (See previous section in this chapter on water.) MSG has been found to improve the flavor of low-salt sausage products.

Sodium Lactate

Sodium lactate is the sodium salt of lactic acid with the formula $CH_3CHOHCOONa$. It occurs naturally in animals and humans, but FSIS regulations at this time do not provide for its use. However, FSIS nevertheless does permit its use at 2 percent as a flavoring as long as it has been approved on an individual basis and declared on the label. Sodium lactate improves product stability and improves shelf life because of its bacteriostatic effects. It is a colorless, syrupy liquid.

Extenders

A number of ingredients are used as extenders to reduce the cost of the product and to provide certain functional properties related to product bind, texture, and flavor. Extenders used in fresh meat processing were discussed in Chapter 18, and those used in sausages will be covered in detail in the following section. Examples of extenders include non-fat dry milk, sodium caseinate, cereal flours, and soy protein, which comes in a variety of forms.

Spices, Seasonings, and Flavorings

Many different spices, seasonings, and flavorings are used in sausage products. Their use levels are primarily dictated by product-identity standards and personal flavor preferences. Combining different levels of the various spices, seasonings, and flavorings available creates infinite variety in the sausage supply. Table 21–4 lists the major spices available to food processors. Data include country of origin, physical description, flavor characteristics and intensity, and examples of use. Anyone interested in learning more about the use of spices in the food industry is encouraged to consult Dziezak's review in *Food Technology*.[6]

By definition, *spices* are any aromatic vegetable substances in whole, broken, or ground form whose function as ingredients in foods is seasoning rather than nutrition and from which none of the flavoring principle has been removed. *Flavorings* are extractives that contain flavoring constituents from fruits, vegetables, herbs, roots, meat, seafood, poultry, eggs, and dairy products whose function as ingredients in foods is flavoring rather than nutrition. *Seasonings* is a comprehensive term that can be applied to any ingredients that improve the flavor of the product in question.

Spices vary greatly in composition, but the aromatic and pungent properties that render them valuable probably reside in volatile oils, resins, or oleoresins. The active principles are usually very small proportions of the spice as a whole. Some success has been achieved in separating this active principle and using it for flavor instead of using the spice from which it is obtained. In addition to having an agreeable effect upon the organs of taste and smell, the principles of spices stimulate the flow of digestive juices (gastric and salivary). Rosemary and sage have antioxidant effects, while marjoram has the opposite effect — it is prooxidative.

Spices may be added as natural spice or as spice extracts. In the latter case, they must be labeled as "flavorings." The USDA–FSIS has definite rules stating exactly which substances must be specifically identified on the label and which can be included under the generic terms *natural flavoring, flavor,* and *flavoring.* Actually, the spice extracts offer the advantages of being easier to control for flavor intensity and of not being visible in the product as spice particles would be. Spice extracts result in fewer microbial contamination problems in the product because the oil-based extracts do not provide a source of spoilage microorganisms like their natural counterparts. The extracts are also easier to store because they are less bulky than natural spices.

---

[6]Dziezak, J. D. 1989. Spices. Food Technol. 40(1):102.

Table 21–4. Some Common Spices: Country of Origin, Description, Flavor and Flavor Intensity, and Typical Uses[1]

Spice and Country of Origin	Description	Flavor and Flavor Intensity[2]	Typical Uses[3]
Allspice Jamaica Honduras Mexico	Reddish brown pimento berries, nearly globular; 1/8- to 5/16-in. diameter. Available: whole and ground.	Pungent, clove-like odor and taste. **250**	Bologna, pork sausage, frankfurters, hamburgers, mince meat, potato sausage, headcheese, and many other meat food products.
Anise seed Spain Netherlands Mexico	Greenish brown, ovoid-shaped seeds; 3/16 in. long. Available: whole and ground.	Pleasant, licorice-like odor and taste. **65**	Dry sausage: mortadella and pepperoni.
Basil U.S. (California) Hungary France Former Yugoslavia	As marketed, small bits of green leaves. Available: whole and ground.	Aromatic; faintly anise-like, mildly pungent taste. **70**	Pizza sausage, certain poultry products.
Bay leaves Turkey Portugal	Elliptical leaves; up to 3 in. long; deep upper surface, paler underneath. Available: whole and ground.	Fragrant, sweetly aromatic; slightly bitter taste. **500**	Pickling spice for corned beef, beef tongue, lamb tongue, pork tongue, and pigs' feet.
Caraway seed Netherlands Poland	Curved, tapered brown seeds; up to 1/4 in. long. Available: whole.	Characteristic odor; warm, slightly sharp taste. **320**	Polish sausage.
Cardamon seed Guatemala India	Small, angular, reddish brown seeds; often marketed in their pods—greenish- or buff-colored (blanched). Available: whole, decorticated, and ground	Pleasant fragrant odor; warm, slightly sharp taste. **125**	Bologna, frankfurters, similar products.
Celery flakes U.S. (California)	Medium to dark green flakes; about 3/8-in. diameter. Available: flakes, granulated, and powdered.	Sweet, strong, typical celery odor and taste. **NA**	Chicken and turkey products.
Celery seed India France	Grayish brown seeds; up to 1/16-in. diameter. Available: whole, ground, and as salt.	Warm, slightly bitter celery odor and taste. **300**	Beef stews, meat loaf, chicken and turkey products.

(Continued)

Table 21–4 (Continued)

Spice and Country of Origin	Description	Flavor and Flavor Intensity[2]	Typical Uses[3]
Chili powder U.S. (California)	Red to very dark red powder. Contains chili pepper, cumin, oregano, garlic, salt, and sometimes other spices.	Characteristic aromatic odor, with varying levels of heat or pungency. NA	Chili con carne, taco meat, some Spanish and Mexican sausages.
Cinnamon Indonesia Seychelles Taiwan Ceylon	Tan to reddish brown quills (stocks) of rolled bark, varying lengths. Available: whole and ground	Agreeably aromatic, with sweet, pungent taste. 400	Ham loaf, other pork loaves, pastrami rub, and sometimes in bologna, mortadella, and blood sausage.
Cloves Madagascar Indonesia Tanzania	Reddish brown; ½ to ¾ in. long. Available: whole and ground.	Strong, pungent, sweet odor and taste. 600	Bologna, frankfurters, headcheese, liver sausage, corned beef, and pastrami. Whole cloves often stuck into hams when the hams are baked.
Coriander seed Morocco Rumania Argentina	Yellowish brown nearly globular seeds; ⅛- to ³⁄₁₆-in. diameter. Available: whole and ground.	Distinctively fragrant; lemon-like taste. 230	Frankfurters, bologna, knackwurst, Polish sausage, and many other cooked sausages.
Cumin seed Iran India Lebanon	Yellowish brown elongated oval seeds; ⅛- to ¼-in. diameter. Available: whole and ground.	Strong, aromatic, somewhat bitter. 290	Chorizo, chili con carne, and other Mexican and Italian sausages. Its principal use is in making curry powder.
Dill seed India	Light brown oval seeds; ³⁄₃₂ to ³⁄₁₆ in. long. Available: whole and ground.	Clean, aromatic odor; warm, caraway-like taste. 160	Headcheese, souse, jellied tongue loaf, and similar products.
Fennel seed India Argentina	Green to yellowish brown oblong or oval seeds; ⁵⁄₃₂ to ⁵⁄₁₆ in. long. Available: whole and ground.	Warm, sweet, anise-like odor and taste. 280	Italian sausages, pizza sausage, pizza salami.
Garlic, dried U.S. (California)	White material, ranging in standard particle size from powdered, granulated, ground, minced, chopped, large chopped, sliced, to large sliced.	Strong, characteristic odor; extremely pungent taste. NA	Polish sausage, most beef sausages, salamis; subtle amounts in bologna, frankfurters, and similar products.

(Continued)

Table 21–4 (Continued)

Spice and Country of Origin	Description	Flavor and Flavor Intensity[2]	Typical Uses[3]
Ginger Nigeria Sierra Leone Jamaica	Irregularly shaped pieces ("hands") 2½ to 4 in. long; brownish- to buff-colored (when peeled and bleached). Available: whole, ground, and cracked.	Pungent, spicy-sweet odor; clean, hot taste. **475**	Pork sausage, frankfurters, knackwurst, and numerous other cooked sausages.
Mace Indonesia Grenada	Flat, brittle pieces of lacy material, yellow to brownish orange in color. Available: whole and ground.	See Nutmeg; but somewhat stronger, less delicate. **340**	Bologna, mortadella, bratwurst, bockwurst, and many other sausages—both fresh and cooked.
Marjoram France Portugal Greece Rumania	As marketed, small pieces of grayish green leaves. Available: whole and ground.	Warm, aromatic, pleasantly bitter, slightly camphoraceous. **85**	Braunschweiger, liverwurst, headcheese, and Polish sausage.
Mustard Denmark Canada U.K. U.S.	Tiny, smooth, nearly globular seeds, yellowish or reddish brown. Available: whole and ground.	Yellow: no odor, but sharp, pungent taste when water is added. Brown: with water added, sharp, irritating odor; pungent taste. Powder **800** Seed **240**	Bologna, frankfurters, salamis, summer sausage, and similar products.
Nutmeg Indonesia Grenada	Large, brown, ovular seeds; up to 1¼ in. long. Available: whole and ground.	Characteristic sweet, warm odor and taste. **360**	Frankfurters, bologna, knackwurst, minced ham sausages, liver sausage, and headcheese.
Onion, dried U.S. (California)	White material ranging in particle size from powdered, granulated, ground, minced, chopped, sliced, to large sliced.	Sweetly pungent onion odor and taste. **60**	Braunschweiger, liver sausage, headcheese, and baked luncheon loaves, including Dutch loaf and old-fashioned loaf.
Oregano Greece Mexico Japan	As marketed, small pieces of green leaves. Available: whole and ground.	Strong, pleasant, somewhat camphoraceous odor and taste. **90**	Most Mexican and Spanish sausages, fresh Italian sausage; sometimes in frankfurters and bologna.

(Continued)

Table 21–4 (Continued)

Spice and Country of Origin	Description	Flavor and Flavor Intensity[2]	Typical Uses[3]
Paprika U.S. (California) Spain Bulgaria Morocco	Powder, ranging in color from bright rich red to brick-red, depending on variety and handling.	Slightly sweet odor and taste; may have moderate bite. 50	Frankfurters, bologna, and many other cooked and smoked sausage products; also fresh Italian sausage.
Pepper, black Indonesia Brazil India Malaysia	From green (unripe) pepper berries, brownish black, wrinkled berries; up to $\frac{1}{8}$-in. diameter. Available: whole, ground, cracked, and decorticated.	Characteristic, penetrating odor; hot, biting taste. 450	Most used of all spices. Frankfurters, bologna, pork sausage, summer sausage, salamis, liver sausage, loaf products, and *most* other sausages.
Pepper, red Japan Mexico Turkey U.S.	Elongated and oblate-shaped red pods of varying sizes; from $\frac{3}{8}$ to $\frac{1}{2}$ in., depending on variety. Available: whole and ground.	Characteristic odor, with heat levels mildly to intensly pungent. NA	Chorizo, smoked country sausage, Italian sausage, pepperoni, fresh pork sausage, and many others.
Pepper, white Indonesia Brazil Malaysia	From ripe, yellowish gray pepper seeds; up to $\frac{3}{32}$-in. diameter. Available: whole and ground.	Like black pepper, but less pungent. 390	Used when black pepper specks are not desired; e.g., pork sausage and deviled ham.
Rosemary France Spain Portugal U.S. (California)	Bits of pine, needle-like green leaves. Available: whole and ground.	Agreeable, aromatic odor; fresh, bittersweet taste. Somewhat like sage in flavor. 95	Chicken stews and a few other poultry products.
Saffron Spain Portugal	Orange and yellow strands; approximately $\frac{1}{2}$ to $\frac{3}{4}$ in. long. Available: whole and ground.	Strong, somewhat medicinal odor; bitter taste. 40	The most expensive of all spices. Used primarily for color, but in very few sausages.
Sage Former Yugoslavia Albania	Oblanceolate-shaped leaves, grayish green; about 3 in. long. Available: whole, cut, rubbed, ground.	Highly aromatic, with strong, warm, slightly bitter taste. 80	Pork sausage, pizza sausage, breakfast sausage, and old-fashioned loaf.
Savory France Spain	As marketed, bits of dried, greenish brown leaves. Available: whole and ground.	Fragrant, aromatic odor. 65	Used primarily in pork sausage, but good in many sausages.

(Continued)

Table 21–4 (Continued)

Spice and Country of Origin	Description	Flavor and Flavor Intensity[2]	Typical Uses[3]
Thyme Spain France	As marketed, bits of gray to greenish brown leaves. Available: whole and ground.	Fragrant, aromatic odor; warm, quite pungent taste. 85	Pork sausage, liver sausage products, headcheese, and bockwurst.
Turmeric India Jamaica	Fibrous roots, orange-yellow in color; $\frac{1}{3}$ in. long. Available: ground.	Characteristic odor, reminiscent of pepper; slightly bitter taste. 220	Used more for color than flavor. Constituent of curry powder.

[1]Adapted from information made available from the American Spice Trade Association, New York.

[2]Flavor intensity on a scale of 1,000 as the strongest flavor and 0 as no flavor. Based on the ranking of Ann Wilder, President and CEO of VANNS Spices, Baltimore, MD, as reported by Meat Business Magazine. March 1997. 58:13.

[3]The seasoning in most sausage and meat products is a blend of several spices.

NA = Not available

Extenders and Binders

Certain sausage products may contain extenders and binders such as non-fat dry milk, dried milk, dried whey, reduced lactose whey, whey protein concentrate, sodium caseinate, calcium lactate, wheat gluten, cereal flours, tapioca dextrin, soy flour, soy protein concentrate, isolated soy protein, and(or) vegetable starch at a level up to 3.5 percent of the finished product (alone or in combination). Isolated soy protein is an exception to the regulations and only may be used up to a level of 2 percent in the finished product. If these ingredients are used in the formulation, their presence must be reflected in the product name label, for example, *"Frankfurter, Cereal Added"* or *"Bologna, Soy Protein Concentrate, and Non-fat Dry Milk Added."*

The functional properties and appropriate uses of various extenders are discussed as follows.

Milk Powder

Non-fat dry milk and similar dried products of milk origin are used primarily as extenders; however, an improvement in product flavor — probably due to a sweetening effect — has been noted. The calcium-reduced form of non-fat dry milk is most commonly used, since high-calcium levels interfere with protein solubility. Dried milk powders and whey powders exhibit limited binding properties.

Cereal Flours

Cereal flours are composed principally of starch; however, they may serve as both extenders and binders. Their function varies with the source, which may be wheat, rice, oats, corn, etc. Cereal flours are generally added to the lower-quality products for economic reasons. Some of the cereal flours help to improve binding qualities, cooking yields, and slicing characteristics.

Soy Flour, Grits, and Texturized Soy Protein

These soy protein ingredients contain 50 percent protein and are used to boost the protein content and help bind water. Flour, grits, and texturized soy differ primarily in their particle size and texture. Soy grits have a larger particle size than soy flour. Texturized soy is very similar to grits except that the texture is changed to more closely duplicate the texture of ground meat. The principal use of soy flour is in nonspecific loaf products, whereas grits are often used in pizza toppings, chili, and sloppy Joe mixes. Texturized soy protein is commonly used in meat patties, meat loaves, and similar items (see Chapter 18).

Soy Protein Concentrates

Soy protein concentrates are made up of 70 percent protein available as flours or coarse granules in a form similar to grits. They are generally used in emulsion-type sausages in which they function to bind water.

One problem sometimes encountered with soy protein concentrates, grits, or flours is the development of a beany flavor. This usually occurs after storage of the product and is more pronounced when higher concentrations are used. Modern soy processing technology has substantially reduced the beany-flavor problem.

Soy Protein Isolate

This 90 percent protein product is useful as both a binder and an emulsifier. It is the only soy product that functions similarly to meat in forming an emulsion. The soy protein isolate should not be considered equal in functional quality to the contractile meat proteins (e.g., actomyosin), but none-theless, it is useful in creating more stable emulsions in marginal formulations.

Unlike soy concentrates and soy flours, isolated soy protein adds no flavor or odor of its own to formulated meat products.

Although the protein content of all these soy products is high as noted, the digestibility, by humans, of soy protein is approximately 90 percent of the digestibility of meat, dairy products, and egg proteins.[7] Also, the amino acid content/balance is not as well-matched to human requirements as are the animal protein sources. Thus, these soy products serve as excellent extenders and binders for meat products, but they are not intended to completely replace meat in the human diet (see Chapter 23). Regulatory agencies around the world are concerned about the illegal use of vegetable proteins in meat products. Researchers have developed an enzyme-linked immunosorbent assay (ELISA) procedure (see Chapter 3) that provides a simple and direct method for detecting and quantitating soy protein.

MEAT EMULSIONS

Finely comminuted sausage batters are often referred to as meat emulsions. However, by strictest definition, they are not true emulsions. An appropriate definition follows: "A meat emulsion is a finely comminuted dispersion of lean and fat particles into a two-phase system which consists of a dispersed phase (fat droplets) and a complex continuous phase composed of water, solubilized proteins, cellular components, and miscellaneous spices and seasonings." A true emulsion, such as mayonnaise, is a heterogeneous mixture of two immiscible liquids (fat and water) stabilized by an emulsifying agent such as the protein fraction albumin. The result is a stable colloidal suspension of the two liquids that do not readily separate because the emulsifying agent acts as a physical barrier between the two phases but is miscible with both. This phenomenon occurs because the emulsifying agent (the protein) will undergo a conformational change in its structure toward a point of maximum stability with the immediate environment. The protein will unfold, orienting *hydrophobic (water-hating)* portions of the protein molecules toward

---

[7]National Research Council Food and Nutrition Board. 1999. Dietary Reference Intakes. http://www.iom.edu/iom/iomhome.nsf/Pages/FNB+DRI

the lipid (fat) phase and *hydrophilic (water-loving)* portions of the protein molecules toward the continuous phase. This is the mechanism of membrane formation around lipid droplets.

A schematic illustration of the structural components present in a typical meat emulsion is shown in Figure 21–1. The scanning electron micrograph illustrated in Figure 21–2 depicts a three-dimensional view of fat droplet entrapment in the emulsion matrix (continuous phase).

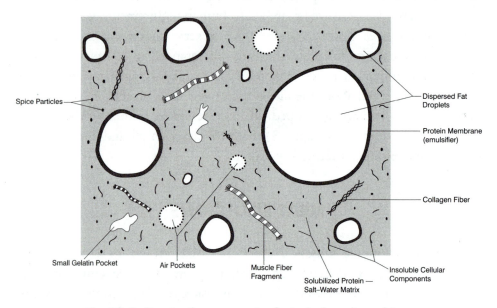

Fig. 21–1. Structural components of a typical meat emulsion.

In order for a protein to be a good emulsifier of fat and water, it must possess both hydrophobic and hydrophilic properties. Of the meat proteins, *myosin* demonstrates the best emulsifying properties and is readily abundant, comprising approximately 45 percent of the myofibrillar proteins in skeletal muscle (see Chapter 22). The protein *collagen* exhibits little or no emulsifying properties due to its unique triple-helical structure, which is extremely stable and not readily solubilized. For this reason, meat emulsions containing a high percentage of collagen are very unstable because phase separation will readily occur.

The role of soy protein concentrate and soy protein isolate as binders and emulsifiers was discussed earlier, indicating that they were useful, but not equal in functionality to myosin. Many familiar food products would not exist without emulsifiers. Because of this demand, the worldwide emulsifier market

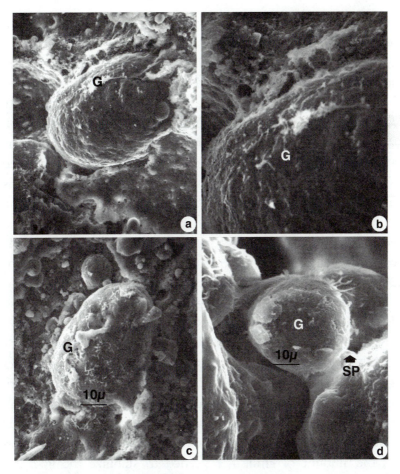

Fig. 21–2. Cooked frankfurter emulsions chopped to 60°F (a and b) and 72°F (c and d). Large encapsulated fat globules are labeled G. Solubilized proteins binding fat particles together are represented by SP. Magnifications: (a) 625 x, (b) 1250 x, (c) 1250 x, and (d) 2500 x. Scale bars represent 10 micrometers. (Scanning Electron Micrographs by Kevin W. Jones)

has been estimated to be worth over $1.6 billion. Researchers in England have designed, synthesized, and characterized a series of synthetic polypeptides with emulsifying properties.[8] Because of the small size and stability of these polypeptides, they are not damaged during processing and can even be reformed if an emulsion is broken. These polypeptides will probably be produced commercially by genetically engineered microorganisms (see Chapter 2) and might be available to the meat industry in the near future. The

---

[8]Bloomberg, G. 1990. Folded protein adds wrinkle to emulsifier technology. Prep. Foods. 159(5):93.

complexity of protein folding and the importance of temperature are explained by Mitraki, et al.[9]

Several factors that must be critically controlled in order for the sausage processor to produce a stable emulsified product are meat temperature during chopping, particle size, protein quality, fat source, fat condition, and thermal processing conditions. Controlling the chopping temperature is critical in the extraction of the contractile proteins within muscle fibers. The contractile protein myosin is known to be the principal emulsifier in comminuted meat systems. Optimal extraction temperatures of myosin have been demonstrated to occur between 35° and 45°F. In Germany, cooling meat emulsions with liquid nitrogen (see Chapter 19) in factories producing 200 tons of sausage per week has resulted in the formation of more stable emulsions with improved texture and color compared to ice-cooled emulsions. Nevertheless, the use of freshly harvested, hot processed, *prerigor* meat results in excellent protein solubility and excellent binding and emulsion strength because of its high pH (>6.0). So, both temperature and pH must be considered.

The particle size of the fat droplets can also have a profound effect on the ultimate stability of a meat emulsion, especially if the emulsifying proteins are of marginal quality (e.g., moderate collagen levels). Particle size is primarily influenced by chopping time or by the degree of comminution that is attained in high-speed emulsion mills. A reduction in particle size of the fat droplets greatly increases the exposed surface area of lipid droplets and, therefore, also increases the amount of emulsifier needed to stabilize the lipid droplets. When the surface area of the lipid droplets exceeds the level of available emulsifying agent, the emulsion will short out, or fat out. *Shorting out* refers to the separation of phases and the subsequent coalescence of large pools of fat. This is seldom a problem in meat emulsions made with high-quality raw meat materials.

The selection of raw materials is important to the sausage processor in determining the desired quality and price of the finished product. For stable emulsions, the processor should limit the use of low-binding raw materials such as tripe to no more than 15 percent of the formulation. Low-binding meats often contain a large percentage of their total protein content in the form of collagen rather than contractile proteins (see Table 21–1). Collagen may bind considerable amounts of water during chopping; however, this

[9]Mitraki, A., B. Fane, C. Hasse–Pettingell, J. Sturtevant, and J. King. 1991. Global suppression of protein folding defects and inclusion body formation. Science. 253:54.

water is lost during thermal processing. Collagen molecules will shrink up to one-third of their original length when heated to 150°F and will be converted to gelatin at higher temperatures. When collagen, as an emulsifier, surrounds a fat droplet, shrinkage and(or) gelatinization of the collagen will occur upon thermal processing. This allows fat droplets to migrate and coalesce into fat pockets. Gelatin will also migrate into gelatin pockets. In severe shorting out of the emulsion, the fat droplets will migrate to the surface to form fat caps. This problem will often occur in poor-quality formulations with a high fat content and a low ratio of contractile proteins to collagen.

The source of fat, as well as the condition of the fat, can influence the stability of meat emulsions to lesser degrees. Rendered fats such as lard will produce unstable emulsions (probably due to a large surface area of fine fat droplets). Beef, pork, and lamb or mutton fats will respond differently, due to differences in their plasticization/hardness properties based on their degree of fatty acid saturation (see Chapter 23). The harder (more saturated) fats from the kidney knob are much less desirable than the softer intermuscular fats from the carcass.

When emulsion products are heated too rapidly, some fatting out will often occur. This problem is generally associated with weaker formulations and is characterized by a greasy coating on the product surface or a small pocket of fat in the smokehouse stick mark (where the sausage hangs over the smokestick). Slower-processing schedules allow a better skin to be formed on the product surface, which minimizes this problem.

SAUSAGE CASINGS

Sausage casings may be categorized as two types: natural and manufactured. Natural casings (also called animal casings) are made from the stomachs, intestines, and bladders of hogs, sheep, and cattle. Packing houses that save casings will flush them thoroughly with water and pack them in salt, whereupon they are generally purchased by casing processors. The casing processors do the final cleaning, scraping, sorting, grading, and salting of the casings. Most animal casings are sold on the basis of a *tierce* (55-gallon barrel). The number of pieces of casing packed per tierce will depend on the size and capacity of the casing.

Anyone who purchases large quantities of natural casings should set up a quality control schedule to check each shipment for some of the important characteristics of casings, such as conformance to size, freedom from defects, salt content, and microbial contamination.

Sometimes meat processors will clean their own hog and sheep casings for sausage production. This practice is generally limited to using only the small intestines of hogs and sheep. The intestinal wall consists of five basic layers of tissue, very clearly described and pictured by Wang.[10] Small-diameter sheep and hog casings primarily consist of the collagen-rich submucosa layer, which is next to the innermost layer of the five tissues. The other four layers, including the circular and longitudinal smooth muscle layers, are removed in cleaning.

In order to be cleaned, the intestine must first be inverted. One end is turned inside out and water is run into the lip that is created. The weight of the water will force the casing to be completely inverted. The inverted casing should then be thoroughly washed in a dilute chlorine solution (0.5 percent) and brushed with a soft-bristled brush to remove excess fat and connective tissues. Next, the casing should be rinsed in clean water (inside and out) and inverted back into its original form and packed in a saturated salt solution (1 g salt/2.8 ml H_20 @77°F) for storage. It is best to put the casings in a freezer for long-term storage (more than two weeks). The casings will not freeze in the saturated salt solution. Halophilic (salt-loving) bacteria do not have an opportunity to grow as they might if the casings were stored in a cooler. The saturated salt solution inhibits the growth of other bacteria that survived the dilute chlorine solution wash.

Casings should be thoroughly flushed prior to being used to remove excess clinging salt. However, they should not be overflushed because of the possibility of removing the natural oils that make the casings easier to work with. Individuals who use a limited quantity of casings at any one time would benefit from purchasing pretubed casings. The plastic sleeve containing a length of casing crimped together to conserve space can be rinsed and slipped onto the stuffing horn without disturbing the larger supply of casings stored in the freezer. Processors who save their own casings can store them this convenient way as well.

Hog Casings

Natural hog casings used in sausage production include bungs, stomachs, middles, small casings, and bladders.

---

[10]Wang, H. 1954. Histology of Beef Casings. Bull. 15, American Meat Institute Foundation.

Hog Bungs

Hog bungs are straight sections of the large intestine, generally cut into 32-inch lengths. They are graded according to size and may hold $1\frac{1}{2}$ to 4 pounds per each 32-inch cut piece (depending on the grade) after being stuffed. The grade names from largest to smallest are export, large prime, medium prime, special prime, small prime, and skip, with a diameter range of over 2 inches down to $1\frac{1}{8}$ inches. Specially selected grades of hog bungs are often used to manufacture sewed hog bungs for stuffing liver sausage, thuringer, and Genoa salami. Sewed hog bungs are manufactured to specified sizes and are generally made of double or triple thicknesses for additional strength (see Figure 21–3).

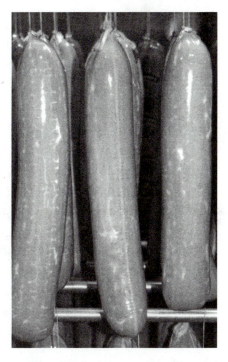

Fig. 21–3. An example of sewed hog bungs used for liver sausage production.

Hog Stomachs

Hog stomachs are principally used for stuffing headcheese and will hold between 5 and 8 pounds. Three hundred pieces are packed to a tierce.

Hog Middles

Hog middles are made from the caecum, or blind gut (internal end of the large intestine), of the hog. They are used for stuffing frizzes, Italian-type sausages, and liver sausage.

They are put up according to three specifications:

- *Hog middle caps* are the closed ends of the hog middles and are sold by the piece in tierces packed to capacity.

- *Cap-on hog middles* are the total length of the middles containing both the cap ends and the open ends. Cap-on middles measure 8 to 10 feet in length with a diameter of 3 to $4\frac{1}{2}$ inches. There are 150 cap-on middles packed to a tierce.

- *Cap-off hog middles* have the cap ends removed and thus are open on both ends. Cap-off middles measure 8 to 10 feet in length and are packed 180 pieces to a tierce.

Small Hog Casings

Small hog casings are made from the small intestines and have three commonly accepted grades: narrow, medium, and wide. These grades may be further subdivided by some of the casing processors, depending on the demands of their customers. The casings are put up in "hanks," with 290 to 355 hanks per tierce, depending on the grade (size). A hank is the equivalent of 100 yards of casing tied up in a bundle. The stuffing capacity of narrow hog casings is approximately 90 pounds per hank; medium hog casings, 115 pounds per hank; and wide hog casings, 135 pounds per hank. Small hog casings are used for stuffing products such as Polish sausage, smoked pork sausage, bratwurst, and many other coarse-cut, small-diameter sausage products.

Hog Bladders

Hog bladders are sold in both salted and dried form and are used primarily as casings for minced luncheon meats. Hog bladders are more fragile than beef bladders, are round in shape, and have a stuffing capacity of 2 to 6 pounds. They are most commonly put up salted and packed in tierces to capacity.

Sheep Casings

Sheep casings are among the most valuable of the animal casings and are produced from the small intestine. The other portions of the digestive tract of sheep are not generally used as sausage casings. The length, diameter, and thinness of tissue make sheep casings ideal for most of the small-diameter sausages. Sheep casings are used primarily for stuffing frankfurters and pork sausages. After they have been cleaned, they are graded for quality, measured for diameter and length, and salted and packed according to the specifications required by the trade.

North American sheep casings are put up in 100-yard hanks and graded according to size. The available sizes and their stuffing capacities per hank are as follows: 16 to 18 mm (37 lbs.), 18 to 20 mm (46 lbs.), 20 to 22 mm (55 lbs.), 22 to 24 mm (64 lbs.), 24 to 26 mm (72 lbs.), and 26 mm and over (77 lbs.).

Sheep casings from South America are generally taken from older animals and thus are stronger casings with a slightly higher stuffing capacity. They also are graded at 2-mm size increments, starting at 17 to 19 mm for the smallest available and ranging up to 27 mm and over for the largest grade.

Other countries that supply sheep casings to the United States are New Zealand, Iran, Iraq, Syria, and the former USSR. Of these, New Zealand is the largest supplier of sheep casings to the United States.

Beef Casings

Beef casings that are used in sausage production include the weasand (esophagus), rounds, middles, bungs, and bladders.

Beef Weasand

A beef weasand casing is made from the tube leading from the throat (esophagus) to the first stomach (rumen). Preparing this tube necessitates removing the weasand meat from the casing, inspecting the casing for grubs, and, if the casing is free of grubs, inflating the weasand with air and then drying it. At this point, the casing is graded for quality and size. Weasand casings are used for stuffing bologna and salami.

Beef Rounds

Beef rounds, made from the small intestine, are used for stuffing ring bologna, holsteiner, and mettwurst. Beef rounds are put up in "sets" containing a minimum of 100 feet and no more than five pieces with a length less than 5 feet. They are separated into "export" (those that have a clear surface) and "domestic" (those that have small nodules of scar tissue left by intestinal parasites) grades and then are further graded according to size. Exports have four size grades: narrow (29 to 36 mm), medium (36 to 38 mm), wide (38 to 44 mm), and extra wide (over 44 mm). Export rounds have a stuffing capacity of 55 to 90 pounds (per 100-foot set), depending on the size grade. Domestic rounds are available in two sizes: wide (over 38 mm) and regular (under 38 mm). Domestic rounds have a stuffing capacity of 65 to 75 pounds (per 100-foot set), depending on the size grade.

Beef Middles

Beef middles obtained from the large intestine are used for stuffing bologna, cervelat, and salamis. Middles are put up in "sets" containing a minimum of 57 feet and with no more than five pieces comprising a set. They are graded according to the following sizes: medium (45 to 50 mm), regular (50 to 60 mm), wide (60 to 70 mm), and extra wide (over 70 mm). Their stuffing capacity ranges from 55 to 95 pounds (per 57-foot set), depending on the graded size.

Beef Bungs

Beef bungs obtained from the caecum or appendix of cattle are used for stuffing bologna and cappicola. Beef bungs are put up by the piece in tierces that may contain 350 to 500 pieces, depending on the size. The diameter of bungs will range from 3 inches to over 5 inches, representing eight different grades. The stuffing capacity per bung will range from 12 to over 20 pounds.

Beef Bladders

Beef bladders are sold in both salted and dried form and are used to stuff mortadella and various luncheon meat specialties. Bladders have a diameter range of less than 5 inches to over $7\frac{1}{2}$ inches and a stuffing capacity of 3 to 10 pounds.

The stuffing capacities given in the preceding discussion are approximate. The actual stuffing capacities will vary, depending on how tight the stuff is, whether or not the sausage is linked, and the natural size variations that occur between animal casings. Anyone who wishes to obtain more information on natural casings should refer to Rust's *Sausage and Processed Meats Manufacturing*[11] from which much of this information was derived.

Manufactured Casings

Manufactured casings account for the vast majority of casings used to stuff sausage products. The principal advantages of manufactured casings over

---

[11]Rust, R. E. 1975. Sausage and Processed Meats Manufacturing. AMI Center for Continuing Education, American Meat Institute, Chicago.

their natural counterparts are price, uniformity, and versatility. There are three classifications of manufactured casings: cellulose, collagen, and casings formed from the meat.

Cellulose Casings

Cellulose casings are inedible, provide high-structural strength, and display good permeability to moisture and smoke when the casings are moist. The three types of cellulose casings manufactured are small cellulose, large cellulose, and fibrous. Small and large cellulose casings are produced from cotton linters (the fuzz from cotton seeds), which are dissolved and regenerated into casings. Fibrous casings are the toughest of all casings and are made from a special paper pulp base that is impregnated with cellulose.

Small cellulose casings are used to produce skinless wieners and similar skinless small-diameter products. These casings are available as clear casings or in a variety of tinted colors that impart a food-grade dye on the product's surface. Small cellulose casings are designed to be easily peeled from the product's surface (see Figure 21–15). They are available in sizes ranging from 15 to 50 mm in diameter and 70 to 160 feet in length. They are generally produced in the "shirred" form, which means they are crimped into short strands. For example, a shirred casing that is 8 inches in length may stuff out at 100 feet. These casings should *not* be allowed to dry out and become brittle in storage, which will make them difficult to work with. Nevertheless, they should *not* be soaked prior to being used.

Large cellulose casings are used for stuffing products such as bologna and braunschweiger. They are often colored but do not impart dye onto the product's surface. They are available as moisture permeable or impermeable. Impermeable large cellulose casings are used for water-cooking processes (as in braunschweiger).

Fibrous casings exhibit the greatest structural strength and thus insure uniform product diameter from end to end. This is an important factor in portion-controlled luncheon meat slicing where uniform slices are essential in automated slicing/packaging lines. Fibrous casings are commonly used for luncheon meat specialties, summer sausage, bologna, and many other products. Large cellulose and fibrous casings should be soaked prior to being used. Casings can be pre-stuck, which consists of puncturing small holes along the entire casing to allow surface air to escape and to hasten drying. If the casings are not pre-stuck, piercing can be done after stuffing with a casing perforator that has eight long, tapered stainless steel needles attached to a round base. This small, handy tool is available from any supplier. Large casings are often pre-drilled in the end(s) to allow entrapped air to escape as the casing is being filled (stuffed out) with meat emulsion.

Collagen Casings

Collagen casings were developed to be an edible casing replacement for natural casings, and at the same time, to have the uniformity of a manufactured product. Collagen casings are made from the corium layer of split beef hides. The collagen from the corium layer is ground, swelled in acid, sieved, filtered, and finally extruded. These casings are generally limited to smaller-diameter products because of their lower structural strength. Palatability of the casing is a function of casing diameter (and subsequent membrane thickness). Larger-diameter collagen casings (which are heavier casings) are generally less palatable than smaller-diameter casings. Unlike large cellulose and fibrous casings, collagen casings should *not* be soaked in water prior to being stuffed. As the product is stuffed into the casings, the casings absorb moisture from within the meat, making them pliable for linking. The product stuffed into collagen casings may be smoked immediately upon entering the smokehouse (no drying period needed). Collagen casings are available in shirred form or as regular flat casings. Small-diameter shirred casings tend to become brittle with age, resulting in breakage problems during stuffing. It is best to store these casings in a 40°F cooler and to use them within six months.

Co-extruded collagen casings are a newer type of casing being used in the sausage industry. The casings are actually formed from a collagen paste that is co-extruded over the sausage batter during stuffing. This co-extrusion process requires very specialized equipment and a very large capital investment. It is designed exclusively for high-volume operations.

Casings Formed from the Meat

A relatively new extrusion system forces the meat emulsion into a porous mold that causes the soluble protein (myosin) to coat the inside of the mold. Acetic acid is then pumped through the mold, which lowers the pH and causes the myosin to coagulate on the surface, forming a natural skin.

This whole process is conducted in a cold environment with a cold product, so the product texture is not changed. Such an extruder in production is pictured in color on the cover of the November, 1990, issue of *Meat Business Magazine*.[12] This process is being used by several companies, each of which produces 50,000 pounds or more of sausage per week.

Casey B. Frye of Burke Corporation, summed up the last two casing concepts in a column in the Institute of Food Technologists' Muscle Foods Division Newsletter.[13]

Sausage Links Without Traditional Casing

Typical sausage link production requires some type of casing to form and encase the sausage for handling, cooking and chilling. Casings of some sort have been used for thousands of years, and today they are used to provide certain characteristics to sausage products, as well as serving as meat containers.

However, casings contribute a significant cost to sausage products. To lessen costs, technology is available to manufacture sausage links without casings. The most common methods for forming "casingless" sausage links are by forming in a modified patty machine and by extrusion. Two methods that are not so common include co-extrusion and sintered mold sausage link manufacturing.

Co-Extrusion Sausage Links

Scientists at Unilever (London, England) developed a cheaper alternative to casing sausage products in the 1960s. This first system was developed with breakfast sausage, but those efforts failed.

Unilever then partnered with the engineering efforts of Protecon (formerly Stork) in Oss, the Netherlands, to develop a production system for smoked link sausages and frankfurters.

In 1980, a license agreement between Unilever and Protecon was established and Protecon began selling co-extrusion systems outside of Unilver. By the late 1970s, the system established manufacturing successes in Spain, Holland, Sweden and Japan, but it has had only limited success in the United States.

---

[12]Admore Publishing Co. Inc., 9701 Gravois Avenue, St. Louis, MO 63123-4389.

[13]Frye, C. B. 1999. Sausage links without traditional casing. Institute of Food Technologists' Muscle Foods Division Newsletter. Fall.

Basically, the system works by extruding a collagen batter around an endless rope of sausage batter. The sausage rope is then briefly bathed in a salt brine, crimped to length, dried, sprayed with liquid smoke, dried again, and then further processed. Currently, Townsend Engineering (Des Moines, Iowa) has refined the process and has made it more cost effective with higher throughput and less variability than the older Protecon system.

Regarding co-extruded products, European manufacturers are experiencing 6.5 days of production with one half day of mainte- nance downtime. Changeover time is short with computerized read- justments.

About 40 percent of the frankfurters in the Spanish markets are those made by co-extrusion. Acceptance is limited, however, because of high capital costs relative to limited production capacity.

The system has been somewhat difficult to cost justify in the United States based on casing and labor savings. The additional cost of retro-fitting current sausage systems is also a deterrent to use.

Sintered Mold Sausage Links

In the late 1980s, engineers at APV Baker (Peterborough, Eng- land) developed a system to make sausage links by denaturing the outer surface of the link, thus creating its own skin without the use of traditional casings.

The system has since found acceptance with several companies in England and France, and limited use in the United States and Japan.

The process involves pumping a meat batter through a rotary valve into portioning cylinders. The portioned meat is then seg- mented into fill nozzles that load into molds made of porous stain- less-steel (sintered molds).

Depending on the product to be manufactured, 8 to 12 such molds are in each mold block. Five mold blocks on the machine are arranged in a pentagon configuration. The pentagon shifts one posi- tion with every fill cycle.

While the portioned meat is in the mold, a modified acetic acid solution is injected through the porous walls of the sintered molds. This acetic acid solution denatures the surface proteins of the raw sausage batter, thus creating a casingless natural protein skin sausage link. With every shift in rotation the pentagon of mold blocks has a specific function for proper link development. After the links are formed they are passed through a water rinse and air knife to rinse off excessive acetic acid and quickly dry the surface. This step is needed for marketing uncooked links but is unnecessary if they are to be cooked soon after forming.

Different link sizes dramatically change throughput. No flavor concerns are noted from acetic acid and the acetic acid does not have to be listed on the ingredient statement.

PRINCIPLES OF
SAUSAGE PRODUCTION

Some students, after reading and studying this chapter, may feel inclined to go into sausage manufacturing in more detail. Individuals who seek additional information on the principles and modern practices of sausage production should refer to *Sausage and Processed Meats Manufacturing,* written by Professor Robert E. Rust, retired but still associated with Iowa State University, which is a very complete compilation of information related to sausage manufacturing practices.

The fundamental processing methods that are important in the successful production of sausage are particle reduction or comminution (chopping, grinding, emulsifying, or flaking), mixing or blending, stuffing, smoking, peeling, and packaging. These manufacturing processes are discussed throughout the remainder of this chapter.

Manufacturing Procedures for Fresh Sausage

Pork sausage comprises the vast majority of fresh sausage produced throughout the world, whether it be in bulk, link, or patty form. Two sources of raw materials are used to produce fresh pork sausage: trimmings from pork cutting operations and boneless pork derived from boning entire hog carcasses (usually sows). If by-products or variety meats are used in fresh sausage, the sausage must be properly labeled. Most recently, fresh turkey sausage has emerged on the marketplace and appears to have established a steadily growing market share. Fresh turkey sausage is generally made from boneless drumsticks, mechanically separated turkey meat, and(or) miscellaneous trimmings from the boning lines (see Chapter 18). Other common fresh sausages include fresh thuringer, fresh bratwurst, fresh bockwurst, and country-style sausage.

The term *pork sausage* generally means the ground and seasoned fresh pork product. There are, however, several kinds or "styles" of fresh pork sausage that differ in texture, seasoning, and meat content.

COUNTRY-STYLE — Country-style usually contains from 10 to 20 percent of beef ground with the fresh pork; it is coarsely ground, with the $\frac{3}{16}$-inch plate, and does *not* contain sage as a seasoning. It is stuffed into hog casings or regenerated collagen casings of different sizes and is unlinked. It is also sold loose (unstuffed).

The term *farm* or *country* cannot be used on labels in connection with products unless such products are actually prepared on the farm or in the country. However, if the product is prepared in the same way as on the farm or in the country, the term *farm* or *country* qualified by the word *style,* in the same size and style of lettering, may be used. Further, the term *farm* may be used as part of a brand designation when it is qualified by the word *brand* in the same size and style of lettering and followed with a statement identifying the locality in which the product was prepared. Sausages containing cereal grains or soy or vegetable proteins *cannot* be labeled *"farm-style"* or *"country-style."*

BREAKFAST-STYLE — Breakfast-style is an all-pork sausage that is finely ground and seasoned with sage, salt, and pepper. It is stuffed into narrow or medium sheep casings, narrow hog casings, or regenerated collagen casings, which are then linked to make the various-sized sausages.

WHOLE HOG SAUSAGE — Whole hog sausage is sausage prepared with fresh and(or) frozen pork meat in such proportions as are normal for a single animal and may be seasoned with condiments as permitted in any sausage product. It *cannot* be made with any lot of product which, in the aggregate, contains more than 50 percent trimmable fat, that is, fat which can be removed by thorough practicable trimming and sorting. To facilitate chopping or mixing, water or ice may be used in an amount not to exceed 3 percent of the total ingredients used.

Modern whole hog sausage is made from 240- to 250-pound butcher hogs. Depending on the fatness (grade) of the hogs, the leaf fat may be pulled out of the raw material for sausage or, conversely, certain lean muscles (ham, loin, tenderloin) may be pulled out if the hogs are too lean. Of course, the carcasses must be boned out before being ground into sausage.

The Code of Federal Regulations (CFR sec. 319.141) stipulates that fresh pork sausage may not contain more than 50 percent fat. However, optimum palatability appears to be found around 30 percent fat. The ratio of fat to lean has a pronounced effect upon the flavor, texture, and tenderness of the sausage. Up to a point, a fatter pork sausage will exhibit more desirable palatability characteristics; however, consumer preferences are creating a greater demand for leaner sausage. Producing leaner pork sausage requires quality raw materials. (Review the section on quality control in Chapter 18.) Pork blade meat trimmings (see Chapter 14) work very well. The finished product should have a bright red appearance of the lean pork mottled with the white of the fat. Good lean and fat particle distinction are paramount in a high-quality fresh pork sausage.

Grinding

The first step in fresh sausage processing is grinding. Prior to grinding, the meat should be chilled to 32°F to minimize fat smearing of the product. The pork trimmings are ground through a $\frac{3}{16}$-inch grinder plate with a bone-chip collection attachment, if available (see Figure 18–5). Some processors prefer to grind the lean pork through a $\frac{1}{8}$-inch plate and the fatter pork through a $\frac{3}{16}$-inch plate before blending them together. It is extremely important that grinder knives always be kept sharp and matched with the same plate. The meat temperature must be maintained at close to freezing. Grinding prepares the meat for easy mixing with the spices.

Fresh sausage prepared in a silent cutter usually has a slightly brighter color that is retained longer; however, the chopper has no effective means of screening out bone or cartilage chips that were missed in the boning operation.

Seasoning Pork Sausage

The most satisfactory way to get the seasoning nearly right for the greatest number of people is to *weigh* the meat and add the following mixture for each 100 pounds of ground pork: 28 to 30 oz. table salt, 6 oz. black pepper, and 2 oz. ground sage.

> ≡ ≡ **NOTE:** Small quantities are more accurately weighed in grams
> (1 oz. = 28 g). ≡ ≡

This imparts an excellent flavor, and the different batches of sausage will always be seasoned the same. Even so, seasonings can be adjusted to cater to the tastes of the largest number of clientele. Many people prefer butcher's pepper (coarsely ground black pepper) instead of table pepper.

Those desiring a more highly seasoned sausage might try the following formula:

Highly Seasoned Pork Sausage

2 lbs. salt	⅜ oz. (10 g) red pepper
6 oz. dextrose (corn sugar)	⅜ oz. (10 g) Jamaica ginger
3 oz. white pepper	⅜ oz. (10 g) thyme
2 oz. (56 g) rubbed sage	

Mix the preceding ingredients with each 100 pounds of ground pork.

Mixing

The ground pork trimmings are placed in a mixer with salt, pepper, and sage (or any other spices that may be used) and are then thoroughly mixed with the seasoning ingredients as well as any antioxidants that may be used.

An alternative method preferred by many processors is to grind the pork through a coarse plate (½ to ¾ inch) and then mix in a blender with the seasonings. After thorough mixing, the sausage is reground through a ³⁄₁₆- or ⅛-inch grinder plate. A combination mixer/grinder lends itself to this method very well and minimizes the cleaning of equipment when the job is completed.

Excessive mixing after the salt has been added can cause too great an extraction of the soluble proteins (largely myosin) and can result in a tough, rubbery product. If fat begins to adhere to the inside of the mixer, it is a sure sign of overmixing.

Stuffing

The majority of the pork sausage produced is stuffed into 1- or 2-pound chubs (bulk-style) in opaque plastic bags. When pork sausage is stuffed into casings and linked, the appearance is much more critical. Many processors prefer to stuff at 28° to 30°F to minimize fat smearing and to get a good "bloom" on the product. Pork sausage is generally stuffed into narrow or medium sheep casings, narrow pork casings, or collagen casings. Natural casings should be flushed with water prior to being used, which also enables them to be placed on the stuffing horn with greater ease. During stuffing, each casing should be held firmly between the fingers on the stuffing horn, with sufficient pressure applied so that the casing fills to maximum capacity (or slightly less) with no air pockets.

Bratwurst

100 lbs. pork trimmings   3.2 oz. (90 g) black pepper
   (70% to 80% lean)   1.6 oz. (45 g) mace
1.9 lbs. salt   1.6 oz. (45 g) coriander
1.9 lbs. non-fat dry milk   1.6 oz. (45 g) hot mustard
1.9 lbs. water

Grind pork through a ⅜-inch grinder plate, mix in the ingredients uniformly, regrind through a 3/16-inch plate, and stuff into hog casings.

Manufacturing Procedures for Cooked and Smoked Sausages

There are many varieties of cooked and smoked sausages, each requiring slightly different processing procedures. Products such as frankfurters, bologna, and many of the loaf products are made from finely comminuted meat emulsions. Other products, such as smoked pork sausage, Polish sausage, and smoked thuringer, are coarser cut in texture and do not require the use of a silent cutter or an emulsion mill. Since frankfurters (wieners/hot dogs) and bologna are the most popular of all sausages, their production will be discussed first.

Frankfurters and Bologna

The manufacture of frankfurters (wieners/hot dogs) is more complex with more critical steps than the production of fresh sausage. For optimum stability of emulsion products, a lean-meat source and a fat-meat source should be handled separately.

GRINDING — The first operation in the production of frankfurters is grinding. Grinding procedures vary with processors. Some prefer to grind the lean trimmings of the formula through a ¾- to 1-inch plate and depend upon the silent cutter to achieve the necessary fineness before stuffing. Others prefer to grind the lean meat through a finer plate (e.g., ⅛ to ¼ inch) and to use an emulsion mill after blending to achieve the desired emulsion texture. The first procedure increases the capacity of the grinder, while the latter procedure is more readily adapted to continuous processing lines. Usually the lean meats (generally beef) are chopped or ground first and to a finer consistency than the fat meats (generally pork).

CHOPPING — After the lean meat has been ground, it is placed in a bowl chopper (silent cutter), and the salt, curing ingredients, and one-half of the water (generally in the form of ice) are added. The temperature of the meat will rise during chopping because of the friction of the knives passing through the meat. The lean meat should be chopped to a temperature of 42° to 45°F before the remaining water, fat meat, and spices are added. Chopping should continue until a temperature of 62° to 63°F is reached. For optimal stability of the final product, temperature is more critical than chopping time. When an emulsion mill is used following chopping, the meat should be chopped to a temperature of 52° to 55°F. When "hard" fats [more saturated with higher melting points (see Chapter 23)] are used as the fat-meat source (e.g., beef or mutton), the end-point chopping temperature should be 2° to 3° higher (65° to 66°F) in order to make the fat plastic (soft/flexible) enough to produce a good emulsion.

Other ingredients, particularly nonmeat proteins such as cereal flours and non-fat dry milk, are added with the last ingredients, since they readily absorb water, making it less available for soluble protein extraction. It is important that half the water (generally in the form of ice) be added initially to produce a brine concentration of sufficient strength to solubilize the meat proteins, which are essential for a stable emulsion. If nonskeletal meats, which are high in collagen (such as tripe), are used in the formulation, they should be added with the fat-meat portion.

BLENDING — Most larger meat processors employ a practice known as "preblending" in the production of frankfurters and other emulsion products. *Preblending* refers to the blending of the salt, cure, and a portion of the water with coarse-ground meats prior to use. Preblends are generally held 24 hours before they are chopped. (The reasons for and advantages of preblending are discussed later in this chapter.) Some processors use a grinder, a mixer, and an emulsion mill to produce emulsion products — skipping the use of the bowl chopper. If this procedure is followed, both the lean and the fat trim must be ground finer (⅛ inch and ¼ inch respectively). The lean meat is then blended with the salt, cure, and one-half of the water for several minutes before the remaining ingredients are added. The meat batter is then emulsified in an emulsion mill.

STUFFING AND LINKING — Home sausage operations or very small processing operations may forego purchasing the chopper, mixer, and stuffer and purchase only a grinder. There are stuffing horns that will fit onto most grinders. Prior to making the final grind, a processor could mount a stuffing horn in front of the grinder plate, under the tightening ring, and then stuff directly out of the grinder. However, in normal sausage production operations, the chopped meat is placed in a stuffer that may be a piston type or a continuous vacuum. Frankfurters and similar products are then stuffed into cellulose casings (for skinless wieners) or sheep, hog, or edible collagen casings of the appropriate diameter. Bologna is generally stuffed into 3½- to 4-inch diameter fibrous casings (except for ring bologna, which is stuffed into 1¼- to 1¾-inch cellulose, collagen, or natural beef round casings).

Cleaned and salted hog or sheep casings should be flushed with water prior to being placed upon the stuffing horn. The water should be expelled from the casing by being stripped between the operator's fingers before the casing is stuffed. As the casing fills with meat, it is allowed to work off the

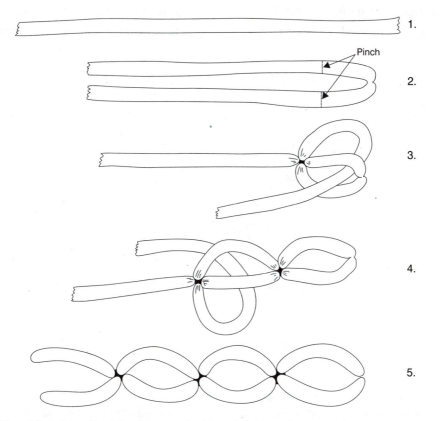

Fig. 21–4. The "braiding" method of hand linking. Stepwise description: (1) start with a 5– to 6–foot section of stuffed casing; (2) fold it in half, twist it at the fold so it holds and then pinch the casing on both halves about 5 inches from the fold; (3) fold one side over the pinched area and then work the end back through the loop that has just been formed; and (4) repeat the process until the ends of the casing have been reached (5).

stuffing horn. The operator should hold it firmly to insure that it is filled to maximum capacity and to eliminate air pockets.

Linking machines link natural casings, collagen casings, and cellulose casings to produce a uniform link portion (see Figures 21–11 and 21–12). Otherwise, natural and collagen casings may be linked manually, using one of two methods. The first method, known as braiding, is accomplished by taking a 5- to 6-foot section of stuffed product, folding it in half, and twisting it at the fold so that it holds. The casing is pinched about 5 inches from the folded end on both halves. One side is folded over the pinched spot on the second side, and the other end is looped through the loop that is created. This process is repeated to the ends of the casing (see Figure 21–4).

The second hand-linking method is accomplished by twisting every other link in opposite directions (see Figure 21–5). With this method, more care is required in hanging the product on smokehouse trucks because the links can become untwisted. However, this method is faster than the braiding method.

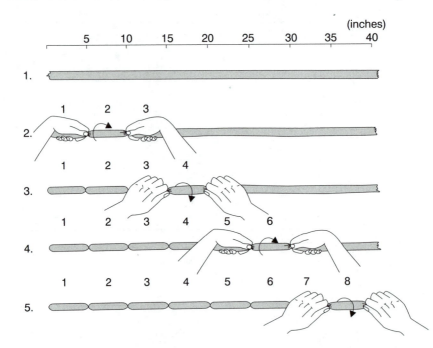

Fig. 21–5. The "alternate twist" method of hand linking. Stepwise description: (1) the casing to be linked can be virtually any length; (2) start on one end, and pinch the casing between thumb and forefinger at points 5 and 10 inches from the end, using both hands, twisting the link between your hands away from your body; (3) continue in the same fashion, pinching the casing at points 15 and 20 inches from the same end, except this time twist the link between your hands towards yourself; and (4 and 5) continue twisting every other link in the opposite direction as the preceding twist until you reach the end of the casing.

SMOKING THE SAUSAGE — See Chapter 20 for a discussion of meat smoking. In smoked sausage manufacture, the smoke should be applied to a surface that is rather sticky or slightly moist but not wet. A short drying cycle is generally completed in the smokehouse before smoking is initiated. This drying cycle conditions the meat surface for optimal adherence of smoke particles to the product. The length of time a product is smoked depends upon the smoke density inside the smokehouse, the surface condition of the product, and the air velocity inside the smoking chamber.

In the modern closed system thermal processing chambers, a good smoke application can be achieved in 30 minutes. Liquid smoke can be uniformly applied in three to five minutes by a pulsing atomization system.

If the product surface is too wet during smoking, an unattractive streaking of smoke will occur on the product. If the product surface is too dry during smoking, little smoke will adhere to the surface, and the sausage will be pale in color. When loaded onto a smokehouse truck, the product should be evenly spaced, with the product not touching any other product. This is essential for uniform surface color.

THERMAL PROCESSING — Thermal processing is the last critical step in producing a stable emulsion product. The highest-quality meat formulations can produce an inferior product when an improper thermal processing schedule is followed. The modern thermal processing chamber serves many functions, including (1) drying, (2) smoking, (3) heating, (4) cooling, and (5) adding humidity. Each of these functions can be carefully controlled independently.

Processed meat products may be processed in "batch-type" or "continuous" thermal processing chambers. The continuous smokehouses are primarily used for small-diameter sausages with continuous frankfurter processing lines capable of producing in excess of 12,000 pounds of franks per hour.

In a typical system, the frankfurter emulsion is stuffed into a cellulose casing and linked with a high-speed mechanical linker (see Figure 21–11). After the franks have been stuffed and linked, the remaining processing time is usually 60 to 80 minutes. The links are then treated with either one or a combination of coagulants (natural smoke, vinegar, citric acid, or malic acid mixed with liquid smoke). The coagulants precipitate the meat proteins at the inner surface of the casing (to aid peelability) and contribute to color and flavor.

After coagulation, the franks are cooked in high air velocity, high relative humidity cooking zones in a very large cooker (see Figure 21–14). The cooked franks are then chilled, peeled, and packaged. Chilling can be accomplished in about 10 minutes by spraying with a 60 percent brine solution (NaCl — see previous discussion about salt in this chapter) and refrigerated at about 28°F. The product is chilled to 38° to 40°F.

High-speed peeling machines can peel franks at a rate in excess of 5,000 pounds per hour with one operator. A knife edge slits the moistened casings as the strand of franks moves through the machine. An airjet assists in blowing the casings off the franks (see Figure 21–15).

Using ascorbate and leaving the product in the oven after the internal meat temperature has reached 150° to 155°F is essential to getting the characteristic cured-meat color in such a short processing time. This extra heating to develop color takes about 20 minutes in a high air velocity 200°F oven.

In batch systems, the franks are generally processed in a graduated, stepwise processing schedule. A typical smokehouse schedule is as follows:

Time	Temperature	Rel. Humidity (%)	Function	Damper
30 min.	125°F	25	Drying	Open
1 hr.	140°F	35	Smoking	Closed
1 hr.	165°F	35	Cooking	Closed
10 min.	180°F	100	Steam cooking	Closed

Heating meat products (cooking) is a very complex subject. Anyone desiring to know about it should consult the very complete explanation given by Robert E. Hanson, Alkar Division of DEC International, in the Proceedings of the Forty-third Reciprocal Meat Conference.[14]

Formulations for Cooked and Smoked Sausages

Quality Deli Franks

50 lbs. fresh regular pork trimmings (50% lean)
30 lbs. bull or cow trimmings (90% lean)
4 oz. white pepper
4 oz. paprika

20 lbs. fresh veal trimmings
26 lbs. crushed ice
2 lbs. 4 oz. salt
5 oz. dextrose
⅞ oz. (24 g) monosodium glutamate

---

[14]Hanson, R. E. 1990. Cooking Technology. Proc. 43rd Reciprocal Meat Conf. 43:109.

4 oz. certified cure (6% nitrite)
2 oz. (56 g) phosphate
1½ oz. (42 g) onion powder
⅞ oz. (24 g) sodium ascorbate

¾ oz. (22 g) mace
¾ oz. (22 g) ginger
½ oz. (14 g) allspice

≡ ≡ NOTE: Small quantities are more accurately weighed in grams
(1 oz. = 28 g). ≡ ≡

Grind bull or cow trimmings through a ¼-inch grinder plate; veal and pork trimmings through a ⅜-inch grinder plate. Place bull or cow trimmings in a bowl chopper, and add one-half (12 pounds) of the moisture (crushed ice), the salt, and the certified cure. (Certified cure may be purchased from one of many spice companies that advertise in the industry publications listed in the Appendix.) Chop to a temperature of 42° to 45°F. While continuing chopping, add the remaining meat ingredients, the remaining ice, the spices, and the other ingredients to the chopper. Chop the emulsion to 62° to 63°F. Stuff into narrow or medium sheep casings, and link at 5-inch intervals. Use the batch smoking schedule described previously and cook until the internal product temperature reaches 155°F.

Bologna

60 lbs. cow meat
  (85% to 90% lean)
40 lbs. regular pork trimmings
  (50% lean)
20 lbs. crushed ice
2 lbs. salt
3 lbs. non-fat dry milk
5 oz. dextrose
4 oz. white pepper
4 oz. certified cure (6% nitrite)

2 oz. (56 g) phosphate
2 oz. (56 g) paprika
1 oz. (28 g) coriander
1 oz. (28 g) onion powder
1 oz. (28 g) monosodium
  glutamate
¾ oz. (22 g) mace
¾ oz. (22 g) ginger
½ oz. (14 g) cloves
¼ oz. (7 g) garlic powder

≡ ≡ NOTE: Small quantities are more accurately weighed in grams
(1 oz. = 28 g). ≡ ≡

Follow the procedures for frankfurters, except stuff into 3½- to 4-inch fibrous casings. If ring bologna is to be made, stuff into 1¼- to 1¾-inch collagen casings or beef rounds 18 inches in length, and tie into a loop. Using the batch smoking schedule described previously, cook until the internal product temperature reaches 155°F.

Polish Sausage

50 lbs. special pork trimmings (80% lean)	4 oz. paprika
30 lbs. boneless cow trimmings	3 oz. black pepper
20 lbs. regular pork trimmings (50% lean)	2 oz. (56 g) white pepper
15 lbs. cold water (crushed ice if a chopper is used)	1 oz. (28 g) monosodium glutamate
2 lbs. 8 oz. salt	1 oz. (28 g) onion powder
5 oz. dextrose	¾ oz. (22 g) mace
4 oz. certified cure (6% nitrite)	¾ oz. (22 g) allspice
	½ oz. (14 g) garlic powder

≡ ≡ NOTE: Small quantities are more accurately weighed in grams (1 oz. = 28 g). ≡ ≡

Grind the lean beef and pork trimmings through a ¼-inch grinder plate; regular pork trimmings through a ⅜-inch grinder plate. Add the lean meat to the blender with the salt, cure, and water (or crushed ice). Mix until the water is absorbed by the meat. Add the remaining meat, dextrose, spices, and other ingredients to the blender, and then mix for two minutes. Regrind through a ⅛- or 3⁄16-inch grinder plate, and then stuff into hog casings. Linking should be done at 5-inch intervals. Using the batch smoking schedule described previously, cook until the internal product temperature reaches 155°F.

Smoked (Country-Style) Pork Sausage

100 lbs. lean pork (20% to 25% fat)	4 oz. certified cure (6% nitrite)
15 lbs. water (crushed ice if chopper is used)	4 oz. red pepper
	4 oz. paprika
2 lbs. 8 oz. salt	2 oz. (56 g) coriander
5 oz. dextrose	1 oz. (28 g) nutmeg

≡ ≡ NOTE: Small quantities are more accurately weighed in grams (1 oz. = 28 g). ≡ ≡

Grind meat through a ⅜-inch grinder plate. Add salt, cure, and water (or crushed ice), and blend until the water is absorbed by the meat. Add the remaining ingredients, and blend until adequate tackiness, or bind, is achieved (usually two additional minutes). Regrind the sausage through a fine ($\frac{1}{8}$- or $\frac{3}{16}$-inch) grinder plate, and then stuff into hog casings. Stuffed sausage should be coiled around smokesticks in 12- to 18-inch loops. *Do not link.* Using the batch smoking schedule described previously, cook until the internal product temperature reaches 155°F.

Braunschweiger

40 lbs. pork liver (semi-frozen)	8 oz. dextrose
30 lbs. pork jowls	4 oz. certified cure (6% nitrite)
20 lbs. special pork trimmings (80% lean)	2 oz. (56 g) marjoram
	2 oz. (56 g) nutmeg
10 lbs. bacon end pieces	2 oz. (56 g) ginger
10 lbs. water (or crushed ice)	1 oz. (28 g) allspice
4 lbs. peeled onions (or 3 to 4 oz. onion powder)	1 oz. (28 g) cloves
	$\frac{7}{8}$ oz. (24 g) sodium erythorbate
3 lbs. 8 oz. non-fat dry milk	
2 lbs. 8 oz. salt	$\frac{1}{2}$ oz. (14 g) liquid smoke (optional)
8 oz. white pepper	

≡ ≡   NOTE: Small quantities are more accurately weighed in grams
(1 oz. = 28 g).   ≡ ≡

Grind the liver, special pork trimmings, pork jowls, bacon ends, and onions separately through a $\frac{3}{8}$-inch grinder plate. After placing the ground liver, salt, and cure in the chopper, chop until bubbles appear on the surface (usually one to three minutes). Add *in order* ground pork and pork jowls, bacon ends, spices, other ingredients, and water (crushed ice) and continue to chop until the temperature reaches 50°F. At approximately 50°F, add the non-fat milk, and chop until the temperature reaches but does not exceed 65°F. Stuff into moistureproof casings (about 3 inches in diameter) or sewed hog bungs. Steam or cook the casings or bungs in hot water (180°F) until the internal temperature reaches 155°F. Immediately chill in an ice bath to 40°F. If the product is to be smoked, use a cool smoke after the product has been chilled.

Principles and Manufacturing Practices for Dry and Semi-dry Sausages

In modern practices, the dry and semi-dry sausages are the most difficult and time-consuming to produce. Ironically, these types of sausages are thought to have been the first produced in early history, dating back to the Babylonian culture around 1,500 B.C.

The key to successful production of dry and semi-dry sausages hinges on the *timely lowering of the pH* of the sausages during processing. The pH of fresh muscle that has proceeded through glycolysis (see Chapter 22) and is suitable for processing is 5.6 to 5.7. However, the pH of some meats may be as high as 6.4. The final pH of fermented sausages will typically range from about 4.8 to 5.4, depending upon the tanginess desired and the product identity characteristics. The pH drop was traditionally accomplished through fermentation by chance, contaminating microorganisms that originated from the meat itself or from the sausage production equipment. As a result, successful products were often achieved in one plant or in one batch, but not in another, when apparently identical procedures were followed. Today, most of the dry and semi-dry sausages are produced with a starter culture, which inoculates the sausage batch with desirable lactic acid–producing bacteria. These bacteria function to lower the pH of the sausage (by producing lactic acid), giving the product its characteristic tangy flavor. Also, reducing the pH of the sausage brings the pH closer to the isoelectric point of the proteins, at which point the proteins most easily give up water. The enhanced drying effect, along with the effect of the acid itself, creates an unfavorable environment for spoilage organisms, thus aiding in the preservation of the product (see Chapter 2).

In order for fermentation to occur, the bacteria must have "food" to grow on. Lactic acid–producing bacteria require simple sugars for optimal growth. The most common simple sugar used in sausage formulations is dextrose or corn sugar. Varying the level of dextrose in the formulation can alter the end point pH. More dextrose is used for lower pH products. Typically, dry and semi-dry sausages will contain 0.5 to 2 percent dextrose.

The actual fermentation process requires specific environmental conditions for optimal product quality. These conditions may be achieved in a "greening room" or in a smokehouse. Larger dry sausage processors typically use a greening room to hang the product during fermentation. This is a room where temperature, humidity, and air velocity can be carefully controlled. Dry sausage is typically fermented at 70° to 75°F, 75 to 80 percent relative humidity (RH), and a slow, steady movement of air until the desired pH is attained (usually one to three days). Semi-dry sausages are normally fermented in a smokehouse for 8 to 12 hours at slightly higher temperatures, 85° to 100°F.

A Good Manufacturing Practice (GMP) guide produced by the American Meat Institute (AMI) (see Appendix) spells out the time–temperature limitations for dry and semi-dry sausage manufacture. The critical aspect of dry and semi-dry sausage manufacture is to lower the pH quickly enough as you raise the temperature so that undesirable microorganisms cannot grow and spoil the sausage. Thus, if the meat temperature is held too long at 60°F or higher and at pH 5.6 or higher, the troublesome microorganisms can get started. It is critical to reach the pH of 5.0 as soon as possible. The GMP from AMI defines the degree/hour limitations, i.e., the maximum number of degrees over 60°F the product should be held before a pH of 5.3 is reached, that are safe, acceptable, and attainable, yet will not infringe upon a processor's use of individualized recipes or uniquely flavored products.

Timely Lowering of pH

Timely lowering of the pH of the product is achieved by the use of any one of four methods: fermentation through the use of starter cultures, backslop, natural fermentation, and direct addition of organic acids (acidification or acidulation).

STARTER CULTURES — The use of starter cultures is by far the most common fermentation method used in the United States. Starter cultures are available from commercial sources in two forms: as a frozen concentrate and a lyophilized (freeze-dried) form. Both forms have produced acceptable results. Starter cultures typically contain a blend of two or more different microorganisms and sometimes different strains of the same microorganism. The microorganisms most commonly used belong to the genera *Pediococcus, Micrococcus,* and *Lactobacillus.* Specific species include *P. cerevisiae, P. acidilactici, M. aurantiacus,* and *L. plantarum,* the last also occurring as the most common organism in natural fermentation methods.

Some dry sausage processors use a custom-developed starter culture. The microorganism content and ratio of these starter cultures is a trade secret. This enables their product to have a unique flavor, different from other products on the market.

The primary advantages of using starter cultures over other fermentation methods are as follows:

- A more uniform product is produced from batch to batch.

- The risk of a batch undergoing spoilage is greatly reduced.

- The fermentation is more controlled, and the time to desired end point pH can be more readily predicted.

- The fermentation takes place faster.

BACKSLOP — The backslop fermentation method is the second most common of the three fermentation methods. Backslopping entails setting aside a small quantity of one day's production in order to inoculate the next day's production with the same microorganisms. Generally, processors who routinely practice backslopping will keep isolated cultures in frozen storage and then periodically revert to their original culture. This is done because "extraneous" microorganisms, which might impart strange flavors to the final product, may be picked up in day-to-day operations. There is a greater chance of undesirable microorganisms predominating during the fermentation process, sometimes resulting in product spoilage.

NATURAL FERMENTATION — When there is no inoculation of a batch with microflora, this method is called natural fermentation. This method is the least practiced, primarily because it produces the most inconsistent results. The end product may display little or no fermentation, it may spoil, or it may occasionally produce an outstanding batch. The success or failure of this method is dependent upon the types of bacteria that happen to be present within the meat and anything with which the meat comes in contact. If there are insufficient lactic acid–producing microorganisms present, spoilage microorganisms have the opportunity to grow and flourish because the pH does not decline.

DIRECT ADDITION OF ORGANIC ACIDS — It is possible and permissible to directly add organic acids to the sausage, which almost immediately causes lowering of the pH. The acids, such as citric acid, lactic acid, and glucono delta lactone, are encapsulated in various vegetable oil coatings such as palm oil or other hydrogenated oils. These coatings melt and release the acid when the appropriate temperature is reached. Two advantages to direct acidification are increased production capacity and energy savings, both of which result from the decreased holding and storage time needed before lowering of the pH occurs. Product uniformity may be increased because the final pH is usually reproducible and predictable. However, due to the shortness of the processing time, more subtle and unique flavors are not allowed to develop. Direct acidification, which gives a characteristic tangy flavor, is most widely used in pepperoni manufactured for the booming pizza trade.

Drying and Cooking

Dry sausages, such as the Italian salamis, are not cooked and do not require refrigeration after manufacture (except for prolonged storage). These sausages are preserved by a low-moisture content of 30 to 40 percent (a_w less than 0.91) and must have a moisture–protein ratio (M:P) of 1.9:1 or less, except when the pH is <5.0, then the M:P may be as high as 3.1. Because dry sausages are not cooked, the U.S. Department of Agriculture has established procedures for certifying the use of dried products containing pork in the Code of Federal Regulations (CFR sec. 318.10c). The sausages must be held at specified temperatures for specified times, which are determined by the sausage diameter. It is preferable to use certified trichinae-free trimmings when possible (see Chapter 3).

Carefully controlled environmental conditions are also critical during the drying process of dry sausage. If the product dries too fast, case hardening will occur. The outside of the sausage will become hard and dry, inhibiting further moisture migration out of the interior part of the sausage, which is still wet. Sausages that have undergone case hardening are prone to internal spoilage by anaerobic bacteria. This problem is caused by the humidity being too low during drying. On the other hand, excessive humidity causes the product to dry too slowly, often resulting in excessive mold growth, yeast growth, and bacterial surface slime on the product surface.

Theoretically, the drying rate at the surface of the sausage should be only slightly greater than that required to remove the moisture, which migrates from the inside of the sausage. These conditions are approached by various combinations of temperature, humidity, and air velocity. There is little agreement as to the most effective combination of these three variables. As a general recommendation, the temperature of the drying room should be maintained between 45° and 55°F at a relative humidity of 70 to 72 percent. The air velocity in the room should be between 15 and 25 air changes per hour. In a 10 foot x 20 foot drying room with 10-foot ceilings, 20 air changes per hour equals air moving 667 cubic feet per minute (cfm), or 7.5 miles per hour (mph).

The drying time will vary considerably, depending on the diameter of the product and the size and type of product. Most dry sausages will require between 10 and 120 days of drying time.

Semi-dry sausages are generally cooked (held) in a smokehouse at a temperature of approximately 100°F with 80 percent relative humidity (RH) for the first 24 hours to allow the friendly microorganisms to grow and produce lactic acid, thus lowering the pH. After that, the smokehouse temperature is raised and the RH is reduced to finish cooking the product rapidly to the internal temperature of 155°F. This rather high temperature will assure that all microbiological activity will stop and provide a safety margin if the product contains uncertified pork (pork that has not been certified as trichinae-free). Semi-dry sausages will generally contain approximately 50 percent moisture (a_w approximately 0.95) and will thus require refrigeration to prevent spoilage. Most semi-dry sausages are smoked during the last 30 to 50 minutes of the cooking cycle, whereas many of the dry sausages are not smoked.

Manufacturing Practices

Maximum particle definition is an important factor in the attractiveness of dry sausages. Particle definition refers to the clear demarcation between fat and lean particles. The raw meat materials should be ground (or chopped) at very low temperatures (20° to 25°F) to achieve the best particle definition. Mixing and unnecessary handling of the meat should be minimized as much as possible. The cold meat temperatures act to reduce protein extraction and to help reduce the deformation in particle shapes. This is important in the reduction of "smearing" of the product. The mixing time should be only enough to allow uniform distribution of spices and other ingredients. Fermentation temperatures for the true dry sausages should be cooler, i.e., approximately 75°F, rather than 100°F for summer sausage, to maintain a more distinct particle definition.

The traditional dry sausages, such as the Italian salamis, are generally stuffed into hog middles, hog bungs, or sewed beef middles (see "Sausage Casings"). Some processors have switched to artificial casings because of the cost of natural casings. Artificial casings used in dry sausage are the fibrous type previously discussed.

Dry sausages are the most expensive sausages to produce (and to buy) because of the lengthy production time and subsequent product shrinkage.

Semi-dry sausages such as thuringer and cervelat, which are both varieties of summer sausage, and Lebanon bologna are generally fermented and cooked in a smokehouse and subsequently can be produced in 12 to 18 hours (see preceding discussion). This shorter process dramatically reduces production costs. Particle definition is less critical in these sausages, and a moderate bind is often desirable. Therefore, the meat in semi-dry sausages is not ground at subfreezing temperatures. The meat batter should also be mixed for a longer period of time in order for adequate protein extraction to occur. Without adequate protein extraction, the sausages may fat out during thermal processing. Fat out occurs when there is insufficient emulsifying protein available to entrap and stabilize the fat particles. The fat will migrate during cooking to form "pockets" of fat inside the product or on the surface, forming undesirable "fat caps."

Formulations for Dry and Semi-dry Sausages

Genoa Salami (Dry)

40 lbs. regular pork trimmings (50% lean, semi-frozen)	6 oz. white pepper
40 lbs. boneless chuck	1 oz. (28 g) fresh garlic (peeled)
20 lbs. pork shoulder butts (80% lean, semi-frozen)	¼ oz. (7 g) nitrate
	⅛ oz. (3.5 g) cloves
3 lbs. 6 oz. salt	⅛ oz. (3.5 g) nitrite
6 oz. sugar (sucrose)	starter culture (follow directions for recommended amount)
6 oz. dextrose	8 oz. red wine (optional)

≡ ≡ NOTE: Small quantities are more accurately weighed in grams
(1 oz. = 28 g). ≡ ≡

Grind pork trimmings and shoulder butts through a ½-inch grinder plate. Grind beef through a ¼-inch grinder plate. Place all the ingredients except salt in a bowl chopper, and then chop for one to two minutes. Add the salt during the last 30 seconds of chopping. Stuff the sausage into hog bung casings; if hog bung casings are unavailable, use fibrous casings of similar size. Place the sausage in a greening room for 36 hours (75°F and 75 percent RH). Genoa is traditionally wrapped tightly with No. 9 Italian hemp 2-ply cord. When wrapping, make a hitch of the twine every ½ inch the entire length of the sausage (see Figure 21–6). The salami should not be smoked but should be dried for 9 to 10 weeks in a drying room (50°F and 70 percent RH).

Fig. 21–6. Genoa salami stuffed in natural casings and hanging in a "drying" room. Each of the traditional dry Italian salamis has a unique and different type of "cording." Twine is wrapped vertically and horizontally for Genoa salami, as shown here.

Pepperoni

50 lbs. special pork trimmings
   (80% lean, semi-frozen)
30 lbs. boneless chuck
   (85% lean)
20 lbs. regular pork trimmings
   (50% lean, semi-frozen)
3 lbs. 6 oz. salt
8 oz. cayenne pepper
8 oz. pimento

6 oz. dextrose
6 oz. sugar (sucrose)
4 oz. crushed red pepper
1 oz. (28 g) whole anise seed
½ oz. (14 g) fresh garlic
   (peeled)
¼ oz. (7 g) nitrate
⅛ oz. (3.5 g) nitrite

≡ ≡ NOTE: Small quantities are more accurately weighed in grams
(1 oz. = 28 g). ≡ ≡

Grind all meat ingredients through a ⅛-inch grinder plate, and then mix with nonmeat ingredients in a blender for two minutes. Cure the meat in 6-inch deep pans for 48 hours at a temperature of 40°F. After this curing period, remix the meat in a blender for approximately one minute, and then stuff into narrow or medium hog casings. Break the casings into 21- to 22-inch lengths, making allowance for enough casing to tie or clip each end where the casing is broken. Twist the casings in the center to form twin links about 10 inches long. The pepperoni should be held in a greening room for 24 hours (75°F and 75 percent RH) before it is ready for drying. If the pepperoni is to be smoked, apply a cold smoke after the sausage has come out of the greening room, but before it starts to dry. Dry at 50°F (70 percent RH) for 21 days. A 10 percent solution of citric acid may be sprayed on the outer surface prior to packaging to aid in maintaining the characteristic cured-meat color.

Summer Sausage (Semi-dry)

40 lbs. special pork trimmings (80% lean)	2 oz. (56 g) coarse-ground black pepper
20 lbs. regular pork trimmings (50% lean)	2 oz. (56 g) intermediate-ground pepper
30 lbs. boneless chuck (80% lean)	1 oz. (28 g) red pepper
10 lbs. beef hearts	¾ oz. (22 g) coriander
2 lbs. 8 oz. salt	½ oz. (14 g) mace
1 lb. dextrose	½ oz. (14 g) allspice
4 oz. certified cure (6% nitrite)	¼ oz. (7 g) fresh garlic (peeled)
3 oz. paprika	starter culture (follow directions for recommended amount)

≡ ≡ NOTE: Small quantities are more accurately weighed in grams (1 oz. = 28 g). ≡ ≡

Grind all meats through a ⅛-inch grinder plate, and then mix thoroughly with nonmeat ingredients; however, *do not overmix.* Three to four minutes is generally sufficient, depending on the speed of the mixer. Regrind the mixture through a ⅛-inch grinder plate, and then stuff into appropriate natural or fibrous casings, usually 2 to 3 inches in diameter.

The following smokehouse schedule should be used with a starter culture in the production of summer sausage:

- Ferment for 24 hours at 100°F (75 to 85 percent RH).

- Smoke for one to two hours near the end of the fermentation cycle.

- Cook for 1½ to 3 hours at 155°F (60 percent RH).

- Cook until done (155°F internal temperature) at 175°F (45 percent RH).

- Shower with cold water for three minutes, and then chill immediately in cooler.

Principles and Manufacturing Practices for Luncheon Meats/Loaves

Meat loaf manufacturing has provided meat processors many excellent opportunities for new product development and diversification over the years. As a result, some of the favorite, most popular luncheon/sandwich loaves, such as pickle and pimento, old-fashioned, ham and cheese, olive, and honey loaves, have been developed. Since other foods are added to the meat loaf formulae, the flavor and the distinctive appearance are changed. Each loaf product thus develops a strong appetite appeal.

Seasoning is an important factor in maintaining the popularity of old favorites and in securing public acceptance of new loaf products. Seasonings especially blended and prepared for meat loaves are obtainable from reputable manufacturers. Essentially the same meat ingredients are used in both sausages and loaves. These ingredients are handled similarly up to the point at which they are cooked or baked. Meat loaf ingredients are ground through plates of various sizes and are sometimes chopped in the silent cutter. Equipment for cubing fat and cheese, etc., will save time. The mechanical mixer is ordinarily used to secure a good mix of materials. When cured-meat flavor and color are desired, pre-cured meats are used. The same emulsion used for frankfurter and bologna manufacture can be used for making loaves. The last items to be added are the pickles, pimentos, olives, cheeses, etc., that make each loaf unique. These must be added very carefully so that they do not break apart. In modern sausage kitchens, meat loaf emulsions are stuffed into plastic bags inside 27-inch-long stainless steel forms (tubes) that have inside cross-section dimensions of 4 × 4 inches. The emulsions are compressed into the tubes and the tubes are capped. The plastic bags permit the loaves to be easily slipped out of the tubes after they have been cooked. Softer emulsions, such as liver–cheese emulsions, are poured into open-top stainless steel troughs of similar size to the tubes, which are lined with waxed paper and a layer of fat prior to being filled. To create an old-fashioned appearance, a limited number of loaves are still prepared in pans similar to bread pans to give the desired shape. In these open pans the emulsion is mounded higher than the sides of the pans, and the top surfaces are brushed with a thin syrup of water and sugar to create an appetizing brown color during heating. The use of both the stainless steel tubes and the bread pans makes loaves that slice nicely, and the slices are used on bread in sandwiches; thus, these products are known as "sandwich meats" as well as luncheon meats.

Loaf products are not normally smoked but are often cooked in a smoke-house. Modern smokehouses equipped with computerized heating and mois-ture (RH) control serve as excellent ovens for cooking loaves and many other meat products. After being cooked and cooled, loaf "logs" are sent through automated slicing/packaging lines to be sliced into portion-controlled lun-cheon meat packages of 4 to 8 ounces each. Uniform slices are possible because of the stainless steel tubes (forms) that were used to form the logs. Loaf products must be clean-slicing, compact, smooth, and meaty in appear-ance. The natural moisture should be held in the loaves until they are eaten. Manufacturers are sometimes troubled with leakage of juices in baking and sometimes with warping of loaves. (Note earlier discussions on binders and phosphates that are permitted in meat products to help retain moisture.) Overheating when chopping causes the emulsion to break down and the product to lose moisture. However, the starting oven temperature is also very important to prevent water loss and warping. Although not applying to all loaves, the starting oven temperature should be at 170° to 200°F. After 1 to 1½ hours, depending on the particular loaf product, the oven temperature should be raised to 250° to 300°F, and the product cooked until the internal temperature reaches at least 155°F. Meat–cheese loaves should contain a type of cheese that will not melt at these temperatures.

Formulations for Luncheon Meat/Loaf Products

Luncheon Loaf Base

70 lbs. beef (85% lean)	1 lb. dextrose
30 lbs. regular pork trimmings (50% lean)	8 oz. ground mustard
24 lbs. water (or crushed ice)	4 oz. ground white pepper
2.5 lbs. non-fat dry milk (NFDM)*	4 oz. monosodium glutamate
	4 oz. speed cure (6.25% nitrate)
	2 oz. (56 g) ground celery

2 oz. (56 g) ground nutmeg   ⅞ oz. (25 g) sodium
1 oz. (28 g) onion powder   erythorbate
1 oz. (28 g) ground coriander

* Milk protein hydrolysates may be substituted for NFDM at the 0.75-lb. level.

≡ ≡ NOTE: Small quantities are more accurately weighed in grams (1 oz. = 28 g). ≡ ≡

Chop one-half the beef with one-half the ice and all the salt and cure to a fine slurry in a bowl chopper. While continuing to chop, add the remaining beef, pork trimmings, water (or ice), non-fat dry milk, and other ingredients. The final chopping temperature should not exceed 65°F. If the chopper has a mixing cycle, add condiments and mix sufficiently for distribution. If the chopper has no mixing cycle, blend condiments into batter in a blender sufficient for distribution. *Do not overmix* — one minute is generally sufficient. Stuff in appropriate casings or loaf pans.

Condiments

- *Pickle and pimento loaf:* Add 5 pounds well-drained pickle and pimento mix or 2.5 pounds diced sweet pickles and 2.5 pounds diced pimentos to the batter at the final blending step.

- *Olive loaf:* Add 6 pounds well-drained stuffed olives to the batter at the final blending step.

- *Macaroni and cheese loaf:* Add 3 pounds cooked and drained macaroni and 5 pounds diced high melt point processed cheddar cheese to the batter at the final blending step.

PRODUCT DEFECTS

There are a number of problems that plague sausage manufacturers, often resulting in unacceptable or undesirable finished products. These problems are often related to poor sanitation practices or the use of poor-quality raw materials.

Table 21–5 is a useful trouble-shooting source of information when product defects or manufacturing problems arise.

Table 21–5. Types of Sausage Product Defects —
Description, Causes, and Remedies

Defect	Description	Cause(s)	Remedy(ies)
Ruptured casings	Casings rupture or split during cooking	Stuffed too tightly	Stuff less tightly.
		Bacterial-gas producer	Check sanitation.
			Change sanitizer.
Color problems (external)	Streaking	Smoked while too wet	Increase drying time.
	Poor smoke color	Smoked while too dry	Shorten drying time.
Color problems (internal)			
Pale color	Lacks desirable color	Insufficient curing time	Use preblends.
Greening	Uniform greenish color	Too much NO_2^- (nitrite burn)	Reduce NO_2^-. Check sanitation.
Green cores	Green center	Bacterial-*Lactobacillus* viridescence	Check cook cycle (153°F critical temperature).
Green rings	Concentric green rings	Bacterial-*Lactobacillus* viridescence	Check cook cycle (153°F critical temperature).
Fading	Light or chemical fading	Display lights	Gas flush.
		Sloppy rinse of sanitizers	Vacuum package.
			Use amber film.
			Rinse equipment better.
Case hardening	Hard, dry surface	Humidity too low	Increase humidity.
	Wet interior	Drying too fast	Increase drying time.
Fatting (shorting) out	Separation of fat	Too much fat in formulation	Reduce fat level.
Fat caps	Fat on surface	Chopping temperature too high	Chop to 60° to 62°F.
Fat caps	Fat on surface between casing and product	Cooked too fast	Cook slower.
Fat pockets	Fat pockets throughout product	Insufficient contractile proteins	Decrease level of high collagen meats.
		Excessive chopping	Sharpen chopper knives, reduce chopping time.
		Too little salt or salt omitted.	Check salt addition.
		Overworking or pumping long distances	Change production layout.
		Too much rework (previously cooked product)	Reduce rework level.
		Insufficient mixing (coarse-cut products)	Increase mixing time.

(Continued)

Table 21–5   (Continued)

Defect	Description	Cause(s)	Remedy(ies)
Gassiness	Vacuum packages bloat	Bacterial gas produced	Check sanitation, thermal-processing schedule.
Gelatin pockets	Gelatin pockets	Too much connective tissue in formulation	Reduce connective tissue level.
			Preblend.
Mold growth	Fuzz on surface	Poor sanitation (outside air in plant)	Check sanitation.
			Use filters in ventilation system.
			Maintain positive room air pressure.
Particle definition poor	Coarse-cut sausages lack clean fat/lean distinction	Dull chopper–grinder blades	Sharpen plates and knives.
		Chopping temperatures too warm.	Chop at 28° to 32°F.
Peelability poor	Inedible casings will not peel easily	Surface drying too severe	Operate smokehouse at higher humidity.
		Poor skin formation	Steam cook last 10 minutes of cycle.
			Reduce collagen level in formulation.
Putrefaction	Undesirable odor	Putrefactive bacteria	Check thermal-processing temperature.
			Check sanitation.
Smearing	Fat smearing, leaving undesirable appearance in fresh sausages	Overworking	Minimize mixing.
		Meat temperature too warm	Stuff at 32°F.

The data contained in Table 21–5 and in Tables 21–6 to 21–24 are courtesy of Least Cost Formulations, Ltd., 827 Hialeah Drive, Virginia Beach, Virginia 23464.

LEAST COST SAUSAGE FORMULATIONS

The best sausage formulation is one that meets the predetermined quality specifications at the lowest possible cost of production. This is a problem that is not easily determined without the assistance of computer formulations; and even then, the results are only as good as the information entered into the computer. The dimensions of the best formulation problem vary widely from company to company, depending on factors such as size of the firm, ingredient availability, operating capacity, storage capacity, and product diversity. For most companies, the ultimate formulation of the product is also influenced by raw material procurement policies, inventory policies, production scheduling, and price strategies.

Much of the following information pertaining to least cost computer formulations and preblending applications was obtained from Robert A. LaBudde, Least Cost Formulations, Ltd.[15] Tables 21–6 to 21–24 are copies of the printouts received from the Least Cost Formulator®, a computer software system designed by the aforementioned company. The information in these tables is an abbreviated form for computer input simplicity. The authors are also indebted to Professor Robert E. Rust, formerly at, and still associated with, Iowa State University, for his useful suggestions and the valuable information that has been published in the Iowa State Sausage Short Course Proceedings. The following information should acquaint the reader with the basic approach to least cost formulation and the use of computers in the meat industry. The detailed application will need to be worked out with each individual's own computer/software specialist.

Computers and Least Cost Formulation

Advances in electronics have brought the cost of computers down from millions of dollars to the thousands (small-business systems are available for under $10,000). Along with the drastic drop in prices has come ease-of-use. The operator does not need an engineering degree to use the computer. Though the art of sausage making is thousands of years old, it is only relatively recently that formulating has begun to make use of twentieth century technology. With the systems available today, even the smallest family-owned meat processor can take advantage of sophisticated techniques to produce consistent quality products and increase the bottom line profit. Systems have been designed assuming the user has not had any computer experience or training and has no familiarity with the intricacies of least cost formulation.

Least cost formulation is a mathematical technique for determining the best use (optimum allocation) of available resources when there are many possible uses for these resources. The limiting resources are the raw materials that are purchased or are available from a harvesting operation from which

[15]824 Timberlake Drive, Virginia Beach, VA 23464. Phone: (757) 467-0954. Fax: (757) 467-2947. ral@lcfltd.com

finished products are made. Sausage processors face the problem of manufacturing many products each day, with fixed amounts and types of raw materials but with chemical analyses that change with each lot. Moreover, the finished products must meet product specifications. Control of the many necessary variables cannot be accomplished by hand in a practical time frame, but it can be done in a matter of seconds on a computer.

Misconceptions About Least Cost Formulation

Linear programming is usually referred to as least cost formulation. With this has come the misconception that only inferior products can be made using least cost formulation — after all they are the "cheapest." Least cost formulation can better be described as *yielding the least expensive formulation that will satisfy the product specifications and make use of the available raw materials.*

The mathematical technique first finds the raw materials that achieve the product specifications, and only then does it find substitutions that still meet the product specifications but reduce the price. Therefore, any quality-level product can be least cost formulated and produced consistently, using either the same or the varying raw materials, provided the specifications are well-defined and the raw materials are accurately characterized. This technique works on different product-quality levels, as considered here in the manufacture of four types of franks — regular frank (#1 — Table 21–3), by-product frank (#3 — Table 21–3), regular frank with isolated soy protein (ISP) (#4 — Table 21–3), and by-product frank with ISP (#5 — Table 21–3).

The Problems with Standard Formulas

It is common practice for meat processors to have a standard formula for making a product. For example, the regular frank might have the following:

Quantity (lbs.)	Raw Materials	Short Name
30	Beef plates	B–plates
70	Pork trimmings (80%)	P–80
25	Water	Water
2¾	Salt	Salt
1	Corn syrup	I–corn
¾	Frank spice and cure	S–frank

Raw meat materials used in manufacturing processes have a wide range of variability that makes it impossible to produce a consistent finished product from a standard formula. The fat content may vary by 5 to 10 percent, and the protein content may vary by 2 to 3 percent between suppliers. Hence, only by chance can the processor repeatedly make the same finished product when fixed amounts of the raw materials are used. Table 21–3 shows the analysis of five frank batches (#s 1 through 5) made according to the standard formula and different lots of raw materials. It is these typical variations that are responsible for product failure, as well as the wide variation in product cost.

Creating Product Standards

Five categories that affect product specifications and their contributions to the product standards for franks are examined as follows:

1. *USDA regulations for franks:* USDA regulations specify what may and may not be included in the product as well as the upper limit of many product attributes: (a) fat content, 30 percent finished weight; (b) USDA added water plus fat, 40 percent finished weight; (c) isolated soy protein, 2 percent finished weight; and (d) non-fat dry milk, 3.5 percent finished weight.

2. *Product label:* The choice of the product label determines the quantity of each raw material and its relation to other raw materials. Hence, a label that reads "pork, beef, water, salt, corn syrup, spices, and nitrite" would have to have more pounds of pork than pounds of beef, more pounds of beef than pounds of water, and so on (see Chapter 3).

3. *Plant equipment:* The type and size of equipment in the plant will have an influence on the product as will the care employees take in carrying out the manufacturing steps. A successful formulation in one plant may fail in another plant in which equipment and technique vary.

   Equipment will affect product specifications in several ways. The first deals with how the batches are handled. The size of each batch will be a function of equipment size. The use of frozen meat and its quantity will depend on freezer space and the ability to temper and(or) flake the boxes of meat. The ability to accurately control the smokehouse will affect the product shrink. The available cooler storage area could determine how many different raw materials are on hand to use in the formulations, as well as the ability to make preblends. The importance of preblends will be covered later.

4. *Market requirements:* Input from salespersons and customers and knowledge of the marketplace will influence the kind and amount of spices in the products as well as the salt content, the color level, and the bite (bind). The method of consumer cooking, the use of a roller grill for instance, will influence the dextrose level and the percent and kind of fat (pork vs. beef). All this must also take into account the price for which the product may be sold.

5. *Company standards:* Company reputation is maintained by choice of labels and the quality level of the products produced. For each kind and type, it must be consistent and at a price that will yield a profit.

The use of a least cost formulation system allows all these parameters to be taken into account, thus producing the identical product, at the specified quality level, day after day.

Getting Started Using Least Cost Formulations

The overall process of setting up a least cost formulation system is flowcharted in Figure 21–7.

Raw Material Attributes

Table 21–6 lists the raw materials that might be available and the attributes that should be controlled in production.

To begin, five attributes are generally sufficient: fat, moisture, protein, bind, and color, but there really is no limit to the possible attributes to measure and control. A comprehensive raw material attribute table is included in Table 21–24. The data are based upon the averages for each material collected over several years from dozens of installed least cost formulation users. The bind and color values are constants by meat type. The values shown were obtained from Professor John Carpenter[16] at the University of Georgia. These

---

[16]Carpenter, J. A., and R. L. Saffle. 1964. A simple method of estimating the emulsifying capacity of various sausage meats. J. Food Sci. 29:774.

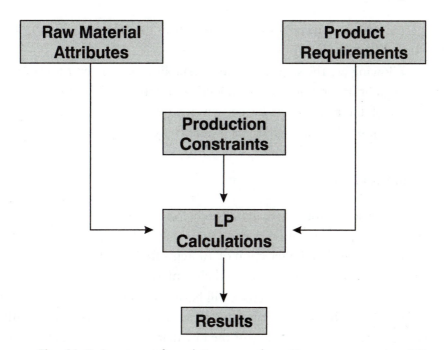

Fig. 21–7. Least cost formulation procedure. Linear programming (LP) calculations provide the final results.

values were updated by J. D. Porteous,[17] but the original values are still widely used. Each meat packer/processor has its own particular system for determining bind and color values. (Review Table 21–1, which lists the values for bind and color in terms of indexes, with bull meat at 100. This table also gives the relationship of various raw meat materials for bind and color.)

The raw material parameter in Table 21–7 shows the price and quantity of each raw material available to the formulations.

Least Cost Formulation: Programming Product Standards

In order for product standards to be created that can be used in least cost formulation, each attribute must be expressed as a constraint with a low limit and a high limit. A standard formula (see Table 21–3) has been designed so that the values of its attributes would most likely fall into the allowable ranges required by the U.S. Department of Agriculture, the product label, and so on,

---

[17]Porteous, J. D. 1979. Some physico-chemical "constants" of various meats for optimum sausage formulation. Can. Inst. Food Sci. Technol. 12:145.

but specific attributes were not given as targets in the table showing the standard formula (see Table 21–3).

The following are the steps required to establish the least cost formulation low and high limits on both the raw materials and the product attributes. The four types of limits that can be chosen are finished weight (FW), green weight (GW), batch weight (BW), and pounds (lbs.).

Raw Material Limits

For each product, the raw materials to be offered to the formulation must be determined. For illustration purposes, assume that the local availability of raw materials has led to the selection of 10 items from the raw materials table (see Table 21–6). These materials and their limits are given in Table 21–7.

The majority of the raw materials will have a low limit of zero and a high limit of Big. In mathematical terms, this means that the least cost formulation system may use as few or as many of the raw materials as it desires to achieve the desired product specifications. Hence, in Table 21–8, the 10 raw materials offered all have low limits of zero and high limits of Big with the limit type of green weight (GW). I–salt (I = Ingredient) has a low limit of zero and a high limit of 3 percent of the green weight.

I–corn has identical values for both the low and the high limits, which require *exactly* 1 percent of the block weight of corn syrup. Similarly, S–frank (S = Standard) limits will cause exactly ¾ pound of spices and cure to be included in the batch.

Attribute Limits

From analysis of the five batches shown in Table 21–3, you can determine a range of limits that the standard formulation yields. Using these data as a base, combined with the USDA regulations and experience, you can set target limits for the attributes of the regular frank. These are given in Table 21–8.

For fat, you may wish to approach the high limit of 30 percent of the finished weight (FW) allowed by the USDA regulations. To give a margin of safety, set the high limit as 29 percent FW. The low limit for fat has been set at 25 percent FW. However, if a low-fat frank were being produced, the low and high limits for fat would be set at 10 percent FW. The least cost formulation system will use as much fat as possible because it is an inexpensive attribute.

Table 21–6. Short Attribute Level Table for Least Cost Formulations

SEQ NO.	SHORT NAME	RAW MATERIAL	GROUP TYPES	MOISTURE	FAT	PROTEIN	SALT	USDA-AN	BIND	COLOR	COLLAGEN
1.	B-BULL	BF FC BULL MEAT	B	0.7000	0.0950	0.2000	0.0000	-0.1000	6.0000	9.4000	0.0400
2.	B-CHEEK	BF CHEEK MEAT	BK	0.6300	0.1950	0.1700	0.0000	-0.0500	2.3800	8.1600	0.1003
3.	B-CHUCK	BF BNLS CHUCKS	B	0.5700	0.2600	0.1600	0.0000	-0.0700	3.8400	13.6000	0.0480
4.	B-COW	BF FC COW MEAT	B	0.6650	0.1500	0.1800	0.0000	-0.0550	4.4100	7.0200	0.0378
5.	B-HEAD	BF HEAD MEAT	BK	0.6100	0.2300	0.1550	0.0000	-0.0100	1.2400	4.0300	0.1131
6.	B-HEART	BF HEARTS	O	0.6650	0.1800	0.1500	0.0000	0.0650	0.9000	6.0000	0.0405
7.	B-LIP	BF LIPS	O	0.6200	0.2200	0.1550	0.0000	0.0000	0.0465	0.0620	0.1395
8.	B-NAVEL	BF BNLS NAVELS	B	0.3800	0.5150	0.1000	0.0000	-0.0200	1.2500	1.9000	0.0420
9.	B-PLATE	BF PLATES	B	0.4000	0.4800	0.1100	0.0000	-0.0400	1.7600	2.7500	0.0462
10.	B-SHANK	BF SHANK MEAT	BK	0.7100	0.1000	0.1850	0.0000	-0.0300	5.1800	8.5100	0.0610
11.	B-50	BF 50% TRIMMINGS	B	0.3750	0.5100	0.1050	0.0000	-0.0450	1.2600	1.9950	0.0441
12.	B-60	BF 60% TRIMMINGS	B	0.4500	0.4100	0.1300	0.0000	-0.0700	2.0800	3.2500	0.0526
13.	B-65	BF 65% TRIMMINGS	B	0.4900	0.3600	0.1500	0.0000	-0.1100	2.7000	4.2000	0.0600
14.	B-75	BF 75% TRIMMINGS	B	0.5700	0.2700	0.1550	0.0000	-0.0500	3.4100	5.2700	0.0589
15.	B-85	BF 85% TRIMMINGS	B	0.6550	0.1550	0.1800	0.0000	-0.0650	4.3200	7.0200	0.0504
16.	B-90	BF 90% TRIMMINGS	B	0.6950	0.1100	0.1850	0.0000	-0.0450	4.8100	7.4000	0.0518
17.	P-BACEND	PK BACON ENDS	OR	0.3050	0.6000	0.0900	0.0000	-0.0550	0.0090	0.2250	0.0774
18.	P-BELLY	PK BELLY STRIPS	P	0.3400	0.5650	0.0900	0.0000	-0.0200	0.5400	0.4500	0.0315
19.	P-BLADE	P-BLADE MEAT	P	0.7150	0.0900	0.1900	0.0000	-0.0450	4.5600	3.8000	0.0437
20.	P-CHEEK	PK CHEEK MEAT	PKZ	0.6900	0.1400	0.1600	0.0000	0.0500	1.4400	4.6400	0.1152
21.	P-DIAPH	PK DIAPHRAGM MEAT	P	0.7000	0.1500	0.1450	0.0000	0.1200	2.3200	2.6100	0.1015
22.	P-FATBCK	PK SKND FATBACKS	P	0.1500	0.8000	0.0400	0.0000	-0.0100	0.1200	0.0400	0.0240
23.	P-HAM	PK FRESH HAM TRIM	P	0.5650	0.3000	0.1300	0.0000	0.0450	2.7300	2.4700	0.0325
24.	P-HAMCUR	PK CURED HAM TRIM	PR	0.5700	0.3000	0.1200	0.0000	0.0900	2.1600	1.2000	0.0360
25.	P-HAMFAT	PK HAM FAT	P	0.0400	0.9500	0.0100	0.0000	0.0000	0.0100	0.0100	0.0040
26.	P-HEAD	PK HEAD MEAT	PK	0.6100	0.2200	0.1600	0.0000	-0.0300	1.2000	2.5600	0.1120
27.	P-HEART	PK HEARTS	OZ	0.7000	0.1400	0.1550	0.0000	0.0800	0.9300	5.1150	0.0418
28.	P-JOWL	PK SKND JOWLS	P	0.3000	0.6150	0.0750	0.0000	0.0000	0.3450	0.1275	0.0322
29.	P-LIP	PK LIPS	O	0.6200	0.2200	0.1550	0.0000	0.0000	0.1550	0.0775	0.1395
30.	P-LIVER	PK LIVERS	OL	0.7400	0.0600	0.1900	0.0000	-0.0200	0.3800	9.3100	0.1805
31.	P-PIC	PK BNLS PICNICS	P	0.6000	0.2400	0.1550	0.0000	-0.0200	3.1000	2.4800	0.0356
32.	P-NECK	PK NECKBONE TRIM	P	0.6000	0.2300	0.1600	0.0000	-0.0400	3.0400	2.5600	0.0400
33.	P-PICHRT	PK PICNIC HEARTS	P	0.7000	0.1100	0.1800	0.0000	-0.0200	4.1400	3.4200	0.0396
34.	P-SNOUT	PK SNOUTS	O	0.5500	0.3000	0.1450	0.0000	-0.0300	0.3625	0.0725	0.1160
35.	P-SPLEEN	PK SPLEENS	O	0.7000	0.1450	0.1550	0.0000	0.0800	0.1085	8.6800	0.1395
36.	P-STOM	PK STOMACHS	O	0.0000	0.0000	0.0000	0.0000	0.0000	0.0000	0.0000	0.0000
37.	P-TNGTRM	PK TONGUE TRIMMINGS	O	0.0000	0.0000	0.0000	0.0000	0.0000	0.0000	0.0000	0.0000
38.	P-TONGUE	PK TONGUES	O	0.0000	0.0000	0.0000	0.0000	0.0000	0.0000	0.0000	0.0000
39.	P-50	PK 50% TRIMMINGS	P	0.3200	0.5900	0.0850	0.0000	-0.0200	1.0200	0.7650	0.0289
40.	P-80	PK 80% TRIMMINGS	P	0.5600	0.2800	0.1500	0.0000	-0.0400	2.8500	2.5500	0.0360
41.	P-95	PK 95% VL TRIMMINGS	P	0.0000	0.0000	0.0000	0.0000	0.0000	0.0000	0.0000	0.0000
42.	C-MDB	CK COMMINUTED MEAT	C	0.6600	0.2050	0.1250	0.0000	0.1600	1.8750	0.6250	0.0437
43.	T-MDB	TK MDB COMMINUTED MT	C	0.0000	0.0000	0.0000	0.0000	0.0000	0.0000	0.0000	0.0000
44.	V-BNLS	VEAL BNLS	V	0.0000	0.0000	0.0000	0.0000	0.0000	0.0000	0.0000	0.0000
45.	S-BEBR	BEBRECZINER SPICE		0.0500	0.0000	0.0100	0.0000	0.0100	0.0000	0.0000	0.0000
46.	S-BOLO	BOLOGNA SPICE		0.0000	0.0000	0.0000	0.0000	0.0000	0.0000	0.0000	0.0000
47.	S-FRANK	FRANK SPICE		0.0800	0.0000	0.1500	0.0000	-0.5200	3.0000	1.5000	0.0000
48.	S-SALAMI	SALAMI SPICE		0.0000	0.0000	0.0000	0.0000	0.0000	0.0000	0.0000	0.0000
49.	S-SAUS	FR SAUSAGE SPICE		0.0000	0.0000	0.0000	0.0000	0.0000	0.0000	0.0000	0.0000
50.	I-CORN	CORN SYRUP		0.2600	0.0000	0.0000	0.0000	0.2600	0.0000	0.0000	0.0000
51.	I-CURE	DRY CURE (W/SALT)		0.0000	0.0000	0.0000	0.0000	0.0000	0.0000	0.0000	0.0000
52.	I-SALT	SALT		0.0000	0.0000	0.0000	1.0000	0.0000	0.0000	0.0000	0.0000
53.	I-CEREAL	CEREAL FLOUR	X	0.0000	0.0000	0.0000	0.0000	0.0000	0.0000	0.0000	0.0000
54.	I-ISP	ISOLATED SOY PROTEIN	X	0.2900	0.0000	0.7100	0.0000	-2.5500	31.9500	0.0000	0.0000
55.	I-NFDM	NON-FAT DRY MILK	X	0.0000	0.0000	0.0000	0.0000	0.0000	0.0000	0.0000	0.0000
56.	I-PREMUL	CASEINATE PRE-EMULS.		0.5030	0.4190	0.0620	0.0100	0.2550	12.3380	0.0000	0.0000
57.	I-SPC	SOY PROTEIN CONCENC.	X	0.2000	0.0000	0.7000	0.0000	-2.6000	28.0000	0.0000	0.0000
58.	I-TVP	TEXTURED VEG PROTEIN	X	0.0000	0.0000	0.5000	0.0000	-2.0000	20.0000	0.0000	0.0000
59.	I-TSPC	TEXTURED SOY PROTEIN	X	0.0000	0.0000	0.7000	0.0000	-2.8000	28.0000	0.0000	0.0000
60.	WATER	WATER		1.0000	0.0000	0.0000	0.0000	1.0000	0.0000	0.0000	0.0000

Decimal parts of 1.0, except for bind and color which are indexes (see text). B = beef, C = chicken, D = domestic, F = fresh, H = hearts, I = imported, K = high collagen (cheek, head), L = lean, M = meat or mutton, O = offal, P = pork, R = cured meat, T = turkey, V = veal, X = extenders, Z = frozen.

Table 21–7. Raw Material Parameter Table: Example Listing

SEQ NO.	SHORT NAME	RAW MATERIAL	MAT. TYPE	GROUP TYPES	COST #1	COST #2	COST #3	QUANTITY #1 (LBS)	QUANTITY #2 (LBS)	ROUNDING UNIT	CURRENT	TO-DATE
1.	B-BULL	BF FC BULL MEAT	B		1.5400	1.5600	0.0000	BIG	BIG	0.000	0.	0.
2.	B-CHEEK	BF CHEEK MEAT	BK		0.9600	0.9800	0.0000	60.0	BIG	1.000	0.	0.
3.	B-CHUCK	BF BNLS CHUCKS	B		1.4400	1.4500	0.0000	BIG	BIG	0.000	0.	0.
4.	B-COW	BF FC COW MEAT	B		1.4700	1.4500	0.0000	600.0	BIG	1.000	0.	0.
5.	B-HEAD	BF HEAD MEAT	BK		0.8100	0.8100	0.0000	BIG	BIG	0.000	0.	0.
6.	B-HEART	BF HEARTS	0		0.4900	0.4900	0.0000	60.0	BIG	0.000	0.	0.
7.	B-LIP	BF LIPS	0		0.5500	0.5500	0.0000	BIG	BIG	0.000	0.	0.
8.	B-NAVEL	BF BNLS NAVELS	B		0.6200	0.6300	0.0000	BIG	BIG	0.000	0.	0.
9.	B-PLATE	BF PLATES	B		0.5200	0.5200	0.0000	.0	BIG	1.000	0.	0.
10.	B-SHANK	BF SHANK MEAT	BK		1.2100	1.2200	0.0000	BIG	BIG	0.000	0.	0.
11.	B-50	BF 50% TRIMMINGS	B		0.6400	0.6450	0.0000	.0	BIG	1.000	0.	0.
12.	B-60	BF 60% TRIMMINGS	B		0.8000	0.8000	0.0000	BIG	BIG	0.000	0.	0.
13.	B-65	BF 65% TRIMMINGS	B		0.8700	0.8700	0.0000	BIG	BIG	0.000	0.	0.
14.	B-75	BF 75% TRIMMINGS	B		1.0200	1.0300	0.0000	BIG	BIG	0.000	0.	0.
15.	B-85	BF 85% TRIMMINGS	B		1.1500	1.1700	0.0000	.0	BIG	1.000	0.	0.
16.	B-90	BF 90% TRIMMINGS	B		1.3000	1.3100	0.0000	BIG	BIG	0.000	0.	0.
17.	P-BACEND	PK BACON ENDS	OR		0.2400	0.2400	0.0000	BIG	BIG	0.000	0.	0.
18.	P-BELLY	PK BELLY STRIPS	P		0.4100	0.4100	0.0000	BIG	BIG	0.000	0.	0.
19.	P-BLADE	P-BLADE MEAT	P		1.1000	1.1200	0.0000	BIG	BIG	0.000	0.	0.
20.	P-CHEEK	PK CHEEK MEAT	PKZ		0.8900	0.9000	0.0000	60.0	BIG	0.000	0.	0.
21.	P-DIAPH	PK DIAPHRAGM MEAT	P		0.6300	0.6300	0.0000	BIG	BIG	0.000	0.	0.
22.	P-FATBCK	PK SKND FATBACKS	P		0.2700	0.2750	0.0000	.0	BIG	1.000	0.	0.
23.	P-HAM	PK FRESH HAM TRIM	P		0.7000	0.7100	0.0000	BIG	BIG	0.000	0.	0.
24.	P-HAMCUR	PK CURED HAM TRIM	PR		0.5700	0.5700	0.0000	BIG	BIG	0.000	0.	0.
25.	P-HAMFAT	PK HAM FAT	P		0.0800	0.0800	0.0000	10000.0	BIG	1.000	0.	0.
26.	P-HEAD	PK HEAD MEAT	PK		0.6600	0.6600	0.0000	BIG	BIG	0.000	0.	0.
27.	P-HEART	PK HEARTS	OZ		0.4800	0.4800	0.0000	.0	BIG	1.000	0.	0.
28.	P-JOWL	PK SKND JOWLS	P		0.3600	0.3650	0.0000	17.0	BIG	1.000	0.	0.
29.	P-LIP	PK LIPS	0		0.3400	0.3500	0.0000	75.0	BIG	0.000	0.	0.
30.	P-LIVER	PK LIVERS	OL		0.1750	0.1700	0.0000	BIG	BIG	0.000	0.	0.
31.	P-PIC	PK BNLS PICNICS	P		0.8300	0.8300	0.0000	BIG	BIG	0.000	0.	0.
32.	P-NECK	PK NECKBONE TRIM	P		0.8100	0.8150	0.0000	BIG	BIG	0.000	0.	0.
33.	P-PICHRT	PK PICNIC HEARTS	P		1.1500	1.1600	0.0000	BIG	BIG	0.000	0.	0.
34.	P-SNOUT	PK SNOUTS	0		0.3100	0.3100	0.0000	25.0	BIG	1.000	0.	0.
35.	P-SPLEEN	PK SPLEENS	0		0.1200	0.1200	0.0000	BIG	BIG	0.000	0.	0.
36.	P-STOM	PK STOMACHS	0		0.2200	0.2100	0.0000	BIG	BIG	0.000	0.	0.
37.	P-TNGTRM	PK TONGUE TRIMMINGS	0		0.2000	0.2000	0.0000	BIG	BIG	0.000	0.	0.
38.	P-TONGUE	PK TONGUES	0		0.5800	0.5800	0.0000	BIG	BIG	0.000	0.	0.
39.	P-50	PK 50% TRIMMINGS	P		0.4200	0.4300	0.0000	375.0	BIG	1.000	0.	0.
40.	P-80	PK 80% TRIMMINGS	P		0.8000	0.8200	0.0000	365.0	BIG	1.000	0.	0.
41.	P-95	PK 95% VL TRIMMINGS	P		1.1000	1.1300	0.0000	BIG	BIG	0.000	0.	0.
42.	C-MDB	CK COMMINUTED MEAT	C		0.2400	0.2400	0.0000	BIG	BIG	0.000	0.	0.
43.	T-MDB	TK MDB COMMINUTED MT	C		0.3200	0.3200	0.0000	BIG	BIG	0.000	0.	0.
44.	V-BNLS	VEAL BNLS	V		1.3500	1.3600	0.0000	BIG	BIG	0.000	0.	0.
45.	S-BEBR	BEBRECZINER SPICE			0.3000	0.3000	0.0000	BIG	BIG	0.000	0.	0.
46.	S-BOLO	BOLOGNA SPICE			0.4500	0.4500	0.0000	BIG	BIG	0.000	0.	0.
47.	S-FRANK	FRANK SPICE			0.3400	0.3400	0.0000	BIG	BIG	0.000	0.	0.
48.	S-SALAMI	SALAMI SPICE			0.5500	0.5500	0.0000	BIG	BIG	0.000	0.	0.
49.	S-SAUS	FR SAUSAGE SPICE			0.3200	0.3200	0.0000	BIG	BIG	0.000	0.	0.
50.	I-CORN	CORN SYRUP			0.3000	0.3000	0.0000	BIG	BIG	0.000	0.	0.
51.	I-CURE	DRY CURE (W/SALT)			0.2500	0.2500	0.0000	BIG	BIG	0.000	0.	0.
52.	I-SALT	SALT			0.0550	0.0550	0.0000	BIG	BIG	0.000	0.	0.
53.	I-CEREAL	CEREAL FLOUR	I		0.2200	0.2200	0.0000	BIG	BIG	0.000	0.	0.
54.	I-ISP	ISOLATED SOY PROTEIN P	I		1.0100	1.0100	0.0000	BIG	BIG	0.000	0.	0.
55.	I-NFDM	NON-FAT DRY MILK	I		1.0200	1.0200	0.0000	BIG	BIG	0.000	0.	0.
56.	I-PREMUL	CASEINATE PRE-EMULS.			0.1620	0.1620	0.0000	BIG	BIG	0.000	0.	0.
57.	I-SPC	SOY PROTEIN CONCENC.	I		0.5000	0.5000	0.0000	BIG	BIG	0.000	0.	0.
58.	I-TVP	TEXTURED VEG PROTEIN	I		0.3500	0.3500	0.0000	BIG	BIG	0.000	0.	0.
59.	I-TSPC	TEXTURED SOY PROTEIN	I		0.5400	0.5400	0.0000	BIG	BIG	0.000	0.	0.
60.	WATER	WATER			0.0000	0.0000	0.0000	BIG	BIG	0.000	0.	0.

B = beef, C = chicken, D = domestic, F = fresh, H = hearts, I = imported, K = high collagen (cheek, head), L = lean, M = meat or mutton, O = offal, P = pork, R = cured meat, T = turkey, V = veal, X = extenders, Z = frozen.

Table 21–8. Product Specifications for Regular Frank

```
PRODUCT SHORT NAME:  ISu-REG          AS OF:        CURRENT DATE
PRODUCT LONG NAME:   ISU REGULAR FRANK PRODUCT TYPE:         GROUPS:

LAST COSTS:     $     0.6556     $    0.6671    $    0.0000
LAST CALCULATION USED COST # 1         DATE LAST CALCULATION: 7/13    CALC. # 3
MIN OR MAX:           MIN

NO. RAW MATERIALS:        14      PRODUCT BLOCK SIZE:   100.0000 (SCALED ON BW)
NO. REQUIREMENTS:        10      SHRINKAGE RATE:         8.0000 %
PRODUCT DISPLAY UNITS: LBS
```

<--- RAW MATERIALS IN PRODUCT ---> (NAMES FROM TABLE: ISU MEAT COURSE)

SEQ NO.	SHORT NAME	RAW MATERIAL DESCRIPTION	IN BLOCK	LO LIMIT	HI LIMIT	LIMIT TYPE	CURRENT VALUE	CURRENT STATUS
1.	B-CHEEK	BF CHEEK MEAT	Y	.0000	BIG	% BW	16.3674	BETWEEN
2.	B-COW	BF FC COW MEAT	Y	.0000	BIG	% BW	8.5269	BETWEEN
3.	B-PLATE	BF PLATES	Y	.0000	BIG	% BW	0.0000	HI-LIMIT
4.	B-50	BF 50% TRIMMINGS	Y	.0000	BIG	% BW	0.0000	HI-LIMIT
5.	B-85	BF 85% TRIMMINGS	Y	.0000	BIG	% BW	0.0000	HI-LIMIT
6.	P-FATBCK	PK SKND FATBACKS	Y	.0000	BIG	% BW	0.0000	HI-LIMIT
7.	P-JOWL	PK SKND JOWLS	Y	.0000	BIG	% BW	0.0000	LO-LIMIT
8.	P-50	PK 50% TRIMMINGS	Y	.0000	BIG	% BW	29.0407	BETWEEN
9.	P-80	PK 80% TRIMMINGS	Y	.0000	BIG	% BW	46.0651	BETWEEN
10.	I-CORN	CORN SYRUP		1.0000	1.0000	% BW	1.0000	BETWEEN
11.	I-SALT	SALT		.0000	3.0000	% BW	2.6769	BETWEEN
12.	S-FRANK	FRANK SPICE		.7500	.7500	LBS	0.7500	BETWEEN
13.	WATER	WATER		.0000	BIG	% BW	24.8942	BETWEEN
14.	P-CHEEK	PK CHEEK MEAT	Y	.0000	BIG	% BW	0.0000	LO-LIMIT

<--- REQUIREMENTS IN PRODUCT --->

SEQ NO.	ITEM-1	ITEM-2	REQ. TYPE	LO LIMIT	HI LIMIT	LIMIT TYPE	LAST CALC. VALUE	CURRENT STATUS
1	MOISTURE		A	-BIG	BIG	% FW	55.4230	BETWEEN
2	PROTEIN		A	-BIG	BIG	% FW	11.6058	BETWEEN
3	FAT		A	25.0000	29.0000	% FW	29.0000	HI-LIMIT
4	BIND		A	180.0000	BIG	% FW	201.4825	BETWEEN
5	COLOR		A	190.0000	BIG	% FW	280.9178	BETWEEN
6	SALT		A	2.2500	2.2500	% FW	2.2500	HI-LIMIT
7	COLLAGEN		A	-BIG	3.7500	% FW	3.7500	HI-LIMIT
8	USDA-AW		A	-BIG	9.0000	% FW	9.0000	HI-LIMIT
9	P	> B	GG	.0000	BIG	% BW	38.8270	BETWEEN
10	B	> WATER	GW	.0000	BIG	% BW	-0.0000	LO-LIMIT

For protein, there are really no requirements. The least cost formulation system will use as little protein as possible because it is an expensive attribute. From the analysis of the standard formula, set the low limit at 10 percent finished weight (FW), and leave the high limit unbounded at Big.

For moisture, leave it completely unbounded, because it will be controlled by the USDA added water (AW) plus fat.

For USDA–AW, the USDA regulation limits the maximum amount so that when it is summed with fat, the total of fat and added water will not exceed 40 percent finished weight (FW). Because water is an inexpensive ingredient, the least cost formulation system will include as much as possible. To give a margin of safety, set the high limit at 9 percent finished weight (FW) for a 29 percent fat frank or at 29 percent for a 10 percent fat frank.

Bind and color are dependent on the protein content of the raw materials. The least cost formulation system will use as little protein as possible because it is an expensive attribute. Therefore, the bind and color will be close to their low limits, and the high limits will be unbounded at Big. The values 180 for bind and 190 for color are based upon the analysis of many hundreds of formulations.

The identical limits for salt will insure that exactly $2\frac{1}{4}$ pounds of salt will be used. This requirement comes from both the choice of label and the marketplace preference.

The limits for collagen were chosen for several reasons, including to:

1. Prevent excessive gelatin formation in the product.

2. Improve peelability by formation of a "skin."

3. Prevent physical shrink (curling).

4. Improve fat binding.

5. Improve "bite" (firm rather than elastic).

An upper limit of 3.75 percent of finished weight was used.

At this point you have taken care of the attributes you wish to control in the regular frank. The next step is to insure that product label constraints are met.

One of the unique features of the least cost formulator is its group capability. The group parameter allows the handling of a number of raw materials as a single item. For example, all beef raw materials may be designated by Group B. In Table 21–6, all the raw materials have been categorized into one or more groups.

To insure that the batch meets the requirements of the chosen label, specify that all the pounds of pork raw materials (P) are greater than the pounds of beef raw materials (B) by Group P>B. Similarly, specify beef greater than water (W) by Group B>W. The other items in the label (salt, corn syrup, and spices) have been fixed by specifying their individual limits. In Table 21–8, the complete product specifications are shown for the regular frank.

Least Cost Formulation: Final Reports

Product Formulation Material Usage Report

Table 21–9 shows the calculated cost of the regular frank and the quantities of raw materials required to achieve these results. Of the 10 meat raw materials offered, only 4 were used. This resulted in a raw material price for regular franks of $0.66 per pound (total cost ÷ finished weight = 78 ÷ 119 = $0.66). When the standard formula (see Table 21–3) was used, the cost per pound was calculated at $0.69, using the same day's raw material prices. This represents a savings of $0.03 per pound. In large operations that produce over 100,000 pounds of franks weekly, this would amount to a savings of $3,000 per week.

Table 21–9. Product Formulation Material Usage Report for Regular Frank

```
PRODUCT NAMES - SHORT: ISU-REG      LONG: ISU REGULAR FRANK    TYPE: MIN
LAST UPDATED:        7/13          LAST CALCULATED:   CURRENT DATE

                    CALC. #    3 BASED ON COST #    1

                    <- CALCULATED COST ($ PER LB) ->
                    COST #1   COST #2   COST #3
                    ---------------------------------
    BLOCK WEIGHT:    0.7800    0.7937    0.0000
    GROSS WEIGHT:    0.6031    0.6137    0.0000
    FINISHED WEIGHT: 0.6556    0.6671    0.0000
```

SEQ NO.	MATERIAL NAME SHORT	LONG	USAGE	COST #1	EXTENDED COST #1	BW	GW	FW	LO-LIMIT	HI-LIMIT	TYPE
1.	B-CHEEK	BF CHEEK MEAT	16.4	0.9600	15.71	16.37	12.66	13.76	.0000	BIG	% GW
2.	B-COW	BF FC COW MEAT	8.5	1.4700	12.53	8.53	6.59	7.17	.0000	BIG	% GW
3.	P-50	PK 50% TRIMMINGS	29.0	0.4200	12.20	29.04	22.46	24.41	.0000	BIG	% GW
4.	P-80	PK 80% TRIMMINGS	46.1	0.8000	36.85	46.07	35.62	38.72	.0000	BIG	% GW
		BLOCK WEIGHT (BW):	100.0 LBS		77.30	100.00	77.33	84.05			
5.	I-CORN	CORN SYRUP	1.0	0.3000	0.30	1.00	0.77	0.84	1.0000	1.0000	% BW
6.	I-SALT	SALT	2.7	0.0550	0.15	2.68	2.07	2.25	.0000	3.0000	% GW
7.	S-FRANK	FRANK SPICE	0.7	0.3400	0.25	0.75	0.58	0.63	.7500	.7500	LBS
8.	WATER	WATER	24.9	0.0000	0.00	24.89	19.25	20.92	.0000	BIG	% GW
		GROSS WEIGHT (GW):	129.3 LBS		78.00	129.32	100.00	108.70			
		SHRINKAGE LOSS:	10.3			10.35	8.00	8.70			
		FINISHED WEIGHT:	119.0 LBS		78.00	118.98	92.00	100.00			

Product Formulation
Requirements Report

This report displays the predicted value of the attributes for the product and is shown in Table 21–10 in the "Value" column. All calculations fall between the low and high limits. As expected, the predicted values for fat and USDA–AW are at the high limits, while the values for protein, bind, and color are near the low limits.

Table 21–10. Product Formulation Requirements Report for Regular Frank

```
PRODUCT NAMES - SHORT: ISU-REG      LONG: ISU REGULAR FRANK    TYPE: MIN
LAST UPDATED:          7/13          LAST CALCULATED:   CURRENT DATE


                             LAST CALCULATION #    3  USING COST #   1
SEQ  (------ REQUIREMENT ------)              (----- LIMITS ON RANGE -----)    PENALTY
NO.  TYPE     DESCRIPTION          VALUE    LOW-LIMIT    HI-LIMIT    TYPE      COST
-------------------------------------------------------------------------------------
 1  A    MOISTURE           55.4230      -BIG        BIG      I FW    0.0000
 2  A    PROTEIN            11.6058      -BIG        BIG      I FW    0.0000
 3  A    FAT                29.0000     25.0000     29.0000   I FW    0.0159
 4  A    BIND              201.4825    180.0000      BIG      I FW    0.0000
 5  A    COLOR             280.9178    190.0000      BIG      I FW    0.0000
 6  A    SALT                2.2500      2.2500      2.2500   I FW    0.0157
 7  A    COLLAGEN            3.7500      -BIG        3.7500   I FW    0.0848
 8  A    USDA-AW             9.0000      -BIG        9.0000   I FW    0.0105
 9  GG   P      )   B       38.8270       .0000       BIG     I GW    0.0000
10  GW   B      )   WATER   -0.0000       .0000       BIG     I GW    0.0064

NOTE: "PENALTY-COST" IS THE AMOUNT IN $ PER LB BY WHICH THE FORMULATION COST COULD BE REDUCED
      IF THE REQUIREMENT LIMIT WAS CHANGED BY 1.0 AS GIVEN. "VALUE" IS GIVEN IN THE SAME UNITS
      AS THE LIMITS.
```

Product Formulation
Price-ranging Report

Table 21–11 is divided into two groups of raw materials: those that were used in the formulation and those that were excluded. Those included offered the right combination of attributes for a given price. For example, item #2, B–Cow, costs $1.47 per pound; if this raw material were to drop to $1.21 per pound, more of it would be included; if B–Cow rose in price to $4.31 per pound, less of it would be used. Notice that P–50 is very close to the price where less would be used. However, the negative number in the "Use–More" column means that no additional P–50 would be included in the formulation regardless of a lower price because of other constraints (fat content).

For those raw materials (#9 through #14) that were not included in the formulation, the "Use–More" column represents the price at which those raw

Table 21-11. Product Formulation Price-ranging Report
for Regular Frank

```
PRODUCT NAMES - SHORT: ISU-REG   LONG: ISU REGULAR FRANK   TYPE: MIN
     LAST UPDATED:       7/13    LAST CALCULATED:   CURRENT DATE

                        CALC. #    3 ... USING COST # 1
```

				<-- CALC. COST ($ PER LB) -->				
				COST #1	COST #2	COST #3		
	BLOCK WEIGHT:			0.7800	0.7937	0.0000		
	GROSS WEIGHT:			0.6031	0.6137	0.0000		
	FINISHED WEIGHT:			0.6556	0.6671	0.0000		
SEQ	<----- MATERIAL NAME ----->		USAGE	<-- COST RANGE -->		<---- CURRENT COSTS ---->		
NO.	SHORT	LONG	(LBS)	USE-MORE	USE-LESS	COST-1	COST-2	COST-3
1.	B-CHEEK	BF CHEEK MEAT	16.367	-20.8668	1.1611	0.9600	0.9800	0.0000
2.	B-COW	BF FC COW MEAT	8.527	1.2068	4.3138	1.4700	1.4500	0.0000
3.	P-50	PK 50% TRIMMINGS	29.041	-.6438	.4335	0.4200	0.4300	0.0000
4.	P-80	PK 80% TRIMMINGS	46.065	.6837	1.1456	0.8000	0.8200	0.0000
5.	I-CORN	CORN SYRUP	1.000	-86.3779	2187.7686	0.3000	0.3000	0.0000
6.	I-SALT	SALT	2.677	-BIG	BIG	0.0550	0.0550	0.0000
7.	S-FRANK	FRANK SPICE	0.750	-115.2305	2916.9653	0.3400	0.3400	0.0000
8.	WATER	WATER	24.894	-21.4985	.8519	0.0000	0.0000	0.0000
9.	B-PLATE	BF PLATES	0.000	-BIG	.9572	0.5200	0.5200	0.0000
10.	B-50	BF 50% TRIMMINGS	0.000	-BIG	.9366	0.6400	0.6450	0.0000
11.	B-85	BF 85% TRIMMINGS	0.000	-BIG	1.3823	1.1500	1.1700	0.0000
12.	P-FATBCK	PK SKND FATBACKS	0.000	-BIG	.3737	0.2700	0.2750	0.0000
13.	P-JOWL	PK SKND JOWLS	0.000	.3452	BIG	0.3600	0.3650	0.0000
14.	P-CHEEK	PK CHEEK MEAT	0.000	.3427	BIG	0.8900	0.9000	0.0000

```
NOTE: "COST-TO-USE-MORE" IS THE PRICE AT WHICH THE FORMULATION WILL START TO USE MORE OF THE MATERIAL.
      "COST-TO-USE-LESS" IS THE PRICE AT WHICH LESS OF THE MATERIAL WILL BE USED IN THE FORMULATION.
```

materials will enter the formulation. Hence, P–Jowl that costs $0.36 per pound will enter at $0.345 per pound. Other than P–Cheek, all the rest of the raw materials will not enter regardless of lower price, due to lack of availability.

With these price ranges, it is possible to improve the advance purchase of the raw materials. If a harvesting operation is involved, it is possible to decide if the raw material is better used in the products or sold in the open market by comparing prices from the pricing services listed in Chapter 13 with the "Use–More" prices.

Other Reports

The Least Cost Formulator® provides a wide range of additional reports that keep a running inventory of raw materials and finished product on hand; raw material usage by day, week, month, and year-to-date; and production formulation tickets.

Other Frank Formulations

For each of the other products to be made, additional raw materials were offered, but the product specifications for moisture, protein, and fat remained the same. The identical reports for each product are shown following their description.

Regular Frank with By-Products

Two additional raw materials — beef hearts and pork lips — are included in the formulation to produce this #3 frank (see Table 21–3). The label will read "pork, water, pork lips, beef hearts, beef, salt, corn syrup, spices, and nitrite." Except for bind and collagen, the low and high limits of the attributes for this product are identical to those of the #1 frank. The reports for this by-product (BP) frank are given in Tables 21–12 to 21–15.

Regular Frank with Isolated Soy Protein

Isolated soy protein (ISP) in the amount of 2 percent of the finished weight of the product is forced into the formulation to produce this frank. The label will read "pork, beef, water, salt, ISP, corn syrup, spices, and nitrite." The limits for the attributes for this product are identical to those of the #1 frank given in Table 21–3. The reports for this product are given in Tables 21–16 to 21–19.

Regular Frank with By-Products and Isolated Soy Protein

All the raw materials that were incorporated into both the third and fourth formulations are used to produce this frank. The label will read "pork, water, pork lips, beef hearts, beef, salt, ISP, corn syrup, spices, and nitrite." The limits for the attributes on this product are identical to those of the #3 frank with by-products indicated in Table 21–13. The reports for this product are given in Tables 21–20 to 21–23.

The results of the least cost formulation examples reveal that considerable cost variations are encountered for the four frankfurter formulations. These cost variations are summarized as follows: regular frankfurter (see Table 21–9) — $65.56/cwt.; by-product (BP) frankfurter (see Table 21–12) — $52.97/cwt.; isolated soy protein (ISP) frankfurter (see Table 21–16) — $60.35/cwt.; and BP–ISP frankfurter (see Table 21–21) — $46.82/cwt.

Table 21–12. Product Specifications for BP Frank

```
PRODUCT SHORT NAME:   ISU-BP            AS OF:        CURRENT DATE
PRODUCT LONG NAME:    ISU BYPRODUCT FRANK   PRODUCT TYPE:   F       GROUPS:

LAST COSTS:      $      0.5297      $     0.5402      $     0.0000
LAST CALCULATION USED COST # 1         DATE LAST CALCULATION: 7/13      CALC. #    3
MIN OR MAX:           MIN

NO. RAW MATERIALS:        16        PRODUCT BLOCK SIZE:    100.0000 (SCALED ON BW)
NO. REQUIREMENTS:         13        SHRINKAGE RATE:          8.0000 %
PRODUCT DISPLAY UNITS: LBS
```

<--- RAW MATERIALS IN PRODUCT --->   (NAMES FROM TABLE: ISU MEAT COURSE)

SEQ NO.	SHORT NAME	RAW MATERIAL DESCRIPTION	IN BLOCK	LO LIMIT	HI LIMIT	LIMIT TYPE	CURRENT VALUE	CURRENT STATUS
1.	B-CHEEK	BF CHEEK MEAT	Y	.0000	BIG	% GW	0.0000	LO-LIMIT
2.	B-COW	BF FC COW MEAT	Y	.0000	BIG	% GW	2.6213	BETWEEN
3.	B-PLATE	BF PLATES	Y	.0000	BIG	% GW	0.0000	HI-LIMIT
4.	B-50	BF 50% TRIMMINGS	Y	.0000	BIG	% GW	0.0000	HI-LIMIT
5.	B-85	BF 85% TRIMMINGS	Y	.0000	BIG	% GW	0.0000	HI-LIMIT
6.	P-FATBCK	PK SKND FATBACKS	Y	.0000	BIG	% GW	0.0000	HI-LIMIT
7.	P-JOWL	PK SKND JOWLS	Y	.0000	BIG	% GW	0.0000	LO-LIMIT
8.	P-50	PK 50% TRIMMINGS	Y	.0000	BIG	% GW	26.8319	BETWEEN
9.	P-80	PK 80% TRIMMINGS	Y	.0000	BIG	% GW	43.1589	BETWEEN
10.	I-CORN	CORN SYRUP		1.0000	1.0000	% BW	1.0000	BETWEEN
11.	I-SALT	SALT		.0000	3.0000	% GW	2.6213	BETWEEN
12.	S-FRANK	FRANK SPICE		.7500	.7500	LBS	0.7500	BETWEEN
13.	WATER	WATER		.0000	BIG	% GW	22.2632	BETWEEN
14.	B-HEART	BF HEARTS	Y	.0000	BIG	% GW	13.6939	BETWEEN
15.	P-LIP	PK LIPS	Y	.0000	BIG	% GW	13.6939	BETWEEN
16.	P-CHEEK	PK CHEEK MEAT	Y	.0000	BIG	% GW	0.0000	LO-LIMIT

<--- REQUIREMENTS IN PRODUCT --->

SEQ NO.	ITEM-1	ITEM-2	REQ. TYPE	LO LIMIT	HI LIMIT	LIMIT TYPE	LAST CALC. VALUE	CURRENT STATUS
1	MOISTURE		A	-BIG	BIG	% FW	55.4038	BETWEEN
2	PROTEIN		A	-BIG	BIG	% FW	11.6009	BETWEEN
3	FAT		A	25.0000	29.0000	% FW	29.0000	BETWEEN
4	BIND		A	130.0000	BIG	% FW	153.3243	BETWEEN
5	COLOR		A	190.0000	BIG	% FW	200.2796	BETWEEN
6	SALT		A	2.2500	2.2500	% FW	2.2500	HI-LIMIT
7	COLLAGEN		A	-BIG	4.2000	% FW	4.2000	BETWEEN
8	USDA-AW		A	-BIG	9.0000	% FW	9.0000	BETWEEN
9	P	> WATER	GW	.0000	BIG	% GW	37.6893	BETWEEN
10	WATER	> P-LIP	MM	.0000	BIG	% GW	6.7669	BETWEEN
11	P-LIP	> B-HEART	MM	.0000	BIG	% GW	0.0000	LO-LIMIT
12	B-HEART	> B	MG	.0000	BIG	% GW	8.7437	BETWEEN

<--- REQUIREMENTS IN PRODUCT --->

SEQ NO.	ITEM-1	ITEM-2	REQ. TYPE	LO LIMIT	HI LIMIT	LIMIT TYPE	LAST CALC. VALUE	CURRENT STATUS
13	B	> SALT	GA	.0000	BIG	% GW	0.0000	LO-LIMIT

Table 21-13. Product Formulation Material Usage Report for BP Frank

PRODUCT NAMES - SHORT: ISU-BP   LONG: ISU BYPRODUCT FRANK   TYPE: MIN
LAST UPDATED: 7/13   LAST CALCULATED: CURRENT DATE

CALC. # 3 BASED ON COST # 1

<- CALCULATED COST (\$ PER LB) ->

	COST #1	COST #2	COST #3
BLOCK WEIGHT:	0.6172	0.6293	0.0000
GROSS WEIGHT:	0.4873	0.4970	0.0000
FINISHED WEIGHT:	0.5297	0.5402	0.0000

SEQ NO.	SHORT	LONG	USAGE	COST #1	EXTENDED COST #1	BW	GW	FW	LO-LIMIT	HI-LIMIT	TYPE
1.	B-COW	BF FC COW MEAT	2.6	1.4700	3.85	2.62	2.07	2.25	.0000	BIG	% GW
2.	P-50	PK 50% TRIMMINGS	26.8	0.4200	11.27	26.83	21.19	23.03	.0000	BIG	% GW
3.	P-80	PK 80% TRIMMINGS	43.2	0.8000	34.53	43.16	34.08	37.05	.0000	BIG	% GW
4.	B-HEART	BF HEARTS	13.7	0.4900	6.71	13.69	10.81	11.75	.0000	BIG	% GW
5.	P-LIP	PK LIPS	13.7	0.3400	4.66	13.69	10.81	11.75	.0000	BIG	% GW
		BLOCK WEIGHT (BW):	100.0 LBS		61.02	100.00	78.97	85.83			
6.	I-CORN	CORN SYRUP	1.0	0.3000	0.30	1.00	0.79	0.86	1.0000	1.0000	% BW
7.	I-SALT	SALT	2.6	0.0550	0.14	2.62	2.07	2.25	.0000	3.0000	% GW
8.	S-FRANK	FRANK SPICE	0.8	0.3400	0.25	0.75	0.59	0.64	.7500	.7500	LBS
9.	WATER	WATER	22.3	0.0000	0.00	22.26	17.58	19.11	.0000	BIG	% GW
		GROSS WEIGHT (GW):	126.6 LBS		61.72	126.63	100.00	108.70			
		SHRINKAGE LOSS:	10.1			10.13	8.00	8.70			
		FINISHED WEIGHT:	116.5 LBS		61.72	116.50	92.00	100.00			

Table 21-14. Product Formulation Requirements Report for BP Frank

PRODUCT NAMES - SHORT: ISU-BP   LONG: ISU BYPRODUCT FRANK   TYPE: MIN
LAST UPDATED: 7/13   LAST CALCULATED: CURRENT DATE

LAST CALCULATION # 3 USING COST # 1

SEQ NO.	TYPE	DESCRIPTION			VALUE	LOW-LIMIT	HI-LIMIT	TYPE	PENALTY COST
1	A	MOISTURE			55.4038	-BIG	BIG	% FW	0.0000
2	A	PROTEIN			11.6009	-BIG	BIG	% FW	0.0000
3	A	FAT			29.0000	25.0000	29.0000	% FW	0.0152
4	A	BIND			153.3243	130.0000	BIG	% FW	0.0000
5	A	COLOR			200.2796	190.0000	BIG	% FW	0.0000
6	A	SALT			2.2500	2.2500	2.2500	% FW	0.0094
7	A	COLLAGEN			4.2000	-BIG	4.2000	% FW	0.0844
8	A	USDA-AW			9.0000	-BIG	9.0000	% FW	0.0158
9	GW	P	>	WATER	37.6893	.0000	BIG	% GW	0.0000
10	MM	WATER	>	P-LIP	6.7669	.0000	BIG	% GW	0.0000
11	MM	P-LIP	>	B-HEART	0.0000	.0000	BIG	% GW	0.0034
12	MG	B-HEART	>	B	8.7437	.0000	BIG	% GW	0.0062
13	GA	B	>	SALT	0.0000	.0000	BIG	% GW	0.0062

NOTE: "PENALTY-COST" IS THE AMOUNT IN \$ PER LB BY WHICH THE FORMULATION COST COULD BE REDUCED IF THE REQUIREMENT LIMIT WAS CHANGED BY 1.0 AS GIVEN. "VALUE" IS GIVEN IN THE SAME UNITS AS THE LIMITS.

Table 21–15.   Product Formulation Price-ranging Report
for BP Frank

PRODUCT NAMES - SHORT: ISU-BP   LONG: ISU BYPRODUCT FRANK   TYPE: MIN
LAST UPDATED: 7/13   LAST CALCULATED: CURRENT DATE

CALC. #   3 ... USING COST # 1

<-- CALC. COST ($ PER LB) -->
COST #1   COST #2   COST #3

	BLOCK WEIGHT:		0.6172	0.6293	0.0000			
	GROSS WEIGHT:		0.4873	0.4970	0.0000			
	FINISHED WEIGHT:		0.5297	0.5402	0.0000			

SEQ NO.	SHORT	LONG	USAGE (LBS)	USE-MORE	USE-LESS	COST-1	COST-2	COST-3
1.	B-COW	BF FC COW MEAT	2.621	.9768	1.4784	1.4700	1.4500	0.0000
2.	P-50	PK 50% TRIMMINGS	26.832	-.7271	.4415	0.4200	0.4300	0.0000
3.	P-80	PK 80% TRIMMINGS	43.159	.7942	1.1377	0.8000	0.8200	0.0000
4.	I-CORN	CORN SYRUP	1.000	-153.8512	10.6458	0.3000	0.3000	0.0000
5.	I-SALT	SALT	2.621	-BIG	BIG	0.0550	0.0550	0.0000
6.	S-FRANK	FRANK SPICE	0.750	-205.1950	14.1344	0.3400	0.3400	0.0000
7.	WATER	WATER	22.263	-.1017	1.5150	0.0000	0.0000	0.0000
8.	B-HEART	BF HEARTS	13.694	-37.3937	.5046	0.4900	0.4900	0.0000
9.	P-LIP	PK LIPS	13.694	-4.1098	.3546	0.3400	0.3500	0.0000
10.	B-CHEEK	BF CHEEK MEAT	0.000	.9516	BIG	0.9600	0.9800	0.0000
11.	B-PLATE	BF PLATES	0.000	-BIG	.9583	0.5200	0.5200	0.0000
12.	B-50	BF 50% TRIMMINGS	0.000	-BIG	.9412	0.6400	0.6450	0.0000
13.	B-85	BF 85% TRIMMINGS	0.000	-BIG	1.3857	1.1500	1.1700	0.0000
14.	P-FATBCK	PK SKND FATBACKS	0.000	-BIG	.3720	0.2700	0.2750	0.0000
15.	P-JOWL	PK SKND JOWLS	0.000	.3361	BIG	0.3600	0.3650	0.0000
16.	P-CHEEK	PK CHEEK MEAT	0.000	.2870	BIG	0.8900	0.9000	0.0000

Header columns: SEQ NO. SHORT | MATERIAL NAME LONG | USAGE (LBS) | <-- COST RANGE --> USE-MORE USE-LESS | <---- CURRENT COSTS ----> COST-1 COST-2 COST-3

NOTE: "COST-TO-USE-MORE" IS THE PRICE AT WHICH THE FORMULATION WILL START TO USE MORE OF THE MATERIAL.
"COST-TO-USE-LESS" IS THE PRICE AT WHICH LESS OF THE MATERIAL WILL BE USED IN THE FORMULATION.

Table 21–16.   Product Specifications for ISP Frank

```
PRODUCT SHORT NAME:  ISU-ISP              AS OF:         CURRENT DATE      CURRENT DATE
PRODUCT LONG NAME:   ISU FRANK WITH ISP   PRODUCT TYPE:  F        GROUPS:

LAST COSTS:      $   0.6035        $    0.6128      $    0.0000
LAST CALCULATION USED COST # 1           DATE LAST CALCULATION: 7/13      CALC. #   2
MIN OR MAX:          MIN

NO. RAW MATERIALS:        15         PRODUCT BLOCK SIZE:      100.0000 (SCALED ON BW)
NO. REQUIREMENTS:         10         SHRINKAGE RATE:            8.0000 %
PRODUCT DISPLAY UNITS: LBS
```

<--- RAW MATERIALS IN PRODUCT --->   (NAMES FROM TABLE: ISU MEAT COURSE)

SEQ NO.	SHORT NAME	RAW MATERIAL DESCRIPTION	IN BLOCK	LO LIMIT	HI LIMIT	LIMIT TYPE	CURRENT VALUE	CURRENT STATUS
				ACCEPTABLE RANGE OF VALUES				
1.	B-CHEEK	BF CHEEK MEAT	Y	.0000	BIG	% BW	23.5800	BETWEEN
2.	B-COW	BF FC COW MEAT	Y	.0000	BIG	% BW	9.5475	BETWEEN
3.	B-PLATE	BF PLATES	Y	.0000	BIG	% BW	0.0000	HI-LIMIT
4.	B-50	BF 50% TRIMMINGS	Y	.0000	BIG	% BW	0.0000	HI-LIMIT
5.	B-85	BF 85% TRIMMINGS	Y	.0000	BIG	% BW	0.0000	HI-LIMIT
6.	P-FATBCK	PK SKND FATBACKS	Y	.0000	BIG	% BW	0.0000	HI-LIMIT
7.	P-JOWL	PK SKND JOWLS	Y	.0000	BIG	% BW	0.0000	LO-LIMIT
8.	P-50	PK 50% TRIMMINGS	Y	.0000	BIG	% BW	40.9518	BETWEEN
9.	P-80	PK 80% TRIMMINGS	Y	.0000	BIG	% BW	25.9207	BETWEEN
10.	I-CORN	CORN SYRUP		1.0000	1.0000	% BW	1.0000	BETWEEN
11.	I-SALT	SALT		.0000	3.0000	% BW	2.9056	BETWEEN
12.	S-FRANK	FRANK SPICE		.7500	.7500	LBS	0.7500	BETWEEN
13.	WATER	WATER		.0000	BIG	% BW	33.1275	BETWEEN
14.	P-CHEEK	PK CHEEK MEAT	Y	.0000	BIG	% BW	0.0000	LO-LIMIT
15.	X-ISP	ISOLATED SOY PROTEIN		2.0000	2.0000	% FW	2.5827	HI-LIMIT

<--- REQUIREMENTS IN PRODUCT --->

SEQ NO.	<-- REQUIREMENT --> ITEM-1	ITEM-2	REQ. TYPE	LO LIMIT	HI LIMIT	LIMIT TYPE	LAST CALC. VALUE	CURRENT STATUS
				ACCEPTABLE RANGE OF VALUES				
1	MOISTURE		A	-BIG	BIG	% FW	55.5938	BETWEEN
2	PROTEIN		A	-BIG	BIG	% FW	11.6484	BETWEEN
3	FAT		A	25.0000	29.0000	% FW	29.0000	HI-LIMIT
4	BIND		A	180.0000	BIG	% FW	231.2576	BETWEEN
5	COLOR		A	190.0000	BIG	% FW	277.2159	BETWEEN
6	SALT		A	2.2500	2.2500	% FW	2.2500	HI-LIMIT
7	COLLAGEN		A	-BIG	3.7500	% FW	3.7500	HI-LIMIT
8	USDA-AW		A	-BIG	9.0000	% FW	9.0000	HI-LIMIT
9	P	> B	66	.0000	BIG	% BW	24.0408	BETWEEN
10	B	> WATER	6W	.0000	BIG	% BW	0.0000	LO-LIMIT

Table 21–17. Product Formulation Material Usage Report for ISP Frank

```
PRODUCT NAMES - SHORT: ISU-ISP      LONG: ISU FRANK WITH ISP   TYPE: MIN
LAST UPDATED:        7/13           LAST CALCULATED:   CURRENT DATE

                    CALC. #    2 BASED ON COST #    1

                    <- CALCULATED COST ($ PER LB) ->
                    COST #1   COST #2   COST #3
                    ------------------------------
        BLOCK WEIGHT:     0.7793    0.7914    0.0000
        GROSS WEIGHT:     0.5552    0.5638    0.0000
        FINISHED WEIGHT:  0.6035    0.6128    0.0000
```

SEQ NO.	SHORT	MATERIAL NAME LONG	USAGE	COST # 1	EXTENDED COST # 1	BW	GW	FW	LO-LIMIT	HI-LIMIT	TYPE
1.	B-CHEEK	BF CHEEK MEAT	23.6	0.9600	22.64	23.58	16.80	18.26	.0000	BIG	% GW
2.	B-COW	BF FC COW MEAT	9.5	1.4700	14.03	9.55	6.80	7.39	.0000	BIG	% GW
3.	P-50	PK 50% TRIMMINGS	41.0	0.4200	17.20	40.95	29.18	31.71	.0000	BIG	% GW
4.	P-80	PK 80% TRIMMINGS	25.9	0.8000	20.74	25.92	18.47	20.07	.0000	BIG	% GW
		BLOCK WEIGHT (BW):	100.0 LBS		74.61	100.00	71.24	77.44			
5.	I-CORN	CORN SYRUP	1.0	0.3000	0.30	1.00	0.71	0.77	1.0000	1.0000	% BW
6.	I-SALT	SALT	2.9	0.0550	0.16	2.91	2.07	2.25	.0000	3.0000	% GW
7.	S-FRANK	FRANK SPICE	0.7	0.3400	0.25	0.75	0.53	0.58	.7500	.7500	LBS
8.	WATER	WATER	33.1	0.0000	0.00	33.13	23.60	25.65	.0000	BIG	% GW
9.	X-ISP	ISOLATED SOY PROTEIN	2.6	1.0100	2.61	2.58	1.84	2.00	2.0000	2.0000	% FW
		GROSS WEIGHT (GW):	140.4 LBS		77.93	140.37	100.00	108.70			
		SHRINKAGE LOSS:	11.2			11.23	8.00	8.70			
		FINISHED WEIGHT:	129.1 LBS		77.93	129.14	92.00	100.00			

Table 21–18. Product Formulation Requirements Report for ISP Frank

```
PRODUCT NAMES - SHORT: ISU-ISP      LONG: ISU FRANK WITH ISP   TYPE: MIN
LAST UPDATED:        7/13           LAST CALCULATED:   CURRENT DATE

                    LAST CALCULATION #    2 USING COST #    1
```

SEQ NO.	TYPE	REQUIREMENT DESCRIPTION	VALUE	LOW-LIMIT	HI-LIMIT	TYPE	PENALTY COST
1	A	MOISTURE	55.5938	-BIG	BIG	% FW	0.0000
2	A	PROTEIN	11.6484	-BIG	BIG	% FW	0.0000
3	A	FAT	29.0000	25.0000	29.0000	% FW	0.0172
4	A	BIND	231.2576	180.0000	BIG	% FW	0.0000
5	A	COLOR	277.2159	190.0000	BIG	% FW	0.0000
6	A	SALT	2.2500	2.2500	2.2500	% FW	0.0170
7	A	COLLAGEN	3.7500	-BIG	3.7500	% FW	0.0921
8	A	USDA-AW	9.0000	-BIG	9.0000	% FW	0.0113
9	GG	P > B	24.0408	.0000	BIG	% GW	0.0000
10	GW	B > WATER	0.0000	.0000	BIG	% GW	0.0070

NOTE: "PENALTY-COST" IS THE AMOUNT IN $ PER LB BY WHICH THE FORMULATION COST COULD BE REDUCED IF THE REQUIREMENT LIMIT WAS CHANGED BY 1.0 AS GIVEN. "VALUE" IS GIVEN IN THE SAME UNITS AS THE LIMITS.

Table 21-19. Product Formulation Price-ranging Report for ISP Frank

```
PRODUCT NAMES - SHORT: ISU-ISP   LONG: ISU FRANK WITH ISP   TYPE: MIN
LAST UPDATED:        7/13    LAST CALCULATED:    CURRENT DATE

              CALC. #    2 ... USING COST # 1

              <-- CALC. COST ($ PER LB) -->
              COST #1   COST #2   COST #3
              -------------------------------
BLOCK WEIGHT:    0.7793    0.7914    0.0000
GROSS WEIGHT:    0.5552    0.5638    0.0000
FINISHED WEIGHT: 0.6035    0.6128    0.0000
```

SEQ NO.	SHORT	LONG	USAGE (LBS)	USE-MORE	USE-LESS	COST-1	COST-2	COST-3
1.	B-CHEEK	BF CHEEK MEAT	23.580	-20.8668	1.1611	0.9600	0.9800	0.0000
2.	B-COW	BF FC COW MEAT	9.548	1.2068	4.3138	1.4700	1.4500	0.0000
3.	P-50	PK 50% TRIMMINGS	40.952	-.6438	.4335	0.4200	0.4300	0.0000
4.	P-80	PK 80% TRIMMINGS	25.921	.6837	1.1456	0.8000	0.8200	0.0000
5.	I-CORN	CORN SYRUP	1.000	-86.3779	2187.7686	0.3000	0.3000	0.0000
6.	I-SALT	SALT	2.906	-BIG	BIG	0.0550	0.0550	0.0000
7.	S-FRANK	FRANK SPICE	0.750	-115.2305	2916.9653	0.3400	0.3400	0.0000
8.	WATER	WATER	33.127	-21.4985	.8519	0.0000	0.0000	0.0000
9.	I-ISP	ISOLATED SOY PROTEIN	2.583	-BIG	3.6153	1.0100	1.0100	0.0000
10.	B-PLATE	BF PLATES	0.000	-BIG	.9572	0.5200	0.5200	0.0000
11.	B-50	BF 50% TRIMMINGS	0.000	-BIG	.9366	0.6400	0.6450	0.0000
12.	B-85	BF 85% TRIMMINGS	0.000	-BIG	1.3823	1.1500	1.1700	0.0000
13.	P-FATBCK	PK SKND FATBACKS	0.000	-BIG	.3737	0.2700	0.2750	0.0000
14.	P-JOWL	PK SKND JOWLS	0.000	.3452	BIG	0.3600	0.3650	0.0000
15.	P-CHEEK	PK CHEEK MEAT	0.000	.3427	BIG	0.8900	0.9000	0.0000

NOTE: "COST-TO-USE-MORE" IS THE PRICE AT WHICH THE FORMULATION WILL START TO USE MORE OF THE MATERIAL. "COST-TO-USE-LESS" IS THE PRICE AT WHICH LESS OF THE MATERIAL WILL BE USED IN THE FORMULATION.

Table 21-20.   Product Specifications for BP-ISP Frank

```
PRODUCT SHORT NAME:  ISU-BPI          AS OF:         7/13/81
PRODUCT LONG NAME:   ISU BYPRODUCT + ISP    PRODUCT TYPE:   F      GROUPS:

LAST COSTS:      $    0.4682      $    0.4776     $    0.0000
LAST CALCULATION USED COST # 1        DATE LAST CALCULATION: CURRENT DATE    CALC. #   2
MIN OR MAX:          MIN

NO. RAW MATERIALS:       17        PRODUCT BLOCK SIZE:     100.0000 (SCALED ON BW)
NO. REQUIREMENTS:        13        SHRINKAGE RATE:           8.0000 %
PRODUCT DISPLAY UNITS:  LBS
```

<--- RAW MATERIALS IN PRODUCT --->   (NAMES FROM TABLE: ISU MEAT COURSE)

SEQ NO.	SHORT NAME	RAW MATERIAL DESCRIPTION	IN BLOCK	ACCEPTABLE RANGE OF VALUES LO LIMIT	HI LIMIT	LIMIT TYPE	CURRENT VALUE	CURRENT STATUS
1.	B-CHEEK	BF CHEEK MEAT	Y	.0000	BIG	% GW	6.3792	BETWEEN
2.	B-COW	BF FC COW MEAT	Y	.0000	BIG	% GW	0.0000	LO-LIMIT
3.	B-PLATE	BF PLATES	Y	.0000	BIG	% GW	0.0000	HI-LIMIT
4.	B-50	BF 50% TRIMMINGS	Y	.0000	BIG	% GW	0.0000	HI-LIMIT
5.	B-85	BF 85% TRIMMINGS	Y	.0000	BIG	% GW	0.0000	HI-LIMIT
6.	P-FATBCK	PK SKND FATBACKS	Y	.0000	BIG	% GW	0.0000	HI-LIMIT
7.	P-JOWL	PK SKND JOWLS	Y	.0000	BIG	% GW	0.0000	LO-LIMIT
8.	P-50	PK 50% TRIMMINGS	Y	.0000	BIG	% GW	37.0056	BETWEEN
9.	P-80	PK 80% TRIMMINGS	Y	.0000	BIG	% GW	27.7887	BETWEEN
10.	I-CORN	CORN SYRUP		1.0000	1.0000	% BW	1.0000	BETWEEN
11.	I-SALT	SALT		.0000	3.0000	% GW	2.8415	BETWEEN
12.	S-FRANK	FRANK SPICE		.7500	.7500	LBS	0.7500	BETWEEN
13.	WATER	WATER		.0000	BIG	% GW	30.1518	BETWEEN
14.	B-HEART	BF HEARTS	Y	.0000	BIG	% GW	14.4133	BETWEEN
15.	P-LIP	PK LIPS	Y	.0000	BIG	% GW	14.4133	BETWEEN
16.	P-CHEEK	PK CHEEK MEAT	Y	.0000	BIG	% GW	0.0000	LO-LIMIT
17.	X-ISP	ISOLATED SOY PROTEIN		2.0000	2.0000	% FW	2.5258	HI-LIMIT

<--- REQUIREMENTS IN PRODUCT --->

SEQ NO.	<-- REQUIREMENT --> ITEM-1	ITEM-2	REQ. TYPE	ACCEPTABLE RANGE OF VALUES LO LIMIT	HI LIMIT	LIMIT TYPE	LAST CALC. VALUE	CURRENT STATUS
1	MOISTURE		A	-BIG	BIG	% FW	55.5606	BETWEEN
2	PROTEIN		A	-BIG	BIG	% FW	11.6402	BETWEEN
3	FAT		A	25.0000	29.0000	% FW	29.0000	BETWEEN
4	BIND		A	130.0000	BIG	% FW	182.3456	BETWEEN
5	COLOR		A	190.0000	BIG	% FW	190.0000	LO-LIMIT
6	SALT		A	2.2500	2.2500	% FW	2.2500	HI-LIMIT
7	COLLAGEN		A	-BIG	4.2000	% FW	4.2000	HI-LIMIT
8	USDA-AW		A	-BIG	9.0000	% FW	9.0000	HI-LIMIT
9	P	> WATER	GW	.0000	BIG	% GW	25.2369	BETWEEN
10	WATER	> P-LIP	MM	.0000	BIG	% GW	11.4655	BETWEEN
11	P-LIP	> B-HEART	MM	.0000	BIG	% GW	0.0000	LO-LIMIT

<--- REQUIREMENTS IN PRODUCT --->

SEQ NO.	<-- REQUIREMENT --> ITEM-1	ITEM-2	REQ. TYPE	ACCEPTABLE RANGE OF VALUES LO LIMIT	HI LIMIT	LIMIT TYPE	LAST CALC. VALUE	CURRENT STATUS
12	B-HEART	> B	MG	.0000	BIG	% GW	5.8528	BETWEEN
13	B	> SALT	GA	.0000	BIG	% GW	2.5772	BETWEEN

Table 21–21.   Product Formulation Material Usage Report
for BP–ISP Frank

```
PRODUCT NAMES - SHORT: ISU-BPI      LONG: ISU BYPRODUCT + ISP   TYPE: MIN
LAST UPDATED:        7/13            LAST CALCULATED:  CURRENT DATE

              CALC. #   2 BASED ON COST #   1

                    <- CALCULATED COST ($ PER LB) ->
                    COST #1   COST #2   COST #3
                    -------------------------------
          BLOCK WEIGHT:     0.5912    0.6032    0.0000
          GROSS WEIGHT:     0.4307    0.4394    0.0000
          FINISHED WEIGHT:  0.4682    0.4776    0.0000
```

SEQ NO.	SHORT	LONG	USAGE	COST #1	EXTENDED COST #1	BW	GW	FW	LO-LIMIT	HI-LIMIT	TYPE
1.	B-CHEEK	BF CHEEK MEAT	6.4	0.9600	6.12	6.38	4.65	5.05	.0000	BIG	% GW
2.	P-50	PK 50% TRIMMINGS	37.0	0.4200	15.54	37.01	26.96	29.30	.0000	BIG	% GW
3.	P-80	PK 80% TRIMMINGS	27.8	0.8000	22.23	27.79	20.24	22.00	.0000	BIG	% GW
4.	B-HEART	BF HEARTS	14.4	0.4900	7.06	14.41	10.50	11.41	.0000	BIG	% GW
5.	P-LIP	PK LIPS	14.4	0.3400	4.90	14.41	10.50	11.41	.0000	BIG	% GW
		BLOCK WEIGHT (BW):	100.0 LBS		55.86	100.00	72.85	79.18			
6.	I-CORN	CORN SYRUP	1.0	0.3000	0.30	1.00	0.73	0.79	1.0000	1.0000	% BW
7.	I-SALT	SALT	2.8	0.0550	0.16	2.84	2.07	2.25	.0000	3.0000	% GW
8.	S-FRANK	FRANK SPICE	0.8	0.3400	0.25	0.75	0.55	0.59	.7500	.7500	LBS
9.	WATER	WATER	30.2	0.0000	0.00	30.15	21.97	23.88	.0000	BIG	% GW
10.	I-ISP	ISOLATED SOY PROTEIN	2.5	1.0100	2.55	2.53	1.84	2.00	2.0000	2.0000	% FW
		GROSS WEIGHT (GW):	137.3 LBS		59.12	137.27	100.00	108.70			
		SHRINKAGE LOSS:	11.0			10.98	8.00	8.70			
		FINISHED WEIGHT:	126.3 LBS		59.12	126.29	92.00	100.00			

Table 21–22.   Product Formulation Requirements Report
for BP–ISP Frank

```
PRODUCT NAMES - SHORT: ISU-BPI      LONG: ISU BYPRODUCT + ISP   TYPE: MIN
LAST UPDATED:        7/13            LAST CALCULATED:  CURRENT DATE

                    LAST CALCULATION #    2 USING COST #   1
```

SEQ NO.	TYPE	DESCRIPTION		VALUE	LOW-LIMIT	HI-LIMIT	TYPE	PENALTY COST
1	A	MOISTURE		55.5606	-BIG	BIG	% FW	0.0000
2	A	PROTEIN		11.6402	-BIG	BIG	% FW	0.0000
3	A	FAT		29.0000	25.0000	29.0000	% FW	0.0093
4	A	BIND		182.3456	130.0000	BIG	% FW	0.0000
5	A	COLOR		190.0000	190.0000	BIG	% FW	0.0013
6	A	SALT		2.2500	2.2500	2.2500	% FW	0.0115
7	A	COLLAGEN		4.2000	-BIG	4.2000	% FW	0.0995
8	A	USDA-AW		9.0000	-BIG	9.0000	% FW	0.0122
9	GM	P)	WATER	25.2369	.0000	BIG	% GW	0.0000
10	MM	WATER)	P-LIP	11.4655	.0000	BIG	% GW	0.0000
11	MM	P-LIP)	B-HEART	0.0000	.0000	BIG	% GW	0.0084
12	MG	B-HEART)	B	5.8528	.0000	BIG	% GW	0.0000
13	GA	B)	SALT	2.5772	.0000	BIG	% GW	0.0000

NOTE: "PENALTY-COST" IS THE AMOUNT IN $ PER LB BY WHICH THE FORMULATION COST COULD BE REDUCED
      IF THE REQUIREMENT LIMIT WAS CHANGED BY 1.0 AS GIVEN. "VALUE" IS GIVEN IN THE SAME UNITS
      AS THE LIMITS.

Table 21–23. Product Formulation Price-ranging Report for BP–ISP Frank

```
PRODUCT NAMES - SHORT: ISU-BPI   LONG: ISU BYPRODUCT + ISP   TYPE: MIN
        LAST UPDATED:        7/13        LAST CALCULATED:   CURRENT DATE

                       CALC. #   2 ... USING COST # 1

                          (-- CALC. COST ($ PER LB) --)
                          COST #1   COST #2   COST #3
                          --------------------------------
           BLOCK WEIGHT:  0.5912    0.6032    0.0000
           GROSS WEIGHT:  0.4307    0.4394    0.0000
        FINISHED WEIGHT:  0.4682    0.4776    0.0000
```

SEQ NO.	SHORT	LONG	USAGE (LBS)	USE-MORE	USE-LESS	COST-1	COST-2	COST-3
1.	B-CHEEK	BF CHEEK MEAT	6.379	.4584	1.0885	0.9600	0.9800	0.0000
2.	P-50	PK 50% TRIMMINGS	37.006	.1570	.4847	0.4200	0.4300	0.0000
3.	P-80	PK 80% TRIMMINGS	27.789	.5932	.8696	0.8000	0.8200	0.0000
4.	I-CORN	CORN SYRUP	1.000	-102.1871	398.2238	0.3000	0.3000	0.0000
5.	I-SALT	SALT	2.841	-BIG	BIG	0.0550	0.0550	0.0000
6.	S-FRANK	FRANK SPICE	0.750	-136.3094	530.9052	0.3400	0.3400	0.0000
7.	WATER	WATER	30.152	-3.9108	1.0072	0.0000	0.0000	0.0000
8.	B-HEART	BF HEARTS	14.413	.2683	.9506	0.4900	0.4900	0.0000
9.	P-LIP	PK LIPS	14.413	.1183	1.0893	0.3400	0.3500	0.0000
10.	I-ISP	ISOLATED SOY PROTEIN	2.526	-BIG	3.4305	1.0100	1.0100	0.0000
11.	B-COW	BF FC COW MEAT	0.000	1.3694	BIG	1.4700	1.4500	0.0000
12.	B-PLATE	BF PLATES	0.000	-BIG	.5941	0.5200	0.5200	0.0000
13.	B-50	BF 50% TRIMMINGS	0.000	-BIG	.7665	0.6400	0.6450	0.0000
14.	B-85	BF 85% TRIMMINGS	0.000	-BIG	1.2762	1.1500	1.1700	0.0000
15.	P-FATBCK	PK SKND FATBACKS	0.000	-BIG	.3220	0.2700	0.2750	0.0000
16.	P-JOWL	PK SKND JOWLS	0.000	.2884	BIG	0.3600	0.3650	0.0000
17.	P-CHEEK	PK CHEEK MEAT	0.000	.4133	BIG	0.8900	0.9000	0.0000

NOTE: "COST-TO-USE-MORE" IS THE PRICE AT WHICH THE FORMULATION WILL START TO USE MORE OF THE MATERIAL. "COST-TO-USE-LESS" IS THE PRICE AT WHICH LESS OF THE MATERIAL WILL BE USED IN THE FORMULATION.

Appearancewise, the frankfurters look similar because each of them has been formulated to the same color index (190). However, the four formulations vary in textural differences because by-products and higher collagen meats have been added. (Note the addition of beef hearts and pork lips to the BP frankfurters.) With the addition of these meats, the product standards (specifications) had to be revised slightly. The bind index lower limit was reduced from 180 to 130, and the upper limit for collagen was increased from 3.75 to 4.20 to accommodate the lower-quality meats in the formulation (see Tables 21–8, 21–13, and 21–21). The end result is that the two cheaper frankfurter formulations that contain by-products have a lower-bind index and a subsequently softer texture (slightly mushy).

The cost of each of the four types of frankfurters represents the least cost to produce, given the constraints placed on each type and the available raw materials.

Table 21-24. Material Attribute Level Table

SEQ NO.	SHORT NAME	ATTRIBUTE LONG NAME	ATT. TYPE	B-BULL	B-CHEEK	B-CHUCK	B-COW	B-HEAD	B-HEART	B-LIP	B-NAVEL	B-PLATE
1.	MOISTURE	MOISTURE		0.7000	0.6300	0.5700	0.6650	0.6100	0.6650	0.6200	0.3800	0.4000
2.	FAT	FAT		0.0950	0.1950	0.2600	0.1500	0.2300	0.1800	0.2200	0.5150	0.4800
3.	PROTEIN	PROTEIN (TOTAL)		0.2000	0.1700	0.1600	0.1800	0.1550	0.1500	0.1550	0.1000	0.1100
4.	SALT	SALT		0.0000	0.0000	0.0000	0.0000	0.0000	0.0000	0.0000	0.0000	0.0000
5.	USDA-AW	USDA ADDED WATER	C	-0.1000	-0.0500	-0.0700	-0.0550	-0.0100	0.0650	0.0000	-0.0200	-0.0400
6.	BIND	CALCULATED BIND	C	6.0000	2.3800	3.8400	4.4100	1.2400	0.9000	0.0465	1.2500	1.7600
7.	COLOR	CALCULATED COLOR	C	9.4000	8.1600	13.6000	7.0200	4.0300	6.0000	0.0620	1.9000	2.7500
8.	COLLAGEN	CALC COLLAGEN	C	0.0400	0.1003	0.0490	0.0378	0.1131	0.0405	0.1395	0.0420	0.0462
9.	M+P+F	MOISTURE+PROTEIN+FAT	C	0.9950	0.9950	0.9900	0.9950	0.9950	0.9750	0.9950	0.9950	0.9900
10.	BIND-NX	UNIV-GA BIND INDEX		30.0000	14.0000	24.0000	24.5000	8.0000	6.0000	0.3000	12.5000	16.0000
11.	COLOR-NX	UNIV-GA COLOR INDEX		47.0000	48.0000	85.0000	39.0000	26.0000	40.0000	0.4000	19.0000	25.0000
12.	COLL-NX	COLLAGEN/PROT RATIO		0.2000	0.5900	0.3000	0.2100	0.7300	0.2700	0.9000	0.4200	0.4200

SEQ NO.	SHORT NAME	ATTRIBUTE LONG NAME	ATT. TYPE	B-SHANK	B-50	B-60	B-65	B-75	B-85	B-90	P-BACEND	P-BELLY
1.	MOISTURE	MOISTURE		0.7100	0.3750	0.4500	0.4900	0.5700	0.6550	0.6950	0.3050	0.3400
2.	FAT	FAT		0.1000	0.5100	0.4100	0.3600	0.2700	0.1550	0.1100	0.6000	0.5650
3.	PROTEIN	PROTEIN (TOTAL)		0.1850	0.1050	0.1300	0.1500	0.1550	0.1800	0.1850	0.0900	0.0900
4.	SALT	SALT		0.0000	0.0000	0.0000	0.0000	0.0000	0.0000	0.0000	0.0000	0.0000
5.	USDA-AW	USDA ADDED WATER	C	-0.0300	-0.0450	-0.0700	-0.1100	-0.0500	-0.0650	-0.0450	-0.0550	-0.0200
6.	BIND	CALCULATED BIND	C	5.1800	1.2600	2.0800	2.7000	3.4100	4.3200	4.8100	0.0090	0.5400
7.	COLOR	CALCULATED COLOR	C	8.5100	1.9950	3.2500	4.2000	5.2700	7.0200	7.4000	0.2250	0.4500
8.	COLLAGEN	CALC COLLAGEN	C	0.0610	0.0441	0.0526	0.0600	0.0589	0.0504	0.0518	0.0774	0.0315
9.	M+P+F	MOISTURE+PROTEIN+FAT	C	0.9950	0.9900	0.9900	1.0000	0.9950	0.9900	0.9900	0.9950	0.9950
10.	BIND-NX	UNIV-GA BIND INDEX		28.0000	12.0000	16.0000	18.0000	22.0000	24.0000	26.0000	0.1000	6.0000
11.	COLOR-NX	UNIV-GA COLOR INDEX		46.0000	19.0000	25.0000	28.0000	34.0000	39.0000	40.0000	2.5000	5.0000
12.	COLL-NX	COLLAGEN/PROT RATIO		0.3300	0.4200	0.4050	0.4000	0.3800	0.2800	0.2800	0.8600	0.3500

SEQ NO.	SHORT NAME	ATTRIBUTE LONG NAME	ATT. TYPE	P-BLADE	P-CHEEK	P-DIAPH	P-FATBCK	P-HAM	P-HAMCUR	P-HAMFAT	P-HEAD	P-HEART
1.	MOISTURE	MOISTURE		0.7150	0.6900	0.7000	0.1500	0.5650	0.5700	0.0400	0.6100	0.7000
2.	FAT	FAT		0.0900	0.1400	0.1500	0.8000	0.3000	0.3000	0.9500	0.2200	0.1400
3.	PROTEIN	PROTEIN (TOTAL)		0.1900	0.1600	0.1450	0.0400	0.1300	0.1200	0.0100	0.1600	0.1550
4.	SALT	SALT		0.0000	0.0000	0.0000	0.0000	0.0000	0.0000	0.0000	0.0000	0.0000
5.	USDA-AW	USDA ADDED WATER	C	-0.0450	0.0500	0.1200	-0.0100	0.0450	0.0900	0.0000	-0.0300	0.0800
6.	BIND	CALCULATED BIND	C	4.5600	1.4400	2.3200	0.1200	2.7300	2.1600	0.0100	1.2000	0.9300
7.	COLOR	CALCULATED COLOR	C	3.8000	4.6400	2.6100	0.0400	2.4700	1.2000	0.0100	2.5600	5.1150
8.	COLLAGEN	CALC COLLAGEN	C	0.0437	0.1152	0.1015	0.0240	0.0325	0.0360	0.0040	0.1120	0.0418
9.	M+P+F	MOISTURE+PROTEIN+FAT	C	0.9950	0.9900	0.9950	0.9900	0.9950	0.9900	1.0000	0.9900	0.9950
10.	BIND-NX	UNIV-GA BIND INDEX		24.0000	9.0000	16.0000	3.0000	21.0000	18.0000	1.0000	7.5000	6.0000
11.	COLOR-NX	UNIV-GA COLOR INDEX		20.0000	29.0000	18.0000	1.0000	19.0000	10.0000	1.0000	16.0000	33.0000
12.	COLL-NX	COLLAGEN/PROT RATIO		0.2300	0.7200	0.7000	0.6000	0.2500	0.3000	0.4000	0.7000	0.2700

SEQ NO.	SHORT NAME	ATTRIBUTE LONG NAME	ATT. TYPE	P-JOWL	P-LIP	P-LIVER	P-PIC	P-NECK	P-PICHRT	P-SNOUT	P-SPLEEN	P-STOM
1.	MOISTURE	MOISTURE		0.3000	0.6200	0.7400	0.6000	0.6000	0.7000	0.5500	0.7000	0.0000
2.	FAT	FAT		0.6150	0.2200	0.0600	0.2400	0.2300	0.1100	0.3000	0.1450	0.0000
3.	PROTEIN	PROTEIN (TOTAL)		0.0750	0.1550	0.1900	0.1550	0.1600	0.1800	0.1450	0.1550	0.0000
4.	SALT	SALT		0.0000	0.0000	0.0000	0.0000	0.0000	0.0000	0.0000	0.0000	0.0000
5.	USDA-AW	USDA ADDED WATER	C	0.0000	0.0000	-0.0200	-0.0200	-0.0400	-0.0200	-0.0300	0.0800	0.0000
6.	BIND	CALCULATED BIND	C	0.3450	0.1550	0.3800	3.1000	3.0400	4.1400	0.3625	0.1085	0.0000
7.	COLOR	CALCULATED COLOR	C	0.1275	0.0775	9.3100	2.4800	2.5600	3.4200	0.0725	8.6800	0.0000
8.	COLLAGEN	CALC COLLAGEN	C	0.0322	0.1395	0.1805	0.0356	0.0400	0.0396	0.1160	0.1395	0.0000
9.	M+P+F	MOISTURE+PROTEIN+FAT	C	0.9900	0.9950	0.9900	0.9950	0.9900	0.9900	0.9950	1.0000	0.0000
10.	BIND-NX	UNIV-GA BIND INDEX		4.6000	1.0000	2.0000	20.0000	19.0000	23.0000	2.5000	0.7000	0.0000
11.	COLOR-NX	UNIV-GA COLOR INDEX		1.7000	0.5000	49.0000	16.0000	16.0000	19.0000	0.5000	56.0000	0.0000
12.	COLL-NX	COLLAGEN/PROT RATIO		0.4300	0.9000	0.9500	0.2300	0.2300	0.2200	0.8000	0.9000	0.0000

(Continued)

Table 21–24 (Continued)

SEQ NO.	SHORT NAME	ATTRIBUTE LONG NAME	ATT. TYPE	P-TNGTRM	P-TONGUE	P-50	P-80	P-95	C-MDB	T-MDB	V-BNLS	S-BEBR
1.	MOISTURE	MOISTURE		0.0000	0.0000	0.3200	0.5600	0.0000	0.6600	0.0000	0.0000	0.0500
2.	FAT	FAT		0.0000	0.0000	0.5900	0.2800	0.0000	0.2050	0.0000	0.0000	0.0000
3.	PROTEIN	PROTEIN (TOTAL)		0.0000	0.0000	0.0850	0.1500	0.0000	0.1250	0.0000	0.0000	0.0100
4.	SALT	SALT		0.0000	0.0000	0.0000	0.0000	0.0000	0.0000	0.0000	0.0000	0.0000
5.	USDA-AW	USDA ADDED WATER	C	0.0000	0.0000	-0.0200	-0.0400	0.0000	0.1600	0.0000	0.0000	0.0100
6.	BIND	CALCULATED BIND	C	0.0000	0.0000	1.0200	2.8500	0.0000	1.8750	0.0000	0.0000	0.0000
7.	COLOR	CALCULATED COLOR	C	0.0000	0.0000	0.7650	2.5500	0.0000	0.6250	0.0000	0.0000	0.0000
8.	COLLAGEN	CALC COLLAGEN	C	0.0000	0.0000	0.0289	0.0360	0.0000	0.0437	0.0000	0.0000	0.0000
9.	M+P+F	MOISTURE+PROTEIN+FAT	C	0.0000	0.0000	0.9950	0.9990	0.0000	0.9900	0.0000	0.0000	0.0600
10.	BIND-NX	UNIV-GA BIND INDEX		0.0000	0.0000	12.0000	19.0000	0.0000	15.0000	0.0000	0.0000	0.0000
11.	COLOR-NX	UNIV-GA COLOR INDEX		0.0000	0.0000	9.0000	17.0000	0.0000	5.0000	0.0000	0.0000	0.0000
12.	COLL-NX	COLLAGEN/PROT RATIO		0.0000	0.0000	0.3400	0.2400	0.0000	0.3500	0.0000	0.0000	0.0000

SEQ NO.	SHORT NAME	ATTRIBUTE LONG NAME	ATT. TYPE	S-BOLO	S-FRANK	S-SALAMI	S-SAUS	I-CORN	I-CURE	I-SALT	X-CEREAL	X-ISP
1.	MOISTURE	MOISTURE		0.0000	0.0800	0.0000	0.0000	0.2600	0.0000	0.0000	0.0000	0.2900
2.	FAT	FAT		0.0000	0.0000	0.0000	0.0000	0.0000	0.0000	0.0000	0.0000	0.0000
3.	PROTEIN	PROTEIN (TOTAL)		0.0000	0.1500	0.0000	0.0000	0.0000	0.0000	0.0000	0.0000	0.7100
4.	SALT	SALT		0.0000	0.0000	0.0000	0.0000	0.0000	0.0000	1.0000	0.0000	0.0000
5.	USDA-AW	USDA ADDED WATER	C	0.0000	-0.5200	0.0000	0.0000	0.2600	0.0000	0.0000	0.0000	-2.5500
6.	BIND	CALCULATED BIND	C	0.0000	3.0000	0.0000	0.0000	0.0000	0.0000	0.0000	0.0000	31.9500
7.	COLOR	CALCULATED COLOR	C	0.0000	1.5000	0.0000	0.0000	0.0000	0.0000	0.0000	0.0000	0.0000
8.	COLLAGEN	CALC COLLAGEN	C	0.0000	0.0000	0.0000	0.0000	0.0000	0.0000	0.0000	0.0000	0.0000
9.	M+P+F	MOISTURE+PROTEIN+FAT	C	0.0000	0.2300	0.0000	0.0000	0.2600	0.0000	0.0000	0.0000	1.0000
10.	BIND-NX	UNIV-GA BIND INDEX		0.0000	20.0000	0.0000	0.0000	0.0000	0.0000	0.0000	0.0000	45.0000
11.	COLOR-NX	UNIV-GA COLOR INDEX		0.0000	10.0000	0.0000	0.0000	0.0000	0.0000	0.0000	0.0000	0.0000
12.	COLL-NX	COLLAGEN/PROT RATIO		0.0000	0.0000	0.0000	0.0000	0.0000	0.0000	0.0000	0.0000	0.0000

SEQ NO.	SHORT NAME	ATTRIBUTE LONG NAME	ATT. TYPE	X-NFDM	X-PREMUL	X-SPC	X-TVP	X-TSPC	WATER
1.	MOISTURE	MOISTURE		0.0000	0.5030	0.2000	0.0000	0.0000	1.0000
2.	FAT	FAT		0.0000	0.4190	0.0000	0.0000	0.0000	0.0000
3.	PROTEIN	PROTEIN (TOTAL)		0.0000	0.0620	0.7000	0.5000	0.7000	0.0000
4.	SALT	SALT		0.0000	0.0100	0.0000	0.0000	0.0000	0.0000
5.	USDA-AW	USDA ADDED WATER	C	0.0000	0.2550	-2.6000	-2.0000	-2.8000	1.0000
6.	BIND	CALCULATED BIND	C	0.0000	12.3380	28.0000	20.0000	28.0000	0.0000
7.	COLOR	CALCULATED COLOR	C	0.0000	0.0000	0.0000	0.0000	0.0000	0.0000
8.	COLLAGEN	CALC COLLAGEN	C	0.0000	0.0000	0.0000	0.0000	0.0000	0.0000
9.	M+P+F	MOISTURE+PROTEIN+FAT	C	0.0000	0.9840	0.9000	0.5000	0.7000	1.0000
10.	BIND-NX	UNIV-GA BIND INDEX		0.0000	199.0000	40.0000	40.0000	40.0000	0.0000
11.	COLOR-NX	UNIV-GA COLOR INDEX		0.0000	0.0000	0.0000	0.0000	0.0000	0.0000
12.	COLL-NX	COLLAGEN/PROT RATIO		0.0000	0.0000	0.0000	0.0000	0.0000	0.0000

PREBLENDING

Preblending is a term used in the sausage industry to describe the grinding and mixing of raw meat materials with salt and(or) water and(or) nitrite for later use in sausage formulations. Preblended meats are typically held in a cooler from 8 to 72 hours before the preblends are used in actual sausage production. During this time, processors use representative sampling techniques to carefully analyze the preblends for fat content. Some processors perform a variety of other laboratory tests on the preblends to insure consistent uniformity in the quality of the finished product (see section on quality control in Chapter 18). By knowing the exact compositional makeup of each preblended batch, processors can combine lean-meat preblends and fat-meat preblends to consistently produce the desired fat content, bind, color, etc., from batch to batch. Precise control of compositional factors such as fat, moisture, and protein content is extremely important for maximizing profit and maintaining compliance with government regulations. Preblending operations are also well-adapted to linear programming techniques, since there is time to collect the necessary data on raw material lots after preblending and before further processing. The preblends become the raw material sources for the Least Cost Formulator®.

Benefits of Preblending

Modern sausage production is plagued by several common problems (see Table 21–5). Preblending helps to eliminate some of them.

Fat and Water Binding

If insufficient time is allowed for salt-soluble protein (myosin) extraction before fat meats and water are added, the bind capability is reduced. The high salt-to-lean content of meat in the preblend improves salt-soluble protein (myosin) extraction. Storage for 8 to 72 hours allows time for optimum protein extraction and fat and water binding. Preblending and storage of lean meats with salt improves the bind characteristics by up to 50 percent.

Raw Material Variations

Individual raw material analyses often vary by more than ±5 percent in fat and moisture from lot to lot, causing similar swings in finished product analyses, particularly in small batches. When raw materials are preblended in large batches, these fluctuations tend to average out and are compensated for by the actual analysis taken during the holding phase and the subsequent correction at the final blend. The preblend is sampled when dumped, and a lab analysis is performed for moisture, fat, and protein (see Chapter 18) during the holding period. Each finished product batch is reformulated to bring fat and water content back into specification. Correction is accomplished by using standard lean and fat meats, which may themselves be preblended or thoroughly characterized. Spices and other ingredients are then added to make individual finished products.

Weighing errors, variations in blending times, and other production variations tend to average out in the larger batches common to preblending and can be corrected after chemical analysis.

Problem Meats

High-collagen meats, such as cheek and head meat, heart meat, and shank meat, are more fully emulsified during the preblending step and present less of a problem in the finished product. Double blending helps guarantee adequate mixing.

Production Scheduling

Since each finished product batch is prepared from materials on hand at a particular time, material flow becomes complex. It is difficult to have the right meats and equipment available at the right time. Preblending formulas account for 80 to 90 percent of the meats used and are typically held constant over a week of production. Production schedules can be made in advance and do not suffer as much from material movement problems that characterize reformulation on a batch-by-batch basis. Variations in the scheduled use of materials are limited to the correction step, where only a few meats are used. Once the materials have been preblended, their storage life is increased to several days, and the preblend can be stored in standard quantities.

Production Efficiency

Small batches and the common use of equipment for different purposes impair the efficient utilization of resources. Preblends are prepared in batch sizes equivalent to 2 to 10 finished-product batches, resulting in good mixing and economy of scale. The preblend is usually prepared in large batch sizes and then poured into tubs for use in several finished-product batches. Economy of scale reduces labor and equipment costs, reduces lab analysis workload, and minimizes material movement. Preblending and correction can be carried out in different shifts. More automation through interconnected scales, grinders, flakers, mixers, and blenders can be used.

Extensive Product Line

The manufacture of many individual finished products, each with its own formula and special instructions, causes confusion and inefficiencies in the sausage kitchen. Preblends can be made in large batches for use in a variety of similar products, reducing the number of manufacturing steps. Blends for a given label, such as "beef" or "loaf base," are commonly prepared, with the individual finished-product requirements handled by different spices and constraints in the correction chop or blend.

Preblending Techniques

Preblending can be carried out by utilizing any of several different strategies to decrease final production variations. These strategies may be based upon: (1) raw material; (2) lean or fat content; (3) species, e.g., beef vs. pork; (4) label, such as "pork and beef," "pork, beef, water," or "beef"; and (5) individual finished product.

The methods used depend on the importance of various factors, including equipment costs, available storage, handling requirements, production efficiency, and simplicity.

The preblending of raw materials is usually associated with a company that prepares a narrow range of products from a limited number of meats in large quantities. Each meat ingredient is segregated and preblended by lot, each with its own analysis, thereby controlling variations and allowing optimization of the final formulation of finished products.

Generally, the industry has opted for preblending by label or finished product as the most effective use of equipment, labor, and storage resources and as the method giving the maximum production efficiency and simplicity. The large batch sizes and fixed instructions of this method give good control of material and production variations.

Formulation Procedures
for Preblending

A typical sequence of events in the purchasing/production cycle is:

> *Tuesday:* Have sales meeting to determine next week's production requirements.

Wednesday: Get availabilities and prices. Standard products, cut-and-harvest, and boning are estimated and accounted for.

Thursday: Procurement department starts purchasing bill-of-materials resulting from the multi-product formulation.

Friday: Raw materials are delivered and quality control checks made on raw materials (i.e., fat content, condition).

Next week: Preblends are made by formula. Correct to finished product on a batch-by-batch basis. Adjust preblend formulas for abnormal events.

Formulating the Preblend

Each preblend will have its own formulation specifications. When preblending to a finished product, these specifications may be as simple as including only the fat and USDA–AW at the finished-product percentage limits. Materials offered to the preblend would include all meats offered to the finished product, salt, cure, and water.

The general principle in formulating a preblend is to arrive at a recipe that will bring the preblend correction steps back into accord with the overall least cost solution, if the preblend is at its target analysis, and to minimize the cost to correct when the preblend is off its target analysis. The Least Cost Formulator® system distributed by Least Cost Formulations, Ltd. (see footnote 15) uses the following procedure to accomplish this purpose:

- Set up preblend product specifications as outlined.

- Identify meats offered to finished product that are suitable for use as corrector meats. Exclude from these cheek, head, and shank meat and offal.

- Inform the formulation system which meats must be used in total in the preblend, which meats can be used as correctors, and which ingredients must not be used in the preblend.

- Set up a basic corrector specification. This would include added ingredients and product requirements on fat, added water, etc. Modify this basic specification at formulation to include the actual preblend and corrector meats.

Once the preblend and corrector specifications have been set up, routine steps would include:

- Performing a special preblend least cost formulation based on the finished-product solution from the multi-product run. The computer will prepare the preblend at least cost and choose the corrector meats to be added at the final blend.

- If the preblend specifications are being adjusted, verifying the corrector will be feasible if the preblend is on or off target analysis in the usual range (typically up or down 2.5 to 3 percent in fat).

- When the actual preblend analysis is obtained from the lab, running the corrector formulation, using the cost and actual analysis of the preblend and offering the corrector meats available.

Making a Preblend

The following is a typical procedure for preparing a preblend:

- Grind meats (coarse grind fat meat to $\frac{5}{16}$ to $\frac{3}{8}$ inch, fine grind lean meat to $\frac{1}{8}$ to $\frac{1}{4}$ inch, and flake frozen meats).

- Sample ground raw materials for chemical analysis (vendor control, see Chapter 18).

- Mix in blender to 80 to 85 percent finished weight of final product. Include all problem meats (cheek, head, shank, etc.), 75 to 80 percent of other meats, all salt (as finely divided and pure as possible), one-half or all of the cure, and sufficient water to work meat (approximately 10 percent of total preblend weight). The materials should be blended adequately (7 to 10 minutes), with a resulting pH of 6 or higher.

- Sample preblend for laboratory analysis. Perform fat, moisture, and protein tests to as high an accuracy as possible. Make sure the sample is representative (see Chapter 18).

- Dump into tubs. Sampling could be done on a grab-sample basis during this step (see Chapter 18). The same preblend may be used for various products.

- Store 8 to 72 hours in cooler. Protein extraction increases 15 percent the first day, 10 percent the second, and 5 percent the third. Longer storage can make up for slight underblending. Shorter storage gives better water bind; longer storage gives better fat bind. The best policy is to plan on holding the preblend 12 to 16 hours, with the shelf life of the preblend used to handle contingencies and Monday start-up needs.

Correcting to Finished Product

- Run corrector formulation based on preblend laboratory analysis.

- Mix in chopper or blender: the preblend (fixed at 80 to 85 percent of finished weight), lean and fat corrector meats, and remaining cure. Add water, spices, and other ingredients and rework (if necessary; rework is failed product from previous runs — use less than 5 percent in quality products).

- Chop or run through emulsifier.

- Stuff.

The corrector meats will vary in usage, depending on the actual chemical analysis of the preblend. Therefore, deviations from projected usages will occur, and the meats chosen should include consideration for increased or decreased needs. The corrector meats should themselves be preblended or well-mixed and characterized by good sampling and lab analysis (see Chapter 18) so that finished-product fluctuations will be minimized.

A MODERN SAUSAGE KITCHEN

The mechanization necessary for efficient sausage production is difficult for the layperson to visualize, especially the majority of consumers who enjoy tasty hot dogs but are prone to complain if they seem to cost too much. Then, too, there are many "nutrition experts" harping about the lack of nutritional value in hot dogs. Most of these consumers probably never take time to think about how the product became transformed from the living matter it once was into the delicious, *nutritious* human food product they enjoy.

Americans **love** hot dogs — to the tune of 581 being eaten every second of every 24-hour day all year in the United States. It has been reported that O'Hare Airport in Chicago sells the most hot dogs of any place in the world — about 2 million per year.

Figures 21–8 to 21–15 illustrate equipment and processes employed in modern, sanitary sausage kitchens.

Fig. 21–8. Portable incline screw conveyor moves meats from a Weiler® grinder (with add–on sides) to a Griffith® blender. (Courtesy, *The National Provisioner*)

Fig. 21–9. When smooth emulsion products are made, the meats are moved to this high–wall batching screw conveyor mounted to a Toledo® scale. This has ample space above the conveyor discs to accumulate a batch before unloading into the silent cutter (colloid mill). (Courtesy, *The National Provisioner*)

Fig. 21–10. A 750–liter Seydelmann® vacuum chopper. Large choppers such as this one can make fine– or coarse–cut emulsion products. During chopping, the lid is closed by hydraulic arms, and a vacuum is drawn. Chopping under vacuum is reported to improve color stability, shelf life, and product uniformity. (Courtesy, Wimmer's, Inc., Westpoint, Nebraska)

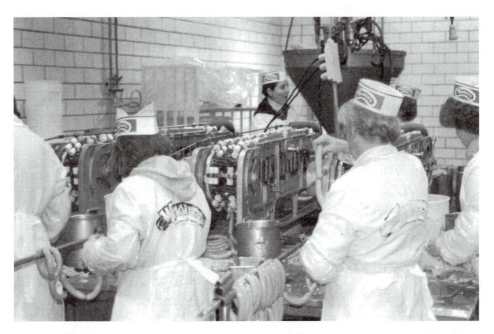

Fig. 21–11. Two Famco® linkers link delicate natural sheep casings to a uniform size. A Vemag® continuous vacuum stuffer can be seen in the background. (Courtesy, Wimmer's, Inc., Westpoint, Nebraska)

Fig. 21-12. At this linking station, natural casing products such as mettwurst are separated into link sizes of approximately the same weight with the aid of a Pratco® clipper. (Courtesy, *The National Provisioner*)

Fig. 21-13. Frankfurters in cellulose casings have been stuffed and linked with this Townsend® machine. (Courtesy, *The National Provisioner*)

Fig. 21–14. Six–station trolley cages are moved into position for cooking and smoking in fully automatic Alkar® cook cabinets. (Courtesy, *The National Provisioner*)

Fig. 21–15. After the weiners have been cooked and then cooled, their cellulose casings are removed by this high–speed peeler to make "skinless wieners." A closely adjusted knife slits the casings, and an air jet blows the casings off the wieners.

22

Structure and Function of Muscle

Most of the edible portion of an animal consists of muscle. This chapter will identify and describe the types of muscle tissue and explain the functions of each. It will also discuss the protein composition of muscle, postmortem conversion of muscle to meat, postmortem quality problems, and postmortem technology.

Muscle tissue is composed of three distinct types: *skeletal, smooth,* and *cardiac.* The identifying characteristics of each type are listed in Table 22–1. Skeletal muscles, the most important of the three types because of quantity and economic value, are directly or indirectly attached to bone. These muscles facilitate movement and/or give support to the body. Smooth muscles are commonly found in systems of tubular or hollow organs. Some examples are the digestive tract, the blood circulatory system (arteries and veins), the urinary tract, and the reproductive organs. Cardiac muscle, as the name would imply, is found in the heart.

Skeletal muscles are covered by a dense connective tissue sheath called the *epimysium* (see Figure 22–1). Each muscle is divided into sections (called bundles) by thinner connective tissue layers known as the *perimysium.* Clusters of fat cells, small blood vessels, and nerve branches are found in borders between the muscle bundles. The cylindrical, multinucleated muscle fibers (cells) of

Table 22–1. Muscle Types and Their Characteristics

	Skeletal	Smooth	Cardiac
Method of control	Voluntary	Involuntary	Involuntary
Banding pattern	Striated	Non-striated	Striated
Nuclei/cell	Multinucleated	Single nucleus	Single nucleus

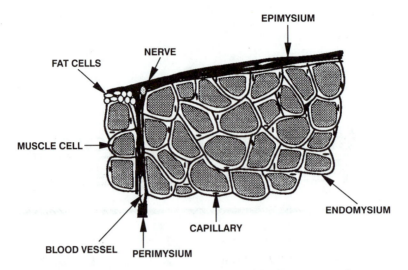

Fig. 22–1. Diagram of a skeletal muscle in cross section. The muscle is surrounded by a dense connective tissue layer called the *epimysium*. A thinner connective tissue layer (the *perimysium*) encircles groups of muscle fibers, dividing them into bundles. Fat cells, nerves, and blood vessels are found in this layer. Each muscle cell is separated from its neighbors by a thin connective tissue sheath called the *endomysium*. A few capillaries are positioned near each muscle cell.

varying lengths are individually wrapped with a thin *endomysium* sheath (see Figure 22–1). Muscle fibers are typically 30 to 50 μm in diameter and several millimeters long. The connective tissue layer forms a network on the surface of each fiber (see Figure 22–2). Several capillaries are also located around each muscle cell.

Not all muscle fibers are the same. Some contract rapidly and depend on *glycolysis* (see discussion later in chapter) for energy production. These are the fast or white fibers. Other fibers contract more slowly and depend on oxidative metabolism. These are the slow or red fibers. All muscles are mixtures of the two major types, with the proportions varying between muscles. The mixing of fiber types occurs even in the same muscle bundle (see Figure 22–3). Usually, the muscles closest to the bones have higher proportions of red fibers and thus appear darker red. This is readily apparent in the cut face of a fresh ham.

Examination of muscle fibers in the light microscope reveals alternating light and dark stripes running perpendicular to their long axes (diagrammed in Figure 22–4). The origin of these bands or striations stems from the muscle cell, which is packed with organelles called *myofibrils*. These too are cylindrical in shape (like the muscle fiber), and they occupy about 80 percent of the cell volume. The myofibrils are approximately 1 μm in diameter, and there are about 1,000 myofibrils in a cross sectional view of a muscle fiber.

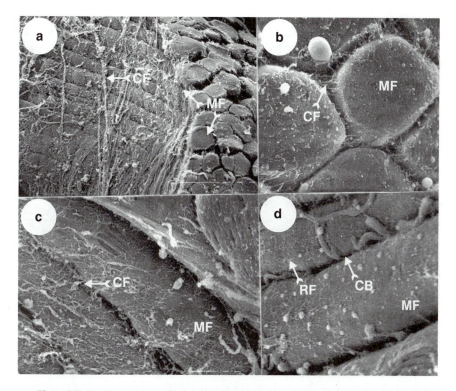

Fig. 22–2. Scanning electron micrographs of bovine skeletal muscle. (a) Muscle fibers (MF) run from the upper left to the lower right. Dense collagen fibers (CF) of the perimysium can be seen surrounding this muscle bundle. Magnification = 175 x. (b) Cross-sectional view of muscle fibers (MF). Collagen fibers (CF) of the endomysium are evident between the muscle fibers. Magnification = 1000 x. (c) Transverse view of muscle fibers (MF) running from upper left to lower right. Collagen fibers (CF) of the endomysium surround each muscle fiber. Magnification = 1000 x. (d) Transverse view of muscle fibers (MF) showing a capillary bed (CB) surrounding a fiber. Fuzzy-appearing reticular fibers (RF) surround the surface of each muscle fiber. Magnification = 1230 x. (Micrographs by Kevin W. Jones and Barbara Schrag)

They contain no membrane boundary in contrast to other organelles such as nuclei and mitochondria.

Myofibrils produce force when skeletal and cardiac muscles contract. A light micrograph of a myofibril is shown in Figure 22–5. Alternating dark (A) and light (I) bands are clearly visible. In addition, a Z line passes perpendicularly through the middle of the I band. The region spanning from one Z line to the next is called a *sarcomere* and is the smallest functional contractile unit. A sarcomere will typically measure 2 to 3 μm in length or about 0.0001 inch.

The reason for the banding patterns becomes apparent when the muscle fiber is viewed with an electron microscope (see Figure 22–6). Two major

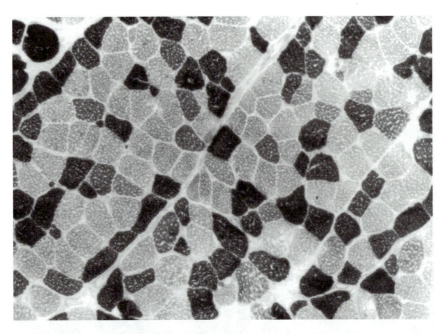

Fig. 22–3. Light micrograph of a cross section from the *Vastus intermedius* muscle of a pig. Magnification = 1000 x. The muscle section was stained for ATPase after alkaline preincubation. Note the differences in staining between different fibers: fast or white fibers are stained dark and slow or red fibers are stained light. (Micrograph provided by Robert Cassens, University of Wisconsin-Madison)

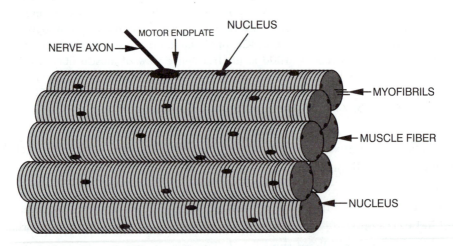

Fig. 22–4. Diagram of a longitudinal view of a group of muscle fibers. The fibers are striated (striped) when viewed in the light microscope. Many nuclei are found in each fiber, and they are located just underneath the muscle cell membrane. A motor end plate is formed from a nerve branch in close contact with a special region on the muscle cell surface. Every muscle fiber has its own motor end plate; only one is shown in the diagram for simplicity.

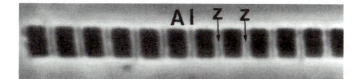

Fig. 22–5. Phase contrast micrograph of a myofibril from pig skeletal muscle. Magnification = 2500 x. Alternating dark (A) and light (I) bands are visible. A Z line splits the I bands in the middle.

types of filaments are present: thick filaments (15 nm diameter and 1.6 μm long) and thin filaments (8 to 10 nm diameter and 1.0 to 1.3 μm long). Thick filaments lie side to side in parallel and define the region of the A band (see Figure 22–6). The middle of the A band (when the muscle fiber is in an extended position) has only thick filaments and thus appears less dense; this region is called the H zone. An M line extends perpendicularly across the middle of the A band (see Figure 22–6).

Small projections (called cross-bridges) are found on the surface of these filaments. Thin filaments attach to the Z lines and interdigitate with the thick filaments (see Figure 22–7).

Thick and thin filaments slide over each other in an accordion-like fashion during muscle contraction and relaxation (see Figure 22–8). Muscle thick and

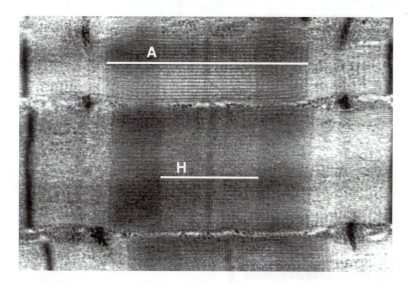

Fig. 22–6. Electron micrograph of a rabbit *psoas* muscle. Magnification = 35,000 x. The A band, I band (only half in this photograph), H zone (the less dense region in the center of the A band), Z line, and M line structures are labeled. There are also parts of the sarcoplasmic reticulum and T tubules located between the myofibrils.

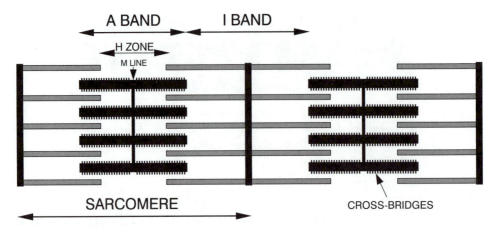

Fig. 22–7. Diagram of the thick and thin filament structure of the myofibril. Only a few filaments are shown; an actual sarcomere contains several hundred filaments. Cross-bridges (consisting of the myosin heads) cover the surface of the thick filaments except near the center.

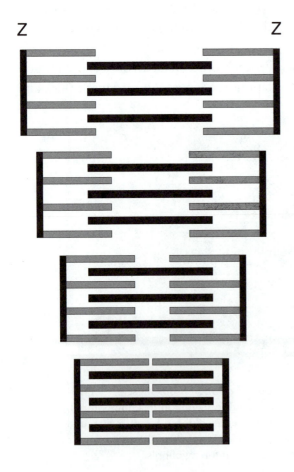

Fig. 22–8. Diagram of the changes in sarcomere structure during contraction. Thick and thin filaments do not shorten but rather slide past each other in an accordion-like fashion. The extent of shortening varies and depends on the load and physical restraints on movement. Note that the I band and H zone regions become narrower as the sarcomere shortens (top to bottom).

thin filaments do not shorten individually, only move past one another. However, the banding pattern changes during muscle contraction with the widths of the I bands and H zones decreasing.

The muscle filaments are arranged so that there are three thick filaments surrounding each thin filament and six thin filaments surrounding each thick filament in the overlap region (see Figure 22–9). Thus, there are several thin filaments that may interact with each thick filament during muscle contraction.

The thick and thin filaments are composed of proteins. Thick filaments contain primarily myosin, and thin filaments have actin, tropomyosin, and troponin. These four proteins are responsible for muscle contraction and relaxation. The cross-bridges consist of the myosin heads. The heads bind to actin, rotate, and release. The concerted effort of millions of cross-bridges produces the muscle shortening.

Muscle contraction is controlled by changing the calcium ion concentration in the muscle cell. Normally, the free calcium ion concentration in the *cytosol* (the soluble material inside the cell) bathing the myofibrils is about 10^{-9}M. It is maintained at this level by an intracellular membrane system called the *sarcoplasmic reticulum*. This system envelops each myofibril. The sarcoplasmic reticulum contains a protein that pumps calcium to its interior in an ATP–dependent process. Muscle contraction is triggered when a message comes from the nerve. A special region where the nerve terminal comes close to the surface of the muscle cell is

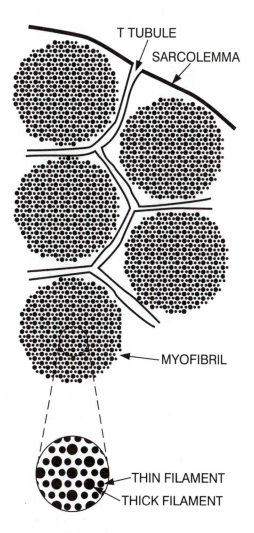

Fig. 22–9. Diagram of a cross section of a portion of a muscle cell. Note that the T tubules connect to the sarcolemma (the cell outer membrane). The myofibrils show the arrangement of the thick and thin filaments in the overlap region (outer part of the A band). Each thick filament is surrounded by six thin filaments, and each thin filament by three thick filaments.

called the motor endplate (see Figure 22–4). The nerve ending releases a small amount of a chemical called acetyl choline that diffuses to the muscle cell membrane and binds to acetyl choline receptors. These receptors cause the muscle cell membrane to become leaky and the membrane becomes "depolarized." (Sodium concentrations usually are high outside the muscle cell and low inside; conversely potassium levels are high inside and low outside. The muscle cell interior is normally negatively charged. When the membrane becomes leaky, the sodium moves across more rapidly than potassium, neutralizing the interior charge. The muscle cell then no longer has a charge; it is depolarized). This depolarization region spreads out over the entire surface of the muscle cell. In addition, it spreads down the T tubules, perpendicular extensions of the outer cell membrane (called the *sarcolemma)* that penetrate though the center of the cell (see Figure 22–9). The T tubules also are physically adjacent to the sarcoplasmic reticulum. The depolarization wave along the T tubules triggers the release of calcium (about 10^{-5} to 10^{-6} M) from the sarcoplasmic reticulum. A special four subunit protein called the *ryanodine receptor* (located at the interface between the T tubule and the sarcoplasmic reticulum) (see Figure 22–10) opens a channel to release the calcium. This protein is also referred to as the "calcium release channel."

Calcium diffuses to the myofibrils and binds specifically to troponin. Troponin moves tropomyosin and allows the myosin heads to bind to actin. Cross-bridges cycle to produce force, filament sliding, and the resultant muscle shortening. The released calcium is quickly moved back into the sarcoplasmic reticulum by the Ca^{++} pump embedded in the membrane (see Figure 22–10). The calcium is then ready for another contraction cycle. Sodium and potassium are also pumped out of and into the muscle cell respectively to reestablish the polarity. The time for one contraction and relaxation cycle is very short, only a couple hundred milliseconds. This rapid time course is due to (1) the quick passage of the signal to the cell interior by way of the T tubules and (2) the short diffusion distance for calcium between the sarcoplasmic reticulum and the myofibrils. The major steps in a muscle contraction cycle are summarized as follows:

1. A message to initiate contraction passes down the nerve.

2. The nerve terminal releases acetyl choline at the motor end plate.

3. Acetyl choline diffuses to the muscle cell membrane and binds to the acetyl choline receptors.

RELAXATION   CONTRACTION

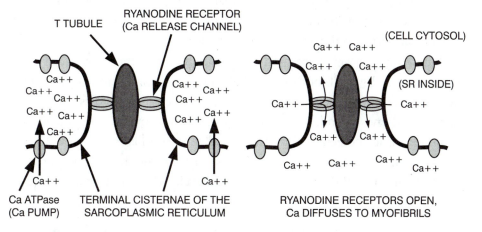

Fig. 22–10. Diagram of a longitudinal view of the interface between the T tubule and the sarcoplasmic reticulum. During the relaxation phase, the calcium is trapped inside the terminal cisternae of the sarcoplasmic reticulum (SR). When the depolarization wave travels down the T tubule, it causes the ryanodine receptor to open and calcium ions to leak into the cytosol. Then, the calcium pump in the SR moves calcium out of the cytosol, using ATP to cause relaxation.

4. The acetyl choline receptors cause the muscle cell membrane to become leaky, allowing sodium to move inside and potassium to move outside.

5. This membrane depolarization spreads across the surface of the cell and down the T tubules.

6. The T tubules cause the ryanodine receptors (calcium release channels) on the sarcoplasmic reticulum to open, releasing calcium into the cell cytosol.

7. Calcium diffuses to the myofibrils and binds to troponin.

8. Troponin changes its shape and moves tropomyosin, allowing the myosin heads to bind to actin.

9. The myosin heads move through a power stroke, filament sliding occurs, the muscle shortens, releasing ADP plus phosphate.

10. The sarcoplasmic reticulum pumps the calcium back inside, the troponin and tropomyosin move back to their original positions, and the muscle relaxes.

11. Myosin splits ATP and remains ready for the next contraction signal.

ATP provides the energy for muscle contraction. It is split into ADP plus phosphate plus energy. Myosin heads break the ADP–phosphate bond during the relaxation part of the cycle. The ADP and phosphate are released after the myosin head binds to the actin during contraction.

Glucose or fatty acids are used to produce ATP in the muscle cell. They come to muscle by the bloodstream. The glucose is stored in the cell as glycogen. Glycogen is a polymer of thousands of glucose molecules (six carbon sugar) linked end to end. The glycolysis pathway converts one glucose into two pyruvic acid molecules with a net yield of two ATP (see Figure 22–11). The pyruvate can be converted to CO_2 and H_2O in the mitochondria to give an additional 36 ATP through the citric acid cycle. This latter process requires oxygen.

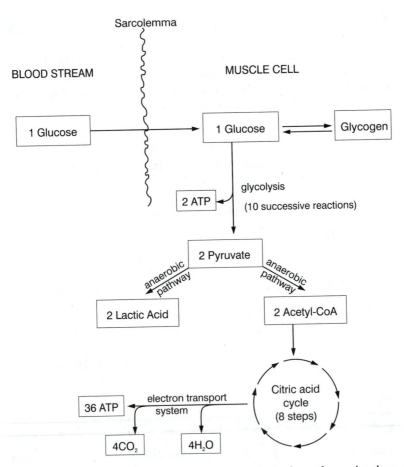

Fig. 22–11. Simplified version of the catabolic fate of muscle glycogen. One glucose molecule following the aerobic pathway can generate 38 molecules of ATP (36 plus 2 from glycolysis). However, one glucose molecule following the anaerobic pathway will yield only two ATP molecules and lactic acid as the end product instead of CO_2 and H_2O.

PROTEIN COMPOSITION OF MUSCLE

Lean muscle of all species consists of approximately 20 percent protein, 72 percent water, 7 percent fat, and 1 percent ash. These proportions change as the animal is fattened: as fat increases, the percentages of protein and water decrease.

Muscle Proteins

Muscle tissue contains many proteins serving many different functions. Muscle proteins can be grouped into three general classifications: (1) *myofibrillar,* (2) *sarcoplasmic,* and (3) *stromal* proteins. Each of these protein classes plays a specific role in the life of the animal and as muscle is converted to meat.

Myofibrillar Proteins

Myofibrillar proteins are also known as salt-soluble proteins because of their ability to be dissolved in salt solutions. These proteins comprise about 11.5 percent of the muscle weight (over half of the total protein). The major myofibrillar proteins are listed in Table 22–2. Myosin and actin are the two most important myofibrillar proteins, since they are the most abundant in this class and provide the structural backbone of the myofibril. Tropomyosin and troponin perform a regulatory role in contraction. Titin is believed to extend from the M line to the Z line in each half sarcomere. It appears to prevent muscle from being stretched too far. Nebulin has been proposed to bind to the thin filaments and may act as a ruler determining the thin filament length. C protein is located in the center of each half A band; its function is not understood. M protein, α–actinin, and desmin are believed to stabilize the sarcomere filament lattice.

Sarcoplasmic Proteins

The sarcoplasmic proteins consist mainly of glycolytic enzymes. These proteins are also known as water-soluble proteins. Myoglobin, the oxygen-binding protein inside the muscle cell, is also a member of this protein class. Hemoglobin is found in the sarcoplasmic protein fraction; it comes from trapped blood that is left in the muscle capillaries, arteries, and veins. Both

Table 22–2. Proteins of the Myofibril

Protein	Molecular Weight	Subunits	Location	% Myo-fibrillar Protein
Contractile				
Myosin	520,000	2 of 220Kd[1], 4 of 20Kd	Thick filaments	43
Actin	42,000		Thin filaments	22
Tropomyosin	68,000	2 of 34Kd	Thin filaments	5
Troponin	69,000	30Kd, 21Kd, 18Kd	Thin filaments	5
Structural				
Titin	2,800,000		Full sarcomere	8
Nebulin	600,000		Thin filaments	3
C protein	140,000		Thick filaments	2
α–actinin	200,000	2 of 100Kd	Z lines	2
M protein	160,000		M lines	2
Desmin	55,000		Z lines	<1

[1]Kilodalton = 1,000 daltons. One dalton is a unit of mass very nearly equal to that of a hydrogen atom.

myoglobin and hemoglobin are responsible for the typical red color of meat. The darker color of beef versus pork (or chicken leg versus chicken breast muscle) is due to larger quantities of these pigment proteins. The heme group also changes in color with oxygen binding, oxidation, and when complexed with nitrite during meat curing (see Chapter 20).

Stromal Proteins

The stromal proteins, or connective tissue proteins, consist primarily of collagen and elastin. *Collagen* is the single most abundant protein found in mammalian species, being present in all tissues with the highest concentrations in bone, skin, tendons, cartilage, and muscle. The primary functions of collagen are to provide strength and support and to act as an impervious membrane (as in skin). The ratio of the amount of connective tissue to the number of muscle cells in a muscle affects the tenderness of meat. Similarly, collagen becomes increasingly cross-linked with age, and this results in greater toughness with meat from older animals. Muscles with postural functions (such as those found in the lower legs) typically have high connective tissue contents, while the back muscles *(longissimus* and *psoas)* have lower proportions of connective tissue. Collagen is not broken down easily except with moist-heat cookery methods (see Chapter 24).

Elastin is found in arterial walls and gives elasticity to those tissues. Elastin is sometimes referred to as "yellow" connective tissue because of its color. The *ligamentum nuchae* or backstrap contains large quantities of elastin. In contrast to collagen, elastin is not degraded by moist-heat cookery methods. Fortunately, muscle tissue contains very small amounts of elastin.

POSTMORTEM CONVERSION OF MUSCLE TO MEAT

When an animal is harvested, its life processes gradually cease, causing dramatic changes in the muscle. These changes can ultimately affect the eating quality of meat. In normal situations when an animal is harvested, the muscle undergoes a gradual decline in pH (from about 7.0 to 5.5) (see Table 22–3). This decline is caused by the depletion of the animal's glycogen reserves held within the muscle and their conversion to lactic acid, the end product of anaerobic glycolysis (see Figure 22–11). When an animal is bled, oxygen is no longer available to the muscle cells, and anaerobic glycolysis takes over. Lactic acid builds up, and the pH drops. The glycogen becomes depleted, and some of the enzymes responsible for glycolysis become inactivated by the low muscle pH. Initially, the creatine phosphate reserves are used to convert ADP back to ATP. However, the creatine phosphate is soon gone and the ATP concentration declines. In the absence of ATP, myosin heads form a tight bond with actin. The muscle filaments can no longer slide over one another, and the muscle becomes stiff and inextensible. This condition is called *rigor mortis.* The rate of ATP depletion closely coincides with the rate of fall in muscle pH. The time for the muscle to reach its final pH varies with species (beef and lamb take longer than pig muscle), cooling rate (metabolism is slowed at lower temperatures), and extent of struggle at death (less activity immediately before harvest prolongs the length of the pH decline period).

Some changes occur in the muscle proteins during the postmortem period. The Z lines develop visible breaks. The proteins desmin and troponin T (one of the three subunits of the troponin complex) are partially degraded. There is also evidence that titin and nebulin are altered postmortem. These changes that occur during cooler storage are referred to as aging. The resulting improvement in tenderness of the meat continues for one to two weeks. Holding carcasses for this time period is an extremely costly procedure because of the amount of cooler space needed and the weight loss of the carcasses during the long-term storage. However, some primal cuts are stored 10 to 14 days in cryovac bags to improve tenderness, particularly for the elite restaurant trade.[1]

The cause of the postmorten tenderization is not known, but indirect evidence suggests that *calpains* (calcium-activated proteins) may be involved. Labeled as the *calcium-activated factor* (CAF), calpains were originally discovered by Drs. William Dayton and Darrel Goll, and their coworkers.[2] First, the proteins that seem to be most susceptible to digestion by calpains in the laboratory (troponin T, desmin, titin, and nebulin) are the same ones that become degraded during postmortem storage. Second, freezing and thawing muscle has been shown to partially inactivate *calpastatins* (proteins that normally inhibit the activity of the calpains), resulting in improved meat tenderness. Third, calcium chloride solutions injected into muscle or used to marinate beef cuts increase the rate of meat tenderization.

POSTMORTEM QUALITY PROBLEMS

Dark, Firm, and Dry (DFD) Meat

The final pH that is attained is the ultimate pH of the meat and has an important influence on the meat's color, texture, water-holding ability, and tenderness. If an animal undergoes vigorous stress or exercise of the muscles before it is harvested, the glycogen content within the muscles may be reduced substantially (see Table 22–3). After this animal has been harvested, a higher ultimate pH will occur in the muscles, and the meat will be dark, firm, and dry (DFD). This is a somewhat common event in beef (especially among bulls), and the resulting carcasses are termed *dark cutters*. The DFD condition is also found in pork carcasses. The dark color of high pH meat is thought to be a result of its greater water-holding capacity, which causes muscle fibers to

---

[1]Parish, F. C., Jr. 1997. Aging of beef. Beef Facts: Meat Science. 11–421, 3977. National Cattlemen's Beef Association.

[2]Dayton, W., et al. 1976. A CA^{2+}-activated protease possibly involved in myofibrillar protein turnover. Purification from porcine muscle. Biochemistry. 15:2150–2158.

Table 22–3. Postmortem Changes in Muscle and Variations Leading to Quality Defects

Condition	At Death				24 Hours After Death			
	Muscle Glyco-gen[1]	Muscle Color	Lactic Acid	Muscle pH	Muscle Glyco-gen	Muscle Color	Lactic Acid	Muscle pH
Normal pigs	60	Purple	Low	7.0	1 to 10	Bright red	High	5.5
DFD pigs	20	Purple	Low	7.0	1 to 10	Dark red	Low	6.0 to 6.5
PSE pigs	20	Dark red	Medium	6.4	1 to 10	Pale	High	5.5
Normal beef	50	Purple	Low	7.0	1 to 10	Bright red	High	5.5
Dark cutters (beef)	20 to 40	Purple	Low	7.0	1 to 10	Shady, dark or black	Low	6.3 to 6.7

[1]Expressed as μmoles of glucose equivalents per gram of muscle.

swell. The swollen state of the fibers results in more incident light being absorbed by the meat surface and therefore a darker color. Dark cutters are severely discounted by the packing industry because of the poor consumer appeal of this meat; thus, it is important to minimize stress and rough handling of animals before harvesting them. Meat from DFD and dark cutters also has reduced shelf life since the higher pH is more conducive to bacterial growth.

Pale, Soft, and Exudative (PSE) Meat

The rate of pH decline is an important factor affecting meat quality. An extremely rapid postmortem drop in muscle pH is associated with the pale, soft, and exudative condition that often occurs in pork. The PSE meat is characterized by a soft, mushy texture, a poor water-holding capacity, and a pale color. These conditions are caused by a rapid pH decline while the muscle temperature is still high. The looser muscle structure associated with a lower water-holding capacity results in a greater reflectance of incidental light and therefore a pale color. The PSE condition is also known to be highly stress-related. Market hogs having the hereditary stress condition known as porcine stress syndrome (PSS) produce carcasses with a high incidence of PSE meat (see Chapter 3). The molecular defect for both PSS and PSE has been shown to be due to a mutation of the ryanodine receptor. Calcium leaks more easily from this protein in affected individuals and results in continued contraction without a nerve signal. A very rapid glycolysis occurs in these muscles in an attempt to replenish the ATP that is being broken down. A similar condition in humans is called *malignant hyperthermia*. People with this condition often will die or become dangerously overheated when they are given an anesthetic.

RN

Another genetic condition that occurs in certain pigs is called RN, Napole gene, or "acid meat" gene. Pigs with this condition have abnormally high muscle glycogen content during life, and their ultimate pH postmortem is lower than usual, often below pH 5.4. The mutation is prevalent in the Hampshire breed and is sometimes referred to as the "Hampshire effect." The rate of pH decline postmortem is normal. The resulting meat has improved tenderness, but the low pH results in decreased water-binding capacity and processed product yield. The mutation has been recently identified to occur in a regulatory subunit of a minor protein kinase found in muscle.[3] This AMP-activated kinase normally inhibits the enzyme glycogen synthetase. In mutant animals the synthetase remains active, with a resulting higher muscle glycogen content.

Callipyge

Some sheep have a genetic condition that results in very heavy muscling. Such muscling is advantageous from a meat yield and growth efficiency standpoint. However, the meat from such animals has been found to be excessively tough. This toughness has been ascribed to higher calpastatin activity in these animals compared with that in normal sheep.

Capillary Breakage

Short-term, violent excitement in cattle (usually from running and(or) fighting) causes blood to fill capillaries in subcutaneous fat and muscle. If these animals are harvested immediately, their muscles may display small blood spots. Holding such animals for one or two days before harvesting

---

[3]Milan, D., et al. 2000. A mutation in PRKAG3 associated with excess glycogen content in pig skeletal muscle. Science. 288:1248–1251.

them usually eliminates this problem. Similar problems may occur in pigs and lambs. If the capillary breakage occurs in the fat, the condition is referred to as "fiery fat."

Cold Shortening

Cold shortening is a problem associated with rapid chilling of the carcass, which results in a dramatic decrease in the tenderness of the muscles. This problem affects beef and lamb carcasses more than pork carcasses, since muscles from these species have a higher percentage of red muscle fibers (slow twitch) as opposed to white muscle fibers (fast twitch). Red muscle fibers are more prone to cold shortening than white fibers. Meat that has undergone cold shortening may have as much as a fivefold increase in toughness. Factors that increase the problem of cold shortening include lower cooler temperatures (20° to 25°F versus 32° to 36°F), smaller carcass size, faster air movement in the cooler, and carcasses with less fat cover. Since the carcass attachments may limit the degree of shortening, the procedures involving hot boning and rapid chilling may cause considerable shortening and toughening of the meat. Two methods may be used to reduce or eliminate cold shortening problems: delayed chilling and electrical stimulation.

Thaw Rigor

Thaw rigor is a condition somewhat similar to cold shortening, but it is much more severe. Thaw rigor is a phenomenon that occurs when muscle is frozen pre-rigor. When such a muscle is thawed, it becomes highly contracted. The sarcoplasmic reticulum becomes damaged during the freezing process and allows large amounts of Ca^{++} ions to move to the myofibrils during thawing. Since high levels of ATP are still present, rapid and severe muscle contraction occurs. Thaw rigor is not a major concern among meat processors because very little meat is frozen in the pre-rigor state. However, it could become a problem in the future.

Heat Ring

Heat ring is a problem associated with beef carcasses that have a relatively thin layer of external fat. The outer portion of the exposed rib eye muscle (after ribbing) appears darker in color and coarser in texture and often sinks, or drops, down. Heat ring is caused by differential chilling rates of the rib eye muscle. The outer portion chills faster, resulting in a slower glycolytic rate, slower pH decline, and longer delay till rigor onset than the better insulated inner portion of the muscle. Electrical stimulation of the carcass dramatically reduces the incidence of heat ring caused by the accelerated rate of pH decline and the subsequent increased speed of rigor onset (see discussion on electrical stimulation).

POSTMORTEM TECHNOLOGY

Better understanding of postmortem biophysical changes in muscle has led to the development of new technology. This technology has capitalized on natural physical events to improve the quality of meat and the efficiency of harvesting and processing. Electrical stimulation, hot boning, and delayed chilling are all relatively new developments in the meat industry.

Electrical Stimulation

Electrical stimulation refers to the passage of an electric current through a carcass immediately after harvest. It is a widely accepted process in the beef packing industry. Electrical stimulation of beef carcasses has been shown to improve carcass quality grade, tenderness, lean color, texture, and firmness. It also reduces heat ring problems and makes the hide pulling process easier (see discussion in Chapter 6). The stimulation process causes the muscles on the carcass to contract violently. This results in an accelerated anaerobic glycolysis to replenish the ATP, thereby increasing the rate of pH decline and reducing the time to rigor mortis. Electrical stimulation has been demonstrated to improve the lean color and enhance the definition of the marbling naturally present within the *longissimus* muscle. Thus, packers often see an increase (up to 5 percent) in the number of carcasses that grade USDA Choice or higher. The mechanisms leading to improved tenderness are not completely understood. Some muscles of the carcass show little or no effect on tenderness due to stimulation, while others show moderate improvements.

Hot Boning

With the exception of the whole-hog sausage industry, hot boning has not been widely accepted in this country, even though tremendous savings in energy have been demonstrated. The reasons for the lack of acceptance in the beef packing industry are primarily related to grading and marketing problems of hot-boned meat. The grading standards (see Chapter 12) have no provisions to deal with such a system. Thus, if or when this technology is adopted by the beef industry, it will likely happen with cow beef first, since a large portion of this is generally further processed without grading.

Hot boning allows the processor to salt the meat before the muscle has gone into rigor mortis. This keeps the muscle pH from falling as low as it would normally, thus improving the water-holding capacity and keeping the color darker. Both of these properties are advantageous in sausage production.

Delayed Chilling

Delayed chilling refers to the holding of carcasses at room temperature (60° to 70°F) for two to four hours following harvest before transferring them to the cooler. The purpose of this practice is to accelerate the glycolytic process and thus speed the onset of rigor. Metabolic activity has been shown to occur at an accelerated rate at higher temperature. Delayed chilling may allow muscles to go into rigor before cold shortening can occur. However, this process may shorten the product's shelf life if adequate sanitation (see Chapters 2 and 3) and USDA–approved bacteriostatic agents are not applied to the carcass (e.g., a dilute chlorine solution is sometimes sprayed onto carcasses). Delayed chilling is not a common practice with large-volume packers because the additional time requirements are not conducive to high chain–speed operations. The advantages of delayed chilling can be accomplished more efficiently by electrical stimulation. Perhaps the greatest potential use of the delayed-chilling concept is in conjunction with hot boning.

23

Meat as a Food

This chapter will discuss the composition of meat and its value as food. Moderate servings of today's leaner meat contribute significant amounts of most nutrients needed for a healthy diet.

Dietary reference intakes for individual components of the diet are established by the Food and Nutrition Board of the National Academy of Sciences, National Research Council (NRC).[1] A well illustrated discussion of this topic is contained in the National Live Stock and Meat Board's *Lessons on Meat*.[2]

The energy value of foods, measured in calories, is of prime importance for two quite different groups of the world's population. In the developed world, most people are trying to limit their calorie intake because consuming too many calories can result in obesity and its related diseases. Conversely, in many parts of the underdeveloped world, people are dying from starvation. Calories in the human diet come from fats, proteins, carbohydrates, and in some cases alcohol. The relative effective "power" (calories/gram) provided by these sources is as follows: fats 9, proteins 4, carbohydrates 4, and alcohol 7. These estimates are rounded from specific calorie factors of 9.02, 4.27, and 3.87 for fats, proteins, and carbohydrates, respectively, which were used in the nutrient value tables (http://www.nal.usda.gov/fnic/foodcomp/) and are based on net physiologic energy remaining after losses in digestion and metabolism have been deducted. Vitamins and minerals are very important cofactors in energy metabolism, but they do not provide calories. There are very few carbohydrates in meat except in liver (see later discussion). If alcohol (wine) is used in cooking, most of it will burn off or evaporate during the cooking process. Thus, in meat the primary concern is calories from fats and proteins.

---

[1]National Research Council Food and Nutrition Board. 1999. http://www.iom.edu/iom/iomhome.nsf/Pages/FNB+DRI

[2]National Live Stock and Meat Board, Chicago (now National Cattlemen's Beef Association, Denver).

The nutrient deficiencies that cause growth-stunting and delayed development in Third World children has been under rather intense study.[3] The general conclusion of parallel studies in Egypt, Kenya, and Mexico was that the *quality* of the diet, in terms of its content of animal products, was more important for the growth, cognitive, and behavioral development of children than the total amount of energy or protein consumed. Available micronutrients, such as iron and zinc, may be more limiting to marginally malnourished children's growth and development than previously recognized. Nutritionists and dietitians use the term *nutrient dense* to describe a food that provides a greater share of the dietary reference intakes for at least four essential nutrients than it provides in calories. Meat is nutrient dense in more than four essential nutrients.

The composition of meat will be discussed in the order of the predominance of its components: water, proteins, fats, carbohydrates, minerals, and vitamins.

WATER

Water is the most abundant chemical component of meat, which is logical because of its importance in the animal body. However, the essential role of water is often overlooked, as explained by Wahlstrom:

> On the basis of the magnitude of animal requirements for water and the variety of its functions in the body, water can be considered as the most important essential nutrient. Animals may live more than ten times longer without food than without water. The body can lose almost all of its fat and over half its protein and yet live, while a loss of one-tenth of its water may result in death. The water content of newborn animals approximates 75 to 80 percent. In adult animals, water content varies inversely with the amount of fat in the body. Expressed on a fat-free basis, the water content of adult animals is approximately 75 percent. Water has many diverse functions in the body. It is intimately involved in basic body metabolism, is an

---

[3]Lindsay, H. A. 1991. Child growth and nutrition. Food and Nutr. News (National Live Stock and Meat Board). 63:15.

important transport medium, serves as a temperature regulator, serves as a solvent and lubricant, transports sound, and cushions the nervous system, to name a few of its most important functions.[4]

Fatty tissue contains little or no moisture; therefore, the higher the fat content (finish) of a carcass or cut, the lower the total water content. Mature beef carcasses that are overfinished may contain as little as 45 percent, while veal may run as high as 80 percent moisture. An important point for the consumer to remember is that as leaner meat is demanded, more water rather than fat is purchased at meat prices.

Water is a key factor in meat processing, as noted in Chapters 18 to 21. Studying the nutrient value tables (http://www.nal.usda.gov/fnic/foodcomp/) will show the effect of the heat of cooking in driving off moisture and thus concentrating the remaining nutrients.

PROTEINS (AMINO ACIDS)

The general classification of muscle proteins was discussed in Chapter 22. The word *protein* comes from the Greek word *protos,* meaning *first,* and *proteios* meaning *primary.* Proteins are essential constituents of all living cells. Amino acids are the basic building blocks of all proteins. Proteins are hydrolyzed by the digestive enzymes into amino acids, the only form in which they can readily be absorbed into the bloodstream.

Some 30 recognized amino acids exist in nature. Of these, 20 are commonly found in most proteins and are required for protein synthesis. Amino acids that cannot be synthesized by a person and(or) animal and must be included in the diet are called *essential amino acids.* Amino acids that can be synthesized from carbon, hydrogen, nitrogen, and sulfur from carbohydrates and proteins in the diet are known as *nonessential amino acids.* Nine amino acids are considered essential for adult humans. When protein is produced, these 9 are supplemented with the other 11, either from dietary sources or from biosynthesis. Table 23–1 lists the known amino acids and identifies which ones are essential for humans and many animals.

---

[4]Wahlstrom, R. C. 1981. Water — the forgotten nutrient. Proc. Second Western Nutrition Conf:152, Canadian Feed Ind. Assoc., Univ. of Alberta.

Table 23–1. Amino Acids

Essential	Nonessential	Less Common, Nonessential
Histidine (His)	Alanine (Ala)	Cystine (Cys)$_2$
Isoleucine (Ile)	Arginine (Arg)	Hydroxyproline (Hyp)
Leucine (Leu)	Asparagine (Asn)	Hydroxylysine (Hyl)
Lysine (Lys)	Aspartic acid (Asp)	Citrulline
Methionine (Met)	Cysteine (Cys)	B-alanine
Phenylalanine (Phe)	Glutamine (Gln)	Aminobutyric acid
Threonine (Thr)	Glutamic acid (Glu)	Diaminopimelic acid
Tryptophan (Trp)	Glycine (Gly)	Dihydroxyphenylalanine
Valine (Val)	Proline (Pro)	Ornithine
	Serine (Ser)	Taurine
	Tyrosine (Tyr)	

All amino acids have the following basic structure:

$$
\begin{array}{c}
\text{COOH} \\
| \\
\text{H}_2\text{N–C–H} \\
| \\
\text{R}
\end{array}
$$

R can be a hydrogen (as in glycine) or a complex chain attached to the basic unit (as in methionine):

Glycine

$$
\begin{array}{c}
\text{COOH} \\
| \\
\text{H}_2\text{N–C–H} \\
| \\
\text{H}
\end{array}
$$

Methionine

$$
\begin{array}{c}
\text{COOH} \\
| \\
\text{H}_2\text{N–C–H} \\
| \\
\text{CH}_2 \\
| \\
\text{CH}_2 \\
| \\
\text{S} \\
| \\
\text{CH}_3
\end{array}
$$

All of the essential amino acids have been found prevalent in muscle, heart, liver, and kidney tissue in quantities considered to be very adequate; thus, meat protein has a *high biological value*. Three of the less common, nonessential amino acids are present in meat, specifically in connective tissue. These are hydroxyproline, hydroxylysine, and cystine.

Supplying sufficient amino acids in the diet to maintain the protein reserves of the body is an important factor in permitting antibody synthesis, thus allowing for acquired immunity to disease. The ability of the human body to resist disease is dependent upon its ability to produce antibodies — substances that tie up specific foreign proteins. The blood globulins and albumins are built up from the amino acids in food.

Another group of substances that are related to proteins (but are not true proteins) are the purines, pyrimidines, and nucleopeptides. These largely water-soluble components have little nutritive value in themselves but are biochemical substances that may excite the flow of gastric juices. Along with fats, they provide a great deal of the aroma and flavor of meat. These components are present to a greater degree in older animals, and they are particularly abundant in the more active muscles that make up the less tender cuts. They impart to game animals that so-called "gamey" flavor. Other examples of this group of substances are creatine and creatinine.

FATS

The total amount of fats in the U.S. food supply has increased since 1900, but the increase has been in fats of vegetable origin, while animal fats have decreased. About one-half of the fat in the human diet comes from plants, and the other half from animals. CAST's Task Force Report entitled *Food Fats and Health*[5] is a very complete and concise treatise on this subject. The major constituent of food fats is triglycerides, which contain a variety of saturated, monounsaturated, and polyunsaturated fatty acids. Animal fats, which are highly digestible, supply the *essential fatty acid* linoleic acid (18:2 n–6; see diagram), carry fat-soluble vitamins (A, D, E, and K), and provide protection and insulation for the human body as well as energy. Fats contribute 2.25 times as much energy per unit of weight to the diet as do carbohydrates and proteins, as noted previously. Aside from these values, fats play a role in add-

---

[5]CAST. 1991. Food fats and health. Task Force Report. Ames, Iowa.

ing palatability to the lean in meat because of the flavor, aroma, and apparent tenderness they generate.

The American Heart Association (AHA) and the U.S. Department of Agriculture recommend that no more than 30 percent of the calories in the human diet come from fats. The average person in this country obtains approximately 37 percent of his or her calories from fats. However, today's lean meat fits well into the AHA recommended diet, and more extensive acceptance should reduce that figure.

The chemical difference between saturated and unsaturated fats lies in the existence of double bonds between certain of the carbon atoms in the fatty acid molecules. When the carbon valencies are satisfied with hydrogen, the fat is saturated. However, if one pair of adjacent carbons in a fatty acid chain each lack a hydrogen and are connected by a double bond, it is a *monounsaturated* fatty acid and when more than one pair of adjacent carbons lack hydrogen and are double-bonded, it is a *polyunsaturated* fatty acid. These are illustrated as:

Saturated

$$\text{HO–C–C–C–C–C–C–C–C–C–C–C–C–C–C–C–H} \quad \text{(Palmitic acid, } C_{16}H_{32}O_2\text{)}$$

Monounsaturated

$$\text{HO–C–C–C–C–C–C–C–C–C = C–C–C–C–C–C–C–C–C–H} \quad \text{(Oleic acid, } C_{18}H_{34}O_2\text{)}$$

Polyunsaturated

$$\text{HO–C–C–C–C–C–C–C–C–C = C–C–C = C–C–C–C–C–C–H} \quad \text{(Linoleic acid, } C_{18}H_{32}O_2\text{)}$$

The most common fatty acids in food fats contain 16 or 18 carbon atoms. The "shorthand" method for indicating the different fatty acids is to indicate the number of carbon atoms to the left of a colon and the number of double bonds to the right of the colon. For instance, linoleic acid is indicated by 18:2, meaning it has 18 carbons and 2 unsaturated bonds. See the nutrient value tables (http://www.nal.usda.gov/fnic/foodcomp/) for a listing of the major fatty acids in meat.

Saturated fats aid in the keeping quality of meats because they are less subject to oxidation. Highly unsaturated fats are soft and oily and may lower the quality of pork and poultry carcasses, thus reducing shelf life, since they are more readily oxidized. Oxidation produces off-flavors.

The degree of unsaturation of a fat can be determined by its reaction with iodine, which quantitatively reacts with the carbons at the double bonds. The *iodine number* that is produced is the number of grams of iodine reacting with 100 grams of lipid. Another method of determining unsaturation is to run polarized light through a prepared sample of fat and then to read its refractive characteristics. Well-established and well-equipped lipid (fat) laboratories use gas–liquid chromatography, high-pressure liquid chromatography (HPLC), and/or thin-layer chromatography (TLC) to separate and quantitate fatty acids.

The excessive intake of saturated fats is associated positively with blood cholesterol concentrations; thus, it is considered a risk factor for heart disease. Some meats have been criticized for having a high content of saturated fatty acids and a shortage of unsaturated fatty acids. However, all saturated fatty acids are not equal in their effect on blood cholesterol levels. Myristic (14:0) and palmitic (16:0) acids seem to increase cholesterol concentrations, while stearic (18:0) and shorter chain fatty acids, like the monounsaturated fatty acids, seem to cause no change. Therefore, stearic acid (18:0) is included in the saturated fatty acid total but also is listed separately in the nutrient value tables so that it may be subtracted from the saturated total when diet/health recommendations are given. Most meats do contain unsaturated fatty acids and stearic acid in substantial quantity.

A special group of unsaturated fatty acids is the omega–3 family (sometimes designated n–3), which derives its name from the location of a double bond on the third carbon from the methyl (CH_3) end of the fatty acid molecule (chain) known as the omega end. *(Omega* is the last letter in the Greek alphabet.) The "front" end of the fatty acid molecule is the carboxyl (COOH) end, which esterifies with (attaches to) a carbon from glycerol to form a mono-, di-, or triglyceride, depending on how many fatty acids esterify with the three glycerol hydroxyls. Nearly all unsaturated fatty acids in meat are of the n–6 family, meaning that the last double bond is on the sixth carbon from the methyl end of the molecule (see diagrams). Alpha–linolenic acid (18:3 n–3) is an omega–3 fatty acid, while gamma linolenic acid (18:3 n–6) is a member of the omega–6 family, which predominates over the omega–3 fatty acids in the U.S. food supply. Only certain plants have the enzymes that can desaturate the omega–3 carbon. Cold water fish can elongate and desaturate α–linolenic acid from algae in their diets to the longer chain, more highly unsaturated omega–3 fatty acids eicosapentaenoic (EPA [22:5 n–3]) and docosahexaenoic (DHA [22:6 n–3]). EPA and DHA have been reported to be beneficial in individuals who have heart disease, by helping to prevent the formation of blood clots. Feeding feedstuffs containing α–linolenic acid can increase EPA and DHA in animal tissues.

Fats may be hardened by the hydrogenation process, a process in which free reactive hydrogen atom(s) are combined with the unsaturated carbon(s) by the use of a catalyst (sodium methoxide or others), which saturates the carbon double bond. This process is now in general use for hardening vegetable oils and some lards. This hardening causes a change in the configuration of the double bonds in relation to the plane of the string of carbon atoms of the fatty acids, resulting in trans fatty acids. A trans fatty acid resembles a saturated fatty acid in that the conformation is straight, allowing for closer packing or alignment of fatty acid chains and thus less mobility and fluidity. The majority of fatty acids in foods are of the cis configuration. These have a greater bond angle or kink and are thus more fluid, because they will not pack as closely as the trans fatty acids. The diagrams of oleic and linoleic acid in this chapter show the cis conformation.

There is a "family" of fatty acids, all isomers of linoleic acid, known as conjugated linoleic acid (CLA). CLA is unique because the two double bonds are closer together — that is, conjugated. Counting carbon atoms from the carboxyl (COOH) end, the two double bonds are located on carbon numbers 9 and 12 and are separated by a methylene group, CH_2 in linoleic acid, as diagrammed in this chapter. In CLA, the two double bonds are located on carbon atoms 9 and 11.

The melting point of a fat depends on the unsaturation and chain length of its component fatty acids. Longer chain and more saturated fatty acids have higher melting points. Melting points vary with the class of domestic animal and the kind of feed the animals received; e.g., fatty acids can be absorbed intact in nonruminant animals and transmitted into the body tissues. Pork and poultry contain 50 percent or more unsaturated fatty acids, largely due to linoleic acid (18:2), most of which comes ultimately from corn (http://www.nal.usda.gov/fnic/foodcomp/). However, bacteria in the rumen of ruminants metabolize dietary fatty acids and in turn synthesize mostly saturated fatty acids for deposition in the body. So less than 50 percent of the fatty acids in beef and lamb are unsaturated.

The following shows the range of melting points of fat from different classes of animals:

Pork

Backfat	86°–104°F
Leaf fat	110°–118°F

Beef

External fat	89°–110°F
Kidney fat	104°–122°F

Lamb

External fat	90°–115°F
Kidney fat	110°–124°F

Poultry

Abdominal fat	80°–110°F

Internal fats are more saturated and have higher melting points than external fats. External fats are more unsaturated than internal fats so that a more liquid state can be maintained in the outer layers of the live animal where the body temperature is lower.

By utilizing the nutrient values (see http://www.nal.usda.gov/fnic/foodcomp/) and making a few calculations, you can determine the percentage of calories contributed from fats in any of the meat products. For instance, multiplying the grams of fat listed in the composite, lean only, cooked column (9.91) in Table NDB No: 13012 times 9 equals 89.19, the number of calories from fat in this composite. The composite provides 216 total calories, so 89.19 ÷ 216 = 41.29 percent of calories from fats. As can be noted from these data, most of the meats show total fats to contribute between 35 and 50 percent of the total calories in the product, whereas the extremes show 19 percent for chicken breasts and up to 65 percent for 73 percent lean hamburger. Total fats are made up largely of triglycerides and to a lesser extent phospholipids. Three fatty acids are joined to each glycerol molecule forming a triglyceride, while two fatty acids, a phosphate and an alcohol, form the phospholipid. Cholesterol is also a fat (lipid). Thus, total fats include more than the fatty acids listed individually in the tables.

Cholesterol contents are generally proportional to fat; however, there are exceptions. For example, veal shows higher values because of the large proportion of cell walls containing cholesterol and the high percent of water and low percent of fat in the cells. *Dietary cholesterol has only a minor effect on blood cholesterol — it is secondary to total caloric consumption and relative saturated fatty acid intakes in influencing blood levels of cholesterol.*

CARBOHYDRATES

The liver is the major carbohydrate reservoir, containing about one-half of all the carbohydrates found in the animal body. Carbohydrates are stored as glycogen and represent 2 to 8 percent of the liver weight. The remaining half is distributed in the muscles, with a limited amount in the bloodstream and other tissues.

Glycogen is converted to lactic acid postmortem. Thus, few carbohydrates remain in meat at the time it is consumed.

MINERALS

Meat is an especially good dietary source of iron, zinc, and phosphorus, but it is not a good source of calcium, iodine, and magnesium.

Muscle meats, and more particularly glandular meats, are exceptionally rich in *iron*. Dietary iron occurs in two forms. The most readily absorbed form is known as *heme iron* because it is contained in the porphyrin ring of hemoglobin and myoglobin (see Chapter 22 and the center color photo section). Heme iron composes more than half of the iron in meats. The remaining iron in meats and iron in grains, fruits, and vegetables is called *nonheme iron*. The absorption of nonheme iron is enhanced by an unidentified "meat factor." Liver contains twice as much iron as muscle meats. Iron is essential in the formation of hemoglobin in red blood cells, a lack of which causes anemia. Because liver is an excellent source of easily assimilated iron, it is often prescribed in the diet of anemia sufferers. Women between the ages of 15 and 44, teenage males, and vegetarians are among the approximately 10 percent of the U.S. population who may not be consuming enough iron.

The presence of *zinc* in the pancreas led to further research, which has shown that zinc is a necessary mineral in the diet. It is essential to the composition or function of nearly 200 enzymes, e.g., it is critical in the latter part of the healing process, and it influences reproduction, growth, brain maturation, functioning of the immune system, gene expression, and appetite through its effect on the senses of taste and smell. Meat, particularly liver; eggs; and shellfish are excellent sources of zinc. Zinc is more readily absorbed from animal sources than vegetable sources. *Phosphorus* is an essential constituent of body cell tissue with important biochemical functions in carbohydrate, protein, and fat metabolism. Phosphorus is necessary for the assimilation of calcium from other sources and along with vitamin D contributes to strong bones and teeth. Meat is a very good source of dietary phosphorus.

Animals also require *molybdenum, nickel, selenium, chromium, copper, fluorine, manganese, cobalt, magnesium,* and *iodine* for normal functioning; therefore, these elements are also found in animal tissues. Only a trace of manganese is found in muscle — liver contains considerably more manganese than muscle does. Muscle contains a small amount of aluminum; liver has a slightly larger amount. Copper, which has been shown to increase iron utilization, is found in small quantities in muscle and in slightly larger quantities in liver. All of the other trace minerals are found in animal tissues, usually in trace amounts, but they are very significant to human dietary needs.

Muscle itself is a poor source of *calcium;* the calcium content of the body is centered in bone tissue. The boneless meat product recovered from mechanical deboning machines may have calcium levels up to 0.75 percent, but because of USDA–FSIS regulations, its use is limited in consumer products (see Chapter 18). Since calcium is often deficient in many people's diets, the use of mechanically separated meat should probably be encouraged.

VITAMINS

It was not until the latter part of the nineteenth century that scientists discovered that dietary factors other than the proteins, carbohydrates, fats, and minerals were vital for health maintenance. In 1912, Casimir Funk, a Polish biochemist, coined the word *vitamine* to cover this group of dietary essentials because he wished to designate a particular one that he believed at the time to be an amine. Since then many new factors (vitamins) have been discovered, all of which have been isolated, identified, and chemically synthesized. These vitamins were not necessarily amines, and so the *e* has been dropped in the terminology. Vitamins are classified as to their solubility in fat solvents (fat-soluble) or water (water-soluble).

Fat-soluble Vitamins

Vitamin A (Anti-xerophthalmia Factor)

Vitamin A or retinol is an alcohol of high molecular weight that is soluble in oils and fats but nearly insoluble in water. It is somewhat stable to heat, acids, and alkalis but is destroyed by light and by mild oxidation. It occurs in animal tissues chiefly in the form of fatty acid esters. Alpha, beta, and gamma carotene ($C_{40}H_{56}$) and cryptoxanthin, the yellow coloring matter in corn, green feeds, vegetables, and fruits, are called precursors or provitamin A, and the animal body is able to convert portions of them into vitamin A. Beta carotene is most prevalent, and each molecule should yield two molecules of vitamin A, but frequently beta carotene is poorly converted, especially in ruminants.

Vitamin A was one of the early vitamins to receive recognition as a required nutrient. It is considered an essential factor in keeping healthy the epithelial tissues, the mucous membranes of the respiratory and genito-urinary tracts, plus the cornea and conjunctiva of the eye, and in preventing night blindness by catalyzing metabolism in the retina. Vitamin A promotes growth, aids in the resistance to infection, tones the nervous system, and is essential for successful reproduction.

Sheep and calf livers are particularly rich in vitamin A, followed in order by *beef, lamb, hog and pig livers, kidneys,* and *chicken liver.* Cod liver oil, butter, cheese, eggs, and *fish roe* (the egg-laden ovary of a fish) are also excellent sources, followed by beef fat, cream, ice cream, and whole milk. Apricots, broccoli, carrots, kale, spinach, pumpkins, yellow squash, sweet potatoes, and turnip greens, along with certain fruits, are sources of carotenoid pigments some of which can be transformed into vitamin A in the animal body.

It is estimated that $1/4$ pound of calf liver or $1/2$ pound of beef liver could supply the daily requirement of vitamin A.

Vitamin D or Cholecalciferol (Anti-rachitic Factor)

Vitamin D is formed in plants following irradiation of ergosterol to form calciferol or vitamin D_2, which may be converted to vitamin D_3. Vitamin D_3

is formed in the bodies of animals and humans when the skin containing 7–dehydrocholesterol is exposed to direct sunlight or ultraviolet light. The resulting cholecalciferol is insoluble in water but soluble in fat solvents. It is further modified for specific functions to 25–hydroxy and 1,25–dihydroxy derivatives by controlled feedback mechanisms in the liver and the kidney. Because these derivatives are so modified in certain animal tissues, they are often referred to as hormones.

The major function of vitamin D metabolites consists of regulating calcium and phosphorus absorption and metabolism, and thus being essential to normal bone growth and tooth development. Vitamin D helps prevent rickets in infants and children and softening of the bones, especially in aging adults, and helps stimulate growth and reproduction. Rich sources of vitamin D are cod liver oil, fish, egg yolk, irradiated foods, and fortified milk. *Pork and beef livers* are considered good sources; calf liver and other meat products contain only fair amounts.

Vitamin E (Anti-encephalomalacia Factor)

A number of different substances having vitamin E activity are known; however, the most common and potent compound is alpha-tocopherol ($C_{29}H_{50}O_2$). It is found in seed germ and green feeds and is a most effective metabolic antioxidant. Animals deficient in vitamin E are sterile or unable to conceive or produce sperm; hence, the vitamin was earlier called the sterility vitamin. Vitamin E and selenium interact closely in metabolism; both are essential, but one may partially compensate for the requirements of the other.

Although *liver* and *some variety meats* may contain limited amounts of vitamin E, plant products, such as alfalfa meal and wheat germ oil or meal, are the most common sources.

Vitamin K (Anti-hemorrhagic Factor)

Vitamin K was discovered in 1935. Only two forms, designated as vitamin K_1 from plants and vitamin K_2 from bacteria, have been isolated from natural sources. Vitamin K is heat and light stable. It is essential for the production of prothrombin, a blood-clotting metabolite, and it aids in the minimization of hemorrhage, as during the birth of infants, and in the prevention of obstructive jaundice. A form of vitamin K that can be injected intramuscularly prior to surgical operations has been developed.

The K vitamins were first isolated from alfalfa leaf meal and from putrefying fish, the former once being the principal practical source. ***Hog liver*** is also a rich source. Cabbage, carrot greens, spinach, soybean oil, tomatoes, hempseed, cauliflower, rice bran, kale, and egg yolks are considered good sources. Today's principal commercial source is a synthetic product known as *menadione,* or primarily its sodium bisulfite, water-soluble derivative, menadione sodium bisulfite.

Water-soluble Vitamins

Vitamin B₁ — Thiamine (Anti-neuritic Factor)

Vitamin B_1 is a thiazol–pyrimidine compound called thiamine that is soluble in water but insoluble in oils and fats. It exists in the pyrophosphate ester form in animal tissue and is an important coenzyme that plays a role in carbohydrate metabolism. Vitamin B_1 or thiamine was isolated in 1926, was chemically synthesized in 1936, and is sold as the salt thiamine hydrochloride.

A deficiency of this vitamin is the cause of a defect in nerve metabolism known as beriberi (polyneuritis in animals). Without sufficient vitamin B_1 in the diet, lactic acid builds up in the body, which can result in death. Symptoms include loss in weight, loss of appetite, slowing of the heart beat, impaired intestinal functioning, impaired reproductive functioning, and lactation failure. Vitamin B_1 promotes growth, stimulates appetite, aids digestion and assimilation, and is essential for normal functioning of nerve tissue.

Pork is an excellent source of vitamin B_1. From 12 to 50 percent of this vitamin may be lost during the cooking process, with some of it being extracted from the meat by the water in which the meat is cooked. The meat juice, or broth, contains the dissolved vitamin and should not be discarded. The loss of this vitamin is less when meat is fried. ***One center-cut pork chop*** contains upwards of 118 milligrams of vitamin B_1, more than the entire daily requirement for women or children. ***One ½-pound*** serving of ***round steak*** or ***two hamburger patties*** will furnish 30 milligrams. ***Liver, muscle meat, bacon, fish,*** eggs, milk, and ***oysters*** are also considered good sources of vitamin B_1, as are fruits and vegetables. Yeast, bran, cereal grains, and legume seeds are rich vegetable sources.

Riboflavin (Vitamin B₂, Formerly G)

Riboflavin, a yellowish-green, fluorescent, water-soluble pigment, a compound of flavin and the pentose sugar ribose, is another of the growth-promoting factors of the B–complex. It was isolated in 1933 and chemically synthesized in 1935. Riboflavin is stable to heat, mineral acids, and oxidizing agents but is rather sensitive to light.

A deficiency of this factor causes stunted growth, premature aging, unwholesomeness of the skin, and a general lessening of the muscle tone of the body. Riboflavin is now known to take part in a number of enzyme systems in the animal body, all of which play important roles in tissue oxidation. It is a valuable agent, in addition to niacin and vitamin B_1, in the treatment of certain cases of pellagra (a condition characterized by skin lesions, gastrointestinal disturbances, and nervous disorders).

Veal and beef liver, followed by beef kidney, lamb liver, pork liver, and *pork kidney,* are rich sources of riboflavin. *Beef heart,* milk, *oysters,* eggs, *sardines,* yeast, whey (dried), crabs, legumes, prunes, and strawberries are also excellent sources. *Ham, bacon, chicken, fish, lamb, beef,* cereals, and certain fruits and vegetables are good sources. Very little loss of the vitamin occurs in cooking. It is estimated that the *meat* in the ordinary diet furnishes about 20 percent of the necessary daily riboflavin requirement.

Niacin (Nicotinic Acid)

Niacin is a simple compound that exists as a white powder in the pure form or in needle crystal form and is soluble in both water and alcohol. Its biological importance was discovered in 1937. It is also called the anti-black tongue factor because it is a cure for the black tongue symptom in dogs. In the form of nicotinamide, niacin also plays an important part in oxidative enzyme systems in body tissues. A deficiency of niacin over an extended period will cause pellagra, dermatitis (cracking of the skin; a chapped appearance), and glossitis (inflamed mouth and tongue). It also contributes to a form of insanity in humans. Therapeutic doses of niacin may reduce blood cholesterol.

Niacin is somewhat heat stable and is found abundantly in *pork, beef, veal, and lamb livers. Liver* (¼ pound) or *veal, pork,* or *beef* (½ pound) per day will furnish the daily human niacin requirement. *Pork and beef kidney* rank next to *liver,* followed by *pork and beef heart, pork meat, veal, chicken, beef,* and *lamb. Salmon,* wheat germ, whey (dried), and yeast are also excellent sources. Other sources are buttermilk, eggs, *haddock,* milk, kale, peas, potatoes, tomatoes, and turnip greens; however, niacin from plant sources is less available. Niacin can be synthesized from tryptophan, but this amino acid is usually in short supply.

Vitamin B₆ (Pyridoxine)

Vitamin B_6 or pyridoxine has been called the rat acrodynia (lesions on the mouth, paws, tail, ears, and nose) factor, having been found to be essential for the maintenance of a healthy skin in rats, for amino acid metabolism, and for the utilization of unsaturated fatty acids. It was isolated in the crystalline state in 1938 and is available commercially. It is stable to heat, acids, and alkali but is destroyed by light and ultraviolet irradiation. It ($C_8H_{11}NO_3$) has a base structure similar to nicotinic acid, but with additions, i.e., 2–methyl, 3–hydroxy, 4,5–dihydroxymethyl pyridine.

Vitamin B_6 deficiency symptoms in the human have largely been found in infants and young children subsisting on certain prepared dietary formulas. It has been established that vitamin B_6 derivatives function in certain enzyme systems that have to do with the transfer of amino (NH_2) groups in the metabolism of nitrogen compounds. **Lean meat** and **kidney** are reported to be slightly more potent sources of vitamin B_6 than **liver,** with **heart** and **brains** furnishing lesser amounts. Egg yolk, wheat germ, and yeast are also excellent sources. **Fish,** milk, legumes, and wheat are good sources. **Meat** furnishes a large share of the daily requirement of this vitamin.

Choline

Choline is usually included as a member of the B–complex group of vitamins. It is B–hydroxyethyl–trimethyl ammonium hydroxide and functions to mobilize fatty acids. It is one of several substances that help to prevent perosis (enlarged and(or) twisted hock joint and misplaced tendons) and fatty livers. **Liver, pancreas,** egg yolk, and **muscle meat** are considered rich sources of choline.

Folic Acid (Pteroylglutamic Acid)

Names formerly used for folic acid include vitamin M, vitamin Bc, factor R, and L. casei factor. This vitamin is essential for the development of both red and white blood cells. Pure folic acid crystals are used medicinally for the treatment of certain macrocytic anemias (enlarged red blood cells). Folic acid was synthesized in 1945, is only moderately heat stable, and is present in green leaves. Pteroylglutamic acid is the name of the basic folic acid molecule. At one end of the molecule is the double ring or pterin group, a previously known yellow compound. The central grouping is a para–aminobenzoic acid ring formerly thought to protect against graying hair. The third component in the folic acid molecule is at least one unit of glutamic acid, a normal constituent of most proteins. *Liver, kidney, beef, veal,* yeast, green and leafy vegetables, wheat, and soybeans are good sources of folic acid.

Vitamin B_{12} (Anti-pernicious Anemia Factor)

The animal protein factor (APF) is largely represented by vitamin B_{12} or cobalamin, a vitamin that contains an inorganic base, namely, cobalt. Gastric juice in the stomach provides the "intrinsic factor" for improving its absorption. A lack of vitamin B_{12} ("extrinsic factor") or a lack of the "intrinsic factor" leads to pernicious anemia. Both vitamin B_{12} and folic acid contribute to the prevention and treatment of human macrocytic anemia (enlarged red blood cells), a condition found in pernicious anemia, sprue, and the anemia that can occur during pregnancy. Even a mild lack of vitamin B_{12} in people over 60 has been shown to cause tingling sensations, muscular coordination problems, weakened limbs, memory loss, and mood changes. *Variety meats* are the best sources of vitamin B_{12}. *Muscle meats* and *fish* are good sources, with eggs contributing a lesser amount.

Pantothenic Acid

Pantothenic acid, also a B–vitamin, plays an important role in the metabolism of fats and other compounds such as cholesterol. It is a combination of a derivative of butyric acid and α–amino propionic acid and is available as a synthetic preparation in the form of dextrorotatory calcium pantothenate. Experiments with chicks, rats, dogs, and pigs indicate that requirements of this vitamin are 5 to 10 times those of thiamine or riboflavin. Clinical evidence indicates that humans on diets low in pantothenic acid show anemias, sore tongues, and foot irritations.

Kidney, liver, and *beef heart* are potent sources, followed by *beef spleen, beef pancreas,* and *beef tongue,* with the major cuts having $\frac{1}{10}$ the potency of liver or kidney. Pantothenic acid is destroyed by prolonged dry heat.

Biotin

Biotin is another of the B–complex group of vitamins. Its formula is based on an imidazole ring with a valeric acid side chain. Biocytin is a naturally occurring form of biotin combined with lysine and converted to biotin in animal tissues. Biotin acts as a cofactor for several enzymes involved in carboxylation and is involved in both gluconeogenesis and lipogenesis. *Gluconeogenesis* is the formation of glucose within the animal body, especially by the liver, from substances (as fats and proteins) other than carbohydrates, and *lipogenesis* is the generation of fat within the animal body. Biotin deficiency in chicks results in altered fatty acid metabolism. Excess fat deposits may occur in the liver, with an accompanying hemorraghic condition. Skin lesions, a peculiar form of dermatitis around the mouth and eyelids, and shortening of long bones may develop. Biotin is found almost universally in feeds and foods, with **animal tissues** being good sources. Some combined forms such as that in the wheat kernel are unavailable to the very young. A protein in egg albumin, avidin, combines with biotin to make it unavailable. The binding by this latter protein is destroyed by cooking.

Vitamin C (Anti-scorbutic Factor)

Vitamin C, also known as ascorbic acid, is a water-soluble product of hexose sugar metabolism. Most animals synthesize adequate amounts, but humans and guinea pigs do not. A deficiency of vitamin C causes scurvy, and under stress conditions, as in disease (including the common cold) or in wound healing, its requirements may greatly increase. Sprouting plants are rich in this vitamin, as are citrus fruits and some vegetables. **Meats** are only a fair source of vitamin C. The greatest concentration in animal tissue exists in the adrenals, corpus lutea, and the thymus. Open-kettle cooking or wilting destroys considerable vitamin C. Vitamin C is used to hasten the development of the characteristic cured-meat color and to retard or prevent nitrosamine formation in cured meats (see Chapter 20).

UNIDENTIFIED FACTORS

Evidence exists that animal tissues contain other unidentified factors that are needed for maximum growth, superior reproduction, and development. These factors do not appear to be any known vitamins, minerals, amino acids, or fatty acids. Animal nutritionists usually recommend using some form of animal protein in diets for very young or reproducing animals. Humans probably need these nutrients also, with animal products being able to provide them, as well as most of the other nutrients, in the best and most efficiently used forms.

NUTRIENT COMPOSITION

The USDA Web site http://www.nal.usda.gov/fnic/foodcomp/ provides a quick, accurate, current reference for the nutritional composition of the edible portion of all meat, poultry, game, and fish products formerly published in the *USDA Handbook*. When utilizing a nutrient value table, be sure that it is not outdated.

As indicated in previous chapters, the composition of meat animals, and thus their carcasses and the meat we eat, has changed and continues to change toward a leaner, higher-protein product.

24

Preparing and Serving Meats

The ultimate satiety or enjoyment that comes from eating meat is largely dependent on how it is cooked. We stress again that eating raw meat (steak tartar, tiger meat), raw poultry, or raw fish (sushi) should not even be considered. First, from the food safety/microbiological concerns (see Chapter 2), eating raw products can be very detrimental to your health and livelihood. Second, palatability of raw products pales in comparison with that of properly cooked products. The very highest-quality steaks/chops can be rendered unpalatable by overcooking or by other improper cooking methods. On the other hand, the lowest-quality cuts of meat can be turned into an exquisite dining experience when proper cooking guidelines have been followed correctly. The complexity of meat and meat products demands that successful chefs and cooks, at home and in business, apply the principles that have been described and explained in the previous chapters.

THE EFFECTS OF HEAT ON MEAT

Tenderness/Texture

Changes in meat tenderness/texture in response to heat are summarized in Table 24–1. A key temperature area is 149°F (65°C), the zone at which muscle (myofibrillar) proteins begin to harden and become less tender (the positive words for *tough*). Collagen begins to become solubilized (tender) in this temperature range. Thus, cuts that have low amounts of connective tissue (e.g., loin and rib steaks/chops) should not be heated (cooked) beyond this temperature because increased heating will toughen them. Loin and rib cuts can normally withstand fast cooking to this temperature. But cuts that have higher levels of connective tissues (e.g., shoulder, leg, and rough cuts) and cuts from older animals should be cooked longer and more slowly to allow time for the connective tissues to break down (see Chapter 22).

Research to determine how meat tenderness might be controlled and improved has been conducted for many years. Meat tenderness can be measured objectively, i.e., by a machine/instrument, or subjectively by humans serving on a taste panel. Of the several machines/instruments used to measure meat tenderness, the most widely and universally used is the Warner-Bratzler Shear Method.[1] In this system 1 or ½-inch cores are removed from steaks/chops and inserted in the machine to be sheared with a blunt "knife." The force required to shear the core is measured on a dynamometer in pounds or kilograms.

Taste panels evaluate flavor, juiciness, mouth feel, and numerous other traits as well as tenderness.[2] This is known as sensory evaluation. Two types of panels may be used, a large (perhaps 50 to 100), untrained panel and a small (6 to 10 member) trained panel. The details and recommended procedures for meat evaluation have been published by The American Meat Science Association.[3] Sensory evaluation has become increasingly important as the food industry produces more new products. These systems have been computerized to shorten the turnaround time between evaluation and application of results.[4]

Juiciness

Juiciness is closely related to tenderness, i.e., more juicy meat items are also most often the most tender. Meat juices are composed of water and fat plus many water- and fat-soluble compounds. In lean, closely trimmed muscles, the only fat present is in the marbling. Moisture is more easily lost during cooking than fat; therefore, low-fat meat is subject to changes in juiciness. Cooking has a great effect on the juiciness of meat (see Table 24–1). Overcooking will dry out a meat product because overcooking causes loss of water and fat, making meat less tender and less juicy.

---

[1]Bratzler, L. J. 1949. Determining the tenderness of meat by use of the Warner-Bratzler method. Proc. 2nd Ann. Recip. Meat Conf. 117.

[2]Brooks, R. I., and A. M. Pearson. 1989. Odor thresholds of the C_{19}-Δ^{16}-steroids responsible for boar odor in pork. Meat Sci. 25:11.

[3]American Meat Science Association. 1995. Research Guidelines for Cookery, Sensory Evaluation and Instrumental Tenderness Measurements of Fresh Meat. American Meat Science Association and National Live Stock and Meat Board. ISBN: 0-88700-018-5.

[4]Billmeyer, B. A., and G. Wyman. 1991. Computerized sensory evaluation system. Food Technol. 45(7):100.

Table 24–1.   Changes in Meat During Heating

Changes	Initial Heating Period Up to 122°F (50°C) Internal Temperature	122°F to 149°F Internal Temperature	149°F (65°C) and over Internal Temperature
Tenderness/ Texture	Muscle fiber width and length gradually decrease. Collagen fibers "buckle," but tenderizing is negligible.	Muscle fiber width rapidly decreases (from 113°F to 144°F) and extensively shrinks in length (from 131°F to 149°F). Proteins coagulated by 144°F begin to break down when tissue is held 30 minutes or more at 140°F	Muscle tissues harden and become tough. Connective tissue shrinks. Solubization of collagen is about half completed at around 143°F (higher for meat from older animals; see text and Chapter 22).
Juiciness	Muscle loses a slight amount of water.	Water content of muscle decreases as proteins are broken down. Tender meat is juicy at this point.	Beef becomes less juicy as temperature increases.
Flavor and Aroma	Desirable aromas and undesirable odors (if present) are accentuated, and specie characteristic flavor development becomes more intense as meat temperature rises.		
Color (see center color section)	Center of beef becomes bright red. Outside becomes grayish brown as temperature rises Veal begins as pink. Pork begins as a grayish pink color. Pork shoulder steak and uncured ham are a dull rose-red when raw. Raw lamb begins as pinkish red.	Beef cooked rare (140°F internal temperature) is red with a thin brown layer on the surface.	Beef at 160°F is medium pink in color. Beef at 170°F is well-done and brown throughout. The final color of veal is brownish gray. Well-done pork is almost white in color. Pork shoulder steak and uncured ham are grayish brown when fully cooked. Well-done lamb is a grayish brown.
Microorganisms	Most microorganisms are still actively reproducing at these temperatures.	Most yeasts and molds are inactivated by moist heat of 140°F for 10 minutes. *Trichinella spiralis* is destroyed at 137°F.	Pathogenic bacteria are destroyed at 149°F for 12 to 15 minutes. Cooking to 160°F (to 170°F in a microwave) allows a margin of safety. (See text.)

Flavor and Aroma

Meat flavor and aroma is a very complex subject. Meat from each species has its own characteristic flavor and aroma. *(Aroma* is a more pleasant and descriptive term for meat than *odor.)* Sex condition and maturity have a great effect on the flavor and aroma of meat from certain species (see Table 24–1). For instance, meat from market weight (220- to 260-pound) boars may have an offensive, definite "sex" or "piggy" odor. However, meat from ram lambs and bulls of the same relative sexual maturity (not the same chronological age) as market weight boars will not normally exhibit such an odor. As male animals of all species mature sexually, the possibility that flavor problems will occur in meat produced from them increases. Pork fat is more unsaturated than beef and lamb fat (see Chapters 19 and 23) and thus more subject to the development of rancidity and the sex odor condition. Dr. A. M. Pearson (deceased) and coworkers at Michigan State University (see footnote 2) identified five steroids as the compounds responsible for the permeating and objectional "perspiration-like" or "urine-like" odors. Heating definitely accentuates the sex/piggy odor. However, there is nothing unsanitary or unhealthful about mildly odoriferous pork, so it can be used in precooked meat products such as luncheon meats that are not normally heated by consumers. In such instances the odor is undetected.

Animal maturity has an effect on meat flavor aside from the effect of sex. Meat from more mature beef animals has the robust, beefy aroma and flavor, while calf and veal is somewhat bland and lacking flavor intensity. Raw aged beef has a characteristic aged aroma. Heating accentuates the mouth-watering aromas and flavor development of all meat products that were not abused before they were cooked. It is better to experience it, than to try to describe it.

The way meat has been handled, i.e., time and temperature (see Chapter 2), and the conditions under which it has been stored (see Chapter 19) will affect the development of flavors and odors. Warmed-over flavor (WOF) is a condition that is oftentimes cited as being the cause of unpleasant flavor developments and quality deterioration in cooked, stored meat products.[5] If meat is abused, microorganisms will grow and produce putrid odors. Products with these characteristics should be discarded.

---

[5]Brewer, S. M. What is "warmed-over flavor"?? Facts. #04373-11.98. National Pork Producers Council. http://www.nppc.org/facts/flavor.html.

Color

Wide variability exists in the degree of doneness preferred by consumers — especially for beef and somewhat for lamb. This variation in consumer preferences created a need for standardization of endpoint cooking temperatures with well-defined degrees of doneness (see Table 24–1). In 1995, the American Meat Science Association and the National Live Stock and Meat Board published a beef steak color guide illustrating six degrees-of-doneness standards (see color section near the center of the book). These degrees of doneness — *very rare, rare, medium rare, medium, well done,* and *very well done* — define a maximum internal temperature to which each degree should be cooked. The following descriptions are also helpful in categorizing each degree of doneness for broiled steaks.

> *Very rare:* 130°F; red color throughout except for narrow reddish pink layer under meat surface.
>
> *Rare:* 140°F; red in center third; reddish pink to outer surface.
>
> *Medium rare:* 150°F; reddish pink in center third; pink to light brown to outer surface.
>
> *Medium:* 160°F; light pink in center; light brown to outer surface.
>
> *Well done:* 170°F; light brown in center; darker brown to outer surface.
>
> *Very well done:* 180°F; darker brown throughout; dry texture; charred surface.

This guide is valid for fresh and frozen meat, although there are some species differences, as noted in Table 24–1. Also it has been shown that high pH (>6.0) meat will not "cook through" and thus will not show the cooked color described above. The chemistry of color development in cured meats was discussed in Chapter 20 and is summarized in a colored diagram in the center section of this textbook. When meat is heated with nitrite present, the characteristic cured meat pink color will result. If the pink color occurs in cooked hamburger or other uncured beef products, consumers feel it is uncooked because this color is not expected. In addition to the possibility that somehow traces of nitrite left in a grinder or mixer, etc., caused this pink color development, it is also possible that heat caused the breakdown of protein-bound nitrogenous compounds that released nitrite to interact with myoglobin, causing the pink color development.

Understanding why and how heating affects the color of meat necessitates understanding some concepts about color. Visible light comprises wavelengths 380 to 780 nanometers (nm), a very small portion of the electromagnetic radiation spectrum (see Appendix — electricity). Each wavelength is associated with a particular color and interacts with meat to create the perception of color. Color results from three factors: the light source; the selective absorption of light by the meat surface; and the detection of light by the red, green, and blue light receptors in the eye leading to the perception of color in the brain. Because of these factors, sensory evaluation of meat is most often done under controlled lighting to mask the effect of doneness. Lighting of retail meatcases is critical to displaying meat most favorably.

Meat cooked in a microwave oven typically does not brown. The browning or Maillard reaction (named for the French scientist who first described browning) involves the reaction of free amino acids with sugars or other carbohydrates. A meat surface temperature of >350°F is necessary for this reaction to take place. Since microwaves penetrate and produce heat energy throughout the product, the surfaces of meat cuts do not reach browning temperature. Commercial browning agents, sprayed on the meat surface at low levels (<1 percent) just prior to packaging and freezing, cause a rich brown color to develop upon microwave heating. (Microwave cookery will be discussed in detail later in this chapter.)

Thus, the effect of cooking on meat color is not always as straightforward as it seems. Color developments that are difficult to explain crop up periodically in the present-day meat industry. For instance, iridescence, the appearance of green, sometimes orange, color when cooked roasts (sometimes cured as well) are sliced, has been extensively studied by Howard J. Swatland of the University of Guelph.

Nutrient Content Loss

The major loss in weight in cooked meat is due to drip loss composed of moisture and melted fat. Additional moisture is evaporated by the heat. This change in weight alters the percentage of protein, fat, and ash of the cooked meat as compared to the fresh meat (see Chapter 23). This will be affected in large part by the degree of doneness.

The work of Leverton and Odell at Oklahoma State University in 1956–1958[6] showed that evaporation loss during cooking varied from 1.5 to

---

[6]Leverton, R. M., and G. V. Odell. 1958. The Nutritive Value of Cooked Meat. Misc. Publ. MP–49, Oklahoma Agric. Exp. Sta., Oklahoma State University.

Table 24–2. Retention of Thiamin, Riboflavin, Niacin, and Vitamin B$_6$ in Cooked Meat, Poultry, and Fish[1]

	Thiamin	Riboflavin	Niacin	B$_6$
	- (%) -			
Beef				
Braised	45	85	60	45
Broiled	70	90	80	75
Roasted	60	100	75	65
Microwaved	55	90	90	—
Veal				
Braised	45	80	65	50
Panfried	65	90	85	65
Roasted	60	90	80	50
Lamb				
Braised	30	60	60	45
Broiled	60	90	80	—
Roasted	60	90	80	75
Microwaved	50	80	70	—
Pork				
Braised	40	75	80	50
Broiled	70	100	80	65
Roasted	60	95	85	65
Microwaved	70	80	90	—
Poultry				
Roasted	70	85	85	75
Stewed	55	95	60	—
Fish				
Baked or broiled	90	100	95	
Breaded deep fat–fried	85	95	100	
Microwaved	95	100	100	
Shrimp				
Baked	95	100	95	
Broiled	90	75	75	
Deep fat–fried	85	95	95	
Microwaved	95	95	95	

[1]McCarthy, M. A., and R. H. Matthews. 1988. Conserving nutrients in foods. Administrative Report No. 384. USDA Human Nutrition Information Service, Nutrition Monitoring Division, Hyattsville, MD 20782.

— = Lack of reliable data.

54.5 percent, with an average range between 15 and 35 percent. The cooked lean meat with all separable fat removed contained from 5 to 10 percent fat and up to 35 percent protein.

USDA nutritionists McCarthy and Matthews have compiled a comprehensive report entitled *Conserving Nutrients in Foods,*[7] which is an excellent reference. A summary of their report covering vitamin retention in meat, poultry, and fish is found in Table 24–2.

Table 24–2 shows that thiamine (vitamin B_1) is the vitamin most sensitive to cooking, and B_6 is quite sensitive as well. Braising (cooking in liquid — see discussion later in this chapter) is especially damaging because the B–vitamins are water-soluble. The protein of meat is not destroyed by cooking, and only small amounts go into the drippings. Even when meats and poultry are stewed in large amounts of water, not more than 10 percent of the protein passes from meat to broth. Mineral retentions of meat cooked by dry heat or braising generally range from 80 to 100 percent. Most (80 to 100 percent) of the minerals are retained in roasted poultry. Baked, broiled, breaded, deep fat–fried, and microwave-cooked fish have iron, potassium, and zinc retentions between 85 and 100 percent.

Safety

Note the general relationships between heat and microorganism growth indicated in Table 24–1. (See the section entitled "Temperature and Time" in Chapter 2 covering this topic.)

USDA-FSIS has published the following guidelines.

Guidelines for Cooked Beef, Roast Beef, and Cooked Corned Beef[8]

1. Cooked beef and roast beef, including sectioned and formed roasts, chunked and formed roasts, and cooked corned beef can be prepared using one of the following time and temperature combinations to meet either a 6.5-log_{10} or 7-log_{10} reduction of *Salmonella*. The stated temperature is the minimum that must be achieved and maintained in all parts of each piece of meat for at least the stated time:

---

[7]McCarthy, M. A., and R. H. Matthews. 1988. Conserving nutrients in foods. Administrative Report No. 384, USDA Human Nutrition Information Service, Nutrition Monitoring Division, Hyattsville, Maryland 20782.

[8]FSIS-USDA. June 1999. Appendix A to Compliance Guidelines. http://www.fsis.usda.gov/oa/fr/95033F-a.htm

Minimum Internal Temperature		Minimum Processing Time, in Minutes or Seconds, After Minimum Temperature Is Reached	
Degrees Fahrenheit	Degrees Centigrade	6.5-log_{10} Lethality	7-log_{10} Lethality
130	54.4	112 min.	121 min.
131	55.0	89 min.	97 min.
132	55.6	71 min.	77 min.
133	56.1	56 min.	62 min.
134	56.7	45 min.	47 min.
135	57.2	36 min.	37 min.
136	57.8	28 min.	32 min.
137	58.4	23 min.	24 min.
138	58.9	18 min.	19 min.
139	59.5	15 min.	15 min.
140	60.0	12 min.	12 min.
141	60.6	9 min.	10 min.
142	61.1	8 min.	8 min.
143	61.7	6 min.	6 min.
144	62.2	5 min.	5 min.
145	62.8	4 min.*	4 min.*
146	63.3	169 sec.	182 sec.
147	63.9	134 sec.	144 sec.
148	64.4	107 sec.	115 sec.
149	65.0	85 sec.	91 sec.
150	65.6	67 sec.	72 sec.
151	66.1	54 sec.	58 sec.
152	66.7	43 sec.	46 sec.
153	67.2	34 sec.	37 sec.
154	67.8	27 sec.	29 sec.
155	68.3	22 sec.	23 sec.
156	68.9	17 sec.	19 sec.
157	69.4	14 sec.	15 sec.
158	70.0	0 sec.**	0 sec.**
159	70.6	0 sec.**	0 sec.**
160	71.1	0 sec.**	0 sec.**

*Past regulations have listed the minimum processing time for roast beef cooked to 145°F as "Instantly." However, due to their large size, most of these roasts dwell at 145°F, or even at higher temperatures, for at least 4 minutes after the minimum internal temperature is reached. FSIS has revised this time/temperature table to reflect this and emphasizes that, to better ensure compliance with the performance standard, establishments should ensure a dwell time of at least 4 minutes if 145°F is the minimum internal temperature employed.

**The required lethalities are achieved instantly when the internal temperature of a cooked meat product reaches 158°F or above.

[**Authors' note:** See Chapter 2 for explanation of 6.5-log_{10} and 7-log_{10} lethality.]

2. Cooked beef, including sectioned and formed roasts and chunked and formed roasts, and cooked corned beef should be moist cooked throughout the process or, in the case of roast beef or corned beef to be roasted, cooked as in paragraph (3) of this compliance guide. The moist cooking may be accomplished by placing the meat in a sealed, moisture impermeable bag, removing the excess air, and cooking; by completely immersing the meat, unbagged, in water throughout the entire cooking process; or by using a sealed oven or steam injection to raise the relative humidity above 90 percent throughout the cooking process.

3. Roast beef or corned beef to be roasted can be cooked by one of the following methods:

- Heating roasts of 10 pounds or more in an oven maintained at 250°F (121°C) or higher throughout a process achieving one of the time/temperature combinations in (1) above;

- Heating roasts of any size to a minimum internal temperature of 145°F (62.8°C) in an oven maintained at any temperature if the relative humidity of the oven is maintained either by continuously introducing steam for 50 percent of the cooking time or by use of a sealed oven for over 50 percent of the cooking time, or if the relative humidity of the oven is maintained at 90 percent or above for at least 25 percent of the total cooking time, but in no case less than 1 hour; or

- Heating roasts of any size in an oven maintained at any temperature that will satisfy the internal temperature and time combinations of the above chart of this compliance guide if the relative humidity of the oven is maintained at 90 percent or above for at least 25 percent of the total cooking time, but in no case less than 1 hour. The relative humidity may be achieved by use of steam injection or sealed ovens capable of producing and maintaining the required relative humidity.

4. Establishments producing cooked beef, roast beef, or cooked corned beef should have sufficient monitoring equipment, including recording devices, to assure that the time (accuracy assured within 1 minute), the temperature (accuracy assured within 1°F), and relative humidity (accuracy assured within 5 percent) limits of these processes are being met. Data from the recording devices should be made available to FSIS program employees upon request.

Guidelines for Cooked Poultry Rolls and Other Cooked Poultry Products

1. Cooked poultry rolls and other cooked poultry products should reach an internal temperature of at least 160°F prior to being removed from the cooking medium, except that cured and smoked poultry rolls and other cured and smoked poultry should reach an internal temperature of at least 155°F prior

to being removed from the cooking medium. Cooked ready-to-eat product to which heat will be applied incidental to a subsequent processing procedure may be removed from the media for such processing provided that it is immediately fully cooked to the 160°F internal temperature.

2. Establishments producing cooked poultry rolls and other cooked poultry products should have sufficient monitoring equipment, including recording devices, to assure that the temperature (accuracy assured within 1°F) limits of these processes are being met. Data from the recording devices should be made available to FSIS program employees upon request.

Compliance Guidelines for Cooling Heat-Treated Meat and Poultry Products (Stabilization)[9]

It is very important that cooling be continuous through the given time/temperature control points. Excessive dwell time in the range of 130° to 80°F is especially hazardous, as this is the range of most rapid growth for the clostridia. Therefore, cooling between these temperature control points should be as rapid as possible.

1. During cooling, the product's maximum internal temperature should not remain between 130° and 80°F for more than 1.5 hours nor between 80° and 40°F for more than 5 hours. This cooling rate can be applied universally to cooked products (e.g., partially cooked or fully cooked, intact or non-intact meat or poultry) and is preferable to (2) below.

2. Over the past several years, FSIS has allowed product to be cooled according to the following procedures, which are based upon older, less precise data: chilling should begin within 90 minutes after the cooking cycle is completed. All product should be chilled from 120°F (48°C) to 55°F (12.7°C) in no more than 6 hours. Chilling should then continue until the product reaches 40°F (4.4°C); the product should not be shipped until it reaches 40°F (4.4°C).

   This second cooling guideline is taken from the former "Requirements for the production of cooked beef, roast beef, and cooked corned beef," CFR 318.17(h)(10). It yields a significantly smaller margin of safety than the first cooling guideline above, especially if the product cooled is non-intact product. If an establishment uses this older cooling guideline, it should ensure that cooling is as rapid as possible, especially between 120° and 80°F, and monitor the cooling closely to prevent deviation. If production remains between 120° and 80°F more than one hour, compliance with the performance standard is less certain.

---

[9]FSIS-USDA. June 1999. Appendix B to Compliance Guidelines.

3. The following process may be used for the slow cooling of ready-to-eat meat and poultry cured with nitrite. Products cured with a minimum of 100 ppm ingoing sodium nitrite may be cooled so that the maximum internal temperature is reduced from 130 to 80°F in 5 hours and from 80 to 45°F in 10 hours (15 hours total cooling time).

This cooling process provides a narrow margin of safety. If a cooling deviation occurs, an establishment should assume that its process has exceeded the performance standard for controlling the growth of *Clostridium perfringens* and take corrective action. The presence of the nitrite, however, should ensure compliance with the performance standard for *Clostridium botulinum*.

Establishments that incorporate a "pasteurization" treatment after lethality and stabilization treatments (e.g., applying heat to the surface of a cooled ready-to-eat product after slicing) and then restabilize (cool) the product should assess the cumulative growth of *C. perfringens* in their HACCP plans. That is, the entire process should allow no more than 1-log_{10} total growth of *C. perfringens* in the finished product. When employing a post-processing "pasteurization," establishments may want to keep in mind that at temperatures of 130°F or greater, *C. perfringens* will not grow.

Thermometers

Note above the requirements for commercial use of accurate recording devices, especially thermometers for measuring meat temperatures. Accurate thermometers are also necessary for home use. Educational efforts are being coordinated between the USDA and industry to encourage consumers to use thermometers when cooking meat, including hamburgers. Some supermarkets are featuring disposable paper probe thermometers with a plastic-coated tip. A color change in the probe indicates that the internal temperature of a patty has reached 160°F. Although necessary for nearly complete accuracy, inserting any type of temperature-measuring device into a hamburger patty is cumbersome, to say the least.

Another method that provides a close approximation of final internal temperature of any broiled patty, chop, or steak is the moisture bubble technique. Simply record the time when the item is put on the grill; then, when moisture bubbles appear clearly on the surface of the item, record the time, turn the item over, and broil for the same length of time that it took for moisture bubbles to appear on the surface of the first side. Regardless of the state of the item (frozen or thawed) or the type and intensity of heat, this method will give a good estimation of 150°F. Check this technique with a thermometer so that you can use it in the future without thermometer verification.

Recommendations for pork cookery endpoint temperatures based on university research have been made by the National Live Stock and Meat Board and the National Pork Producers Council. Fresh (uncured) chops and roasts from the loin can be cooked to either 160°F (71°C) or 170°F (77°C) internal temperature, depending on personal choice, for optimum palatability. Leg and shoulder roasts and ground pork loaves should be cooked to the higher of these two temperatures to avoid the pink appearance that sometimes remains when these cuts are cooked to the lower temperature. Cooking pork to either 160° or 170°F provides a comfortable safety margin against the remote possibility of trichina infection (see later discussion in this chapter and Chapter 3). Reheating fully cooked hams [by law they must have been heated to 158°F (70°C) during processing, thus making them safe to eat], to an internal temperature of 140°F (60°C) results in optimum palatability (see Tables 24–5, 24–7, 24–9, 24–12, and 24–14 for pork cookery timetables).

Two Cardinal Principles to Remember

1. When meat is being cooked and(or) cooled, it should pass through a temperature range of from 40°F (4°C) to 140°F (60°C) and vice versa in four hours, or preferably less. For instance, it is recommended that cooked turkey be cooled within two hours. *Meat must be kept either cold or hot to be safe.*

2. *After handling raw, uncooked products, never handle fully cooked, ready-to-eat (RTE) meat products with the same equipment and utensils or before washing your hands and changing your apron and other apparel that may contaminate the RTE product. Furthermore, RTE products must be kept completely separate (in another area or room) from fresh, raw products.*

THE COOKING PROCESS

Specific Heat

The unit of heat measurement in the *SI* (System International) is the *calorie*. A *calorie* is the quantity of heat that is necessary to raise the temperature of 1 gram of water from 15° to 16°C. The exact temperature is specified, because the amount of heat required changes with changes in temperature. In the English system, the unit of heat is known as the British thermal unit *(B.t.u.)*, which is the quantity of heat needed to raise the temperature of 1 pound of water from 59° to 60°F. The B.t.u. is much larger than the calorie.

The amount of heat required to raise the temperature of 1 gram of any substance 1°C may be compared with the amount of heat required to raise the temperature of 1 gram of water 1°C. Such a comparison determines the *specific heat* of the substance. The specific heat of a material is a very important basic property that governs its heating characteristics. Thus, the smaller the numerical specific heat rating, the more rapidly it will cook. The specific heat of various cooking media and the elements that compose meat are listed in Table 24–3.

Table 24–3. Specific Heat of Various Media and Elements

Media and Elements	Specific Heat (cal/g)
Hydrogen	3.41
Water	1.00
Oil	0.50
Steam	0.50
Air	0.25
Nitrogen	0.24
Oxygen	0.219
Carbon	0.17
Copper	0.09

By observing the specific heat of the cooking media listed, you can see that cooking in air requires less heat than cooking in steam and in oil, both of which require less heat than cooking in water. Water requires twice as much heat (calories) as fat does to heat the same amount. Another way of expressing this relationship is to consider that if an equal amount of heat is applied to an equivalent weight of fat and water, the fat will heat twice as rapidly as the water. Note that hydrogen has the highest specific heat, nearly 3½ times that of water, while the specific heat of oxygen is only one-fifth that of water. Carbon has the lowest specific heat of the major chemical components of meat. Meat is composed largely of carbon, hydrogen, oxygen, and nitrogen (see Chapter 23). Fats and oils have less hydrogen and more carbon and oxygen per mass than water and protein, so cooking in oil requires less heat, and fatter meat will cook faster. Also, bone-in cuts will cook faster, because the bone is high in calcium and phosphate, both of which conduct heat more rapidly into the meat. Copper has a very low specific heat (0.09), which makes it ideal for use in cooking utensils. An explanation of the basis for computing specific heat and a listing of the specific heats of various foods, including meat, is provided by Mohsenin.[10]

Methods of Heat Transfer

There are basically three methods of heat transfer: convection, conduction, and radiation. Usually, the process of cooking meat involves more than one of these methods.

Convection

This is the simplest method of heat transfer. As gas (air) or liquid (water) is heated in one area of the pan or oven, the hot air or water expands and rises, while the colder air or water sinks, setting up circulating currents. Meat is cooked by the currents of air or water circulating around it. Hot air furnaces and trade winds are examples of convection heat transfer.

Conduction

In this process, the heat, which consists of the vibrations of the molecules of the object being heated, is "handed on" from one molecule to the next. The layer of molecules in contact with the gas fire or electric element is heated first and thus made to vibrate more rapidly. This layer hands on the motion to the adjacent layer because each layer is bound to the adjacent layer by certain cohesive forces. Meat is cooked when it is in direct contact with the heated surface. There are no moving currents set up when heat is transferred by conduction.

Radiation

If you have ever been seated in front of a campfire on a chilly evening and found your face burning while your backside was freezing, you were experiencing heat transfer by radiation. Radiation is the basis for the solar heating

---

[10]Mohsenin, N. N. 1980. Thermal Properties of Foods and Agricultural Materials. Gordon and Breach, Science Publishers, Inc., New York.

of homes with sunlight throughout the year, in spite of freezing temperatures outside. The interesting relationship that exists between solar radiation and convection is described in *Science* magazine.[11] Radiant waves also transfer heat to meat. In conventional ovens and most broilers, infrared electromagnetic waves of radiant energy (see Appendix — electricity) pass from the heat source to the pan and into the meat. Microwaves are another form of radiant energy that comprise a very popular cooking method which will be discussed in detail.

Types of Heating Equipment

Conventional Gas or Electric Ranges

The conventional gas or electric ranges have been used for many years. Nowadays, many fine additions are available for this type of heating equipment. They include self-cleaning, programmable automatic controls, grills, and microwave combinations. Regardless of whether the power is electric or gas, the oven will cook largely by convection, with some heat being transferred by radiation. The burners on top of the stove cook by conduction. Gas will usually heat food quicker than electric power.

Forced Air Convection Ovens

Convection ovens are similar to conventional electric or gas ovens or ranges except that the hot air within is circulated with blowers or fans. This effectively reduces the cooking time because the heat is transferred more quickly. Thus, when compared to regular ovens, convection ovens cook faster and need less energy to operate. Meat will brown uniformly just as it does in regular ovens. Convection ovens have become more popular in recent years for these reasons.

Impingement Ovens

To *impinge* means "to strike or dash especially with a sharp collision." With impingement cooking, jet-like high-pressure streams of hot water or air

---

[11]Chin, G. (ed.). 2000. This week in science. Science. 287:2373. (March 31.)

are applied directly to the meat through nozzles that are fixed perpendicular to the meat surface. Since impingement heating occurs within the confines of an oven, cooking time is lessened considerably and yields are improved. Impingement cooking is becoming quite widely used commercially, an example being the prebrowned, precooked pork breakfast sausage (see later discussion).

Microwave Ovens

Regular ovens are heated by gas or electricity, and the heated air inside cooks the food. In an electronic (microwave) oven, a magnetron (think of a vacuum tube) produces microwaves that are absorbed by the food, causing the molecules within the food to vibrate against each other. These microwaves are a low-level form of radiant energy, just as are radio waves, visible light, and infrared heat (see Appendix — electricity); they all have long wave lengths, so their radiant energy is nonionizing, meaning that it has no cumulative harmful effect on humans. Friction is created by the vibrating molecules, causing a chain reaction from the outside where the microwaves initially hit the meat toward the center, resulting in heat penetration by conduction throughout the food, thus cooking it. Microwaves are reflected by metal (the oven walls), transmitted through glass, paper, pottery, and plastic (the materials the food is cooked in) and absorbed by the food. This explains why only the food gets hot, leaving the oven walls and pan cool.

The primary advantage of microwave cooking is speed; the cooking time is usually cut about in half. This includes the actual cooking time plus the "standing time" that most foods require for heat equalization after being cooked — conventionally cooked food continues to cook for a while after it has been removed from the oven. This standing time is especially important in microwave cooking where uneven heating is likely. (See Table 24–1 and Chapter 2.)

Because there is no heat in the oven itself, meat and other foods that require the hot air of a conventional oven for browning and crisping may not be as satisfactory — unless the speed factor is more important than a crisp, brown exterior. One alternative is to cook the meat in a microwave oven until it is halfway done, then use a regular oven or broiler for quick browning and crisping. Stoves that combine microwave with conventional or convection oven capabilities for automatic browning are being manufactured. Partial cooking in a microwave oven before grilling shortens the grilling and overall cooking time, which besides helping busy cooks, lessens the chance of carcinogens developing from excessive grilling. A microwave oven can also be used to thaw frozen foods quickly (see later discussion).

Microwave cookery is not yet the panacea for the home or the institutional cook, but it does offer tremendous advantages that are recognized by many people. The energy efficiency of microwave ovens is often overlooked by the homemaker. Microwave ovens will typically use 40 to 50 percent less energy to cook a given quantity of food than conventional ovens.

Continuous Cooking Ovens

"There has been a tremendous growth in the development and production of precooked, portion control meat products for foodservice, fast food, and vending machine markets. These products may be whole muscle, ground or finely comminuted red meat, poultry and seafood items. Continuous cooking methods have been developed to achieve superior economy, product quality and uniformity, safety, hygiene, and shelf life. The continuous cooking methods include fryers, broilers, grills and ovens.

Continuous ovens may be classed according to the air flow design. Counter flow ovens have the heat source in one end of the oven and the cold product enters at the other end. The air flow moves in opposite directions to the product flow. In these ovens, high humidities are typically used to achieve a maximum cook yield or minimum cook loss. Browning of the surface may be accomplished by passing the products through or near a direct flame or heating elements at the beginning or end of the process.

Impingement ovens cause browning of products because of very high heat transfer coefficients, on the order of 80 to 150 watts/m²°K. [See Appendix for units.] To put this in perspective, natural convection has heat transfer coefficients on the order of 5 to 10 watts/m²°K. Forced convection over flat surfaces such as in counter flow ovens would be 15 watts/m²°K at air velocities of 150 m/minute to 30 watts/m²°K at air velocities of 1,000 m/minute."[12]

---

[12]Richert, S., Ralston Purina Co., 900 Checkerboard Square Plaza 4RN, St. Louis, MO 63164

Other Types of Heating Equipment

Frying pans are perhaps the most common and oldest type of heating equipment. Iron skillets have retained their popularity over the years, and it has been shown[13] that continued use of iron cookware to cook acetic foods (applesauce and spaghetti sauce, which are often combined with meat) resulted in more iron in the food, a nutritional plus. There are many types and sizes of frying pans, some with coatings to make them easier to clean. Electric frying pans are popular too. A variation of the frying pan is the *wok*, a bowl-shaped cooking utensil used in the preparation of Chinese food, which has become popular in the United States for preparing stir-fry meals. *Griddles* are a form of frying utensil, flat and usually larger with more capacity than frying pans or woks. Cooking takes place by heat conduction in all types of frying appliances.

Crock pots and *Dutch ovens* as well as *kettles* are used to cook meat slowly, which will increase the tenderness of meat from lower-quality animals and less tender cuts. Moisture is often added. It is very important to bring the meat through the dangerous temperature range between 40°F (4°C) and 140°F (60°C) within four hours, or preferably less, to prevent the growth of harmful microorganisms that might be present (see earlier discussion and Chapter 2). Cooking takes place by heat conduction and convection.

Charcoal, gas, and(or) electric *grills* are popular, especially for outdoor cookery or simulated outdoor cookery. Cooking occurs by radiant heat and also by convection. Charring the meat is not recommended because charred meat is not as healthful or as palatable as properly grilled meat.

METHODS OF COOKING MEAT

A national committee has simplified and standardized the cooking of meat into two fundamental methods: *dry heat,* in which the meat is surrounded by dry air in the oven or under the broiler, a method that is adaptable to the preparation of the more tender cuts of meat, and *moist heat,* in which the meat is surrounded by hot liquid or steam, a method suitable for the preparation of the less tender cuts of meat.

---

[13]Cheng, Y. J., and H. C. Brittin. 1991. Iron in food: Effect of continued use of iron cookware. J. Food Sci. 56(2):584

Dry-Heat Methods

Broiling

Broiling is the preferred method for cooking the more tender steaks and chops and cured pork. The meat is directly exposed to heat, either from above or from below, as with outdoor charcoal broiling, or from an oven broiler. (See the discussion on barbecuing/grilling, a form of broiling, later in this chapter.) Panbroiling is a faster and more convenient method than oven broiling for cooking thinner steaks and chops. The seasoning may be applied before or after broiling, the latter being preferable. (See Table 24–4 for oven broiling steps, Table 24–5 for oven broiling times, Table 24–6 for panbroiling steps, and Table 24–7 for panbroiling times.)

When controlled temperatures are possible, a constant broiler temperature of 350° to 400°F gives the best results. Without controls, a "high" setting or continuously "on" will have to be regulated, with the oven door being opened periodically and(or) the height of the rack being adjusted below the heat source.

Table 24–4. Steps for Oven Broiling[1]

---

1. Set the oven regulator for broiling. The broiler may be preheated, but this is not essential.

2. Place meat in a broiler pan on the rack of the broiler. For steaks, chops, or patties ¾ to 1 inch thick, position the pan so the surface of the meat is 2 to 3 inches from the heat. Place thicker cuts 3 to 6 inches away from the heat. Broiler heat may vary, causing a modification of the recommended distances.

3. Broil until the top side is brown. Cured meat and smoked pork should be cooked until lightly brown. The meat should be approximately half-done by the time it is browned on top.

4. Turn and brown the other side. Table 24–7 offers a guide for broiling times. For the most accurate determination of the degree of doneness of a thick steak or chop, a meat thermometer (quick-recovery) should be inserted horizontally into the center of the cut. The steak or chop should be tested shortly before the end of the estimated total broiling time. *Do not leave the thermometer in the meat during broiling.* To test for doneness of a steak or a chop, cut a small slit in the center or close to a bone and observe the color of the meat.

5. Season, if desired, and serve at once.

---

[1]National Live Stock and Meat Board. 1991. The Meat Board's Lessons on Meat. p. 106.

Table 24–5. Timetable for Oven Broiling[1]

Cut	Approx. Thickness	Approx. Weight	Distance from Heat	Approx. Cooking Time		
				Rare	Medium	Well
	(in.)	*(lbs.)*	*(in.)*	- - - - - - - - - *(total min.)* - - - - - - - - -		
Beef						
Chuck shoulder steak[2]	¾	¾ to 1	2 to 3	12	14	—
(boneless)	1	1 to 1¼	3 to 4	14	18	—
Rib steak	¾	11 to 14 oz.	2 to 3	8	12	—
	1	1 to 1½	3 to 4	10	15	—
	1½	1½ to 2	4 to 5	20	25	—
Rib eye steak	¾	7 to 8 oz.	2 to 3	8	12	—
	1	9 to 10 oz.	3 to 4	10	15	—
	1½	12 to 14 oz.	4 to 5	20	25	—
Top loin steak	¾	11 to 14 oz.	2 to 3	8	12	—
	1	1 to 1½	3 to 4	10	15	—
	1½	1½ to 2	4 to 5	20	25	—
Sirloin steak	¾	1¼ to 1¾	2 to 3	10	15	—
	1	1½ to 3	3 to 4	16	21	—
	1½	2¼ to 4	4 to 5	21	25	—
Porterhouse steak	¾	12 to 16 oz.	2 to 3	8	12	—
	1	1½ to 2	3 to 4	10	15	—
	1½	2 to 3	4 to 5	20	25	—
Tenderloin (filet mignon)		4 to 8 oz.	2 to 4	10	15	—
Ground beef patties (4)	½	4 oz. each	3 to 4	—	10[3]	—
	1	5⅓ oz. each	3 to 4	—	14[3]	—
Top round steak[2]	1	1¼ to 1¾	3 to 4	15	18	—
	1½	1½ to 2	4 to 5	20	25	—
Flank steak[2]		1 to 1½	2 to 3	12	14	—
Veal						
Loin/rib chop	1	8 oz.	4	—	14 to 16	15 to 17
	1½	11 oz.	5	—	21 to 23	23 to 25
Arm/blade steak[2]	¾	16 oz.	4	—	14 to 15	15 to 16
Ground veal patties (4)	½	4 oz. each	4	—	8 to 10	10 to 12
Pork						
Loin/rib chop	¾		4	—	8 to 11	11 to 14
	1½		4	—	19 to 22	23 to 25
Boneless loin/	1		4	—	11 to 13	13 to 15
butterflied chop	1½		4	—	16 to 18	18 to 20

(Continued)

Table 24–5 (Continued)

Cut	Approx. Thickness	Approx. Weight	Distance from Heat	Approx. Cooking Time		
				Rare	Medium	Well
	(in.)	(lbs.)	(in.)	- - - - - - - - (total min.) - - - - - - - - -		
Pork (continued)						
Butterflied single loin roast (boneless)		3	4	—	22 to 24	26 to 28
Blade chop (bone-in)	¾		4	—	—	13 to 15
	1½		4	—	—	26 to 29
Shoulder chop (bone-in)	¾		4	—	—	16 to 18
	1		4	—	—	18 to 20
Cubes for kabobs, loin	1		4	—	9 to 11	11 to 13
leg tenderloin	1		4	—	12 to 14	16 to 18
Tenderloin		½ to 1	4	—	16 to 21	20 to 25
Ground pork patties (4)	½	4 oz. each	4	—	—	7 to 9
Country-style ribs	1		5	cook until tender		45 to 60
Back ribs/spareribs			5	cook until tender		45 to 55
Lamb						
Shoulder chop[2]	¾ to 1	5 to 9 oz.	3 to 4	—	7 to 11	—
Loin/rib chop	1	3 to 5 oz.	3 to 4	—	7 to 11	—
	1½	4½ to 7½ oz.	4 to 5	—	15 to 19	—
Sirloin chop	¾ to 1	6 to 10 oz.	3 to 4	—	12 to 15	—
Butterflied leg (sirloin removed, boneless)		4	5 to 7	40 to 46	47 to 53	54 to 60
Leg steak	¾ to 1	11 to 18 oz.	3 to 4	—	14 to 18	—
Cubes for kabobs	1 to 1½		4 to 5	—	8 to 12	—
Ground lamb patties (4)	½	4 oz. each	3	—	5 to 8	—

Cut	Approx. Thickness		Total Cooking Time		
	- - - - - - - - - - (in.) - - - - - - - - - -		- - - - - - - - - (min.) - - - - - - - - -		
Pork, smoked					
Ham slice	½		—	8 to 10	—
	1		—	14 to 16	—
Ham kabobs	1 to 1½		—	8 to 12	—
Loin chop	½ to 1		—	15 to 20	—
Canadian-style bacon	½		—	6 to 8	—

[1]National Live Stock and Meat Board. 1991. The Meat Board's Lessons on Meat. p. 107.

[2]Marinate 6 hours or overnight, if desired.

[3]USDA–FSIS recommends that ground beef pattties be cooked to 160°F.

Table 24–6. Steps for Panbroiling[1]

1. Place the meat in a preheated heavy frying pan or on a griddle. Most meat cuts have enough fat to prevent sticking; however, the cooking surface may be lightly brushed or sprayed with oil for panbroiling very lean cuts. No oil is needed when a nonstick frying pan is used.

2. *Do not add oil or water and do not cover.*

3. For cuts thicker than $\frac{1}{4}$ inch, cook at medium heat, turning occasionally. Cuts less than $\frac{1}{4}$ inch should be cooked at medium-high. Turn more than once to assure even cooking.

4. Remove fat as it accumulates. If fat is permitted to collect, the meat will be fried instead of panbroiled.

5. Cook until brown on both sides. *Do not overcook.* Count on about one-third to one-half the time for panbroiling or griddle broiling as required for regular broiling. To test for doneness of a steak or a chop, cut a small slit in the center or close to the bone and observe the color of the meat.

6. Season, if desired, and serve at once.

[1]National Live Stock and Meat Board. 1991. The Meat Board's Lessons on Meat. p. 110.

Table 24–7. Timetable for Panbroiling[1]

Cut	Approx. Thickness	Range Temp.	Approx. Cooking Time		
			Rare	Medium	Well
	(in.)		- - - - - - - - - *(total min.)* - - - - - - - - -		
Beef					
Rib eye steak	$\frac{1}{2}$	Med.-high	3 to 5	—	—
Top loin steak	$\frac{1}{4}$	Med.-high	2 to 3	—	—
Eye round steak	$\frac{1}{2}$	Med.-high	2 to 4	—	—
	1	Med.	8 to 10	—	—
Tenderloin	$\frac{3}{4}$ to 1	Med.	6 to 9	—	—
Round tip steak	$\frac{1}{8}$ to $\frac{1}{4}$	Med.-high	1	—	—
Sirloin steak (boneless)	$\frac{3}{4}$ to 1	Med.-low to med.	10 to 12	—	—
Top round steak	1	Med.	13 to 16	—	—
Ground beef patties (4)	$\frac{1}{2}$	Med.	—	7 to 8[3]	—

(Continued)

Table 24–7 (Continued)

Cut	Approx. Thickness	Range Temp.	Approx. Cooking Time		
			Rare	Medium	Well
	(in.)		- - - - - - - - - (total min.) - - - - - - - - -		
Veal					
Loin/rib chop	¾ to 1	Med.-low to med.	—	10 to 12	12 to 14
Blade/arm steak[2]	¾	Med. to med.-high	—	13 to 14	14 to 15
Ground veal patties (4)	½	Med.-low to med.	—	6 to 7	8 to 9
Pork					
Loin/rib chop (bone-in)	½	Med.	—	7 to 8	9 to 10
	1	Med.	—	12 to 14	15 to 17
Loin chop (boneless)	½	Med.	—	7 to 8	9 to 10
	1	Med.	—	10 to 12	12 to 14
Butterflied chop	½	Med.	—	8 to 9	10 to 11
	1	Med.	—	12 to 14	15 to 17
Ground pork patties (4)	½	Med.	—	—	7 to 9

[1]National Live Stock and Meat Board. 1991. The Meat Board's Lessons on Meat. p. 110.

[2]Marinate 6 hours or overnight, if desired.

[3]USDA–FSIS recommends ground beef patties be cooked to 160°F.

Panfrying

Panfrying is suitable for small or thin cuts of meat, ground meat, thin strips, and pounded, scored, or otherwise tenderized cuts that do not require prolonged heating for tenderization. The meat is placed in a heavy fry pan and browned on both sides. After browning, the temperature is lowered, and the meat is turned occasionally until done. The fat that accumulates remains in the pan. (See Table 24–8 for panfrying steps and Table 24–9 for panfrying times.)

Bacon is more desirable when panbroiled below its smoke point (290° to 300°F) in its own grease until limp, but not quite crisp, and a light golden brown in color. Draining off the grease as the bacon fries may cause scorching or burning; bacon broiled on a rack about 4 inches below the flame retains more of its original thiamine (vitamin B_1).

Table 24–8. Steps for Panfrying[1]

1. Brown the meat on both sides in a small amount of oil. Some cuts will cook in the fat that comes from the meat. Lean cuts such as cubed steak and liver, and floured or breaded cuts require additional oil on the surface of the frying pan to prevent sticking.

2. If the meat is cooked with a coating, seasoning may be added to the coating ingredients. Otherwise, the meat may be seasoned after it has browned.

3. *Do not cover the meat.* If covered, the meat is braised and will not have the crisp texture inherent to frying. In frying, there may be some sacrifice of tenderness for crispness and flavor.

4. Cook at medium heat, turning occasionally. The temperature should be kept below the smoke point of the fat. The object in frying is to cook the meat through while it is browning. Turning occasionally is necessary to promote even (uniform) cooking.

5. Serve at once.

[1]National Live Stock and Meat Board. 1991. The Meat Board's Lessons on Meat. p. 111.

Table 24–9. Timetable for Panfrying[1]

Cut	Approx. Thickness	Approx. Cooking Time		
		Rare	Medium	Well
	(in.)	- - - - - - - - - - (total min.) - - - - - - - - - -		
Beef				
Flank strips	⅛ to ¼	1	2 to 3	—
Sirloin strips	⅛ to ¼	1	2 to 3	—
Top round strips	⅛ to ¼	1	2 to 3	—
Veal				
Cutlets	⅛	—	3 to 4	—
	¼	—	5 to 6	—
Ground veal patties	½	—	5 to 7	—
Pork				
Loin/rib chop (bone-in)	¼	tender	4	—
	½	—	5 to 6	7 to 8
	¾	—	8 to 10	11 to 13
	1	—	13 to 15	15 to 18

(Continued)

Table 24–9 (Continued)

Cut	Approx. Thickness	Approx. Cooking Time		
		Rare	Medium	Well
	(in.)	- - - - - - - - (total min.) - - - - - - - -		
Pork (continued)				
Boneless loin/butterflied chop	½	—	5 to 6	7 to 8
	¾	—	10 to 13	14 to 16
	1	—	17 to 19	19 to 22
Sirloin chop	¾	—	—	14 to 15
Tenderloin	¼	tender	3 to 4	—
	½	—	4 to 5	6 to 7
Ground pork patties	½	—	—	7 to 9
Loin chop (boneless)	¼	tender	4	
Sirloin cutlet	¼	tender	4	
Cubed steak		tender	6 to 7	
Lamb				
Leg	⅛ to ¼	2	3	—

[1]National Live Stock and Meat Board. 1991. The Meat Board's Lessons on Meat. p. 111.

Stir Frying

Stir frying is similar to panfrying except that the food is almost constantly stirred, and it is usually cooked in a wok. (See Table 24–10 for stir-frying steps.) Because of the smaller, thinner pieces of meat and vegetables and the high temperature, this is a fast cooking method.

Deep Fat Frying

Deep fat frying is generally used for only very tender meat cuts and for brains, sweetbreads, and liver, which are usually covered with a batter composed of eggs, bread crumbs, etc., and then completely immersed in fat and cooked at a temperature of 300° to 325°F. However, several of the many meat snack foods are designed to be deep fat–fried. A popular version of deep fat frying is the *fondue,* in which small individual tender meat pieces (loin or sirloin cuts) are individually skewered onto fondue forks and then immersed

Table 24–10.   Steps for Stir Frying[1]

1. Partially freeze the meat to facilitate slicing.

2. Slice the meat across the grain into thin, uniform slices, strips, or pieces.

3. If desired, marinate the meat while the other ingredients are being prepared.

4. Place the meat in a small amount of hot oil in a wok or a large frying pan. Stir fry about half a pound at a time.

5. Cook at high temperature. Cook the vegetables separately.

6. Slide the spatula under the meat and turn continuously with a scooping motion.

7. Combine the meat and vegetables.

8. Season, if desired, and serve at once.

[1]National Live Stock and Meat Board. 1991. *The Meat Board's Lessons on Meat.* p. 112.

in oil until cooked to the desired doneness. The Institute of Food Technologists published a very complete overview of a symposium on the chemistry and technology of deep fat frying.[14]

Roasting

Roasting is adapted to the preparation of the larger, more tender cuts such as the beef rib, beef sirloin, top round, sirloin tip, veal leg, veal rump, veal loin, veal shoulder, pork loin, pork shoulder, leg of lamb, sirloin lamb roll, loin lamb roll, rolled shoulder of lamb, and fresh or cured pork. The cut (preferably not less than $2\frac{1}{2}$ inches thick) is placed in an open roasting pan with the fat side up so that it will be self-basting. No water is added, and the roast is left uncovered. Lamb roasts are enclosed in the natural fell (see Chapters 7 and 16), which helps retain moisture during roasting. The fell (thin connective tissue layer) should be removed after the roast has been cooked and before it is served. Smoked pork and fresh beef, veal, and lamb are roasted at an oven temperature of 300° to 325°F, whereas fresh pork is roasted at an oven temperature of 325° to 350°F. Beef Wellington is a beef tenderloin that has been rare roasted, spread with *pâté de foie gras* (goose liver

[14]Institute of Food Technologists. 1991. The chemistry and technology of deep-fat frying. Food Technol. 45:67.

paste), covered with pastry, and then baked until brown. (See Table 24–11 for roasting steps and Table 24–12 for roasting times and temperatures.) The most accurate method for determining the proper degree of doneness is the use of the meat thermometer, which is inserted into the center of a roast in such a way that it does not touch a bone and it does not rest in a seam of fat. Boneless or rolled cuts require from 5 to 10 minutes more time per pound to cook than do bone-in cuts. Roasts with long bones require less time than do thick, chunky cuts.

Cooking longer at lower temperatures in roasting has been found to cut down considerably on the shrinkage that occurs at higher oven temperatures. Placing the fat side up or placing loose fat or bacon strips on the top of lean cuts eliminates basting. Searing does not materially assist in keeping the meat juices from escaping, but it does give the meat color and aroma.

Table 24–11. Steps for Roasting[1]

1. Season the meat with spices, herbs, salt, or other seasonings before, during, or after cooking, as desired. The meat should be taken directly from the refrigerator and placed in a cold oven.

2. Place the meat with the fat side up, on a rack (to hold the roast out of the drippings) in an open, shallow roasting pan. The fat on top lets the meat baste itself as it cooks, so further moistening of the surface is unnecessary. In a roast such as a beef rib, a pork loin, or a rack of lamb, the ribs form a natural rack.

3. Insert a meat thermometer so that the tip is in the center of the roast. The tip should not touch bone or rest in fat.

4. *Do not add water and do not cover.* Roasting is a dry-heat method of cooking, and if the pan is covered or water is added, the meat will be cooked by moist heat rather than roasted.

5. Roast in a slow oven — 300°F to 325°F. Preheating the oven is not necessary.

6. Remove the meat from the oven when the thermometer registers five degrees lower than the desired level of doneness. The meat will continue to cook as it stands. The use of a meat thermometer will help avoid overcooking or undercooking. Roasts are easier to carve if they are allowed to "set" for 15 to 20 minutes after they have been removed from the oven so juices can set up.

7. Roast a less tender cut of meat, such as a bottom round roast, for a longer time and at a lower temperature — 250°F to 300°F, although surface drying may occur if there is little fat covering.

8. Serve at once.

[1]National Live Stock and Meat Board. 1991. The Meat Board's Lessons on Meat. p. 104.

Table 24–12. Timetable for Roasting[1]

Cut	Approx. Weight	Oven Temp.	Approx. Cooking Time		
			Rare	Medium	Well
	(lbs.)	*(°F)*	- - - - - - - - - - - *(min./lb.)* - - - - - - - - - - -		
Beef					
Rib roast	6 to 8	300 to 325	23 to 25	27 to 30	32 to 35
	4 to 6	300 to 325	26 to 32	34 to 38	40 to 42
Rib eye roast	4 to 6	350	18 to 20	20 to 22	22 to 24
Boneless rump roast	4 to 6	300 to 325	—	25 to 27	28 to 30
Round tip roast	3½ to 4	300 to 325	30 to 35	35 to 38	38 to 40
	6 to 8	300 to 325	22 to 25	25 to 30	30 to 35
Top round roast	4 to 6	300 to 325	20 to 25	25 to 28	28 to 30
Tenderloin roast, whole	4 to 6	425	45 to 60 (total)		
half	2 to 3	425	35 to 45 (total)		
Ground beef loaf (9" × 5")	1½ to 2½	300 to 325	1 to 1½ hrs. (total)		
Veal					
Loin roast (bone-in)	3 to 4	300 to 325	—	34 to 36	38 to 40
(boneless)	2 to 3	300 to 325	—	18 to 20	22 to 24
Rib roast	4 to 5	300 to 325	—	25 to 27	29 to 31
Crown roast (12 to 14 ribs)	7½ to 9½	300 to 325	—	19 to 21	21 to 23
Rib eye roast	2 to 3	300 to 325	—	26 to 28	30 to 33
Rump roast (boneless)	2 to 3	300 to 325	—	33 to 35	37 to 40
Shoulder roast (boneless)	2½ to 3	300 to 325	—	31 to 34	34 to 37
Pork					
Loin roast					
Center (bone-in)	3 to 5	325	—	20 to 25	26 to 31
Blade loin or sirloin					
(boneless, tied)	2½ to 3½	325	—	—	33 to 38
Top (double)	3 to 4	325	—	29 to 34	33 to 38
Top (single)	2 to 4	325	—	23 to 33	30 to 40
Crown roast	6 to 10	325	—	—	20 to 25
Leg roast					
Whole (bone-in)	12	325	—	—	23 to 25
Top (inside)	3½	325	—	—	38 to 42
Bottom (outside)	3½	325	—	—	40 to 45
Blade Boston roast (boneless)	3 to 4	325	—	—	40 to 45
Lamb					
Leg roast	7 to 9	325	15 to 20	20 to 25	25 to 30
Whole (bone-in)	5 to 7	325	20 to 25	25 to 30	30 to 35
(boneless)	4 to 7	325	25 to 30	30 to 35	35 to 40
Shank half	3 to 4	325	30 to 35	40 to 45	45 to 50
Sirloin half	3 to 4	325	25 to 30	35 to 40	45 to 50

(Continued)

Table 24–12 (Continued)

Cut	Approx. Weight	Oven Temp.	Approx. Cooking Time		
			Rare	Medium	Well
	(lbs.)	(°F)	- - - - - - - - - - - (min./lb.) - - - - - - - - - - -		
Lamb (continued)					
Shoulder roast					
boneless	3½ to 5	325	30 to 35	35 to 40	40 to 45
bone-in, pre-sliced	3½ to 5	325	35 to 40	40 to 45	45 to 50
			- - - - - - - - - - - (total min.) - - - - - - - - - - -		
Pork					
Tenderloin roast, whole	½ to 1	425	—	27 to 29	30 to 32
Ground pork loaf	1 to 1½	350	—	—	55 to 65
Back ribs		325	cook until tender		1½ to 1¾ hrs.
Country-style ribs, 1-inch slices		325	cook until tender		1½ to 1¾ hrs.
Spareribs		325	cook until tender		1½ to 1¾ hrs.

Cut	Approx. Weight	Oven Temp.	Final Thermometer Reading	Approx. Cooking Time
	(lbs.)	(°F)	(°F)	(min./lb.)
Pork, smoked				
Ham (cook before eating)				
Whole (boneless)	8 to 12	300 to 325	160	17 to 21
(bone-in)	14 to 16	300 to 325	160	18 to 20
Half (bone-in)	7 to 8	300 to 325	160	22 to 25
Portion (bone-in)	3 to 5	300 to 325	160	35 to 40
Loin	3 to 5	300 to 350	140	20 to 25
Ham (fully cooked)				
Whole (boneless)[2]	8 to 12	300 to 325	140	13 to 17
(bone-in)[3]	14 to 16	300 to 325	140	12 to 14
Half (boneless)[2]	6 to 8	300 to 325	140	17 to 20
(bone-in)[3]	6 to 8	300 to 325	140	14 to 17
Portion (boneless)	3 to 4	300 to 325	140	20 to 23
Canadian-style bacon (fully cooked)	2 to 4	300 to 350	140	20 to 30

[1]National Live Stock and Meat Board. 1991. The Meat Board Lessons on Meat. p. 105.

[2]Add ½ cup water, cover tightly, and cook as directed.

[3]Cover tightly and cook as directed.

Note: Smaller roasts require more minutes per pound than larger roasts.

Moist-Heat Methods

Braising

Braising (in some regions of the United States referred to as *fricassee*) is the preferred cooking method for the less tender cuts such as the blade and arm roast of beef or steak from the same cuts; the heel of the round of beef; round and flank steak of beef; the steaks of veal, such as round, sirloin, blade, and arm veal steak; veal loin and rib chops; pork chops (both loin and rib); blade and arm pork steak from the pork shoulder; breast of lamb; neck slices of lamb; and lamb trotters.

The meat to be braised is seasoned, dredged with flour (if desired), and browned. Water, meat or vegetable stock, or sour cream (for beef stroganoff) or milk is then added. The kettle or cooking utensil is then covered, and the cut is cooked at a simmering temperature. It may be cooked in the oven, on top of the range, or in a crock pot. This method is commonly called *pot roasting.* (See Table 24–13 for braising steps and Table 24–14 for braising times.)

Table 24–13.   Steps for Braising[1]

1. Brown the meat slowly on all sides in a heavy pan in just enough oil to prevent the meat from sticking. Cuts with sufficient fat require no added fat unless they are coated with flour or crumbs. The browning develops flavor and color. Slow browning adheres to the meat better than quick browning at a high temperature. Drain drippings after browning.

2. Season the meat with salt, herbs, or spices, unless seasoning has been added to the coating.

3. Add a small amount of liquid [as little as 2 tablespoons may be used; however, $\frac{1}{4}$ to $\frac{1}{2}$ cup (maximum) is recommended], such as water, tomato juice, wine, meat stock, or other seasoned liquid.

4. Cover tightly to provide a moist atmosphere for cooking.

5. Simmer at low heat until the meat is tender. This may be done on top of the range or in an oven at 300°F to 325°F. Use the timetable in Table 24–16 as a guide to total cooking time.

6. If desired, make sauce or gravy from the liquid in the pan.

7. Serve at once.

[1]National Live Stock and Meat Board. 1991. The Meat Board's Lessons on Meat. p. 113.

Table 24–14.   Timetable for Braising[1]

Cut	Approx. Thickness	Approx. Weight	Approx. Cooking Time
	(in.)	(lbs.)	(total hrs.)
Beef			
Blade pot roast		3 to 5	1¾ to 2¼
Arm pot roast		3 to 5	2 to 3
Chuck roast (boneless)		3 to 5	2 to 3
Short ribs	2 × 2 × 4		1½ to 2½
Flank steak		1½ to 2	1½ to 2½
Round steak	¾ to 1		1 to 1½
Swiss steak	1½ to 2½		2 to 3
Veal			
Boneless breast (stuffed)		2 to 2½	1¼ to 1½
		4 to 4½	2 to 2½
Boneless breast (rolled & tied)		2 to 3	1½ to 2½
Riblets			50 to 70 min.
Arm/blade steak	¾ to 1		45 to 60 min.
Round steak	¼		30 min.
	½		45 min.
Shoulder roast (boneless)		3½ to 4	2 to 2½
Loin/rib chop	½		8 to 10 min.
	¾ to 1		20 to 25 min.
Pork			
Chops			
Rib, loin	¾		30 min.
Boneless loin	1½		45 min.
Spareribs/back ribs			1½
Country-style ribs			1½ to 2
Tenderloin			
Whole		½ to 1	40 to 45 min.
Slices	½		25 min.
Shoulder steak	¾		40 to 50 min.
Cubes	1 to 1¼		45 to 60 min.

(Continued)

Table 24–14 (Continued)

Cut	Approx. Thickness	Approx. Weight	Approx. Cooking Time
	(in.)	*(lbs.)*	*(total hrs.)*
Pork (continued)			
Leg steak[2] (inside)	⅛ to ¼		5 to 7 min.
Blade Boston (boneless)		2½ to 3½	2 to 2½
(bone-in)		3 to 4	2¼ to 2¾
Sirloin (boneless)		2½ to 3½	1¾ to 2¼
Arm picnic shoulder (boneless)		2 to 4	2½ to 3
(bone-in)		4 to 8	2¼ to 2¾
Lamb			
Breast (stuffed)		2 to 3	1½ to 2
(rolled)		1½ to 2	1½ to 2
Riblets		¾ to 1 each	1½ to 2
Neck slices	¾		1
Shanks		¾ to 1 each	1 to 1½
Shoulder chop	¾ to 1		45 to 60 min.
Lamb for stew	1½		1½ to 2

[1]National Live Stock and Meat Board. 1991. The Meat Board's Lessons on Meat. p. 114.
[2]Marinate 6 hours or overnight, if desired.

Cooking in Liquid

Cooking in liquid can be used for preparing both small (stewing) and large (simmering) pieces of meat. It is suitable for cuts such as beef shank (soup bones), beef plate and brisket, veal shank and breast, lamb shank and breast, pork spareribs, and fresh or smoked pork shoulder (butts and picnics). Beef that is simmered in a small amount of water until tender, then finely chopped, is used to make mincemeat for pies, cookies, and bars, etc.

When small pieces of meat are stewed or large pieces are simmered, the seasoning is added, and the meat is browned (this is optional) in its own or added fat and then covered with hot or cold water; in some cases tomato juice is added. The kettle is then covered, and the meat is cooked at a simmering temperature. The temperature of the liquid should not exceed 195°F because boiling (212°F) toughens meat protein. If vegetables are to be added, they should be added just long enough before the meat is tender so that they will not be overdone. The liquid can be thickened so that it may be served separately or with the stew.

Very tender cuts of beef may be poached, a third method of cooking in liquid traditionally used to cook poultry and fish. The beef cuts are first browned and then placed in a poaching liquid, which may be composed of beef broth or consommé (clarified broth), red wine and herbs, etc. The roast is gently simmered to 130°F because it will continue cooking after it is removed from the heat. (See Table 24–15 for poaching steps, Table 24–16 for cooking in liquid times, and Table 24–17 for variety meats cooking times.)

Table 24–15.   Steps for Cooking in Liquid—Poaching[1]

1. Rub the beef roast with seasonings, if desired. Tie the roast with heavy string at 2-inch intervals, if necessary. Brown on all sides in a small amount of oil.

2. Pour off excess drippings.

3. Cover with liquid. Season with additional ingredients if desired.

4. Bring to a boil. Reduce the heat, cover and simmer until the internal temperature of the roast registers 10 degrees below the doneness desired.

5. Remove the roast and place on a carving board. Cover tightly with plastic wrap or aluminum foil and allow it to "stand" 10 minutes before carving. Remove the string from the roast and then carve into thin slices.

6. Serve at once.

[1]National Live Stock and Meat Board. 1991. The Meat Board's Lessons on Meat. p. 115.

Table 24–16.   Timetable for Cooking in Liquid[1]

Cut	Approx. Thickness	Approx. Weight	Approx. Cooking Time
	(in.)	(lbs.)	(total hrs.)
Beef			
Fresh or corned beef brisket		2½ to 3	2½ to 3
Shank crosscut		¾ to 1¼	2 to 3
Beef for stew			1¾ to 2¼

(Continued)

Table 24–16 (Continued)

Cut	Approx. Thickness	Approx. Weight	Approx. Cooking Time
	(in.)	*(lbs.)*	*(total hrs.)*
Veal			
Boneless breast	1		1¼ to 1½
Shank crosscuts	1½		1 to ¼
Veal for stew	1 to 1½		45 to 60 min.
Pork			
Spareribs			2 to 2½
Country-style ribs			2 to 2½
Cubes	1 to 1¼		45 to 60 min.
Lamb			
Lamb for stew	1 to 1½		1½ to 2
Pork, smoked			
Ham, country or		10 to 16	4½ to 5
country-style, half		5 to 8	3 to 4
Arm picnic shoulder		5 to 8	3½ to 4
Shoulder roll		2 to 4	1½ to 2
Hocks			2 to 2½

[1]National Live Stock and Meat Board. 1991. The Meat Board's Lessons on Meat. p. 116.

Table 24–17. Timetable for Cooking Variety Meats[1]

Kind	Broiled	Braised[2]	Cooked in Liquid
	(min.)		
Liver			
Beef			
3- to 4-pound piece		2 to 2½ hrs.	
Sliced		20 to 25 min.	
Veal (calf), sliced	8 to 10		
Pork			
Whole (3 to 3½ pounds)		1½ to 2 hrs.	
Sliced		20 to 25 min.	
Lamb, sliced	8 to 10		

(Continued)

Table 24–17 (Continued)

Kind	Broiled	Braised[2]	Cooked in Liquid
	(min.)		
Kidney			
Beef		1½ to 2 hrs.	1 to 1½ hrs.
Veal (calf)	10 to 12	1 to 1½ hrs.	¾ to 1 hr.
Pork	10 to 12	1 to 1½ hrs.	¾ to 1 hr.
Lamb	10 to 12	¾ to 1 hr.	¾ to 1 hr.
Heart			
Beef			
Whole		3 to 4 hrs.	3 to 4 hrs.
Sliced		1½ to 2 hrs.	
Veal (calf), whole		2½ to 3 hrs.	2⅓ to 3 hrs.
Pork		2½ to 3 hrs.	2½ to 3 hrs.
Lamb		2½ to 3 hrs.	2½ to 3 hrs.
Tongue			
Beef			3 to 4 hrs.
Veal (calf)			2 to 3 hrs.
Pork			
Lamb			
Tripe			
Beef	10 to 15[3]		1 to 1½ hrs.
Sweetbreads	10 to 15[3]	20 to 25 min.	15 to 20 min.
Brains	10 to 15[3]	20 to 25 min.	15 to 20 min.

[1]National Live Stock and Meat Board.

[2]On top of range or in a 300°F to 325°F oven.

[3]Time required after precooking in water.

Microwave Cookery

When meat is prepared using microwave cookery methods, the recommended practice is to cook the cut (steak, chop, or roast) in a plastic cooking bag at a low to medium power setting. If the microwave oven is not equipped with a rotating turnstile device, the cooking dish should be rotated 180°F halfway through the cooking process. The use of a non-melting cooking bag is especially important in cooking pork in a microwave oven. The bag creates a steamy cooking environment that helps insure greater temperature uniformity of the meat cut being cooked. Without the cooking bag, "hot" and "cold" spots can occur in the steak, chop, or roast. It has been demonstrated that viable trichinae and other harmful microorganisms can survive in these "cold" pockets when pork cuts register an overall internal temperature of 160°F. Research at Iowa State University[15] has shown that *no viable trichinae survive when pork is microwave cooked to an internal temperature of 160°F or higher in a cooking bag (not sealed) and at a low to medium power setting.*

Table 24–18, from the International Microwave Power Institute,[16] provides guidelines for the microwave cooking of meat. The National Live Stock and Meat Board[17] also has many excellent colorful materials that have up-to-date recipes and recommendations for cooking meats in microwave ovens.

HANDLING FROZEN MEAT IN THE HOME AND THE INSTITUTION

Below 0°F temperatures are best for holding frozen meat, the lower the better. Frozen meat that has been thawed under refrigeration need not be used immediately as is commonly recommended. Repeated tests have shown that such meat will keep as long as fresh meat if it is properly refrigerated.

Refreezing meat does not materially affect its quality. Tests were made in which beef, properly wrapped in a good grade of locker paper, film, or aluminum foil, was thawed in the unopened package, was refrozen, and later was rethawed and held in that condition at 38°F in a household refrigerator for an additional week — the meat was still in excellent condition when it was cooked. Obviously, the results of this extreme test do not mean that you should become careless; what they do suggest is that you need not become panicky about using all the meat in a package that has been properly thawed if it is more than is needed for a meal. Rewrap it, refreeze it, and use it at another time.

---

[15]Microwave Cooking of Pork. 1984. Proc. of the Meat Industry Res. Conf. National Live Stock and Meat Board, Chicago.

[16]International Microwave Power Institute. 1987. The Microwave Cooking Handbook. Clifton, Virginia 22024.

[17]National Live Stock and Meat Board. (Contact the American Meat Science Association, Savoy, IL 61874 www.meatscience.org)

Table 24–18. Microwave Chart for Meat[1]

Meat Item	Power Level	Time[2]	Comments
Beef			
Ground, crumbled	High (100%)	5 min.	Arrange meat in a ring in an all plastic sieve or colander. Place the sieve in a bowl to collect drippings. After 5 minutes, stir and break up. Let stand, covered, 2 to 3 minutes.
Ground, patties			
1, 4 oz.	High (100%)	1¼ min.	Form into doughnut shapes. Top with browning agent, if desired. Place patties on a microwaveable roast rack. Cover with waxed paper. Invert patties during cooking. Let stand 1 minute.
4, 4 oz.	High (100%)	3½ min.	
Ground, meat loaf			
1 lb.	High (100%)	10 min.	Form loaf into round or doughnut shape. Place loaf in a microwaveable baking dish. Top with waxed paper. Rotate dish during cooking. Let stand 10 minutes.
1½ lbs.	High (100%)	14½ min.	
1½ lbs.	Med. (50%)	24 min.	
Ground, meatballs			
16, ½-inch balls (1 lb.)	High (100%)	10 min.	Place meatballs in a circle in a microwaveable baking dish, leaving the center empty. To add color, top with a sauce. Cover meatballs with waxed paper. Rotate dish during cooking.
Roasts, pot roasts (2 to 3½ lbs.)			
Arm	Med.-low (30%)	27 to 33 min./lb.	Place meat in a microwaveable dish. Cover with browning agent if desired. Add ¼ cup water (onion if desired) and cover. Invert meat midway through cooking and rotate dish two or three times. Let stand 10 to 15 minutes.
Bone-in blade	Med.-low (30%)	23 to 29 min./lb.	
Boneless blade	Med.-low (30%)	32 to 39 min./lb.	
Boneless shoulder	Med.-low (30%)	41 to 48 min./lb.	
Roasts, high-quality (3 to 4 lbs.)			
rib eye; boneless top round, top loin, tip (cap off), rump and cross rib; rib roast cut from small end	Med. (50%)	11 to 13 min./lb. (rare–med.)	Place meat on a microwaveable rack. Cover with waxed paper. Shield edges if needed. Rotate rack one-quarter turn every 20 minutes. Invert meat halfway through cooking. Let stand 15 to 20 minutes.
	Med.-low (30%)	18 to 22 min./lb. (rare–med.)	

(Continued)

Table 24–18 (Continued)

Meat Item	Power Level	Time[2]	Comments
Beef (continued)			
Roasts, corned beef (3 to 4 lbs.)	Med.-low (30%)	2 to 2½ hrs.	Place meat in a microwaveable baking dish. Add ½ cup water (onion if desired) and cover. Invert meat halfway through cooking and rotate dish one-quarter turn every 30 minutes. Let stand 10 minutes.
Pork			
Bacon			
1 slice	High (100%)	1 min.	Arrange slices in a single layer on a microwaveable roast rack. Cover slices with paper towel to prevent spattering. For crisper bacon, let stand a few minutes.
4 slices	High (100%)	3 min.	
Canadian-style bacon			
whole, 1 lb.	Med. (50%)	12 to 15 min.	Place bacon in a microwaveable baking dish. Add 2 tablespoons water, cover with plastic wrap, and vent. Invert meat and rotate dish one-quarter turn during cooking. Let stand 10 minutes.
1 lb., ¼-in. slices	Med. (50%)	5 to 7 min.	Overlap or stack slices in a microwaveable baking dish. Rotate one-quarter turn during cooking. Let stand a few minutes. Cover with waxed paper.
Ground, crumbled (75% lean), 1 lb.	High (100%)	6 min.	Arrange in a ring in a microwaveable sieve or small colander. Place sieve in a microwaveable bowl. Cover meat with plastic wrap. (Covered microwaveable utensil may also be used.) Stir to break up pork halfway through cooking period; stir again upon removal from oven.
Cubes, boneless shoulder			
½-in. pieces	Med. (50%)	24 min./lb.	Place in 10- × 16-inch oven cooking bag or covered microwaveable utensil with 1 cup water. Close bag loosely with string or ½-inch strip cut from open end of bag. Place bag and contents in microwaveable baking dish. Rearrange (without opening bag) halfway through cooking period.
1-in. cubes	Med. (50%)	28 min./lb.	

(Continued)

Table 24–18 (Continued)

Meat Item	Power Level	Time[2]	Comments
Chops, fresh (5 to 7 oz.; ¾- to 1-in. thick), boneless top loin, center-cut rib, center-cut loin			
1 chop 4 chops	Med.-low (30%) Med.-low (30%)	20 min./lb. 18 min./lb.	Place in microwaveable dish; cover tightly. Invert chops halfway through cooking period and rotate dish one-half turn.
Country-style ribs (3 to 3½ lbs.; 1-in. thick portions)	Med. (50%)	14 min./lb.	Place in 14- × 20-inch oven cooking bag with 1 cup water. Close bag loosely with string or a ½-inch strip cut from open end of bag. Place bag and contents in a microwaveable baking dish. Invert or turn ribs, rearranging carefully, halfway through cooking period and rotate dish one-half turn, if desired.
Back ribs or spareribs (3 to 3½ lbs.; 3 or 4 ribs/portion)	Med. (50%)	14 min./lb.	Place ribs, bones down, in a 14- × 20-inch oven cooking bag. Add 1 cup water. (Continue same as for country-style ribs.)
Roasts, boneless (3 to 3½ lbs.; 4-in. diameter), loin blade, loin sirloin, Boston shoulder	Med.-low (30%)	22 min./lb.	Place in a 10- × 16-inch cooking bag in a microwaveable baking dish. Sprinkle meat with browning agent, if desired. Close bag loosely with string or a ½-inch strip cut from open end of bag. Invert or turn roast halfway through cooking and rotate dish one-half turn. Cook until roast temperature reaches 170°F. Wrap bag tightly in foil. Let stand 10 minutes.
Roasts, bone-in (3 to 3½ lbs.), loin center (rib or loin), shoulder, blade Boston	Med.-low (30%)	20 min./lb.	Same as boneless roasts.
Pork sausage links, smoked (2)	High (100%)	1½ to 2 min.	Add ½ cup water, cover with plastic wrap, and vent.
Pork sausage links, precooked (2 or 3)	High (100%)	2 min.	Cook in preheated browning skillet.
Pork sausage, continuous link, fully cooked, smoked, 1 lb.	High (100%)	3 min.	Uncovered.

(Continued)

Table 24–18 (Continued)

Meat Item	Power Level	Time[2]	Comments
Frankfurters (2, in buns)	High (100%)	1½ min.	Wrap in paper napkin or paper towel to absorb moisture.
Smoked rib or loin chops			
2, 1 in.	High (100%)	7 to 8 min.	Place chops on a microwaveable roast
4, 1 in.	High (100%)	12 to 13 min.	rack. Top with sauce or glaze to hold
2, ¾ in.	High (100%)	5½ to 6½ min.	in moisture. Cover with waxed paper.
4, ¾ in.	High (100%)	8½ to 10 min.	Invert chops halfway through cooking and rotate dish one-quarter turn.
Ham, fully cooked whole, 3 lbs.	Med. (50%)	10 to 15 min./lb.	Place ham on a microwaveable roast rack. Place plastic wrap over cut surface of ham. Shield edges. Invert ham halfway through cooking. Rotate rack one-quarter turn two or three times during cooking. Let stand 15 minutes.
slices, 2-in. thick (3 to 4 lbs.)	Med. (50%)	20 to 35 min.	Place slices in a microwaveable baking dish. Cover with plastic wrap; vent. Rotate dish one-quarter turn several times during cooking.
slices, ¾ to 1 in. thick	Med. (50%)	14 min.	Place slices in a microwaveable baking dish. Cover with plastic wrap; vent. Invert ham slices halfway through cooking. Rotate dish one-quarter turn twice during cooking. Let stand 3 minutes.
slices, ¼ in. thick (1½ lbs.)	Med. (50%)	10 min.	Overlap slices in a microwaveable baking dish. Add ¼ cup water. Cover with plastic wrap; vent. Rotate dish one-quarter turn during cooking. Let stand 10 minutes.
Lamb			
Roast, bone-in center leg (3 to 5 lbs.)	Med.-low (30%)	14 to 17 min. (rare) 18 to 21 min. (med.) 22 to 26 min. (well)	Place roast on a microwaveable roast rack. Cover with waxed paper. Rotate rack one-quarter turn during cooking. Shield edges if needed. Let stand 10 minutes.

[1]The Microwaveable Cooking Handbook. 1987. International Microwave Power Institute. Clifton, VA 22024, pp. 38–42.

[2]Cooking times are based upon a full power output of approximately 650 watts.

Every time frozen meat is thawed it will lose some of the meat juices. If the position of the thawed meat package is reversed (turned over) when the package is put back in the zero compartment for refreezing, these juices will be reabsorbed to a large extent.

Freezer burn on meats is caused by considerable dehydration (moisture loss) of the meats or parts thereof, due to a poor grade of wrapping paper, improper wrapping, or holes in the paper (see Chapter 19). In badly dehydrated meat, water and other condiments can be added in the cooking process to replace that which was lost, but the meat has lost considerable flavor and tends to be tough and stringy. It is not the lean meat but the fat that changes in flavor in 0°F storage. The oxygen in air will combine with unsaturated fats and break them down into free fatty acids, ketones and aldehydes, giving the fats a stale, rancid flavor. This flavor is, in part, absorbed by the lean. That is the reason for using a good grade of wrapping paper, one that is moisture–vapor proof, and employing the drugstore method of wrapping or vacuum-packaging to exclude the air. It is rather foolish to go to all this trouble, using expensive paper and tape and taking valuable time to do a good job, and then fling the package into a basket and rip it.

When an unthawed cut is roasted, additional time that is equal to one-third to one-half the recommended time for unfrozen cuts should be allowed (see Table 24–19). If a meat thermometer is used, it should be inserted after the meat is partially cooked and the frost is out of the center.

If thawing meat before cooking is preferred, refer to Table 24–19 for a timetable for defrosting frozen meat. Thawing in a refrigerator is the recommended practice. However, a microwave oven can be used to temper or thaw frozen foods quickly, and later they can be cooked on a regular range. The microwave oven is especially good for large, slow-thawing roasts and poultry. Tempering, rather than completely thawing, is raising the temperature of solidly frozen food to just below the freezing point of water, e.g., to 25° to 28°F. Compared to conventional tempering rooms used for large volumes of products in an industrial situation, the use of microwave energy to temper frozen food decreases the time requirement from days to minutes. Because the product will not have to be unpacked, microbial growth and product spoilage on the outside of each package is minimized. In tempering and thawing, the external surfaces of the food product should not be allowed to remain at temperatures above freezing for sufficient time to encourage microbial growth. It is better to plan ahead. Precooked frozen foods can be quickly thawed or reheated in a microwave; this method has been successfully used for years in many restaurants.

Table 24–19. Timetable for Defrosting Frozen Meat[1]

Meat	In Refrigerator
Large roast	4 to 7 hrs./lb.
Small roast	3 to 5 hrs./lb.
1-inch steak	12 to 14 hrs.

[1]National Live Stock and Meat Board.

In a study of cooking times, yields, and temperatures of frozen roasts conducted at the University of Illinois and cosponsored by the National Association of Meat Purveyors and the National Live Stock and Meat Board, the final conclusion was that "Roasts cooked from the frozen state yield as much as roasts partially or completely thawed prior to cooking."[18] It was further observed, after 860 roasts weighing almost 4 tons had been cooked, that roasting from a frozen state required between 1.3 and 1.45 times as long to cook as from a chilled state.

COOKING POULTRY

Preparation

Frozen birds should be thawed to remove the giblets and the internal fat. The inside of such a bird should be sprinkled with salt. Stuffing should be placed in the body cavity of the bird (avoid packing), and the ends of the drumsticks should be drawn down against the opening and tied with a cord that laps over the back of the tail and the ends of the legs. (The legs of all birds must be tied close to the body to avoid overcooking and drying.) Ducks, pheasants, and guineas will require stitching to close the opening. The loose skin over the crop may be filled with the stuffing and the edge fastened with a cord.

The wing tips should be folded back on the wings. The breast and legs are rubbed with butter or margarine, sprinkled with salt, preferably celery salt, and if desired, dusted lightly with flour. Ducks and geese need no added fat.

---

[18]Shoemaker, B. E., B. C. Breidenstein, and D. S. Garrigan. 1969. The Study of the Different Handling Methods for Frozen Roasts in Institutional Food Service. National Live Stock and Meat Board, Chicago.

Roasting

The procedure will vary with the age of the bird. An old bird should be steamed or braised for 1½ to 2 hours. The bird is placed breast up on the rack in the roaster, the bottom of the roaster is covered with hot water, the lid is placed on the roaster, and then the roaster is put in an oven to braise at a temperature of 250° to 275°F. It may be necessary to add water several times during the braising process. Some cooks prefer to stuff the bird before braising it, while others would rather add the stuffing after the braising period. After this steaming period, the lid and water are removed, and the oven temperature is adjusted to 325°F for the remainder of the roasting.

Young birds or those having a flexible tip on the rear end of the breastbone are placed on the roasting rack, with no lid and no water added. The position of the bird in the conventional method of roasting is with the breast up. Another practice is to place the bird on its side or squarely on its breast, using a V-shaped rack for support. Basting at 45-minute intervals with pan drippings or using an aluminum foil tent (loosely covered) will help to minimize drying. The tent should be removed near the end of the roasting time to facilitate browning and to allow undesirable volatiles to escape. Ducks and geese are self-basting; the skin should be pricked with a fork during the roasting process to allow some of the fat to drain. Time guidelines for roasting young birds are given in Table 24–20.

Table 24–20. Timetable for Roasting Young Birds

Kind	Weight	Oven Temperature	Approximate Total Cooking Time
	(lbs.)	(°F)	(hrs.)
Chicken	4 to 5	350	1½ to 2
Duck	5 to 6	350	2 to 2½
Goose	10 to 12	325	3 to 4
Guinea	2 to 2½	350	1½
Turkey	6 to 9	325[1]	2½ to 3[1]
	10 to 13	325	3 to 4
	14 to 17	325	4 to 5
	18 to 25	325	6 to 8

[1]Temperature recommended by the National Turkey Federation. Usually, 20 minutes per pound is required.

A bird cooked according to a time schedule may or may not be done because of various conditions such as age, weight, oven temperature, and air circulation. Some of the indications of doneness are slightly shrunken flesh beneath the skin, flexibility of the leg joint, and absence of any pink juice when the flesh of the thigh is pricked with a fork or a skewer. A meat thermometer placed in the most fleshy portion of the breast will probably give the most accurate indication of doneness. Large birds with thick thighs may have the drumsticks released from the cord that binds them to the body when the roasting period is about three-fourths completed. This permits the heat to circulate more readily around the thick, meaty thighs.

The giblets are simmered to tenderness before they are added to the gravy. The neck may be cooked with the giblets. Chicken giblets require about 1 to 1½ hours to cook and turkey giblets 2 to 3 hours. The liver needs only about 15 minutes of cooking and should be added during the last period. If the giblets are to be incorporated with the stuffing, they should be cooked the previous day.

Stuffing

The ingredients that furnish the bulk of a stuffing are starchy in nature and consist of bread crumbs, boiled rice, or mashed potatoes. To get added richness of flavor, melted butter or some melted fat taken from the body of the bird is added. The seasoning vegetables consist of celery, parsley, and onions. The spices or herbs that are favored are thyme, sweet marjoram, pepper, and sage. Other ingredients that add variety to a stuffing are oysters, nuts, mushrooms, dried apricots or prunes, sausage, raisins, diced salt pork that has been fried crisp, and sliced apples.

Dry stuffing is made of medium dry crumbs without milk or water added. Moist stuffing is made with crumbs with milk or water added or with a base of boiled rice or potatoes.

A 4- to 5-pound bird will require about 4 cups of crumbs, while a 14- to 15-pound turkey will require 10 to 12 cups of crumbs. An ordinary 1-pound loaf of bread (two to four days old) will make approximately 4 to 5 cups of crumbs. If boiled rice is used, it should be 1 cup less than the bread crumbs because the rice will swell.

Oyster Stuffing

(12-lb. turkey)

1½ pt. oysters	1 tbsp. chopped onion
8–10 cups bread crumbs	1 to 2 tsp. celery salt
¾ cup butter or other fat	½ tsp. savory seasoning
⅛ cup chopped parsley	

Heat the oysters at a boiling temperature for several minutes, then drain. Cook the parsley and onion for several minutes in the melted butter or fat, and then add the cooked mixture, the drained oysters, and the seasonings to the bread crumbs.

Sausage Bread Stuffing

(12–lb. turkey)

1 lb. sausage	1 cup diced celery
7 cups bread crumbs	4 tbsp. chopped parsley
2 eggs	2 tbsp. diced celery
1 cup milk	

Panfry the sausage until brown, and then drain off the fat. Beat the eggs slightly, and then add hot milk to the egg mixture. Pour it over the remaining ingredients.

Savory Stuffing

(12-lb. turkey)

8–10 cups bread crumbs	¾ cup butter or other fat
1 pt. chopped celery	1–2 tsp. salt
1 small onion chopped	1–2 tsp. savory seasoning
½ cup chopped parsley	pepper to taste

Cook the celery, parsley, and onion in the melted butter or fat for several minutes. Add the cooked mixture, the dry seasoning, and the condiments to the bread crumbs and mix. Add nuts, if desired. If chestnuts are added, boil them in water for 15 minutes, and then remove the shell and brown skin while they are still hot.

Broiling

Young, plump birds split into halves or quarters are the only ones suitable for this purpose. The distance of the broiler rack from the flame or heating element will vary with different ovens, but a temperature of 375° to 400°F is most desirable. This necessitates having the broiler rack 3 inches from the heat in some ovens and 4 to 7 inches away in others. The speed at which the bird browns will govern the distance to use. A 2-pound broiler should cook in 35 to 45 minutes.

The bird should be coated with melted fat, seasoned with salt and pepper, and sprinkled with flour if desired. If a barbecue sauce is to be used, it should be applied during the last half of the cooking period to avoid excessive browning before the meat is fully cooked. The skin side should be away from the broiler heat, and the bird should be turned several times as it browns. A good practice is to partly cook the bird in a 350°F conventional oven or in a microwave oven and then broil. Small 3- to 5-pound turkeys, squabs, guineas, and ducklings are broiled in the same manner.

Frying

Panfrying is widely practiced. Various methods are used and all have enthusiastic supporters. One method consists of steaming the disjointed bird in a 300°F oven until practically tender and then dipping each piece in a beaten egg and rolling it in bread, cracker crumbs, or corn meal. The steamed bird is then placed in a thick skillet that contains melted butter or ½ inch of melted fat and browned quickly. Salt and pepper are then added.

Another method is to bread each piece, put the thickest pieces in the pan first, and add sufficient fat to come up around each piece. The pan is then covered to avoid spattering, and the pieces are cooked at moderate or medium heat until they are brown. As each piece browns, it is turned over. This requires about 20 to 25 minutes for chicken. Final cooking may be done in a moderate oven (325°F).

Stewing

Stewing produces a more tender product. The flavor is due in large part to the type of bird used for stewing. The old bird is high in flavor, as is the case with all meat animals. Flavor intensity increases with the age of the animal or bird.

CARVING POULTRY

The conventional method of carving pursued by most people is done with the bird on its back. Another method is to carve the bird as it rests on its side. With the bird on its back, turn the platter so that the legs of the bird point toward you. Grasp the end of a leg with the fingers of your left hand and cut between the leg and the body (see Figure 24–1). Pull the point of the knife through the joint, and sever the skin between the leg and the back. Lift the leg to a second plate, if the platter space is limited, and separate the drumstick from the second joint or thigh. Slice the dark meat from the second joint and also from the drumstick if it is too large for a single serving. Remove the wing by cutting around the area where it appears to join the body, and force it toward the back.

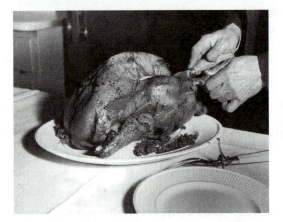

Fig. 24–1. Carving poultry (breast up). Removing the leg. (Photo by Peter Killian; courtesy USDA)

If the bird is on its side, remove the wing between the first and second joint, leaving the second joint attached to the bird. Remove the drumstick, leaving the thigh attached to the body. This ends the most difficult part of the carving operation. Disjointing a bird can be an embarrassing experience for the carver.

If you stand to carve a bird resting on its back, slicing the breast meat by slicing down and away from you is the most comfortable method. Most experts recommend placing the fork squarely across the breastbone toward the end of the keel (see Figure 24–2). This places the left hand, which is steadying the bird with the fork, in a position that does not interfere with the right hand.

If you are sitting, the breast of the bird must be next to you, in which case your right hand is working under your left arm. To avoid this unnatural posi-

Fig. 24-2. Carving poultry (breast up). Slicing the breast meat. (Photo by Peter Killian; courtesy USDA)

tion, point the front of the bird toward you so you can place the fork into the opposite breast several inches below the keel, and slice as you did in the standing position. The slices of white breast meat can be arranged opposite the cuts of dark meat on a separate plate. When sufficient servings have been made or when one side of the bird has been carved, shove the platter away from you, and put the plates in its place. Drop a spoonful of stuffing on each plate and a portion of white and dark meat on top of or beside it. Some carvers prefer to serve as they remove a slice of breast meat, but this slows up the serving because it means extra handling of tools between each operation.

Carving a bird on its side makes the breast easier to carve and eliminates handling the second wing joint and the thigh as separate pieces (see Figures 24-3 to 24-10).

The breast of duck or goose is too shallow to be carved in the same manner as turkey, chicken, or guinea. Instead, long, thin slices are cut with the grain and parallel to the ridge or keel bone and then into portions across the grain if they are too large. Another method consists of lifting the entire breast

Fig. 24-3. When carving a bird resting on its side, remove the wing portion first. Grasp the wing tip firmly between the thumb and fingers, lift up, and sever between the first and second joints. Drop the wing tip and first-joint portion to the side platter. Leave the second joint attached to the bird.

Fig. 24–4. Remove the drumstick. Grasp the end of the drumstick and lift it up and away from the body, disjointing it at the thigh; then transfer the drumstick to the side platter for slicing the meat. Leave the thigh attached to the bird.

Fig. 24–5. Slice the drumstick meat. Hold the drumstick upright, cut down parallel with the bone, and turn the leg to get uniform slices.

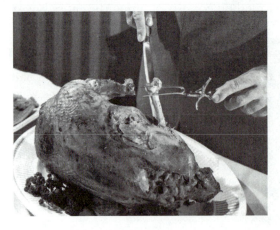

Fig. 24–6. Slice the thigh meat. Anchoring the fork where it is most convenient to steady the bird, cut slices parallel to the body until the bone is reached, and then transfer the slices to the side platter. Run the point of the knife around the thigh bone, lift up with the fork, and using either the fork or your fingers, place the bone on the side platter. Then slice the remaining thigh meat.

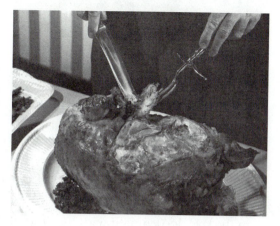

Fig. 24–7. Remove the "oyster," the choice dark muscle above the thigh and adjoining backbone that resembles the shape of an oyster. Use the point of the knife to lift it out of its spoon-shaped cradle.

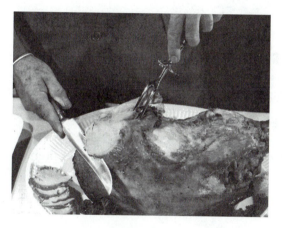

Fig. 24–8. Cut short breast slices until the wing socket is exposed. Sever the second joint of the wing and transfer it to the side platter. Slice the meat in the same manner as you did the drumstick meat.

Fig. 24–9. Continue slicing the breast meat. Steady the bird with a fork. Cut thin slices of breast meat until enough slices have been provided or until the breast bone has been reached.

Fig. 24–10. Remove the dressing. Slit the thin tissues of the thigh region with the tip of the knife, and then make an opening large enough for a serving spoon to enter. Lay the skin back onto the platter with the dressing in the breast end uppermost for ease of serving.

from the keel, loosening it with the point of the knife, and then placing it on a separate plate. Portions for serving are made by cutting across the grain of the meat.

BARBECUING

Barbecuing is a popular means of preparing meat for those small backyard gatherings as well as for very large community picnics or "barbecues." The National Barbecue Association's Web site is http://www.ribman.com for interested readers. Following are some helpful suggestions on how to prepare meat for large groups of people.

Open Fire

The cut (generally ham or beef round) is attached to a metal rod that is mechanically rotated close to a layer of burning charcoal, which glows through the grates. Steaks, chops, and kabobs are grilled on the grate.

Indirect Heat

The cut is coated with a ½-inch layer of dough and placed in an oven (400°F) to roast. This is the least wasteful method since there is no charred meat, and the product is very tasty and juicy.

Although entire hindquarters of beef are barbecued by the open-fire method, a more tasty product will result if it is prepared according to the *open-trench method,* a method originally prescribed by the American Hereford Association.

The Open-Trench (Pit) Method

Barbecuing meat by this method has been a very popular means of preparing meat for large rural gatherings, particularly of the livestock interests. The cooking principle involved in this method of barbecuing is a combination of dry heat roasting and steaming. The steam is formed from the moisture in the meat and held in the sealed pit.

Building the Trench

A heavy soil containing plenty of clay should be used; sandy soil will require a brick lining to prevent a cave-in. The trench should be made 3½ feet deep and 3 to 3½ feet wide. The length will depend upon the number of people to be served. A liberal serving is considered to be ½ pound (on a fresh meat basis) per person, or 50 pounds for 100 people. To barbecue 100 pounds of meat requires 3 feet of pit length; 200 pounds — 5 feet; 400 pounds — 10 feet; 600 pounds — 15 feet; 800 pounds — 20 feet; etc.

Covers for the pit may consist of pieces of corrugated sheet metal or rough boards. If the latter are used, they should be covered with tarpaulins to keep the dirt from sifting through, since the final seal will be about a foot of dirt placed over the top. If steam leaks occur, they should be plugged with more dirt.

Making the Bed of Hot Coals

Dry oak or hickory wood, measuring 4 to 5 inches in diameter and cut in 2- to 3-foot lengths, is best in producing the 15- to 18-inch bed of hot coals. Apple wood is satisfactory, but the soft and resinous woods are not. Four or five hours should be allowed for producing the bed of coals. Any chunks of wood that do not burn to coals should be removed from the pit or moved to one end with a long rod that has a hooked end. It requires twice the volume of the pit in wood to make the desired bed of coals, or 1 cord (1 ton) per 7 feet of pit length. Inefficient cooking, to say the least!

Coating with Sand

The hot coals must have an overcoating of dry sand or fine gravel to a depth of 1 inch. If the sand or gravel is moist, sheet iron should be placed over part of the pit, with the damp sand placed on it and stirred occasionally to dry while the wood is burning. Wet sand will produce too much smoke.

The Meat

Any of the better grades of meat, poultry, or game are suitable for barbecuing, although beef is the most popular. The boneless cuts are a decided advantage for speed in carving, which is necessary when large groups are served. It is important to have each cut as nearly the same thickness as possible so that all the cuts will cook uniformly.

The meat should be liberally seasoned with salt and pepper before being partially wrapped in aluminum foil of 0.0015 gauge, with a lengthwise drugstore lock wrap, but with the ends left partly open to form a tube that will not scoop up sand. A three-tined fork, with the tines bent into a right-angled curve to hold the roast, should be used to place the tube-style wrapped meats on the hot sand. Place the creased fold part of the aluminum foil down on the fork and roll it off into position on the hot sand with the crease up. When all the meat is in position, the cover should be placed over the pit and then sealed with 8 to 12 inches of dirt. The framework or sheet iron covering should be strongly reinforced to prevent the top from falling into the pit.

About 12 hours should be allowed for barbecuing and 4 to 5 hours for building the bed of coals. The pit should not be opened until shortly before serving is to begin.

Barbecue Sauce

A sauce (pickle marinade) recommended by authorities with experience in its use (it makes any slow-cooked roast mouth watering) is made as follows:

Ingredients	oz./lb. Meat	g/lb. Meat	g/kg Meat
Catsup	1.80	50	110
Worcestershire sauce	0.45	13	29
Prepared mustard	0.15	4	9
Prepared barbecue sauce (on sale in stores)	0.80	22	48
Salt	0.20	6	13

1 lb. = 454 g; 1 kg = 2.2 lb.

Mix together with approximately 20 percent of the ingredient volume of water, which is used to rinse out ingredient bottles.

One gallon of sauce should be sufficient for 40 pounds of meat, which is almost enough for 100 hungry people. The roasts should be pumped to 110 percent of their starting weight, and then covered with the remainder. After being held at 32° to 40°F for two or three days, they should be cooked in a smokehouse, in a pit, or on a rotisserie. The cover sauce should be warmed and served over the sliced meat or beside it.

Barbecue Menu for 100 People

The late Professor J. W. Cole of the University of Tennessee reported the following needs:

Meat — 50 pounds (boned and rolled)

Buns — 200 (sliced almost through and buttered)

Potatoes — 6 pounds potato chips, or 30 pounds scalloped potatoes, or 100 pounds baked potatoes

Beans — 30 pounds, baked

Salad — 30 pounds potato salad, with pickles, eggs, etc., or 20 pounds cabbage salad, with dressing, or 15 to 20 pounds lettuce salad, with dressing

Pickles — 1 gallon

Coffee — 7 to 8 gallons

Dessert — 100 cups of ice cream, cup cakes, or fruit in season

Serving

Separate tables should be provided for those doing the carving. The serving tables, each 3 feet by 10 feet, set on trestles, should be covered with clean wrapping paper, and the paper plates, paper napkins, and plastic forks and spoons placed at the head ends. A systematic arrangement, such as buns, meat, potatoes, salad, relishes, dessert, and beverage, should be followed. The serving may be run as self-service, or attendants may fill each plate completely (except for relish and beverage) before handing it to the guests.

Homemade Barbecue Grills

Regular-size concrete building blocks can be used for the walls, placed end to end from two to three tiers high. The width of space between the lateral walls varies from 3 to 5 feet. The ends may be open or closed. Pressed charcoal briquettes are lodged in piles on the ground or gravel base and lighted with lighter fluid. Depending on the size of the undertaking, the fire should be started from 30 minutes to 1 or 2 hours before serving time.

When the briquettes show gray areas (15 to 20 minutes), the piles can be leveled (see Figure 24–11). To check the temperature of the coals, cautiously hold the palm of your hand about 4 inches above the coals. Count the number of seconds before the heat forces you to remove your hand: two seconds = hot (high); three seconds = medium hot; four seconds = medium; five seconds = low; six to seven seconds = very low. Medium to low is the best temperature for grilling. Spreading the coals apart will lower the temperature; moving them together and knocking off the ash will make them hotter. The coals should be arranged in a single layer directly under the food to cook burgers, steaks, ham slices, chops, kabobs, and other quickly cooked foods directly. Larger cuts (roasts) are more uniformly cooked indirectly from burning charcoal that has been piled at both ends or sides of the grill but not directly under the meat. A drip pan may be placed under the meat and the roast covered (with damper open to allow oxygen to enter). Usually it is not necessary to turn the roast in indirect grilling.

Metal-reinforced grills are made with a frame of 1-inch pipe, with a grill surface 3 feet wide and 4 feet long, and of #9 gauge wire with 2-inch mesh. All sections of the grill have long handles. The cuts of steaks, chops, half-chickens, burgers, kabobs, etc., are placed on the grill, and when it is time to turn the cuts (indicated by the appearance of moisture droplets on the upper surface of the patties, chops, or steaks), a second grill is placed over the meat.

Fig. 24–11. Grill walls made of concrete blocks.

Fig. 24–12. The double flip-over grill.

This permits two persons, one on each side of the grill, to turn and baste the cuts (see Figure 24–12).

Hot butter containing some additional salt and some pepper is a good basting for practically all meats. It can be brushed on or sprayed on hot. For chicken, the sauce consists of ½ pint of water, 1 pint of vinegar, ½ pound of butter, and 1 ounce of salt (sufficient for 10 chicken halves). If you want to spice up your pork chops while they are on the grill, try this tasty pork barbecue spice: 1 pound salt, 3 ounces garlic salt, 3 ounces onion salt, 2 ounces chili powder, 4 ounces paprika. A moderate amount should be sprinkled on each side of the chops as they are grilling. Many pork producer groups sell this or a similar spice mixture. There are also several marinades commercially available that will impart a grilled flavor to meat.

A slightly different grill, with the added advantage of portability, obviously lacking in a concrete block grill, is shown in Figure 24–13. This trailer is made of heavy gauge steel and has held up well during several years of use and travel around Illinois.

Hog Roast

A hog roast is a popular and economical means of feeding large groups of people. Roasting hogs that weigh 100 to 150 pounds are ideal, due to their greater trimness at lighter weights than butcher hogs in the 220- to 260-pound range, which are nevertheless often used. In calculating the meat needs of the group to be fed, figure 1 pound of carcass weight per person. An innovative portable arrangement for roasting whole hogs, beef rounds, or any large cut on a rotisserie is shown in Figure 24–14.

The whole hog should be cooked to 165°F, measured at the innermost portion of the shoulder. When it is cooked on a rotisserie, the whole hog

Fig. 24–13. A portable grill. (Courtesy, University of Illinois Hoof and Horn Club)

Fig. 24–14. Portable hog roaster with rotisserie.

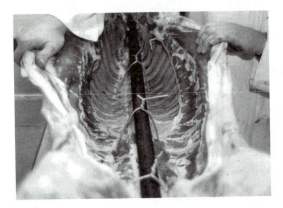

Fig. 24–15. Securing roasting hog to rotisserie bar.

Fig. 24–16. Wrapping new chicken wire around roaster.

should be securely fastened. This is accomplished by looping #9 gauge wire around the loin and backbone (see Figure 24–15). Chicken wire is then tightly wrapped around the carcass (see Figure 24–16).

It will require approximately 60 to 80 pounds of charcoal and 10 to 12 hours to cook a 200-pound roasting hog, using these methods.

CARVING RED MEAT

Carving should not be an objectionable task; it should be a proud accomplishment. Demonstrating carving dexterity will invariably generate commendations from the guests, which is certainly not objectionable. There is a definite technique or way of carving different cuts (see Figures 24–18 to 24–24).[19] In addition to knowing how to carve, remember that:

- The carving knife must be sharp (it should not be sharpened at the table).

- Whenever possible, roasts should be carved across the grain of the meat.

- The carving platter should be of ample size (it is embarrassing to serve cuts from the tablecloth or the lap).

- The purpose of the fork is to hold the cut, not to dull the knife.

- Small, loose, striated pieces should be ignored.

---

[19]All points on carving were courtesy of the National Live Stock and Meat Board, Chicago.

If a piece of meat will not hold together, or if it desires to run a race around the platter, the carver should continue to work swiftly and quietly, without condemning the meat, or the cook, or his or her own shortcomings as a carver.

In order to do a commendable job, the carver must have elbow room and plenty of platter space. Some carvers find it is more convenient to stand while they are carving.

Figure 24–17 shows an arrangement of the carver's place with one tantalizing tumbler placed in the danger zone. The carving task should be finished before the salad, sherbet, water, coffee, or whatever other food is placed before the carver. If the carver is to serve the vegetables, the dishes containing them should be conveniently grouped to your right or left in some sensible pattern that has practicality for its theme rather than artistic effect.

Fig. 24–17. Arrangement of plates and utensils for carving. The tumbler should be moved out of the danger zone before carving begins.

Serving can be hastened and more carving space can be made available if someone else will serve the vegetables and divert the guests' attention from the carving by injecting them with a conversational hypo. (This will not "take" on those who came to learn how or how not to carve.)

A carver should always appear at ease, which is not possible when the hand holding the fork is crossed over the hand doing the carving. It is not bad form to serve small slices, but it is rather embarrassing to serve large, straggly pieces with trailers.

A familiarity with anatomy is of course of inestimable value in efficient carving. By efficient carving is meant the greatest number of neat slices. The homemaker who takes pride in neatness and gastronomic appeal will not be satisfied with the arguments that meat is meat; it all goes to the same place; it has to be cut and mangled anyway; and the small brown pieces are the best.

(Note: The following carving directions are given for right-handed people. If you are left-handed, please reverse the directions.)

Rolled Rib Roast

A rolled rib roast (or any boneless roast) is placed on the platter with the larger cut surface down (see Figure 24–18).

- Use the standard carving set or the slicer and carver's helper.

- With the guard up, push the fork firmly into the roast on the left side an inch or two from the top.

- Slice across the grain toward the fork from the far right side (first illustration). Uniform slices of ⅛- to ⅜-inch thick make desirable servings.

- As each slice is carved, lift it to the side of the platter or to another hot serving platter (second illustration).

Cut first slice.

Lift slice.

Fig. 24–18. Carving a rolled rib roast of beef (or any boneless roast).

Standing Rib Roast

When a standing rib roast is purchased, the short ribs and the backbone most likely will have been removed (see Chapter 15). If this has not been done previously, the meat retailer will, on request, remove the short ribs and loosen the backbone, which can then be removed in the kitchen after roasting. This makes the carving much easier, as only the rib bones remain.

The roast is placed on the platter with the small cut surface up and the rib side to the carver's left (see Figure 24–19).

- Either the standard carving set or the roast meat slicer and carver's helper can be used on this roast.

Cut first slice.

Cut close to bone.

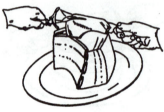

Lift slice.

Fig. 24–19. Carving a standing rib roast of beef.

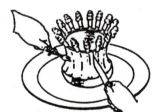

Slice between ribs.

Lift out slice.

Fig. 24–20. Carving a crown roast of lamb.

- With the guard up, insert the fork firmly between the two top ribs. From the far outside edge, slice across the grain toward the ribs (first illustration). Make the slices ⅛- to ⅜-inch thick.

- Release each slice by cutting closely along the rib with the knife tip (second illustration).

- After each cut, lift the slice on the blade of the knife to the side of the platter (third illustration). If the platter is not large enough, have another hot platter nearby to receive the slices.

- Make sufficient slices to serve all guests before transferring the servings to individual plates.

Lamb Crown Roast

A lamb crown roast is made from the rack, or rib section, of the lamb (see Chapter 16). A pork crown roast may be made from the rib sections of two or more loins of pork (see Chapter 14). Both cuts are carved in a similar manner (see Figure 24–20).

- Use a standard carving set.

- Move to the side of the platter any garnish in the center that may interfere with carving. Dressing can be cut and served along with the slices.

- Steady the roast by placing the fork firmly between the ribs.

- Cut down between the ribs, allowing one rib to each slice (first illustration).

- Lift the slice on the knife blade, using the fork to steady it (second illustration).

Leg of Lamb

The leg of lamb should be placed before the carver so that the shank bone is to the right, and the thick, meaty section, or cushion, is on the far side of the platter. Different roasts will not always have the same surface uppermost because of the difference in right and left legs; however, this does not affect the method of carving. The illustrations show a right leg of lamb resting on the large, smooth side (see Figure 24–21).

- A standard carving set is a convenient size for this roast.

- Insert the fork firmly in the large end of the leg, and carve two or three lengthwise slices from the near thin side (first illustration).

- Turn the roast so that it rests on the surface just cut. The shank bone now points up from the platter.

- Insert the fork in the left side of the roast. Starting at the shank end, slice down to the leg bone. Parallel slices may be made until the aitch bone is reached (second illustration). A desirable thickness is $\frac{1}{4}$ to $\frac{3}{8}$ inch.

- With the fork still in place, run the knife along the leg bone, and release all the slices.

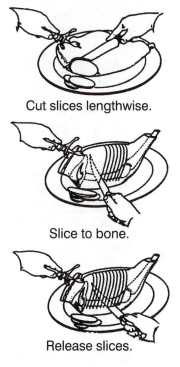

Cut slices lengthwise.

Slice to bone.

Release slices.

Fig. 24–21. Carving a leg of lamb.

Bone-in Ham

This procedure is almost identical to carving a leg of lamb, except that a bone-in ham is usually larger than a leg of lamb. The ham is placed on the platter with the fat or decorated side up. The shank end should always be to the carver's right. The thin side of the ham, from which the first slices are made, will be nearest or farthest from the carver, depending on whether the ham is from a right or a left side of pork. The illustration shows a left ham with the first slices cut nearest the carver. The diagram shows the bone structure and direction of the slices (see Figure 24–22).

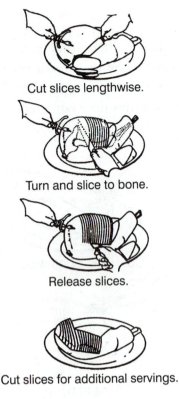

Cut slices lengthwise.

Turn and slice to bone.

Release slices.

Cut slices for additional servings.

Fig. 24–22. Carving a roast ham.

- Use a standard carving set or the slicer and carver's helper on the baked ham.

- Insert the fork, and cut several slices parallel to the length of the ham on the nearest side (first illustration).

- Turn the ham so that it rests on the surface just cut. Hold the ham firmly with the fork and cut a small wedge from the shank end (second illustration). Removing this wedge will make the succeeding slices easier to cut and to release from the bone.

- Keep the fork in place to steady the ham and then cut thin slices down to the leg bone (second illustration).

- Release the slices by cutting along the bone at right angles to the slices (third illustration).

- For more servings, turn the ham back to its original position, and slice at right angles to the bone (fourth illustration).

Pork Loin Roast

It is much easier to carve a pork loin roast if the backbone is separated from the ribs. This is done by the retailer or the processor by sawing across the ribs close to the backbone (see Chapter 14). The backbone becomes loosened during roasting (note in the first illustration that it has fallen away from the ribs [see Figure 24–23]).

- Use the standard carving set for carving the pork loin. A smaller size may also be used.

- Before the roast is brought to the table, remove the backbone by cutting between it and the rib ends (second illustration).

- Place the roast on the platter so that the rib side faces you. This makes it easy to follow the rib bones, which are the guides for slicing. Make sure of the slant of the ribs before proceeding to carve, as all the ribs are not perpendicular to the platter.

- Insert the fork firmly in the top of the roast. Cut closely against both sides of each rib. Alternately make one slice with a bone and one without. Roast pork is more tempting when sliced fairly thin. In a small loin, each slice may contain a rib; if the loin is large, it is possible to cut two boneless slices between ribs.

- Serve each person two slices, which is the usual serving.

Pot Roast

For a pot roast from the chuck, carving is simplified by cutting out solid chunks and turning them in position to make possible the carving of neat slices across the grain. The blade pot roast may contain at least part of one rib and a portion of the blade bone. The long cooking process softens the tissues attached to the bones; therefore, the bones can be slipped out easily before the roast is placed on the table (see Figure 24–24).

- Use the steak set or the standard carving set for carving the pot roast.

- Hold the pot roast firmly with the fork inserted at the left and separate a section by running the knife between two muscles, then close to the bone, if the bone has not been removed (first illustration).

- Turn the section just separated so that the grain of the meat is parallel with the platter

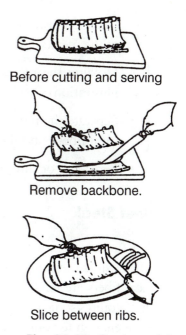

Before cutting and serving

Remove backbone.

Slice between ribs.

Fig. 24–23. Carving a loin roast of pork.

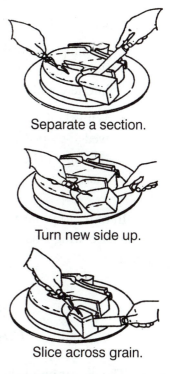

Separate a section.

Turn new side up.

Slice across grain.

Fig. 24–24. Carving a chuck or blade roast.

(second illustration). This enables you to cut the slices across the grain of the meat.

- Holding the piece with the fork, cut slices ¼- to ⅜-inch thick (third illustration).

- Separate the remaining sections of the roast (note the direction of the meat fibers, and then carve across the grain).

- Serve two or three slices, depending on size, to each person.

Beef Steak

Contrary to most carving rules, a steak is carved with the grain (except for a flank steak). A loin or rib steak need not be cut across the grain because the meat fibers are tender and already relatively short. Porterhouse steak is a cut that requires some thought before it is carved because the tenderloin muscle is large enough for only a single serving. A good way to handle this situation is to remove the T-shaped bone and then carve across the tenderloin and loin muscle, giving a piece of each as a serving.

- Use the steak set with a knife blade of 6 to 7 inches.

- Holding the steak with the fork inserted at the left, cut closely around the bone. Then lift the bone to the side of the platter where it will not interfere with the carving.

- With the fork in position, cut across the full width of the steak. Make wedge-shaped portions, widest at the large top-loin muscle side. Each serving will be a piece of the tenderloin and a piece of the large top-loin muscle.

In order to protect the cutting edge of the knife, as well as the platter, use a small cutting board, which is almost a necessity when you are carving a steak.

SOME INSTITUTIONAL
MANAGEMENT PROBLEMS

The operation of dining commons in institutions presents problems dealing with grades of meat, the form in which they should be purchased, their cutting, and their preparation. When economy, because of budget limitations, is the paramount issue, the quantity and quality of meat purchases will naturally suffer. This does not mean that hash from cutter and canner stock, as delicious as it is, must always be the main dish. It is a challenge to the person in charge to make the kitchen a laboratory in which to discover:

- What grades of meat best suit special needs from the standpoint of complete utilization, consumer satisfaction, and cost per serving.

- What advantages there may be in buying meat in the carcass as compared to wholesale and boneless cuts, fresh as compared to frozen, and from the packer as compared to the jobber (wholesaler).

- What cuts furnish satisfactory roasts, steaks, etc., at the lowest cost per serving (for example, chucks as compared to rounds for roasts).

- What method of cutting is best adapted to the utilization of all the meat.

- What new and different ways there may be in preparing and serving the same cuts so as to relieve the monotony so often prevalent.

The Cutting and Cooking Tests

The most businesslike approach to the solution of these problems is to make cutting tests on carcasses and cuts of different grade and price levels, and thereby determine the actual costs of servings (see Chapter 13 for examples). What will servings cost and how do the different roasts compare in palatability and ease in serving? The answers are found by cutting and preparing the meat for the oven and then dividing the usable weight into the total cost. The cooked product when served will give further information as to shrinkage and actual number of servings secured per pound of fresh meat. (See Table 24–21 for estimates of the number of servings per pound and Table 24–22 for the cost per serving.) Food yields are available from the USDA's Food and Consumer Service (http://schoolmeals.nal.usda .gov:8001/). A handy reference to determine a 3½-ounce serving of meat by sight is to compare it to a full deck of cards. They are approximately equal in weight.

Table 24–21. Number of Servings per Pound to Expect from a Specific Cut of Meat[1, 2]

Beef			
Cut	Serving	Cut	Serving
Steaks		**Pot roasts**	
Chuck (arm or blade)........	2	Arm (chuck)...............	2
Club.....................	2	Blade (chuck)	2
Cubed	4	Chuck, boneless	2½
Filet mignon	3	Cross rib.................	2½
Flank	3	**Other cuts**	
Porterhouse...............	2	Brisket...................	3
Rib......................	2	Cubes....................	4
Rib eye (Delmonico)........	3	Loaves	4
Round....................	3	Patties...................	4
Sirloin...................	2½	Short ribs	2
T-bone	2	**Variety meats**	
Top loin	3	Brains...................	5
Roasts		Heart	5
Rib, standing	2	Kidney	5
Rib eye (Delmonico)........	3	Liver....................	4
Rump, rolled	3	Sweetbreads	5
Sirloin tip	3	Tongue..................	5

Pork			
Cut	Serving	Cut	Serving
Chops and steaks		Ham, smoked, bone-in	3½
Blade chops or steaks	3	Ham, smoked, boneless......	5
Boneless chops	4	Ham, smoked, canned.......	5
Fresh ham (leg) steaks	4	Boston shoulder (rolled),	
Loin chops	4	boneless.................	3
Rib chops	4	Loin blade................	2
Smoked (rib or loin) chops	4	Loin (rolled), boneless	3½
Smoked ham (center slice)		Loin, center..............	3
steaks	5	Picnic shoulder (bone-in)	
Roasts		fresh or smoked	2
Ham (leg), fresh, bone-in	3	Sirloin...................	2
Ham (leg), fresh, boneless	3½	Smoked shoulder roll (butt)....	3

(Continued)

Table 24–21 (Continued)

Pork

Cut	Serving	Cut	Serving
Other cuts		Spareribs	1½
Back ribs.	1½	Tenderloin (whole)	4
Back ribs, country-style	1½	Tenderloin (fillets)	4
Bacon (regular), sliced	6	**Variety meats**	
Bacon, Canadian-style	5	Brains	5
Cubes, fresh or smoked	4	Heart	5
Hocks, fresh or smoked	1½	Kidney	5
Pork sausage.	4	Liver	4

Lamb

Cut	Serving	Cut	Serving
Chops and steaks		Shoulder, boneless.	3
Leg chops (steaks)	4	**Other cuts**	
Loin chops	3	Breast.	2
Rib chops	3	Breast (riblets)	2
Shoulder chops	3	Cubes.	4
Sirloin chops	3	Shanks	2
Roasts		**Variety meats**	
Leg, bone-in	3	Heart	5
Leg, boneless	4	Kidney	5
Shoulder, bone-in	2½		

[1]National Live Stock and Meat Board.

[2]The number of servings per pound is only a guide to the average amount to buy to provide 3 to 3½ ounces of cooked lean meat. The cooking method and cooking temperature, the degree of doneness, the difference in the size of bone in the bone-in cuts, and the amount of fat trim are some of the factors that vary and will affect the yield of cooked lean meat.

Table 24–22. Cost of Various Servings of Meat

Approximate Cost ($)

Cost per Pound	1½ Servings per Pound	2 Servings per Pound	2½ Servings per Pound	3 Servings per Pound
0.79	0.53	0.40	0.32	0.26
0.89	0.59	0.45	0.36	0.30
0.99	0.66	0.50	0.40	0.33
1.09	0.73	0.55	0.44	0.36
1.19	0.79	0.60	0.48	0.40
1.29	0.86	0.65	0.52	0.43
1.39	0.93	0.70	0.56	0.46
1.49	0.99	0.75	0.60	0.50
1.59	1.06	0.80	0.64	0.53
1.69	1.13	0.85	0.68	0.56
1.79	1.19	0.90	0.72	0.60
1.89	1.26	0.95	0.76	0.63
1.99	1.33	1.00	0.80	0.66
2.09	1.39	1.05	0.84	0.70
2.19	1.46	1.10	0.88	0.73
2.29	1.53	1.15	0.92	0.76
2.39	1.59	1.20	0.96	0.80
2.49	1.66	1.25	1.00	0.83
2.59	1.73	1.30	1.04	0.86
2.69	1.79	1.35	1.08	0.90
2.79	1.86	1.40	1.12	0.93
2.89	1.92	1.45	1.16	0.97
2.99	1.99	1.50	1.20	1.00
3.09	2.05	1.55	1.24	1.03
3.19	2.13	1.60	1.28	1.06
3.29	2.19	1.65	1.32	1.10
3.39	2.26	1.70	1.36	1.13
3.49	2.33	1.75	1.40	1.16

per Pound[1]

per Serving

3½ Servings per Pound	4 Servings per Pound	5 Servings per Pound	6 Servings per Pound
0.23	0.20	0.16	0.13
0.25	0.22	0.18	0.15
0.28	0.25	0.20	0.17
0.31	0.27	0.22	0.18
0.34	0.30	0.24	0.20
0.37	0.32	0.26	0.22
0.40	0.35	0.28	0.23
0.43	0.37	0.30	0.25
0.45	0.40	0.32	0.27
0.48	0.42	0.34	0.28
0.51	0.45	0.36	0.30
0.54	0.47	0.38	0.32
0.57	0.50	0.40	0.33
0.60	0.52	0.42	0.35
0.63	0.55	0.44	0.37
0.65	0.57	0.46	0.38
0.68	0.60	0.48	0.40
0.71	0.62	0.50	0.42
0.74	0.65	0.52	0.43
0.77	0.67	0.54	0.45
0.80	0.69	0.56	0.46
0.83	0.72	0.58	0.48
0.86	0.74	0.60	0.50
0.88	0.77	0.62	0.51
0.91	0.80	0.64	0.53
0.94	0.82	0.66	0.55
0.97	0.85	0.68	0.57
1.00	0.87	0.70	0.58

(Continued)

Table 24–22

				Approximate Cost ($)
Cost per Pound	1½ Servings per Pound	2 Servings per Pound	2½ Servings per Pound	3 Servings per Pound
3.59	2.39	1.80	1.44	1.20
3.69	2.46	1.85	1.48	1.23
3.79	2.53	1.90	1.52	1.26
3.89	2.59	1.95	1.56	1.30
3.99	2.66	2.00	1.60	1.33
4.09	2.73	2.05	1.64	1.36
4.19	2.79	2.10	1.68	1.40
4.29	2.86	2.15	1.72	1.43
4.39	2.93	2.20	1.76	1.46
4.49	2.99	2.25	1.80	1.50
4.59	3.06	2.30	1.84	1.53
4.69	3.13	2.35	1.88	1.56
4.79	3.19	2.40	1.92	1.60
4.89	3.26	2.45	1.96	1.63
4.99	3.33	2.50	2.00	1.66
5.09	3.39	2.55	2.04	1.70
5.19	3.46	2.60	2.08	1.73
5.29	3.53	2.65	2.12	1.76
5.39	3.59	2.70	2.16	1.80
5.49	3.66	2.75	2.20	1.83
5.59	3.73	2.80	2.24	1.86
5.69	3.79	2.85	2.28	1.90
5.79	3.86	2.90	2.32	1.93
5.89	3.93	2.95	2.36	1.96
5.99	3.99	3.00	2.40	2.00
6.09	4.06	3.05	2.44	2.03

[1]National Live Stock and Meat Board and senior author.

(Continued)

per Serving

3½ Servings per Pound	4 Servings per Pound	5 Servings per Pound	6 Servings per Pound
1.03	0.90	0.72	0.60
1.05	0.92	0.74	0.62
1.08	0.95	0.76	0.63
1.11	0.97	0.78	0.65
1.14	1.00	0.80	0.67
1.17	1.02	0.82	0.68
1.20	1.05	0.84	0.70
1.23	1.07	0.86	0.72
1.25	1.10	0.88	0.73
1.28	1.12	0.90	0.75
1.31	1.15	0.92	0.77
1.34	1.17	0.94	0.78
1.37	1.20	0.96	0.80
1.40	1.22	0.98	0.82
1.43	1.25	1.00	0.83
1.45	1.27	1.02	0.85
1.48	1.30	1.04	0.87
1.51	1.32	1.06	0.88
1.54	1.35	1.08	0.90
1.57	1.37	1.10	0.92
1.60	1.40	1.12	0.93
1.63	1.42	1.14	0.95
1.65	1.45	1.16	0.97
1.68	1.47	1.18	0.98
1.71	1.50	1.20	1.00
1.74	1.52	1.22	1.02

Grade of Meat Purchased

The grade of meat purchased depends upon whether paying or nonpaying clients are being served. Choice and Select grades should be bought for the former. Because appropriations may be limited for nonpaying clients or wards, the more economical Standard, Commercial, and Utility grades of meat may have to be procured. A good policy in preparing these economical, but tasty meals is to use the better grades for chops, steaks, and roasts and to have some of the Utility grade on hand to incorporate with the more wasty cuts, thereby allowing most of the excess fat to be used. Whether hindquarters or forequarters are used for this fat-saving purpose will depend upon the difference in the price between the two, with hindquarters being preferred by many. Hindquarters will usually average about 2 percent less bone than forequarters. The tables on the retail cut yields of the various carcasses (see Chapters 13 to 16) should be helpful in estimating the yield of edible meat from different grades of carcasses and cuts.

25

Meat Judging and Evaluation[1]

> This chapter will provide basic principles of meat judging and selection. This should be ample information for students who wish to participate in local contests and a good background for those interested in advanced meat judging. The chapter includes a few of the general rules of intercollegiate contests to whet the appetite of students who become interested in further competition.
>
> Students seriously interested in competitive meat judging should contact the Intercollegiate Meat Judging Coaches Association through the American Meat Science Association.[2] They can also obtain a copy of the *Meat Evaluation Handbook* from the American Meat Science Association.

Intercollegiate meat judging contests were inaugurated and sponsored by the National Live Stock and Meat Board in the fall of 1926, when the first contest was held in connection with the International Livestock Exposition in Chicago. This contest was later moved to Madison, Wisconsin, and more recently to Dakota City, Nebraska. A contest was begun at the American Royal Livestock Show in 1927 and later moved to Emporia, Kansas. The Meat Board no longer exists, but funding support for meat judging is provided by the National Cattlemen's Beef Association and the National Pork Producers Council. In addition to these two contests, similar contests are held

[1]Special thanks go to Dr. Tom Carr, University of Illinois, for his advice and guidance on this chapter. The authors are greatly appreciative of Dr. H. Dwight Loveday, University of Tennessee, and Dr. Roger Johnson, Triumph Pork Group, LLC, for their critiques of Figure 25–1 and the associated terminology. The authors are also grateful to Dr. Dell Allen, Excel Corporation, for his original work from which Figure 25–1 was modified in the preparation of this chapter. Susan Specht-Overholt, Banner Pharmacaps, provided the information on veal color.

[2]1111 North Dunlap Avenue, Savoy IL 61874. Phone (217) 356-3182. http://www.meatscience.org

throughout the calendar year at Greeley, Colorado; Dallas, Texas; and Wyalusing, Pennsylvania. These contests, in many instances, are concurrent with the livestock expositions held at those places. Several regional contests are also held.

The success of intercollegiate livestock judging contests, plus the fact that more agricultural colleges were teaching meat courses, prompted R. C. Pollock, then general manager of the Meat Board, to get intercollegiate meat judging on its way. Ten teams competed that first year, as compared to 26 teams in 1960 and 15 teams in 1999. The contests have given college students who meet the eligibility rules set up by a rules committee the opportunity to put to a test the meat knowledge they have acquired in their respective institutions and to gain a wider knowledge of the meat industry. As was to be expected, the contests have done more than that. Students have met other students, as well as leaders in the industry. The meat industry (packers) has become interested in these young students. As a result, many college-trained individuals are now holding responsible positions in the industry.

UNITED NATIONAL COLLEGIATE MEAT ANIMAL EVALUATION CONTEST

A contest designed to allow students to compete in an overall program was begun in 1964. The program includes meat judging, breeding animal judging, and market animal evaluation. The following explanation of this contest was written by R. G. Kauffman, University of Wisconsin, who was influential in starting the contest.

> The coordinated approach to meat animal evaluation was initiated to assist and encourage students of animal science to be more aware of the relationships and limitations that exist when evaluating breeding and market animals, and to help them more fully appreciate the importance of carcass excellence as it related to production, as well as meat processing, merchandising and consumption. This program was specifically designed to stimulate college teaching and to motivate students to seek a more complete understanding of meat animal evaluation — from conception to consumption.
>
> The basic idea took roots April 21, 1955, in Chicago, Illinois, when the National Live Stock and Meat Board, through encouragement by concerned educators, sponsored the first of several clinics to

provide students an opportunity to evaluate market livestock, before and after slaughter. From this beginning, there has been a continued growth of interest, support and participation.

The idea for the United National Collegiate Meat Animal Evaluation Contest developed when educators and businesspeople of the livestock and meat industry designed an exercise that would emphasize all aspects of meat animal evaluation. It was decided that breeding livestock, market livestock and carcasses should be included and that such a program be educational, stimulating and competitive. They organized the first one through the cooperation of the Rath Packing Company, Waterloo, Iowa, in 1964. Forty students representing six universities competed through the cooperation of Farmbest, Inc., and Iowa Beef Processors, Denison, Iowa. In 1968, the contest was moved to Omaha, Nebraska, where 117 students from 11 universities competed under the sponsorship of the Knights of Ak–Sar–Ben, and with the cooperation of Armour and Co., Hormel Foods Corporation, and the Omaha Livestock Market, Inc. In 1990, Con Agra, Inc., joined the Knights of Ak–Sar–Ben as sponsors of this activity. In 1995, the contest moved to St. Joseph, Missouri, under the sponsorship of United Livestock Marketing Services and IBP, Inc.

In 1999, students representing 13 universities competed in three divisions of the contest to compare their knowledge.

MEAT JUDGING CONTESTS

The rules and regulations governing intercollegiate meat judging contests undergo constant change as meat industry procedures evolve. The changes must be adopted by the Intercollegiate Meat Coaches Association, which also includes contest official judges and supervisors. Following subsequent approval by the American Meat Science Association, the changes are incorporated into the annual Meat Judging Contest Rules and Regulations and published. The regulations identify eligible participants as:

- Any college or university having adequate instruction in meats is eligible to enter a team composed of four members. Individual educational institutions may compete in either the Senior Division or the A Division of the Intercollegiate Meat Judging Program. Eligibility for each division is as follows:

1. Senior Division: The Senior Division is open to all Bachelor of Science granting agricultural universities and colleges where instruction in meat grading and judging is included as a regular part of the curriculum.

2. A Division: The A Division is open to all post-secondary educational institutions where instruction in meat grading and judging is regularly offered. There is no A Division in the American Royal and International contests.

Contest Sequence and Procedures

First, the contestants are divided into groups that will remain together under the direction of a non-contestant group leader throughout the contest. The group leader is to enforce the rules of the contest, guide the group through the contest, and collect the judging materials at the appropriate times. The contestants must abide by the regulations and may be disqualified by the group leader for any violation upon warning and continued violation. The rules prohibit the contestants from talking among themselves; using aids such as grade-guide cards, photographs, or measuring rulers; handling beef and pork cuts; and touching the rib eye in the carcass-grading class. These rules are necessary to prevent changes and(or) damage to the product that may occur during the contest by the large number of contestants working over the limited number of exhibits. The judges must use their eyes, not their hands, to make their evaluations.

Table 25–1 lists the classes and scoring bases for a standard intercollegiate meat judging contest. Contests, of necessity, are held in meat packing plants or other facilities that have the coolers and equipment for handling meat carcasses and wholesale cuts. The first portion of the contest consists of a series of 15-minute periods during which the contestants will place five carcasses or cuts and record notes on each of the five classes (see Table 25–1) requiring written reasons or questions. During the first four minutes at each placing class, all contestants in the group will stand back, away from the class, at a distance specified by the group leader to observe general appearance. The contestants will be permitted to "walk by" each beef quality class exhibit (five seconds per exhibit) during the initial standback period to observe the muscle quality. After the standback, the contestants may move at will around the

Table 25–1. Classes for a Standard Intercollegiate Meat Judging Contest

Class	Possible Placing Score[1]	Possible Reasons/ Questions Score
Beef carcasses	50	50
Beef carcasses	50	—
Beef cuts	50	50
Beef cuts	50	—
Pork carcasses	50	50
Pork carcasses	50	—
Fresh skinned hams	50	50
Lamb carcasses	50	50
Lamb carcasses	50	—
Beef carcass quality grading (15 carcasses)	150	
Score per carcass		
Correct 10 points		
⅓ grade off 8 points		
⅔ grade off 5 points		
Full grade off 0 points		
Beef carcass yield grading (15 carcasses)	150	
Score per carcass		
Correct 10 points		
Off 0.1 YG 9 points		
Off 0.2 YG 8 points		
Off 0.3 YG 7 points		
Off 0.4 YG 6 points		
Off 0.5 YG 5 points		
Off 0.6 YG 4 points		
Off 0.7 YG 3 points		
Off 0.8 YG 2 points		
Off 0.9 YG 1 point		
Off 1.0 YG 0 points		
Total	750	250
Specifications classes (Senior Division only)		
(10 cuts)	100	
Score per specification cut		
Correct 10 points		
1 incorrect rejection 7 points		
2 incorrect rejections 3 points		
3 incorrect rejections 0 points		
Rejected/accepted when		
correct to accept/reject 0 points		
No correct answer 0 points		
Grand totals Division A		1,000 points
Senior Division		1,100 points

[1]After the judging committee has reached agreement on the points between each of the four exhibits in a class, the scores for each of the 24 possible placings will be secured by calculating by hand, by using the Hormel computing slide, or by using a computer.

class, keeping in mind that the other contestants also have an equal right to see the exhibit. The second standback will occur during the last two minutes at each reasons or questions placing class, following which the placing cards will be collected by the group leader. During one of the 15-minute periods of this segment in the coolers, the contestants will quality grade 15 beef carcasses, with no standback required.

After approximately two hours in the coolers, the contestants are given 30 minutes to warm up and review their reasons notes. Reasons, defending their placings, are written for each of the five classes during consecutive 15- minute timed intervals. When questions are used instead of written reasons, a 15-minute warm-up/review period is permitted, followed by 5-minute intervals between 3-minute answering periods per class. *Using notes is not permitted during the question period, but it is allowed during the writing of reasons.*

The groups return to the coolers for a series of 10-minute stops during which they will place four non-reasons classes (see Table 25–1), yield grade 15 beef carcasses (20 minutes), and evaluate 10 specifications cuts (20 minutes). Non-reasons placing classes have initial two-minute and final one-minute standbacks.

The Official Committee

A seven-person judging committee and a contest superintendent will be selected for each contest by the American Meat Science Association/Intercollegiate Meat Coaches Association. The superintendent, committee chairperson, grader member of the committee, and(or) sponsor are responsible for cooperating with the host plant to select contest material, to prepare the area where the contest will be conducted, to assure that the contest is properly conducted, and to work with the committee. The judging committee makes the final selection of the classes, grading carcasses, and specification cuts. The committee agrees upon the official placing and scoring of each class, the questions (if used) to be asked on each class and the official answers, the final grades for carcasses on the grading rails, and the official responses for each of the specification cuts prior to the beginning of the contest. Five members of the committee are individually assigned to read and grade all the reasons written by the contestants from one of the five reasons classes. The committee is also asked to give oral reasons or written notes supporting the official placing of each class. One member of the committee is an official USDA meat grader who is responsible for providing grade parameters on each of the carcasses graded to justify the official final grades and official specifications responses. At contests, the committee members work as hard as the contestants do. Sponsors locate the host facilities, make local arrangements, register the participants, provide judging materials, supply the materials needed to identify the classes and the exhibits, score the contests, assemble the results, and conduct the awards presentations.

Awards

The regulations of the American Meat Science Association/Intercollegiate Meat Coaches Association, contest sponsors, and most others indicate that:

- Each year the winning team is given the custody of a perpetual trophy, which must be won three times for permanent possession. A plaque is awarded as the permanent property of the winning team, and a place ribbon is presented to each of the 10 high teams.

- An appropriate emblem is presented annually to the highest ranking individual in total points and reasons score and to the highest ranking individual in each of the major classes — beef, pork, and lamb; beef carcass grading; and combined beef and IMPS (Institutional Meat Purchase Specifications) classes.

- A ribbon is given to each of the 10 highest contestants in total points and to the highest contestant in each of the five divisions.

MEAT JUDGING

For the serious student of meat judging, the *Meat Evaluation Handbook* is highly recommended. This 70-page manual, containing more than 190 full-color pictures, is published by the American Meat Science Association in cooperation with the National Cattlemen's Beef Association and the National Pork Producers Council.

Judging carcasses differs from grading in several respects. The carcasses must be rated or placed in the order in which the one surpasses the other on the basis of muscling, trimness, and quality. Since the four specimens may fall in the same grade, or with an assured spread of not more than one full grade, a more critical examination and evaluation of factors such as muscling and trimness as they affect yield must be emphasized. Quality factors to be considered include marbling, maturity, lean color, and texture.

Meat Judging in Three Easy Steps

1. Three important factors in meat are:

    a. *Muscling* (conformation)

    b. *Trimness*

    c. *Quality* ≈ *Palatability*

2. What tells you

    a. *Muscling?*

        (1) Rib eye area.

        (2) *Bulge and plumpness* in round (leg) and chuck (shoulder) and through loin and rib; generally *width, thickness,* and *depth.*

    b. *Trimness?*

        (1) *Trimness* over rib eye, lower rib, flank, brisket, round, chuck.

        (2) *Blue* indicates muscle in the round, not a thick covering of fat. *Penalize outside* fat and *intermuscular* fat (between the muscles), i.e., *seam* fat.

    c. *Quality?*

        (1) Color — *beef:* bright cherry red; *pork:* bright grayish pink; *lamb:* bright reddish pink

        (2) Firmness of lean and fat

        (3) Smooth texture

        (4) Marbling — fine specks or strands *within* the muscle

        (5) Youth — cartilage *between* or *on ends of bones;* blood in bones; bright, light muscle color

3. Put 2a, 2b, and 2c together! The leanest carcass or cut with the most muscling and quality wins, and you place the others accordingly.

    More emphasis on *muscle, leanness (quantity):* chucks, rounds, legs, pork carcasses, lamb carcasses

    More emphasis on *quality:* ribs, loins

    Emphasis on quantity and quality about *equal:* Balance them off, depending on current market prices, when judging beef carcasses.

This judging routine may apply to other classes as well. The first impression is more often right than wrong, so take a good look at the carcasses from a distance in the four minutes allotted. The first general impression is likely to be correct because it takes into consideration the outstanding points of excellence or inferiority as far as overall muscling and trimness are concerned.

Beef Carcass Judging

Muscling

Ribbing between the twelfth and thirteenth ribs, the area where the rib eye *(longissimus)* is exposed, is the best singular indicator of overall muscling. Overall superior muscling is demonstrated in carcasses that are very thickly muscled; are very full and thick in relation to their length; and have a very plump, full, and well-rounded appearance. Inferior muscling indicates a low proportion of meat to bone and a low proportion of the weight of the carcass in the more valuable parts. It is reflected in carcasses that are very thinly muscled; are very narrow and thin in relation to their length; and have a very angular, thin, and sunken appearance.

Trimness

The most accurate estimate of the total amount of external fat on a carcass is made by taking a visual measurement perpendicular to the outside fat surface at a point three-fourths the length of the rib eye from its chine bone end. Unusual amounts of fat on other parts of the carcass, on areas such as the brisket, plate, flank, cod or udder, inside round, rump, and hips, in relation to the actual thickness of the fat over the rib eye, must also be considered.

The amount of kidney, pelvic, and heart fat including the kidney knob (kidney and surrounding fat), the lumbar and pelvic fat in the loin and round, and the heart fat in the chuck and brisket area, which are all removed in making closely trimmed retail cuts, must also be determined.

Quality

The main indicator of quality in beef carcasses is marbling development in relation to maturity development. In a judging class, all cattle will be "young"; nevertheless, some of them may be approaching the most mature age permitted in the A maturity grouping (A+ or A^{90}) or even over into the B^0 (B–) category. In such a case, more marbling is required to qualify for the

Points for Consideration in Judging Beef Carcasses

MUSCLING

1. Round cusion — length, width, thickness, plumpness
2. Sirloin — width, fullness, prominence
3. Loin — fullness, width, depth of chine
4. Rib eye — size, shape
5. Rib — fullness, width

QUALITY

1. Marbling — amount, fineness, distribution
2. Color — youthfulness, brightness of lean
3. Firmness of lean
4. Texture of lean — firmness

5. External fat — whiteness, firmness
6. Bone — youthfulness

TRIMNESS

1. Rib eye
2. Round cushion
3. Rump
4. Sirloin
5. Loin edge
6. Plate
7. Rib
8. Chuck
9. Kidney, heart, and pelvic
10. Cod or udder
11. Brisket

USDA Choice grade. Depending on the four carcasses in the class, and the current market differential between USDA Choice and USDA Select carcasses compared to the market differential between USDA Yield Grade 2's, 3's, and 4's, you may wish to place a Select 2 over a Choice 4. Other indications of more acceptable quality are fineness and even distribution of marbling, a bright cherry red color, and a firm, fine muscle texture.

USDA yield and quality grades should be determined for each carcass before a final placing is made. (See the *Meat Evaluation Handbook,* published by the American Meat Science Association in cooperation with the National Cattlemen's Beef Association and the National Pork Producers Council, for points to consider in beef carcass judging.)

Pork Carcass Judging

Trimness

In judging all classes of carcasses, with the exception of quality grading beef carcasses, contestants should consider trimness first and muscling a close second. That which is most obvious to the eye should come first. In pork judging, trimness definitely should be evaluated first because it is the main factor used in grading. Therefore, determine from the backfat thickness the grade into which the carcass falls. To do this, you must be familiar with the backfat thickness that is associated with the muscling of a carcass designated for the particular grade (see Chapter 12).

It becomes a matter of judgment as to the merits of one carcass over another carcass of the same or a near grade.

- Look for an even distribution of backfat. Many hogs have a tendency to lay the fat on heavily over the shoulder. On ribbed carcasses, consider fat depth over the eye at the tenth rib.

- Look for a minimum belly thickness of approximately 1½ inches. A certain minimum thickness of belly is necessary for a uniform bacon slab; however, it is fat that makes the belly thicker. Any additional is waste fat.

- Look for a trim, neat jowl.

- Look for a firm, white fat that is not greasy to the touch. Extremely soft pork must be discounted rather severely because soft, oily fat makes processing all pork cuts difficult.

The final decision between two specimens may hinge on the general distribution of the external finish. An overfat pork carcass is wider through the belly than at either the leg (ham) or the shoulder, due to excess backfat and belly fat. In such a carcass, the leg (ham) and the shoulder blend smoothly into the loin and the belly from both ends.

Muscling

Note the form or shape of the carcass in respect to its length, width, and thickness through the leg (ham), loin, and shoulder. All these characteristics give balance and uniformity to a carcass and are reflected in the yield of lean and fat cuts.

The length of the carcass is not very important, provided it is in balance and conforms to the standard measurements for the grade. Most of the carcasses that conform to the top-grade specifications are between 29 and 33 inches long (from the first rib to the forward end of the aitch bone) and weigh between 140 and 185 pounds.

Discount long ham hocks, tapering or banjo hams, flat hams, and flat carcasses (those that lack loin development) (see Figure 12–54).

Quality

A carcass of superior quality is one that is bright in appearance, as evidenced by the bright pink color of the flesh on the inside of the belly, and in the ham face and in the loin eye (when exposed). Dark muscle is associated with maturity, intact male sex condition, excessive drying, bacterial action, and other factors, all of which are undesirable.

Look for indications of marbling on the inside of the carcass. Feathering between the ribs is the best indicator of marbling in unribbed pork carcasses. Marbling in the exposed lumbar lean also may be observed. If the carcass is ribbed, look at the eye itself.

Most pork has lower levels of marbling because hogs are marketed at a young age. Excessive backfat thickness is no guarantee that the lean is marbled. Actually, the only sure method of determining marbling in a pork carcass is to see a cross section of the loin eye muscle. Since the size (area) of this muscle and the thickness of the fat covering it are very important in predicting carcass composition, pork carcasses are often ribbed for judging purposes. They must be ribbed for certified litter testing in the development of breeding stock for meat-type hogs. (See the *Meat Evaluation Handbook,* published by the American Meat Science Association in cooperation with the National Cattlemen's Beef Association and the National Pork Producers Council, for points to consider in pork carcass judging.)

Points for Consideration in Judging Pork Carcasses

MUSCLING

1. Length of side
2. Ham — length, width, thickness, plumpness
3. Sirloin — fullness, prominence
4. Lumbar lean — size
5. Loin — fullness, width
6. Chine — depth
7. Shoulder — thickness, fullness
8. Loin eye — size (ribbed carcasses only)

QUALITY

1. Feathering — amount
2. Color — grayish pink
3. Belly — firmness, thickness
4. Loin eye — color, texture, firmness, marbling (ribbed carcasses only)

TRIMNESS

1. Backfat — first rib, last rib, last lumbar
2. Sirloin-loin juncture
3. Ham collar
4. Loin edge
5. Shoulder —clear plate
6. Belly
7. Sternum
8. Jowl
9. Tenth rib fat depth (ribbed carcasses only)

Lamb Carcass Judging

Trimness

The amount of finish necessary to make a Choice quality lamb is very small. Consumers generally are not very tolerant of fats, particularly lamb and mutton fat. The great difficulty with highly finished lamb or mutton carcasses is that the deposition of intermuscular masses of fat outstrips marbling. This is particularly true of the shoulder.

Sufficient white, brittle fat to cover the back with $\frac{1}{8}$ to $\frac{1}{4}$ inch of fat with a lighter covering over the leg and shoulder is adequate finish for a quality lamb carcass. A papery (no fat under the fell) back on a carcass shows lack of finish. A fiery color to the fat is slightly objectionable. When the carcass has been ribbed, the fat cover over the rib eye and the lower rib gives an excellent indication of the overall trimness of the carcass. Special attention should be given to fat deposits in the breast, flank, and cod/udder areas and in the internal kidney and pelvic areas.

Muscling

Thickness and meatiness in lamb carcasses are important because the economic value of cuts from small carcasses depends on the percentage of lean to bone to a large degree. The carcasses should be thick and uniformly wide. Carcasses that are slightly rangy are not objectionable, provided they are uniformly wide and thick. Neat, smooth shoulders, well-fleshed over the blades and covered with a thin layer of white fat, are preferred to narrow shoulders. The neck should be short and thick, rather than long and thin.

The legs should be plump. Tapering legs are *not* characteristic of a valuable lamb carcass. A slight crease over the backbone is indicative of a well-fleshed back, but a prominent backbone reflects shallow muscling and a small rib eye. Flat lamb shoulders with prominent blades at the top are *not* characteristic of excellent conformation. The loin and rib rack should be broad, thick, full, and well turned in the rib to give the carcass a neat, trim appearance. When the carcass has been ribbed, the area of the rib eye itself gives a good indication of the overall muscling in the carcass.

Pot-bellied carcasses are objectionable because they increase the amount of cheap flank and breast meat.

Quality

The break joint must show four well-defined red ridges (see Figures 7–6 and 7–7), indicating youth, as does redness in the ribs.

If the carcass is ribbed, which is very desirable, marbling, a major determinant of quality, can readily and accurately be observed. Feathering between the ribs in the chest cavity and fat streaks in the flank are indices of marbling; however, feathering has been eliminated as a quality grading factor. Firmness is associated with finish, and thin carcasses are naturally soft because the hard fat is absent. The fat should be firm, white, and waxy and evenly distributed over the entire carcass; however, if the fat is oily and soft, the carcass is lacking in firmness, even if it is well-finished. The flank should be firm and dry, and the inside of the flank should show a few fat streaks and have a bright reddish pink color to the flesh.

The practice of ribbing has been followed for years in the Commercial carcass contests and is the only way to do a complete, accurate job of carcass evaluation and judging. Ribbing would allow for more accurate student appraisal of meat quantity and quality in lamb carcasses, which involves too

---

Points for Consideration in Judging Lamb Carcasses

MUSCLING

1. Leg cushion — length, width, thickness, plumpness
2. Sirloin — fullness, prominence
3. Loin — fullness, width
4. Rib eye — size, shape (ribbed carcasses only)
5. Rack — fullness, width
6. Shoulder — thickness, fullness

QUALITY

1. Flank streaking — primary and secondary flanks, amount, fineness
2. Color — reddish pink, youthfulness, brightness
3. Firmness
4. Feathering — amount, fineness (eliminated in 1983 as a quality-grading factor)

5. Marbling — amount, fineness, distribution (ribbed carcasses only)
6. Texture — fineness (ribbed carcasses only)

TRIMNESS

1. Rib eye (ribbed carcasses only)
2. Lower rib (ribbed carcasses only)
3. Leg cushion
4. Sirloin
5. Dock
6. Loin
7. Loin edge
8. Rack
9. Shoulder — top and lower portions
10. Kidney, heart, and pelvic
11. Cod or udder
12. Flank
13. Breast

much guesswork without ribbing. Contest lambs for FFA and 4–H should definitely be ribbed.

USDA yield and quality grades should be determined for each carcass before a final placing is made. (See the *Meat Evaluation Handbook,* published by the American Meat Science Association in cooperation with the National Cattlemen's Beef Association and the National Pork Producers Council, for points to consider in lamb carcass judging.)

Judging Wholesale (Primal and Sub-primal) Beef Cuts

Judging Beef Rounds

Beef rounds are the source of the very popular round steaks and roasts; popular because they contain so little bone and fat. More emphasis should be placed on cutability than quality in beef rounds. (See the *Meat Evaluation Handbook,* published by the American Meat Science Association in cooperation with the National Cattlemen's Beef Association and the National Pork Producers Council, for points to consider in judging beef rounds.)

MUSCLING — Muscling is very important, since it determines the poundage of round steak that can be cut from the area between the rump and the

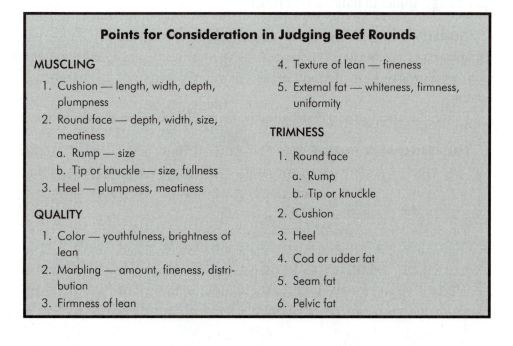

Points for Consideration in Judging Beef Rounds

MUSCLING

1. Cushion — length, width, depth, plumpness
2. Round face — depth, width, size, meatiness
   a. Rump — size
   b. Tip or knuckle — size, fullness
3. Heel — plumpness, meatiness

QUALITY

1. Color — youthfulness, brightness of lean
2. Marbling — amount, fineness, distribution
3. Firmness of lean

4. Texture of lean — fineness
5. External fat — whiteness, firmness, uniformity

TRIMNESS

1. Round face
   a. Rump
   b. Tip or knuckle
2. Cushion
3. Heel
4. Cod or udder fat
5. Seam fat
6. Pelvic fat

stifle joint. An ideal round is plump, wide, and deep and carries the muscling down well toward the hock (full at the heel). Flat, tapering, and dished rounds are heavily discounted.

TRIMNESS — The external fat covering is generally rather sparsely distributed over the round but tends to be heavy or patchy over the rump in the more highly finished beef. Select for smoothness and trimness in this area. The amount of cod/udder fat under the round and the amount of seam fat in the round face are important in evaluating trimness.

QUALITY — Firmness, marbling, and acceptable color are quality considerations for desirable rounds.

Judging Beef Ribs

Beef ribs constitute the highest-priced cuts in the forequarter. Ribs are suitable for dry-heat roasting, and the steaks cut from the top grades are very popular. Acceptable quality is very important in beef ribs and should have equal or greater emphasis than muscling and trimness. Therefore, the judge should estimate the yield and quality grades on each rib in the class early in the decision-making process. (See the *Meat Evaluation Handbook,* published by the American Meat Science Association in cooperation with the National Cattlemen's Beef Association and the National Pork Producers Council, for points to consider in judging beef ribs.)

QUALITY — Inspect the bone to determine age, particularly the presence or absence of the buttons on the ends of the feather bones. Hard bone, flat ribs, a yellow cast to the fat, and a dark color to the lean must be severely discounted.

Determine the degree of marbling, the firmness, the color, and the texture.

MUSCLING AND TRIMNESS — The external finish should be firm and dry and evenly distributed over the entire cut. Excessive external fat covering should be discounted, as should excessive intermuscular fat in the blade end or in the lower rib on the loin end.

The rib eye muscle should be proportionally large, in relation to the size of the cut, and oval in shape. A kidney-shaped eye muscle is undesirable. A large rib eye muscle will make a meaty-appearing rib. The blade end of the rib should be deep and meaty, not flat. The contour of the external part of the rib down to the short-rib section should be gradual, not dipped or dished. A

Points for Consideration in Judging Beef Ribs

MUSCLING

1. Rib eye — size, shape
2. Back — width, length, fullness
3. Blade — width, depth, size, meatiness

TRIMNESS

1. Rib eye
2. Lower rib
3. Rib ends or short ribs
4. Lower blade
5. Back
6. Blade end
7. Seam fat in blade

QUALITY

1. Rib eye (loin end)
   a. Marbling — amount, fineness, distribution
   b. Color — youthfulness, brightness of lean
   c. Firmness of lean
   d. Texture of lean — fineness
2. Blade end: Same as loin end
3. External fat — whiteness, firmness, uniformity

combination of these desirable features of conformation results in a thick, well-balanced cut that will yield a high proportion of the desirable rib eye *(longissimus)*.

Judging Beef Loins and Shortloins

Quality is very important in these wholesale cuts because from them are secured the most tender and most expensive steaks in the entire beef carcass. Quality should have more emphasis than cutability; however, the latter should not be ignored. (See the *Meat Evaluation Handbook,* published by the American Meat Science Association in cooperation with the National Cattlemen's Beef Association and the National Pork Producers Council, for points to consider in judging beef loins.)

QUALITY — Marbling is highly desirable in any cut of meat, but it is doubly so in the loin cuts. When the choice to be made is between a fine weblike marbling and a coarse-type marbling, if the total amount appears to be the same, give preference to the former. Texture is also very important because it affects the tenderness of the steaks. Color is probably more important in these cuts than in any other cuts because consumers see such a large area of exposed meat, and if they do not like the color of these high-priced items, they are less likely to purchase them.

The meat should be firm. As marbling increases, firmness increases and moisture decreases. As stated previously, the moisture content of fat is 8 percent and of lean it is 60 to 70 percent. Another reason for desiring firmness in any wholesale cut, and particularly those cuts that are cut into steaks, is that it is easier to cut a firm steak of even thickness without having it flop over the knife or ooze away from the knife or edge.

Do not fail to inspect the chine bone to determine if it can qualify for the top grades for age, or whether a white, hard, fused bone in the *sacral* region will reduce the quality grade and expected palatability.

MUSCLING — The thicker and heavier the muscling on a loin, the greater the yield of steak. Look for a full, rounded, meaty sirloin end blending well into the shortloin. The shortloin should show fullness with a large, oval eye muscle on the rib end. Flat sirloin ends, prominent hips, and depressed and shallow shortloins should be degraded.

TRIMNESS — Highly finished cattle have heavy external fat deposits on loin cuts and particularly over the edge of the shortloin. A dip or depression in front of the hip indicates trimness. Select for smoothness and a minimum amount of patchiness. In the trimmed loin (flank off, kidney and suet out), the remainder of the kidney fat should be hard and brittle. The external fat should have similar qualities. A lack of external and internal kidney fat is associated with excellent cutability.

Points for Consideration in Judging Beef Loins

MUSCLING

1. Loin eye — size, shape
2. Back or shortloin — fullness, width
3. Sirloin end — depth, width, size, meatiness
   a. Top sirloin — size
   b. Bottom sirloin (knuckle) — size

QUALITY

1. Loin eye (rib end)
   a. Marbling — amount, fineness, distribution
   b. Color — youthfulness, brightness of lean
   c. Firmness of lean
   d. Texture of lean — fineness

2. Sirloin end: Same as rib end
3. External fat — whiteness, firmness, uniformity

TRIMNESS

1. Eye
2. Loin edge
3. Flank edge
4. Back or shortloin
5. Sirloin end
   a. Top sirloin
   b. Botton sirloin
6. Seam fat in sirloin end
7. Channel fat (pelvic fat)

Judging Beef Chucks (Square-cut)

Square-cut beef chucks are no longer used for intercollegiate judging classes because they are not traded heavily anymore. Instead, chucks are divided into their sub-primals inside and outside (clod) for the trade. Nevertheless, opportunities still exist to merchandise square-cut chucks, and the chuck is the heaviest wholesale cut in a beef carcass, so this discussion is included.

Beef chucks are utilized primarily for pot roasts, Swiss and braised steaks, boiling beef, stew beef, dried beef, and ground meat. The blade and arm ends of the chucks present a considerable cut area for judging trimness and muscling differences. Quality should have less emphasis than muscling and trimness in judging beef chucks. (See Figure 25–1 for points to consider in judging beef chucks.)

MUSCLING — Uniformity of depth is important. The arm end should be rounded and heavily muscled and should not fall away too rapidly into the cross-rib region (arm end of the fourth and fifth chuck ribs, formerly called the English cut). The blade end should be deep and give the appearance of plumpness as opposed to flatness. A very good indication of meatiness and plumpness is the prominence of the shoulder joint. The neck should be short and blend in with the rest of the chuck. A long, flat neck is objectionable.

TRIMNESS — Discount soft, oily fats. A tendency toward an undesirable heavy fat deposit over the clod muscle in the center of the shoulder is evident in highly finished chucks. Large fat deposits between muscles (intermuscular fat) must be discounted.

QUALITY — The color of the lean should be a bright cherry red. Degrade the darker colors of red. Marbling should be slight to modest on the blade end. The same degree of marbling will not be evident on the arm end, since the muscles are attached to the much exercised shank muscles. The surface of the meat should present a smooth, velvety appearance and should be firm and not watery.

Judging Fresh Hams

Trimness

Commercial fresh hams are trimmed into skinned hams, thus removing the major part of the fat from about two-thirds of the surface area of the fresh

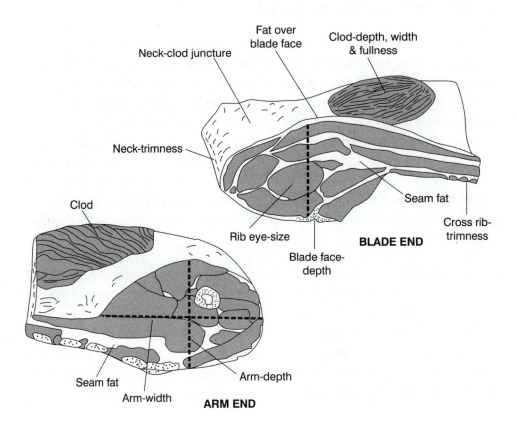

Fig. 25–1. Points for consideration in judging beef chucks.

MUSCLING

1. Clod — width, depth, fullness
2. Neck–clod juncture — depth, fullness
3. Neck — fullness, meatiness
4. Blade end — depth, size, meatiness
5. Arm end — depth, width, size
6. Eye — size

QUALITY

1. Blade end
   a. Color — youthfulness, brightness of lean
   b. Marbling — amount, fineness, distribution
   c. Firmness of lean
   d. Texture of lean — fineness

2. Arm end: Same as blade end
3. External fat — whiteness, firmness, uniformity

TRIMNESS

1. Clod
2. Neck–clod juncture
3. Neck
4. Arm end
5. Blade end
6. Throat region
7. Cross rib or English cut
8. Seam fat in arm end
9. Seam fat in blade end

ham; however, variations in trimming could result in excess fat remaining under the rump (butt) face, which should be penalized. A heavy layer of heel and collar fat and prominent amounts of intermuscular fat showing in the rump (butt) face also should be discounted.

Muscling

Basically, a fresh ham may be considered in three parts: rump (butt), center-cut, and shank/hock. Pricewise, the center-cut is the profitable item, the rump (butt) is a moderately profitable item, and the shank/hock is a loss item.

This price picture reflects the pattern of what a fresh ham should have in its conformation. A short, slim shank/hock with a moderate bulge at the heel means less weight in this loss item. A long (from aitch bone to 1 inch above stifle joint) center-cut that is deep (the distance through the ham from inside to outside) and has a good proportionate depth will increase the profit. A ham rump (butt) that is full-fleshed rather than pointed will throw more weight into this cut and will more nearly cover the loss in the shank/hock. The overall appearance of a fresh ham with the desired conformation features meatiness, plumpness, depth, and general trimness, with as much weight as possible represented in the more valuable center section.

Quality

The quality of the lean is an important factor to consider in the judging of fresh hams because it relates to the final quality and smokehouse yield of the cured cuts. Quality in fresh hams is reflected in the firmness and color of the lean and fat. Soft, oily, and off-color fat should be discounted.

The texture of the lean is very similar for hogs in the same age bracket. The grain and color of the lean change with the increased age of the animal. The desirable color for pork is bright pink, but it is more often a grayish pink tinged with red, and in many cases, the muscles next to the *ilium* bone *(iliopsoas, gluteus profundus, gluteus accessorius)* may be dark, giving a *two-toned condition.* Other things being equal, the brighter, more evenly colored lean is preferred. Hams that are very pale in color and soft and watery in texture should be severely discounted (e.g., see PSE pork, Chapter 22). (See the *Meat Evaluation Handbook,* published by the American Meat Science Association in cooperation with the National Cattlemen's Beef Association and the National Pork Producers Council, for points to consider in judging fresh hams.)

Points for Consideration in Judging Fresh Hams

MUSCLING

1. Ham — length, width, depth, plumpness

2. Butt face — depth, width, size

3. Knuckle or forecushion — depth, fullness

4. Heel — plumpness, meatiness

QUALITY

1. Firmness of butt face

2. Color of butt face — grayish pink, brightness, uniformity

3. Marbling — amount, fineness, distribution

4. Texture of butt face — fineness

TRIMNESS

1. Butt face — underneath

2. Knuckle or forecushion

3. Collar

4. Cushion

5. Rump portion

6. Heel

7. Seam fat in butt (rump) face

SPECIFICATION CLASSES

In an effort to improve the usefulness of meat judging as an educational activity, the intercollegiate coaches association elected to include Institutional Meat Purchase Specifications (IMPS) classes in the 1982 contests on a trial basis. This idea was very well received by the beef fabrication industry because it provided a meaningful training experience for college graduates entering the meat field. At the 1982 Reciprocal Meats Conference, a committee was established to finalize rules and guidelines for IMPS classes in the 1983 and subsequent contests.

Rules and Guidelines

- No rulers or copies of the specifications will be provided or used during the contest.

- There will be a total of 10 cuts per contest. The cuts will be divided into two classes of five each, with 10 minutes allowed per class.

- Scorecards have the defect codes listed on the left side and the exhibit numbers across the top (see Figure 25–2). Contestants should place a mark on the "Acceptable, meets all specifications" line, or they should mark up to three defects. A defect code number can only be used one time per exhibit.

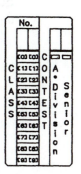

INTERCOLLEGIATE MEAT JUDGING CONTEST

SPECIFICATIONS CLASS

DEFECT	1	2	3	4	5		6	7	8	9	10	
EXHIBIT NUMBER												
1. Bone, cartilage, connective tissue and backstrap removal	▭	▭	▭	▭	▭	1	▭	▭	▭	▭	▭	1
2. Breast flap removal	▭	▭	▭	▭	▭	2	▭	▭	▭	▭	▭	2
3. Chine/feather bone removal	▭	▭	▭	▭	▭	3	▭	▭	▭	▭	▭	3
4. Deckle and associated fat removal	▭	▭	▭	▭	▭	4	▭	▭	▭	▭	▭	4
5. Diaphragm and/or hanging tender removal	▭	▭	▭	▭	▭	5	▭	▭	▭	▭	▭	5
6. Fat exceeds recommended depth	▭	▭	▭	▭	▭	6	▭	▭	▭	▭	▭	6
7. Fat/skin not trimmed or tapered to proper length	▭	▭	▭	▭	▭	7	▭	▭	▭	▭	▭	7
8. Fat not entirely removed	▭	▭	▭	▭	▭	8	▭	▭	▭	▭	▭	8
9. Foot/shank removal location	▭	▭	▭	▭	▭	9	▭	▭	▭	▭	▭	9
10. Hip bone or cartilage exposed or removed	▭	▭	▭	▭	▭	10	▭	▭	▭	▭	▭	10
11. Length of cut	▭	▭	▭	▭	▭	11	▭	▭	▭	▭	▭	11
12. (False) Lean exposure	▭	▭	▭	▭	▭	12	▭	▭	▭	▭	▭	12
13. Lymph gland and associated fat present or removed	▭	▭	▭	▭	▭	13	▭	▭	▭	▭	▭	13
14. Muscle number incorrect	▭	▭	▭	▭	▭	14	▭	▭	▭	▭	▭	14
15. Muscle relative size incorrect	▭	▭	▭	▭	▭	15	▭	▭	▭	▭	▭	15
16. Muscle(s) are not firmly attached	▭	▭	▭	▭	▭	16	▭	▭	▭	▭	▭	16
17. (Not) Perpendicular or parallel at juncture or length of cut	▭	▭	▭	▭	▭	17	▭	▭	▭	▭	▭	17
18. Quality grade/Yield grade	▭	▭	▭	▭	▭	18	▭	▭	▭	▭	▭	18
19. Rib mark incorrect	▭	▭	▭	▭	▭	19	▭	▭	▭	▭	▭	19
20. Rib number incorrect	▭	▭	▭	▭	▭	20	▭	▭	▭	▭	▭	20
21. Split tails too long/not perpendicular	▭	▭	▭	▭	▭	21	▭	▭	▭	▭	▭	21
22. Sacral/caudal vertebrae number or removal	▭	▭	▭	▭	▭	22	▭	▭	▭	▭	▭	22

ACCEPTABLE

	1	2	3	4	5		6	7	8	9	10	
23. Meets all specifications	▭	▭	▭	▭	▭	23	▭	▭	▭	▭	▭	23

	1	2	3	4	5	6	7	8	9	10
ACCEPT / DEFECT 1										
DEFECT 2										
DEFECT 3										

SCORE ☐☐☐☐☐   ☐☐☐☐☐

TOTAL SCORE _____

Fig. 25–2. Example of a meat judging contest scorecard. (Courtesy, *Intercollegiate Meat Judging Contest Official Handbook*)

- Each exhibit will be worth 10 points, for a total of 100 points per individual for the contest. (Team points total 400 per contest.) Scoring is as follows:

Situation		Points
(a)	Correct	10
(b)	1 incorrect answer	7
(c)	2 incorrect answers	3
(d)	3 or more incorrect answers or no correct answers	0
(e)	If cut is acceptable and student marks a defect	0
(f)	If cut is unacceptable and student marks acceptable	0
(g)	If no correct answers are given for a cut	0
(h)	More than 3 answers given for a cut	0

- Each exhibit will be labeled with the correct IMPS number and name.

- Exhibits may not be touched or moved.

- Each contestant should assume that unexposed areas of a cut are acceptable.

- On cuts that have been fabricated to be used as rejects, the defect in each should be obvious in order to eliminate any questions as to the reason for rejecting the cut.

- Factors that would be cause for rejection because of conditions, such as ragged edges, bone dust, etc., will not be considered in the contest.

REASONS

A contestant's ability to tell by oral or written word the reasons why one carcass or cut is superior to another depends on training and experience, knowledge of meat terminology, and method of presenting the reasons. Written reasons are unlike oral reasons in that the writer must be much more concise and efficient in presenting arguments to defend the placing in question. In most intercollegiate meat judging contests, the difference between first and fifth place teams is often found in the reasons scores. Like other competitive activities, high scoring reasons in meat judging contests are the result of the writers' dedication of time, effort, and thought to the reasons writing practice.

The card upon which meat reasons are written is divided into four equal parts, headed respectively by "First," "Second," "Third," and "Fourth." The first section is designated for presenting the reasons for placing the top exhibit over the second place exhibit. The second section of the card is used to support placing the second choice over the third, and the third section contains the comparison of the third and fourth place exhibits. The fourth section presents evidence as to why the exhibit that placed last is the least desirable in the class.

If the No. 2 carcass or cut is placed first in the class over the No. 1, in second place, it is suggested that the pair placing be listed as 2/1, meaning 2 over 1, at the beginning of the first section of the reasons card. Listing the pair of exhibits to be discussed at the start of each section helps the writer construct the paragraph without being confused, as well as helping the person who is reading and grading the reasons.

For each of the five classes in a meat judging contest, a 15-minute period is allowed for the contestants to write their set of reasons. An intermission of three minutes is given between each of the writing sessions for the participants to review and organize their notes for the next class. Five sets of reasons are written by each contestant during the reason-writing segment of the competition. Being able to achieve the same level of work quality and concentration when writing the fifth set as the first set necessitates practice. Reasons writing is tiring work.

An official committee member will rapidly read and grade the reasons written by every contestant for that particular class. Reasons that are well-organized, clear, concise, and neatly written in legible form will catch the grader's eye and score better than those that make the same points in a less readable manner. Hard work to improve penmanship, legibility, punctuation, spelling, and neatness will increase reasons scores as effectively as developing meat judging terminology and learning to see trimness, muscling, and quality differences.

The most important aspect of reasons writing is accuracy. The writers should double check their placings to verify that they are writing about the same cut or carcass that they are thinking about.

Giving the reasons in a simple, easy-to-follow manner will favorably impress the scorer of the reasons. Inventing differences, unseen by the official judge, to strengthen the argument for a placing is usually more costly in points than leaving space blank in a paragraph. Meat judging reasons are an example of *comparative and descriptive, **not** creative, writing.*

It is important for most meat judges to develop their own style of reasons writing. Reasons for different classes may be organized the same, use many of the same terms, and even look the same. Since the reasons for different classes are not read and scored by the same person, there may be no harm in that similarity. However, within each set of reasons, it is important to use some variety in the organization; to avoid repeating terms, especially connectors, verbs, and modifiers; and to vary the sequence of categories (trimness, muscling, quality) discussed, depending upon the situation. Grading reasons can be tedious, and adding variety makes the task easier for the grader, which may enhance the score. Developing a style is not the same thing as writing the same set of reasons for each class and simply varying a few terms to fit the species or cut. Such *stereotyped* or *canned* reasons should be avoided because they will not identify the important differences in a specific class. Understanding meat judging terminology is essential, and being able to generate a logical flow of reasoning using accurate terminology takes practice. No two people express themselves alike, so each reasons writer must work to develop an individual style that is sufficiently flexible to accommodate variation between and within the different classes.

Comparisons that identify the superior qualities of one cut or carcass over another are more convincing and effective arguments to defend a placing than *descriptions* of the merits and faults of the carcasses or cuts in the class. Whatever the style, the opening and the closing must be strong, since these are the first and last statements the official scorer reads. The most important factors should be stressed first as the primary basis for placing one carcass or cut over another. In *close placings,* if differences in small details or less important traits will contribute to the placing decision, they should be mentioned in the reasons; otherwise, they may not be worth the time and space required to include them. It is preferable to cite all the advantages of a particular trait that support the placing, e.g., all the muscling advantages, before moving to the next trait rather than mixing statements about the various traits (trimness, muscling, and quality) throughout the paragraph.

Judging usually involves placing one carcass or cut over another because the first had a greater balance of more desirable traits than the second; however, different carcasses and cuts vary in the degree of emphasis on cutability and quality, as indicated in the following comparison. Awareness of the relative emphasis not only will influence the placing of the class but also may alter the sequence of factors discussed in the reasons writing of a pair comparison. Even though placing the class correctly is important, a high reasons score on an incorrect placing may result if the class is correctly analyzed and the important points are emphasized in legible, well-organized reasons.

Quality Classes	*Equal*	*Cutability Classes*
Beef ribs	Beef carcasses	Beef rounds
Beef loins		Hams
		Pork carcasses
		Lamb carcasses

General Rules for Reasons Writing

1. Use the past tense throughout your writing.

2. Use a variety of connecting words.

3. Use a variety of verbs.

4. Use a variety of words denoting *degrees of difference*.

5. In 2, 3, and 4, remember the meaning of the words you use. Do not plug in words indiscriminately.

6. Try to write something worthwhile and meaningful. Since you must write a full page of words, put forth a little more effort.

7. Make opening and closing comparisons especially strong, because the opening comparison will influence the judge's thinking through the reasons, and the final comparison will be made immediately before the judge records the reasons score.

Grants

Many times *grants* are necessary in the thorough analysis of a class. A grant is an admission that an exhibit that has been placed below another has its merits also. In an extremely close placing, the grants may take as much consideration in your reasons as the defense of the placing; however, in most cases, the grants are brief and cover only a few points.

Do not neglect grants if they are important in the comparison.

Penmanship and Grammar

Reasons writing will give you training in neatness and good penmanship. Through the corrections made, correct spelling and good sentence structure will also be a part of the training. Just as oral reasons give you training in speaking, written reasons give you training in writing.

Terminology

Before reasons writing can even be attempted, a knowledge of the terminology used for carcasses and cuts is essential. If you do not know the accepted carcass nomenclature, how can you accurately convey an idea to anyone else?

General and specific terms or statements may be used singly or several may be used in sequence; however, if you make a broad general statement, never forget to ask yourself where, how, or why to describe this statement more specifically.

Quality

General

1. More youthful
2. Higher quality
3. Firmer

Specific

1. A higher degree
2. A greater amount of
3. More extensive
4. More evenly dispersed
5. More evenly distributed
6. Redder rib bones
7. Lighter, brighter cherry red beef color
   Lighter, brighter grayish pink pork color
   Lighter, brighter reddish pink lamb color
8. Smoother, finer-textured
9. More finely dispersed marbling
10. More evenly dispersed feathering
11. Softer, more pearly white buttons (beef)
12. Whiter, flakier fat
13. More uniformly covered
14. Firmer, thicker flank
    Firmer, thicker side

Last Place

1. Underfinished
2. Lacked quality
3. Wasty

4. Dark-colored
5. Soft, oily
6. Hard-boned

Muscling

General

1. More symmetrical
2. Meatier

3. Heavier muscled
4. Thicker fleshed

Specific

1. Bulging
2. Thicker
3. Deeper
4. Fuller
5. Plumper
6. Wider

7. Shorter
8. Longer
9. Shorter shanked
10. Shorter necked
11. Larger
12. Deeper chined

Last Place

1. Ill-shaped
2. Small
3. Angular
4. Lacked meatiness

5. Long-shanked, thin, tapering, round
6. Long, thin-fleshed ham (leg)
7. Poorly balanced

Useful Terms

Comparative Verbs

1. As shown, showed
2. Displayed
3. Possessed
4. Exhibited

5. Indicated
6. Demonstrated
7. Lacked

Grants

1. Realize
2. Although
3. Grant

4. However
5. Recognize
6. Admit

Degrees of Comparison

1. Unsurpassed
2. Somewhat
3. Much
4. Great
5. Distinctly
6. Large
7. Extreme
8. Superior
9. Excessive
10. Higher degree
11. Little
12. Slightly
13. Smaller
14. Limited
15. Lesser
16. Lower

Connectives

1. Furthermore
2. In addition
3. Also
4. Carrying into
5. Along with
6. Resulting in
7. As evidenced by
8. Blending into
9. Characterized by
10. For being
11. Contributing to
12. Coupled with
13. Shown by

Commonly Misspelled Words

1. Blade
2. Bulging
3. Carrying
4. Chine
5. Conformation
6. Desirably
7. Exudative
8. Feathering
9. Heel
10. Length
11. Loin
12. Meatier
13. Muscling
14. Quality
15. Quantity
16. Symmetrical
17. Thoracic
18. Trimmer
19. Value

Notes

A good share of successful reasons writing may be attributed to accurate and thorough note taking. Since reasons must be written in a limited time, the notes taken during judging need to be organized and clear. They should contain enough phrase terms that can easily be incorporated into the reasons, and they should remind you of the class you are writing about.

Before recording a single note, make certain you are on the correct side of the class and that the placing is the one you want. Take notes *systematically* in a logical order; use abbreviations and short sentences to save time. Take the notes in pairs just as the class is placed; i.e., if the class is placed 1–2–3–4, start with the first pair 1/2; then 2/3; followed by 3/4, and finally 4. Record the merits of one carcass or cut over another on the left side of the page and the grants concerning the pair on the right side, separated by a ruled line for accuracy. Divide your note cards into sections for muscling, trimness, and quality of each pair; however, wait until you see the class before making these divisions, because you want obvious characteristics listed first. By doing this, you can later write your reasons in order directly from your note cards. Underline important facts and double underline those notes and reasons you wish to emphasize more. *Remember, there is no substitute for good notes.*

Sample Reasons

Sample reasons are found in the *Meat Evaluation Handbook,* published by the American Meat Science Association in cooperation with the National Cattlemen's Association and the National Pork Producers Council.

HAM AND BACON SHOWS

At each annual convention, the American Association of Meat Processors (AAMP)[3] holds a national cured meats show known as the American Cured Meat Championships. Many states have similar competition prior to the national show. Any meat processor belonging to AAMP is eligible to enter a ham, bacon, or sausage product.

Classes of Entries

Several classes of products are open for competition among AAMP members, as follows:

> Country Ham — Smoked
> Country Ham — Unsmoked

---

[3]P.O. Box 269, Elizabethtown, PA 17022. E-mail: aamp@aamp.com

Bone-in Ham — Lightweight
Bone-in Ham — Heavyweight
Boneless Ham
Semi-boneless Ham
Sectioned and Formed Ham
Country Bacon — Dry Cured
Bacon
Summer Sausage — Cooked
Summer Sausage — Uncooked
Non-fermented Semi-dried Sausage
Cooked Ring Bologna
Dried Beef
Jerky — Whole Muscle
Jerky — Restructured
Meat Snack Stick
Smoked Turkey
Small Diameter Smoked & Cooked Sausage
Frankfurters/Wieners — Emulsified
Frankfurters/Wieners — Coarse Ground
Innovative Beef Product
Innovative Pork Product

Standards of Evaluation

Since judges may change from year to year and there may be considerable difference of opinion as to the characteristics of the ideal or average product, entries are judged according to standards (a total of 1,000 points is possible for each class) established by AAMP or any state or local organization that is sponsoring a contest.

MEAT CONTESTS FOR ALL INTERESTED STUDENTS

Competition is undoubtedly one of the greatest instruments for arousing interest among young people. It has proved so effective that it has actually become a so-called final examination for many courses offered in schools and colleges.

This method of fostering interest in the youth of the United States has been applied by the FFA and the 4–H Club Congress to various phases of their work. Contests of national importance that have to do with meat are the

Meat Judging Contest and the Meat Identification Contests held for the FFA and for the National 4–H organization.

Meats Evaluation and Technology Career Development Event

The Meats Evaluation and Technology Career Development Event is a special project of the National FFA Foundation, Inc.[4] Members complete a written test; judge quality, yield, and grade of beef carcasses; and evaluate and place beef, lamb, and pork carcasses. Members also complete a meat formulation problem and cooperatively fulfill a beef carcass order. This textbook is used as a reference.

An example of one meat identification scorecard is illustrated in Figure 25–3.

Coaching a Team for Meat Identification

Charts that show the location of the wholesale and retail cuts of a carcass are of value to students in memorizing the location of the cuts in the carcass. The study of photographs alone for identification of cuts is of doubtful value. Whether purchaser or student, it is absolutely necessary that one come in direct contact with the product through actual purchasing, by learning the cuts through meat displays, or by doing the actual cutting. It is rather useless for vocational schools to enter meat judging or meat identification teams in national contests with no coaching other than from photographs.

Suggested Coaching Methods

One of the most effective ways of helping students obtain the necessary training is to place them as helpers for certain hours during the week, with competent and interested meat retailers, with the understanding that the students be given every opportunity to learn the names of the wholesale and retail cuts and to help make those cuts whenever convenient. The students work with or without pay, according to agreement. Several weeks or a month

---

[4]http://www.FFA.org/activities/cde/general/about.htm

Contestant Number: _____

4-H Meat Retail Cut Identification

Species	Primal Cut	
B – Beef	A. Breast	J. Rib (Rack)
L – Lamb	B. Brisket	K. Round
P – Pork	C. Chuck	L. Shank
	D. Flank	M. Shoulder
	E. Ham/Leg	N. Side (Belly)
	F. Jowl	O. Spareribs
	G. Leg	P. Variety Meats
	H. Loin	Q. Various
	I. Plate	

Retail Names
Chops, Roasts/Pot Roasts, Steaks, Slices

1. American-Style	31. Round
2. Arm	32. Round, Boneless
3. Arm, Boneless	33. Rump, Boneless
4. Arm Picnic, Whole	34. 7-Bone
5. Bacon	35. Shoulder, Boneless
6. Blade	36. Sirloin
7. Blade Boston	37. Sirloin, Boneless
8. Bottom Round	38. Sirloin, Flat Bone
9. Butterfly	39. Sirloin, Half
10. Center	40. Sirloin, Pin Bone
11. Chuck Eye, Boneless	41. Sirloin, Round Bone
12. Cubed	42. Sirloin, Wedge Bone
13. Double	43. Skirt, Boneless
14. Eye Round	44. Smoked Center
15. Flank	45. Smoked Ham, Boneless
16. Frenched-Style	46. Smoked Loin
17. Fresh, Center	47. Smoked Rib
18. Fresh, Rump Portion	48. Smoked Rump Portion
19. Fresh, Shank Portion	49. Smoked Shank Portion
20. Fresh Side	50. Smoked Shoulder Picnic, Whole
21. Heel of Round	51. Square Cut
22. Loin	52. T-Bone
23. Mock Tender	53. Tenderloin
24. Neck	54. Tip
25. Porterhouse	55. Tip, Cap Off
26. Rib	56. Top Loin
27. Rib Eye	57. Top Loin, Boneless
28. Rib, Large End	58. Top Loin (Double), Bnls
29. Rib, Small End	59. Top Round
30. Rib, Small End, Bnls	60. Top Sirloin

Variety Meats

61. Brains	65. Oxtail
62. Heart	66. Sweetbreads (beef thymus)
63. Kidney	67. Tongue
64. Liver	68. Tripe

Miscellaneous Other Cuts

69. Back Ribs	81. Ground Beef
70. Bacon, Slab	82. Ground Pork
71. Beef for Stew	83. Point Half, Boneless
72. Breast	84. Riblets
73. Brisket, Whole, Bnls	85. Sausage
74. Canadian Style Bacon	86. Sausage Links
75. Corned	87. Shank
76. Country Style Ribs	88. Short Ribs
77. Cross Cuts	89. Sirloin Cutlets
78. Cross Cuts, Boneless	90. Smoked Hock
79. Flat Half, Boneless	91. Smoked Jowl
80. Fresh Hock	92. Spareribs

Type of Cut	Cookery Method
Ch – Chop(s)	D – Dry Heat
Ro – Roast/Pot Roast	M – Moist Heat
Sl – Slice(d)	D/M – Dry or Moist Heat
St – Steak	

Cut No.	Species (1 pt)	Primal Cut (1 pt)	Retail Name (1 pt)	Type (1 pt)	Cookery Method (1 pt)
Ex.	Lamb **L**	Leg **G**	Center **10**	Slice **Sl**	Dry **D**
1.	_____	_____	_____	_____	_____
2.	_____	_____	_____	_____	_____
3.	_____	_____	_____	_____	_____
4.	_____	_____	_____	_____	_____
5.	_____	_____	_____	_____	_____
6.	_____	_____	_____	_____	_____
7.	_____	_____	_____	_____	_____
8.	_____	_____	_____	_____	_____
9.	_____	_____	_____	_____	_____
10.	_____	_____	_____	_____	_____
11.	_____	_____	_____	_____	_____
12.	_____	_____	_____	_____	_____
13.	_____	_____	_____	_____	_____
14.	_____	_____	_____	_____	_____
15.	_____	_____	_____	_____	_____
16.	_____	_____	_____	_____	_____
17.	_____	_____	_____	_____	_____
18.	_____	_____	_____	_____	_____
19.	_____	_____	_____	_____	_____
20.	_____	_____	_____	_____	_____
21.	_____	_____	_____	_____	_____
22.	_____	_____	_____	_____	_____
23.	_____	_____	_____	_____	_____
24.	_____	_____	_____	_____	_____
25.	_____	_____	_____	_____	_____
26.	_____	_____	_____	_____	_____
27.	_____	_____	_____	_____	_____
28.	_____	_____	_____	_____	_____
29.	_____	_____	_____	_____	_____
30.	_____	_____	_____	_____	_____
Number Incorrect	_____	_____	_____	_____	_____
	×1	×3	×4	×1	×1
Points Off	_____	_____	_____	_____	_____

300 – _____ **=** _____

Total Points Off   Final Score

Fig. 25–3. Example of a meat identification scorecard.

of such training for students will teach them more than is possible by any other method.

Another method that is more convenient, but not as fruitful, is one in which arrangements are made with one or more meat retailers to give the students practice in identifying available cuts one evening a week.

Still another method that can be employed with some degree of satisfaction is one in which the coach makes frequent shopping tours with one or two members of the class to engage in identifying showcase display cuts. Under this method, it is, of course, a prerequisite that the instructor must know the meats. More than two students in a group is unwieldy and may crowd the otherwise already busy shop.

Under any system of coaching, a list of the cuts as published by the group that is responsible for the contest should be used so that there will be no time lost in identifying cuts other than those indicated on the chart.

When there are more trained contestants than the number required to constitute a team, an elimination contest can be arranged by the meat department of the state agricultural university and(or) college.

Identification Features

Chapters 13 to 17 provide the basis for gaining a foundation in the knowledge of cut identification and should be read before a great deal of direct contact is attempted. Variety meat pictures appear in Figure 11–1.

The first task that confronts a contestant is that of determining whether the cut is beef, pork, veal, or lamb. These four kinds of meat are recognized by:

- *Color of the lean:* Beef varies from bright to dark red. Pork is grayish pink to grayish red. Veal is light grayish pink. The older the veal, the more it borders on dark grayish pink. Lamb is light, reddish pink. Mutton is brick red.

- *Size of the cut:* Beef cuts are large in size. Pork and veal cuts run similar in size. Lamb cuts are small in size.

- *Type of fat:* Beef has a white or cream-white (yellow in the lower grades), firm, and rather dry fat. Pork has a characteristic white, greasy fat. Veal is readily recognized by the absence of fat. Lamb has a chalk-white, brittle, rather dense fat, usually covered with the fell, a colorless connective tissue membrane.

Having identified the cut as to kind, the next task is to identify the cut, both as to name and as to the wholesale cut from which it was derived. (Refer to the color section near the center of this text for retail cut identification assistance.) This requires a familiarity with anatomy to determine location by the shape of the bone and the shape and contour of the muscles. The contestant must remember that the difference between a roast and a steak of the same name is merely one of thickness. Steaks are generally from $\frac{1}{2}$ to $1\frac{1}{2}$ inches thick, whereas roasts are over 2 inches thick.

Appendix

ELECTRICITY

Definitions

Ampere — the unit of electrical current strength; such a current as would be given with an electromotive force of 1 *volt* through a wire having a resistance of 1 *ohm;* the rate of flow of 1 *coulom* per second.

$$= \frac{\text{volts}}{\text{ohms}}$$

Coulom — the amount of electricity conveyed by 1 *ampere* in one second; the quantity of electricity required to liberate 0.001118 gram of silver from a solution of an ionizable silver salt = 6.3×10^{18} electrons.

Dyne — the fundamental unit of force in the centimeter, gram, second (cgs) system that if applied to 1 gram would give it an acceleration of 1 centimeter per second.

eV — an **electron volt** (eV) is equivalent to the kinetic energy that an electron would acquire while moving through a potential gradient of 1 volt (an extremely small energy unit).

Erg — in the cgs system, a unit of work and of energy; the work done in moving a body 1 centimeter against the force of 1 dyne. To convert eV to ergs, multiply eV by 1.60×10^{-12}.

Faraday — the quantity of electricity required to liberate 1 gram-equivalent weight of an ion = 96,494 couloms = gram equivalent weight of silver =

$$\frac{107.88}{0.001118}$$

Hertz — a unit of electromagnetic wave frequency equal to one cycle per second.

Joule — the meter, kilogram, second *(mks)* system unit of work equivalent to the force of 1 newton acting through a distance of 1 meter and equal to 10^6 ergs or 0.737324 foot-pound.

Newton — a unit of force in the mks system, which equals 100,000 dynes.

Ohm — the unit of electrical resistance equal to the resistance of a conductor carrying a current of 1 ampere at a potential difference of 1 volt between the terminals; the resistance at 0°C of a uniform column of mercury having a mass of 14.4521 grams and a length of 106.3 centimeters.

Volt — the unit of electromotive force, or that difference of potential that when steadily applied against the resistance of 1 ohm, will produce a current of 1 ampere.

Watt — a unit of power, especially electrical power, which is equivalent to 1 joule per second.

THE ELECTROMAGNETIC RADIATION SPECTRUM[1,2]

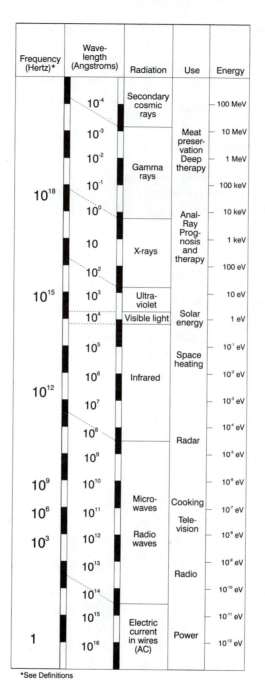

Frequency (Hertz)*	Wave-length (Angstroms)	Radiation	Use	Energy
	10^{-4}	Secondary cosmic rays		100 MeV
	10^{-3}		Meat preser-vation	10 MeV
	10^{-2}	Gamma rays	Deep therapy	1 MeV
	10^{-1}			100 keV
10^{18}	10^{0}		Anal-Ray Prog-nosis and therapy	10 keV
	10	X-rays		1 keV
	10^{2}			100 eV
10^{15}	10^{3}	Ultra-violet		10 eV
	10^{4}	Visible light	Solar energy	1 eV
	10^{5}		Space heating	10^{-1} eV
	10^{6}	Infrared		10^{-2} eV
10^{12}	10^{7}			10^{-3} eV
	10^{8}		Radar	10^{-4} eV
	10^{9}			10^{-5} eV
10^{9}	10^{10}			10^{-6} eV
	10^{11}	Micro-waves	Cooking	10^{-7} eV
10^{6}	10^{12}	Radio waves	Tele-vision	10^{-8} eV
10^{3}	10^{13}			10^{-9} eV
	10^{14}		Radio	10^{-10} eV
	10^{15}	Electric current in wires (AC)		10^{-11} eV
1	10^{16}		Power	10^{-12} eV

*See Definitions

[1]Electromagnetic waves reflect, diffract, and interfere like acoustic waves (i.e., ultrasound waves). However, longitudinal acoustic waves differ from electromagnetic waves in that acoustic waves cannot be polarized, and when they hit a boundary of a solid medium, their velocity is changed.

[2]Electron-beam (e-beam) irradiation generates 4 to 10 MeV (see page 720).

METRIC SYSTEM

LENGTH

Millimeter (mm) . . = Meter ÷ 1,000

Centimeter (cm) . . . = Meter ÷ 100

Decimeter (dm). . . . = Meter ÷ 10

Basic Unit . . Meter (m)

Decameter (dkm). . . . = Meter × 10

Hektometer (hm). . . = Meter × 100

Kilometer (km). . . = Meter × 1,000

Myriameter (mym) = Meter × 10,000

1 mm . . . . 0.0394 in.

1 cm. . . . . 0.3937 in.

1 dm . . . . 3.937 in.

1 m . . . . 39.37 in.

1 dkm . . . . 32.81 ft.

1 hm . . . . 328 ft. 1 in.

or 109 yd. 13 in.

1 km . . . . 3280 ft. 1 in.

or 0.62 mi.

1 mym . . . . 6.2 mi.

2.540 cm. . . . . 1 in.

30.480 cm . . . . 1 ft.

0.914 m . . . . 1 yd.

1.609 km . . . . 1 mi.

CAPACITY

Milliliter (ml). . . . . = Liter ÷ 1,000

Centiliter (cl) . . . . . . = Liter ÷ 100

Deciliter (dl) . . . . = Liter ÷ 10

Basic Unit . . Liter[1] (l)

Decaliter (dkl) . . . . . = Liter × 10

Hektoliter (hl) . . . . = Liter × 100

Kiloliter (kl) . . . . . = Liter × 1,000

Cubic Capacity

1 ml . . . . 0.06 cu. in.

1 cl. . . . . 0.6 cu. in.

1 dl . . . . 6.1 cu. in.

1 l. . . . . 61.02 cu. in.

1 dkl . . . . 0.35 cu. ft.

1 hl . . . . 3.53 cu. ft.

1 kl . . . . 1.31 cu. yd.

1.67 cl. . . . . 1 cu. in.

2.85 dkl . . . . 1 cu. ft.

0.76 kl . . . . 1 cu. yd.

Liquid Capacity

1 ml. . . . . 0.27 fluidram

or 0.0338 fl. oz.

1 cl . . . . 0.338 fl. oz.

1 dl . . . . 3.3815 fl. oz.

or 0.21 pt.

1 l . . . . 2.1134 pt.

or 1.0567 qt.

or 0.2642 gal.

1 dkl . . . . 9.081 qt.

or 2.64 gal.

0.02957 l . . . . 1 oz.

0.473 l . . . . 1 pt.

0.946 l . . . . 1 qt.

3.785 l . . . . 1 gal.

---

[1]Liter = 1,000 cu cm (approximately 1 qt.). Liter implies cubic capacity, so is *not* prefixed by *cubic*. In the metric system, dry and liquid measures are identical.

Dry Capacity

1 dl . . . . 0.18 pt.

1 l . . . . 1.816 pt.

or 0.9081 qt.

or 0.1135 pk.

or 0.028 bu.

1 dkl. . . . . 18.162 pt.

or 9.0810 qt.

or 0.14 pk.

or 0.28378 bu.

1 hl. . . . . 2.84 bu.

0.550 l . . . . 1 pt.

1.101 l . . . . 1 qt.

8.809 l . . . . 1 pk.

35.238 l . . . . 1 bu.

AREA

Sq. centimeter (sq cm)

. . . . = 0.0001 centare (ca)

or 1 ca ÷ 10,000

Basic Unit . . . Centare (ca)
which is 1 Square Meter (sq m)

Are (a)² . . . . = 1 ca × 100

Hectare (ha) . . . . . = 1 ca × 10,000

Sq. kilometer (sq km)

. . . . . = 1 ca × 1,000,000

1 sq mm . . . . 0.00155 sq. in.

1 sq cm . . . . 0.155 sq. in.

1 sq m or ca . . . . 10.764 sq. ft.

or 1.196 sq. yd.

1 a . . . . 119.60 sq. yd.

1 ha . . . . 2.47 acres

1 sq km . . . . 0.3861 sq. mi.

6.451 sq cm . . . . 1 sq. in.

0.093 sq m . . . . 1 sq. ft.

0.836 sq m . . . . 1 sq. yd.

4047 sq m . . . . 1 acre

2.590 sq km. . . . . 1 sq. mi.

or 640 acres

VOLUME

Cu. centimeter (cu cm)

. . . . . = 0.000001 cu m

or 1 cu m ÷ 1,000,000

Decistere (ds)

. . . . = 0.10 cu m or cu m ÷ 10

Basic Unit . . . Stere (s)
which is 1 Cubic Meter (cu m)

Decastere (dks)

. . . = 10 steres or 1 cu m × 10

1 cu mm . . . . 0.000061 cu. in.

1 cu cm . . . . 0.0610 cu. in.

1 cu dm . . . . 61.02 cu. in.

or 0.035 cu. ft.

1 ds . . . . 3.53 cu. ft.

1 cu m or 1 s . . . . 35.31 cu. ft.

or 1.31 cu. yd.

1 dks or 10 cu m . . . . . 353.1 cu. ft.

or 13.10 cu. yd.

16.387 cu cm . . . . 1 cu. in.

0.028 cu m . . . . 1 cu. ft.

0.765 cu m . . . . 1 cu. yd.

WEIGHT & MASS

Milligram (mg) . . . = Gram ÷ 1,000

Centigram (cg) . . . . = Gram ÷ 100

Decigram (dm) . . . . . = Gram ÷ 10

---

²A metric unit of area, **not** *acre*.

Basic Unit . . . Gram (g)

Decagram (dkg). . . . . = Gram × 10
Hectogram (hg) . . . . = Gram × 100
Kilogram (kg) . . . . = Gram × 1,000
Quintal (q) . . . . = Gram × 100,000
Metric ton (MT) . . . = Quintal × 10

1 mg . . . . 0.015 grain
1 cg . . . . 0.154 grain
1 dg . . . . 1.543 grains
1 g . . . . 0.035 oz.
1 dkg . . . . 0.353 oz.
1 hg . . . . 3.527 oz.
1 kg or 1,000 g . . . . . . 35.274 oz.
or 2.2046 lbs.
1 q or 100,000 g . . . . . 220.46 lbs.

Ton

0.907 MT . . . . 1 short ton
2,000 lbs.
1.016 MT   1 long ton
2,240 lbs.

Hundredweight (cwt)
45.359 kg. . . . . . . 1 short cwt
100 lbs.
50.802 kg . . . . 1 long cwt
112 lbs.

28.349 g. . . . . 1 oz.
453.00 g. . . . . 1 lb.
0.453 kg . . . . 1 lb.
0.907 MT. . . . . 1 short ton
1.016 MT . . . . 1 long ton

MASS & LENGTH

Mass

1 dalton

. . . . . = mass of one hydrogen atom

or 1.67×10^{-24} g

1 picogram . . . . = 1×10^{-12} g

Length

1 nanometer (nm) . . . . . . = 10^{-9} m

= 10 angstroms (Å)

1 micrometer (mm) . . . . . = 10^{-6} m

= 1,000 nm

= 10,000 angstroms (Å)

PREFIXES FOR POWERS OF 10 FOR USE WITH SI UNITS[1]

Value	Prefix	Abbreviation
10^6	mega	M
10^3	kilo	k
10^{-1}	deci	d
10^{-2}	centi	c
10^{-3}	milli	m
10^{-6}	micro	μ
10^{-9}	nano	n
10^{-12}	pico	p
10^{-15}	femto	f
10^{-18}	atto	a

[1]Combinations of prefixes are no longer allowed, so that n- is used instead of mμ and p- instead of μμ-.

Quick Reference for Conversions Between

Volume			Weight		Weight	
qt.	L	gal.	lbs.	kg	g	ounce
1	1	0.3	0.2	0.1	1	0.04
2.1	2	0.5	0.4	0.2	2	0.07
3.2	3	0.8	0.7	0.3	3	0.1
4.2	4	1.1	0.9	0.4	4	0.14
5.3	5	1.3	1.1	0.5	5	0.18
6.3	6	1.6	1.3	0.6	6	0.21
7.4	7	1.8	1.5	0.7	7	0.25
8.5	8	2.1	1.8	0.8	8	0.28
9.5	9	2.4	2	0.9	9	0.32
11	10	2.6	2.2	1	10	0.35
21	20	5.3	4.4	2	20	0.7
32	30	7.9	6.6	3	30	1.1
42	40	11	8.8	4	40	1.4
53	50	13	11	5	50	1.8
63	60	16	13	6	60	2.1
74	70	18	15	7	70	2.5
85	80	21	18	8	80	2.8
95	90	24	20	9	90	3.2
106	100	26	22	10	100	3.5
211	200	53	44	20	200	7.1
317	300	79	66	30	300	11
423	400	106	88	40	400	14
528	500	132	110	50	500	18
634	600	159	132	60	600	21
740	700	185	154	70	700	25
846	800	211	176	80	800	28
951	900	238	198	90	900	32
1,057	1,000	264	220	100	1,000	35
			441	200		
			661	300		
			882	400		
			1,102	500		
1.057*		0.264*	2,205*			0.0353*

*Multiply the SI units by this factor to convert to English units for this column.

System International (SI) and English Units

Area		Temperature		Pressure		Length		
cm²	in.²	°C	°F	kPa	psi	in.	cm	feet
10	1.5	100	212	50	7	0.4	1	
20	3	90	194	75	11	0.8	2	
30	5	80	176	100	14	1.2	3	0.1
40	6	70	158	125	18	1.6	4	
50	8	60	140	150	22	2	5	
60	9	50	122	175	25	2.4	6	0.2
70	11	40	104	200	29	2.8	7	
80	12	35	95	225	33	3.1	8	
90	14	30	86	250	36	3.5	9	
100	16	25	77	275	40	3.9	10	0.3
200	31	20	68	300	43	7.9	20	0.6
300	46	15	59	325	47	12	30	1
400	62	10	50	350	51	16	40	1.3
500	77	5	41	375	54	20	50	1.6
600	93	0	32	400	58	24	60	2
700	108	−5	23	500	73	28	70	2.3
800	124	−10	14	600	87	32	80	2.6
900	139	−15	5	700	101	36	90	3
1,000	145	−20	−4	800	116	39	100	3.3
		−25	−13	900	130	79	200	6.6
		−30	−22	1,000	145	118	300	9.8
		−40	−40			157	400	13
						197	500	16
						236	600	20
						276	700	23
						315	800	26
						354	900	30
						394	1,000	33
	0.155*		(9/5 C) + 32		0.145*	0.3937*		0.0328*

THERMOMETRIC EQUIVALENTS[1]

Temperature Scales

Symbol	Designation	Zero Point	Freezing Point of Water	Boiling Point of Water at Standard atm Pressure
°C	degree Celsius or Centigrade	Freezing point of water	0°C	100°C
°K	degree Kelvin or absolute temperature in degrees Centigrade	Absolute zero	273.15°K	373.15°K
°F	degree Fahrenheit	−17.8°C	+32°F	212°F
°Rank	degree Rankine or absolute temperature in degrees Fahrenheit	Absolute zero	491.4°Rank	671.4°Rank

The following formulas may be used to convert temperatures from one scale to another:

Temperature Given in	Temperature Wanted in			
	°C	°K	°F	°Rank
°C	°C	°C + 273.15	1.8°C + 32	1.8°C + 491.4
°K	°K − 273.15	°K	1.8°K − 459.4	1.8°K
°F	0.556°F − 17.8	0.556°F + 255.3	°F	°F + 459.4
°Rank	0.556°Rank − 273.1	0.556°Rank	°Rank − 459.4	°Rank

[1]Budavari, Susan, et al. (eds.). The Merck Index: An Encyclopedia of Chemicals, Drugs, and Biologicals (11th Ed.). Merck & Co., Inc., Rahway, NJ. 1989. MISC–114.

Conversion Table (Thermometric Equivalents)[1]

To convert a temperature from one scale to the other, find the desired temperature (Fahrenheit or Celsius) in one of the columns of boldface figures. For a Fahrenheit value, the Celsius equivalent will appear in the column to the left. For a Celsius value, the Fahrenheit equivalent will be listed in the column to the right. For example, the Celsius equivalent of 100°F is 37.78°C; the Fahrenheit equivalent of 100°C is 212°F.

To Convert Degrees			To Convert Degrees			To Convert Degrees		
To C	←F or C→	To F	To C	←F or C→	To F	To C	←F or C→	To F
−40	**−40**	−40	−23.33	**−10**	14	−6.67	**20**	68
−39.44	**−39**	−38.2	−22.78	**−9**	15.8	−6.11	**21**	69.8
−38.89	**−38**	−36.4	−22.22	**−8**	17.6	−5.56	**22**	71.6
−38.33	**−37**	−34.6	−21.67	**−7**	19.4	−5	**23**	73.4
−37.78	**−36**	−32.8	−21.11	**−6**	21.2	−4.44	**24**	75.2
−37.22	**−35**	−31	−20.56	**−5**	23	−3.89	**25**	77
−36.67	**−34**	−29.2	−20	**−4**	24.8	−3.33	**26**	78.8
−36.11	**−33**	−27.4	−19.44	**−3**	26.6	−2.78	**27**	80.6
−35.56	**−32**	−25.6	−18.89	**−2**	28.4	−2.22	**28**	82.4
−35	**−31**	−23.8	−18.33	**−1**	30.2	−1.67	**29**	84.2
−34.44	**−30**	−22	−17.78	**0**	32	−1.11	**30**	86
−33.89	**−29**	−20.2	−17.22	**1**	33.8	−0.56	**31**	87.8
−33.33	**−28**	−18.4	−16.67	**2**	35.6	0	**32**	89.6
−32.78	**−27**	−16.6	−16.11	**3**	37.4	0.56	**33**	91.4
−32.22	**−26**	−14.8	−15.56	**4**	39.2	1.11	**34**	93.2
−31.67	**−25**	−13	−15	**5**	41	1.67	**35**	95
−31.11	**−24**	−11.2	−14.44	**6**	42.8	2.22	**36**	96.8
−30.56	**−23**	−9.4	−13.89	**7**	44.6	2.78	**37**	98.6
−30	**−22**	−7.6	−13.33	**8**	46.4	3.33	**38**	100.4
−29.44	**−21**	−5.8	−12.78	**9**	48.2	3.89	**39**	102.2
−28.89	**−20**	−4	−12.22	**10**	50	4.44	**40**	104
−28.33	**−19**	−2.2	−11.67	**11**	51.8	5	**41**	105.8
−27.78	**−18**	−0.4	−11.11	**12**	53.6	5.56	**42**	107.6
−27.22	**−17**	1.4	−10.56	**13**	55.4	6.11	**43**	109.4
−26.67	**−16**	3.2	−10	**14**	57.2	6.67	**44**	111.2
−26.11	**−15**	5	−9.44	**15**	59	7.22	**45**	113
−25.56	**−14**	6.8	−8.89	**16**	60.8	7.78	**46**	114.8
−25	**−13**	8.6	−8.33	**17**	62.6	8.33	**47**	116.6
−24.44	**−12**	10.4	−7.78	**18**	64.4	8.89	**48**	118.4
−23.89	**−11**	12.2	−7.22	**19**	66.2	9.44	**49**	120.2

(Continued)

Conversion Table (Thermometric Equivalents) (Continued)

To Convert Degrees			To Convert Degrees			To Convert Degrees		
To C	←F or C→	To F	To C	←F or C→	To F	To C	←F or C→	To F
10	50	122	29.44	85	185	48.89	120	248
10.56	51	123.8	30	86	186.8	49.44	121	249.8
11.11	52	125.6	30.56	87	188.6	50	122	251.6
11.67	53	127.4	31.11	88	190.4	50.56	123	253.4
12.22	54	129.2	31.67	89	192.2	51.11	124	255.2
12.78	55	131	32.22	90	194	51.67	125	257
13.33	56	132.8	32.78	91	195.8	52.22	126	258.8
13.89	57	134.6	33.33	92	197.6	52.78	127	260.6
14.44	58	136.4	33.89	93	199.4	53.33	128	262.4
15	59	138.2	34.44	94	201.2	53.89	129	264.2
15.56	60	140	35	95	203	54.44	130	266
16.11	61	141.8	35.56	96	204.8	55	131	267.8
16.67	62	143.6	36.11	97	206.6	55.56	132	269.6
17.22	63	145.4	36.67	98	208.4	56.11	133	271.4
17.78	64	147.2	37.22	99	210.2	56.67	134	273.2
18.33	65	149	37.78	100	212	57.22	135	275
18.89	66	150.8	38.33	101	213.8	57.78	136	276.8
19.44	67	152.6	38.89	102	215.6	58.33	137	278.6
20	68	154.4	39.44	103	217.4	58.89	138	280.4
20.56	69	156.2	40	104	219.2	59.44	139	282.2
21.11	70	158	40.56	105	221	60	140	284
21.67	71	159.8	41.11	106	222.8	60.56	141	285.8
22.22	72	161.6	41.67	107	224.6	61.11	142	287.6
22.78	73	163.4	42.22	108	226.4	61.67	143	289.4
23.33	74	165.2	42.78	109	228.2	62.22	144	291.2
23.89	75	167	43.33	110	230	62.78	145	293
24.44	76	168.8	43.89	111	231.8	63.33	146	294.8
25	77	170.6	44.44	112	233.6	63.89	147	296.6
25.56	78	172.4	45	113	235.4	64.44	148	298.4
26.11	79	174.2	45.56	114	237.2	65	149	300.2
26.67	80	176	46.11	115	239	65.56	150	302
27.22	81	177.8	46.67	116	240.8	66.11	151	303.8
27.78	82	179.6	47.22	117	242.6	66.67	152	305.6
28.33	83	181.4	47.78	118	244.4	67.22	153	307.4
28.89	84	183.2	48.33	119	246.2	67.78	154	309.2

(Continued)

Conversion Table (Thermometric Equivalents) (Continued)

To Convert Degrees			To Convert Degrees			To Convert Degrees		
To C	←F or C→	To F	To C	←F or C→	To F	To C	←F or C→	To F
68.33	155	311	87.78	190	374	107.22	225	437
68.89	156	312.8	88.33	191	375.8	107.78	226	438.8
69.44	157	314.6	88.89	192	377.6	108.33	227	440.6
70	158	316.4	89.44	193	379.4	108.89	228	442.4
70.56	159	318.2	90	194	381.2	109.44	229	444.2
71.11	160	320	90.56	195	383	110	230	446
71.67	161	321.8	91.11	196	384.8	110.56	231	447.8
72.22	162	323.6	91.67	197	386.6	111.11	232	449.6
72.78	163	325.4	92.22	198	388.4	111.67	233	451.4
73.33	164	327.2	92.78	199	390.2	112.22	234	453.2
73.89	165	329	93.33	200	392	112.78	235	455
74.44	166	330.8	93.89	201	393.8	113.33	236	456.8
75	167	332.6	94.44	202	395.6	113.89	237	458.6
75.56	168	334.4	95	203	397.4	114.44	238	460.4
76.11	169	336.2	95.56	204	399.2	115	239	469.4
76.67	170	338	96.11	205	401	115.56	240	464
77.22	171	339.8	96.67	206	402.8	116.11	241	465.8
77.78	172	341.6	97.22	207	404.6	116.67	242	467.6
78.33	173	343.4	97.78	208	406.4	117.22	243	469.4
78.89	174	345.2	98.33	209	408.2	117.78	244	471.2
79.44	175	347	98.89	210	410	118.33	245	473
80	176	348.8	99.44	211	411.8	118.89	246	474.8
80.56	177	350.6	100	212	413.6	119.44	247	476.6
81.11	178	352.4	100.56	213	415.4	120	248	478.4
81.67	179	354.2	101.11	214	417.2	120.56	249	480.2
82.22	180	356	101.67	215	419	121.11	250	482
82.78	181	357.8	102.22	216	420.8	121.67	251	483.8
83.33	182	359.6	102.78	217	422.6	122.22	252	485.6
83.89	183	361.4	103.33	218	424.4	122.78	253	487.4
84.44	184	363.2	103.89	219	426.2	123.33	254	489.2
85	185	365	104.44	220	428	123.89	255	491
85.56	186	366.8	105	221	429.8	124.44	256	492.8
86.11	187	368.6	105.56	222	431.6	125	257	494.6
86.67	188	370.4	106.11	223	433.4	125.56	258	496.4
87.22	189	372.2	106.67	224	435.2	126.11	259	498.2

(Continued)

Conversion Table (Thermometric Equivalents) (Continued)

To C	←F or C→	To F	To C	←F or C→	To F	To C	←F or C→	To F
126.67	260	500	146.11	295	563	165.56	330	626
127.22	261	501.8	146.67	296	564.8	166.11	331	627.8
127.78	262	503.6	147.22	297	566.6	166.67	332	629.6
128.33	263	505.4	147.78	298	568.4	167.22	333	631.4
128.89	264	507.2	148.33	299	570.2	167.78	334	633.2
129.44	265	509	148.89	300	572	168.33	335	635
130	266	510.8	149.44	301	573.8	168.89	336	636.8
130.56	267	512.6	150	302	575.6	169.44	337	638.6
131.11	268	514.4	150.56	303	577.4	170	338	640.4
131.67	269	516.2	151.11	304	579.2	170.56	339	642.2
132.22	270	518	151.67	305	581	171.11	340	644
132.78	271	519.8	152.22	306	582.8	171.67	341	645.8
133.33	272	521.6	152.78	307	584.6	172.22	342	647.6
133.89	273	523.4	153.33	308	586.4	172.78	343	649.4
134.44	274	525.2	153.89	309	588.2	173.33	344	651.2
135	275	527	154.44	310	590	173.89	345	653
135.56	276	528.8	155	311	591.8	174.44	346	654.8
136.11	277	530.6	155.56	312	593.6	175	347	656.6
136.67	278	532.4	156.11	313	595.4	175.56	348	658.4
137.22	279	534.2	156.67	314	597.2	176.11	349	660.2
137.78	280	536	157.22	315	599	176.67	350	662
138.33	281	537.8	157.78	316	600.8	177.22	351	663.8
138.89	282	539.6	158.33	317	602.6	177.78	352	665.6
139.44	283	541.4	158.89	318	604.4	178.33	353	667.4
140	284	543.2	159.44	319	606.2	178.89	354	669.2
140.56	285	545	160	320	608	179.44	355	671
141.11	286	546.8	160.56	321	609.8	180	356	672.8
141.67	287	548.6	161.11	322	611.6	180.56	357	674.6
142.22	288	550.4	161.67	323	613.4	181.11	358	676.4
142.78	289	552.2	162.22	324	615.2	181.67	359	678.2
143.33	290	554	162.78	325	617	182.22	360	680
143.89	291	555.8	163.33	326	618.8	182.78	361	681.8
144.44	292	557.6	163.89	327	620.6	183.33	362	683.6
145	293	559.4	164.44	328	622.4	183.89	363	685.4
145.56	294	561.2	165	329	624.2	184.44	364	687.2

(Continued)

Conversion Table (Thermometric Equivalents) (Continued)

To C	←F or C→	To F	To C	←F or C→	To F	To C	←F or C→	To F
	To Convert Degrees			To Convert Degrees			To Convert Degrees	
185	365	689	204.44	400	752	223.89	435	815
185.56	366	690.8	205	401	753.8	224.44	436	816.8
186.11	367	692.6	205.56	402	755.6	225	437	818.6
186.67	368	694.4	206.11	403	757.4	225.56	438	820.4
187.22	369	696.2	206.67	404	759.2	226.11	439	822.2
187.78	370	698	207.22	405	761	226.67	440	824
188.33	371	699.8	207.78	406	762.8	227.22	441	825.8
188.89	372	701.6	208.33	407	764.6	227.78	442	827.6
189.44	373	703.4	208.89	408	766.4	228.33	443	829.4
190	374	705.2	209.44	409	768.2	228.89	444	831.2
190.56	375	707	210	410	770	229.44	445	833
191.11	376	708.8	210.56	411	771.8	230	446	834.8
191.67	377	710.6	211.11	412	773.6	230.56	447	836.6
192.22	378	712.4	211.67	413	775.4	231.11	448	838.4
192.78	379	714.2	212.22	414	777.2	231.67	449	840.2
193.33	380	716	212.78	415	779	232.22	450	842
193.89	381	717.8	213.33	416	780.8	232.78	451	843.8
194.44	382	719.6	213.89	417	782.6	233.33	452	845.6
195	383	721.4	214.44	418	784.4	233.89	453	847.4
195.56	384	723.2	215	419	786.2	234.44	454	849.2
196.11	385	725	215.56	420	788	235	455	851
196.67	386	726.8	216.11	421	789.8	235.56	456	852.8
197.22	387	728.6	216.67	422	791.6	236.11	457	854.6
197.78	388	730.4	217.22	423	793.4	236.67	458	856.4
198.33	389	732.2	217.78	424	795.2	237.22	459	858.2
198.89	390	734	218.33	425	797	237.78	460	860
199.44	391	735.8	218.89	426	798.8	238.33	461	861.8
200	392	737.6	219.44	427	800.6	238.89	462	863.6
200.56	393	739.4	220	428	802.4	239.44	463	865.4
201.11	394	741.2	220.56	429	804.2	240	464	867.2
201.67	395	743	221.11	430	806	240.56	465	869
202.22	396	744.8	221.67	431	807.8	241.11	466	870.8
202.78	397	746.6	222.22	432	809.6	241.67	467	872.6
203.33	398	748.4	222.78	433	811.4	242.22	468	874.4
203.89	399	750.2	223.33	434	813.2	242.78	469	876.2

(Continued)

Conversion Table (Thermometric Equivalents) (Continued)

To C	←F or C→	To F	To C	←F or C→	To F	To C	←F or C→	To F
	To Convert Degrees			**To Convert Degrees**			**To Convert Degrees**	
243.33	470	878	260	500	932	276.67	530	986
243.89	471	879.8	260.56	501	933.8	277.22	531	987.8
244.44	472	881.6	261.11	502	935.6	277.78	532	989.6
245	473	883.4	261.67	503	937.4	278.33	533	991.4
245.56	474	885.2	262.22	504	939.2	278.89	534	993.2
246.11	475	887	262.78	505	941	279.44	535	995
246.67	476	888.8	263.33	506	942.8	280	536	996.8
247.22	477	890.6	263.89	507	944.6	280.56	537	998.6
247.78	478	892.4	264.44	508	946.4	281.11	538	1,000.4
248.33	479	894.2	265	509	948.2	281.67	539	1,002.2
248.89	480	896	265.56	510	950	282.22	540	1,004
249.44	481	897.8	266.11	511	951.8	282.78	541	1,005.8
250	482	899.6	266.67	512	953.6	283.33	542	1,007.6
250.56	483	901.4	267.22	513	955.4	283.89	543	1,009.4
251.11	484	903.2	267.78	514	957.2	284.44	544	1,011.2
251.67	485	905	268.33	515	959	285	545	1,013
252.22	486	906.8	268.89	516	960.8	285.56	546	1,014.8
252.78	487	908.6	269.44	517	962.6	286.11	547	1,016.6
253.33	488	910.4	270	518	964.4	286.67	548	1,018.4
253.89	489	912.2	270.56	519	966.2	287.22	549	1,020.2
254.44	490	914	271.11	520	968	287.78	550	1,022
255	491	915.8	271.67	521	969.8	288.33	551	1,023.8
255.56	492	917.6	272.22	522	971.6	288.89	552	1,025.6
256.11	493	919.4	272.78	523	973.4	289.44	553	1,027.4
256.67	494	921.2	273.33	524	975.2	290	554	1,029.2
257.22	495	923	273.89	525	977			
257.78	496	924.8	274.44	526	978.8			
258.33	497	926.6	275	527	980.6			
258.89	498	928.4	275.67	528	982.4			
259.44	499	930.2	276.11	529	984.2			

[1]Budavari, Susan, et al. (eds.). The Merck Index: An Encyclopedia of Chemicals, Drugs, and Biologicals (11th Ed.). Merck & Co., Inc., Rahway, NJ. 1989. MISC–115–118.

RELATIVE HUMIDITY TABLES[1]

These values are correct for air velocity of not less than 600 feet per minute.

Dry Bulb Temperature °F	Wet Bulb Temperature °F	Relative Humidity
134°F	92°F	20%
134°F	100°F	30%
134°F	106°F	39%
134°F	110°F	46%
140°F	100°F	25%
140°F	106°F	34%
140°F	110°F	38%
140°F	112°F	41%
140°F	116°F	47%
150°F	106°F	25%
150°F	110°F	28%
150°F	114°F	35%
150°F	120°F	41%
150°F	124°F	46%
150°F	128°F	53%
150°F	132°F	60%
160°F	114°F	25%
160°F	120°F	31%
160°F	122°F	33%
160°F	126°F	38%
160°F	128°F	40%
160°F	130°F	43%
160°F	132°F	46%
170°F	126°F	29%
170°F	128°F	31%
170°F	130°F	33%
170°F	132°F	35%
170°F	134°F	38%

---

[1]Modern Curing and Processing with Griffith's Prague Powder. 1970. Griffith Laboratories, Inc., One Griffith Center, Alsip, IL 60658.

Dry Bulb Temperature °F	Wet Bulb Temperature °F	Relative Humidity
170°F	136°F	40%
170°F	140°F	45%
170°F	144°F	50%
180°F	134°F	30%
180°F	138°F	33%
180°F	142°F	37%
180°F	146°F	42%
180°F	150°F	47%
180°F	154°F	52%
200°F	150°F	30%
200°F	154°F	34%
200°F	158°F	38%
200°F	162°F	42%
200°F	166°F	46%
200°F	170°F	51%
210°F	158°F	30%
210°F	162°F	34%
210°F	166°F	38%
210°F	170°F	42%
210°F	174°F	46%
210°F	178°F	50%

Recall from Chapter 2, that

$$\text{Relative humidity} = \text{Water activity } (a_w) \times 100$$

CARE OF WET BULB THERMOMETERS

1. Remove and wash wick daily. Replace with new wick when it becomes saturated with tars.

2. Check water level in pan. Be sure bulb is well above water level. Wick must have water supply at all times when operating.

3. Calibrate thermometers regularly to be sure they are registering correctly.

4. Clean bulbs regularly, using fine emery paper or fine steel wool.

WINDCHILL[1]

t_{app} = apparent temperature due to windchill

V = wind velocity in miles/hour

t_a = air temperature (°F)

$$t_{app} = 59.42 - (0.52\sqrt{V} + 0.81 - \frac{V}{28.91}) \left[33 - \frac{5}{9}(t_a - 32)\right] + 32$$

Example: $V = 10$ $t_a = 32$

$$= 59.42 - (1.64 + 0.81 - 0.35)(33) + 32$$

$$59.42 - 69.30 + 32 = 22$$

---

[1]Siple, P. A., and C. F. Passel. 1945. Measurements of dry atmosphere in subfreezing temperatures. Proc. Am. Phil. Soc. 89:177–199.

This formula may at first appear complicated, but it replaces a 10-page table that weather reporters use.

DECIMAL SYSTEM

Decimal Equivalents of Common Fractions

$\frac{1}{64}$ — 0.0156	$\frac{5}{12}$ — 0.4167
$\frac{1}{32}$ — 0.0313	$\frac{3}{7}$ — 0.4286
$\frac{3}{64}$ — 0.0469	$\frac{4}{9}$ — 0.4444
$\frac{1}{16}$ — 0.0625	$\frac{5}{11}$ — 0.4545
$\frac{1}{12}$ — 0.0833	$\frac{15}{32}$ — 0.4688
$\frac{1}{11}$ — 0.0909	$\frac{1}{2}$ — 0.5000
$\frac{3}{32}$ — 0.0938	$\frac{17}{32}$ — 0.5313
$\frac{1}{10}$ — 0.1000	$\frac{6}{11}$ — 0.5455
$\frac{7}{64}$ — 0.1094	$\frac{5}{9}$ — 0.5556
$\frac{1}{9}$ — 0.1111	$\frac{9}{16}$ — 0.5625
$\frac{1}{8}$ — 0.1250	$\frac{4}{7}$ — 0.5714
$\frac{9}{64}$ — 0.1406	$\frac{7}{12}$ — 0.5833
$\frac{1}{7}$ — 0.1429	$\frac{3}{5}$ — 0.6000
$\frac{1}{6}$ — 0.1667	$\frac{5}{8}$ — 0.6250
$\frac{11}{64}$ — 0.1719	$\frac{7}{11}$ — 0.6364
$\frac{2}{11}$ — 0.1818	$\frac{2}{3}$ — 0.6667
$\frac{3}{16}$ — 0.1875	$\frac{11}{16}$ — 0.6875
$\frac{1}{5}$ — 0.2000	$\frac{7}{10}$ — 0.7000
$\frac{13}{64}$ — 0.2031	$\frac{45}{64}$ — 0.7031
$\frac{7}{32}$ — 0.2188	$\frac{5}{7}$ — 0.7143
$\frac{2}{9}$ — 0.2222	$\frac{8}{11}$ — 0.7273
$\frac{15}{54}$ — 0.2344	$\frac{47}{64}$ — 0.7344
$\frac{1}{4}$ — 0.2500	$\frac{3}{4}$ — 0.7500
$\frac{3}{11}$ — 0.2727	$\frac{7}{9}$ — 0.7778
$\frac{9}{32}$ — 0.2813	$\frac{51}{64}$ — 0.7969
$\frac{2}{7}$ — 0.2857	$\frac{4}{5}$ — 0.8000
$\frac{19}{64}$ — 0.2969	$\frac{13}{16}$ — 0.8125
$\frac{3}{10}$ — 0.3000	$\frac{9}{11}$ — 0.8182
$\frac{1}{3}$ — 0.3333	$\frac{5}{6}$ — 0.8333
$\frac{11}{32}$ — 0.3438	$\frac{6}{7}$ — 0.8571
$\frac{4}{11}$ — 0.3636	$\frac{55}{64}$ — 0.8594
$\frac{3}{8}$ — 0.3750	$\frac{7}{8}$ — 0.8750
$\frac{2}{5}$ — 0.4000	$\frac{8}{9}$ — 0.8889
$\frac{13}{32}$ — 0.4063	$\frac{9}{10}$ — 0.9000

$^{10}/_{11}$ — 0.9091 $^{61}/_{64}$ — 0.9531

$^{11}/_{12}$ — 0.9167 $^{63}/_{64}$ — 0.9844

$^{59}/_{64}$ — 0.9219   1 — 1.0000

$^{15}/_{16}$ — 0.9375

To convert to percentage, carry the decimal point two places to the right. Thus, 63/64, or 0.9844, equals 98.44%.

ENERGY ALLOWANCES

Energy allowances in kilocalories (kcal) are defined as follows[1]: 1 kilocalorie is the amount of heat necessary to raise 1 kg of water from 15°C to 16°C. The accepted international unit of energy is the *joule (J)*. To convert energy allowances from kilocalories to kilojoules (kJ), the factor 4.2 may be used (1 kcal equals exactly 4.184 kJ). Because the energy content of diets is usually greater than 1,000 kJ, the preferred unit is the megajoule (mJ), which is 1,000 kJ. In this textbook, *calories,* the popular term, which actually means *kcal,* is used.

---

[1]Subcommittee on the Tenth Edition of the RDAs, Food and Nutrition Board, Commission of Life Sciences, National Research Council (NRC). 1989. Recommended Dietary Allowances (10th Ed.). National Academy Press, Washington, DC.

POPULAR MEAT INDUSTRY PERIODICALS AND ONLINE SERVICES

Periodicals and online services from the

American Association for the Advancement of Science

1200 New York Avenue, N.W.
Washington, DC 20005

Tel: (202) 326-6400
http://aaas.org

- *GrantsNet* (research funding database)
  http://www.grantsnet.org
- *NeuroAIDS* (experimental Web site)
  http://www.sciencemag.org/NAIDS
- *Next Wave* (weekly career updates)
  http://www.nextwave.org
- *Science* (the journal) (ISSN 0036-8075)
  http://www.sciencemag.org
- *ScienceNOW* (daily news service)
  http://www.sciencenow.org
- *Science Online*
  http://www.scienceonline.org

Food Engineering (ISSN 0193-323X)

Cahners Business Information
201 King of Prussia Road
Radnor, PA 19089

Tel: (610) 964-4000
Fax: (610) 964-2915
http://www.foodexplorer.com

Food Processing (ISSN 0015-6523)

Putnam Publishing
555 West Pierce Road, Suite 301
Itasca, IL 60143

Tel: (630) 467-1300
Fax: (630) 467-1179
http://www.foodprocessing.com

Meat & Poultry (ISSN 0892-6077)

4800 Main Street, Suite 100
Kansas City, MO 64112

Tel: (816) 756-1000
Fax: (816) 756-0494
Sales: cjolley@sosland.com
Editorial: knunes@sosland.com
http://www.meatpoultry.com

Meat Business Magazine (ISSN 1049-5908)

Printing & Publishing, Inc.
109 West Washington Street
Millstadt, IL 62260-1155

Tel: (800) 451-0914
bizmag@stimo.com

Meat International

Elsevier International Business Information
The Netherlands
Editor: Cedric Porter

Tel: +31-314-349678
Fax: +31-314-340515
C.Porter@ebi.nl

Meat Marketing & Technology (ISSN 1079-1604)

Marketing & Technology Group, Inc.
1415 North Dayton Street
Chicago, IL 60622

Tel: (312) 266-3311
Fax: (312) 266-3363
http://www.meatingplace.com

National Provisioner, The (ISSN 0027-996X)

Stagnito Communications, Inc.
1935 Shermer Road, Suite 100
Northbrook, IL 60062-5354

Tel: (847) 205-5660
Fax: (847) 205-5680
info@stagnito.com
http://www.nationalprovisioner.com

Poultry (ISSN 1096-3057)

Marketing & Technology Group, Inc.
1415 North Dayton Street
Chicago, IL 60622

Tel: (312) 274-2201
Fax: (312) 266-3363
http://www.meatingplace.com

Prepared Foods (ISSN 0747-2536)

Cahners Publishing Company
1350 East Touhy Avenue
Des Plaines, IL 60018-3358

Tel: (847) 635-8800
Fax: (847) 390-2445
http://www.preparedfoods.com

Periodicals from

Watt Publishing Co.

122 South Wesley Avenue
Mt. Morris, IL 61054-1497

Tel: (815) 734-4171
Fax: (815) 734-9091
http://www.wattnet.com

- *Broiler Industry* (ISSN 0007-2176)
- *Meat Processing* (ISSN 0025-6390)
- *Poultry Digest* (ISSN 0032-5724)
- *Poultry International* (ISSN 0032-5767)
- *Poultry Processing* (ISSN 0898-4565)
- *Turkey World* (ISSN 0041-4271)

FEDERAL AGENCIES[1]

Economic Research Service

Tel: (202) 694-5050
    (Information Center)
http://www.econ.ag.gov

Environmental Protection Agency

401 M Street, S.W.
Washington, DC 20460

Tel: (202) 260-2090
http://www.epa.gov

Federal Trade Commission

600 Pennsylvania Avenue, N.W.
Washington, DC 20580

Tel: (202) 326-2021
    (202) 326-2000 (employee/
    department locator)
http://www.ftc.gov

Food and Drug Administration

5600 Fishers Lane, #1471
Rockville, MD 20857

Tel: (301) 443-4177
    (888) INFO FDA or
    (888) 463-6332 (consumer
    information hotline)
http://www.fda.gov

FDA's Center for Food Safety and Applied Nutrition

200 C Street, S.W.
Washington, DC 20204

Tel: (202) 205-4850 (director)
    (800) 332-4010 (consumer
    inquiries)
http://www.fda.gov

Government Printing Office

Superintendent of Documents
P.O. Box 371954
Pittsburgh, PA 15250-7954

Tel: (202) 512-1800
Fax: (202) 512-2250
http://www.access.gpo.gov

Hazard Analysis Critical Control Point (HACCP) System

Tel: (800) 233-3935, press 2 (hotline)
    (202) 720-2709 (technical
    inquiries: Office of Program
    Policy Development and
    Evaluation)
http://www.fsis.usda.gov

National Archives and Records Administration

700 Pennsylvania Avenue, N.W.
Washington, DC 20408

http://www.nara.gov

OSHA

Tel: (202) 219-6091 (Assistant
    Secretary)
    (202) 693-2043 (Directorate of
    Safety Standards Programs)
http://www.osha.gov

U.S. Department of Agriculture

1400 Independence Avenue, S.W.
Washington, DC 20250

Tel: (202) 720-2791 (general
    information)
    (202) 720-8732 (employee/
    department locator)
http://www.usda.gov/usda.htm

---

[1]Updated and selected from a supplement to Food Processing, Vol. 60, No. 9, September 1999.

U.S. Department of Commerce

14th Street and Constitution
   Avenue, N.W.
Washington, DC 20230

Tel: (202) 482-2000
http://www.doc.gov

U.S. Department of Labor

Occupational Safety and
   Health Administration
200 Constitution Avenue, N.W.
Washington, DC 20210

Tel: (202) 576-7100
http://www.dol.gov

NATIONAL ASSOCIATIONS

Alaska Seafood Marketing Institute

1111 West Eighth Street, Suite 100
Juneau, AK 99801

Tel: (800) 478-2903
     (907) 465-5572
alaska_seafood@commerce.stateak.us
http://www.alaskaseafood.org

**American Association of
Cereal Chemists**

3340 Pilot Knob Road
St. Paul, MN 55121-2097

Tel: (651) 454-7250
Fax: (651) 454-0766
aacc@scisoc.org
http://www.scioc.org/aacc

**American Association of
Meat Processors**

One Meating Place
P.O. Box 269
Elizabethtown, PA 17022

Tel: (717) 367-1168
Fax: (717) 367-9096
aamp@aamp.com
http://www.aamp.com

American Dairy Science Association

1111 North Dunlap Avenue
Savoy, IL 61874

Tel: (217) 356-3182
Fax: (217) 398-4119
http://www.adsa.uiuc.edu

American Dietetic Association, The

216 West Jackson Boulevard
Suite 800
Chicago, IL 60606-6995

Tel: (312) 899-0040
Fax: (312) 899-0008
http://www.eatright.org

**American Feed Industry
Association, Inc.**

1501 Wilson Boulevard, Suite 1100
Arlington, VA 22209

Tel: (703) 524-0810
Fax: (703) 524-1921
http://www.afia.org

American Frozen Food Institute

2000 Corporate Ridge, Suite 1000
McLean, VA 22102

Tel: (703) 821-1350
http://www.affi.com

American Meat Institute

1700 North Moore Street, Suite 1600
Arlington, VA 22209

Tel: (703) 841-2400
Fax: (703) 527-0938
http://www.meatami.org

American Meat Science Association

1111 North Dunlap Avenue
Savoy, IL 61874

Tel: (217) 356-3182
Fax: (217) 398-4119
http://www.meatscience.org

American Oil Chemists' Society

1608 Broadmoor Drive
P.O. Box 3489 (61826-3489)
Champaign, IL 61821-5930

Tel: (217) 359-2344
Fax: (217) 351-8091
http://www.aocs.org

American Society of Agricultural Engineers

2950 Niles Road
St. Joseph, MI 49085

Tel: (616) 429-0300
Fax: (616) 429-3852
http://www.asae.org/

American Spice Trade Association, Inc.

560 Sylvan Avenue
P.O. Box 1267
Englewood Cliffs, NJ 07632

Tel: (201) 568-2163
Fax: (201) 568-7318

AOAC International

481 North Frederick Avenue
Suite 500
Gaithersburg, MD 20877-2417

Tel: (301) 924-7077
Fax: (301) 924-7089
aoac@aoac.org
http://www.aoac.org

Association of Food & Drug Officials

P.O. Box 3425
York, PA 17402

Tel: (717) 757-2888
Fax: (717) 755-8089
http://www.foodsafety.org/afdo

Center for Science in the Public Interest (CSPI)

1875 Connecticut Avenue, N.W.
Suite 300
Washington, DC 20009-5728

Tel: (202) 332-9110
Fax: (202) 265-4954
cspi@cspinet.org
http://www.cspinet.org

Conference for Food Protection

110 Tecumseh Trail
Frankfort, KY 40601

Tel: (502) 695-0253
Fax: (502) 695-0253
leontown@dcr.net
http://www.uark.edu.cfpncims

Council for Agricultural Science and Technology (CAST)

4420 Lincoln Way
Ames, IA 50014-3447

Tel: (515) 292-2125
http://www.cast-science.org

Council for Responsible Nutrition

1875 Eye Street, N.W., Suite 400
Washington, DC 20006-5409

Tel: (202) 872-1488
Fax: (202) 872-9594
http://www.crnusa.org

Flexible Packaging Association

1090 Vermont Avenue, N.W.
Suite 500
Washington, DC 20005-3880

Tel: (202) 842-3880
Fax: (202) 842-3841
http://www.flexpack.org

Food and Nutrition Board

Institute of Medicine,
National Academy of Sciences
2101 Constitution Avenue, N.W.
Washington, DC 20418

Tel: (202) 334-1732
Fax: (202) 334-2316
http://www2.nas.edu/fnb

Food Distributors International

201 Park Washington Court
Falls Church, VA 22046-4521

Tel: (703) 532-9400
Fax: (703) 538-4673
http://www.fdi.org

Food Industry Suppliers of Canada

P.O. Box 152
Apple Hill, ON Canada K0C 1B0

Tel: (613) 525-2775
Fax: (613) 525-4328
http://www.fisc.ca

Food Institute, The

28-12 Broadway
Fair Lawn, NJ 07410-3913

Tel: (201) 791-5570
Fax: (201) 791-5222
http://www.foodinstitute.com

Food Institute of Canada

1600 Scott Street #415
Ottawa, ON, Canada K1Y 4N7

Tel: (613) 722-1000
Fax: (613) 722-1404

Food Marketing Institute

800 Connecticut Avenue, N.W.
Washington, DC 20006-2701

Tel: (202) 452-8444
Fax: (202) 429-4559
http://www.fmi.org

**Food Processing Machinery &
Supplies Association**

200 Daingerfield Road
Alexandria, VA 22314

Tel: (703) 684-1080
Fax: (703) 548-6563
http://www.fmpsa.org &
http://www.iefp.org

Food Processors Institute, The

1350 I Street N.W., Suite 300
Washington, DC 20004

Tel: (202) 393-0890
Fax: (202) 639-5991
http://www.nfpa-food.org

Foodservice & Packaging Institute

1550 Wilson Boulevard, Suite 701
Arlington, VA 22209

Tel: (703) 527-7505
Fax: (703) 527-7512
fpi@fpi.org
http://www.fpi.org

**Foodservice Equipment Distributors
Association**

223 West Jackson Boulevard
Suite 620
Chicago, IL 60606

Tel: (312) 427-9605
Fax: (312) 427-9607
fede@earthlink.net

Institute of Food Technologists

221 North LaSalle Street, Suite 300
Chicago, IL 60601-1291

Tel: (312) 782-8424
Fax: (312) 784-8348
http://www.ift.org

Institute of Packaging Professionals

481 Carlisle Drive
Herndon, VA 20170-4823

Tel: (703) 318-8970
Fax: (703) 814-4961
iopp@pkgmatters.com
http://www.packinfo-world.org

Institute of Shortening & Edible Oils, Inc.

1750 New York Avenue, N.W.
Suite120
Washington, DC 20006

Tel: (202) 783-7960
Fax: (202) 393-1367
http://www.iseo.org

International Association of Food Industry Suppliers

1451 Dolley Madison Boulevard
McLean, VA 22101-3850

Tel: (703) 761-2600
Fax: (703) 761-4334
http://iafis.org

International Association of Milk, Food & Environmental Sanitarians (IAMFES)

6200 Aurora Avenue, Suite 200W
Des Moines, IA 50322

Tel: (800) 369-6337
    (515) 276-3344
Fax: (515) 276-8655
iamfes@iamfes.org
http://www.iamfes.org

International Association of Refrigerated Warehouses

7315 Wisconsin Avenue, N.W.
Suite 1200 N
Bethesda, MD 20814

Tel: (301) 652-5674
Fax: (301) 652-7269
http://www.iarw.org

International Foodservice Manufacturers Association

180 North Stetson Avenue
Chicago, IL 60601-6710

Tel: (312) 540-4400
Fax: (312) 540-4401
ifma@prodigy.com
http://www.foodserviceworld.com/ifma

International Institute of Ammonia Refrigeration

1200 19th Street, N.W., Suite 300
Washington, DC 20036

Tel: (202) 857-1110
Fax: (202) 223-4579
http://www.iiar.org

Maine Sardine Council

470 North Main Street
P.O. Box 337
Brewer, ME 04412-0337

Tel: (207) 989-2180
Fax: (207) 989-2154
msardine@mint.net
http://www.mint.net/sardine

Midwest Food Processors Association, Inc.

P.O. Box 1297
Madison, WI 53701-1297

Tel: (608) 255-9946
Fax: (608) 255-9838
http://www.mwfpa.org

National Cattlemen's Beef Association

P.O. Box 3469
Englewood, CO 80155

Tel: (303) 694-0305
Fax: (303) 694-2851
http://www.beef.org

National Chicken Council

1015 15th Street, N.W., Suite 930
Washington, DC 20005-2605

Tel: (202) 296-2622
Fax: (202) 293-4005
wroenigk@chickenusa.org

National Fisheries Institute

1901 North Fort Myer Drive
Arlington, VA 22209

Tel: (703) 524-8880
Fax: (703) 524-4619
http://www.nfi.org

**National Food Distributors
Association**

401 North Michigan Avenue,
Suite 2200
Chicago, IL 60611-4267

Tel: (312) 644-6610
Fax: (312) 527-6783
nfda@sba.com

**National Food Processors
Association**

1350 I Street, N.W.
Washington, DC 20005

Tel: (202) 639-5900
Fax: (202) 639-5932
nfpa@nfpa-food.org
http://www.nfpa-food.org

National Frozen Food Association

4755 Linglestown Road, Suite 300
Harrisburg, PA 17112-0069

Tel: (717) 657-8601
Fax: (717) 657-9862
info@nffa.org
http://www.nffa.org

National Grocers Association

1825 Samuel Morse Drive
Reston, VA 20190-5317

Tel: (703) 437-5300
Fax: (703) 437-7768
natlgrocer@ad.com
http://www.nationalgrocers.org

**National Hot Dog and
Sausage Council**

Post Office Box 3556
Washington, DC 20007

Tel: (703) 841-2400
Fax: (703) 527-0938
http://www.hot-dog.org

National Meat Association

1970 Broadway, Suite 825
Oakland, CA 94612

Tel: (510) 763-1533
Fax: (510) 763-6186
http://www.hooked.net/users/nma/

National Meat Canners Association

1700 North Moore Street, Suite 1600
Arlington, VA 22209

Tel: (703) 841-3680
Fax: (703) 841-9656
http://www.meatami.org/svc02.htm

National Paperbox Association

801 North Fairfax Street, Suite 211
Alexandria, VA 22314

Tel: (703) 684-2212
Fax: (703) 683-6920
boxmaker@paperbox.org
http://www.paperbox.org

National Pork Producers Council

1776 N.W. 114th Street
Clive, IA 50325

Tel: (515) 223-2600
Fax: (515) 223-2646
http://www.nppc.org

National Restaurant Association

1200 17th Street, N.W.
Washington, DC 20036-3097

Tel: (800) 424-5156
       (202) 331-5995
Fax: (202) 331-2429
http://www.restaurant.org

**North American Association of
   Food Equipment Manufacturers
   (NAFEM)**

401 North Michigan Avenue
Chicago, IL 60611-4267

Tel: (312) 644-6610
Fax: (312) 527-6658
http://www.nafem.org

**North American Meat Processors
   Association**

1920 Association Drive, Suite 400
Reston, VA 20191-1547

Tel: (703) 758-1900
Fax: (703) 758-8001
http://www.namp.com

**Northwest Food Processors
   Association**

6950 S.W. Hampton Street, Suite 340
Portland, OR 97223-8329

Tel: (503) 639-7676
Fax: (503) 639-7007
http://www.nwfpa.org

**Packaging Machinery Manufacturers
   Institute**

4350 North Fairfax Drive, Suite 600
Arlington, VA 22203

Tel: (703) 243-8555
Fax: (703) 243-8556
matt@pmmi.org

Pet Food Institute

1200 19th Street N.W., Suite 300
Washington, DC 20036

Tel: (202) 857-1120
Fax: (202) 223-4579
http://www.petfoodinstitute.com

**Process Equipment Manufacturer's
   Association**

111 Park Place
Falls Church, VA 22046

Tel: (703) 538-1796
Fax: (703) 241-5603
http://www.webmasters.net/pema/

Refrigerated Foods Association

2971 Flowers Road, S., Suite 266
Atlanta, GA 30341

Tel: (770) 452-0660
Fax: (770) 455-3879
judy@refrigeratedfoods.org
http://www.refrigeratedfoods.org

Research & Development Associates

16607 Blanco Road, Suite 1506
San Antonio, TX 78232

Tel: (210) 493-8024
Fax: (210) 493-8036
http://www.rdajff@flash.net

Salt Institute, The

700 North Fairfax Street, Suite 600
Alexandria, VA 22314-2040

Tel: (703) 549-4648
Fax: (703) 548-2194
http://www.saltinstitute.org

Snack Food Association

1711 King Street, Suite One
Alexandria, VA 22314

Tel: (703) 836-4500
Fax: (703) 836-8262
http://www.sfa.org

Society for Foodservice Management

304 West Liberty Street, Suite 201
Louisville, KY 40202

Tel: (502) 583-3783
Fax: (502) 589-3602
http://www.sfm-online.org

Soy Protein Council

1255 23rd Street, N.W., Suite 200
Washington, DC 20037

Tel: (202) 467-6610
Fax: (202) 833-3636
http://www.spcouncil.org

U.S. Meat Export Federation

1050 17th Street, Suite 2200
Denver, CO 80265

Tel: (303) 623-MEAT (6328)
Fax: (303) 623-0297
http://www.usmef.org

U.S. Poultry & Egg Association

1530 Cooledge Road
Tucker, GA 30084-7303

Tel: (770) 493-9401
Fax: (770) 493-9257
http://www.poultryegg.org

Western Hemisphere Association of Meat Marketers

1700 North Moore, Suite 1600
Arlington, VA 22209

Tel: (703) 841-3690
Fax: (703) 841-9656
jbreiter@meatami.org
http://www.meatami.org

World Food Logistics Organization

7315 Wisconsin Avenue, Suite 1200N
Bethesda, MD 20814

Tel: (301) 652-5674
Fax: (301) 652-7269
http://www.iarw.org

PROFESSIONAL ORGANIZATIONS

American Meat Science Association (AMSA)

AMSA Mission

The mission of the American Meat Science Association is to contribute to the betterment of the human life by leading the discovery and application of sound scientific and technological principles of the meat sciences through education and research.

Purposes/Activities

To: promote the application of science and technology to production, processing, packaging, distribution, preparation, evaluation, and utilization of meat and meat products.

stimulate the exchange, discussion, and dissemination of information concerning meat research.

encourage sound and useful research and educational techniques.

coordinate educational, research, development, and service activities in meat science and related areas.

encourage recognition of those engaged in meat science.

Membership Classifications

Professional: Any person who is active in any aspect of meat science and who evidences interest in supporting the objectives of the Association.

Emeritus: Any AMSA member retired from active professional life, provided that person held "Professional" or "Member" status for a minimum of 10 years.

Student: Any person actively pursuing candidacy for a degree (associate or higher) with a major or majors in a science or technology associated with meat science.

Benefits of Membership in AMSA

- Direct, personal contact at AMSA meetings with meat scientists from academia, industry, and government.

- Opportunities to learn about current meat research, teaching, extension, industry, and government activities in the United States and worldwide.

- A distinguished forum to express opinions on matters of professional interest, including teaching programs in the meat sciences.

- Attendance at various exclusive short courses on meat subjects and at the International Congress of Meat Science and Technology (ICoMST).

- The annual *Proceedings of the Reciprocal Meat Conference* and the annual *Proceedings of the Meat Industry Research Conference.*

- The *AMSA Directory of Members,* which includes individual addresses and phone numbers.

- The quarterly *AMSA Newsletter* of activities, news affecting meat science, and job opportunities.

- Discounted subscription rates *Meat Science,* the official journal of AMSA.

> American Meat Science Association
> 111 North Dunlap Avenue
> Savoy, IL 61874
> Tel: (217) 356-3182
> Fax: (217) 398-4119
> http://www.meatscience.org

Institute of Food Technologists (IFT)

Purpose/Activities

Food science, as defined by IFT, employs biological, chemical, and physical research to help individuals develop the basic understanding necessary to provide a nutritious, safe, and acceptable diet. *Food technology* applies this basic knowledge to the practical development of new and improved food sources, products, and processes.

The areas of interest reach from agriculture (at the growing end), to nutrition and safety (at the point of consumption) and everything in between.

Objectives

To: stimulate basic and applied research and investigations into technological problems.

present, discuss, and publish the results of such research and investigations.

stimulate the free interchange of information and ideas.

raise the educational standards of food scientists.

promote recognition of the food scientist's role in today's food supply.

promote the application of science and technology for the improvement of production, processing, packaging, distribution, preparation, evaluation, and utilization of foods, in order to provide better and more adequate foods for humankind.

IFT publishes two journals: *Food Technology* (monthly) and the *Journal of Food Science* (bimonthly). *Food Technology* specializes in general interest articles and news of the industry and IFT. The *Journal of Food Science* publishes the results of original basic and applied food research. Subscriptions to both are available to non-members.

Membership

Student membership is encouraged.

Institute of Food Technologists
221 North LaSalle Street, Suite 300
Chicago, IL 60601-1291
Tel: (312) 782–8424
Fax: (312) 782–8348
E-mail: info@ift.org
http://www.ift.org

American Society of Animal Science (ASAS)

Purpose/Activities

ASAS is a professional organization for animal scientists that is designed to help members provide effective leadership through research, extension, teaching, and service for the dynamic and rapidly changing livestock and meat industries.

Objectives

To: obtain more effective research results.

publish the *Journal of Animal Science* as a means of disseminating scientific and educational information.

promote professional advancements in the field of Animal Science.

cooperate with other organizations having similar objectives.

Membership Classifications

Membership is open to individuals interested in research, instruction, or extension in Animal Science or associated with the production, processing, marketing, and distribution of livestock and livestock products.

Professional member: To apply for membership, an individual must be sponsored by a member who is in good standing.

Student affiliate member: To apply for membership, an individual must be a regularly enrolled college student who does not hold a full-time job. A professional member of ASAS must certify eligibility for this membership.

Retired member: A professional member with 10 or more years in good standing may become a retired member at no cost. By paying one-half of the annual professional membership dues, a retired member can still receive the *Journal of Animal Science*.

> American Society of Animal Science
> 309 West Clark Street
> Champaign, IL 61820
> Tel: (217) 356–3182
> Fax: (217) 398–4119
> http://www.asas.org

Poultry Science Association (PSA)

Objectives

To: stimulate the discovery, application, and dissemination of knowledge.

create a forum for the exchange of information among various segments of the poultry industry.

publish original research, reviews, and timely information in the official PSA publication, *Poultry Science*.

To recognize outstanding personal achievement.

Membership Eligibility

To be eligible for membership, individuals must have an interest in research, instruction, or extension in Poultry Science or be associated with the production, processing, marketing, and distribution of poultry and poultry products.

Membership Classifications

Active: Available to professionals in the field with certification from two current members.

Student: Available to any full-time undergraduate or graduate student in the field or a closely related field with certification from his or her college or university advisor.

Emeritus: Offered to any active member in good standing in the Association who has reached the age of 70 or who has attained a retired status with the organization with which the member is associated and who is no longer in receipt of a regular salary based on work in the poultry area.

Corporate: Available to corporations who wish to further their contribution to poultry science.

> Poultry Science Association
> 309 West Clark Street
> Champaign, IL 61820
> Tel: (217) 356–3182
> Fax: (217) 398–4119
> http://www.psa.org

Federation of Animal Science Societies (FASS)

Purpose

FASS serves the American Dairy Science Association, the American Society of Animal Science, and the Poultry Science Association as an advocate for animal agriculture. Animal agriculture is a foundation of society and encompasses all aspects of animal production and the use of animals for food that contribute to the nutritional well-being of humankind.

Membership

Individual memberships in FASS are paid as a portion of membership dues paid to the American Dairy Science Association, the American Society of Animal Science, or the Poultry Science Association.

> Federation of Animal Science Societies
> 1111 North Dunlap Avenue
> Savoy, IL 61874
>
> Tel: (217) 356-3182
> Fax: (217) 398-4119
> fass@assochq.org
> http://www.fass.org

American Registry of Professional Animal Scientists (ARPAS)

ARPAS — The Symbol of Today's Professional Animal Scientist

All successful certification and licensing programs are targeted to serve and protect the public's interest. Modern agriculture, like all segments of society, is becoming more complex. More government regulations and controls require that practicing professionals establish accountability by means of registry and certification programs. In today's business climate, producer and industry clients want assurance that they are getting advice from certified professionals who stay on the cutting edge.

By completing the requirements for registration, maintaining your continuing education units, and adhering to the code of ethics, ARPAS registration provides you with a new level of recognition to help you distinguish yourself to your clients as a Professional Animal Scientist.

What Is ARPAS?

The American Registry of Professional Animal Scientists is a self-supporting, self-governing organization with five affiliated societies:

- American Dairy Science Association

- American Meat Science Association

- American Society of Animal Science

- Equine Nutrition and Physiology Society

- Poultry Science Association

ARPAS approves events that qualify for CEUs and thus provide appropriate education to help members stay current in their field.

Why Was ARPAS Established?

- To register professionally competent animal scientists.

- To provide a mechanism to sustain their professional competence.

- To enhance their professional stature and credibility.

- To strengthen and promote any profession of the animal sciences.

- To provide a means for those in industry, companies, and consulting to establish and maintain professional recognition.

Levels of Certification Available

Professional Scientist (PAS) — based on degree, experience, examination and CEUs. (Can begin as an Affiliate Member and then advance to full membership.)

Registered Animal Specialist (RAS) and *Registered Animal Product Specialist* (RAPS) — based on experience, examination, and CEUs. Ideal for individuals who have not completed a college degree program but have established their credentials by significant experience, examination, and continuing education.

Board Certification (Diplomate) — highest requirements. Available to professionals with an M.S. or Ph.D. degree who can benefit from the highest level of certification.

Note: Meat scientists would join the **American College of Animal Food Science (ACAFS),** as Dr. Robert Rogers (page 1082) did when he achieved Diplomate status.

PAS and RAS/RAPS registration can be in one or more of the following species or product areas: aquaculture, beef cattle, companion animals, dairy cattle, horses, laboratory animals, poultry, sheep and goats, swine, meat science, dairy product science, and poultry products.

Board Certification is in a specific discipline of animal science and includes the American Colleges of: Animal Behavior; **Animal Food Science;** Animal Genetics; Animal Nutrition; and Animal Physiology.

Benefits of Being a Registered Professional with ARPAS

(Some reasons present ARPAS members have given)

1. Provides professional credentials for animal scientists who must increasingly verify their professional competence and uniqueness to legislative, legal, and industry groups by some mechanism in addition to formal education.

2. Documents that the professional has the know-how to apply relevant cutting-edge knowledge to real world problems.

3. Brings recognition of abilities and knowledge to the one being served. Provides another certification title, which creates more credibility on the farm and among other professionals (i.e., veterinarians, attorneys, CPAs).

4. ARPAS provides a variety of mechanisms to enable members to stay current in their field and maintains documentation of the CEUs.

5. Provides assurance to those you serve that they are getting advice from a professional who keeps up-to-date.

6. Proactively positions animal scientists for a time when some form of licensing may be required to practice certain animal sciences.

7. Provides a pool of credentialed animal scientists to be called upon in settings involving public policy and law making, especially where new laws and decisions are made.

8. The journal, *The Professional Animal Scientist,* provides a source of useful information relevant from research as well as field studies.

9. Provides certification and diplomate status when dealing with other governments in international affairs.

10. Provides appropriate certification and titles, after examination, for individuals working at all levels of animal science, from on-farm production (Registered Animal Scientist), to expertise gained by academic degrees and experience (Professional Animal Scientist), to Diplomate status (Board Certification).

ARPAS Programs Are Growing to Serve You Better

The **journal** of ARPAS is *The Professional Animal Scientist,* which is published four times a year. It provides an opportunity for publishing topics in applied animal science from research and field studies as well as issue-oriented articles, and review articles.

The ARPAS **Web page** (http://www.arpas.uiuc.edu) is being totally upgraded as a source of continuing education opportunities, newsletters, meetings and conferences with CEUs, other information relevant to members, and reference materials.

Special **seminars and courses** are being planned to provide convenient access to first-class continuing education opportunities.

Efforts are being undertaken to acquaint government agencies and other public policy bodies with the opportunities they have to access the ARPAS membership for high-quality information on key policy issues from certified professionals.

What ARPAS Members Say

"ARPAS helps me project a clear signal that I am a nutritionist. The ACAN Diplomate certification provides credentials to help our customers understand that I have met the highest standards for our profession. Our customers like dealing with a professional.

"ARPAS also helps me maintain my technical knowledge through its CEU program. This program provides one more incentive to encourage me to stray abreast of the latest technical information."

Gary T. Lane, Ph.D., PAS, Dipl. ACAN
Burkmann Feeds
Danville, Kentucky

"ARPAS gives us a level of qualification that is recognized. When most other professions (i.e., engineers, pharmacy, medical, accountants, real estate brokers, etc.) have to be certified in order to be 'professionals,' people expect no less from those in our profession. Also, in dealing in international affairs, certification and diplomate status are very important."

Robert Rogers, Ph.D., PAS, Dipl. ACAFS
Mississippi State University

"ARPAS has provided the necessary credibility and credentials I need to be on a par with other board certified professionals in court cases involving questions of proper nutrition and feeding practice when I was required to appear as an expert witness."

Gary D. Potter, Ph.D., PAS, Dipl. ACAN
Texas A&M University

"As a certified animal scientist my clients can be assured that my professional and ethical integrity are being monitored by the American Registry of Professional Animal Scientists. By displaying the Diplomate title of ARPAS on my business card, the public is aware that I have conformed to a code of ethical conduct and am maintaining my professional training."

Lee A. Shell, Ph.D., PAS, Dipl. ACAN
Omaha Nutritional Services
Omaha, Nebraska

American Registry of Professional Animal Scientists
1111 North Dunlap Avenue
Savoy, IL 61874

Tel: (217) 356-3182
Fax: (217) 398-4119
arpas@assochq.org
http://www.arpas.uiuc.edu

PLACEMENT

Don't just wait for good job offers to find you!

Dreaming about the perfect job and finding the perfect job are not the same. While the job market may look favorable, receiving the most desirable job offers continues to be a job in itself. Occasionally a graduate is blessed with a job offer "out of the blue," but generally invitations come as the result of a well-planned job search.

What are the characteristics of a good campaign?[1]

1. *Make numerous contacts.* Finding a job is a "numbers" game. The competition is tough for the best positions. It's not uncommon for an organization to interview 20 or 25 graduates for every offer. Thus, it is important to make enough contacts to ensure that you make the right one.

2. *Be realistic.* Direct your letters, phone calls, and résumés to companies that have positions for which you are qualified. It's fine to aim high, but expectations that are beyond your limitations are doomed to failure.

3. *Be in the right place at the right time.* Sometimes it pays to go back to a company that wasn't interested in you a few weeks or months ago because the situation may have changed in the meantime.

4. *Be persistent.* State in the cover letter that you will call in a reasonable period of time to determine if there is opportunity for an interview. Then follow up, and keep calling until you reach the person who can make a hiring decision. Don't say, "I'll call at your convenience." You have no way of making that determination.

5. *Organize your job search.* Collect company addresses and phone numbers, the names of key people and their titles, and put each on a card. Use the card to make notes when you do your research about a particular company. Keep a record of every contact on that card. Then when you receive a job offer, you can remember the entire sequence of events and be ready to give an enthusiastic "Yes."

What are the characteristics of the ideal job candidate?

Everyone has fantasies, and that includes campus recruiters who dream of finding "ideal students" to fill job openings in their companies. In a survey conducted at Michigan State University, 151 organizations were asked, "What would you ideally like to consider when pre-screening candidates from the Placement Service for possible interviews?"

---

[1]Career Concerns. 1989. News Letter. Career and Academic Planning Center. South Dakota State University, Brookings.

See also Gardner, P. D., S. W. J. Koslowski, and A. Broadus. 1988. The Ideal Job Candidate: What Recruiters Would Like to See. Collegiate Employment Research Institute, Michigan State University. (November)

An average of eight characteristics was placed on each recruiter's list, and 56 different characteristics were mentioned.

Characteristics mentioned most frequently were outside activities, personal traits, grade point averages, work experiences, and other communication skills. Other important "most desirable" included interview preparation and flexibility. On the lists of personality traits, initiative and self-esteem were ranked as the two most important characteristics.

The researchers concluded that recruiters were seeking individuals with both quantifiable and non-measurable factors difficult to place on résumés. Consequently, they recommended that students need to use multiple strategies in the job search process in order to emphasize their strengths from both categories.

Advice from the Experts: Interviewing Tips from On-Campus Recruiters[2]

Research organizations in advance of interviews — Since most on-campus interviews are relatively short, it is important that you use this time to sell yourself to an employer. Don't waste this opportunity be spending too much time on issues that could have been answered by reading the company's literature and/or viewing its videotape. Displaying your knowledge about a potential employer will greatly enhance your chances of interview success.

Define your career goals and the opportunities you want — One of the keys to making a successful sale is product knowledge. In the case of job interviews, that product is *you*. You need to perform a thorough self-evaluation well in advance of your interviews. Know what your strengths, weaknesses, skills, and abilities are and be prepared to discuss them in the interview.

Be enthusiastic and sincere during your interviews — It is important for you to convey a genuine sense of interest during your interviews. You must appear eager and flexible, but not too rehearsed. Don't fixate on being nervous. Even seasoned pros can have the "interview jitters." Above all, *never* be late for an interview appointment.

Be honest — Don't claim interest in an employer if you really do not intend to work for that organization. Don't lie on your résumé or during the interview. While you should never draw attention to your weaknesses, don't attempt to hide a shortcoming by being untruthful. Learn how to deal with perceived (or real) weaknesses *before* your interviews by talking to a campus career services professional and/or reading books on job interviewing techniques.

Be realistic — Carefully evaluate what an employer has to offer you ... and what you have to offer the employer. Don't accept a position that isn't suited to you "just because you need a job." Although most entry-level salaries have been on the rise, don't set your starting salary expectations too high. If a starting salary seems inordinately low, but is for a position that you really want, you might be able to arrange for an early salary review.

---

[2]Some of this material is adapted from Scheetz, L. P. Recruiting Trends. Collegiate Employment Research Institute. © Michigan State University.

Index

A

A band 891, 893, 894

Abdominal cavity
  beef 190
  hog 166
  pork leaf fat 290
  sheep and lamb 207–208
  veal and calf 220–221

Abdominis internus muscle 583

Abscesses 27, 67, 100–102
  beef 102, 107
  cattle liver 102
  swine 100, 170

Accelerated processing (*see* Hot boning *and* Hot fat trimming)

Accidental death 95

Acetylated monoglycerides (food-grade Dermatex®) 730

Acetyl choline 896

Achilles tendon 568, 604

ACTH (adrenocorticotropic hormone, corticotropin) 341

Actin 800, 895, 899, 900

Adam's apple (epiglottis)
  beef 182
  pork 159

Added water (AW)
  in curing 739
  in sausages 797–800

Addison's disease 334

Additives 14, 54

Adductor muscle
  beef 566, 569

Adenosine triphosphate (ATP) 895, 897, 898, 901

ADH (antidiuretic hormone, vasopressin) 342

Adipic acid 679

Adrenals 334

Aerobic microorganisms 22

Age classification of veal and calf 629

Aging
  beef 544, 902, 907
  deer 250
  meat tenderness and flavor 699–700

Agricultural Marketing Service (AMS) 91, 472

Air-powered equipment 193–196, 199, 200, 221

Air (used in hide removal) 155, 182, 185

Aitch bone
  beef 188–190, 353, 354, 355, 564, 565, 572
  deer 249, 254
  lamb leg 207, 603
  pork leg (ham) 161, 172, 505, 510, 512, 515, 516
  veal and calf 639

Aldehydes 699

Alginate 678–679

Aluminum foil 723, 724

Alum tanning 327

American Heart Association 3, 914

American Lamb Council 596, 627

American Meat Institute (AMI)
  and beef carcass retail cut yield 588
  and pork carcass retail cut yield 534

American Meat Science Association 502

Amino acids 1, 663, 683, 746, 911–913
  structure 912

AMS (Agricultural Marketing Service) 91, 472

AMS, USDA, IMPS standards 500–501, 1024

Anaerobic microorganisms 22

Analytical and scientific support, FSIS 86

Analytical procedures for fat determination 648–649
  (*see also* Quality control)

Anatomy
  avian 238
  bovine 542, 631
  ovine 597
  pices 269
  porcine 505, 507
  veal 631

Anemia 338

Animal by-products 275–346

Animal food/feed 300, 301, 304–308

Anisakiasis 114

Anisakis marina 114

Antelope 247, 262

Antemortem inspection 64

Antioxidants 14, 291–292, 300, 803–804

Anyl–Ray® 651–653

Appert, Nicolas 707

Aquaculture 7–8

Archaeology 133

Armbone chuck 552

Aroma 914, 931, 932
  lard 293

Artery cure 170, 510, 749, 755–757

Arthritis 102, 334

Ascorbate 802–803

Ascorbic acid 802–803

Aseptic canned foods 710

Asthma 334

Astronaut diets 720

Atlas joint
  beef 182
  deer 252

hog 157–159
lamb 206
rabbit 267
veal and calf 218

a_w (water activity) 18, 21–22,
669, 713–716, 734, 843
minimum values for microbial
growth, etc. 714, 716

B

Baby veal (bob veal) 629
Bacillus cereus 31
Backfat (fatback), pork
cutting test 481–484
separating from loin
518–519, 524, 525, 536
(*see also* Fat, external, *and* Fat
thickness)
Backfat measurement
beef 379–380
lamb 412–413
pork 423, 427–428
Backs
chicken 241
veal and calf 633
Backslop fermentation 842
Backstrap (*ligamentum nuchae,*
neck leader) 548, 554, 616
Bacon
and Utility pork grade
425–426
beef 561
curing regulations 736–737
evaluation of 1033–1034
Bacteria 17
hides 312
identification 24–34
pork leg (ham) 509
prevalence 17
size 17
Bacterial fermentation 841–842
"Bald" tip, beef 567
Barbecue
back ribs, pork 523–524
country-style ribs, pork 523
neck bones, pork 526–527
spareribs, lamb 622
spareribs, pork 525–526
Barbecue sauce 982–983
Barbecues for large groups
981–987
Barbecuing 980–987
Bating hides 316
Bear meat 110–113
Bed pads 325
Beef carcass (*see* Carcass, beef)

Beef casings 821–822
beef bladders 822
beef bungs 822
beef middles 822
beef rounds 821
beef weasand 821
Beef consumption 469
Beef fabrication 541–593
Beef jerky 715, 772
Beef measles 108
Beef patties
defined 664–665
mechanically separated meat
permitted 683
Beef patty industry 661–670
low-fat ground beef and patties
662–665
patty machine 667
patty size 661
sources of raw materials 662
Beef stew 709
Beef triangle 547, 583
Belly (side), pork
cutting test 481–484
flank 525
percent of live and carcass
weight 525
seeds 525
separating from loin 518,
524–525
Bevel 511
BHA (butylated hydroxyanisole)
292, 700, 803
BHT (butylated hydroxytoluene)
392, 700, 803
Biceps femoris muscle
beef 566, 570, 576, 577, 578
lamb 609, 610
pork 515
retail cut identification
497–498
Big Horn sheep 247, 263
Bile 67, 167, 190, 208, 234,
336
Binders 243, 664, 674,
811–813
Binding meats 794, 795,
796–797
Biodegradable 304
Biological oxygen debt (BOD)
140
Biological value 683
Biotechnology 13–16, 339
Birds
condemned 49, 53, 68, 69
game 264–266
Bird, veal 637

Bladder
beef 55, 190, 822
hog 820
uses 275, 820, 822
Blade Boston, pork 528–530
boneless roast, fresh or cured
529
cutting test 481–484
percent of live and carcass
weight 527
Blade meat 551
Bleeding
cattle 178–179
deer 248
hogs 139–140
poultry 229–230
sheep and lambs 201–202
vealers and calves 217
Blood 27, 51, 53, 54, 93,
95–96, 110, 208
clotting, hog skin collagen for
342
composition of 284, 346
handling of 140–141
meal 307
splash 137–139
spots 137–139
uses 284–285, 306–307,
335, 346
yield 141, 201, 346
Bob veal 215
BOD (biological oxygen debt)
140
Body composition 658–659
Boiling
beef 371
Haggis 788
veal and calf breast 636
Bologna
definition of 782, 785–786
formulation of 837
procedures 831–836
raw material, beef 372
Bone chip collector 666
Bone darkening, poultry 245
Boneless
beef 551, 558–559,
567–570, 573, 584–585
lamb 607–610, 611, 615,
616–617, 624, 626
pork 512–516, 531–532,
760–761
venison 256–258
Bone meal 306, 308, 346
Bones 5, 116
beef 544, 562, 569
cartilage from 344
"ham bone" 513
lamb 597, 625
maturity of 357–361

pork 505
processing of 346
use in retail cut identification
497–498, 1038
uses for 275, 283, 346
veal and calf 631

Boning
beef chuck 556
beef loin 582
beef round 568
beef rump 572
Boston shoulder, pork 530
lamb leg 607–610
lamb shoulder 616
mechanical 560, 627
pork ham (leg) 512–516,
760–761
pork loin 521–522
pork picnic 531–532
poultry 238–242
tunnel 512–516, 760–761
veal chuck (shoulder) 635
veal shanks and breasts 636

Boston shoulder (butt) (*see* Blade Boston, pork)

Bottom round (beef) 570
yield based on wholesale round
576

Botulism 30

Box curing (pressure) 751

Boxed beef 464, 590
IMPS 500–501
sub-primals 568, 577
tallow production 298

Boxed lamb 599, 625–627

Brains
cooking 283
freezing 282
medula oblongata 230
pharmaceuticals from 335,
336
shipping 282

Braising 959–961
beef 369
lamb shanks 623
liver 281
tongue 282

Branded meat products 469

Brands, hide 308, 310, 312

Bratwurst
definition of 786
formulation of 831
procedures 827, 829, 830,
833–834

Braunschweiger (liverwurst)
definition of 786, 790
formulation of 839
procedures 839

Breakfast-style pork sausage 828

Break joint
beef 184–185
deer 251
lamb and mutton 203–204,
403–404, 604

Breast
chicken 240
pheasant 265
poultry 228, 240–243 (*see
also* Meat grading, federal)
sheep and lamb 204, 412,
622 (*see also* Meat judging,
lamb carcasses *and* Meat
grading, federal)
veal and calf 635, 636
percent of carcass weight
638

Breast, flank, and shank, lamb 612, 622
breast separation from the rib
612, 618, 620
cutting test 485–488
percent of carcass weight 622

Brine concentration 734, 800–801

Brisket
beef 51, 184
corned beef 770–771
cutting test 476–480
prices 471–472, 495
wholesale 561–562 (*see
also* Meat judging, beef
carcasses *and* Meat
grading, federal)
veal and calf 217–218

Broiler or fryer
chicken 438
duck 439
rabbit 447
turkey 438

Broiling 948–952
liver 281
poultry 975

Brokers, livestock 455
distributors 463
wholesalers 463

Bromelin 672

Brucellosis 103–104

Bruises 95–99, 216
poultry 225, 441, 442, 443

Buffalo (bison) 263

Buffalo wings 238

Bullock class 352–355

Bunging
cattle 188–189
hogs 162–163
sheep and lambs 206
veal and calves 220

Burgers
buffalo 263
deer 254
turkey 243

Burn dressings 324

Businesses, meat industry 457–464

**Butcher's string 609, 621, (*see
also* Netter)**

Buyers, livestock 455

By-products, edible 277–285, 449
acetylated monoglycerides
730
blood 277, 284
brains 277, 283
bungs and middles 819, 822
casings 817–822
chitterlings (pork intestines)
277, 284
composition of 280
cracklings 293–294
fats 277
fries 277, 283
gelatin 323
heart 277, 282
imports and exports of 281
in frankfurter formula 866
intestines 277–284
kidney 277, 283
liver 277, 281
liver extract 337–338
oleomargarine 296–297
oxtail 277, 283
pancreas 283
pepsin 342
rendering 286
rennin 343
rounds, intestine 821
stomachs 283, 819
sweetbreads 277, 283
testicles 277, 283
thymus 277, 283, 343
tongue 277, 282
tripe 277, 283
variety meats 277–285
yields, beef and hog 278–279

By-products, inedible
adhesives 305
adrenal 334
animal food/feed 300–301,
304–308
bile 336
blood 305, 335, 346
bone 305, 346
brains 335–336
collagen 345
cortisone 334
fat 299, 304
fatty acids 301
feed 305–308
fertilizer 305

gall 336
gelatin 345–346
gland extracts 333–344
glue 345
glycerin 303
greases 294–304
hair 324–325
heart valves 333, 336
hides 308–324
horns and hooves 346
imports and exports of 281, 300
insulin 339
lanolin 285, 326
lard oil 291, 302
leather (*see* Leather)
meat meal 300–301, 304–308
neat's-foot oil 303
ovaries 338
pancreas 338–340
parathyroid 340
pelts 325–329
pericardial tissue 337
pet foods 301
pharmaceuticals 333–344
pigskin 323–324, 345
pituitary 341–342
rendering of 286–288, 304–308
skins (*see* Skins)
soap 303–304
stick 286
sutures 337
tallows 294–304
tankage 286, 306
thromboplastin 336
thyroid 167, 343–344
wool 326
By-products, meat 1
composition of 280
definition of 275, 474
export of 279, 281
hide and offal
defined 474
value of 332
history of 275
income from 276
prices, beef 471–473, 493, 495
processed meat 277
used for food 277–285
yield from 278–279

C

Caecum 167, 819, 822
Cala 530
Calcium 334, 678, 682–683, 694, 895–897

Calf
carcass (*see* Carcasses, veal and calf)
skins 322–323
Callipyge 904
Calories (energy) 909, 1061
definition 941
Calpains (calcium activated proteins) 902
Calves (classification) 629
Campylobacter jejuni 26–27
Canadian bacon 524
Canning 22, 707–713
history of 707
Capillary rupture/breakage 139, 904–905
Caping, deer 251–252
Cap muscles 549, 551
Cap off tip roast, beef 567
Capon 438
Carbohydrates 19, 918
Carbon dioxide (CO_2)
liquid for freezing 694
MAP (modified atmosphere packaging) 706
stunning 138–139, 229
supercritical fluid extraction 656–657
Carcass, beef 542, 543, 586
cutting 541–593
cutting test 476–480
price 471–473, 491–496
sprays 55–56
value (*see* price)
yield 379
Carcasses passed for cooking 105
Carcass evaluation
beef 350–397
lamb 402–421
pork 422–436
loin eye area measurement 536
ultrasonics 658
yield of four lean cuts 427
veal and calf 397–401
Carcasses, veal and calf 222, 401, 631–632, 634
Carcass, lamb
carcass certification program 627
cutting 595–627
cutting test 485–488
yield 412
Carcass, pork
cutting 503–534
cutting test 481–484

evaluation 422–436, 535–540
weight 509
yield of cuts 427–428, 533–534
Carcass sprays 55–56, 171, 705–706
CARDS (Computer Assisted Retail Decision Support) 472–473
Carotid arteries, severing
cattle 178, 179
hogs 140
sheep 202
vealers and calves 217
Carpal tunnel syndrome (CTS) 116
Carrageenan 663
Cartilage, used in surgery 344
Carving
poultry 976–980
red meat 987–994
Case hardening
drying sausage 843
steel 346
vegetable tanning 320
Case-ready meat cuts 465
Casings, sausage 817–827
Catalase 338
Cattle harvested, number per state and rank 174–175
Caudal (coccygeal) vertebrae (tail)
beef rounds 192
lamb 597, 602, 604
Caul fat 277
Ceilings 51
Cells 17, 29, 30
Cellulose casings 823–824
Centers for Disease Control 25
Cereal flours 805, 811, 812
Cervelot 782, 786–787
Cervical vertebrae (neck) 540, 553, 560
Cesium-137 720
CGMP (Current Good Manufacturing Practices) 31
Changes in meat during heating 931
Characteristics of food-borne diseases (*see* Food-borne diseases, characteristics of)
Cheek meat 182, 664
Chemists, FSIS 87
Chenodeoxycholic acid 336

Chicago round 562–564
  boneless retail yield 566
  percent of live and carcass
    weight 566
Chicken parts 237–242
Child Nutrition Program 80
Chilled meat preservation
  705–706
Chiller, poultry 227
Chilling, of carcasses 706
  beef 192–193, 590
  deer and other game 250
  delayed 907
  lamb and mutton 209, 212
  pork 171
  poultry 233
  veal and calf 220
Chine bones
  beef 355, 358, 359–360, 548
  pork 521
  veal and calf 637
Chitterlings (pork intestines) 54
  cooking 284
Chlorine 171
Cholesterol 335, 344, 657, 915,
  918
Cholic acid 338
Choline 924
Chopped and formed restructured meat products 675–680
Chopping 78, 641, 677
Chops, lamb
  arm 614
  blade 614
  double 611, 619
  double boneless loin 611
  English 611
  Frenched 621
  kidney 611
  loin 612
  rack (rib) 619
  Saratoga 615
  shoulder 614
  sirloin 606
  (see also center color section and
    page 598)
Chops, origin of the name 612
Chops, pork
  blade 519, 521
  boneless 522, 524
  butterfly 522, 524
  center loin
    loin 520, 521, 524
    rib 520, 521, 524
  cured and smoked loin 524
  double 522
  pocket 521, 524
  sirloin 520, 521
  Windsor 524

Chops, veal and calf
  Frenched 637
  kidney 638
  loin 638
  pocket chops (bird) 637
  rack (rib) 637
Chops, venison 255–258
  butterfly 256
  loin 254, 255
  rib 254, 255
  tenderloin 256
Chrome-tanned leather 320
Chromium impurities 678
Chuck, beef 552–560, 592
  armbone 552
  cutting test 476–480
  percent of live and carcass
    weight 552
  prices 471–473, 491–496
  rib separation 592
  two-piece 552
  yield of roasts and steaks 552,
    559
Chuck, square cut, veal and calf
  635
  name preferred instead of
    shoulder 635
  percent of carcass weight 635
  separation from rib 635
Chunked and formed restructured meat products 676
Chunking 676
Chymotrypsin 339
Citric acid 802
  cycle 898
City chicken 636
Classes, of carcasses
  beef 352–355
  lamb and mutton 403–404
  pork 424–425
  poultry 437–440
  veal and calf 398–399
Classification
  veal and calf 629
  zoological xi–xiii
Clavicle 240
Cleaning agents (FSIS approved)
  130
Cleanliness (basic requirement
  for successful meat processing)
  34, 465, 470
Clear plate, pork 528
  cutting test 481–484
Clod
  beef 556, 1022
  veal and calf 635
Clostridium botulinum 17, 19,
  30, 686, 708, 732, 741, 801
Clostridium perfringens 29

Clothing 52, 293
Cobalt–60 720
Coccygeal (caudal) vertebrae (tail)
  beef 572, 574
  lamb 597, 602, 604
  veal and calf 638
Cock, or rooster 438
Cod
  beef 190, 352, 353
  sheep and lamb 204, 412
  veal and calf 218
Codex Alimentarius, The 718
Co-extruded casings 824–827
Cold shortening 192, 233, 905
Collagen 900–901
  and tenderness 672
  casings 824
  hide 316–317, 320
Color 14, 20, 35, 933–934
  beef 180, 354, 358, 361–362
  differences due to pigment
    proteins 899–900
  fat 296
  feed 301
  frankfurters 856, 859, 862,
    875
  lamb, yearling mutton, and
    mutton 404
  lard 290, 292, 293
  pork 425–426
  poultry 441, 442, 443, 444
  rabbits 447
  restructured meat products
    673–680
  retail meat display 471, 725
  tallows and greases 295, 296
  veal and calf 214, 398, 629
  (see also center color section)
Combination cure 757
Comitrol® 677
Commercial processing
  beef 193–196, 282, 590–593
  hides 314, 315, 316, 318
  lamb 210–212, 625–627
  pork 155–157, 172, 758–762
  poultry 223–225, 226–228
  veal and calf 213, 220–222
Comminution, defined 641
Complexus muscle 551
Compliance 82–83, 449, 501
Composition
  hides 313
  knife handle 117
  muscle (myofibrillar) protein
    899–900
  part of quality control 31, 39,
    79
  variety meats 280

Composition and Quality Assessment Procedures 426

Compulsory grading 350

Computer Assisted Retail Decision Support (CARDS) 472–473

Computerized
grading 453
inspection 65–66
least cost formulation 852–883
pricing 472–473
processing 172, 304, 591, 593

Condemned animals, carcasses, meat, and meat products 49, 65, 68–69

Conformation
beef 351–352, 373–378, 392–396
lamb, yearling mutton, and mutton 403, 404, 408–411, 417–421
poultry 440, 442
rabbit 448
swine 103, 422, 424, 431–435
veal and calf 215, 397–401

Connective tissue 606, 617, 672, 675, 900–901

Consumer expenditures
amount of food and beverage dollar spent for meat 464
amount spent on food service meals 467
menu item growth rate 464
species composition of the meat dollar 464

Consumer grades, poultry 437

Consumer lifestyles 465, 466, 467

Consumer preference 82, 468–470
for lamb 595
for pork leg (ham) portions 512
influence on price 472

Consumer traits 468–470, 567

Consumption of meat 4–12, 223, 267, 271, 464–467, 469, 595, 629, 883
away from home 12, 467

Contract buying of livestock 455

Convection ovens 943

Convenience foods 465–466
frozen 672–673

Cooked meat products
bacon 72
cured pork products 740

sausages 782, 787, 831–836
manufacturing procedures 831–836
recipes 836, 837, 838, 839

Cookery, servings per pound 996–997

Cooking
food safety 19–21, 83
garbage 110
in liquid 961–963
natural products 78
rendering 286–288
traditional retail operation 88
to destroy tapeworm cysts 108
to destroy *Toxoplasma gondii* 114
to destroy trichina 111–112
variety meats 963

Cooking instructions
beef 543, 544, 570
label 76
lamb 598
patties 669–670
pork 506, 929–1002
(*see also center color section*)
spareribs 525
veal 632

Cooking losses 9, 662, 696, 712

Cooking poultry 971–975
broiling 975
frying 975
preparation 971
roasting 972–973
stuffing 973–974
stewing 975

Cooking process 941–947

Cooking temperatures 85, 949–964, 966–969

Cooking times 949–964, 966–969

Cook-in processing 712

Cold shortening 192–193, 233, 905

Collagen 316–317, 320, 342, 345, 814, 816, 817, 929

Cooling
for food safety 19–21
(*see also* Chilling, of carcasses)

Copolymer films 725

Copper 293, 678, 699, 942

Corned beef 562, 709, 770–771

Corning 770–771

Cortisone 334

Cost
brucellosis 103

frankfurters compared 875
labels 75
liver abscesses 102
meat packing 459–462
processing rough (thin) lamb cuts 627
quality products and good service 468
USDA grading service 449
USDA meat inspection 40

Country cure, dry 750–751

Country hides and skins 308

Country-style
pork ribs 523
pork sausage 828

Couscous (and lamb shanks) 623

Cover pickle 749

Cracklings 294

Crest 354

Crock pot 947

Crown roast
lamb 619–621
veal 637

Cryogenic freezing 666, 668

Cryovac® 703

Cubing 551, 571, 622, 670, 671

Cured-meat color 733, 746, 747–748, 762 (*see also center color section*)

Cured-meat competition and classes 1033–1034

Cured-meat products 769–778
bacon 769
beef bacon 770
beef jerky 772
blade Boston (Boston shoulder, Boston butt) 776
Canadian bacon 770
Cappicola 777
corned beef 770–771
Reuben sandwich 771
cured beef 771
dried beef 771
Buendnerfleisch 771
moisture–protein ratio regulations 771
game 778
hams 754–762, 772–773
canned hams 773
country hams 773
imported canned hams 774
Prosciuto (Parma) hams 775
Westphalian (German) hams 775–776
ham hocks 777
jowl 778

lean or formed bacon 770
loins (pork) 776
neck bones 777
pastrami piece, beef 561
picnic shoulder 777
poultry 778
shoulder (and ham) hock
513, 530, 777
shoulders (pork) 776
spareribs 777
Curing and smoking 88–89,
731–778
basic methods 748–758
containers 753
cured-meat shrinkage
757–758
FSIS labeling rules and
limitations 739, 740,
758
length of cure 753–754
presmoke soaking 757
temperature 753
functions of 733–734
history of 731–733
to destroy trichina 113
Curing hides 312–315
Curing ingredients 734–739
alkaline phosphates 737–738
monosodium glutamate
(MSG) 663
nitrite and nitrate 735–737
FSIS regulations and lim-
its 736
functions 735
salt 734
sodium ascorbate and
erythorbate 737
spices and flavorings 738
sugar 735
water 738–740
Curing terms glossary 749
Current developments in beef
processing 589
Current Good Manufacutring
Practices (CGMP) 31
Cushion 1016, 1017, 1024
Custom exemption 87–88
Cutaneous omobrachialis muscle
194, 558
Cuts, types of
primal 496
retail 497
sub-primal 497
wholesale 496
Cutting and cooking tests 995
Cutting method
calf carcass 629–639
effect on pricing 491, 492,
494
veal carcass 630–639

Cutting steaks across the
grain/perpendicular to muscle
fibers 570, 584, 987
Cutting tests
forms 476–488
pricing 471–473
Cutting yield
definition of 473
effect on pricing 475
Cut-up poultry 237–243
Cysticercosis 66, 108

D

Dark-cutting beef 362, 902,
903
Dark meat 237, 902–903
Deacon skins 215
Dead animals 65, 288, 309
Death loss 95–99
Deboners 227, 228, 239–240
Debraining 230
Decimal system 1059–1060
Deckle (beef brisket) 562
Deer processing 247–262
Deerskins 252
Defeathering 230–233
Defrosting meat 701–703, 965,
970–971
Degrees of doneness 933
Dehairing hogs 142–147
Dehydrated meats 699
Delaney Clause 739–740, 743
Delayed chilling (high-
temperature aging) 907
Demand, seasonal 494–496
Denaturing 65
Denuding 675
Depilating hog carcasses 148
Dermatex®, food-grade 730
Desinewed 794, 795
Dextrose 803, 840
DFD (dark, firm, and dry) meat
95, 426, 902–903 *(see also cen-
ter color section)*
Diabetes 338–339
Diagnosis of food-borne disease
32–33
illness and death 718–720
Diamond round 564
Diamond® sharpener 124–127
Diaphragm (skirt)
beef 191, 560–561
hanging tenderloin 191, 564
hog 165–166

lamb 610, 620
veal and calf 220–221
Diced veal 636, 638
Dimethylnitrosamine (DMNA)
741
Direct addition of organic acids
842
Direct Product Profit (DPP)
spreadsheet 472
Disabled animals 64–65
Discolorations, poultry
441–442, 443–444, 446
Diseased animals 64, 100–101
Diseases 40, 94–114
Disjointed
poultry 441, 443
rabbit 448–449
Distribution 31, 455–502
beef 590–593
general 31, 277, 456, 503
shortening 298
Distributors 463
Docosahexaenoic acid (DHA)
916
Dog food 301
Dorsal aorta, lamb 601
Dorsal process, vertebra, lamb
603, 620
Double rail live animal con-
veyer/restrainer 137, 193,
210, 221
Doves 265
Down animals 64–65
DPP (Direct Product Profit)
spreadsheet 472
Dressing
deer 249
for crown roast filling 621
for heart filling 282
for Scotch roast 622
pheasants 264–265
poultry 225–229
procedures (FSIS) 54–57
rabbits 267–269
rooms 52
veal and calf 216, 218
Dressing percent
defined 473
not to be confused with Yield
Grade 379, 412
of cattle 176–177
of hogs 136, 169
of poultry 226
of sheep and lambs 200
of veal and calves 216
(see also Yield, dressing percent)
Dried beef, evaluation of 1034
manufacture 771–772

Drift (fast)
    cattle 176
    hogs 135
    poultry 226
    sheep and lambs 198
    vealers and calves 216
Drop credit
    defined 474
    value 332
Drumsticks 238, 239, 241
Dry and semi-dry sausages 715, 782, 839–848
    drying conditions 843–844
    formulations 845–848
    manufacturing procedures 839–845
    timely lowering of pH 840–841
Dry bulb temperature 1055–1056
Dry-heat cookery 953–958
Drying 21–22, 27, 29, 190, 713–716, 765, 843–844
    history 713
Dry-picking, poultry 230
Dry rendering 286
Dry salt cure 749–750
Dry sugar cure (dry country cure) 750–751
Ducks 232, 243–244, 266
    Long Island style ducklings 244
Durand, Peter 707
D value 20

E

Ears, hog 145–146
Edible yield
    beef cutting fats 298
    effects on price 475–496
    poultry muscle and skin 243
Effluent 14, 108, 276, 313
E_h 22, 27
Eicosapentaenoic acid 916
Elastin 901
    enzymatic tenderization 672
    neckstrap
        beef 548, 554
        lamb 616
Electrical hazards 130
Electrical harvest 137–138, 200–201
Electrical stimulation 179–181, 229, 236, 906
Electricity definitions 1039–1041

Electromagnetic radiation spectrum 1041
Electrons 22
    radiation 720
ELISA test 84, 813
Elk 247, 262
EMME (electronic meat measuring equipment) 657–658
Emulsifying agent 813
Emus xiii, 245
Endocrine glands 333
Endomysium 890
Endospores 707
Endotoxin 17
Energy (calories) 909, 913–914, 1061
Enterotoxin 17, 28
Environment 275–276, 304, 321, 691, 707, 713, 720
Enzyme
    by-products 333
    catalyzed chemical changes in meat palatability 690
    tenderization 672
EPH (endangering public health) 89
Epidemiology 83
Epiglottis (Adam's apple)
    beef 181
    pork 158–159
Epimysium 576, 889–890
Epinephrine 334
Epiphyseal plate, or break joint
    beef 184–185
    lamb and mutton 202–204, 208, 604 bottom
Equipment
    commercial 155–156, 193–196, 210–212, 221–222, 226–228, 665–668, 671, 677, 883–887
    condition and cleanliness 50–52
    hide and skin processing 314, 315, 318, 323
    least cost formulation planning 857
    processing game 250–251
    rendering 299
    safety 115+
Ergonomics 116
ERH (equilibrium relative humidity) 21
Erysipelas 103
Erythorbate 802–803
Escherichia coli (E. coli) 26

Esophagus
    beef 106, 181–182, 191, 279, 821
    deer 249
    hog 158, 279
    sheep and lamb 205
    thyroid location 343–344
    veal and calf 217–218
Essential
    amino acids 1, 304, 911–913
    cholesterol 344, 918
    fatty acids 290, 913
    pharmaceuticals 337
    steroids 334
    vitamins 1
Estrogen 94
Ether 364, 649–651
Evaluation
    carcass 649
        (*see also* Meat grading, federal)
    of foreign meat plants 84–86
    pork carcass 535–540
Evisceration 56, 283
    beef 56, 190–191
    hog 159–167
    poultry 69, 227, 233–237
    sheep and lamb 206–208
    veal and calf 220, 221
Exemptions from meat inspection 87–89
Exotic animals 40–41
Exotoxins 17, 29
Exports to Japan 705–706
Exposed flesh, poultry 441, 443
Exsanguination
    beef 178–179
    hog 139–141
    poultry 229–230
    sheep and lamb 201–202
    veal and calf 216–217
Extenders 664, 679, 805, 811–813
Extrusion 680–681
"Eye" loin muscle 497
    beef 563
    lamb 610
    pork loin 519, 520, 536
    veal and calf 638
    (*see also Longissimus* muscle)
EZE–LAP Diamond® sharpener 124–127

F

FAC (Fat Analysis Committee) color 296
Facia 256, 412, 582

Facilities and equipment 50–53, 70–71, 74–75, 88, 93

Facing hams 169–170, 507, 512

Factors that stimulate oxygen absorption 699

Facultative microorganisms 22

Fajitas 560, 561

False lean 558

Fast (drift)
cattle 176
hogs 135
poultry 225–226
sheep and lambs 198
vealers and calves 216

Fatback (see Backfat)

Fat caps 851

Fat covering, poultry 441, 443

Fat cuts 508

Fat, dietary 913–918

Fat, external (subcutaneous) 277
beef 379–380, 392–396
lamb 412–415, 417–421, 599, 603, 604, 610, 611
pork 422, 423, 428, 429–435
poultry 441, 443
Product Examination Service 451

Fat, intermuscular
arm, beef 557, 558
cross-rib, beef 553
pork leg (ham) 514–515
rib, beef 550
shoulder, lamb 617

Fat, kidney, pelvic, and heart 277
beef 190, 379–380, 389 392–396, 563, 589
lamb 207, 402, 403, 602, 611
pork 165–166, 169, 290, 507
veal and calf 399

Fat, maximum permitted in ground beef 664

Fat oxidation 291–293

Fats
carry fat-soluble vitamins 913
defined 285
dietary 913–918
edible 277
effect on meat palatability 913–914
functions in the body 913
inedible 51, 53, 299–304
in feeds 300, 301
rendering of 286–288

supplies and use of 288–290, 296, 297, 298, 299, 300
trimming and oxidation 700
uses for, historic 285
why we eat fat 662

Fat streaking
veal and calf 399–400

Fat thickness 277
beef 379–380, 392–396
lamb 412–415, 417–421
pork 422, 423, 428, 429–435
poultry 441, 443
ultrasonics 658

Fatting out 851

Fatty acids 285, 301–303, 699
essential 913
free 291, 293, 295
methyl esters 302
monounsaturated 913–914
omega–3 family 915–916
polyunsaturated 913–914
saturated 285, 291, 817, 913–915
structure 914
unsaturated 285, 291, 817, 914–916
uses for 302

Fatty alcohols 302

Fatty amines 302

Feather bones
beef 192, 358, 361, 548
lamb 620
pork 521

Feathering
pork 425
veal and calf 399–400

Feather meal 307

Feathers, uses for 275

Federal–state relations 84

Feed 300–301, 304–308

Feet
beef 55, 182, 184–185
lamb 202–204
pig 136, 144–145, 509
veal and calf 218

Fell 205–206, 609, 610, 618, 621

Femoral artery
lamb 601
pork 169–170, 510, 755–757

Femorotibial joint 238–239

Femur (thigh bone)
beef 542, 565, 566, 568, 572, 574
lamb 597, 602, 605, 608, 609
pork 512–515

poultry 238–239
veal and calf 631, 638

Fermented meat 13, 78, 840–842

Fibrinolysin 335

Fibrous casings 823

Fibula (shank bone)
lamb 605
pork 513

Ficin 672

Field dressing
deer 249–250
pheasants 264–265

Fiery carcass (fat) 176, 905

Fill
beef 176
hog 136
poultry 225–226
sheep and lamb 198, 200
veal and calf 216

Filler (vs. binder) meats 796

Films 725–730

Finish 65, 215, 589–590, 599, 603, 610
(see also Fat, external [subcutaneous])

Fires 131

Firmness
beef 180, 362, 364
lamb and mutton 405–406
pork 425–426
veal and calf 399

Fish 269–273, 285, 302, 681–683
divided into two groups based on oil content of flesh 703
freezing 703–705

Fish meal 307

Fisting
deer 251, 253
sheep and lamb 204–206
veal and calf 218–219

Flaked and formed restructured meat products 673–680

Flaking 676–677
paint 51

Flank
beef 545, 563–564, 565
cutting test 476–480
percent of live and carcass weight 563
prices 471–473
lamb 600, 611
pork 510
ribbing, beef 545
veal and calf 638

Flap meat 583

Flavor   39, 913–914, 932
  beef   363, 369
  effect of salt   678, 700
  "gamey"   913
  ground beef   663
  heart   282
  lamb   595, 602
  lamb shanks   623
  lard   288–291
  liver   279
  oxtail   283
  veal   629, 630

Flavoring   75, 463, 806
  defined   806

Flaying   312, 317

Fleshing, poultry   440, 442

Flexor digitorium superficialis
  muscle   576

Floors   51, 130

Folic acid (pteroylglutamic acid)
  924–925

Food and Nutrition Board of the
  National Academy of Sciences,
  National Research Council
  (NRC)   909

Food-borne diseases, characteristics of   32–33

Food-borne infection   17, 25,
  110

Food-borne intoxication   17, 25,
  28, 29, 30

Food Safety and Inspection Service (FSIS)   31, 39, 277, 284

Food service industry   463–464
  chucks   559
  frozen meat items   696
  retailers   463–468
  volume   464–468

Foreign programs, inspection
  84–86

Forequarter, beef   544–562
  cutting test   476–480
  prices   471–473, 495
  ribbing   545–546
  shipping   545, 590
  yield   495

Foresaddle
  lamb   612–623
    yield   600
  veal and calf   635–637

Foreshank
  beef   562
  lamb   612, 623

Forming structured meats
  679–680

Forms
  for loaf products   848
  pricing   476–496

retail cutting test   476–488,
  490

Formulations, inspection controls
  on   72

Four d's of meat inspection   64

Four lean cuts, pork   508
  expected quality and yields for
    grades   427
  specific gravity to estimate
    composition   655–656

F$_o$ value   708

Fox food   301

Frankfurters
  company standards   856
  comparative costs   865, 869,
    872, 875
  consumption   883
  definition of   782, 787
  formulation of   836–837, 861
  manufacturing procedures for
    831–836, 855
  market   856
  raw material   856–857,
    793–797
    beef   372, 837
  regulations, USDA   72, 82,
    855
    label   855

Freeze drying   715

Freezer
  burn   699
  space   568, 697
  storage of seasoned meats
    700

Freezing   27, 35, 88, 171
  defects, poultry   231, 245,
    442, 443–444
  fish   703–706
    storage recommendations
      704–705
  meat   19, 78, 694–705
    packaging   701
      butcher's wrap   701
      drugstore
        (apothecary) wrap
        701
      vacuum packaging
        701
    physical effects   696–698
    rate and amount   697
    temperature, length of
      storage, and thawing
      701–703
  poultry   78, 245–246, 703
  recommendations   694–698
  to destroy *Anisakis marina*
    114
  to destroy tapeworm cysts
    108
  to destroy trichina   112

Frenching   621, 637

Fresh sausage(s)   827–831
  identification   781
  manufacturing procedures
    827–831
  formulations   830, 831
  regulations, FSIS   828–829
  venison   259

Fries (mountain oysters, testicles)
  cooking   283

"Frosting over the rose"   558

Frozen convenience foods
  672–673

Frozen meat
  cooking losses   696
  cutting method   698
  handling in the home and insti-
    tution   965, 970–971
  freezer burn   970
  refreezing   965
  roasting frozen meat
    970–971
  thinking ahead and thawing
    (defrosting) meat in the
    refrigerator or cooler
    970–971
  palatability   696
  purchase options   466

Frozen storage   691–705

Fryer
  chicken   236, 437, 438
  rabbit   447

Fryer–roaster turkey   438

Frying   952–955
  liver   281
  poultry   975

FSIS (Food Safety and Inspection
  Service)   31, 39, 277, 284

Further-processed turkey
  242–243

Further processing, defined   641

Futures, livestock   455

G

Gall bladder   336
  beef   190
  hog   167
  sheep and lamb   208
  veal and calf   220

Gallstones (price)   336

Game birds   264–266

Game conservation   247

Game processing   247+

Gamma radiation   720

Garbage treatment
  greases   288
  trichina control   110

Gastrocnemius muscle   571, 605

GDL (glucono delta lactone) 803, 804, 805

Geese 225, 243–244, 266, 439

Gelatin 323, 345–346

Genetic engineering 341, 815

GES (*E. streptococcus*) 100

GH (growth hormone) 341

Gizzard 235

Gland extracts 333–344
    adrenal 334–335
    collection specifications 340
    liver 337–338
    ovaries 338
    ox gall 336
    pancreas 338–340
    parathyroid 340
    pepsin 342–343
    pineal 340
    pituitary 341–342
    rennin 343
    thromboplastin 336
    thymus 283, 343
    thyroid 167, 343–344

Glandular meats 277–285

Glucono delta lactone (GDL) 803, 804–805

Glue 345
    horse 43

Glutaraldehyde tanned shearling 325

Gluteus medius muscle
    beef 354, 577, 580, 583
    lamb 603, 606, 610
    pork 512, 520

Glycerin 303

Glycerol 285, 291, 917–918
    uses 302

Glycine 292

Glycogen 362, 898, 918

Glycolysis 362, 898

GMP (good manufacturing practices) 31, 840

Goldfisch® fat extractor 650

Gonad-stimulating hormone 341

Good manufacturing practices (GMP) 31, 840

Gooseneck round, beef 575

Gracilis muscle, beef 353–354, 569

Grader 208, 451–452
    qualifications 451
    salaries ranking 451–452

Grades
    beef 350–397
        percent graded 397
        computerized grading 453
        cost 452

definition 348–349, 473
    lamb, yearling mutton, and mutton 402–421
        percent graded 416
    Meat Certification Service 449–450
    pork 422–436
        calculation of grade 428
    poultry 436–447
    Product Examination Service 450–451
    rabbit 447–449
    stamp application 388–390, 450, 452–453, 590
    tallows and greases 295
    to purchase by the institutional food buyer 1002
    veal and calf 397–401
        percent graded 401

Grading
    chart for beef 365
    not to be confused with inspection 39, 390

Grainy lard 293

Granulomas 101

Greases 294–304
    as oils 302–303
    as soap 303–304
    classification of 294–296
    fatty acids from 301–302
    production of 300
    utilization of 300

Green cores 851

Greening room 840

Green rings 851

Grinder knives 666, 677

Grinding 78, 88
    knives 120–121
    meat 641, 665–666

Ground
    beef
        additional uses 670
        composition of 664
        consumption 469
        definition of 664
        manufacturing 661–670
        knives 666
        percent of total beef consumption 469
        raw material 555, 556, 557, 583, 662
    lamb 623
    veal and calf 636, 638

Grouse 264–266

Grubs 67, 106–107, 310

Guinea 244, 439–440

Gullet 106, 234, 249, 252

H

HACCP (Hazard Analysis Critical Control Points) 31, 43–44, 58–64, 88, 90, 227–228, 464–465, 940

Haggis 284

Hair
    all species 85, 275, 324–325
    calves 56, 66, 215
    growth in pigs 323
    hog 55, 142–148, 323
    horse 43
    human 28, 52
    preparations 324–325
    removal 142–148
    uses 275, 324–325

Half turkey 243

Halophilic organisms 734

Ham 772–773, 754–762
    bruises 95–96
    canned 773
    chopped 683
    commercial production 758–762
    cook-in 712
    country 773
        Smithfield 774
        Virginia 774
    definition 509–510
    evaluation of 1034
    facing 169–170, 507, 512
    imported 774
    Prosciutto (Parma) 775
    regulations, FSIS 72, 738–740
    sectioned and formed 676, 760–762
    Westphalian (German) 775–776
    (*see also* Leg [ham], pork)

Ham and bacon shows 1033–1034

Hamburger
    composition of 664
    definition of 664
    manufacture 665–668
    raw material for, beef 371

Handling
    acids during fat extraction 651
    beef in industry 590–593
    by-products 275–276, 287
    calves 56, 99, 216–217, 220
    equipment and material 129
    fish 271–272
    food 16, 25, 28, 29, 50, 53–54, 83
    instructions on food label 76
    livestock 95–99
    meat meals 304

poultry 225–226
rabbits 267
raw meat 71–72
sewage 74
sheep 198, 210
solvents during fat extraction
649

Hanging tenderloin
beef 564
lamb 610, 620, 622

Hank 820

Harvest cost, defined 474

Harvesters, poultry 227

Harvest numbers, by state and rank
cattle 174–175
hogs 134
poultry 224
sheep and lambs 199
veal and calves 214

Head, inspection of 66

Head removal
beef 182–183
deer 252
hog 157–159, 172
rabbit 267
sheep and lamb 206
veal and calf 218

Head trimmings 277

Head weights
beef 182
hog 159
lamb 206

Heart
American Heart Association
(month, prudent diet) 3
cooking 282, 284
electrical harvest and stunning
137, 177, 200–201
epinephrine 334
fiery carcass 176
inspection 67
listeriosis effects 27
poultry 234–235
puncture as cause of death 95
raw material for surimi 685
valves 336

Heating 19–21, 29, 111, 112

Heating equipment 944–947
continuous cooking ovens
946
conventional gas or electric
ranges 944
forced air convection ovens
944
frying pan, wok, griddle, crock
pot, Dutch oven, grills 947
impingement ovens 944–945
microwave ovens 945–946

Heat ring 180, 905–906

Heat transfer methods 943-944
conduction 943
convection 943
radiation 943–944

Heel flies 106

Heme group 745

Heme iron 279, 745

Hemoglobin 279, 899

Hen 242, 264, 437–438

Heparin 337–338

Herbs 36

Hermetic seal (bond) 728

Hide pattern, beef 310–312

Hides 308–322
bating 316–317
blue stock 320
brining 314
classification and grading
308, 309, 315
composition of 313
costs 313
curing 313–315
glue from 345
grubs 106
handling 310
horse 43
murrain/renderer 309
packer curing of 313, 317
plumping 317
post-tan processing 321–322
preservation 312–313
processing 312–322
production and export
329–332
salt pack curing of 313
tanning 319–322
chrome 320–321
commercial 319
home 326–329
miscellaneous 321
setting out 321
splitting 317, 321
staking 322
vegetable 319–320
wet blue 317
trim standard for 310–312
value 276, 332
weights 308, 309

**High-temperature conditioning
(delayed chilling) 907**

High-temperature drying 714

Hindquarter, beef 562–585
cutting on the rail 573–575
cutting test 476–480
prices 471–473, 495
round separation 562–564
shipping 545, 589–593
streamlined 573
yield 562

Hindsaddle
lamb 601–612
yield 600
long cut 633
veal and calf 638–639

Hipbones
lamb 597, 601
veal and calf 631, 633

Hog casings 818–820
hog bladders 820
hog bungs 819
hog middles 819–820
hog stomachs 819
small hog casings 820

Hog dressed
lambs 209–210
veal 218

Hog roasts 985–987

Hogs harvested
composition change 288–289
number in one plant in 24
hours 128
number per state and rank of
134

Honing 120, 121–123

Hooves, uses for 346

Hormones 94, 334, 336, 354

Horns, uses for 346

Horse meat 41–43, 110

Hot boning 504, 589, 906–907

Hot carcass weight
beef 379, 381
definition 473

Hot dogs 780, 782, 831–837
consumption 883

Hot fat trimming 277, 589

**Hotel–restaurant–institutional
(HRI) purveyors 463–464,
627, 670**
(*see also* Purveyors)

Hothouse lambs 209–210

Hot processing 504, 589

Hot salt cure 751–752

**HRI (hotel–restaurant–institu-
tional) purveyors 463–464,
627, 670**

Humane harvest 93

Humerus (armbone)
beef 542, 552, 554, 556
lamb 612
pork 531, 532
poultry 238
veal 635

Humulin 339

**Hurdle concept 35–36, 669,
686**

**HVP (hydrolyzed vegetable pro-
teins) 663**

Hyaluronidase 343

Hydrogenated lard 290

Hydrogenated vegetable oils 293, 916

Hydrolyzed vegetable proteins (HVP) 663

Hydrophilic and hydrophobic portions of protein molecules 813–814

Hydroxides 303

H zone 893

I

I band 891, 893, 894

Ice

    crystal formation 697, 706

    effects on poultry grades 442–444

    from potable water 73

    glaze for storing fish in freezer 704

    to cool early refrigerated rail-road cars 456

ICMISC (Industrial Cooperative Meat Identification Standards Committee) 498–499, 603

Identification, of cuts 1035–1036 *(see also center color section)*

    beef 541–593

    Guide to Identifying Meat Cuts 502

    lamb 595–627

    pork 503–540

    poultry 237–242

    seven types, retail 497–498

    species 497

    Uniform Retail Meat Identity Standards 498–500

    veal and calf 632, 634–639

Identification, of microorganisms 24

IFN (International Feed Number) 305–308

Ilium (pelvic bone)

    beef 542, 573–574, 577, 583

    deer 254

    lamb 597, 601, 602, 603, 606, 607

    pork 505, 512, 520

    poultry 238, 241

    veal and calf 631, 639

Imitation ground beef 664

Immature veal 215

Immersion cure 749, 752–754, 757

Immobilization 93

    cattle 177–178

hogs 136–139

poultry 229

sheep and lambs 200–201

vealers and calves 216–217

Imperfections in hides 308

Imports, meat 84–86, 456–457

    inspection of 84–86, 457

IMPS (Institutional Meat Purchase Specifications) 449–450, 500–501, 596, 1024

Income 12

Industrial Cooperative Meat Identification Standards Committee (ICMISC) 499, 556, 603

Industry beef handling and distribution 590–593

Inedible material

    effects on consumption data 5

    fats 299–300

    inspection regulations 53, 69

Infection

    definition 17

    Salmonella 25

    swine abscesses 100–101

    swine arthritis 102–103

    trichinosis 111

Infraspinatus muscle 551

Injection

    cure 758–760

    site 101–102

Injury 115

Ink 42, 453

Inspection 31, 37–93

    and Meat Certification Service 449–450

    and Product Examination Service 450–451

    definition 45–50

    not to be confused with grading 39, 390

    rabbits 267

    requirements for grading 349

    stamps 92, 171, 610

Inspector classification 91

Institutional management problems 994–1002

Institutional Meat Purchase Specifications (IMPS) 449–450, 500–501, 596, 1024

Institutional method, chuck boning 556

Insulin 338–340

Intercollegiate meat judging contests 1003–1004, 1005–1009

Intercostal muscles, lamb 621

Intermediate-moisture foods 715–716

Intermuscular (seam) fat 550, 557, 558

Intestines

    beef 190–191, 821

    chitterlings 54, 284

    deer 249

    hog 135, 163–167, 820

    inspections 54, 66–67

    pharmaceuticals from 337

    poultry 227, 234–236

    sheep 206–208, 820–821

    uses for 275, 818–822

        as sutures 337

    veal and calf 221

Intoxication 17, 28, 29, 30

Intramuscular fat

    beef 180, 362–365

        plotted against marbling 365

        (see also center color section)

    lamb, yearling mutton, and mutton 405

    pork 426

        (see also center color section)

    veal and calf 399

Iodine number 915

Iron 292, 678, 699, 745, 910, 918

Irradiation 79, 312, 316, 717–722

    background and status 717–718

    cost of 721–722

    dose requirements 721

    energy comparisons 722

    methods/processes 720–722

    to render pork safe from trichina 113

    safety 720

    (see also Radiation)

Ischium 542, 565

Isoelectric point of proteins in sausage 840

J

Jellied products 284, 797

    Haggis 284, 788

    headcheese 284, 788

    scrapple 284, 792

    tongue 796

    veal and calf breasts 636

Jerky

    beef 715, 772

    venison 261

Jowl, pork 527

    cutting test 481–484

Judging sausages and cured meats 1033–1034

Jugular veins, severing
cattle 178
hogs 139–140
poultry 229
sheep 201–202
vealers and calves 217

Juiciness
all meat 930–931
frozen meat 696
ground beef 663
marbling 363
quality grades 39, 363, 368, 370
veal 630
(see also center color section)

Jump muscle 354

Jungle, The 37

K

Kabobs 583, 606, 615, 618, 623, 681

Kallikrein 339

Kidney, pelvic, and heart fat 277
beef 190, 379–380, 389 392–396, 563, 589
lamb 207, 402, 403, 602, 611
pork 165–166, 169, 290, 507
veal and calf 399

Kidneys
beef 190, 379, 380, 389, 392–396, 541–594
cooking 283
hog 68, 165, 169, 507
human 334
sheep and lamb 68, 207, 334, 602, 611
veal and calf 638
(see also Kidney, pelvic and heart fat)

Kidney worms 68, 107–108

kiloGray (kGy) 717

Kneecap (patella)
beef 542, 567
lamb 605, 609
pork 505, 514
poultry 238–239

Knives 52
boning 118
for sheep harvest 200
grinder 666, 677
Rockwell number 118
safe handling 127, 128
safety 117, 127
selection 117–120
skinning 118
slicing 118
tang 118

Knuckle
beef 562, 564, 567
pork 762
veal and calf 639

Kosher
calves 217
cattle 178
cut of hides 311
stamp 93

Kraft paper 724

L

Labeling
fresh meat 498–500
Guide to Identifying Meat Cuts 502
ham 738–740
irradiation symbol 79
recommended in Uniform Retail Meat Identity Standards 498–500
regulations, USDA 75–83
lard 290–291
sausage 794
Universal Product Code (UPC) system 501–502

Labels
components 76
frankfurter 855

Lactic acid 23, 24, 170, 898, 901, 902–904

Lactobacillus plantarum 841

Lamb carcass *(see* Carcass, lamb)

Lamb, fabrication of 595–627

Lamb merchandising, developments and history 595

Laminated foils 730

Lanolin 285, 326

Lard 288–294, 298, 299
antioxidants 291–292
aroma 285, 291, 293
chilling 293
color 290, 293
consumption 288–290
definitions 290, 291
export 299
flavor 290, 291, 292, 293
from backfat 519
hardening 290
overcooking 292
pressing, cooling 293
production 288–290
regulations 290–291
rendering 292–293
steam 286
temperature 292, 293

shortening value 290, 293
smoke point 293
smoothing or creaming 293
stabilizing 293
value 290
yield 293

Lard oil 291, 293

Latissimus dorsi 551

LCI (Livestock Conservation Institute) 96

Leaf fat 169–170, 277, 290, 507
rendered 290

Lean cuts, pork 426–428, 508

Least cost formulations 852–878
final reports 863–865
product standards 855–858
raw material attributes 856–862

Leather
effect of hide brands 308, 310
export and import 329, 330, 331
past and future 329–332
production 130, 186, 215
quality 315, 322–323
shoddy 321–322
uses 315–316, 329–330

Leg (ham), pork
cutting test 481–484
facing 169–170, 507, 512
percent of live and carcass weight 510

Leg, lamb 601–610
cutting test 485–488
percent of live and carcass weight 602
retail yield 602

Leg (round), veal and calf 638–639
percent of carcass weight 639
veal cutlet 639

Legs, poultry 237–239

Leviticus 17:14 93

Lifter muscles 551

Ligamentum nuchae (neck leader/backstrap) 899
beef 548, 554
lamb 616

Light fading 747

Lighting requirements 70–71, 590

Linking
alternate twist method 834
braiding method 833

Linoleic acid 290, 699, 913, 914

Linolenic acid 699

Lipid oxidation 291–292, 300

Lip, rib steaks 544, 551

Lips
  pork 146–147

Liquid CO$_2$ 694

Liquid nitrogen 694

Liquid smoke 766

Listeria monocytogenes 17, 27–28

Liver
  beef 190
  carbohydrate reservoir 918
  condemned 68
  cooking 281–282
  inspection 67
  nutritive value 277, 279, 280
  pharmaceuticals from
    337–338
  pork 165–167
  poultry 234
  sheep and lamb 208
  veal and calf 220

Live–retail price spreads
  491–493

Liver sausage 31, 819, 839

Livestock Conservation Institute
  (LCI) 96

Livestock production 455–456

Livestock Weather Safety Index
  97

Loading rates for livestock
  97–99

Loaf products 848–850
  formulations 849–850
  liver 281–282

Loin, beef 577–585
  cutting test 476–480
  percent of live and carcass
    weight 577
  prices 471–473, 495
  separating from round
    562–566, 572–574
  steak yield 577

Loin, lamb 610–612
  cutting test 485–488
  eye 603, 612
  yield of tender, juicy loin chops
    610

Loin, pork 517–524
  blade chop 519, 529
  cutting test 481–484
  loin eye area measurement
    536
  percent of live and carcass
    weight 517

Loin, veal and calf 633, 638
  percent of carcass weight 638

Longissimus costarum muscle
  551

Longissimus muscle
  beef 380, 547, 550, 551,
    552, 554
  blade Boston 529
  lamb 412, 599, 603, 610,
    613, 614, 615
  pork loin 519–524
  retail cut identification 497
  veal and calf 637
  (*see also* "Eye" loin muscle)

Longus colli muscle 616

Loose side (kidney fat location in
  beef carcasses) 563

Losses, in shipping livestock
  95–98

Lovibond color 296

Low-fat ground beef and patties
  662, 665

Low-moisture foods 713–716

Low-temperature drying 714

Lumbar lean, pork carcass 510

Lumbar vertebrae
  beef 192, 358–360, 542, 545
  lamb 597, 602, 603,
    611–612
  pork 517, 520
  retail cut identification
    497–498
  veal and calf 638

Luncheon (specialty) meats
  782–783, 848–849
  formulations 849–850

Lungs
  cooking 284
  inspection of 66–67
  pharmaceuticals from 338

Lymph nodes 16, 68
  atlantal 66
  bronchial 66, 67
  iliac 67, 68
  inguinal 67
  inspection of 66–68
  internal *iliac* 67
  lumbar 67
  mandibular 66, 159
  mediastinal 66, 67
  mesenteric 67
  parotid 66
  popliteal 68, 515, 570, 575,
    609
  portal 67
  prefemoral 68, 604
  pre-scapular 555, 558, 616
  renal 67
  superficial inguinal 67, 68
  super-mammary 67
  suprapharyngeal 66

M

Macerator 571, 670, 671

Management, meat business
  470–471

Mandibular lymph nodes 66,
  158–159

Manganese 699

Manufactured casings 822–827
  casings formed from the meat
    824–826
  cellulose 823–824
  collagen 824

MAP (Modified Atmosphere
  Packaging) 706, 725–726

Marbling
  beef 180, 362–365
    plotted against intramus-
      cular fat 365
    (*see also center color section*)
  lamb, yearling mutton, and
    mutton 405
  pork 426
    (*see also center color section*)
  veal and calf 399

Margarine 296–299

Margin
  calculation 489–496
  defined 474
  effect on pricing 489–491
  live/carcass, beef 491–496
  live/quarters, beef 491–496
  meat packing profit 461–462
  problem 489–491

Market classes
  of beef 352–355
  of hogs 424–425
  of sheep and lambs 403–404
  of veal and calves 398–399

Marketing 23, 39, 455
  auction 455
    cattle 174
    hogs 133–135
    sheep and lambs 198
    veal and calves 213
  beef prices 471–472, 495
  boxed beef 456, 464
  buyers, livestock 455
  carcass weight and grade 459
    cattle 175
    hogs 135
    sheep and lambs 198
    veal and calves 213
  chicken 242
  direct 455
    cattle 174–175
    hogs 133
    sheep and lambs 198
    veal and calves 213
  frankfurters 856

futures   455
hides   310
imports   4–6, 84–86, 456–457
livestock   95–99, 455, 458
Packers and Stockyards Act   455, 458–459
seasonal variations in   494–496
speed   494
stockyards   455
Markup
calculations   489–496
defined   474
Massaging (tumbling) pork, leg (ham)   515, 762
Mass and length; prefixes for powers   1045
Masseter muscle   108 (*see also* Cheek meat)
Materials handling   129
Mature (old) poultry   439
Maturity
beef   357–361
lamb, yearling mutton, and mutton   403–404
pork   424–425, 435
trichina   109
veal and calf   398–399
Maws   283
Meals away from home   466, 467
Measles, beef   108
Meat Buyers Guide, The   501
Meat Certification Service   449–450
Meat color (*see* Color)
Meat consumption   4–12, 223, 267, 271, 464, 466, 467, 468, 469
Meat cookery   929–1002
Meat curing, history of   731–733
Meat emulsions, or batters   813–817
Meat grading, federal   347–453
Meat Import Act of 1979   456
Meat industry
and meat inspection exemptions   87–89
annual sales   3, 459–462
businesses   457–468
comparisons   459–462
cooperates with FSIS   73
disassembly   457–458
history and status of the horse meat industry   41–43
investing money in worker safety   116

life quality enhanced with by-products   275
sanitation throughout   16, 50–58
Meat judging
beef carcasses   1011–1012
lamb carcasses   1015–1017
pork carcasses   1012–1014
specification classes   1024–1026
wholesale cuts   1017–1024
written reasons   1026–1033
Meat judging, vocational
FFA and 4–H contests   1034–1038
retail cut identification   1035–1038
Meat meal   301, 304–308
Meat, number of servings per pound   996–997
Meat, nutrient composition of   927
Meat Nutri-Facts   468
Meat packing economics   459–462
sales and costs   461–462
Meat preservation   22
defined   689
history   689
reasons for   689–690
Meat processing
defined   463, 641
functions   641
Meat processors   23, 463–464
Meat production   4, 456
Meat retailing   464–468
take-out food   466
Mechanical deboning   560, 627, 674
Mechanically desinewed meat   794
Mechanically separated meat, poultry, and fish   681–684, 794
defined   681
poultry   681, 794
regulations   682–684
separated tissue   560
Mechanical tenderization   571, 670–672
Medicinal products   333–344
Melatonin   340
Melting points, fats   285, 291, 916–917
Melts   208
Merchandising, meat   464–468
Mesophiles   19
Metacarpus   238–239

Metatarsus   238–239
Metmyoglobin   745–746
(*see also center color section*)
Metric system   1042–1054
Microbial counts for clean equipment   51–52
Microbiology   16–19, 91, 707
monitoring   51
Micrococcus aurantiacus   841
Microwave cooking   294, 653–654, 672, 934, 964–965, 966–969
Microwave reheatables   713
Milk powder   805, 811, 812
Minerals   19, 699, 918–919, 927
Mink food   301
Minor cuts, pork   508, 532
Mitochondria   891
MIU (moisture, impurities, and unsaponifiables)   295
M line   893, 894, 899
Mock
chicken   636
duck   623
tender   557
Modified atmosphere packaging (MAP)   706, 725–726
Moist-heat cookery   947, 959–965
Moisture   18, 21–22
Moisture–protein (M:P) ratios   797
Mold inhibitors   715, 804
potassium sorbate   804
Molds   18, 21, 34
Monoclonal antibodies   14
Monosodium glutamate (MSG)   663, 805
Moose   247, 262
Motor endplate   892, 896
Mountain oysters   277
Mrad   717
MRE (Meal, Ready-to-Eat) rations   710–711
MSG (monosodium glutamate)   663, 805
Multifidus dorsi muscle
beef   551
lamb   603
Murrain hides   309
Muscle
abscesses caused by injections   100–102
blood splashes   139, 904, 905
contraction   895–898

density 136
grub damage 106–107
location 504–507
pH 901, 902–904, 906
physical characteristics 889–894
protein composition 899–901
sterile 16
structure 890, 891, 892–895
tenosynovitis 116
tremors with PSS 95
trichina 109–113
types 889
cardiac 889
skeletal 889
smooth 889
(see also pH)
Muscling, pork 424
Mutilated and/or bloodshot areas in game 258–259
Mycoplasmas 102
Myofibril 890, 891, 892, 893, 894, 895
Myofibrillar protein 762, 899, 929
Myoglobin 279, 736, 744–748, 902
chemical structure of 745–746
content in different species and muscles 744
pigment states of 746
Myosin 515, 678, 800, 814, 816, 895–897
Myxedema 344

N

Napole 904
National Cattlemen's Beef Association resources 278, 499, 501, 534, 673
National Pork Producers Council (NPPC) 426, 502, 535–540
National Safety Council 115–117
Natural casings 817–822
Natural fermentation 842
Natural food 78, 468
Nature veal 629
Navel 560
(see also Plate [short plate], beef)
Neat's-foot oil 303
Neck bones
beef chuck 560
cutting test 481–484

pork 526–527
poultry 241
Neck leader (backstrap, ligamentum nuchae) 548, 554
Neck slices, lamb 613, 618
Needle tenderizing 670–672
Nerve 30, 116, 344
Netter, roast-tying
beef chuck 559
lamb 609, 617
pork blade, Boston and picnic 532
pork leg (ham) 515–516
rump 573
New York
pork shoulder 528
round 564
strip (steak) 585
Nicotinic acid (niacin) 923
Nitrate 72, 735–737, 801–802
regulations 736, 802
Nitric oxide 736
Nitrite 30, 72, 735–737, 739, 741–743, 801–802
functions of 735, 801
other sources 743–744
regulations, FSIS 736, 744, 802
Nitrogen
liquid 694
MAP (modified atmosphere packaging) 706, 725–726
Nitrosamines 741, 743–744
Nitrosohemochrome 746
Nitrosomyoglobin 746
N-nitrosomorpholine (NMOR) 741
N-nitrosopyrrolidine (NPYR) 741
Nonfat dry milk 805
Nonmeat ingredients, in restructured products 678–679
"Nontraditional" products 739
North American Meat Processors Association 501
NPPC (National Pork Producers Council) 426, 502, 535–540
Nutrient content loss during cooking 934–936
Nutrient density 277, 279, 910
Nutrition 277, 909–927
effect on hide development 310
labeling 79, 468
lard 290

O

Obturator foramen (hole in pelvic bone) 575
Obturatorius internus (oyster muscle) 575
Occupational Safety and Health Administration (OSHA) 117
Odor 932
lard 290–292
rancidity 698–699
soap 303
Offal 34, 54
defined 275, 474
horse 43
value 276, 331–332
Office of Price Administration (OPA) 350
Oil 302–303
defined 285
lard 291, 293, 302–303
neat's-foot 302–303
vegetable 293, 303
Oleic acid 291, 699, 914
Oleomargarine 296–299
Oleoresin 806
Omega-3 family of fatty acids 915–916
Online services and periodicals 1062–1064
Online shopping 466
OPA (Office of Price Administration) 350
Organic food 78, 80, 468–469
Organizations 502, 1074–1082
OSHA (Occupational Safety and Health Administration) 117
Ossification 355, 357–361, 398, 403–404, 439
Osso bucco 636, 639
Ostriches xiii, 245
Ovaries 338
Overcooking 930
Overhaul 749
Ox gall 336, 338
Oxidative rancidity 291–292, 293, 675, 690, 698–705
Oxtail
cooking 283
Oxygen 22, 28–30, 291, 747–748
transmission through packaging films 725–729
Oxymyoglobin 746
Oxytocin 94, 341

Oyster muscle (*Obturatorius internus*)
  beef 575
  chicken 979
  turkey 241

P

Packaging 22, 23, 35, 723–730
  cost as a portion of manufacturing operating expense 462
  films 725–730
    formation of 725, 729–730
    sealants 727
  lamb 627
  machine 728
  materials characteristics 723–724
Packers 457–462
Packers and Stockyards Act 455, 458–459
Packing industry
  and John Pynchon 455, 458
  functions 457–458
  history 455–456, 503–504
  inventory control 471
  kill and chill 458
  locker plant 458
  multi-plant packers 458
  origin of name of 458
  purchase options 455
  small packers 459–462
Palatable food 469, 909–927
  lamb 595, 606
  venison 250
Palmitic acid 914, 915
Panbroiling 948, 951
Pancreas 338–340
Pancreatin 339
Panfrying 952, 953
Pantothenic acid 925
Papain 672
Parasites, internal 107+
  Anisakis marina 18
  kidney worms 107–108
  roundworms 107–108
  tapeworms 108–109
  Toxoplasma gondii 18
  trichinae 18, 109–113
Parathyroids 340
Partially defatted beef fatty tissue (PDBFT) 287
Partially defatted pork fatty tissue (PDPFT) 287
Particle-size reduction, methods of 675–677

Partridge 264–266
Passed for cooking, carcasses 105
Pasteurization 20, 25, 27, 73, 105, 709–710
  by irradiation (radurization) 718–719
Patella (knee cap)
  beef 542, 567
  lamb 605, 609
  pork 505, 514
  poultry 238–239
Pathology and epidemiology 83
Patties 661–670
  lamb 622
  portion control of 641
  produced by purveyors 463
  veal and calf 636
  (*see also* Ground beef, definition of, *and* Hamburger, definition of)
Patty machine 667–668
Payment for livestock 458–459
Pectoral muscles 614
Pediococcus acidilactici 841
Pediococcus cerevisiae 841
Pelting 202–206, 211
Pelts 308–332
  classification of sheep 325
  curing 312–315
  deer 252
  preservation of 312–313
  rabbit 268–269
  tanning 319–321, 326–329
  uses for 325
Pelvic bone
  beef 187–188, 353–354, 574, 583
  deer 249, 254
  lamb 206, 601–602, 606, 607, 608
  pork 161–162, 172, 505, 512–516
  poultry 238, 241
  veal and calf 639
Pemmican 732
Pepperoni 846–847
  Caserta 786
Pepsin 342–343
Pericardium 166, 337
Perimysium 889–890
Periodicals and online services 1062–1064
Peroxide value 291, 699
Pest control 74
Pet food 301
PFF (protein, fat-free) regulations 739–740

pH 19, 22–23, 27, 28, 30, 53, 72, 735, 898, 901, 903, 906
  binding and solubility 816
  dark-cutting beef 362
  hides 316–317
  patties 669
  soap 304
  tanning 320
  timely lowering of in sausage manufacture 840–842
Phalanges 238
Pharmaceuticals, by-product 1, 333–344
Pheasants 245, 264–265
Phosphates 72, 678, 804
  FSIS regulations and limits 738
Phospholipids 649, 656, 917–918
Phosphorus 918–919
Physical properties of meat 23, 1009–1027
Picker, poultry 227
Pickle 72, 734, 749
Pickled pigs' feet 509, 709
Pickling hides 317
Picnic, pork 526, 528–532
  cutting test 481–484
  percent of live and carcass weight 527
Pigeons 440
Pig roasts 985–987
Pigskins 323–324, 342
  weight 141, 155
Pinbone 579, 583
Pineal 340
  melatonin 340
Pit barbecues 981–982
Pituitary, pharmaceuticals from 341–342
  ACTH (adrenocorticotropic hormone, corticotropin) 341
  ADH (antidiuretic hormone, vasopressin) 342
  growth hormone (GH) 341
  oxytocin 341
  prolactin 341
  TSH (thyroid-stimulating hormone, thyrotropic hormone, thyrotropin) 341
Pizza topping 670
Pizzle
  beef 188–189, 311, 352–354
  hog 160–161
  sheep and lamb 207
Placement 1083–1084
Planimeter 380

Plasmin   335

Plate (motor endplate)   892, 896

Plate (short plate), beef   560–561
   cutting test   476–480
   price   495
   ribbing   545
   separating from rib   547

Pleural membrane   524

Pluck
   beef   191
   deer   249, 252
   hog   166
   sheep and lamb   208
   veal and calf   220–221

P_o (vapor pressure of pure water)   713

Polish sausage formulation   838
   venison   260

Popliteal lymph node   68, 515, 570, 575, 609

Population   1–3, 5, 42, 113, 348, 455–456, 627

Porcine dressings   324

Porcine stress syndrome (PSS)   95

Pork carcass (*see* Carcass, pork)

Pork fabrication   507–534

Pork grades   422–436
   for sow carcasses   435–436
   U.S. No. 1   429, 431
   U.S. No. 2   429, 432
   U.S. No. 3   429–430, 433
   U.S. No. 4   430, 434
   U.S. Utility   430

Pork processing, history of   503–504

Pork sausage   827–831
   breakfast-style   828
   definition of   791

Porphyrin ring   745

Porterhouse steak   580, 581
   history of name   581

Portion control   463

Postmortem inspection   65–69

Potable water   73–74, 89

Potassium–40 liquid scintillation detection   658–659

Potassium sorbate   804

Potted meat   709

Poultry
   by-product meal   307
   cut-up   241, 703

Poultry grading   436–447

Poultry processing equipment, automatic   226–229

Poultry harvested, number per state and rank   224

Poultry stuffing or dressing   973–974

Power tools and equipment   129, 155–156, 193–196, 210–212, 221–222, 226–229, 323

Preblending   877–883
   defined   877

Precooked meat (retail)   465
   char–marked   670
   fully cooked   669
   partially cooked   669
   patties, precautions and regulations   669

Pregnancy   94, 177, 355, 424

Preliminary beef yield grade (PYG)   381–382

Preliminary pork carcass grades   428

Pre-scapular lymph node   555, 558, 619

Preservation   689–730

Price
   animal–carcass   491–496
   by-product, beef   332
   consumer price index   89
   definition   475
   fluctuations influence yield grade value differences   388, 416
   gallstones   336
   horse meat   43
   muscle value comparison   241
   quarters, beef   493, 495
   reporting   471–473
   spareribs   525
   spreads   496
   summer and winter   494–496

Pricing   471–496
   alternatives   467, 470, 489–491
   and cutting tests   475, 489, 491
   and diet changes   468, 469
   and livestock production costs   472
   as a tool to "control the flow"   471
   competition   472, 489
   effect on demand   494
   margin   475–491
   muscle value comparison   241
   problem   489–494
   profitable   461–462
   relationships   492–496
   seasonal effect on   494–496
   supply and demand   472
   terms   473–475

Primal cuts
   boners sell   463
   defined   496
   IMPS   496, 500–501
   list of   501
   pork   508
   prices of   471–473, 495
   standards   501–502

Processed meats
   pork   504, 731–778, 779–888
   specialty shops   466
   variety meats   277, 279, 281–282, 284

Processing
   controls   31, 34–37, 72
   definition of   463
   rate
      beef   590
      pork   504, 535

Processors   463–464

Procurement grades, poultry   446–447

Product Examination Service   450–451

Production
   broilers   223–224
   by-product and livestock efficiency   279
   captive game   247
   clothing   329–330
   livestock   455–456, 535–540, 642
   patties   665–670
   pork production efficiency   535–540
   tapeworm eggs   108
   volume of a single pork plant   459
   wool for one suit   326
   world meat   4

Professional organizations   502, 1074–1082

Prolactin   341

Propyl gallate   292, 700, 803

Propylparaben   804

Protective equipment   127

Protein, fat-free (PFF) regulations   739–740

Proteins, groups 1 and 2 in sausage   799

Proteins, meat, classification of   899–901, 911

PSE (pale, soft, exudative) meat   95, 138, 426, 903–904
   (*see also* center color section)

Pseudomonas   19, 23, 691

Psoas muscle
   beef   563, 578, 580, 582
   lamb   610
   pork   520–521

rabbit (electron micrograph) 893

retail cut identification 497

veal and calf 638

PSS (porcine stress syndrome) 95, 903–904

Psychrophiles 19

Pulled wool 326

Pulling

beef hides 195–196

deer skins 251–253

gizzards apart 235

hog skins 148–157

leaf fat 169

meat through macerating blade 671

veal and calf skins 219

versus crisp cutting of meat when grinding 666

Pump pickling (stitch or spray pumping) 749, 754–755

Purchasing method 455–456

of cattle 174–175

of hogs 133–135

of sheep and lambs 198

of veal and calves 213

Purge 706

Purveyors 463

Putrefaction 852

PYG (see Preliminary beef yield grade [PYG])

Pynchon, John 455, 458

Pyramid, food/nutritional 8

Q

Quadriceps muscles (four)

beef 566, 567

retail cut identification 498

Quail 264–266

Quality

assurance 31, 34–37

beef 180, 357–367

definition 39

game 247

lamb, yearling mutton, and mutton 405–407

pork 95, 425–426

poultry 440–446

protein 304–306

rabbit 447–449

veal and calf 214–216, 399–400

(see also center color section)

Quality control 31, 34–37, 38, 80, 304, 643–661

analytical procedures for fat determination 648–659

Anyl–Ray® 651–653

Babcock 651

cooking 653

ether extraction 649–650

laboratory/research methods for complete lipid extraction 656

modified AOAC 651

rapid microwave moisture and fat analysis system 653–654

specific gravity 654–656

blending samples to get a composite sample 648, 659

electrical conductivity and impedance 657–658

EMME (electronic meat measuring equipment) 657–658

TOBEC (total body electrical conductivity) 658

final blending of fat and lean blends 659–661

algebraic method 660

method of squares 660–661

potassium–40 liquid scintillation detection 658–659

sampling 643–648

accuracy 644, 646–648

number of samples 647

range 647

standard deviation 644–646

variation 644

supercritical fluid extraction and steam stripping 656–657

ultrasonics 658

Quarter turkey 243

Quick cure 749

Quick reference for conversions between SI and English units 1046–1047

R

Rabbi 93

Rabbit pelts 268–269

Rabbits 266–289

grading 447–449

Rack (rib, hotel rack), veal and calf 633, 637

percent of carcass 637

separating from shoulder 635

Rack (rib) lamb 618–621

cutting test 485–488

ICMISC recommended name for rib 618

percent chops of live weight 618

separating from shoulder 613

Radappertization (sterilization) 719–720

Radiation 29, 312–313, 652

(see also Irradiation and Electromagnetic radiation spectrum)

Radius and ulna (foreshank)

beef 542, 562

lamb 597, 612, 623

pork 505, 530–531

poultry 238

veal and calf 635

Radurization (pasteurization) 718–719

Rail height 74–75

Rams 405

Rancidity 291–292, 300

Range produced (fed) 468

Ready-to-eat (RTE) 35, 466, 768

processed meats 466

Rectal temperatures 64–65

Rectus abdominis muscle

beef, flank steak 565

lamb 602

pork 525

Rectus femoris muscle (one of the quadriceps) 567

Refrigeration and freezing equipment 691–694

cryogenic freezing 693–694

meat storage units 692–693

home storage units 692–693

walk-in storage units 693

mechanical refrigeration 691–692

Regulations

game hunting and processing 247, 250–251

Reinspection 70–74

Rejection, imported meat 84–86

Relationship between marbling, maturity, and quality 365–367

Relative humidity (RH) 21, 1055, 1056

Rendering

definition 286, 291

methods

home 292–293

inedible fats 299–300

low-temperature, continuous systems 287

Rennet 343

Residue Avoidance Program (RAP) 84

Residue monitoring 84

Resin guaiac 292

Restaurants
    as a part of the food service industry 463, 467
    greases 288
    sous vide 685–686

Restricted ingredients 69, 70, 72

Restructured meat products 673–680
    comminution method and particle size 675–677
        chopping 677–678
        chunking 676
        flaking 676, 677
        grinding 666–667, 677
        sectioning 675–676
        slicing 676
    nonmeat ingredients 678–679
    processing procedures 679–680
    raw materials source and quality 674–675

Retail cuts 23, 35, 88, 497–498
    beef 543, 547–588
    consumers buy fresh and freeze 694–695
    definition of 497, 602
    guidelines for using 502
    ICMISC 499
    lamb 598
    list of 476–488, 497–498, 1036
    pork 506, 508
    seven types of 497–498
    Uniform Retail Meat Identity Standards 498–499
    Universal Product Code 501–502
    veal and calf 632
    (*see also center color section*)

Retailers, meat
    and beef cutting 544–589
    and lamb availability 595
    cutting test 475–496
    description of types of 464–468
    margin 471–496
    pricing 471–475
    Uniform Retail Meat Identity Standards 498–500
    Universal Product Code 501–502

Retail exemptions from inspection 88–89

Retail yield based on live and carcass weights
    beef 587–589
    lamb 623–625
    pork 533–540

Retained carcasses 68

Reticulum 283
    cooking 283

Retortable pouches 710–711

Retort processing 708

Rheumatic fever 336

Rhomboideus muscle 551

Rib, beef 547–551, 592
    chuck separation 547
    cutting test 476–480
    prices 471–473, 489–491, 495
    yield of steaks and roasts 547

Ribbing
    beef carcasses 356–357, 380, 544–545
    lamb carcasses 599–600
    pork carcasses 536
    veal and calf carcasses 633

Rib bones
    beef 358–360, 542
    fish 273
    lamb 597
    pork 505, 517–518, 525–526
    poultry 238, 240
    veal 631, 635, 637

Rib eye area
    beef 380
        carcass weight schedule 383
    lamb 600
    ultrasonics 658

Rib eye muscle
    beef 356–357, 380, 547, 550, 551, 552, 554
    lamb 412, 618

Rib fingers (*intercostal* muscles, lamb) 621

Riblets, lamb 622

Riboflavin (vitamin B₂, formerly G) 922–923

Rib (rack), lamb 618–621
    cutting test 485–488
    percent chops of live weight 618
    separating from shoulder 613

Rib (rib rack, hotel rack), veal and calf 633, 637
    percent of carcass 637
    separating from shoulder 635

Rigor mortis 179, 230, 901

Ritual harvest 93, 178

RN 904

Roast beef 709

Roaster
    poultry 438–439
    rabbit 447

Roasting 955–958
    beef 368–369
    heart 282
    hogs 985–987
    lamb 610
    poultry 236–237, 242–243

Roasts, beef
    components in carcass, summary of types of cuts, identification and fabrication of 541–589
        arm 554–555
        blade 554–555
        bottom (outside) round 570, 575
        brisket, boneless 562
        chuck eye 559
        clod 558
        cross-rib pot roast 553
        Delmonico 551
        English cut 553
        eye of round 570
        gooseneck 575
        heel 571
        inside chuck 559
        neck pot 556
        outside chuck 558
        Pikes Peak 571
        rib, boneless 550
        rib, standing 549
        rump 572, 573, 576
        seven-bone 555
        shoulder pot 558
        supraspinatus 559
        tip 567
        top (inside) round 566, 575
        (*see also center color section*)

Roasts, lamb
    types of cuts, identification and fabrication of 595–627
        American leg 605
        boneless blade roll 615
        boneless leg 609
        boneless shoulder 617
        center of leg 607
        crown 621
        cushion shoulder 617
        Frenched leg 605
        leg of lamb 602
        leg-o-lamb 602
        leg, long cut 602
        leg, shank half 607
        leg, sirloin half 607
        loin 611
        Saratoga roll 615
        Scotch 622
        square-cut shoulder 613
        (*see also center color section*)

Roasts, pork
    types of cuts, identification and
        fabrication of 503–532
        arm 526–527, 530
        blade 529
        blade Boston butt 529
        country-style ribs 523
        ham, cured and smoked
            516
        leg, boneless 515, 516
        leg, center section 516
        leg, rump portion 516
        leg, shank portion 516
        loin, boneless 521–522
        loin, center 519, 520,
            521
        picnic 526–527, 530
        shoulder 528
        shoulder, boneless fresh
            532
        sirloin 521
        (see also center color section)
Roasts, veal
    types of cuts, identification and
        fabrication of 629–639
        arm 635, 636
        blade 635, 636
        crown 637
        loin 638
        rump 639
        shoulder clod 635
        sirloin 638
        square-cut chuck, clod out
            635
        standing rib 637
Roasts, venison
    types of cuts, identification and
        fabrication of 254–259
        arm 255
        neck 255
        rib 255
        round 255, 257
        round tip 255, 257
        rump 255
        shoulder 255
        sirloin 255, 257
Robotics 130, 172
Rock Cornish game hen 437
Rockwell number 118
Rodding the weasand
    cattle 181–182, 187
    lambs 205
    veal and calves 217–218
Romans 731
Rooster 264–265, 438
Rose muscle 558
Rough cuts
    beef 560–562, 565
        percent of live and carcass
            weight 560
        yield 560

lamb 622, 626
    percent of carcass weight
        622
Round
    beef 562–576
        cutting test 476–480
        prices 471–473, 495
        yield 566
    venison 255, 257
Roundworms 107–108
RTE (ready-to-eat) 35
Ruffle fat 54
Rumen
    bacteria 102
    cooking 283
    pH 102
    relationship to fill 176–177,
        198, 200, 216, 391
    relationship to kidney fat loca-
        tion 563
Ruminants 1
Rump
    beef 572–573
        percent of live and carcass
            weight 571
        percent of wholesale
            round weight 576
    veal and calf 639
Ryanodine receptor 896–897

S

Sacral vertebrae (sacrum)
    beef 192, 358–360, 542,
        562, 564, 572, 577, 578,
        579
    lamb 597, 603, 607
    pork 505, 510
    veal and calf 638
Sacroiliac joint
    beef 542, 578
    lamb 597, 607
    pork 505
Sacrosciatic ligament 576
"Safe cook" 708
Safe handling 16–24, 31–36, 81
    cooking thoroughly 19–21
    keeping hot foods hot 20
    keeping raw meat and poultry
        separate from other foods
        54
    keeping refrigerated or frozen
        20, 54, 81
    refrigerating leftovers immedi-
        ately or discarding 29
    thawing in refrigerator or
        microwave 703, 946, 970
    washing working surfaces
        (including cutting boards),

utensils, and hands after
    touching raw meat or poul-
    try 34–35, 52, 53
Safety
    food 13, 16, 31, 34–37, 669,
        689–690, 936–941
    knife 117
        handling instructions
            127–128
    worker 115–131, 138, 180
Salami
    definition of 791–792
    formulation of 845
    principles 839–845
Salimeter 749, 752
Salivary glands 159
Salmonella 17, 25, 719
Salsus 779
Salt 731
    and rancidity 292
    a_w 715–716
    in carcass rinse 170
    in sausages 800–801
    saturated solution 818
Salting 731–732
Sanitation
    by-product processing 277
    canning 707
    exports to Japan 705
    freezing systems 673
    FSIS requirements 50–58
    knife 117
    preservation 690–691
    program 31, 34–37
    quality control 31–37
    to prevent swine arthritis
        102–103
Saponification 303
Saran wrap 726
Sarcolemma 895, 896
Sarcomere 891, 894
Sarcoplasmic proteins 899–900
Sarcoplasmic reticulum 893,
    895, 897
Saturated fat
    definition 285, 291, 913–915
Sausages 779–887
    batters 813–817
    casings 817–827
        quality control 817
    classification/identification of
        781–783
    equipment 883–887
    evaluation of 1033–1034
    glossary 784–793
    history of 779–781
    ingredients 793–813
    least cost formulations
        852–877

nonmeat ingredients 798–813

per capita consumption 9, 464, 781

product defects 850–852

production 781

raw material attributes 856–862

smoking 835–836

terminology (glossary) 783–793

thermal processing 835–836

variety meats in 281–282, 284, 794

veal and calf 636

venison 259–262

Sautéing

liver 281

Scabbards 52

Scalding

hogs 141–142, 323

poultry 231–232

Scanning, UPC (Universal Product Code) 501–502

Scapula (blade bone)

beef 542, 547, 552, 557, 558

lamb 597, 614, 616

pork 505, 518, 519, 528

poultry 238, 241–242

spine of 555

veal and calf 636

Scotch tender 557

Scrapple 284

Scudding hides 317

Seam fat 550, 557, 558, 614, 617

Seasonal effects on prices 494–496

Seasoning 664, 806

defined 806

Sectioned and formed restructured meat products 673–674

Sectioning 675, 676

Seedless bellies 525

Semimembranosus muscle

beef 566, 569

lamb 607, 610

retail cut identification 497–498

Semitendinosus muscle

beef (eye of round) 566, 570

lamb 607, 609

pork 515

retail cut identification 498

Serratus dorsalis muscle 551

Serratus ventralis muscle 561, 676

Serving a crown roast 621

Serving costs per pound 998–1001

Servings per pound, from various cuts 996–997

Sewage control 74

Sex condition

beef 352–355

lamb 405

pork 424–425

Shank

beef 561

cross cuts 562

veal and calf 636

cross cuts 636

osso bucco 636

Shanking, beef 182, 184–185

Sharpening knives 120–123

Shearling bed pads 325

Sheep casings 820–821

Sheep and lambs harvested, number per state and rank of 199

Shelf life 471

defined 641

Shipping

livestock 95–99

meat imports 4, 84–86, 456–457

poultry 225–226

Shirred casings 823

Shish kabobs, lamb 606, 615, 618, 623

Shoddy leather 321–322

Shohet 93

Shortening 296, 298

composition 296, 298

fats and oils used in U.S. manufacture 296, 298

lard 296–299

Shorting out 816, 851

Short ribs 561

Shoulder, lamb 612–618

cutting test 485–488

separation from rack 613

Shoulder, pork

removing clear plate from 528

removing jowl from 527

removing neck bones from 526–527

separating from loin 517

(*see also* Blade Boston, pork, *and* Picnic, pork)

Shoulder, veal and calf 635

percent of carcass weight 635

Shrink

beef 176

collagen 817

cooler 171, 192–193

Cryovac® bag 703

cured meat 757–758

freezer 724

knife handles 117

marketing 96

meat 699

pork 136

poultry 225

smoking 72

tissue, lamb 198

Shrinkage, definition of 474–475

Shrouding 192

Side, beef 544

Siding 185–186, 317

Sinclair, Upton 37

Skeleton

beef 542

fowl 238

lamb 597

pork 505

veal/calf 631

Skinning

beef 185–186

deer 251–254

hogs 148–157

lamb 202–206

sanitation requirements (FSIS) 55–56

veal and calves 218–219

Skins 28

as bed pads 325, 332

as burn dressing 324, 332

calf 322–323, 328

classification of 308–309, 325

export and import of 329–332

for blood clotting and collagen 342

foxes 322

goat 322

in gelatin 323

kip 308–309, 322

mink 322

pig 136, 155, 158, 323–324

pork for cracklings 294

sheep 325–326

slunk 323

snacks 294

uses 322–326

Skirt

beef 191, 560–561

hanging tenderloin 191, 564

hog 166

lamb 610, 620, 622

veal and calf 220–221, 636, 637

Slicing 88, 570, 573, 639

top round 698

Slunk skins 215, 323

Smoked sausage formulation 836–839
   pork 838–839
   procedure 831–836
   venison 262

Smoked turkey, evaluation of 1033–1034

Smoking
   equipment 763–764
   FSIS regulations against 52
   FSIS regulations for meat products 72, 78, 768
   procedures 764–768
      cooking 767
      drying 765
      showering and chilling 768
      smoking 765–766

Soap
   from by-products 43, 285, 303
   manufacture 303–304

Sodium acid pyrophosphate 803, 804

Sodium ascorbate 737, 802–803

Sodium erythorbate 737, 802–803
   FSIS regulations and limits 737, 744, 802–803

Sodium lactate 805

Sorbitol 803

Sources of imported meat 6, 456–457

Sous vide 685–686, 706

Sow carcass grades 435–436

Soxhlet® fat extraction apparatus 650–651

Soy flour 811, 812

Soy grits 812

Soy protein concentrate 662, 811, 812

Soy protein isolate 681, 811–813, 866

Space recommendations for shipping livestock 95–99

Spareribs
   beef 561
   lamb 622
   pork 524–526
      cooking instructions 526
      cutting test 481–484
      percent of carcass weight 525
      separating loin from 524–525

Specht-Overholt, Susan Marie 633, 636

Special fed veal 629

Special label claims 76–80

Special orders, service counter 465

Specialty (luncheon) meats 466

Species, meat, identification of 84, 497, 499

Specifications 31, 500–501
   collecting pharmaceuticals 333, 340
   cuts 500–501
   cutting tests 475–496
   facilities 74
   frankfurters 875
   IMPS 449–450, 500–501
   purveyor 463
   Uniform Retail Meat Identity Standards 498–500

Specific heat 941–943
   defined 942
   of various media and elements 942

Sphincter muscle 162

Spices
   defined 806
   effect on microbial contamination and growth 36, 806
   in cured meats 738
   irradiated 718
   used in sausages 807–811

Spinalis dorsi (cap muscle) 549

Spinous processes, beef 192, 357–361

Spirulator, cryogenic 668

Splash, blood 138

Spleen
   inspection of 67
   lamb 208
   pharmaceuticals from 342

Splitting
   beef carcass 191–192, 196
   deer carcass 254
   fatty acids 301–302
   hides 317, 321
   lamb double loin 611
   lamb double rack 619
   pork carcass 52, 165, 168–169
   veal and calf carcasses 220

Spool joint 203–204

Spots, blood 139, 904–905

Sprays, carcass 55–56, 169–170, 193

Sprue 338

Squab, or pigeon 440

Staking hides 322

Stamps
   grading 388–390, 450, 610, 618
   ink 42

inspection 92, 610
   Kosher 93

Standard deviation 364, 644–646

Staphylococcus aureus 17, 19, 24, 28–29

Starter culture 841–842

State meat inspection 84

Steaks, beef
   components in carcass, summary of types of cuts, identification and fabrication of 541–589
      boneless strip 585
      bottom (outside) round 570
      club 580, 581
      cubed (minute) 557, 571
      Delmonico 551
      double-bone sirloin 579
      flank 565
      flat-bone sirloin 579
      Kansas City 585
      minute (cubed) 557, 571
      New York strip 585
      pinbone sirloin 578
      pin-wheel 561
      porterhouse 581
      rib, blade (large) end 550
      rib, boneless 551
      rib eye 551
      rib, loin (small) end 549
      round-bone sirloin 578
      round steak 568
      skirt-steak rolls 560, 561
      strip loin, bone-in 585
      T-bone 581
      tenderloin (filet mignon) 582
      tip 567
      top (inside) round 569
      top-loin 580, 581, 584
      top-loin, boneless 585
      top-sirloin, boneless 584
      wedge-bone sirloin 577
      (see also center color section)

Steaks, fish 272, 273

Steaks, lamb leg 606, 607
   identification and fabrication of 606–607

Steaks, pork
   types of cuts, identification and fabrication of 503–532
      arm 531–532
      blade 529
      ham, center slices 516
      (see also center color section)

Steaks, turkey 242–243

Steaks, veal
types of cuts, identification and
fabrication of 632–639
arm 636
blade 636
cutlet 639
round 639

Steaks, venison 255, 257
round 255–257

Steam stripping 656–657

Stearic acid 291

Stearin 291

Steeling 120, 123–124

Sterilization 707–709
radappertization 719–720

Sternum
beef 187, 542, 547, 552, 562
lamb 205, 597, 622
pork 139–140, 163–164,
505, 525
poultry 238, 240

Stewing
beef 371, 556
poultry 975–976

Stew meat
lamb 615
veal 636

Sticking
cattle 178–179
hogs 139–141
sheep and lambs 201–202
veal and calves 216–217

Stick, tankage 286

Stifle joint
beef 542, 567, 568
lamb 597, 605, 608
veal and calf 631, 639

**Stitch (or spray) pumping (pump
pickling) 754–755**

Stomach 1, 96, 110, 166, 283
pepsin 342
pharmaceuticals 342–343
uses 283

Storage
poultry 71, 225, 245–246
raw meat 31, 71

Streaking, fat
lamb, yearling mutton, and
mutton 405–406
veal and calf 399–400

Streptococcus 31, 100

Streptococcus faecalis 31

Streptococcus pyogenes 31

Streptococcus viridans 31

Stress, swine 95, 903–904

Stromal proteins 900–901

Structured meat products 673
(*see also* Restructured meat
products)

Stunner, poultry 227

Stunning
cattle 177–178
hogs 136–139
poultry 229–230
sheep and lambs 200–201
veal and calves 55, 216–217
(*see also* Immobilization)

Sub-primal cuts
beef chuck 556
beef loin 577
beef round 569
boners sell 463
defined 497, 602
IMPS 500–501
lamb 606–607
venison 256

Subscapularis muscle 551

Sugars
energy 898
meat curing 735
sausage 803, 840

Summer sausage 847–848
microbiology 13, 23,
841–842

**Supercritical fluid extraction
656–657**

Supermarket chain ranking 467

Supply
cutting tests 475–496
home-raised pork 503
imports 84–86, 456–457
meat in the United States
455–502
poultry 223–224
source 455
urban shoppers 468–470
vealers 213, 215
water 73

Supraspinatus muscle 557

Surimi 9, 684–685

Sutures 337

Sweetbreads 220, 221, 338
cooking 283

Sweet pickle cure 749, 752–754

Systemic insecticide 107

T

Taenia saginata (human tapeworm) 108

Tail
beef 188–189, 311
steak 588

lamb 602
ribs 618
pork 275, 511
poultry 240
veal and calf 637
ribs 635

Tallows and greases 294–304
as feeds 300–301
as soap 303–304
classification of 294–295
edible 296–299
inedible 299–303
fatty acids from 301–302
production of 296–300
supply, exports, and disposition 299
trading standards for 295
utilization of 296
value 285, 300

Talmadge-Aiken meat plants 38

Tang (knife) 118

Tankage 286, 306

Tank houses 287

Tank water 286

Tanning 318–322, 326–329
alum 327–328
chrome 320
commercial 319–322
deer 252
home 326–329
salt acid 328
salt alum 327
vegetable 319

Tapeworms 66, 108

Taste
lard 290
rancidity 698–699

**TBHQ (tertiary
butylhydroquinone) 292**

Teeth, uses for 275

Temperatures
carcass sprays 170
chilling beef 192–193
conversion F/C 1048–1054
curing 753
dry bulb 1055–1056
effects on microbial growth
19–21
for livestock shipping 97,
225
game handling 249
myosin extraction 816
normal rectal 64–65
pork carcasses 504
rendering 285, 286–287,
291, 292, 293, 302
required by USDA-FSIS to
achieve certain finished
product characteristics
completely cooked 768
ham, cooked 768

poultry, fully cooked, ready-to-eat, baked, roasted 768
restructuring 679–680
water for scalding
hogs 141–142
poultry 231–232
wet bulb 1055–1056

Tenderization 670–672, 902

Tenderizing
enzymatic 672
papain, bromelin, and ficin 672
mechanical 574, 670–671

Tenderloin muscle *(psoas)*
beef 563, 578, 579, 582
deer 256
lamb 610, 612
pork 520, 521
poultry 240
veal and calf 638
(see also center color section)

Tenderness 39, 904, 929–930, 931
beef 180, 568, 571
deer 250
lamb 595, 610
veal 630

Tendons, uses for 275

Tenosynovitis 116

Tensor fasciae latae **muscle** 567, 583

Testicles (mountain oysters, fries) 277
pharmaceuticals from 343

Tests
contamination/adultration 51, 65, 83, 84
cutting, yield 475–496
forms for 475–496

Texture
all meat 929–930, 931
beef 180, 354–355, 359–360, 361–362, 582
feed 300–301
frankfurters 875
ground beef 666
lamb, yearling mutton, and mutton 404, 405, 602
lard 293
oxtail 283
pork 426
rabbit 447–448
veal and calf 398

Texturized soy protein 812

Texturized vegetable protein 664

Thaw rigor 905

Thermal processing 835–836

Thermometer requirements 940–941

Thermophile 19

Thiamine (vitamin B$_1$) 922

Thigh, chicken 238–239, 241

Thoracic cavity
beef 191
vertebrae in 358–362, 553
hog 165–167, 524
pleural membrane 524
sheep 205

Thoracic **vertebrae (rib)**
beef 358–360, 542, 545, 548
lamb 597, 599, 619, 620
pork loin 505, 506, 517–521
veal and calf 637

Thrombin 335

Thromboplastin 336

Thuringer 793

Thymus 220–221, 343
sweetbreads 220–221, 283

Thyroid 167, 343, 344
pharmaceuticals 344
calcitonin 344
thyroglobulin 344
thyroxin 344

Tibia **(shank bone)**
beef 542, 568
lamb 597, 605
pork 505, 513
poultry 238, 239

Tierce 817

Tight side (kidney fat location in beef carcasses) 563

Time and microbial growth 19–21

Tip, sirloin/round/knuckle 562–564, 567

Titer, fat 294–295

Tobacco 52

TOBEC (total body electrical conductivity) 658

Tocopherols 292

Tongue 66, 159, 206, 278, 280, 282
cooking 282
freezing 282
shipping 282

Top round, beef 566, 569
yield based on total round 566

Total quality management (TQM) (*see* Quality control *and* Quality, assurance)

Toxoplasma gondii 113–114

Toxoplasmosis 113

TQM (total quality management) (*see* Quality control *and* Quality, assurance)

Trachea
beef 181–182
deer 249, 252
hog 158
lamb 205
poultry 234, 439
thyroid location 343–344
veal and calf 217–218

Training in knife safety 128

Transportation
deer 250
history and development of 455–456
livestock 95–99

Transverse **vertebral process** 505, 542, 580, 612

Trapezius **muscle**
beef 551, 558
pork 528

Triangle, beef 547, 583

Triceps brachii **muscle**
beef 554
lamb 614

Trichinae
control 90, 111–113, 719
detection by a digestion system 111
life cycle 110

Trichinoscope 111

Triglycerides 285, 917–918

Trim (lean), boneless
beef 585–589
boners sell 463
lamb 623, 625, 627
pork 533–540

Tripe
cooking 283
sausage 816

Tri-tip 583

Trypsin 339–340

TSH (thyroid-stimulating hormone, thyrotropic hormone, thyrotropin) 341

T-tubules 893, 895, 896, 897

Tuber coxae **(hipbone)** 578

Tuberculosis 66, 104–106, 159

Tularemia 266

Tumbling (massaging) pork, leg (ham) 515, 762

Tunnel boning 514–515, 607–609, 760–761

Turkey 31, 224–225
burger 243
formed, cured, and smoked (simulated ham) 243

frankfurters, bologna, and
breakfast sausage 794
fryer-roaster 438
ham 243
mature (old) 439
mechanically deboned 681
roll 243
steak 243
wild 264
yearling 439
young 438
Two cardinal principles to follow
when cooking 941
Two-piece chuck 552

U

Udder
beef 55, 190, 311, 352–355
lamb 207, 412
veal and calf 218
Ulna
lamb 597, 612
pork 505, 531
poultry 238
Ultrasonics 658
Uncooked smoked sausage
781–782
Ungraded beef 370
Uniform Retail Meat Identity
Standards 498–500, 596
United National Collegiate Meat
Animal Evaluation Contest
1004–1005
United Nations 3
Universal Product Code (UPC)
501–502, 593
Unsaturated fat
definition 285, 291–292
Ureters
beef 190–191
hog 108, 165–166
sheep and lamb 207–208
USDA Canner 372
USDA Choice
beef 368–369, 374
lamb 406, 409
USDA Commercial 370–371,
377, 670
USDA Cutter 372
USDA Good 406–407, 410
USDA Handbook 927
USDA lamb carcass certification
program 627
USDA *Livestock, Meat, Wool Mar-*
ket News 472

USDA Prime
beef 368, 373
lamb 405–406, 408
USDA Select 369, 375
USDA Standard 370, 376, 670
USDA Utility
beef 371, 378, 670
lamb 407, 411
pork 430
USDA Yield Grade (YG) 1
beef 386–387, 392
lamb 413, 417
USDA Yield Grade (YG) 2
beef 387, 393
lamb 414, 418, 626
USDA Yield Grade (YG) 3
beef 387, 394
lamb 415, 419, 624, 626
USDA Yield Grade (YG) 4
beef 388, 395
lamb 415, 420, 626
USDA Yield Grade (YG) 5
beef 388, 396
lamb 415, 421
U.S. No. 1 429, 431
U.S. No. 2 429, 432
U.S. No. 3 429–430,
433
U.S. No. 4 430, 434

V

Vacuum breakers 73
Vacuum package 23, 590–593,
599, 625–627, 685, 705–706,
727
Value
added by meat packing
457–458
-based marketing 470
by-products 276
comparison 241
definition of 475
differences, beef yield grades
388
differences, lamb yield grades
416
differences, rib steak 550
effect of dark-cutting beef
362
effect of hide pattern on leather
186, 310
for inspection exemption 89
muscle value comparison 241
of beef carcass conformation
357
of beef quarters 493, 495
of labels 75
of lamb loin 610
of meat grading 349

of spareribs 532
passed for cooking loss 105
pet food industry 301
ranking of poultry parts 242
Variety meats 277–285
proximate composition 277,
279, 280
sausages 794, 866
(*see also* By-products, edible)
Vastus intermedius muscle (one of
the *quadriceps*) 567
Vastus lateralis muscle (one of the
quadriceps) 567
Vastus medialis muscle (one of the
quadriceps) 567
Veal
classification 629
definition 629
Veal and calf grades 397–401
Veal and calves harvested, number per state and rank of 214
Vealers 215, 629
Vegetable(s)
for cooking with meat 623
oil 293, 303
protein 663–664, 681,
812–813, 866
Vending machines 466
Venison processing 247–262
Venison sausage 259–262
Verified animal production control 80
Vibrio parahaemolyticus 31
Vienna-style wieners 709
Violation of meat inspection laws
89–90
Viruses 18, 102
Viscera inspection 52, 66–67
Vitamin(s) 1, 919–926
A 920
B_1 922
B_2 (riboflavin, formerly G)
922–923
B_6 (pyridoxine) 924
B_{12} 925
biotin 925–926
C (ascorbic acid) 802–803,
926
choline 924
D and D_3 920–921
E 921
folic acid (pteroyglutamic acid)
924–925
K 921–922
niacin (nicotinic acid) 923
pantothenic acid 925

W

Walls 51

Warbles 106

Warner-Bratzler Shear Method 930

Washing
  carcasses 55–57
    beef 192
    pork 169–170
  fish 271
  heads 55
  workers 34, 52

Water
  content in meat 910–911
    essential functions in animal body 910–911
  for cleaning 34–35
    carcasses 55–57, 73
  in cured meats 73, 738–739, 752
    FSIS regulations, limits and labeling requirements 739–740
    functions 738
  infection and 100
  in sausage 798–800
  potable 73, 89
  preharvest 96
    and shrink 96
    lamb 198
    poultry 225–226
  poultry chilling (FSIS regulations) 69
  scalding temperature
    hogs 141–142
    poultry 231–232

Water activity (a_w) 18, 21, 669, 713–716, 843–844

Waterfowl 266

Wax- or paraffin-treated Kraft papers 724

Wax picking, poultry 232–233

Weasand
  beef 51, 821
  lamb 205
  veal and calf 217–218

Wedge muscles 551

Weights
  beef carcass 544
  beef live 544
  beef side 544
  carcass (definition) 473

chilled carcass (definition) 473
  hog live 509
  lamb carcass 596, 599, 600
  lamb live 596, 599, 600
  live (definition) 473
  livestock for shipping 97–99
  pork carcass 509
  pork cuts 534

Wet bulb temperature 1055–1056

White meat 237

Whole
  hog sausage 828–829
  muscles 673

Wholesale cuts 23, 88, 503
  and locomotion-support muscle systems 496
  beef 544, 546, 562
  boners sell 463
  cutting tests 475–496
  defined 496, 602
  guidelines for making 496, 499
  IMPS 500–501
  lamb 600
  list of 476–488
  marketing efficiency 472–473
  pork 171, 503, 506, 508
  system 496
  value 493, 495
  veal and calf 633

Wholesalers 463

Wholesome Meat Act 37, 87

Wholesomeness 34, 37, 39, 94, 171

WHO (World Health Organization) 717

Wiltshire side 172

Windsor chop 524, 776

Wings, chicken 238

Wishbones, chicken 240, 241

Withdrawal periods 101

Wool
  length for classification 325
  pulled 326

World population 2, 3

Worms 107–113

X

Xenotransplantation 333–344

Y

Yearling turkey 439

Yersinia enterocolitica 31

Yield
  beef
    carcass based on live weight 176–177
    retail cuts based on live and carcass weights 585–589
  cutting 473
  cutting tests 476–480, 491–496
  definition of 475
  dressing percent 135–136, 473
  effect on price 491–494
  lamb
    carcass based on live weight 200
    retail cuts based on live and carcass weights 625–626
  pork 135–136
  worker safety 117

Yield grades 39
  beef 379–397, 589
    boneless roast and steak yield equivalents 386
    calculation of 386
  lamb 412–421, 626
    by fat thickness chart 414
    calculation of 413
    closely trimmed primal cut yield equivalents 412
    yield at various trim levels 626

Young turkey 438

Z

Zinc 910, 918–919

Z line 893, 894, 899, 901

Zooarchaeologists 133

Zoo foods 301